Next-Generation Plant Breeding Approaches for Stress Resilience in Cereal Crops

Mallana Gowdra Mallikarjuna •
S. Chandra Nayaka • Tanushri Kaul
Editors

Next-Generation Plant Breeding Approaches for Stress Resilience in Cereal Crops

Springer

Editors
Mallana Gowdra Mallikarjuna
Genetics
Indian Agricultural Research Institute
New Delhi, India

S. Chandra Nayaka
Applied Botany and Biotechnology
University of Mysore
Mysore, Karnataka, India

Tanushri Kaul
International Centre for Genetic
Engineering and Biotechnology
New Delhi, India

ISBN 978-981-19-1444-7 ISBN 978-981-19-1445-4 (eBook)
https://doi.org/10.1007/978-981-19-1445-4

This Springer imprint is published by the registered company Springer Nature Singapore Pte Ltd.
The registered company address is: 152 Beach Road, #21-01/04 Gateway East, Singapore 189721, Singapore

Dr. Hari Shanker Gupta

Dr. H. S. Gupta was born in a village of Rudrapur on 1 July 1953, Uttar Pradesh, India. He obtained a B.Sc. degree in 1971 from National Degree College, Barhalganj Gorakhpur University (now Deen Dayal Upadhyaya Gorakhpur University), Gorakhpur; postgraduate degree from G.B. Pant University of Agriculture and Technology, Pantnagar, Uttarakhand; and Ph.D. from the Indian Institute of Technology, Kharagpur (1974–1978). Dr. Gupta completed his postdoctoral fellowship from the University of Nottingham, United

Kingdom (1987–1988), and Washington State University, Pullman, United States of America (1993–1994).

Dr. Gupta is a renowned geneticist and plant breeder who contributed immensely to cereals genetics and breeding during his 35 years of a scientific and administrative carrier at various institutions under the Indian Council of Agricultural Research (ICAR), Government of India, and Borlaug Institute for South Asia, New Delhi. He is behind the release of various conventional and genomics-assisted bred varieties, including India's first marker-assisted biofortified hybrid "Vivek-QPM-9." During his tenure as director at ICAR-VPKAS, Almora, and ICAR-IARI, New Delhi, the institutes were conferred with the prestigious Sardar Patel Outstanding ICAR Institution Award by ICAR, New Delhi. Dr. Gupta served as director general, BISA, New Delhi, from 2014 to 2016.

Dr. Gupta has been conferred with various awards for his outstanding contributions in agricultural sciences, such as the ICAR Team Award (1994–1996, 2006–2007 and 2008–2009), NRDC's Meritorious Invention Award (2006), Hari Om Asharam Trust Award (2007), NRDC's Societal Invention Award (2007), WIPO Gold Medal (2008), Dr. Amrik Singh Cheema Award (2010), and Doctor of Science (Honoris Causa) from Sardar Vallabhbhai Patel University of Agriculture & Technology, Meerut, and Assam Agricultural University, Jorhat.

Dr. Gupta has been elected for various fellowships, such as the National Academy of Agricultural Science, India; Indian Society of

Agricultural Biochemists; Indian Society of Genetics & Plant Breeding; Indian Society of Agricultural Engineering (Hon); and Rockefeller Foundation's Career Fellow. He also served as president of the Indian Society of Genetics and Plant Breeding and vice president of the Society for Plant Biochemistry and Biotechnology, India.

Most importantly throughout his career, Dr. Gupta mentored and inspired several young researchers as a scientific and research administrator in various capacities. He is known for guiding and moulding the younger generations in science. This book is dedicated to Dr. H. S. Gupta for his innumerable contributions to cereals and agricultural science during his tenure as a researcher and scientific administrator.

Foreword

Cereals are an essential source of food and nutrition to mankind. The group of crops contributes immensely to achieve food security, improve nutrition and promote sustainable agriculture in the developing world. The staple cereal crops, such as rice, wheat and maize, supply more than 50% of the calorie requirement of the global population. This was possible by employing traditional breeding over the past several decades and delivery of improved cultivars in different agro-ecological systems to meet the diverse demands of various stakeholders. However, the progress was not very satisfactory in improving stress-resilience traits, which have complex inheritance. At the moment, plant breeders are equipped with several breeding tools ranging from basic "selection" to "genome editing." The advent of next-generation sequencing technologies and advanced breeding informatics has enabled genome-wide selection and prediction approaches to address the challenges of complexity in quantitative traits improvement in crops. Further, getting rapid genetic gain per unit time has been a dream of the plant breeders to deliver the improved cultivars within a short time. The latest additions to breeding tools, such as speed breeding and CRISPR/Cas based genome editing, have opened tremendous opportunities to achieve the breeding objectives more efficiently.

At this juncture, we are witnessing adverse effects of climate change on the production of cereals and affecting the food security regionally and globally. Therefore, the most effective and sustainable approaches to manage various climate change-induced stresses are to develop climate-resilient cereal cultivars that yield better. The book provides extensive coverage on genome-wide association studies, genomic selection, rapid generation advancement and genome editing approaches in cereals for enhanced stress tolerance. Moreover. I feel that the information compiled on the genes, germplasm, genomic databases and bioinformatics tools is also of great importance for stress resilience breeding programmes. I believe that the book will serve as a valuable reference for students, faculty, researchers and industry. I congratulate the editors, Dr. MG Mallikarjuna, Dr. S. Chandra Nayaka and Dr. Tanushri Kaul, for conceptualizing and compliment all the authors for their efforts.

Department of Agricultural Research and Education T. Mohapatra
Indian Council of Agricultural Research
New Delhi, Delhi, India
14 February 2022

Preface

Climate change is now a reality, especially in increased CO_2 levels and temperature and extensive snow melting. Forecasts of climate change impacts on agriculture reveal the drastic fluctuations in agroclimatic parameters, such as temperature, rainfall, soil health, pests and diseases incidences. Furthermore, global warming and unwarranted variation in weather parameters cause the intermittent occurrence of drought, flood and rapid emergence of pathogen races and insect biotypes, leading to a striking reduction in crop yields. Therefore, ensuring food and nutritional security is the most challenging task in the present and forthcoming climate-change era. Cereals are the major contributors to world food and nutrition security and provide more than 60% of the global calories' requirement. Therefore, breeding climate-resilient cereals is the sustainable and practical approach to ensure food and nutritional security to an ever-increasing population.

The plant breeders have several tools in their basket, from basic "selection" to the latest "genome editing." Many of these stress-resilient and adaptive traits show complex genetics. The application of next-generation breeding approaches like genomic selection, genome-wide association mapping, and genome editing coupled with rapid generation advancement methods has enormous potential in delivering climate-resilient cereal cultivars. This book provides a snapshot of genome-wide association studies, genomic selection, genome editing and rapid generation advancement methods in improving the climate-resilient and stresses adaptive traits

The book includes chapters contributed by various researchers on cereal crops and breeding aspects regarding next-generation breeding methods/techniques. The introduction portion comprises the first chapter on targeted next-generation breeding methods/technologies targeted in this book for enhancing the climate resilience in cereals. The second group of chapters are dedicated to genome-wide association studies and genomic selection in cereals for climate resilience and stress tolerance traits, which include three crop-specific chapters, i.e. one each for rice (Chap. 2), wheat (Chap. 3) and maize (Chap. 4), and one chapter on the application of genome-wide approaches for enhancing nutrient use efficiencies in cereals (Chap. 5). The third group includes three chapters dedicated to accelerated generation advancement methods with particular emphasis on stress resilience in cereals. In this category, we have a chapter on doubled haploidy (Chap. 6), rapid generation advancement in wheat (Chap. 7) and speed breeding in rice (Chap. 8). The fourth group is exclusively dedicated to the application of genome editing methods for abiotic and biotic stress tolerance with five chapters. Chapters 9 and 10 discuss the methods, principles and applications of genome editing for stress resilience and nutritional enhancement in cereals. Chapters 11 and 13 mainly focus on applying CRISPR/Cas9 for sustainable disease management in cereals, and Chap. 12 is exclusively confined to the application of genome editing in rice for stress tolerance. The last chapter included in the book (Chap. 14) is mainly dedicated to genomic and bioinformatic resources for next-generation breeding approaches towards stress resilience.

Noticeably there are various books and reviews on the individual aspects of genomics-assisted breeding and next-generation breeding approaches on cereals. However, there are no comprehensive books on applying next-generation breeding approaches for climate resilience in cereals. Therefore, this book is compiled to deliver the latest updates and most comprehensive information on the application of next-generation breeding methods/technologies to enhance stress tolerance and climate resilience in cereals. We sincerely feel that this compilation will be greatly useful for students, research scholars and scientists engaged in crop improvement, genomics, biotechnology and molecular biology in generic and in specific stress-resilient cereals breeding at universities, public and private research institutes, and NGOs with R&Ds, for conducting research and funding and policy agencies for planning future strategies in cereal improvement towards climate resilience.

We are very grateful to all the learned contributors, and without each of their contributions, the task would have been impossible. The contributors have strived hard to update the scientific information of their respective domains and have spared their valuable time and knowledge to come up with quality chapters. We apologize sincerely for any exclusions, mistakes or failure to acknowledge fully.

We are thankful to our families, Smt. Gowramma Ramanagowdru, Smt. Sridevi Suresh and Ms. Jayashri Patil (mother, elder and younger sisters of MG Mallikarjuna); Smt. Chaithra and Master Daivik Chandra (wife and son of Chandra Nayaka) and Ananyanandini Kaul, Shambhunath Kaul and Usha Kaul (Daughter, Father and Mother of Tanushri Kaul) for their continuous encouragement and indirect supports to maintain the academic ambience to complete this editing process. We are also extending our sincere thanks to Dr. Prashant Hanjagi, Dr. Digvinder Pal and Dr. Rajesh Kumar for assisting during compilation. Finally, we highly appreciate all the cooperation and support of Springer Nature for their careful and speedy publication of this book.

New Delhi, India — Mallana Gowdra Mallikarjuna
Mysore, Karnataka, India — S. Chandra Nayaka
New Delhi, India — Tanushri Kaul

Contents

Editors and Contributors

About the Editors

Mallana Gowdra Mallikarjuna is a scientist (senior-scale) in genetics and plant breeding, currently working at Division of Genetics, ICAR-Indian Agricultural Research Institute, New Delhi, India. He holds a graduation in agricultural sciences from the University of Agricultural Sciences, Dharwad, India. He completed the post-graduation from Acharya NG Ranga Agricultural University, Hyderabad, and a doctoral degree from Indian Agricultural Research Institute, New Delhi, with university and institute medals for outstanding academic performance. He was awarded several research grants from various funding agencies. His current area of interest includes the application of genomics and systems biology to understand the stress responses in maize and model plants, deciphering the evolutionary snapshots of stress-responsive gene families and stress-resilient breeding. To date, he has 20 research articles in peer-reviewed journals, five book chapters and four maize hybrids. Besides, he has guided postdoc fellows and research scholars for their project work.

S. Chandra Nayaka is an agricultural biotechnologist specialized in the application of biotechnological tools in plant protection, pathogen detection and plant–pathogen interaction. He is serving as principal investigator of ICAR-AICRP on Pearl Millet and is a professor at the University of Mysore. He has 20 years of experience in research, including visiting scientist at the Danish Government Institute of Seed Pathology, Denmark, and China Agricultural University, Beijing, China. His main activities pertain to pathogen biology, host–pathogen interaction, epidemiology and disease management. His group has pioneered full-length genome sequencing of highly virulent strain of the biotrophic pathogen *Sclerospora graminicola*. He is a recipient of the Japan International Award for Young Agricultural Researchers and The Millennium Plaques Honor by the Indian Science Congress Association. He holds 4 national patents and has published more than 100 research articles, 08 books, 15 book chapters, 03 application notes, and six technical bulletins. Additionally, he regularly guides postdoc fellows and research scholars.

Tanushri Kaul is a renowned senior scientist and group leader, specialized in plant molecular biotechnology at ICGEB, New Delhi, with pioneering work in nutritional improvement of crops via CRISPR gene editing. She has been awarded DBT-IISc-Post Doctoral Fellowship, Indo-Israel; Indo-German-DST Research Associateship, RGYI-Award; BIOCARE-DBT grant cum fellowship; ASEAN Bio-tech Award Malaysia; and Global Certificate Course on Data Science & AI. She has significantly contributed to the field of plant molecular biology and biotechnology for 20 years of her research career. She developed phytase-rich tomatoes, wheat grains with low phytic acid, herbicide-resistant maize and rice, and iron- and zinc-enriched rice. She spearheaded the whole genome sequencing and transcriptomics of rice bean and saffron. She has received four patents for her exceptionally innovative research contribution in the scientific community. She has over 70 publications in international journals, book chapters and articles. Moreover, she has frequently guided international and national postdoctoral fellows, predoctoral fellows and MSc students.

Contributors

V. Mohan Murali Achary Crop Improvement Group, International Centre for Genetic Engineering and Biotechnology, New Delhi, India

Divya Ambati ICAR-Indian Agricultural Research Institute, Regional Station, Indore, India

C. Anilkumar ICAR-National Rice Research Institute, Cuttack, Odisha, India

Srinivas Ankanagari Department of Genetics, Osmania University, Hyderabad, India

H. Prashanth Babu Indian Agricultural Research Institute, New Delhi, India

Alexander Balamurugan Division of Plant Pathology, ICAR-Indian Agricultural Research Institute, New Delhi, India

Santhosh Kumari Banakara Department of Genetics and Plant Breeding, University of Agricultural Sciences, GKVK, Bangalore, India

Shilpi Bansal Division of Plant Pathology, ICAR-Indian Agricultural Research Institute, New Delhi, India

P. S. Basavaraj ICAR-National Institute of Abiotic Stress Management, Baramati, Maharashtra, India

Hemalata Bharti Indian Agricultural Research Institute, New Delhi, India

Jyotsna Bharti Nutritional Improvement of Crops Group, International Centre for Genetic Engineering and Biotechnology (ICGEB), New Delhi, India

Jayant S. Bhat ICAR-IARI's Regional Research Centre, Dharwad, India

Pratik I. Bhoyar ICAR-Indian Agricultural Research Institute, Regional Station, Indore, India

Gaurav N. Chaudhari Dr. Panjabrao Deshmukh Krishi Vidyapeeth, Akola, India

Suvarna Rani Chimmili ICAR-Indian Institute of Rice Research, Hyderabad, Telangana, India

Viswanathan Chinnusamy Division of Plant Physiology, ICAR-Indian Agricultural Research Institute, New Delhi, India

Kapil Deswal ICAR—Indian Institute of Wheat and Barley Research, Karnal, India
CCSHAU, Hisar, India

E. Lamalakshmi Devi ICAR-Research Complex for NEH Region, Sikkim Centre, Tadong, Sikkim, India

Murugesh Eswaran Nutritional Improvement of Crops Group, International Centre for Genetic Engineering and Biotechnology (ICGEB), New Delhi, India

Anil Gaddameedi International Crops Research Institute for the Semi Arid Tropics, Pantancheru, Hyderabad, India

N. C. Gahtyari ICAR-Vivekananda Parvatiya Krishi Anusandhan Sansthan, Almora, India

Kiran B. Gaikwad Indian Agricultural Research Institute, New Delhi, India

Mallikarjuna Garladinne Plant Molecular Biology Laboratory, Agri Biotech Foundation, Hyderabad, India

Vikas Gupta ICAR-Indian Wheat and Barley Research Institute, Karnal, India

Prashantkumar S. Hanjagi ICAR-National Rice Research Institute, Cuttack, Odisha, India

Mohammad Mahmood Hasan Department of Microbiology and Cell Science, University of Florida, Gainesville, FL, USA

Khondokar Nowshin Islam Department of Microbiology, Miami University, Oxford, OH, USA

Mohammad Nurul Islam Plant Breeding and Biotechnology Laboratory, Department of Botany, University of Dhaka, Dhaka, Bangladesh

Shailendra Jha Division of Genetics, ICAR-Indian Agricultural Research Institute, New Delhi, India

Kranti Kamre Indian Agricultural Research Institute, New Delhi, India

Sruthi Kanneboina ICAR-Indian Institute of Rice Research, Hyderabad, Telangana, India

Lakshmi Kant ICAR-Vivekananda Parvatiya Krishi Anusandhan Sansthan, Almora, India

Chikkappa G. Karjagi Unit Office, ICAR-Indian Institute of Maize Research, New Delhi, India

Rashmi Kaul Nutritional Improvement of Crops Group, International Centre for Genetic Engineering and Biotechnology (ICGEB), New Delhi, India

Tanushri Kaul Nutritional Improvement of Crops Group, International Centre for Genetic Engineering and Biotechnology (ICGEB), New Delhi, India

R. K. Khulbe ICAR-Vivekananda Parvatiya Krishi Anusandhan Sansthan, Almora, India

Suneetha Kota ICAR-Indian Institute of Rice Research, Hyderabad, Telangana, India

Aundy Kumar Division of Plant Pathology, ICAR-Indian Agricultural Research Institute, New Delhi, India

Manjeet Kumar Indian Agricultural Research Institute, New Delhi, India

Naresh Kumar Indian Agricultural Research Institute, New Delhi, India

Rakesh Kumar Tel Aviv University, Tel Aviv-Yafo, Israel

Ravindra Kumar Indian Institute of Wheat and Barley Research, Karnal, Haryana, India

Sudhir Kumar ICAR-Indian Agricultural Research Institute, New Delhi, India

Hirenallur Chandappa Lohithaswa Department of Genetics and Plant Breeding, University of Agricultural Sciences, GKVK, Bangalore, India

B. M. Lokeshkumar ICAR-Central Soil Salinity Research Institute, Karnal, Haryana, India

Bingi Pujari Mallikarjuna ICAR-IARI's Regional Research Centre, Dharwad, India

Mallana Gowdra Mallikarjuna Division of Genetics, ICAR-Indian Agricultural Research Institute, New Delhi, India
ICAR—Indian Institute of Wheat and Barley Research, Karnal, India

Khaled Fathy Abdel Motelb Nutritional Improvement of Crops Group, International Centre for Genetic Engineering and Biotechnology (ICGEB), New Delhi, India

Chandra Nayak Department of Studies in Applied Botany and Biotechnology, University of Mysore, Mysore, India

Mamta Nehra Nutritional Improvement of Crops Group, International Centre for Genetic Engineering and Biotechnology (ICGEB), New Delhi, India

R. S. Pal ICAR-Vivekananda Parvatiya Krishi Anusandhan Sansthan, Almora, India

Dharmateja Palaparthi Indian Agricultural Research Institute, New Delhi, India

B. S. Patil ICAR-IARI's Regional Research Centre, Dharwad, India

Rahul M. Phuke ICAR-Indian Agricultural Research Institute, Regional Station, Indore, India

Ganesan Prakash Division of Plant Pathology, ICAR-Indian Agricultural Research Institute, New Delhi, India

Dattatraya Hegde Radhika University of Horticultural Sciences, Bagalkote, Karnataka, India

Malireddy K. Reddy Crop Improvement Group, International Centre for Genetic Engineering and Biotechnology, New Delhi, India

Akshay S. Sakhare ICAR-Indian Institute of Rice Research, Hyderabad, Telangana, India

P. Senguttuvel ICAR-Indian Institute of Rice Research, Hyderabad, Telangana, India

Devender Sharma ICAR-Vivekananda Parvatiya Krishi Anusandhan Sansthan, Almora, India

Parul Sharma Punjab Agricultural University, Ludhiana, India

Rinku Sharma Division of Genetics, ICAR-Indian Agricultural Research Institute, New Delhi, India

Niveditha Shettigar ICAR-Research Complex for NEH Region, Umiam, Meghalaya, India

Sowmya Muntagodu Shreekanth Department of Genetics and Plant Breeding, University of Agricultural Sciences, GKVK, Bangalore, India

G. P. Singh ICAR—Indian Institute of Wheat and Barley Research, Karnal, India

Rajender Singh ICAR—Indian Institute of Wheat and Barley Research, Karnal, India

Sonia Khan Sony Nutritional Improvement of Crops Group, International Centre for Genetic Engineering and Biotechnology (ICGEB), New Delhi, India

K. V. Sripathy Regional Centre, ICAR-Indian Institute of Seed Science, Bangalore, India

N. C. Sunitha University of Agricultural Sciences, Bangalore, Karnataka, India

Ashwini Talakayala Plant Molecular Biology Laboratory, Agri Biotech Foundation, Hyderabad, India
Department of Genetics, Osmania University, Hyderabad, India

Swapnil D. Tayde Dr. Panjabrao Deshmukh Krishi Vidyapeeth, Akola, India

Arulprakash Thangaraj Nutritional Improvement of Crops Group, International Centre for Genetic Engineering and Biotechnology (ICGEB), New Delhi, India

Ratan Tiwari ICAR—Indian Institute of Wheat and Barley Research, Karnal, India

Rakhi Tomar Division of Genetics, ICAR-Indian Agricultural Research Institute, New Delhi, India

Sripada Udupa International Centre for Agricultural Research in the Dry Areas, Rabat, Morocco

Palanisamy Veeraya Division of Genetics, ICAR-Indian Agricultural Research Institute, New Delhi, India

Rachana Verma Nutritional Improvement of Crops Group, International Centre for Genetic Engineering and Biotechnology (ICGEB), New Delhi, India

Zeenat Wadhwa ICAR—Indian Institute of Wheat and Barley Research, Karnal, India

Rajbir Yadav Indian Agricultural Research Institute, New Delhi, India

K. R. Yathish Winter Nursery Centre, ICAR-Indian Institute of Maize Research, Hyderabad, India

Next-Generation Breeding Approaches for Stress Resilience in Cereals: Current Status and Future Prospects

1

Mallana Gowdra Mallikarjuna, Palanisamy Veeraya, Rakhi Tomar, Shailendra Jha, S. Chandra Nayaka, H. C. Lohithaswa, and Viswanathan Chinnusamy

Abstract

Cereals have predominantly been used as a staple food since time immemorial and contribute more than 50% caloric requirement of the global population. By 2050, an increase of 70–100% in the cereal food supply is needed to feed the predicted 9.8 billion world population. However, globally, cereal productivity is adversely affected by numerous stresses, viz. droughts, waterlogging, cold and heat waves, insects, pests, and diseases. Further, climate change exacerbates biotic and abiotic stresses in cereal production systems. Depending on the crop growth stage and stress sensitivity, these stresses result in yield losses up to a cent per cent in cereals. Therefore, feeding and nourishing the generations in the era of climate change demands the development of stress-resilient cereal cultivars. Though numerous attempts were made in the pre-genomic era to improve cereal yield potential through conventional breeding techniques, the degree of success was less due to genetic instability, narrow genetic base, and non-availability of genes for tolerance/resistance in the germplasm. In the twenty-first century, we have now reached the 'genomics and editing' stage from the 'Mendelian era' of the nineteenth century. Additionally, with the integration of novel genomics, the

M. G. Mallikarjuna (✉) · P. Veeraya · R. Tomar · S. Jha
Division of Genetics, ICAR-Indian Agricultural Research Institute, New Delhi, India
e-mail: MG.Mallikarjuna@icar.gov.in

S. C. Nayaka
Department of Studies in Applied Botany and Biotechnology, University of Mysore, Mysore, India

H. C. Lohithaswa
Department of Genetics and Plant Breeding, University of Agricultural Sciences, GKVK, Bangalore, India

V. Chinnusamy
Division of Plant Physiology, ICAR-Indian Agricultural Research Institute, New Delhi, India

M. G. Mallikarjuna et al. (eds.), *Next-Generation Plant Breeding Approaches for Stress Resilience in Cereal Crops*, https://doi.org/10.1007/978-981-19-1445-4_1

next-generation plant breeding approaches are changing the course of plant breeding via understanding the genetics of traits and accelerating the genetic gain. Newly developed next-generation breeding tools, viz. genome-wide association studies, genomic prediction, genome editing, and accelerated generation advancement methodologies, showed promising results by enhancing the stress resilience in cereals with high yield potential. This introductory chapter aims to provide an overview of targeted next-generation breeding tools for enhancing stress resilience in cereals in the present context of climate change, pests, diseases, and abiotic stresses.

Keywords

Abiotic and biotic stresses · Cereals · Climate change · Next-generation breeding · NAM · MAGIC · Genomic selection · Genome editing · GWAS · Speed breeding

1.1 Introduction

Plant breeding is one of the important areas of crop sciences and has extensively contributed to food and nutritional security for a century. Plant breeding continuously alters the genetic architecture of crop plants to meet the current needs and demands of the population, viz. food, fodder, and industrial products. The ever-increasing human population is one of the major pressures on food and nutritional security. Additionally, the world population is estimated to reach about nine billion by 2050 (Keating et al. 2014), and the growth is expected to be denser in the regions of sub-Saharan Africa and South Asia, where the population income mainly depends on agriculture, but food insecurity is >20% (Lipper et al. 2014).

Cereals are a significant portion of dietary intake, and food security is directly and mainly associated with cereal production and supply. The dependency on cereals for food and nutritional requirements in developing countries is more pronounced than in developed countries. Nearly 60–80% of calories in developing and underdeveloped countries are directly derived from cereals compared to 30% in the developed world (Awika 2011). In general, globally, cereals share >65% of the total calorie and protein supply. Therefore, ensuring the cereals supply to a growing population is of prime importance in the changing climatic scenario. Approximately 70–100% increase over the existing cereals production is required to feed the predicted world population of nine billion by 2050 (Godfray et al. 2010). However, the existing and upcoming cereal production is relentlessly challenged by climate change effects like drought, heat waterlogging, frost, diseases, and pest outbreaks (Porter and Reay 2016). The ever-changing climate is expected to produce more devastating effects on the production of wheat, maize, and rice. The declining wheat yield is reported in many parts of the world, and prediction showed a 6% grain yield decline for every 1 °C increase in global mean temperature (Asseng et al. 2014; Porter and Reay 2016). Further, 10–25% yield decline for every 1 °C rise in temperature was reported in major staple crops, including wheat, maize, and rice (Deutsch et al. 2018). The increase in mean global temperature can alter the metabolic activities of insects and pathogens and subsequently increase the pest's food consumption rate (Zavala et al. 2008). On the other hand, for major cereals, the price hikes are estimated at 25–50% without considering climate change impacts (Rosegrant et al. 2013), and, it is going to be 60–97% with change in climate by 2050 (Hubert et al. 2010). These alarming projections could result in food and nutritional insecurity in the developing world and broaden nutritional disparity among the population. Many of these uncertain projections on the impact of climate change demand resilient cereal cultivars and management practices suitable to a diverse range of stresses (Arneth et al. 2019).

The development of stress-resilient cereal cultivars is the most sustainable and economical approach to overcome the above climate change-mediated hurdles to ensure the nutritional and food security to the growing population. The traditional breeding approaches significantly succeeded in delivering improved cereal cultivars with a focus on enhanced grain yield. However, traditional approaches have often failed to deliver the required genetic gain for complex traits like stress tolerance. In the last three decades, advancements in breeding science came with new-generation breeding tools to deal with complex inherited stress tolerance traits and through achieving higher genetic gain per unit time (Barabaschi et al. 2015; Barrangou and Dudley 2016; Sanchez-Garcia 2019). Next-generation plant breeding uses cutting-edge approaches precisely to develop sustainable crop varieties with high yield and resistance or tolerance to biotic and abiotic stresses, thus leading to the development of climatic resilience in crop plants. It encompasses various high-throughput approaches to map the genomic regions in developing the cultivars with enhanced genetic gain per unit time. The genome-wide association analysis (GWAS) resulted in the identification of genomic regions associated with stress-resilient quantitative traits, and their subsequent utility through appropriate breeding pipelines produced promising results (Xiao et al. 2017). Recently, rapid generation advancement tools like doubled haploids and speed breeding were employed in cereals to understand the genetic basis of various stresses and deliver the stress resilience genomic resources and cultivars in a very short period (Collard et al. 2017). The genomic selection (GS) or genome-wide selection (GWS) is a rapid selection approach and is essentially based on genomic estimated breeding values (GEBVs) calculated from genome-wide markers data. The GEBVs are considered for judging the worth of genotypes for subsequent selection. The GS has an advantage over QTL-based breeding approaches in capturing all the minor and major effect loci for target trait variation (Crossa et al. 2011; Shikha et al. 2017; Sweeney et al. 2019). Genome editing is one of the latest additions to next-generation plant breeding approaches in cereals. Genome editing manipulates gene functions associated with various biotic and abiotic responses in cereals (Ansari et al. 2020). Further, integrating these novel genomics-based and next-generation breeding techniques will accelerate stress-resilient cereals breeding and increase the genetic gain under different production systems.

1.2 Mapping of Genomic Regions for Stress Tolerance in Cereals: Genome-Wide Association Analysis (GWAS) Approaches

1.2.1 GWAS Approaches

The genome-wide association mapping (GWAS) relies on the significant association of a phenotypic trait and marker locus (Soto-cerda and Cloutier 2012). It is an approach that accounts for thousands of polymorphisms to evaluate the effect of quantitative trait loci (QTL). The GWAS overcomes various constraints of traditional QTL/gene mapping through high resolution, i.e. an ability to detect single nucleotide polymorphism (SNP) within gene responsible for phenotypic change (Brachi et al. 2011; Soto-cerda and Cloutier 2012). Additionally, the GWAS approach allows associating commonly occurring genetic variations with a phenotypic variation using samples from previously well-studied populations (Brachi et al. 2011). In plants, the concept of association analysis was applied in maize as a candidate gene-based association study (Thornsberry et al. 2001) and on a genome-wide scale (Beló et al. 2007). Since the invention, GWAS has been improved in terms of various statistical algorithms and target population structure to enhance the reliability of marker-trait associations. Presently, GWAS mapping is regularly employed in the genetic dissection of various complex traits in plants, including cereals. In addition to classical/general GWAS, several approaches were developed based on association principles to detect the marker-trait associations, viz. nested association mapping (NAM), multi-parent advanced generation intercross (MAGIC) and other approaches that relate with population structure such as genomic control (GC) (Soto-cerda and Cloutier 2012) and structured association (SA) (Pritchard et al. 2000) etc.

1.2.2 General/Classical Association Mapping

The general/classical association mapping approach is one of the original and efficient forms of GWAS and revealed various stress-associated genomic regions in cereals (Cullis et al. 1998; Xiao et al. 2017). Here, the natural population of a crop is scanned to identify the marker-trait associations using linkage disequilibrium (LD) present between the alleles of several loci in that natural population. Thus, LD is a non-random association of alleles at various loci and deciphered as unequal haplotype frequencies in the population. The LD is population shaped by various genetic (recombination, mutations, population structure [Q], kinship [K], etc.) and demographic factors ((Flint-Garcia et al. 2003; Gupta et al. 2005; Stich et al. 2005, 2007; Oraguzie et al. 2007). Occurrence of new mutations, population stratification, self-pollination, genetic isolation, populations with few founders, epistasis, selection, and kinships are significantly increasing the LD, whereas, significant LD reduction was observed with high recombination and mutation rates, outcrossing, and gene conversion (Flint-Garcia et al. 2003; Gupta et al. 2005). Successful and

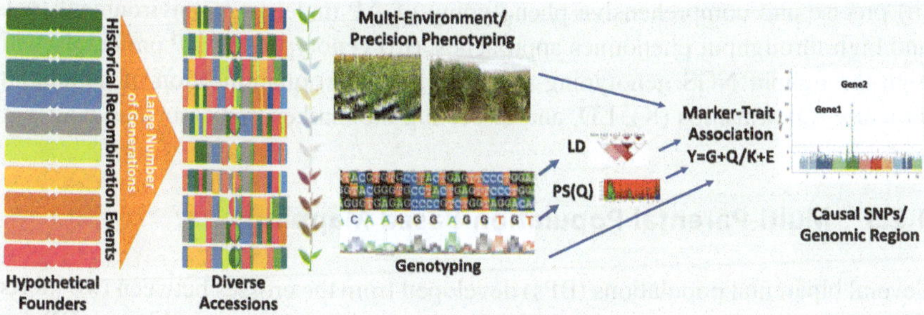

Fig. 1.1 The general pathway for one-step/classical GWAS mapping in plants. The diverse accessions of a plant species subjected to stringent precision phenotyping using multilocation trials/high-throughput phenomics and high-throughput genotyping. Subsequently, the appropriate statistical models, multiple-testing analyses, and software/packages are employed to detect the marker(s)-trait(s) association(s)

practical association for trait improvement arises through genetic linkage (Uitterlinden et al. 2005). Ideally, the kinship creates LD between linked loci, which also generate the LD among unlinked loci in a population with predominant parents (Stich et al. 2005). Additionally, the significant LD between the loci of the same or different chromosomes also arises because of spurious associations. Furthermore, the LD between unlinked loci arises through the 'hitchhiking effect' owing to population stratification and admixture, epistasis, selection of co-adapted loci, etc. (Cannon 1963; Wang et al. 2002; Stephan et al. 2006; Oraguzie et al. 2007).

In GWAS, the marker-trait associations based Q and K are superior as compared to ANOVA based associations in both self and cross-pollinated species (Yu et al. 2005; Stich et al. 2008). A major problem in GWAS mapping is false-positive results, which can be controlled by incorporating covariates for Q and K in mixed linear models. The mixed linear model (MLM) approach has been used to detect multiple levels of relatedness by utilizing random genetic markers. This method proved to control type I and II errors (Yu et al. 2005). Furthermore, in mixed model association-mapping approaches, the kinship matrix estimated by REML proved to be more appropriate than the Q-K method with respect to nominal alpha-level and adjusted power for detection of quantitative trait loci (Stich et al. 2008). Additionally, various studies have demonstrated the effectiveness of MLM over the general linear model (GLM) approach (Yu et al. 2005; Yu and Buckler 2006; Zhao et al. 2007; Raman et al. 2014). Eight different statistical models for association mapping have been compared for three traits ranging from single locus to multi-locus. Recently, the fixed and random model circulating probability unification (FarmCPU) method performed better than other methods in controlling false positives and false negatives in marker-trait association (Kaler et al. 2020).

The general steps of GWAS mapping in crops using germplasm accessions are summarized in pictorial form (Fig. 1.1). The major GWAS steps include (i) creation of association panel (AP) with diverse accessions, viz. landraces, elite cultivars, wild relatives, exotic accessions, etc., to capture the maximum variation in that crop,

(ii) precise and comprehensive phenotyping of AP under multi-environment trials and high-throughput phenomics approaches, (iii) genotyping of AP preferably with high-throughput NGS genotyping platforms, and (iv) quantification of population structure (Q), kinships (K), LD, and dissecting the marker-trait associations.

1.2.3 Multi-Parental Population-Based Mapping

Several biparental populations (BPs) developed from the crosses between two inbred lines have been widely used for QTL/gene mapping in the crops. Although BPs are simple to constitute and show lower LD decay, they lack mapping precision due to limited recombination events and diversity (Scott et al. 2020). Therefore, geneticists recently came up with multi-parent populations (MPPs) to overcome these limitations. The MPPs have been successfully developed and employed in gene mapping and breeding in various crops that include cereals (Huang et al. 2015; Cockram and Mackay 2018; Scott et al. 2020). The multi-parent population mainly constitutes two types of experimental designs: nested association mapping (NAM) and multi-parent advanced generation intercross (MAGIC) populations. The genetic population derived from NAM designs is subjected to joint inclusive composite interval mapping (JICIM). JICIM represents high phenotypic variance and high efficiency of QTL detection showing a 70% likelihood of detecting two distinct QTLs (Li et al. 2011). In maize, high-resolution mapping of the genomic regions associated with leaf architecture and quantitative resistance to leaf blight was accomplished with the development of the NAM population (McMullen et al. 2009; Kump et al. 2011).

1.2.3.1 Nested Association Mapping

NAM is an integrated approach that involves multi-parents and combines the advantages of both linkage mapping and association mapping approaches. NAM constitutes fixed lines (RILs/NILs/DHs) that have been developed by a combination of several diverse and a common founder parents. Individual RIL/NIL/DHs population from each cross of the donor-common founder parents together constitute the NAM population (Gireesh et al. 2021). The NAM design nests the historical LD within the new recombination and uses both historical LD present in a large number of diverse founder parents and recombination-derived LD in the process of population development (Yu et al. 2008; Nice et al. 2016). Systematic reshuffling of genomes of parental inbred lines and underlying mapping strategies during NAM-RIL/NAM-NILs development allows the detection of QTLs with high accuracy and efficiency (Buckler et al. 2009). Additionally, NAM allows capturing of rare alleles governing the traits of interest, which is otherwise difficult in the case of general or classical association mapping. On the other hand, NAM has greater precision and accuracy of mapping similar to association mapping but contrasting to biparental population-based QTL mapping approaches.

The selection of common and donor founders in NAM development is the most crucial step in developing the NAM population. In many cases, the elite or well-characterized inbred line is used as a common founder to derive the NAM. For instance, maize line B73 was used as a common founder parent owing to the availability of its reference genome sequence along with that B73 was well characterized in terms of genetics and basic research (Yu et al. 2008). On the other hand, donor founders should depict the maximum genetic diversity for target traits in the crop. Therefore, donor founders should be as diverse as possible and represent the maximum genetic diversity in the crop. The NAM population is more advantageous when created from a large number of parental inbreds (Stich 2009). Generally, ten diverse founder lines with large phenotypic diversity are preferred in NAM population development. The recombination events of a common founder with the diverse founders favour high-resolution QTL mapping (Xu et al. 2017). Additionally, the diverse nature of founder lines allows the NAM population for conducting multiple studies. At least dozen-plus studies were reported using the first NAM population developed in maize (Gireesh et al. 2021). NAM also allows the genetic improvement of a common founder through incorporating important traits from various diverse founders. In NAM, the maintenance of pedigree of each crosses and lines throughout population development and advancement to the next generation is another crucial step. Moreover, the unequal size of NAM subpopulations (separate NAM-RILs or NILs) during mapping results in uneven allelic distributions across the subpopulations which reduce the accuracy of QTL detection (Li et al. 2016b; Bu et al. 2021). Further, the population size is another important criterion as it is directly proportional to the power of QTL detection owing to increased recombination events with the number of individuals (Stich 2009; Cockram and Mackay 2018). The general NAM population development and mapping strategies are depicted in Fig. 1.2.

1.2.3.2 MAGIC Population

The MAGIC population is a collection of RILs generated with a complex crossing program with multi-parental lines. Basically, these populations are an extension of advanced intercrossed inbred lines. The MAGIC populations combine the advantages of both biparental population and large germplasm collection. The MAGIC is advantageous over AM in autogamous species (example: rice and wheat) where LD is extensive. In such autogamous species, LD based GWAS may not give higher precision; thus, these crops required advanced recombinant lines generated through broad variations. Thus, the major advantage of the MAGIC population is the creation of abundant genetic variation and fast linkage disequilibrium (LD) decay resulting in high efficiency and efficient QTL exploration (Huang et al. 2018). Additionally, MAGIC populations are also ideal for assessing the interactions of QTL $\times$ environment and epistatic effects. Further, the MAGIC population allows breeding lines with the combined genomic regions for multiple traits of interest (Cockram and Mackay 2018).

Development of MAGIC population includes four major steps: (1) founder selection, (2) mixing, (3) advanced intercrossing, and (4) inbreeding/selfing. The founder lines selection for MAGIC population development is based on genetic and/or phenotypic diversity among the lines or geographical origin of the materials

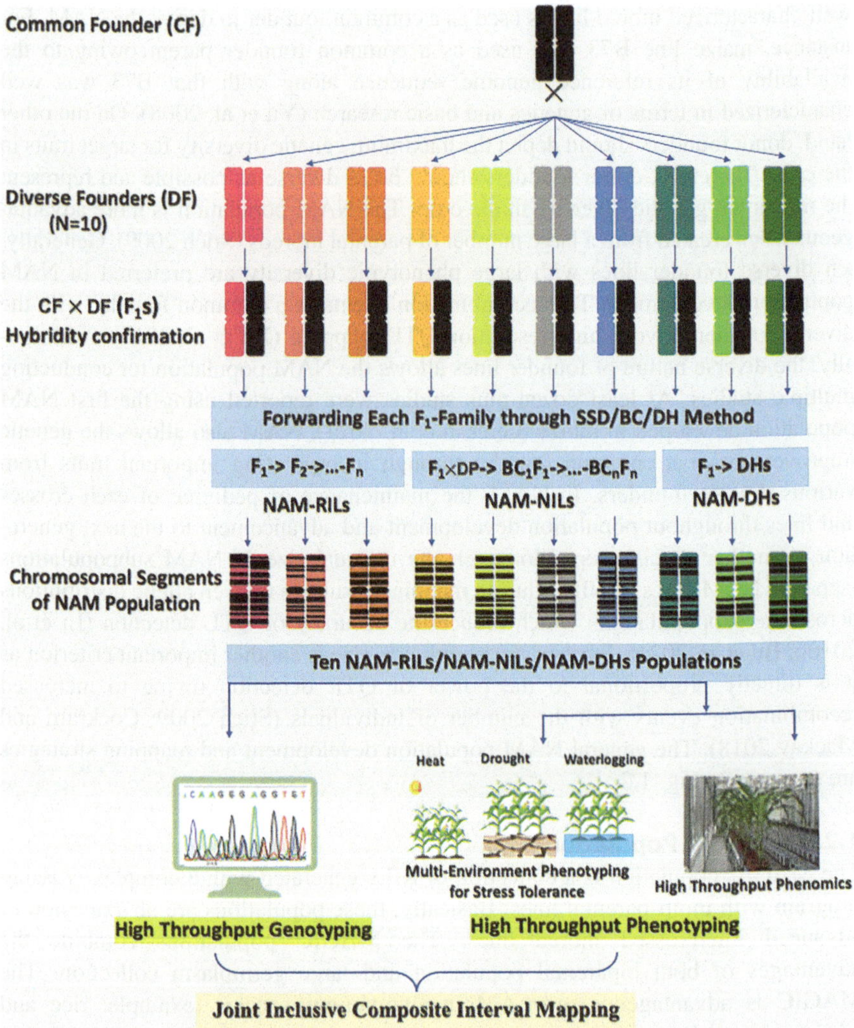

Fig. 1.2 General flow of NAM population design and mapping in cereals. The figure shows the development of the NAM population using ten separate diverse founder lines (DF) crossed with a common founder (CF), and the confirmed F_1s (CF × DF) are advanced via SSD, backcross, and DH technique to generate 10 NAM-RILs, NAM-NILs, and NAM-DHs, respectively. The ten NAM-RILs/NAM-NILs/NAM-DHs populations are subjected to precise multi-environment phenotyping and high-throughput genotyping for subsequent joint interval mapping analysis towards trait mapping

in diverse regions. The first stage of mixing of inbred lines is carried by intercrossing of multiple parents to form a broad genetic base. In advanced intercrossing, the first stage mixed lines from different funnels are randomly and sequentially intercrossed to enhance the number of recombinations in the population. Finally, in the inbreeding/selfing stage, lines resulting from the advanced intercrossing stage are

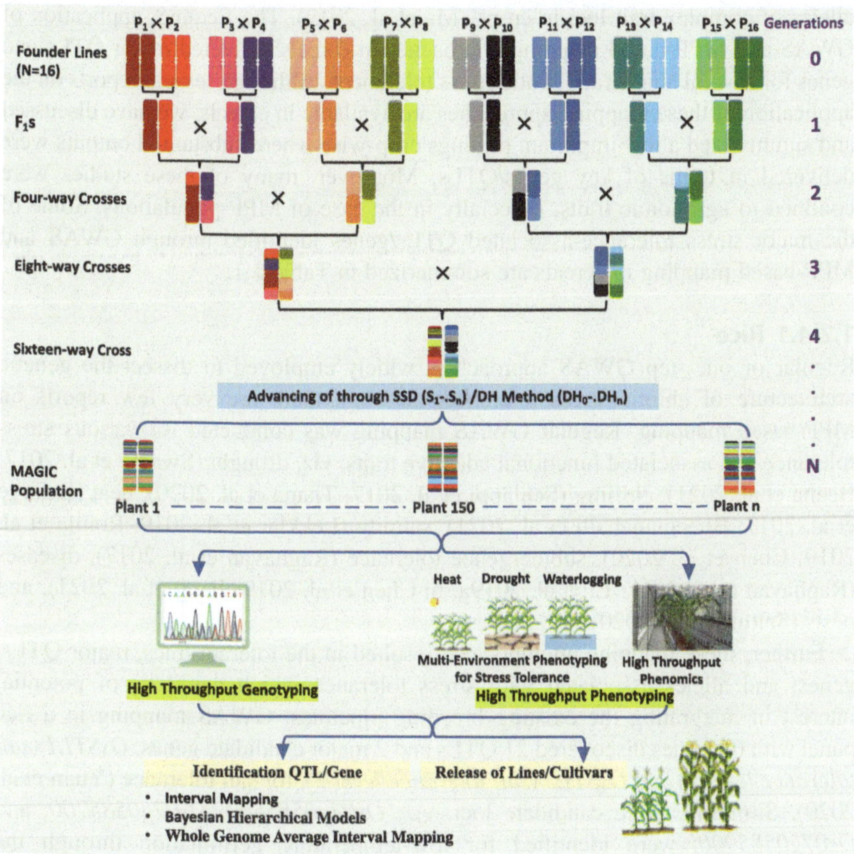

Fig. 1.3 MAGIC population development and mapping: The picture depicts the steps involved in MAGIC population development with sixteen founder lines followed by its utility for mapping of genomic regions and release of line or cultivars. P_n, parents; F_1, first filial generation; SSD, single seed descent method; DH, doubled haploid breeding; S_n, selfing generations; DH_n, DH generations

advanced to create homozygous lines through SSD/DH method (Huang et al. 2015). The representative general flow of MAGIC population development and mapping is provided in Fig. 1.3.

1.2.4 Mapping of Genomic Regions for Stress Tolerance through GWAS and Multi-Parental Population Approaches in Cereals

GWAS and MPP-based mapping approaches are the most powerful genetic tools to dissect multiple or complex trait loci related to biotic and abiotic stress tolerances and many other agronomic traits in plants (Challa and Neelapu 2018). These methods facilitated the discovery of critical stress-related genes and their favourable alleles of complex trait loci in crops (Ma et al. 2012). The frequent application of GWAS and MPP-based mapping approaches in cereals delivered major QTLs and genes for several abiotic and biotic stress tolerances. Although several reports on the application of these mapping approaches are available in cereals, we have discussed and summarized a few important findings crop-wise where substantial outputs were delivered in terms of key genes/QTLs. Moreover, many of these studies were confined to agronomic traits, especially in the case of MPP populations. Some of the major stress tolerance-associated QTLs/genes identified through GWAS and MPP-based mapping in cereals are summarized in Table 1.1.

1.2.4.1 Rice

Regular or one-step GWAS approach is widely employed to dissect the genetic architecture of abiotic stress tolerance, although there are very few reports on MPP-based mapping. Regular GWAS mapping was conducted for various stress tolerance and associated functional adaptive traits, viz. drought (Swamy et al. 2017; Beena et al. 2021), chilling (Schläppi et al. 2017; Thapa et al. 2020), heat (Lafarge et al. 2017; Bheemanahalli et al. 2021), salinity (Lekklar et al. 2019; Rohila et al. 2019; Chen et al. 2020), submergence tolerance (Raghavan et al. 2017), diseases (Raghavan et al. 2017; Li et al. 2019a, b; Chen et al. 2019; Jiang et al. 2021), and pests (Satturu et al. 2020).

Further, these mapping attempts also resulted in the tolerant lines, major QTLs, genes, and alleles associated with stress tolerance which could be of potential interest in integrating the existing breeding pipelines. GWAS mapping in a rice panel with 664 lines discovered 21 QTLs and 2 major candidate genes, *OsSTL1* (*salt tolerance level 1*) and *OsSTL2* (*salt tolerance level 2*), for salt tolerance (Yuan et al. 2020). Similarly, three candidate loci, viz. *Os07g0585500*, *Os07g0585700*, and *Os07g0585900*, were identified for low-temperature germination through the GWAS approach in a panel of 200 indica rice lines (Yang et al. 2020a). For disease tolerance, GWAS studies identified a new allele *Pikx* of *Pik* locus for blast resistance (Li et al. 2019a) and two LRR-containing loci (*Os01g0601625; Os01g0601675*) for Bakanae resistance (Chen et al. 2019). For drought tolerance, Xiong et al. (2018) mapped and characterized an ERF family TF *OsLG3* through GWAS. Additionally, GWAS mapping also reported two candidate genes, viz. *OsTCP19* and $OsNPF6.1^{HapB}$, for adaptation to soil nitrogen (Liu et al. 2021) and utilization efficiency (Tang et al. 2019), respectively.

Further, the MAGIC panels were employed to decipher the genomic regions associated with submergence tolerance and brown spot disease resistance (Raghavan et al. 2017) and important rice insect pest brown planthopper (*Nilaparvata lugens*) (Satturu et al. 2020). In the MAGIC panel of 1316 lines, a major QTL with >20% phenotypic variation for submergence tolerance was detected on chromosome 9, coinciding with *Sub1* QTL, whereas for brown spot disease tolerance, a major QTL with 34.42% phenotypic variation was reported on chromosome 12 (Raghavan et al. 2017). Multiple genome-wide association mapping in a MAGIC panel of 391 lines revealed 13 candidate genes for BPH resistance, including *NB-ARC*

Table 1.1 List of selected genes/alleles detected for various stress tolerance in cereals through GWAS and MPP-based mapping approaches

Crop	Gene/allele	Gene/allele description	Stress	Approach	Reference
Maize	*ZmNAC080308*	*NAC gene*	Drought	GWAS	Wang et al. (2021a)
	ZmRR1	*Response regulator 1*	Chilling	GWAS	Zeng et al. (2021)
	ZmFBL41	*F-box protein*	BLSB	GWAS	Li et al. (2019b)
	ZmPP2C-A	*Clade A PP2C phosphatases*	Drought	GWAS	Xiang et al. (2017)
	ZmCCoAOMT2	*Caffeoyl-CoA O-methyltransferase*	SLB, GLS	GWAS	Yang et al. (2017)
	ZmVPP1	*Vacuolar-type H^+-pyrophosphatase*	Drought	GWAS	Wang et al. (2016c)
	ZmNAC111	*NAC* gene	Drought	GWAS	Mao et al. (2015)
	ZmDREB2.7	*Dehydration responsive element binding proteins*	Drought	GWAS	Liu et al. (2013)
Rice	*OsTCP19*	*Member of TCP gene family*	ASN	GWAS	Liu et al. (2021)
	OsSTL1, OsSTL2	*Salt tolerance level 1 and 2*	Salt	GWAS	Yuan et al. (2020)
	Os10g22484; Os10g22520	–	LTG	GWAS	Yang et al. (2020a)
	Os01g0601625; Os01g0601675	LRR-containing genes	Bakanae	GWAS	Chen et al. (2019)
	OsNPF6.1^{HapB}	*Nitrate transporter*	NUE	GWAS	Tang et al. (2019)
	Pikx	Allele of R gene, *Pik* locus	Blast	GWAS	Li et al. (2019a)
	OsCD1	*Cadmium transporter*	CA	GWAS	Yan et al. (2019)
	OsSAP16	*Stress-associated protein 16*	LT	GWAS	Wang et al. (2018)
	OsLG3	ERF family TF	Drought	GWAS	Xiong et al. (2018)
	OsUGT706D1; OsUGT707A2	*Flavone -glucosyltransferase*	UV	GWAS	Peng et al. (2017)
Wheat	*TaRN1; TaRN2*	*Root number 1* and *root number 2*	Salinity	GWAS	Li et al. (2021)
	QPmlfl-1A (Pm3a)[a]	*Pm3a*	PM	MAGIC	Stadlmeier et al. (2018)
Barley	*Rrs1*[a], *Rrs17*[a], *Rrs18*[a]	Resistance to *Rhynchosporium*	Scald	NAM	Büttner et al. (2020)

(continued)

Table 1.1 (continued)

Crop	Gene/allele	Gene/allele description	Stress	Approach	Reference
	BOPA2_12_30822	*Alpha-glucosidase*	Salinity	NAM	Saade et al. (2016)

Note: *BLSB* banded leaf and sheath blight, *SLB* southern leaf blight, *GLS* gray leaf spot, *LT* low temperature, *LTG* low-temperature germination, *ASN* adaptation to soil nitrogen, *PM* powdery mildew, *UV* ultraviolet ray tolerance
[a]QTL for known gene(s)

domain-containing protein, *NHL repeat-containing protein*, *LRR containing protein*, and *WRKY70* (Satturu et al. 2020).

1.2.4.2 Wheat

The application of regular GWAS approaches is very complicated in wheat owing to genomic complexity and a high proportion of repetitive sequences as compared to other major cereal crops like rice (Gupta et al. 2019; Pang et al. 2020). However, advances in the mapping algorithms and availability of genome sequence data are encouraging geneticists for mapping novel genes in wheat. Recent GWAS with 90 K SNP-chip genotyping revealed four pleiotropic adult plant resistance QTLs, viz. Lr46/Yr29, QLr-2AL.1/QYr-2AL.1, QLr-2AL.2/QYr-2AL.2, and QLr-5BL/QYr-5BL.1, for leaf and yellow rust diseases (Zhang et al. 2021). Further, recent GWAS studies also identified potential QTLs for various other biotic stress tolerance in wheat, viz. spot blotch (Tomar et al. 2021), stem rust (Megerssa et al. 2020), tan spot (Galagedara et al. 2020), wheat blast (Juliana et al. 2020), etc. For abiotic stress tolerance, significant marker-trait associations with major effects were reported for drought tolerance (Schmidt et al. 2020b; Alahmad et al. 2020; Muhu-Din Ahmed et al. 2020; Maulana et al. 2020; Abou-Elwafa and Shehzad 2021), heat tolerance (Schmidt et al. 2020a; Abou-Elwafa and Shehzad 2021), salt tolerance (Chaurasia et al. 2020; Li et al. 2021), etc. Further, a recent GWAS report in wheat revealed two root trait associate candidate genes, viz. *TaRN1* and *TaRN2* assigning salinity tolerance (Li et al. 2021).

The wheat MAGIC populations are available with four (Huang et al. 2012; Rebetzke et al. 2014; Milner et al. 2016)- and eight-way (Huang et al. 2012; Mackay et al. 2014) crosses. The NAMs in wheat are available in both bread wheat (Bajgain et al. 2016; Ren et al. 2017; Jordan et al. 2018) and durum wheat (Kidane et al. 2019) background. The MPPs were also employed to dissect the genetic basis of stress tolerance and adaptive traits in wheat. The dissection of genomic regions for a stay-green trait in the wheat NAM population revealed the QTL parent-specific alleles in the target genomic regions and context-specific expression patterns (Christopher et al. 2021). The mapping for powdery mildew resistance in the MAGIC panel showed five genomic regions collectively explaining >70% phenotypic variations (PV) with a major QTL *QPmlfl-1A* (34% PV) coinciding with candidate gene *Pm3a* (Stadlmeier et al. 2018).

1.2.4.3 Maize

Maize is an ideal crop for GWAS owing to rapid LD decay, but very limited GWAS studies were conducted on dissecting the genetic basis of stress tolerance-associated traits. Although numerous general/classical GWAS studies were reported for various stress tolerance traits in maize, very few studies succeeded in identifying major QTLs, genes, or alleles associated with stress tolerance. For drought tolerance, toe *NAC* genes, viz. *ZmNAC080308* (Wang et al. 2021a) and *ZmNAC111* (Mao et al. 2015), *Clade A PP2C phosphatase* (*ZmPP2C-A*) (Xiang et al. 2017), *Vacuolar-type* H^+*-pyrophosphatase* (*ZmVPP1*) (Wang et al. 2016c), and a *dehydration responsive element binding protein* (*ZmDREB2.7*) (Liu et al. 2013), were identified through GWAS approach. Similarly, for biotic stress tolerance, GWAS studies resulted in *f-box protein* (*ZmFBL41*) for banded leaf and sheath blight (BLSB) (Li et al. 2019b) and *caffeoyl-CoA O-methyltransferase* (*ZmCCoAOMT2*) and for both southern leaf blight (SLB) and gray leaf spot (GLS) diseases (Yang et al. 2017). Compared to other cereals, in maize, MPPs were first developed and widely used for trait mapping. However, many MPP-based mappings were restricted to agronomic traits. Presently three NAM populations, viz. US-corn NAM with 25 diverse founders (Yu et al. 2008), Chinese-NAM with 11 founders (Li et al. 2015), and two European-NAM with 11 founders for each of dent and flint type (Bauer et al. 2013), are available for maize researchers. However, only 6 out of 20 plus mapping reports are related to stress tolerance (Gage et al. 2020). The recent mapping studies for drought tolerance in a Chinese-corn NAM population did not show any novel major genes (Li et al. 2016a). The MAGIC populations were developed in maize with eight (Dell'Acqua et al. 2015) and four founders (Anderson et al. 2018) lines. Interestingly, Dell'Acqua et al. (2015) showed major QTL for grain yield with pleiotropic effects for plant and ear height, which suggested the suitability of the MAGIC population for detecting stress-associated QTL with high precision. Similarly, the major genomic regions for plant and car height and flowering time in maize were detected on the reported candidate genes in four parents-based MAGIC populations (Mahan et al. 2018; Anderson et al. 2018).

1.3 Genomic Selection for Stress Resilience in Cereals

1.3.1 Genomic Selection

Many stress tolerance traits are genetically complex and are governed by multiple genes. Thus, the limited genetic gain has been achieved through marker-assisted selection (MAS). Further, the success of MAS is limited to a few major QTL and does not consider the minor QTLs during the selection process, which hampers the realized genetic gain (Dekkers 2004; Shikha et al. 2017). To overcome this limitation of MAS, a new marker-based 'genomic selection (GS)' was proposed to capture all the allelic variations distributed throughout the genome. Thus, the GS is described as a kind of MAS that concurrently assesses all the genome-wide distributed markers, markers effects, and haplotypes in order to determine genomic

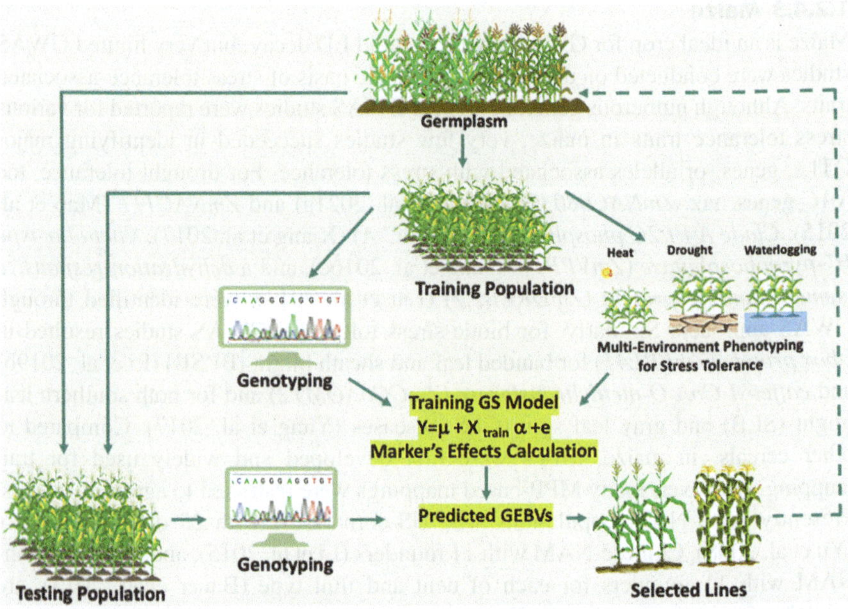

Fig. 1.4 General overview of genomic selection principle in crop plants. There are two steps in genomics selection. In the first stage, the marker effects/breeding values are estimated based on genotyping and phenotyping data points of the training population. In the second stage, the testing population, which consists of members of untested populations, is only genotyped. The individual's selection of testing population is based on their expected phenotypes predicted on the marker effects calculated in the training population

estimated breeding values (GEBVs) (Meuwissen et al. 2001; Dekkers 2007). The subsequent selections are completely based on these GEBVs (Nakaya and Isobe 2012). GS allows the rapid selection of superior genotypes and accelerates the breeding cycle. It's well-proven breeding technology in animal breeding and recently expanded in global plant breeding programmes, at a larger scale especially in the private sector.

The general outline of GS is depicted in Fig. 1.4. GS strategy uses two kinds of populations such as training population (TP) and testing/candidate population (T/CP). The TP encompasses the breeding lines for which detailed genotyping and high-quality phenotyping data for target trait(s) is available. On the other hand, T/CP might be part of TP or derived from the parental lines which are part of TP. The GEBVs are calculated for the training population based on the phenotyping and genotyping data with best-fitting statistical models. Later, the genotyping data generated on the T/CP population is used with the fitted GS models of TP to calculate the GEBVs to select the individuals in the T/CP population. The exclusion of phenotyping in GS reduces the selection time by almost half per cycle compared to the phenotypic selection (Lorenzana and Bernardo 2009). Thus, gain per unit cycle can be increased by replacing the phenotypic selection with the GEBVs (Wong and Bernardo 2008). The GS even work smoothly with a modest number of molecular markers and variable environments (Crossa et al. 2010). The effective use of GS in plant breeding programmes depends upon various factors such as breeding methodology, the number of target traits, genetic architecture and heritability of targeted traits, statistical models, availability of genotyping and phenotyping facilities, and the budget of the breeding program (Heffner et al. 2009; Jannink et al. 2010; Sweeney et al. 2019; Rahim et al. 2020). GS has the capacity to predict the complex traits associated with growth, yield, and biotic and abiotic stress tolerance and also allow breeders to use genome profile or phenotype independent of the underlying trait biology (Cabrera-Bosquet et al. 2012).

1.3.2 Application of GS for Stress Tolerance in Cereals

In cereals, GS is witnessing wider adaptation in enhancing the genetic gain for complex traits such as grain yield and stress tolerance-associated traits. Capturing variations of small-effect QTL-associated with stress tolerance makes GS the best selection method for traits like biotic and abiotic stress tolerance. Further, GS is amenable for both line selection and hybrid breeding (Cui et al. 2020; Xu et al. 2021). Therefore, the utility of GS is expanding to a greater extent in cereal breeding. In forthcoming crop-based chapters, the applications of GS for various stress tolerance are discussed in detail. Here, we have highlighted a few GS studies employed in major cereals for selected stresses.

In rice, the genomic prediction has been performed for various stress adaptive quantitative traits which resulted in moderate to high prediction accuracies (Xu et al. 2021). Under drought stress, employing reproducing kernel Hilbert space (RKHS) model with G × E interaction showed enhanced predictive ability up to 32% higher than single environment GS models in rice (Bhandari et al. 2019). Further, rice blast resistance prediction with the GBLUP model showed the prediction accuracies from 0.15 to 0.72 across the isolates (Huang et al. 2019). The arsenic concentration in rice is an important global concern in many rice-growing regions. The GS was employed to predict the arsenic tolerance in rice showed the prediction accuracies of 0.654 and 0.707 for grain arsenic content and grain yield, respectively. Further, the prediction accuracies with different weights to trait-specific markers in the genomic relationship matrix of single-environment models enhanced baseline performance by 32% (Ahmadi et al. 2021). Similarly, the moderate prediction accuracies of 0.43 and 0.48 were detected for flag leaf and dehulled grain arsenic content, respectively (Frouin et al. 2019).

In wheat, genomic prediction for *Fusarium* head blight (FHB) resistance showed moderate to high prediction accuracies, indicating the suitability of GS as a very promising breeding strategy for FHB resistance in wheat (Arruda et al. 2015; Dong et al. 2018). Further, GS was expanded for both FHB and *Septoria tritici* blotch (STB) in winter wheat which revealed better prediction accuracies for FHB (0.72); however, the low prediction accuracies (0.15) were observed for STB severity, which was also characterized by high genotype × environment variance (Herter et al. 2019). The GS was applied for leaf, stripe, and stem rusts resistance in wheat which resulted in low to high mean prediction accuracies. The prediction accuracies at the seedling stage were 0.31–0.74 and 0.70–0.78 for leaf and stripe rust, respectively, whereas, for adult plant resistance, prediction accuracies were 0.12–0.56, 0.31–0.65, and 0.34–0.71 for leaf, stem, and stripe rust, respectively (Juliana et al. 2017). Further, the GS with population historic datasets for stem rust resistance revealed that including historical data of close relatives increases and decreases the prediction accuracies by 11.9 and 12%, respectively, depending on the heritability of the target trait (Rutkoski et al. 2015). With these results, GS routes for molecular-based resistance breeding through capturing more of the variation due to small effect QTL for disease resistance (Heffner et al. 2009; Jannink et al. 2010).

GS for frost tolerance in bread wheat revealed prediction accuracies of 0.588 with basic GBLUP and 0.592 with weighted effects of frost-tolerant QTL with the GBLUP model. In the same study, the prediction accuracy for winter hardiness was found slightly better with the WBLUP model (0.410) compared to GBLUP (0.398). Further, the combined predictions for both frost tolerance and winter hardiness WBLUP models resulted in the highest prediction accuracy (0.596) among the investigated models (Michel et al. 2019). Similarly, GS in wheat for terminal drought stress revealed prediction accuracies from low (−0.32) to moderate (0.52) range (Shabannejad et al. 2021). The GS in four RILs population of durum wheat under drought tolerance revealed the use of training and validation populations in full sibs relationships as an effective strategy for enhancing prediction accuracies (0.35–0.47) of grain yield and incorporating target QTL as fixed effect in the models resulted in higher significant prediction accuracy in all the four RILs populations (Zaïm et al. 2020).

Maize is another important cereal crop where GS is most effectively employed in contemporary breeding programmes. GS in eight bi-parental mapping populations of maize under drought stress resulted in an average genetic gain of 0.086 Mg ha^{-1} grain yield per cycle of selection (Beyene et al. 2015). The comparative studies on the seven GS models, namely, ridge regression, LASSO, elastic net, random forest, reproducing kernel Hilbert space, Bayes A, and Bayes B, for their prediction accuracies of SNPs for drought-tolerant traits in maize revealed superior prediction accuracies of Bayes B (Shikha et al. 2017). Rapid cycle genomic selection was performed on two multi-parent yellow synthetic populations, MYS1 and MYS2, for drought and waterlogging stress tolerance. The study showed higher realized genetic gain for GS under drought stress (110 and 135 kg ha^{-1} $year^{-1}$) compared to waterlogging (38 and 113 kg ha^{-1} $year^{-1}$) in both MYS-1 and MYS-2 populations, respectively (Das et al. 2021). Further, across different growth stages and environments, GS for stalk strength in maize revealed better prediction accuracies with a multivariate model over univariate model and model with rind penetrometer resistance-associated loci as fixed effects (Liu et al. 2020b).

GS for maize common rust (*Puccinia sorghi*) in GWAS panel and DH population of tropical maize revealed genomic prediction accuracies of 0.61 and 0.51, respectively (Ren et al. 2021). Similarly, for Fusarium ear rot infection and mycotoxin fumonisin content, the maximum prediction accuracies for untested lines were 0.46 and 0.67, respectively (Holland et al. 2020).

1.4 Genome Editing for Stress Tolerance in Cereals

1.4.1 Genome Editing: A New Tool in Cereal Breeding

Genome editing refers to the precise alteration of genomic sequence(s) or gene (s) with the help of group of technologies called 'genome editing tools' to achieve the required changes in the phenotypic expression of target traits in an individual. Genome editing tools involve three general steps to achieve the desired changes in targeted genomic regions. The first step includes the engineering of exogenous nuclease with recognition module and nuclease domain to precisely recognize the target sequence in the genome. Secondly, the binding of exogenous nuclease to target sequence creates double-strand breaks (DSBs). Thirdly and finally, the mutations, viz. insertions and deletions, are inserted in the target regions during the process of DSB repair process by endogenous non-homologous end-joining (NHEJ) or homology-directed repair (HDR) pathways (Osakabe and Osakabe 2015; Wang et al. 2016b; Wada et al. 2020).

In plants, zinc finger nuclease (ZFN), transcription activator-like effector nuclease (TALEN), and clustered regularly interspaced short palindromic repeat (CRISPR)/CRISPR-associated protein9 nucleases are the three important genome editing tools (Mishra et al. 2021; Matres et al. 2021). The first two editing tools, viz. ZFNs and TALENs, use non-specific nuclease domain FokI. In ZFNs, an engineered array of zinc finger motifs with ~30 amino acids in a conserved ββα configuration which adds specificity (Carroll 2011), whereas TALENs are engineered with transcription activator-like effector (TALE) DNA-binding domains (Joung and Sander 2012). In both ZFN and TALEN tools, the target specificity is governed by pairs of ZFNs or TALENs to two closely spaced DNA sequences (Osakabe and Osakabe 2015, 2017). This space allows the Fok I nuclease domains linked to each of the ZFN/TALEN to create the DSBs in the 'spacer region' (Urnov et al. 2005; Christian et al. 2010; Wang and Qi 2016; Wang et al. 2016b). Compared to ZFNs, TALENs show high target specificity and low off-target edits. However, TALENs utility is limited owing to extensive repeat structure DNA-binding domains of TALENs, which create difficulty in designing and multiplexing (Mishra et al. 2021). The latest and third one is the CRISPR-*Cas9* system which uses two components, viz. *Cas9* protein and guide RNA (gRNA). The gRNA is a small RNA of ~20 nucleotides complementary to target sequences in the genome. The *Cas9* is an RNA-guided nuclease whose sequence specificity largely arises from Watson-Crick base pairing between its gRNA and the target DNA site, in addition to a direct interaction between *Cas9* and a short protospacer adjacent motif (PAM) of DNA in the genome (Wang and Qi 2016; Wang et al. 2016b). The discovery and recent improvements in CRISPR/Cas editing approach accelerated the basic research in plant genetics and opened new avenues in breeding climate-resilient cereal cultivars with enhanced grain yield and tolerance to biotic and abiotic stresses. The proof of concept, advanced breeding lines, and cultivars are being developed in various cereal crops (Xing et al. 2014; Wang et al. 2016a; Ahmad et al. 2020; Kuang et al. 2020). The general procedure of CRISPR/*Cas9* editing in cereals is summarized in Fig. 1.5.

1.4.2 Editing Cereals for Stress Tolerance

Several genome editing tools have been used to speed up the crop breeding in various crop plants. Genome editing shows advantages over traditional editing tools, being simple, efficient, highly specific, and amenable to multiplexing (Osakabe and Osakabe 2017; Ahmad et al. 2020; Wada et al. 2020; Matres et al. 2021). Some of the successful applications of CRISPR/*Cas9* editing tools for the creation of tolerant cereals to various biotic and abiotic stresses are tabulated in Table 1.2.

For drought tolerance, several genes were edited in rice and maize through CRISPR/*Cas9* approach. In rice, CRISPR/*Cas9* was employed to enhance the primary root growth and sensitivity to abscisic acid (ABA) treatment through the creation of frameshift mutations in *OsERA1* (Ogata et al. 2020), to induce the curled leaf phenotypes through knocking out of *SRL1* and *SRL2* genes (Liao et al. 2019). Similarly, CRISPR/*Cas9* targeted mutagenesis of the *Pyrabactin resistance-like9* (*OsPYL9*) gene enhanced drought tolerance and grain yield by regulating abiotic stress-responsive proteins and circadian rhythm (Usman et al. 2020). Mutant rice lines of *OsGA20ox-2* gene created through the CRISPR/*Cas9* tool also showed a tolerant response to drought stress under dehydration conditions (Mubarok et al. 2019). In maize, editing of *ARGOS8* improved maize grain yield under the drought stress environment (Shi et al. 2017). Further, CRISPR/*Cas9* edited lines for *ZmPARP2* gene exhibited higher leaf growth rates and biomass under water deficit stress (Njuguna et al. 2017).

For salinity stress tolerance, few of the genes were targeted to create salinity-tolerant cereal cultivars. The CRISPR-*Cas9*-induced drought and salt-tolerant (*OsDST*) gene mutants of MTU1010 rice cultivar exhibited high-level tolerance to drought and salt stresses by downregulating stomatal developmental genes, viz. *SPCH1*, *MUTE*, and *ICE1* (Santosh Kumar et al. 2020). Similarly, the CRISPR/*Cas9*-mediated knockout of *OsRR22* enhanced the salinity tolerance in rice (Zhang et al. 2019a). Additionally, the simultaneous editing of three genes, OsPIN5b, GS3, and OsMYB30, through CRISPR/*Cas9* system improved the panicle length, grain size, and cold tolerance, respectively (Zeng et al. 2020). Furthermore, decreased cold tolerance was reported in the mutant lines generated through CRISPR–*Cas9* system for *OsAnn3* gene, which suggests the key role of *OsAnn3* in cold tolerance (Shen et al. 2017). Additionally, herbicide-tolerant lines were generated in rice (Kuang et al. 2020) and wheat ((Zhang et al. 2019b) by editing the *Acetolactate synthase* (*ALS*) gene.

The CRISPR-*Cas9* tools also delivered appreciable results in the development of biotic stress-resilient cereals. In rice, bacterial blight tolerant lines were created

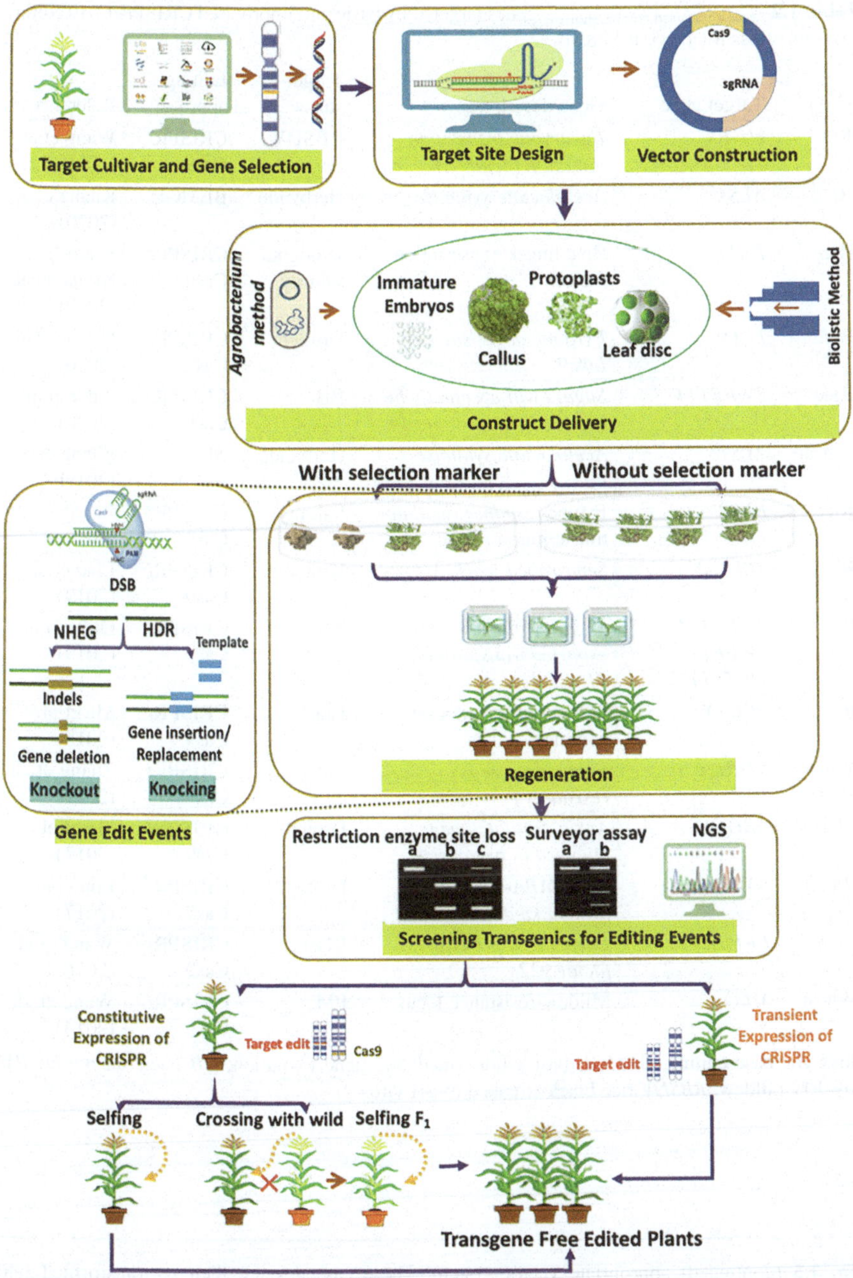

Fig. 1.5 A schematic representation of CRISPR/*Cas9* genome editing approach in cereals. The first step is to identify the cereal cultivar and target trait to be improved. In the next step, the most vulnerable target sites in the gene(s) for target traits are selected specifically using online available web resources for designing primers for complementary 20 nucleotides in the target gene/ sequences. The target sequence-specific sgRNA and *Cas9* cassettes are constructed in an

Table 1.2 List of selected studies showing genes targeted by basic and improvised CRISPR/*Cas9* editing tools for stress tolerance in cereals

Crop	Target gene	Gene description	Target stress	Editing approach	Reference
Rice	*eIF4G*	*Eukaryotic translation initiation factor 4G*	RBSDV	CRISPR/Cas9	Wang et al. (2021b)
Rice	*ALS1*	*Acetolactate synthase*	Herbicide	BEMGE	Kuang et al. (2020)
Rice	*DST*	Zinc finger transcription factor	Drought, salinity	CRISPR/Cas9	Santosh Kumar et al. (2020)
Rice	*PYL9*	*Pyrabactin resistance1-Like9*	Drought	CRISPR/Cas9	Usman et al. (2020)
Rice	*SWEET14*	*Sugars will eventually be exported transporters 14*	BB	CRISPR/Cas9	Zafar et al. (2020)
Wheat	*ALS1*	*Acetolactate synthase*	Herbicide	BE	Zhang et al. (2019b)
Rice	*RR22*	B-type *response regulator* transcription factor	Salinity	CRISPR/Cas9	Zhang et al. (2019a)
Rice	*SRL1, 2*	*Semi-rolled leaf1, 2*	Drought	CRISPR/Cas9	Liao et al. (2019)
Rice	*SWEET11, SWEET13, SWEET14*	*Sugars will eventually be exported transporters 11, 13, 14*	BB	CRISPR/Cas9	Oliva et al. (2019)
Rice	*SEC3A*	Subunit of the exocyst complex	Blast	CRISPR/Cas9	Ma et al. (2018)
Wheat	*EDR1*	*Enhanced disease resistance1*	PM	CRISPR/Cas9	Zhang et al. (2017)
Maize	*ARGOS8*	*Auxin-regulated gene involved in organ size*	Drought	CRISPR/Cas9	Shi et al. (2017)
Rice	*SAPK2*	*Stress/ABA–activated protein kinase 2*	Drought	CRISPR/Cas9	Lou et al. (2017)
Rice	*ERF922*	*Ethylene responsive factor 922*	Blast	CRISPR/Cas9	Wang et al. (2016a)
Wheat	*MLO*	Mildew-resistance locus	PM	CRISPR/Cas9	Wang et al. (2014)

Note: *BE* base-editing, *BEMGE* base-editing-mediated gene evolution, *BB* bacterial blight, *PM* powdery mildew, *RBSDV* rice black-streaked dwarf virus

Fig. 1.5 (continued) appropriate vector system. These cassettes are then co-transformed into embryo or protoplast or callus or leaf discs of target cultivar/genotype employing a suitable transformation method, viz. *Agrobacterium*-mediated or biolistic transformation. Identify the events with constitutive or transient expression of CRISPR. If the mutant plants are showing constitutive expression of CRISPR, self the events or cross with the wild plant to select the transgene-free edited plants among the segregants

through the editing of *SWEET* transporter genes (Oliva et al. 2019; Zafar et al. 2020). Similarly, for fungal diseases, *SEC3A* (Ma et al. 2018) and *ERF922* (Wang et al. 2016a) were edited to generate blast-resistant rice cultivars. In wheat mildew resistance locus (MLO) (Wang et al. 2014) and enhanced disease resistance1 (*EDR1*) gene (Zhang et al. 2017) were edited to generate powdery mildew resistant lines. Plant virus resistance is another important target in cereals editing. The recent report on the editing of *eukaryotic translation initiation factor 4G* (*eIF4G*) resulted in resistance to rice black-streaked dwarf virus (Wang et al. 2021b).

1.5 Rapid Generation Advancement Techniques for Stress Tolerance Breeding in Cereals

The development of pure or inbred lines to fix the additive genetic variation in crops is a fundamental requirement for the development of new cultivars in various breeding methods. Conventional approaches like single seed descent (SSD) and pedigree methods require 6–7 generations of inbreeding and selection from the first filial generations (F_1). Further, the additional requirement of the evaluation process to various biotic, abiotic, and agronomic traits adds up to 3–5 seasons. Therefore, to accelerate the genetic gain per unit time, various methods were suggested to quicken the generations/unit time, viz. shuttle breeding, mutation breeding, transgenic breeding, doubled haploid breeding and marker-assisted selection, speed breeding, etc. (Forster 2014). The principles and proof of concept for many of these rapid generation advancement (RGA) methods were discussed and demonstrated (Christopher et al. 2015; Song et al. 2017; Collard et al. 2017; Patial et al. 2019; Abdul Fiyaz et al. 2020; Rahim et al. 2020; Seguí-Simarro et al. 2021). However, practically utility in regular breeding was limited due to various constraints. Presently, two RGA methods, viz. doubled haploid (DH) technique and speed breeding (SB), are gaining importance and being integrated with major cereal breeding programmes in the world. Thus, DH and SB are discussed in the coming section with relevance to stress tolerance breeding in cereals.

1.5.1 Double Haploidy

Doubled haploids (DHs) are plants derived by doubling chromosomes of haploid plants, which are generated from single immature pollen grain or egg or crossing with haploid inducer lines. Various methods employed to create DHs are grouped into in vitro and in vivo methods. In vitro methods for DH production include tissue culture-based anther and microspore cultures and chromosome elimination followed by embryo rescue methods, whereas the present-day in vivo methods employ inducer lines to generate the haploid plants and subsequently follow doubling of chromosomes with colchicine treatment. The adaptation of each of the techniques is determined by crop type, genetic background, regeneration efficiency, lab facilities, and skilled human resources (Asif 2013). Many of the in vitro methods for DH production are discussed in detail in the forthcoming section on DHs. Here, we are limiting to widely adopted in vivo DH methods which are proved to be more reliable and efficient for high-throughput DH lines generation in maize and wheat (Niu et al. 2014; Chaikam et al. 2019).

1.5.1.1 In Vivo DH Production in Maize

The development of inducer stocks, identification of candidate genes associated with haploid induction, and refinement of DH production procedure are made in vivo DH production in maize as a popular RGA method in global maize breeding of public and private sectors (Liu et al. 2015, 2017; Kelliher et al. 2017; Gilles et al. 2017; Zhong et al. 2019; Chaikam et al. 2019; Meng et al. 2021). The in vivo production of maize lines involve mainly four major steps: (1) induction of haploids by crossing source with inducer line; (2) haploid identification at seed or seedling stage through visible markers; (3) doubling of chromosome in the selected haploids; and (4) production of DH lines by selfing of fertile doubled haploid plants (Chaikam et al. 2019). The general procedure of in vivo DH maize production is summarized in Fig. 1.6. In maize, the DH technology has been employed to generate various mapping populations in addition to line development.

The in vivo DH technology is being extensively utilized for the development of mapping populations for various stresses and line development for stress-resilient hybrid breeding. The QTL mapping study was conducted for grain yield and agronomic traits with five DH populations selected from Improved Maize for African Soils (IMAS) and the Water Efficient Maize for Africa (WEMA) panels of CIMMYT which identified major QTLs for various target traits with phenotypic

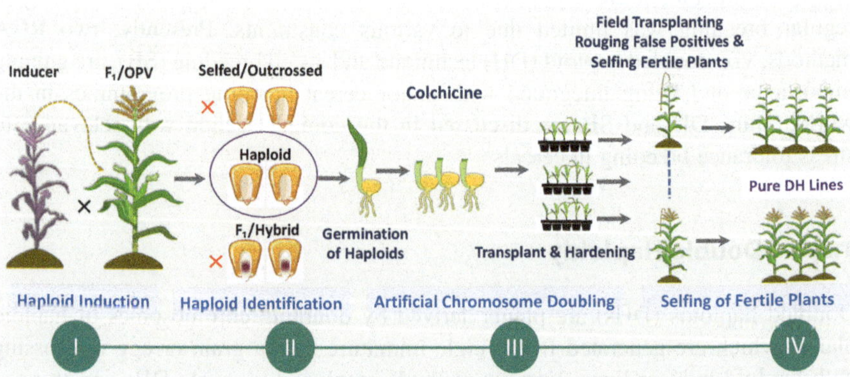

Fig. 1.6 General overview of in vivo maize doubled haploid technology. The sources to produce DH lines are mostly F1s produced from targeted crosses from selected parental lines. Firstly, the source/F_1s are crossed with an inducer line. Secondly, the putative haploid seeds are selected based on a visible phenotypic marker on seeds. In the third stage, the selected putative haploid plants are germinated and subjected to chromosome doubling through treating the seedlings with colchicine. Later, the colchicine-treated seedlings are planted in the greenhouse for hardening. In the fourth stage, plants are transplanted into the field. In the field, false positives are roughed out, and uniform DH plants are produced by selfing of each fertile plant

variations of 8.05–71.31% (Ertiro et al. 2020). Similarly, the QTL mapping in two DH populations (CML495 × LPSC7F64; CML451 × DTPYC9F46) under nitrogen starvation identified candidate genes for nitrogen stress-adoptive traits, viz. *pet1*, *hcf102,* and *spt2* for chlorophyll fluorescence, and *sweet15a* and/or *spt2* affecting chlorophyll content, auxin levels, and senescence (Liu et al. 2020c). Besides development of mapping populations, in vivo DH technology in maize is also used to develop and isolate the genetic stocks and lines showing resistance to various stresses. For instance, DH lines with *msv1* QTL were generated for maize streak virus resistance (Semagn et al. 2015).

Additionally, there are reports on the successful application of maize DH technology for the development of stress-resilient maize hybrids. The DH lines derived from BC_1F_1 of eight tropical maize populations belonging to different heterotic groups were crossed to generate stress-tolerant maize hybrids. Further, the ten best hybrids showed a grain yield of 1–1.4 t/ha under drought stress and 1.6–2.2 t/ha under optimum moisture conditions above the mean yield of the commercial checks (Beyene et al. 2013). Under optimum-moisture and random-drought conditions, two hybrids CKDHH1097 and CKDHH1090 derived from doubled haploid inbred lines showed 23% and 43% of higher grain yield over commercial checks, respectively (Sserumaga et al. 2018). Further, the testcross performance of DH lines showed grain yield of 8.15–8.85 t/ha under optimum moisture condition and 4.53–5.67 t/ha under drought stress conditions, while the best commercial variety showed the grain yield of 7.67 t/ha and 3.43 t/ha under optimum and drought stress conditions, respectively (Odiyo et al. 2014).

1.5.1.2 Wheat × Maize DH Technology in Wheat

In wheat, in vitro androgenesis (anther culture and microspore culture) and intergenic wheat × maize wide hybridization-based embryo culture are the most widely used procedures for the production of DH wheat lines. Among these, the intergenic wheat × maize wide hybridization method is more successful and popular among the wheat-breeding community owing to the rapid generation of DH homologous lines (Santra et al. 2017). The importance of wide hybridization in haploid induction was reported in 1984 (Zenkteler and Nitzsche 1984), followed by the production of haploid wheat plants through embryo rescue approach (Laurie and Bennett 1988). Pollination of wheat plants with maize pollens results in wheat egg fertilization and zygote formation. However, during the initial phase of zygote development, the haploid set of maize chromosomes is so unstable owing to failure to get attached with wheat spindle fibres, which makes the rapid loss of maize chromosomes rapidly after few cell division of embryo with haploid wheat chromosome set (Laurie and Bennett 1989). Further wheat × maize method for wheat DH lines production was responsible for refinement in the in vitro androgenic haploid induction (Bitsch et al. 1998; Sadasivaiah et al. 1999) and *Hordeum bulbosum*-based haploid induction methods (Suenaga 1994). The intergenic wheat × maize method for DH production in wheat comprises the following major seven steps: (1) selection of target wheat genotypes with segregating gametes, (2) emasculation of the selected wheat genotype flower, (3) manual pollination of the emasculated wheat flower with

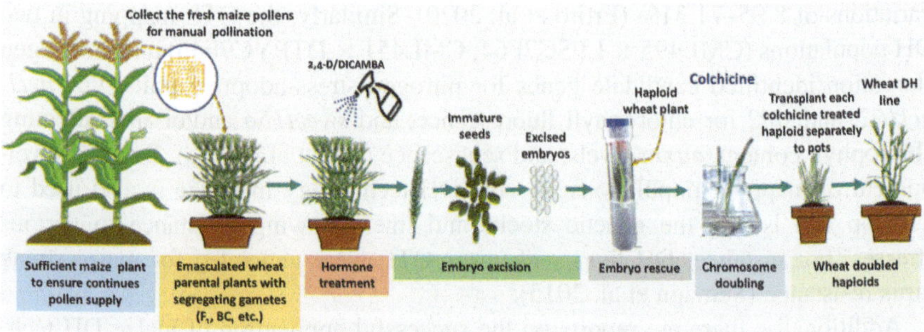

Fig. 1.7 General procedure of wheat × maize in vivo DH technology for wheat DH production

maize pollens, (4) hormone treatment of 2,4-D or Dicamba, (5) collection of embryos from immature seeds, (6) haploid plant regeneration through embryo culture, and (7) haploid plants regeneration and doubling of chromosomes (Niu et al. 2014) (Fig. 1.7).

The wheat × maize hybridization DH technique resulted in several mapping populations and wheat varieties for various basic researches and farming communities (Depauw et al. 2011; Niu et al. 2014). Few recent case reports of potential DH applications are mentioned here. Many stress-resilient wheat DH lines developed from wheat × maize hybridization were released for commercial cultivation in various countries (Depauw et al. 2011; Sanchez-Garcia 2019; Patial et al. 2019). In Canada, various disease-resistant cultivars, viz. Prevail (Kumar et al. 2017), Magnet (Kumar et al. 2019), Viewfield (Cuthbert et al. 2019), Durafield (Singh et al. 2016), and Raymore (Singh et al. 2014) are the few wheat × maize system-based wheat DH lines released recently for commercial cultivation. Similarly, in Romania, four wheat cultivars, viz. Faur F, Glosa, Litera, and Miranda, with improved grain yield and stress resistance traits were released for commercial cultivation and were developed through wheat × maize DH system (Săulescu et al. 2012).

The mapping for chlorophyll content and fluorescence kinetics in a wheat DH population (Opata × SH223) developed through wheat × maize DH technology revealed a major QTL *QTc.wwc-1B-S11* on chromosome 1B with 10.09% phenotypic variation (Ilyas et al. 2014). Mapping of QTL for flag leaf senescence in DH population of Beaver × Soissons cross revealed concurrence of QTL for senescence on chromosomes 2B and 2D under both drought and optimum moisture environments (Verma et al. 2004). Under drought, mapping for drought-induced abscisic acid production in a DH population of Chinese Spring × SQ1 revealed underlying genomic region and marker loci on the long arm of chromosome 5A of wheat (Quarrie et al. 1994).

The wheat × maize DH technology was also employed to develop various wheat DH mapping populations to map the genomic regions for biotic stresses. Thirteen QTLs were identified for *Fusarium* head blight resistance in the DH population derived from AGS2060 × AGS2035 cross. Further, the study also revealed the consistently expressing new QTL on linkage groups 5A, 6B, and 7A (Aviles et al. 2020). The resistance for multiple races of loose smut was mapped in the DH population derived from loose smut-resistant Blackbird and susceptible Strongfield line. The study revealed the major QTL (QUt.spa-6B.2), explaining 74% of the phenotypic variation on chromosome 6B and two other QTL on 7A (QUt.spa-7A.2) and 3A (QUt.spa-3A.2) (Kumar et al. 2018). In durum wheat, four QTLs were mapped for *Claviceps purpurea* resistance in the DH population developed from Greenshank_RIL3 × AC Avonlea with phenotypic variations of 2.9–90% for various target traits in different test locations (Gordon et al. 2020). Besides diseases, the root-lesion nematode (*Pratylenchus thornei*) resistance was mapped in the DH population (Sokoll × Krichauff) which revealed eight QTL including the three major QTLs (*QRlnt.sk-2B.1*; *QRlnt.sk-2B.2*; *QRlnt.sk-6D.1*) with the phenotypic variation of >10% (Linsell et al. 2014).

1.5.2 Speed Breeding

Speed breeding is a set of technique to manipulate the environmental conditions to advance the crops to the next breeding generation as quickly as possible through accelerating the flowering and seed set process (Wanga et al. 2021). The concept of speed breeding evolved with the use of artificial light to addendum the inadequate sunlight in protected cultivation and in in vitro cultures (Siemens 1880; Pfeiffer 1926; Mpelkas 1980; Nakamura et al. 2000). Subsequently, the collaborative efforts by the National Aeronautics and Space Administration (NASA) and Utah State University (USU) on growing wheat in space station have led to the development of dwarf wheat variety 'USU-Apogee' (Bugbee and Koerner 1997). The success of NASA and USU inspired the scientists at the University of Queensland and the University of Sydney of Australia and John Innes Centre, United Kingdom, to improve the technique further and come up with a protocol for the rapid advancement of generations called 'speed breeding' (Watson et al. 2018). Speed breeding does not demand sophisticated in vitro conditions for plants growth. Here, the main principle of speed breeding is to grow the plants in controlled growth chambers, biotrons, or greenhouses with optimum light quality and intensity, day length, and temperature to accelerate physiological processes, viz. photosynthesis and flowering to shorten the seed-to-seed generation time (Ghosh et al. 2018; Chiurugwi et al. 2019).

Three methods, viz. speed breeding I, II and III, have been proposed to implement the speed breeding methods based on need and resources availability (Ghosh et al. 2018). The speed breeding I method uses a controlled environment chamber with a light supply of 360–380 μmol m^{-2} s^{-1} during vegetative stage and 490–500 μmol m^{-2} s^{-1} during adult stage for 22 h of photoperiod. The temperature regimes of 22 °C (photoperiod) and 17 °C (2 h dark period) were set with 70% humidity. The speed breeding II method is based on a temperature-controlled glasshouse fitted with high-pressure sodium vapour lamps. Here the temperatures are maintained at 22 °C during daytime, photoperiod duration of 22 h with light intensity 440–650 μmol m^{-2} $s^{-1}$of and 17 °C during the night. The speed breeding III method is based on low-cost insulated growth room of 3 m × 3 m × 3 m fitted with seven lightboxes and a 1.5 horsepower domestic air conditioner. Here the photoperiod of 12 h for 4 weeks and thereafter 18 h photoperiod with the temperature of 21 °C and light intensity of 210 to 260 μmol m^{-2} s^{-1} and at 50 cm above the pot from 340 to 590 μmol m^{-2} s^{-1} (Ghosh et al. 2018; Abdul Fiyaz et al. 2020).

Speed breeding reduces the cost and space required in the development of a large number of inbred lines. With the advent of speed breeding, a new concept of 'speed DUS testing' has come into the light, where it allows the combining of phenotyping with DNA markers for characterization to reduce the timeline for variety registration (Jamali et al. 2020; Yang et al. 2021). Speed breeding also facilitates the rapid transfer of genes for multiple target attributes into adapted cereal cultivars or pyramiding desirable traits in the background of elite cultivars (Hickey et al. 2017). In addition to cultivar development, the mapping population developed from speed breeding preserves the recessive alleles owing to the absence of natural and artificial forces. Therefore, the phenotyping and genotyping of the populations developed from speed breeding facilitate capturing a full spectrum of variations in target traits.

Speed breeding is presently employed as solo or in combination with other advanced accelerated breeding methods such as genomics−/marker-assisted selection, genome editing, and genomic predictions to develop advanced cultivars. Biotron-based speed-breeding technique was employed to transfer the *hst1* (*OsRR22*) gene from Kaijin into high-yielding Yukinko-mai background through marker-assisted selection within 17 months (Rana et al. 2019). A similar approach was also employed to transfer the low-amylose allele *Wx1–1* from Oborozuki to Akidawara (Tanaka et al. 2016). In barley, a very good proof of concept is demonstrated for combining multiple disease resistance with the help of speed breeding. Resistance for leaf rust, net, and spot forms of net blotch and spot blotch were successfully introgressed to scarlet cultivar from four donors, viz. NRB090683–1, NRB091033, NRB09108, and NRB091092, with eight generations in 2 years of time and modified backcrossing (Hickey et al. 2017). Similarly, in durum wheat, improvement of multiple quantitative stress adaptive traits such as seminal root angle, seminal root number, tolerance to crown rot, resistance to leaf rust, and plant height was undertaken with six generations per year (Alahmad et al. 2018). Integration of speed breeding with a rapid phenotyping approach resulted in the development of a NAM population of ~1000 wheat RILs (F_5) within 18 months and QTL for stay-green and root adaptive traits (Christopher et al. 2015).

1.6 Prospects for Next-Generation Breeding Approaches for Stress Resilience Breeding

The present global efforts are channelized to dissect the genetic architecture of agronomic and stress adaptive traits, analyse the impact of allelic variation on target traits, and catalogue allele variants to enable the cereal breeders to attain the required genetic gain. The advances in genomics and breeding platforms resulted in various genomic technologies to understand the genetic architecture of stress-resilient traits. Furthermore, these technologies contribute towards the characterization of cereal germplasm for various stresses. Further, the developments in statistical algorithms and breeding techniques are easing the metanalysis of various genomic regions and reducing the linkage drag for effective introgression of stress-resilient traits. These proven innovations are being popularly employed in many global and regional cereal breeding pipelines (Zhang et al. 2015; Nice et al. 2016; Aravind et al. 2017; Shikha et al. 2017; Pham et al. 2019; Satturu et al. 2020).

1.6.1 Development of MPPs and Panels for High-Resolution Mapping in Cereals

MPPs and association panels are very effective in mapping various stress-resilient genes in cereals. The MPPs are most useful in genetic analysis, construction and refining of linkage maps, and association analysis (Huang et al. 2015). However, efforts were limited in developing MPPs for various stresses in cereals. The future association mapping for climate resilience should be based on both global and regional diversities existing in cereals with next-generation genotyping to increase the mapping precision and utility in the breeding programmes. In rice and maize, three global NAM populations and only two in wheat were reported (Gireesh et al. 2021). Similarly, in the case of MAGIC populations, four were developed in rice, six in wheat, and two in maize (Huang et al. 2015; Satturu et al. 2020).

Furthermore, most MPPs were developed with founder lines showing broad variation for agronomic traits. Therefore, there is a need to develop global- and regional-specific MPPs populations in cereals with special emphasis on climate-resilient traits in addition to grain yield attributes. These panels and populations are needed to be made available to the global breeding community for extensive and intensive phenotyping for stress-resilient traits. Secondly, the generation of high-density genotyping with NGS for each population and making it available to the global research community will enhance the utilization of MPPs.

1.6.2 Creation and Expansion of High-Throughput Phenotyping and Genotyping Facilities for Stress Tolerance Breeding

The genetic architecture of many of the stress-resilient traits is complex and governed by several genes with significant G × E interactions. Therefore, precise phenotyping is the most essential and integral component of any advanced gene mapping and selection methods. Presently, the phenomics approaches allow non-destructive phenotyping of target traits under various stress environments such as drought, heat, salinity, etc., facilitating the precise genetic dissection of complex stress tolerance mechanisms (Yang et al. 2013, 2020b). Recently, quite a good number of phenomics reports for stress-resilient traits are being published in various cereals (Yang et al. 2013, 2020b; Rouphael et al. 2018; Schmidt et al. 2020a; Saade et al. 2020; Mertens et al. 2021; Kim et al. 2021). Therefore, integrating these precise phenomics approaches with the advanced mapping and selection methods greatly facilitates stress-resilient cereals breeding.

The advancement in genotyping platforms in the last two decades shifted the breeding programmes in cereal crops from simple phenotype-based selection to genomics-assisted selection. Although the cost of NGS is greatly reduced, still the cost of NGS-based genotyping is not affordable in many underdeveloped and developing countries. Therefore, creating centralized no-profit based genotyping facilities on a cluster basis and their effective functioning will speed up the adaptation of next-generation breeding approaches in cereals.

1.6.3 Integration of Next-Generation Plant Breeding Tools in Stress Tolerance Breeding Programmes

The improved genetic gain per unit time and resources by next-generation plant breeding tools was demonstrated in various crops, including cereals (Workshop and Breeders; Desta and Ortiz 2014; Watson et al. 2018; Hickey et al. 2019; Li et al. 2019b, 2021; Cui et al. 2020; Xu et al. 2021). Recently many authors showed that integration of next-generation breeding approaches complements each other in product delivery with rapid genetic gain for target traits. For example, the modified biotron-based speed breeding is integrated with MAS to introgress the salt-tolerant *hst1* gene from highly salt-tolerant Kaijin into Yukinko-mai (Rana et al. 2019). Further, it is proposed that integration of speed breeding with genomic selection or genome editing approach reduces the time required to deliver the cultivars compared to speed breeding alone (Hickey et al. 2019). DH technology could also be of potential interest in rapidly fixing the edited heterozygous genotypes. With the above proof of concept, there is a huge scope to employ the integrated-next-generation breeding methods for resource-efficient and rapid delivery of climate-resilient cereal cultivars.

1.6.4 Genome Editing for Stress-Tolerant Quantitative Traits

Editing plants for stress-tolerant quantitative traits is not a straightforward strategy like qualitative traits. The targeting of abiotic stress governed by multiple genes and adaptive signalling pathways necessitates the identification of master regulator gene (s) for stress tolerance. The knowledge generated on functional genomics and systems biology for various stress adaptive traits in cereals could be utilized to identify the master regulator gene(s) to edit for stress tolerance in cereals (Wasaki et al. 2006; Johnson et al. 2014; Mallikarjuna et al. 2016, 2020a, b; Aravind et al. 2017; Arora et al. 2017a, b; Mittal et al. 2017; Thirunavukkarasu et al. 2017; Liu et al. 2020a; Sun et al. 2020). Further, there is a need to improve the multiplexing efficiency of editing methods for simultaneous editing of targeted genes to alter the adaptive pathways for quantitative traits.

1.6.5 Policy Supports for Next-Generation Plant Breeding Tools

Policy and economic support by governments, especially in developing and under-developed countries, are most important for creating high-throughput phenotyping and genotyping facilities to accelerate the climate-resilient breeding of cereals. Additionally, there is an urgent need to define policies to regulate the release of cultivars developed through next-generation plant breeding approaches. In the case of genome-edited products, regulatory approaches are still evolving in many countries. The countries like the USA and Australia consider gene-edited crops as non-GMOs unless they contain foreign DNA, which means the edited crops entering the release pipeline are as the same as conventionally bred crops. On the other hand, in 2018, the European Union Court of Justice decided the gene-edited crops as GMOs. Thus, rules and regulatory procedures for releasing edited crops are like transgenic cultivars (Holme et al. 2019; Qaim 2020). Further, the policy decision needs to consider the minimum level of tolerance to important biotic and abiotic stress tolerance at least on a regional basis. Including minimum stress tolerance level in newly releasing cultivars not only enhances the stress tolerance breeding of cereals and reduces farmers' cost of cultivation but also ensures food and nutritional security during unwarranted occurrences of biotic and abiotic stresses in the region.

Acknowledgements MGM is grateful to the 'Early Career Research Award scheme' (ECR/2017/000675), Science and Engineering Research Board (SERB), Government of India, and the network project on 'Computational Biology and Agricultural Bioinformatics (Agril.Edn.14(44)/2014-A&P)' for financial support to research work on the stress genomics of maize. The funding agencies had no role in designing the study, data collection and analyses, the decision to publish, or the preparation of the manuscript.

References

Abdul Fiyaz RA, Ajay BC, Ramya KT, Kumar JA, Sundaram RM, Rao LS (2020) Speed breeding: methods and applications. In: Accelerated plant breeding, vol 1. Springer, Cham, pp 31–49

Abou-Elwafa SF, Shehzad T (2021) Genetic diversity, GWAS and prediction for drought and terminal heat stress tolerance in bread wheat (*Triticum aestivum* L.). Genet Resour Crop Evol 68:711–728. https://doi.org/10.1007/S10722-020-01018-Y/FIGURES/5

Ahmad S, Wei X, Sheng Z, Hu P, Tang S (2020) CRISPR/Cas9 for development of disease resistance in plants: recent progress, limitations and future prospects. Brief Funct Genomics 19:26–39. https://doi.org/10.1093/BFGP/ELZ041

Ahmadi N, Frouin J, Norton GJ, Price AH (2021) Genomic prediction of arsenic tolerance and grain yield in rice: Contribution of trait-specific markers and multi-environment models. Rice Sci 28: 268–278. https://doi.org/10.1016/J.RSCI.2021.04.006

Alahmad S, Dinglasan E, Leung KM, Riaz A, Derbal N, Voss-Fels KP, Able JA, Bassi FM, Christopher J, Hickey LT (2018) Speed breeding for multiple quantitative traits in durum wheat. Plant methods 14:1–5. https://doi.org/10.1186/S13007-018-0302-Y/TABLES/3

Alahmad S, Kang Y, Dinglasan E et al (2020) Adaptive traits to improve durum wheat yield in drought and crown rot environments. Int J Mol Sci 21:5260. https://doi.org/10.3390/IJMS21155260

Anderson SL, Mahan AL, Murray SC, Klein PE (2018) Four parent maize (fpm) population: effects of mating designs on linkage disequilibrium and mapping quantitative traits. Plant Genome 11: 170102. https://doi.org/10.3835/PLANTGENOME2017.11.0102

Ansari WA, Chandanshive SU, Bhatt V, Nadaf AB, Vats S, Katara JL, Sonah H, Deshmukh R (2020) Genome editing in cereals: approaches, applications and challenges. Int J Mol Sci 21: 4040. https://doi.org/10.3390/IJMS21114040

Aravind J, Rinku S, Pooja B, Shikha M, Kaliyugam S, Mallikarjuna MG, Kumar A, Rao AR, Nepolean T (2017) Identification, characterization, and functional validation of drought-responsive micrornas in subtropical maize inbreds. Front Plant Sci 8:941. https://doi.org/10.3389/FPLS.2017.00941

Arneth A, Denton F, Agus F et al (2019) Framing and context. In: Shukla P, Skea J, Calvo Buendia V et al (eds) Climate change and land: an IPCC special report on climate change, desertification, land degradation, sustainable land management, food security, and greenhouse gas fluxes in terrestrial ecosystems. IPCC, Geneva, pp 77–129

Arora K, Panda KK, Mittal S, Mallikarjuna MG, Thirunavukkarasu N (2017a) In silico characterization and functional validation of cell wall modification genes imparting waterlogging tolerance in maize. Bioinform Biol Insights 11:1177932217747277. https://doi.org/10.1177/1177932217747277

Arora K, Panda KK, Mittal S, Mallikarjuna MG, Rao AR, Dash PK, Thirunavukkarasu N (2017b) RNAseq revealed the important gene pathways controlling adaptive mechanisms under waterlogged stress in maize. Sci Rep. 7:1–12. https://doi.org/10.1038/S41598-017-10561-1

Arruda MP, Brown PJ, Lipka AE, Krill AM, Thurber C, Kolb FL (2015) Genomic selection for predicting head blight resistance in a wheat breeding program. Plant Genome 8:1–12. https://doi.org/10.3835/PLANTGENOME2015.01.0003

Asif M (2013) Progress and opportunities of doubled haploid production. Springer, Cham, pp 55–71

Asseng S, Ewert F, Martre P, Rötter RP, Lobell DB, Cammarano D, Kimball BA, Ottman MJ, Wall GW, White JW, Reynolds MP (2014) Rising temperatures reduce global wheat production. Nat Clim Change 5:143–147. https://doi.org/10.1038/NCLIMATE2470

Aviles AC, Harrison SA, Arceneaux KJ, Brown-Guidera G, Esten Mason R, Baisakh N (2020) Identification of QTLs for resistance to *fusarium* head blight using a doubled haploid population derived from Southeastern United States soft red winter wheat varieties AGS 2060 and AGS 2035. Genes 11:699. https://doi.org/10.3390/GENES11060699

Awika JM (2011) Major cereal grains production and use around the world. In: ACS symposium series. Am Chem Soc:1–13

Bajgain P, Rouse MN, Tsilo TJ, Macharia GK, Bhavani S, Jin Y, Anderson JA (2016) Nested association mapping of stem rust resistance in wheat using genotyping by sequencing. PLoS One 11:e0155760. https://doi.org/10.1371/JOURNAL.PONE.0155760

Barabaschi D, Tondelli A, Desiderio F, Volante A, Vaccino P, Valè G, Cattivelli L (2015) Next generation breeding. Plant Sci 242:3–13. https://doi.org/10.1016/J.PLANTSCI.2015.07.010

Barrangou R, Dudley EG (2016) CRISPR-based typing and next-generation tracking technologies. Annu Rev Food Sci Technol 7:395–411. https://doi.org/10.1146/ANNUREV-FOOD-022814-015729

Bauer E, Falque M, Walter H, Bauland C, Camisan C, Campo L, Meyer N, Ranc N, Rincent R, Schipprack W, Altmann T (2013) Intraspecific variation of recombination rate in maize. Genome Biol 14:1–7. https://doi.org/10.1186/GB-2013-14-9-R103

Beena R, Kirubakaran S, Nithya N et al (2021) Association mapping of drought tolerance and agronomic traits in rice (*Oryza sativa* L.) landraces. BMC Plant Biol 21:1–21. https://doi.org/10.1186/S12870-021-03272-3

Beló A, Zheng P, Luck S et al (2007) Whole genome scan detects an allelic variant of fad2 associated with increased oleic acid levels in maize. Mol Gene Genom 279:1–10. https://doi.org/10.1007/S00438-007-0289-Y

Beyene Y, Mugo S, Semagn K et al (2013) Genetic distance among doubled haploid maize lines and their testcross performance under drought stress and non-stress conditions. Euphytica 192: 379–392. https://doi.org/10.1007/S10681-013-0867-5

Beyene Y, Semagn K, Mugo S et al (2015) Genetic gains in grain yield through genomic selection in eight bi-parental maize populations under drought stress. Crop Sci 55:154. https://doi.org/10.2135/CROPSCI2014.07.0460

Bhandari A, Bartholomé J, Cao-Hamadoun TV et al (2019) Selection of trait-specific markers and multi-environment models improve genomic predictive ability in rice. PLoS One 14:e0208871. https://doi.org/10.1371/JOURNAL.PONE.0208871

Bheemanahalli R, Knight M, Quinones C et al (2021) Genome-wide association study and gene network analyses reveal potential candidate genes for high night temperature tolerance in rice. Sci Rep 11:1–17. https://doi.org/10.1038/S41598-021-85921-Z

Bitsch C, Gröger S, Lelley T (1998) Effect of parental genotypes on haploid embryo and plantlet formation in wheat x maize crosses. Euphytica 103:319–323. https://doi.org/10.1023/A:1018654000521

Brachi B, Morris GP, Borevitz JO (2011) Genome-wide association studies in plants: the missing heritability is in the field. Genome Biol 12:1–8. https://doi.org/10.1186/GB-2011-12-10-232/FIGURES/2

Bu S, Wu W, Zhang YM (2021) A multi-locus association model framework for nested association mapping with discriminating QTL effects in various subpopulations. Front Genet 11:1709. https://doi.org/10.3389/FGENE.2020.590012/BIBTEX

Buckler ES, Holland JB, Bradbury PJ et al (2009) The genetic architecture of maize flowering time. Science 325:714–718. https://doi.org/10.1126/SCIENCE.1174276

Bugbee B, Koerner G (1997) Yield comparisons and unique characteristics of the dwarf wheat cultivar 'USU-Apogee'. Adv Space Res 20:1891–1894. https://doi.org/10.1016/S0273-1177(97)00856-9

Büttner B, Draba V, Pillen K et al (2020) Identification of QTLs conferring resistance to scald (*Rhynchosporium commune*) in the barley nested association mapping population HEB-25. BMC Genom 21:1–12. https://doi.org/10.1186/S12864-020-07258-7/FIGURES/2

Cabrera-Bosquet L, Crossa J, von Zitzewitz J et al (2012) High-throughput phenotyping and genomic selection: the frontiers of crop breeding converge. J Int Plant Biol 54:312–320. https://doi.org/10.1111/J.1744-7909.2012.01116.X

Cannon GB (1963) The effects of natural selection of linkage disequilibrium and relative fitness in experimental populations of Drosophila melanogaster. Genetics 48:1201

Carroll D (2011) Genome engineering with zinc-finger nucleases. Genetics 188:773–782. https://doi.org/10.1534/GENETICS.111.131433

Chaikam V, Molenaar W, Melchinger AE, Boddupalli PM (2019) Doubled haploid technology for line development in maize: technical advances and prospects. Theor Appl Genet 132:3227–3243. https://doi.org/10.1007/S00122-019-03433-X/FIGURES/4

Challa S, Neelapu NRR (2018) Genome-wide association studies (GWAS) for abiotic stress tolerance in plants. In: Biochemical, physiological and molecular avenues for combating abiotic stress in plants, pp 135–150. https://doi.org/10.1016/B978-0-12-813066-7.00009-7

Chaurasia S, Singh AK, Songachan LS et al (2020) Multi-locus genome-wide association studies reveal novel genomic regions associated with vegetative stage salt tolerance in bread wheat (*Triticum aestivum* L.). Genomics 112:4608–4621. https://doi.org/10.1016/J.YGENO.2020.08.006

Chen SY, Lai MH, Tung CW, Wu DH, Chang FY, Lin TC, Chung CL (2019) Genome-wide association mapping of gene loci affecting disease resistance in the rice-*Fusarium fujikuroi* pathosystem. Rice 12:1–12. https://doi.org/10.1186/S12284-019-0337-3

Chen C, Norton GJ, Price AH (2020) Genome-wide association mapping for salt tolerance of rice seedlings grown in hydroponic and soil systems using the Bengal and Assam AUS panel. Front Plant Sci 11:1633. https://doi.org/10.3389/FPLS.2020.576479

Chiurugwi T, Kemp S, Powell W, Hickey LT (2019) Speed breeding orphan crops. Theor Appl Genet 132:607–616. https://doi.org/10.1007/S00122-018-3202-7

Christian M, Cermak T, Doyle EL et al (2010) Targeting DNA double-strand breaks with Tal effector nucleases. Genetics 186:757–761. https://doi.org/10.1534/GENETICS.110.120717

Christopher J, Richard C, Chenu K et al (2015) Integrating rapid phenotyping and speed breeding to improve stay-green and root adaptation of wheat in changing, water-limited, Australian environments. Procedia Environ Sci 29:175–176. https://doi.org/10.1016/J.PROENV.2015.07.246

Christopher M, Paccapelo V, Kelly A et al (2021) QTL identified for stay-green in a multi-reference nested association mapping population of wheat exhibit context dependent expression and parent-specific alleles. Field Crops Res 270:108181. https://doi.org/10.1016/J.FCR.2021.108181

Cockram J, Mackay I (2018) Genetic mapping populations for conducting high-resolution trait mapping in plants. Adv Biochem Eng Biotechnol 164:109–138. https://doi.org/10.1007/10_2017_48

Collard BCY, Beredo JC, Lenaerts B et al (2017) Revisiting rice breeding methods–evaluating the use of rapid generation advance (RGA) for routine rice breeding. Plant Prod Sci 20:337–352. https://doi.org/10.1080/1343943X.2017.1391705

Crossa J, Campos G, Perez P et al (2010) Prediction of genetic values of quantitative traits in plant breeding using pedigree and molecular markers. Genetics 186:713–724. https://doi.org/10.1534/GENETICS.110.118521

Crossa J, Pérez P, de los Campos G et al (2011) Genomic selection and prediction in plant breeding. J Crop Improv 25:239–261. https://doi.org/10.1080/15427528.2011.558767

Cui Y, Li R, Li G et al (2020) Hybrid breeding of rice via genomic selection. Plant Biotechnol J 18:57–67. https://doi.org/10.1111/PBI.13170

Cullis B, Gogel B, Verbyla A, Thompson R (1998) Spatial analysis of multi-environment early generation variety trials. Biometrics 54:1. https://doi.org/10.2307/2533991

Cuthbert RD, DePauw RM, Knox RE et al (2019) AAC viewfield hard red spring wheat. Can J Plant Sci 99:102–110. https://doi.org/10.1139/CJPS-2018-0147

Das RR, Vinayan MT, Seetharam K et al (2021) Genetic gains with genomic versus phenotypic selection for drought and waterlogging tolerance in tropical maize (*Zea mays* L.). Crop J 9(6): 1438–1448. https://doi.org/10.1016/J.CJ.2021.03.012

Dekkers JCM (2004) Commercial application of marker- and gene-assisted selection in livestock: strategies and lessons. The online version of this article, along with updated information and services, is located on the World Wide Web at: Commercial application of marker. J Anim Sci 82:E313–E328

Dekkers JCM (2007) Prediction of response to marker-assisted and genomic selection using selection index theory. J Anim Breed Genet 124:331–341. https://doi.org/10.1111/J.1439-0388.2007.00701.X

Dell'Acqua M, Gatti DM, Pea G et al (2015) Genetic properties of the MAGIC maize population: A new platform for high definition QTL mapping in *Zea mays*. Genome Biol 16:1–23. https://doi.org/10.1186/S13059-015-0716-Z/FIGURES/8

Depauw RM, Knox RE, Humphreys DG et al (2011) New breeding tools impact Canadian commercial farmer fields. Czech J Genet Plant Breed 47:28–34

Desta ZA, Ortiz R (2014) Genomic selection: genome-wide prediction in plant improvement. Trends Plant Sci 19:592–601. https://doi.org/10.1016/J.TPLANTS.2014.05.006

Deutsch CA, Tewksbury JJ, Tigchelaar M et al (2018) Increase in crop losses to insect pests in a warming climate. Science 361:916–919. https://doi.org/10.1126/SCIENCE.AAT3466

Dong H, Wang R, Yuan Y et al (2018) Evaluation of the potential for genomic selection to improve spring wheat resistance to fusarium head blight in the pacific northwest. Front Plant Sci 9:911. https://doi.org/10.3389/FPLS.2018.00911

Ertiro BT, Olsen M, Das B et al (2020) Genetic dissection of grain yield and agronomic traits in maize under optimum and low-nitrogen stressed environments. Int J Mol Sci. 21:543. https://doi.org/10.3390/IJMS21020543

Flint-Garcia SA, Thornsberry JM, Edward SB IV (2003) Structure of linkage disequilibrium in plants. Annu Rev Plant Biol 54:357–374. https://doi.org/10.1146/annurev.arplant.54.031902.134907

Forster BP (2014) Accelerated plant breeding. CAB Rev Perspect Agric Vet Sci Nutr Nat Resour 9: 1–16. https://doi.org/10.1079/PAVSNNR20149043

Frouin J, Labeyrie A, Boisnard A et al (2019) Genomic prediction offers the most effective marker assisted breeding approach for ability to prevent arsenic accumulation in rice grains. PLoS One 14:e0217516. https://doi.org/10.1371/JOURNAL.PONE.0217516

Gage JL, Monier B, Giri A, Buckler ES (2020) Ten years of the maize nested association mapping population: impact, limitations, and future directions. Plant Cell 32:2083–2093. https://doi.org/10.1105/TPC.19.00951

Galagedara N, Liu Y, Fiedler J et al (2020) Genome-wide association mapping of tan spot resistance in a worldwide collection of durum wheat. Theor Appl Genet 133:2227–2237. https://doi.org/10.1007/S00122-020-03593-1

Ghosh S, Watson A, Gonzalez-Navarro OE et al (2018) Speed breeding in growth chambers and glasshouses for crop breeding and model plant research. Nat Protoc 13:2944–2963. https://doi.org/10.1038/S41596-018-0072-Z

Gilles LM, Khaled A, Laffaire J-B et al (2017) Loss of pollen-specific phospholipase NOT LIKE DAD triggers gynogenesis in maize. EMBO J 36(707–717):10.15252/EMBJ.201796603

Gireesh C, Sundaram RM, Anantha SM et al (2021) Nested association mapping (NAM) populations: present status and future prospects in the genomics era. CRC Criti Rev. Plant Sci 40:49–67. https://doi.org/10.1080/07352689.2021.1880019

Godfray HCJ, Beddington JR, Crute IR et al (2010) Food security: the challenge of feeding 9 billion people. Science 327:812–818. https://doi.org/10.1126/SCIENCE.1185383

Gordon A, Mccartney C, Knox RE et al (2020) Genetic and transcriptional dissection of resistance to Claviceps purpurea in the durum wheat cultivar Greenshank. Theor Appl Genet 133:1873–1886. https://doi.org/10.1007/S00122-020-03561-9

Gupta PK, Rustgi S, Kulwal PL (2005) Linkage disequilibrium and association studies in higher plants: Present status and future prospects. Plant Mol Biol 57:461–485. https://doi.org/10.1007/S11103-005-0257-Z

Gupta PK, Kulwal PL, Jaiswal V (2019) Association mapping in plants in the post-GWAS genomics era. Adv Genet 104:75–154. https://doi.org/10.1016/BS.ADGEN.2018.12.001

Heffner EL, Sorrells ME, Jannink J-L (2009) Genomic selection for crop improvement. Crop Sci 49:1–12. https://doi.org/10.2135/CROPSCI2008.08.0512

Herter CP, Ebmeyer E, Kollers S et al (2019) An experimental approach for estimating the genomic selection advantage for *fusarium* head blight and *Septoria tritici* blotch in winter wheat. Theor Appl Genet 132:2425–2437. https://doi.org/10.1007/S00122-019-03364-7

Hickey LT, Germán SE, Pereyra SA et al (2017) Speed breeding for multiple disease resistance in barley. Euphytica 213:1–14. https://doi.org/10.1007/S10681-016-1803-2

Hickey LT, N Hafeez A, Robinson H et al (2019) Breeding crops to feed 10 billion. Nature Biotechnol 37:744–754. https://doi.org/10.1038/S41587-019-0152-9

Holland JB, Marino TP, Manching HC, Wisser RJ (2020) Genomic prediction for resistance to Fusarium ear rot and fumonisin contamination in maize. Crop Sci 60:1863–1875. https://doi.org/10.1002/CSC2.20163

Holme IB, Gregersen PL, Brinch-Pedersen H (2019) Induced Genetic variation in crop plants by random or targeted mutagenesis: convergence and differences. Front Plant Sci 10:1468. https://doi.org/10.3389/FPLS.2019.01468

Huang BE, George AW, Forrest KL et al (2012) A multiparent advanced generation inter-cross population for genetic analysis in wheat. Plant Biotechnol J 10:826–839. https://doi.org/10.1111/J.1467-7652.2012.00702.X

Huang BE, Verbyla KL, Verbyla AP et al (2015) MAGIC populations in crops: current status and future prospects. Theor Appl Genet 128:999–1017. https://doi.org/10.1007/S00122-015-2506-0

Huang C, Shen C, Wen T et al (2018) SSR-based association mapping of fiber quality in upland cotton using an eight-way MAGIC population. Mol Genet Genom 293:793–805. https://doi.org/10.1007/S00438-018-1419-4

Huang M, Balimponya EG, Mgonja EM et al (2019) Use of genomic selection in breeding rice (Oryza sativa L.) for resistance to rice blast (Magnaportheoryzae). Mol Breed 39:1–16. https://doi.org/10.1007/S11032-019-1023-2

Hubert B, Rosegrant M, van Boekel MAJS, Ortiz R (2010) The future of food: scenarios for 2050. Crop Sci 50:S33–S50. https://doi.org/10.2135/CROPSCI2009.09.0530

Ilyas M, Ilyas N, Gul A, Arshad M (2014) QTL mapping of wheat doubled haploids for chlorophyll content and chlorophyll fluorescence kinetics under drought stress imposed at anthesis stage. Pak J Bot 46(5):1889–1897

Jamali SH, Cockram J, Hickey LT (2020) Is plant variety registration keeping pace with speed breeding techniques? Euphytica 216:1–13. https://doi.org/10.1007/S10681-020-02666-Y

Jannink J-L, Lorenz AJ, Iwata H (2010) Genomic selection in plant breeding: from theory to practice. Brief Funct Genom 9:166–177. https://doi.org/10.1093/BFGP/ELQ001

Jiang N, Fu J, Zeng Q et al (2021) Genome-wide association mapping for resistance to bacterial blight and bacterial leaf streak in rice. Planta 253:1–16. https://doi.org/10.1007/S00425-021-03612-5

Johnson SM, Lim FL, Finkler A et al (2014) Transcriptomic analysis of *Sorghum bicolor* responding to combined heat and drought stress. BMC Genom 15:1–19. https://doi.org/10.1186/1471-2164-15-456

Jordan KW, Wang S, He F et al (2018) The genetic architecture of genome-wide recombination rate variation in allopolyploid wheat revealed by nested association mapping. Plant J 95:1039–1054. https://doi.org/10.1111/TPJ.14009

Joung JK, Sander JD (2012) TALENs: a widely applicable technology for targeted genome editing. Nat Rev. Mol Cell Biol 14:49–55. https://doi.org/10.1038/NRM3486

Juliana P, Singh RP, Singh PK et al (2017) Genomic and pedigree-based prediction for leaf, stem, and stripe rust resistance in wheat. Theor Appl Genet 130:1415–1430. https://doi.org/10.1007/S00122-017-2897-1

Juliana P, He X, Kabir MR et al (2020) Genome-wide association mapping for wheat blast resistance in CIMMYT's international screening nurseries evaluated in Bolivia and Bangladesh. Sci Rep 10:1–14. https://doi.org/10.1038/S41598-020-72735-8

Kaler AS, Gillman JD, Beissinger T, Purcell LC (2020) Comparing different statistical models and multiple testing corrections for association mapping in soybean and maize. Front Plant Sci 10:1794. https://doi.org/10.3389/FPLS.2019.01794

Keating BA, Herrero M, Carberry PS et al (2014) Food wedges: framing the global food demand and supply challenge towards 2050. Glob Food Sec 3:125–132. https://doi.org/10.1016/J.GFS.2014.08.004

Kelliher T, Starr D, Richbourg L et al (2017) MATRILINEAL, a sperm-specific phospholipase, triggers maize haploid induction. Nature 542(7639):105–109. https://doi.org/10.1038/NATURE20827

Kidane YG, Gesesse CA, Hailemariam BN et al (2019) A large nested association mapping population for breeding and quantitative trait locus mapping in Ethiopian durum wheat. Plant Biotechnol J 17:1380–1393. https://doi.org/10.1111/PBI.13062

Kim M, Lee C, Hong S et al (2021) High-throughput phenotyping methods for breeding drought-tolerant crops. Int J Mol Sci 22:8266. https://doi.org/10.3390/IJMS22158266

Kuang Y, Li S, Ren B et al (2020) Base-editing-mediated artificial evolution of *Osals1* in planta to develop novel herbicide-tolerant rice germplasms. Mol Plant 13:565–572. https://doi.org/10.1016/J.MOLP.2020.01.010

Kumar S, Fox SL, Humphreys DG et al (2017) AAC prevail Canada western red spring wheat. Can J Plant Sci 98:475–482. https://doi.org/10.1139/CJPS-2017-0193

Kumar S, Knox RE, Singh AK et al (2018) High-density genetic mapping of a major QTL for resistance to multiple races of loose smut in a tetraploid wheat cross. PLoS One 13:e0192261. https://doi.org/10.1371/JOURNAL.PONE.0192261

Kumar S, Fox SL, Mitchell Fetch J et al (2019) AAC magnet Canada western red spring wheat. Can J Plant Sci 99:988–996. https://doi.org/10.1139/CJPS-2019-0180

Kump KL, Bradbury PJ, Wisser RJ et al (2011) Genome-wide association study of quantitative resistance to southern leaf blight in the maize nested association mapping population. Nat Genet 43:163–168. https://doi.org/10.1038/NG.747

Lafarge T, Bueno C, Frouin J et al (2017) Genome-wide association analysis for heat tolerance at flowering detected a large set of genes involved in adaptation to thermal and other stresses. PLoS One 12:e0171254. https://doi.org/10.1371/JOURNAL.PONE.0171254

Laurie DA, Bennett MD (1988) The production of haploid wheat plants from wheat x maize crosses. Theor Appl Genet 76:393–397. https://doi.org/10.1007/BF00265339

Laurie DA, Bennett MD (1989) The timing of chromosome elimination in hexaploid wheat × maize crosses. Genome 32:953–961. https://doi.org/10.1139/G89-537

Lekklar C, Pongpanich M, Suriya-Arunroj D et al (2019) Genome-wide association study for salinity tolerance at the flowering stage in a panel of rice accessions from Thailand. BMC Genom 20:1–18. https://doi.org/10.1186/S12864-018-5317-2

Li H, Bradbury P, Ersoz E et al (2011) Joint QTL linkage mapping for multiple-cross mating design sharing one common parent. PLOS ONE 6:e17573. https://doi.org/10.1371/JOURNAL.PONE.0017573

Li C, Li Y, Bradbury PJ et al (2015) Construction of high-quality recombination maps with low-coverage genomic sequencing for joint linkage analysis in maize. BMC Biol 13:1–12. https://doi.org/10.1186/S12915-015-0187-4

Li C, Sun B, Li Y et al (2016a) Numerous genetic loci identified for drought tolerance in the maize nested association mapping populations. BMC Genom 17:1–11. https://doi.org/10.1186/S12864-016-3170-8

Li J, Bus A, Spamer V, Stich B (2016b) Comparison of statistical models for nested association mapping in rapeseed (Brassica napus L.) through computer simulations. BMC Plant Biol 16:1–17. https://doi.org/10.1186/S12870-016-0707-6

Li C, Wang D, Peng S et al (2019a) Genome-wide association mapping of resistance against rice blast strains in South China and identification of a new Pik allele. Rice 12:1–9. https://doi.org/10.1186/S12284-019-0309-7

Li N, Lin B, Wang H et al (2019b) Natural variation in ZmFBL41 confers banded leaf and sheath blight resistance in maize. Nat Genet 51:1540–1548. https://doi.org/10.1038/S41588-019-0503-Y

Li L, Peng Z, Mao X et al (2021) Genetic insights into natural variation underlying salt tolerance in wheat. J Exp Bot 72:1135–1150. https://doi.org/10.1093/JXB/ERAA500

Liao S, Qin X, Luo L et al (2019) CRISPR/Cas9-induced mutagenesis of semi-rolled leaf1,2 confers curled leaf phenotype and drought tolerance by influencing protein expression patterns and ROS scavenging in rice (*Oryza sativa* L.). Agronomy 9:728. https://doi.org/10.3390/AGRONOMY9110728

Linsell KJ, Rahman MS, Taylor JD et al (2014) QTL for resistance to root lesion nematode (Pratylenchusthornei) from a synthetic hexaploid wheat source. Theor Appl Genet 27:1409–1421. https://doi.org/10.1007/S00122-014-2308-9

Lipper L, Thornton P, Campbell BM et al (2014) Climate-smart agriculture for food security. Nat Clim Change 4:1068–1072. https://doi.org/10.1038/NCLIMATE2437

Liu S, Wang X, Wang H et al (2013) Genome-wide analysis of *ZMDREB* genes and their association with natural variation in drought tolerance at seedling stage of *Zea mays* L. PLoS Genet 9:e1003790. https://doi.org/10.1371/JOURNAL.PGEN.1003790

Liu C, Li W, Zhong Y et al (2015) Fine mapping of qhir8 affecting in vivo haploid induction in maize. Theor Appl Genet 128:2507–2515. https://doi.org/10.1007/S00122-015-2605-Y

Liu C, Li X, Meng D et al (2017) A 4-bp insertion at *ZmPLA1* encoding a putative phospholipase a generates haploid induction in maize. Mol Plant 10:520–522. https://doi.org/10.1016/J.MOLP.2017.01.011

Liu H, Able AJ, Able JA (2020a) Integrated analysis of small RNA, transcriptome, and degradome sequencing reveals the water-deficit and heat stress response network in durum wheat. Int J of Mol Sci 21:1–28. https://doi.org/10.3390/IJMS21176017

Liu X, Hu X, Li K et al (2020b) Genetic mapping and genomic selection for maize stalk strength. BMC Plant Biol 20:1–16. https://doi.org/10.1186/S12870-020-2270-4

Liu X, Yuan Y, Martinez C et al (2020c) Identification of QTL for early vigour and leaf senescence across two tropical maize doubled haploid populations under nitrogen deficient conditions. Euphytica 216:1–14. https://doi.org/10.1007/S10681-020-2577-0/

Liu Y, Wang H, Jiang Z et al (2021) Genomic basis of geographical adaptation to soil nitrogen in rice. Nature 590:600–605. https://doi.org/10.1038/S41586-020-03091-W

Lorenzana RE, Bernardo R (2009) Accuracy of genotypic value predictions for marker-based selection in biparental plant populations. Theor Appl Genet 120:151–161. https://doi.org/10.1007/S00122-009-1166-3

Lou D, Wang H, Liang G, Yu D (2017) OsSAPK2 confers abscisic acid sensitivity and tolerance to drought stress in rice. Front Plant Sci 8:993. https://doi.org/10.3389/FPLS.2017.00993

Ma Y, Qin F, Tran LSP (2012) Contribution of genomics to gene discovery in plant abiotic stress responses. Mol Plant 5:1176–1178. https://doi.org/10.1093/MP/SSS085

Ma J, Chen J, Wang M et al (2018) Disruption of *OsSEC3A* increases the content of salicylic acid and induces plant defense responses in rice. J Exp Bot 69:1051–1064. https://doi.org/10.1093/JXB/ERX458

Mackay IJ, Bansept-Basler P, Bentley AR et al (2014) An eight-parent multiparent advanced generation inter-cross population for winter-sown wheat: Creation, properties, and validation. G3: Genes. Genom Genet 4:1603–1161. https://doi.org/10.1534/G3.114.012963

Mahan AL, Murray SC, Klein PE (2018) Four-parent maize (FPM) population: Development and phenotypic characterization. Crop Sci 58:1106–1117. https://doi.org/10.2135/CROPSCI2017.07.0450

Mallikarjuna MG, Nepolean T, Mittal S et al (2016) In-silico characterisation and comparative mapping of yellow stripe like transporters in five grass species. Indian J Agric Sci 86:721–727

Mallikarjuna MG, Bhat J, Hossain F et al (2020a) Genetic enhancement of heat tolerance in maize through conventional and modern strategies. In: Heat stress in food grain crops: plant breeding and omics research. Bentham Science Publishers, pp 28–66. https://doi.org/10.2174/9789811473982120010004

Mallikarjuna MG, Thirunavukkarasu N, Sharma R et al (2020b) Comparative transcriptome analysis of iron and zinc deficiency in maize (*Zea mays* L.). Plants 9:–1812. https://doi.org/10.3390/PLANTS9121812

Mao H, Wang H, Liu S et al (2015) A transposable element in a NAC gene is associated with drought tolerance in maize seedlings. Nat Commun 6:1–13. https://doi.org/10.1038/NCOMMS9326

Matres JM, Hilscher J, Datta A et al (2021) Genome editing in cereal crops: an overview. Transgenic Res 30:461–498. https://doi.org/10.1007/S11248-021-00259-6

Maulana F, Huang W, Anderson JD, Ma XF (2020) Genome-wide association mapping of seedling drought tolerance in winter wheat. Front Plant Sci 11:1626. https://doi.org/10.3389/FPLS.2020.573786

McMullen MD, Kresovich S, Villeda HS et al (2009) Genetic properties of the maize nested association mapping population. Science 325:737–740. https://doi.org/10.1126/SCIENCE.1174320

Megerssa SH, Ammar K, Acevedo M et al (2020) Multiple-race stem rust resistance loci identified in durum wheat using genome-wide association mapping. Front Plant Sci 11:1934. https://doi.org/10.3389/FPLS.2020.598509

Meng D, Liu C, Chen S, Jin W (2021) Haploid induction and its application in maize breeding. Mol Breed 41:1–9. https://doi.org/10.1007/S11032-021-01204-5

Mertens S, Verbraeken L, Sprenger H et al (2021) Proximal hyperspectral imaging detects diurnal and drought-induced changes in maize physiology. Front Plant Sci 12:240. https://doi.org/10.3389/FPLS.2021.640914

Meuwissen THE, Hayes BJ, Goddard ME (2001) Prediction of total genetic value using genome-wide dense marker maps. Genetics 157:1819–1829. https://doi.org/10.1093/GENETICS/157.4.1819

Michel S, Löschenberger F, Hellinger J et al (2019) Improving and maintaining winter hardiness and frost tolerance in bread wheat by genomic selection. Front Plant Sci 10:1195. https://doi.org/10.3389/FPLS.2019.01195

Milner SG, Maccaferri M, Huang BE et al (2016) A multiparental cross population for mapping QTL for agronomic traits in durum wheat (Triticum turgidum ssp. durum). Plant Biotechnol J 14:735–748. https://doi.org/10.1111/PBI.12424

Mishra R, Zheng W, Joshi RK, Kaijun Z (2021) Genome editing strategies towards enhancement of rice disease resistance. Rice Sci 28:133–145. https://doi.org/10.1016/J.RSCI.2021.01.003

Mittal S, Mallikarjuna MG, Rao AR et al (2017) Comparative analysis of CDPK family in maize, Arabidopsis, rice, and sorghum revealed potential targets for drought tolerance improvement. Front Chem 5:115. https://doi.org/10.3389/FCHEM.2017.00115

Mpelkas CC (1980) Light Sources for Horticultural Lighting. IEEE Trans Ind Appl 16:557–565. https://doi.org/10.1109/TIA.1980.4503829

Mubarok H, Basunanda P, Santoso TJ (2019) Tolerance of T2 generation 'Kitaake' Rice (Oryza sativa L.) CRISPR/Cas9-OsGA20ox-2 mutant strains to drought condition. Ilmu Pertan Agric Sci 4(123):10.22146/IPAS.37032

Muhu-Din Ahmed HG, Sajjad M, Zeng Y et al (2020) Genome-wide association mapping through 90 k SNP array for quality and yield attributes in bread wheat against water-deficit conditions. Agriculture 10:392. https://doi.org/10.3390/AGRICULTURE10090392

Nakamura S, Senoh M, Nagahama S et al (2000) Blue InGaN-based laser diodes with an emission wavelength of 450 nm. Appl Phys Lett 76:22–24. https://doi.org/10.1063/1.125643

Nakaya A, Isobe SN (2012) Will genomic selection be a practical method for plant breeding? Ann Bot 110:1303–1316. https://doi.org/10.1093/AOB/MCS109

Nice LM, Steffenson BJ, Brown-Guedira GL et al (2016) Development and genetic characterization of an advanced backcross-nested association mapping (AB-NAM) population of wild × cultivated barley. Genetics 203:1453–1467. https://doi.org/10.1534/GENETICS.116.190736

Niu Z, Jiang A, Abu Hammad W et al (2014) Review of doubled haploid production in durum and common wheat through wheat × maize hybridization. Plant Breed 133:313–320. https://doi.org/10.1111/PBR.12162

Njuguna E, Coussens G, Aesaert S et al (2017) Modulation of energy homeostasis in maize and Arabidopsis to develop lines tolerant to drought, genotoxic and oxidative stresses. Afrika. Focus 30(66–76):10.21825/AF.V30I2.8080

Odiyo O, Njoroge K, Chemining G, Beyene Y (2014) Performance and adaptability of doubled haploid maize testcross hybrids under drought stress and non-stress conditions. Int Res J Agric Sci Plants 4:150–158. https://doi.org/10.14303/IRJAS.2014.055

Ogata T, Ishizaki T, Fujita M, Fujita Y (2020) CRISPR/Cas9-targeted mutagenesis of OsERA1 confers enhanced responses to abscisic acid and drought stress and increased primary root growth under nonstressed conditions in rice. PLoS One 15:1–12. https://doi.org/10.1371/JOURNAL.PONE.0243376

Oliva R, Ji C, Atienza-Grande G et al (2019) Broad-spectrum resistance to bacterial blight in rice using genome editing. Nature Biotechnol 37:1344–1350. https://doi.org/10.1038/S41587-019-0267-Z

Oraguzie NC, Rikkerink EHA, Gardiner SE, de Silva NH (2007) Association mapping in plants, 1st edn. Springer, New York, NY

Osakabe Y, Osakabe K (2015) Genome editing with engineered nucleases in plants. Plant Cell Physiol 56:389–400. https://doi.org/10.1093/PCP/PCU170

Osakabe Y, Osakabe K (2017) Genome editing to improve abiotic stress responses in plants, 1st edn. Elsevier Inc.

Pang Y, Liu C, Wang D et al (2020) High-resolution genome-wide association study identifies genomic regions and candidate genes for important agronomic traits in wheat. Mol Plant 13: 1311–1327. https://doi.org/10.1016/J.MOLP.2020.07.008

Patial M, Pal D, Thakur A et al (2019) Doubled haploidy techniques in wheat (*Triticum aestivum* L.): an overview. Proc Natl Acad Sci India Sect B Biol Sci 89:27–41. https://doi.org/10.1007/S40011-017-0870-Z

Peng M, Shahzad R, Gul A et al (2017) Differentially evolved glucosyltransferases determine natural variation of rice flavone accumulation and UV-tolerance. Nat Commun 8:1–12. https://doi.org/10.1038/S41467-017-02168-X

Pfeiffer NE (1926) Microchemical and morphological studies of effect of light on plants. Bot Gaz 81:173–195. https://doi.org/10.1086/333584

Pham AT, Maurer A, Pillen K et al (2019) Genome-wide association of barley plant growth under drought stress using a nested association mapping population. BMC Plant Biol 19:1–16. https://doi.org/10.1186/S12870-019-1723-0

Porter SD, Reay DS (2016) Addressing food supply chain and consumption inefficiencies: potential for climate change mitigation. Reg Environ Change 16:2279–2290. https://doi.org/10.1007/S10113-015-0783-4

Pritchard JK, Stephens M, Donnelly P (2000) Inference of population structure using multilocus genotype data. Genetics 155:945. https://doi.org/10.1093/GENETICS/155.2.945

Qaim M (2020) Role of new plant breeding technologies for food security and sustainable agricultural development. Appl Econ Perspect Policy 42:129–150. https://doi.org/10.1002/AEPP.13044

Quarrie SA, Gulli M, Calestani C et al (1994) Location of a gene regulating drought-induced abscisic acid production on the long arm of chromosome 5A of wheat. Theor Appl Genet 89: 794–800. https://doi.org/10.1007/BF00223721

Raghavan C, Mauleon R, Lacorte V et al (2017) Approaches in characterizing genetic structure and mapping in a rice multiparental population. G3: genes. Genom Genet 7:1721–1730. https://doi.org/10.1534/G3.117.042101

Rahim MS, Bhandawat A, Rana N et al (2020) Genomic selection in cereal crops: methods and applications. In: Gosal S, Wani S (eds) Accelerated plant breeding, vol 1. Springer, Cham, pp 51–88

Raman H, Raman R, Kilian A et al (2014) Genome-wide delineation of natural variation for pod shatter resistance in brassica napus. PLoS One 9:e101673. https://doi.org/10.1371/JOURNAL.PONE.0101673

Rana MM, Takamatsu T, Baslam M et al (2019) Salt tolerance improvement in rice through efficient snp marker-assisted selection coupled with speed-breeding. Int J Mol Sci 20:2585. https://doi.org/10.3390/IJMS20102585

Rebetzke GJ, Verbyla AP, Verbyla KL et al (2014) Use of a large multiparent wheat mapping population in genomic dissection of coleoptile and seedling growth. Plant Biotechnol J 12:219–230. https://doi.org/10.1111/PBI.12130

Ren Y, Hou W, Lan C et al (2017) QTL analysis and nested association mapping for adult plant resistance to powdery mildew in two bread wheat populations. Front Plant Sci 8:1212. https://doi.org/10.3389/FPLS.2017.01212

Ren J, Li Z, Wu P et al (2021) Genetic dissection of quantitative resistance to common rust (puccinia sorghi) in tropical maize (*Zea mays* L.) by combined genome-wide association study, linkage mapping, and genomic prediction. Front Plant Sci 12:1338. https://doi.org/10.3389/FPLS.2021.692205

Rohila JS, Edwards JD, McClung AM et al (2019) Identification of superior alleles for seedling stage salt tolerance in the USDA rice mini-core collection. Plants 8:472. https://doi.org/10.3390/PLANTS8110472

Rosegrant MW, Tokgoz S, Bhandary P (2013) The new normal? A tighter global agricultural supply and demand relation and its implications for food security. Am J Agric Econ 95:303–309. https://doi.org/10.1093/AJAE/AAS041

Rouphael Y, Spíchal L, Panzarová K et al (2018) High-throughput plant phenotyping for developing novel biostimulants: from lab to field or from field to lab? Front Plant Sci 9:1197. https://doi.org/10.3389/FPLS.2018.01197

Rutkoski J, Singh RP, Huerta-Espino J et al (2015) Efficient use of historical data for genomic selection: a case study of stem rust resistance in wheat. Plant Genom 8(09):0046. https://doi.org/10.3835/PLANTGENOME2014.09.0046

Saade S, Maurer A, Shahid M et al (2016) Yield-related salinity tolerance traits identified in a nested association mapping (NAM) population of wild barley. Sci Rep 6:32586. https://doi.org/10.1038/SREP32586

Saade S, Brien C, Pailles Y et al (2020) Dissecting new genetic components of salinity tolerance in two-row spring barley at the vegetative and reproductive stages. PLoS One 15:e0236037. https://doi.org/10.1371/JOURNAL.PONE.0236037

Sadasivaiah RS, Orshinsky BR, Kozub GC (1999) Production of wheat haploids using anther culture and wheat x maize hybridization techniques. Cereal Res Commun 27:33–40. https://doi.org/10.1007/BF03543916

Sanchez-Garcia M (2019) Genetic gains in wheat breeding and its role in feeding the world. Crop Breed Genet Genom 1(e190005):10.20900/CBGG20190005

Santosh Kumar V, Verma RK, Yadav SK et al (2020) CRISPR-Cas9 mediated genome editing of drought and salt tolerance (*OsDST*) gene in indica mega rice cultivar MTU1010. Physiol Mol Biol Plants 26:1099–1110. https://doi.org/10.1007/S12298-020-00819-W

Santra M, Wang H, Seifert S, Haley S (2017) Doubled haploid laboratory protocol for wheat using wheat-maize wide hybridization. Methods Mol Biol 1679:235–249. https://doi.org/10.1007/978-1-4939-7337-8_14

Satturu V, Vattikuti JL, Durga Sai J et al (2020) Multiple genome wide association mapping models identify quantitative trait nucleotides for *Brown Planthopper* (*Nilaparvata lugens*) resistance in MAGIC indica population of rice. Vaccines 8:608. https://doi.org/10.3390/VACCINES8040608

Săulescu NN, Ittu G, Giura A et al (2012) Results of using *Zea* method for doubled haploid production in wheat breeding at Nardi Fundulea - Romania. Rom Agric Res 29:3–8

Schläppi MR, Jackson AK, Eizenga GC et al (2017) Assessment of five chilling tolerance traits and GWAS mapping in rice using the USDA mini-core collection. Front Plant Sci 8:957. https://doi.org/10.3389/FPLS.2017.00957

Schmidt J, Claussen J, Wörlein N et al (2020a) Drought and heat stress tolerance screening in wheat using computed tomography. Plant Methods 16:1–12. https://doi.org/10.1186/S13007-020-00565-W

Schmidt J, Tricker PJ, Eckermann P et al (2020b) Novel alleles for combined drought and heat stress tolerance in wheat. Front Plant Sci 10:1–14. https://doi.org/10.3389/FPLS.2019.01800

Scott MF, Ladejobi O, Amer S et al (2020) Multi-parent populations in crops: a toolbox integrating genomics and genetic mapping with breeding. Heredity 125:396–416. https://doi.org/10.1038/S41437-020-0336-6

Seguí-Simarro JM, Jacquier NMA, Widiez T (2021) Overview of in vitro and in vivo doubled haploid technologies. Methods Mol Biol 2287:3–22. https://doi.org/10.1007/978-1-0716-1315-3_1

Semagn K, Beyene Y, Babu R et al (2015) Quantitative trait loci mapping and molecular breeding for developing stress resilient maize for Sub-Saharan Africa. Crop Sci 55:1449–1459. https://doi.org/10.2135/CROPSCI2014.09.0646

Shabannejad M, Bihamta MR, Majidi-Hervan E et al (2021) A classic approach for determining genomic prediction accuracy under terminal drought stress and well-watered conditions in wheat landraces and cultivars. PLoS One 16:e0247824. https://doi.org/10.1371/JOURNAL.PONE.0247824

Shen C, Que Z, Xia Y et al (2017) Knock out of the annexin gene OsAnn3 via CRISPR/Cas9-mediated genome editing decreased cold tolerance in rice. J Plant Biol 60:539–547. https://doi.org/10.1007/S12374-016-0400-1

Shi J, Gao H, Wang H et al (2017) ARGOS8 variants generated by CRISPR-Cas9 improve maize grain yield under field drought stress conditions. Plant Biotechnol J 15:207–216. https://doi.org/10.1111/PBI.12603

Shikha M, Kanika A, Rao AR et al (2017) Genomic selection for drought tolerance using genome-wide SNPs in maize. Front Plant Sci 8:550. https://doi.org/10.3389/FPLS.2017.00550

Siemens CW (1880) III. On the influence of electric light upon vegetation, and on certain physical principles involved. Proc R Soc Lond 30:210–219. https://doi.org/10.1098/RSPL.1879.0108

Singh AK, Clarke JM, Knox RE et al (2014) AAC Raymore durum wheat. Can J Plant Sci 94:1289–1296. https://doi.org/10.4141/CJPS-2014-048

Singh AK, Depauw RM, Knox RE et al (2016) AAC Durafield durum wheat. Can J Plant Sci 96:719–725. https://doi.org/10.1139/CJPS-2015-0262

Song J, Carver BF, Powers C et al (2017) Practical application of genomic selection in a doubled-haploid winter wheat breeding program. Mol Breed 37:1–15. https://doi.org/10.1007/S11032-017-0715-8

Soto-cerda BJ, Cloutier S (2012) Association mapping in plant genomes. Genetic diversity in plants 14:29–54

Sserumaga JP, Beyene Y, Pillay K et al (2018) Grain-yield stability among tropical maize hybrids derived from doubled-haploid inbred lines under random drought stress and optimum moisture conditions. Crop Pasture Sci 69:691–702. https://doi.org/10.1071/CP17348

Stadlmeier M, Hartl L, Mohler V (2018) Usefulness of a multiparent advanced generation intercross population with a greatly reduced mating design for genetic studies in winter wheat. Front Plant Sci 871:1825. https://doi.org/10.3389/FPLS.2018.01825

Stephan W, Song YS, Langley CH (2006) The hitchhiking effect on linkage disequilibrium between linked neutral loci. Genetics 172:2647–2663. https://doi.org/10.1534/GENETICS.105.050179

Stich B (2009) Comparison of mating designs for establishing nested association mapping populations in maize and *Arabidopsis thaliana*. Genetics 183:1525–1534. https://doi.org/10.1534/GENETICS.109.108449

Stich B, Melchinger AE, Frisch M et al (2005) Linkage disequilibrium in European elite maize germplasm investigated with SSRs. Theor Appl Genet 111:723–730. https://doi.org/10.1007/S00122-005-2057-X

Stich B, Melchinger AE, Piepho HP et al (2007) Potential causes of linkage disequilibrium in a European maize breeding program investigated with computer simulations. Theor Appl Genet 115:529–536. https://doi.org/10.1007/S00122-007-0586-1

Stich B, Möhring J, Piepho HP et al (2008) Comparison of mixed-model approaches for association mapping. Genetics 178:1745–1754. https://doi.org/10.1534/GENETICS.107.079707

Suenaga K (1994) Doubled haploid system using the intergeneric crosses between wheat (*Triticum aestivum*) and maize (*Zea mays*). Bull Nat Inst Agrobiol Res 9:83–139

Sun M, Huang D, Zhang A et al (2020) Transcriptome analysis of heat stress and drought stress in pearl millet based on Pacbio full-length transcriptome sequencing. BMC Plant Biol 20:1–15. https://doi.org/10.1186/S12870-020-02530-0

Swamy BPM, Shamsudin NAA, Rahman SNA et al (2017) Association mapping of yield and yield-related traits under reproductive stage drought stress in rice (*Oryza sativa* L.). Rice 10:1–13. https://doi.org/10.1186/S12284-017-0161-6/TABLES/4

Sweeney DW, Sun J, Taagen E, Sorrells ME (2019) Genomic selection in wheat. In: Applications of genetic and genomic research in cereals, pp 273–302. https://doi.org/10.1016/B978-0-08-102163-7.00013-2

Tanaka J, Hayashi T, Iwata H (2016) A practical, rapid generation-advancement system for rice breeding using simplified biotron breeding system. Breed Sci 66:542–551. https://doi.org/10.1270/JSBBS.15038

Tang W, Ye J, Yao X et al (2019) Genome-wide associated study identifies *NAC42*-activated nitrate transporter conferring high nitrogen use efficiency in rice. Nature Commun 10:1–11. https://doi.org/10.1038/S41467-019-13187-1

Thapa R, Tabien RE, Thomson MJ, Septiningsih EM (2020) Genome-wide association mapping to identify genetic loci for cold tolerance and cold recovery during germination in rice. Front Genet 11:22. https://doi.org/10.3389/FGENE.2020.00022

Thirunavukkarasu N, Sharma R, Singh N et al (2017) Genome-wide expression and functional interactions of genes under drought stress in maize. Int J Genom 2017:2568706. https://doi.org/10.1155/2017/2568706

Thornsberry JM, Goodman MM, Doebley J et al (2001) Dwarf8 polymorphisms associate with variation in flowering time. Nat Genet 28:286–289. https://doi.org/10.1038/90135

Tomar V, Singh D, Dhillon GS et al (2021) New QTLs for spot blotch disease resistance in wheat (*Triticum aestivum* L.) using genome-wide association mapping. Front Genet 11:1740. https://doi.org/10.3389/FGENE.2020.613217

Uitterlinden AG, Fang Y, van Meurs JBJ, Pols HAP (2005) Genetic vitamin D receptor polymorphisms and risk of disease. Vitamin D 2:1121–1157. https://doi.org/10.1016/B978-012252687-9/50071-1

Urnov FD, Miller JC, Lee YL et al (2005) Highly efficient endogenous human gene correction using designed zinc-finger nucleases. Nature 435:646–651. https://doi.org/10.1038/NATURE03556

Usman B, Nawaz G, Zhao N et al (2020) Precise editing of the ospyl9 gene by rna-guided cas9 nuclease confers enhanced drought tolerance and grain yield in rice (*Oryza sativa* L.) by regulating circadian rhythm and abiotic stress responsive proteins. Int J Mol Sci 21:7854. https://doi.org/10.3390/IJMS21217854

Verma V, Foulkes MJ, Worland AJ et al (2004) Mapping quantitative trait loci for flag leaf senescence as a yield determinant in winter wheat under optimal and drought-stressed environments. Euphytica 135:255–263. https://doi.org/10.1023/B:EUPH.0000013255.31618.14

Wada N, Ueta R, Osakabe Y, Osakabe K (2020) Precision genome editing in plants: state-of-the-art in CRISPR/Cas9-based genome engineering. BMC Plant Biol 20:234. https://doi.org/10.1186/S12870-020-02385-5

Wang F, Qi LS (2016) Applications of CRISPR Genome engineering in cell biology. Trends Cell Biol 26:875–888. https://doi.org/10.1016/J.TCB.2016.08.004

Wang W, Thornton K, Berry A, Long M (2002) Nucleotide variation along the *Drosophila melanogaster* fourth chromosome. Science 295:134–137. https://doi.org/10.1126/SCIENCE.1064521

Wang Y, Cheng X, Shan Q et al (2014) Simultaneous editing of three homoeoalleles in hexaploid bread wheat confers heritable resistance to powdery mildew. Nat Biotechnol 32:947–951. https://doi.org/10.1038/NBT.2969

Wang F, Wang C, Liu P et al (2016a) Enhanced rice blast resistance by CRISPR/Cas9-targeted mutagenesis of the erf transcription factor gene *OsERF922*. PLoS One 11:e0154027. https://doi.org/10.1371/JOURNAL.PONE.0154027

Wang H, la Russa M, Qi LS (2016b) CRISPR/Cas9 in genome editing and beyond. Annu Rev. Biochem 85:227–264. https://doi.org/10.1146/ANNUREV-BIOCHEM-060815-014607

Wang X, Wang H, Liu S et al (2016c) Genetic variation in *ZmVPP1* contributes to drought tolerance in maize seedlings. Nat Genet 48:1233–1241. https://doi.org/10.1038/NG.3636

Wang X, Zou B, Shao Q et al (2018) Natural variation reveals that *OsSAP16* controls low-temperature germination in rice. J Exp Bot 69:413–421. https://doi.org/10.1093/JXB/ERX413
Wang N, Cheng M, Chen Y et al (2021a) Natural variations in the non-coding region of *ZmNAC080308* contributes maintaining grain yield under drought stress in maize. BMC Plant Biol 21:1–13. https://doi.org/10.1186/S12870-021-03072-9
Wang W, Ma S, Hu P et al (2021b) Genome editing of rice *eIF4G* loci confers partial resistance to rice black-streaked dwarf virus. Viruses 13:2100. https://doi.org/10.3390/V13102100
Wanga MA, Shimelis H, Mashilo J, Laing MD (2021) Opportunities and challenges of speed breeding: a review. Plant Breed 140:185–194. https://doi.org/10.1111/PBR.12909
Wasaki J, Shinano T, Onishi K et al (2006) Transcriptomic analysis indicates putative metabolic changes caused by manipulation of phosphorus availability in rice leaves. J Exp Bot 57:2049–2059. https://doi.org/10.1093/JXB/ERJ158
Watson A, Ghosh S, Williams MJ et al (2018) Speed breeding is a powerful tool to accelerate crop research and breeding. Nature Plants 4:23–29. https://doi.org/10.1038/S41477-017-0083-8
Wong CK, Bernardo R (2008) Genome-wide selection in oil palm: increasing selection gain per unit time and cost with small populations. Theor Appl Genet 116:815–824. https://doi.org/10.1007/S00122-008-0715-5
Xiang Y, Sun X, Gao S et al (2017) Deletion of an endoplasmic reticulum stress response element in a zmpp2c-a gene facilitates drought tolerance of maize seedlings. Mol Plant 10:456–469. https://doi.org/10.1016/J.MOLP.2016.10.003
Xiao Y, Liu H, Wu L et al (2017) Genome-wide association studies in maize: praise and stargaze. Mol Plant 10:359–374. https://doi.org/10.1016/J.MOLP.2016.12.008
Xing HL, Dong L, Wang ZP et al (2014) A CRISPR/Cas9 toolkit for multiplex genome editing in plants. BMC Plant Biol 14:1–12. https://doi.org/10.1186/S12870-014-0327-Y
Xiong H, Yu J, Miao J et al (2018) Natural variation in OsLG3 increases drought tolerance in rice by inducing ROS scavenging. Plant Physiol 178:451–467. https://doi.org/10.1104/PP.17.01492
Xu Y, Li P, Yang Z, Xu C (2017) Genetic mapping of quantitative trait loci in crops. Crop J 5:175–184. https://doi.org/10.1016/J.CJ.2016.06.003
Xu Y, Ma K, Zhao Y et al (2021) Genomic selection: a breakthrough technology in rice breeding. Crop J 9:669–677. https://doi.org/10.1016/J.CJ.2021.03.008
Yan H, Xu W, Xie J et al (2019) Variation of a major facilitator superfamily gene contributes to differential cadmium accumulation between rice subspecies. Nat Commun 10:1–12. https://doi.org/10.1038/S41467-019-10544-Y
Yang W, Duan L, Chen G et al (2013) Plant phenomics and high-throughput phenotyping: accelerating rice functional genomics using multidisciplinary technologies. Curr Opin Plant Biol 16:180–187. https://doi.org/10.1016/J.PBI.2013.03.005
Yang Q, He Y, Kabahuma M et al (2017) A gene encoding maize *caffeoyl-CoA O-methyltransferase* confers quantitative resistance to multiple pathogens. Nat Genet 49:1364–1372. https://doi.org/10.1038/NG.3919
Yang T, Zhou L, Zhao J et al (2020a) The candidate genes underlying a stably expressed qtl for low temperature germinability in rice (*Oryza sativa* L.). Rice 13:1–15. https://doi.org/10.1186/S12284-020-00434-Z
Yang W, Feng H, Zhang X et al (2020b) Crop phenomics and high-throughput phenotyping: past decades, current challenges, and future perspectives. Mol Plant 13:187–214. https://doi.org/10.1016/J.MOLP.2020.01.008
Yang CJ, Russell J, Ramsay L et al (2021) Overcoming barriers to the registration of new plant varieties under the DUS system. Commun Biol 4:1–10. https://doi.org/10.1038/S42003-021-01840-9
Yu J, Buckler ES (2006) Genetic association mapping and genome organization of maize. Curr Opin Biotechnol 17:155–160. https://doi.org/10.1016/J.COPBIO.2006.02.003

Yu J, Pressoir G, Briggs WH et al (2005) A unified mixed-model method for association mapping that accounts for multiple levels of relatedness. Nat Genet 38:203–208. https://doi.org/10.1038/NG1702

Yu J, Holland JB, McMullen MD, Buckler ES (2008) Genetic design and statistical power of nested association mapping in maize. Genetics 178:539–551. https://doi.org/10.1534/GENETICS.107.074245

Yuan J, Wang X, Zhao Y et al (2020) Genetic basis and identification of candidate genes for salt tolerance in rice by GWAS. Sci Rep 10:1–9. https://doi.org/10.1038/S41598-020-66604-7

Zafar K, Khan MZ, Amin I et al (2020) Precise CRISPR-Cas9 mediated genome editing in super basmati rice for resistance against bacterial blight by targeting the major susceptibility gene. Front Plant Sci 11:575. https://doi.org/10.3389/FPLS.2020.00575

Zaïm M, Kabbaj H, Kehel Z et al (2020) Combining QTL analysis and genomic predictions for four durum wheat populations under drought conditions. Front Genet 11:316. https://doi.org/10.3389/FGENE.2020.00316

Zavala JA, Casteel CL, DeLucia EH, Berenbaum MR (2008) Anthropogenic increase in carbon dioxide compromises plant defense against invasive insects. Proc Natl Acad Sci 105:5129–5133. https://doi.org/10.1073/PNAS.0800568105

Zeng Y, Wen J, Zhao W et al (2020) Rational improvement of rice yield and cold tolerance by editing the three genes *OsPIN5b*, *GS3*, and *OsMYB30* with the CRISPR–Cas9 system. Front Plant Sci 10:1663. https://doi.org/10.3389/FPLS.2019.01663

Zeng R, Li Z, Shi Y et al (2021) Natural variation in a type-A response regulator confers maize chilling tolerance. Nat Commun 12:1–13. https://doi.org/10.1038/S41467-021-25001-Y

Zenkteler M, Nitzsche W (1984) Wide hybridization experiments in cereals. Theor Appl Genet 68: 311–315. https://doi.org/10.1007/BF00267883

Zhang N, Gibon Y, Wallace JG, Lepak N, Li P, Dedow L, Chen C, So YS, Kremling K, Bradbury PJ, Brutnell T (2015) Genome-wide association of carbon and nitrogen metabolism in the maize nested association mapping population. Plant Physiol 168:575–583. https://doi.org/10.1104/PP.15.00025

Zhang Y, Bai Y, Wu G et al (2017) Simultaneous modification of three homoeologs of TaEDR1 by genome editing enhances powdery mildew resistance in wheat. Plant J 91:714–724. https://doi.org/10.1111/TPJ.13599

Zhang A, Liu Y, Wang F et al (2019a) Enhanced rice salinity tolerance via CRISPR/Cas9-targeted mutagenesis of the OsRR22 gene. Mol Breed 39:1–10. https://doi.org/10.1007/S11032-019-0954-Y

Zhang R, Liu J, Chai Z et al (2019b) Generation of herbicide tolerance traits and a new selectable marker in wheat using base editing. Nat Plants 5:480–485. https://doi.org/10.1038/S41477-019-0405-0

Zhang P, Yan X, Gebrewahid TW et al (2021) Genome-wide association mapping of leaf rust and stripe rust resistance in wheat accessions using the 90 K SNP array. Theor Appl Genet 134: 1233–1251. https://doi.org/10.1007/S00122-021-03769-3

Zhao K, Aranzana MJ, Kim S et al (2007) An arabidopsis example of association mapping in structured samples. PLoS Genet 3:e4. https://doi.org/10.1371/JOURNAL.PGEN.0030004

Zhong Y, Liu C, Qi X et al (2019) Mutation of ZmDMP enhances haploid induction in maize. Nat Plants 5:575–580. https://doi.org/10.1038/S41477-019-0443-7

Genome-Wide Association Mapping and Genomic Selection Approaches for Stress Resilience in Rice

2

C. Anilkumar, B. M. Lokeshkumar, N. C. Sunitha, P. S. Basavaraj, and Prashantkumar S. Hanjagi

Abstract

In the era of climate change, continuous breeding efforts are required to dissect and detect genomic regions responsible for stress resilience and achieve a rapid genetic gain in rice. Bi-parental linkage mapping significantly contributed to identifying major genomic regions associated with several agronomic and stress tolerance traits. However, many of the causative alleles responsible for biotic and abiotic stress resistance controlled by minor genes were unidentified in rice. The advancements in molecular marker and genome analysis technologies which evolved from cutting-edge research resulted in new breeding tools, viz. genome-wide association and prediction. The genome-wide association (GWAS) study utilizing LD mapping from diverse panels unravels several loci and alleles for resistance to various stresses in natural populations. Similarly, genomic selection (GS), the upgraded version of MAS, helps in selecting the genotypes based on genomic-estimated breeding values for target traits. The GS approach provides an opportunity to increase the genetic gain per unit time and cost for complex traits. Presently in rice improvement, GWAS and GS were successfully employed to identify causative alleles for resistance to various stresses and predict the genetically appropriate genotypes for resistance breeding in rice, respectively. These genetic tools are proved as the most promising

C. Anilkumar (✉) · P. S. Hanjagi
ICAR-National Rice Research Institute, Cuttack, Odisha, India

B. M. Lokeshkumar
ICAR-Central Soil Salinity Research Institute, Karnal, Haryana, India

N. C. Sunitha
University of Agricultural Sciences, Bangalore, Karnataka, India

P. S. Basavaraj
ICAR-National Institute of Abiotic Stress Management, Baramati, Maharashtra, India

M. G. Mallikarjuna et al. (eds.), *Next-Generation Plant Breeding Approaches for Stress Resilience in Cereal Crops*, https://doi.org/10.1007/978-981-19-1445-4_2

approaches in rice improvement for stress resilience. The present book section summarized the updates and prospects on GWAS and GS in stress resistance rice breeding.

Keywords

Rice genomics · Stress tolerance · GWAS and GS · Genomics-assisted breeding

2.1 Introduction

Half of the world's population rely on rice for staple food, and it serves as a source of income around the world, including 100 million households in Asian and African countries (Khush and Jena 2009; Khan et al. 2009; Basavaraj et al. 2021). The global population is increasing, coupled with diminishing land, water and other resources. With the increased rice-consuming population and living standards, rice consumption is also increasing (Mueller et al. 2012; Patra et al. 2020). Hence, rice production must scale up to 771 million tonnes by 2030 (Khush and Jena 2009; Badawi 2004). However, targeted production can be achieved only by resolving the factors inducing the gap between expected and realized yields of rice cultivars (Chakraborti et al. 2021). Among several constraints reducing rice production, biotic and abiotic stresses have a direct impact by causing severe yield loss (Ramegowda and Senthil-Kumar 2015).

Rice is prone to various biotic stress such as insects, nematodes, diseases, weeds and rats throughout its crop growth period (Das et al. 2017; Prakash et al. 2014). About 70 diseases and > 100 species of insect pests are known to attack rice. However, rice blast caused by *Magnaporthe grisea* Barr, sheath blight caused by *Rhizoctonia solani* and sheath rot *Fusarium fujikuroi* complex, bacterial blight *Xanthomonas oryzae* pv. *oryzae* and tungro caused by the complex of two viruses are regarded as major diseases. At the same time, hoppers (leaf and plant), stem borers and other defoliator insects are major insect pests to Asian ecology where nearly 50% of rice is produced (Gnanamanickam 2009; Pathak and Khan 1994). Globally, pests and diseases cause annual crop losses ranging between 24% and 41%, with an average of 37% (Sparks et al. 2012). However, the dynamics of rice pests and diseases infection have shown changes since the introduction of input-responsive high-yielding varieties in Asia, which, further, transformed minor pests into major pests in rice production systems (Pathak and Khan 1994). Likewise, changing climate and agroecosystem influence chances of occurrence of biotic and abiotic stresses (Hasan et al. 2015; Spindel et al. 2015).

Abiotic stresses like heat, cold, drought, salinity, submergence (intermittent) and oxidative stresses significantly affect rice production and productivity(Das et al. 2017; Lafitte et al. 2004). As rice is grown in four diverse agroecosystems (Koohafkan and Furtado 2004), abiotic constraints differ with the production site or agroecosystem of rice. For example, rice withstands submergence and waterlogging in the deep-water ecosystem; however, it is sensitive to intermittent submergence in the tropics. Thus, despite rice being superior to other crops in stress response, added tolerance level is demanded in different ecosystems better than that found in improved germplasm (Lafitte et al. 2006).

Significant progress has been achieved in developing tolerant cultivars to different stresses using conventional approaches (Das et al. 2017; Werner et al. 2005). Since the expressions of these stresses are complex, identifying stress-tolerant genotypes in the early-stage of the crop is preferred. Thus, conventional approaches are being supplemented with molecular tools to develop stress-tolerant cultivars at a faster pace and precision (Hasan et al. 2015). However, despite utilizing molecular tools, progress is restricted to identify a few major QTL for stress tolerance (Gregorio et al. 2013; Singh et al. 2016). In general, QTL for complex traits is identified either by linkage mapping or association mapping. The bi-parental mapping populations like F_2, $F_{2:3}$, $F_{2:4}$, NILs, RILs, doubled haploid (DH), BILs, reciprocal introgression lines and advanced backcross BC_3F_5 introgression lines can be used for linkage mapping purpose. In contrast, association mapping utilizes diverse natural populations like germplasm or landraces or commercial crop cultivars. Association mapping is a population statistics approach that exploits linkage disequilibrium (LD) in highly diverse natural populations adapted to different natural habitats to investigate the genetic architecture of complex traits. The low resolution and background noise encountered in bi-parental QTL mapping could be addressed using GWAS. The GWAS is a relatively new method of investigating complex traits in the diverse panel of genotypes by genome-wide marker information and precise phenotyping for stress tolerance to identify associated QTL (Fig. 2.1). Recent advances in high-throughput next-generation SNP platforms coupled with computational advancements facilitated the rapid dissection of complex genetic traits employing GWAS and fast track the breeding process through genomics-assisted breeding strategies (Lipka et al. 2015).

Rice has vast diverse germplasm and genomic resources. Since the late 1990s, several candidate genes or major QTL conferring stress tolerance have been identified in rice like *Sub1* locus for submergence tolerance (Xu and Mackill 1996), $>$100 genes or loci for blast resistance (Fang et al. 2016) and other stresses. Genome-wide association analysis (GWAS) is one of the best methods to identify allelic variations associated with target traits in a diverse population considering historical recombination events (Zhou and Huang 2019; Manolio 2010; Bush and Moore 2012). Due to lower LD in diverse GWAS panels, allelic diversity is captured at a higher resolution than bi-parental QTL mapping (Pantalião et al. 2016; Rebolledo et al. 2015). Diverse rice germplasm comprises numerous genomic variants for stress tolerance and can be explored more efficiently through GWAS (Han and Huang 2013). Going deeper into functional genomics, causative genes underlying stress tolerance could be identified using GWAS results (Katara et al. 2021). However, GWAS results could be directly used in rice stress breeding as significant SNPs are most likely to be tightly linked to causative genes responsible for tolerance to many stresses (Zhou and Huang 2019). DNA markers linked to large-effect QTL/genes identified through GWAS have been used to implement marker-assisted selection (MAS) in rice stress breeding programs (Gregorio et al.

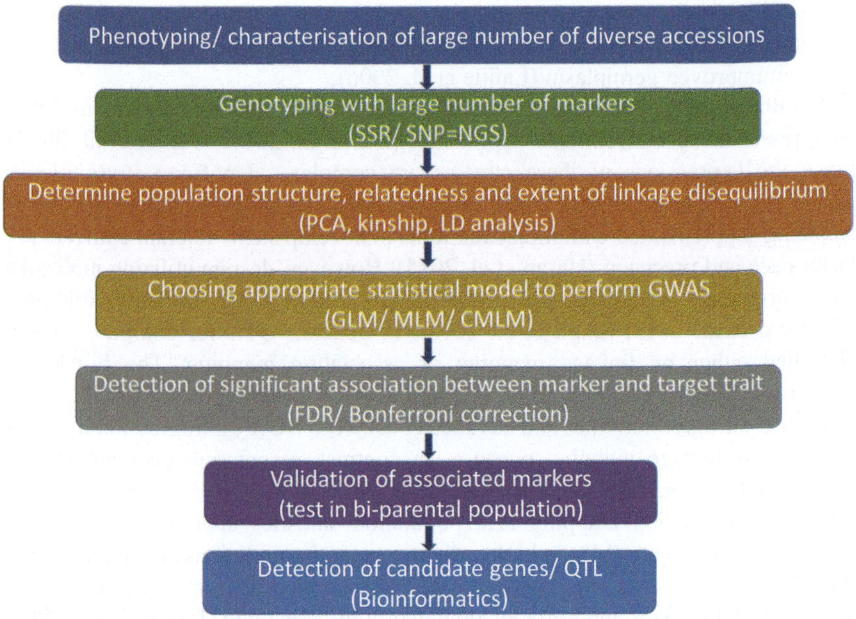

Fig. 2.1 Most important steps in a successful GWAS experiment

2013). However, there are few disadvantages in GWAS like difficulty in detection of rare and small-effect alleles, which could be solved through another advanced breeding tool called genomic selection (GS);. In particular, most of the biotic and abiotic stress tolerances are complex traits controlled by a mixture of small- and large-effect QTL/genes (Flowers 2004; Holland 2007; Spindel et al. 2015) and sometimes show low heritability (Verulkar and Verma 2014), which may go undetected using GWAS.

Of late, genomic selection is emerging as a promising approach to address complex traits with low heritability. A genotyped-only individual with high GEBV is selected in genomic selection. Unlike marker-assisted selection, both linked and unlinked markers are utilized in genomic prediction to achieve improved genetic gains (Cooper et al. 2014; Spindel et al. 2015). Several features associated with rice such as self-pollination, availability of diverse germplasm and ease of genotyping (sequencing) make genomic selection a feasible tool to address stress tolerance (Spindel et al. 2015). However, genomic selection studies in rice are still budding. Thus, we briefly review the GWAS methodology, applications and achievements in rice stress breeding and present possible strategies to implement GS in rice stress resistance breeding.

2.2 Genomic Resources in Rice for GWAS and GS

Crop improvement program mainly depends on the existence of available genetic variation and diversity in the respective crops. Rice has rich genetic diversity consisting of more than 120,000 germplasm comprising traditional varieties, landraces, genetic stocks, breeding lines, wild relatives and 21 wild species (Gur and Zamir 2004; Kovach and McCouch 2008). Therefore, rice breeders have ample opportunities to exploit and utilize these gene pools to develop high potential breeding lines or varieties over the existing ones. Rice has 22 wild and 2 cultivated species, with about 773,948 rice accessions conserved in gene banks worldwide (Wambugu and Ndjiondjop 2018). Utilizing the natural variation within these wild and/or domesticated and cultivated accessions helps mining novel genes for tolerance to many stresses. Rice is one of the major crops in the world, occupying very broad geographic distribution, adapted and cultivated in many ecologies with different agronomic conditions. Accordingly, the genetic diversity of rice is abundant (Zhao et al. 2018b), and the rich genetic diversity is beneficial to utilize in GWAS. All molecular markers, including hybridization-based markers like RFLP, PCR-based markers like RAPD and SSR and sequence-based markers like SNP, are available extensively in rice. Enormous genomic data is available since it was sequenced (International Rice Genome Sequencing Project 2005), which made the availability of rice genome sequences and SNP resources (Verdeprado et al. 2018). With the advancements achieved through next-generation sequencing platforms, the cost of sequencing has drastically reduced. Thus, extensive utilization of next-generation sequencing is being practised in rice breeding programs. Thus, enormous genetic and genomic resources along with high-throughput genotyping and phenomics platforms have rendered GWAS and GS suitable for rice breeding. GWAS utilize a set of diverse germplasm accessions, which is regarded as association mapping panel or diversity panel to pool all possible genetic variants (Wang et al. 2020); Zhou and Huang 2019). GWAS helps to determine multiple genetic factors related to various molecular mechanisms from phenotypic variations of multiple accessions (Portwood et al. 2019: soybean; Song et al. 2013; Bauchet et al. 2017). The existence of unstructured but diverse panel and high-throughput genotyping is necessary for the detection of more associated loci considering historical recombination, i.e. higher resolution with fewer false positives (Wang et al. 2020). Many unstructured natural rice populations have been used for GWAS on different production-related traits, including biotic and abiotic stress tolerance (Huang et al. 2010, 2012; reviewed by Verdeprado et al. 2018; reviewed by Zhang et al. 2016). The self-pollinated property of rice ensured a strong population structure in rice. Therefore, different species of rice exhibits intact population structure and allelic fixation index. Hence, to avoid the influence of strong population structure of different species in rice, GWAS analysis must be carried separately on panels of different species.

Genomic selection is a kind of MAS where desirable genotypes are predicted based on genomic estimated breeding values. Once the prediction model and other factors are optimized using a training population, GS can be implemented for

3–4 cycles of selection, saving time and resources, thus escalating genetic gain (Bernardo 2010). QTL mapping is suitable to resolve high heritable traits and GWAS for high and moderately heritable traits, while low heritable traits are targeted using GS (Alqudah et al. 2020; Verdeprado et al. 2018). GS is gaining importance in rice breeding because of the quantitative traits being governed by many small-effect and low heritable genes.

2.3 Factors Affecting GWAS and GS Application in Rice Stress Resilience Breeding

GWAS is superior to QTL mapping in resolving small-effect genes associated with markers. However, its precision is greatly influenced by the size of a diverse panel of genotypes, marker density, extent of LD and population structure. Large diversity panel and marker density are being used for rice GWAS, which is facilitated by high-throughput genotyping at low cost. Large population size and marker density are critical in GWAS analysis as they both influence the power of GWAS. The panel size ranging from 100 to 500 with sufficient markers covering the entire genome of rice is needed to perform GWAS in rice. Population structure depicts the relatedness and correlation of individuals within the chosen panel, which must be considered during analysis and interpretation of results. The self-pollinated behaviour of rice species exhibits a strong population structure among the subpopulations. Cultivated rice might include various subpopulations such as indica, basmati, temperate japonica, tropical japonica, etc. The population stratification adds to population structure problems leading to a spurious association. Therefore, species or sub-population-wise separate GWAS analysis is required (Zhao et al. 2011; Zhou and Huang 2019; Wang et al. 2020). Despite using large sample sizes and statistical corrections, it is challenging to identify rare alleles within one sub-population (Marouli et al. 2017). Hence, multi-parent recombinant populations like NAM and MAGIC were developed using diverse accessions (within a clade, e.g. indica) in rice (Bandillo et al. 2013). On the other hand, GWAS mapping relies on linkage disequilibrium (LD), a non-random association between two or more loci in a specific population. In other words, few chromosomal regions do not recombine and are inherited as linkage blocks over generations (Flint-Garcia et al. 2003). LD between individuals of a population is assessed prior to association mapping (Slatkin 2008). The strength of LD depends on the extent of historical recombination events that occurred over several generations. Ignoring non-random associations among alleles from different loci leads to the spurious association. LD acts as an indicator to define the distance between loci and find the number of markers to be added for covering the entire genome; for example, high LD means lower number of markers are sufficient to cover the genome (Semagn et al. 2010; Sallam and Martsch 2015).

The success of genomic selection in terms of selection response or genetic gain relies on the prediction accuracy of the model followed. The prediction accuracy of GS model is influenced by various factors, viz. training population size, structure of training population, relatedness between training and test populations, precise

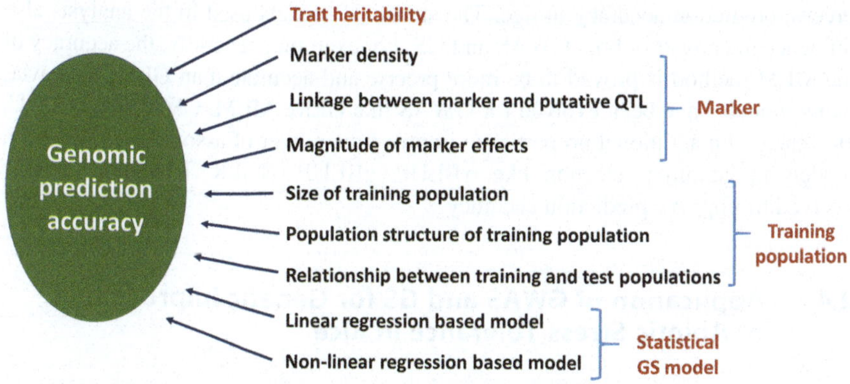

Fig. 2.2 Factors influencing accuracy in genomic selection experiments

phenotypic data of trait and its heritability, marker density and statistical method used to build the model (Fig. 2.2). Thus, GS models are designed (optimization of GS components) to attain higher prediction accuracy, validated (mostly through *n*-fold cross-validation) and implemented in the breeding population to achieve decent genetic gains (Xu et al. 2020). Although there are several reports on optimising GS components to achieve higher accuracy in livestock and maize, there are few such studies in rice. The size and composition of the training population depend on trait heritability and population structure. In rice, a moderate training population size of 100–500 is optimum for genomic prediction of high heritable trait (Bhandari et al. 2019; Ben Hassen et al. 2018a; Spindel et al. 2015), while training population size of <50 is sufficient if the population is highly diverse (Onogi et al. 2015). Having understood reduction in population size with its diversity, highly diverse individuals in a population could affect relatedness between training and test populations during cross-validation. Hence, population structure and relatedness between training and test population are interconnected and influence prediction accuracy (Guo et al. 2014; Grenier et al. 2015; Ben Hassen et al. 2018a). Prediction accuracy is higher when the training and test population are closely related (Desta and Ortiz 2014; Grenier et al. 2015; Ben Hassen et al. 2018b). Full-sib and half-sib families derived from the training population were used as test populations. Training populations with full-sib families gave higher prediction accuracy than half-sib test families (Guo et al. 2012; Riedelsheimer et al. 2013). Designing training population using individuals within a sub-population could resolve the problem of population structure (Guo et al. 2014), while the inclusion of individuals with low genotype-environment interaction (GEI) helps in the selection of stable individuals across environments (Hoffstetter et al. 2016; Ben Hassen et al. 2018b).

For both GWAS and GS, uniformly distributed genome-wide markers are chosen based on the assumption of LD between QTL and at least one marker. Hence, along with all other factors, appropriate and precise genotyping of association panel in GWAS and training and test population of GS is most important. Bhandari et al. (2019) opined that at least 27 SNPs per Mb of the genome are required to achieve precise prediction accuracy in rice. The statistical models used in the analysis also influence the power of both GWAS and GS. For example, generally, the accuracy of the MLM method is proved to be more precise and accurate than GLM. However, many models have been evolved for GWAS like GLM, MLM, CMLM, FarmCPU, etc., each with additional properties to improve the power of association. Similarly, models in genomic selection like rrBLUP, gBLUP, BGLR, RKHS, etc., have evolved to improve prediction accuracy.

2.4 Application of GWAS and GS for Genetic Improvement of Abiotic Stress Tolerance in Rice

The economically important complex traits like grain yield and quality of rice are highly influenced by external factors such as the availability of nutrients and favourable environmental conditions during growth and development. On the other hand, rice grain yield and quality are significantly affected by unfavourable factors like stresses imposed by biotic and abiotic factors. Abiotic stresses like drought, salinity or alkalinity, high or low temperature, submergence or flooding and metal toxicity or nutrient deficiency influence rice grain production and quality at various levels in different regions. Therefore, exploiting existing natural diverse populations to detect the different genes/QTLs responsible for abiotic stress tolerance mechanisms is a needful approach to develop tolerant varieties. Several researchers have used the association mapping approach for the identification of QTL linked with the abiotic stress tolerance in rice presented in Table 2.1. In this section, we will briefly discuss abiotic stress-tolerant QTL identified through GWAS.

2.4.1 Drought Tolerance

Rice is a drought-sensitive crop, and the occurrence of severe drought stress results ingrain yield loss up to 100% (Sahebi et al. 2018). Moreover, drought significantly affects pollen fertility and embryo development after pollination during the reproductive stage, resulting in low grain yield (Ozga et al. 2017). Therefore, understanding the genetic basis of drought tolerance is imperative and the primary factor to develop drought-tolerant rice varieties. Genes or QTLs associated with drought resistance traits like leaf water status maintenance, stomatal closure regulation and root morphology have been reported by several researchers (Price and Tomos 1997; Courtois et al. 2000; Yue et al. 2006; Li et al. 2017). Al-Shugeairy et al. (2015) exploited 328 accessions of the Rice Diversity Panel (http://www.ricediversity.org/) for QTL mapping by GWAS and found only one SNP on chromosome 2, which is significantly related to drought recovery traits. They also analysed the position of candidate genes underneath the QTL and recognized three candidate genes. The first one, *LOC_Os02g40530,* is an MYB family transcription factor coding gene responsible for fundamental factors in regulatory networks governing biotic and abiotic

Table 2.1 Abiotic stress tolerance QTL/MTAs identified through genome-wide association studies (GWAS) in rice

Stress	Genotyping platform/markers	Population/panel	Outcome	Reference
Drought	Affymetrix array consist comprises of 44,100 SNPs	A total of 328 accessions from RDP	SNP position at 24559374 bp associated with drought recovery trait. Three genes encoding for MYB family transcription factor *LOC_Os02g40530*, *LOC_Os02g40510* and *LOC_Os02g40700* were candidates for the QTL associated with drought recovery traits	Al-Shugeairy et al. (2015)
	150,325 SNPs through the GBS approach	175 upland japonica rice accessions	GWAS detected 13 SNPs linked with yield under drought environment. Besides, the study identified 50 genes, of which 30 annotated genes are relevant to drought tolerance TF, such as WRKY, Apetala2 and protein kinases	Pantalião et al. (2016)
	GBS method employing Illumina HiSeq2000. A total of 10,19,883 SNPs were utilized for MTA	A panel of 270 accessions from Asia, African, and American continent	Six associated loci for drought-tolerant along with two functional genes (*OsPYL2* and *OsGA2ox9*) were identified	Ma et al. (2016)
	About a total of 6.4 million SNPs from genotyping data set	529 rice accessions	143 significant associations were identified for 21 root architectural characters under drought stress and non-stress environments	Li et al. (2017)
	125 SSR markers	75 Malaysian genotypes	Seven MTA were identified for grain yield under moisture stress, and 4 MTA were consistent across the season	Swamy et al. (2017)
	43,58,600 (*Indica* and Japonica population), 28,63,169 (*Indica* subpopulation), and 19,59,460 (*Japonica* subpopulation) SNPs were used	507 rice accessions	GWAS analysis identified 470 associated loci, of which some are associated with known drought tolerance genes. Of these 470 loci, 443 loci were detected for image-based traits	Guo et al. (2018)

(continued)

Table 2.1 (continued)

Stress	Genotyping platform/markers	Population/panel	Outcome	Reference
	21,623 SNP markers	180 Vietnamese rice landraces	The study identified 14 QTLs for leaf RWC, 9 associated with the slope of RWC, 12 associated with drought tolerance scale, 3 for drought recovery, and 1 for relative crop growth rate	Hoang et al. (2019)
Salinity and alkalinity	EcoTILLING approach	392 rice accessions	GWAS revealed 11 significant SNPs for salt tolerance. Of which 5 nsSNPs. These 5 nsSNPs are related to salinity-tolerant traits	Negrão et al. (2013)
	Array-Based 6 k SNP chip	220 rice accessions	Detected 20 significant SNPs for Na^+/K^+ ratio, 44 SNPs for related traits. The genomic region associated with *Saltol* was associated with Na^+/K^+ ratio. Besides *Saltol*, novel QTLs were detected	Kumar et al. (2015)
	26,258 SNPs	378 diverse rice genotypes	Fluorescence imaging detected four loci associated with salinity-induced fluorescence responses. Besides, a region on chromosome 1 which controls ionic stress was also identified	Campbell et al. (2015)
	Rice 3 K genome data (SNP data)	A panel of 478 genotypes	GWAS identified 11 loci containing 22 significant SNPs for salinity tolerance. *OsNRT2.1* and *OsNRT2.2* transporter family genes associated with salinity tolerance were detected	Shi et al. (2017)
	30,000 SNP markers	A set of 235 accessions belong to the temperate japonica group	GWAS revealed a total of 27 QTLs validated by subsampling. The position of identified QTLs was compared with 300 genes that play a crucial role in calcium signalling and metabolism	Frouin et al. (2018)

	A high-density array-based 700,000 SNPs	306 rice accessions	GWAS detected 1200 candidate genes for cation transporters and transcription factors associated with a role in salinity tolerance	Patishtan et al. (2018)
	162,529 SNPs	478 rice accessions	56 QTNs and 66 candidate genes were detected, and 2 of these genes (*LOC_Os01g45760* and *LOC_Os10g04860*) associated with auxin biosynthesis	Culi et al. (2018)
	SNPs from Rice 3 K RGP	708 rice accessions	A total of 2255 MTA were detected. The SNPs are dispersed in 903 genes. In addition, 5 known and 2 novel genes identified are related to yield and salinity tolerance traits	Liu et al. (2019)
	Resequencing yielded 788,396 SNPs	295 varieties belong to the japonica group	Eight QTLs positively associated with the score of alkalinity tolerance, the concentration of Na^+ in the shoots and Na^+/K^+ ratio of shoots were detected	Li et al. (2019a)
	Exome sequencing (112,565 SNPs)	104 Thai rice accessions	GWAS identified 200 loci harbouring 448 SNPs on exons. Of the 200 loci, 146 loci co-localized with earlier reported salinity-tolerant QTL	Lekklar et al. (2019)
	32,315 SNPs	181 Rice cultivars	A total of 54 loci related to salt tolerance were identified. Of which, 17 QTLs are associated with dry weight ratio	An et al. (2020)
	SNP data from 3 K RGP	664 genotypes from 3 K RGP	The study identified 21 QTLs and 2 candidate genes. Sequence and haplotype analysis found *OsSTL* gene was a homolog of salinity-tolerant gene *SRP1*in *Arabidopsis*	Yuan et al. (2020)

(continued)

Table 2.1 (continued)

Stress	Genotyping platform/markers	Population/panel	Outcome	Reference
	Two million SNPs	A panel of 204 rice accessions comprised of collected Bengal and Aus rice	A total of 97 and 74 QTLs under hydroponic and soil systems were identified, of which 11 QTLs were detected in both conditions. Further, 65 candidate genes and a major gene *OsHKT1;5* were also identified	Chen et al. (2020)
	> 33,000 SNP	155 rice varieties	GWAS revealed 151 markers linked to 29 genomic regions. Candidate genes *Os01g0304100*, *Os01g0624700*, *Os01g0812000*, *Os01g0966000* and *Os01g0963000* are associated with different transcription factors such as WRKY 12, GAMyb, ABI/VP1 and retinoblastoma-related protein (RBR), Lox, F-box and Na+/H+ antiporter	Nayyeripasand et al. (2021)
Low temperature	Total of 273 SSR primers	174 Chinese accessions from mini core	The study detected 51 QTLs for cold tolerance. Of which 22 are associated with tolerance at the germination stage and 33 at the booting stage	Pan et al. (2015)
	4,358,600 SNPs	529 rice accessions	GWAS revealed 132 significant loci, of which 12 were found common for both chilling and cold tolerance	Lv et al. (2016)
	Genotypic data set from 44 K SNP chip	295 rice cultivars in the RDP1	The study revealed 67 QTLs for cold tolerance at the seedling stage	Wang et al. (2016)
	700,000 SNP markers	400 accessions from RDP1	The study detected 42 QTLs for low-temperature tolerance at the seedling stage. Of these, 22 QTLs are co-localized with previously reported QTLs	Shakiba et al. (2017)

	1672 SNP markers	200 cultivars	31 significant MTA for various cold tolerance traits	Sales et al. (2017)
	A total of 157 markers including, 148 SSRs 3 InDels &6 SNPs	202 rice accessions belong to a mini core collection	A total of 48 QTLs at 39 regions spanning across 2 new low temperature seedling survivability (LTSS)–QTL, *qLTSS3*–4 and *qLTSS4*–1, were identified	Schlappi et al. (2017)
	351,124 SNP markers	1033 accession diversity panel	At the seedling stage, 8 loci were reported for cold tolerance. Besides, the study also revealed *LOC_Os10g34840, LOC_Os10g34840* as the candidate gene for cold tolerance	Xiao et al. (2018)
	44 k SNP	187 rice accessions	A total of 53 loci responsible for low-temperature germination identified	Wang et al. (2018a)
	5 K rice array	249 *indica* rice varieties	A total of 47 significant SNP related to the severity of damage and seed survival rate were identified	Zhang et al. (2018)
	7 K SNP array	257 rice accessions	A total of 51 QTLs were detected, and in addition, 21 novel genomic regions were identified from the whole set, 11 from the indica group and 10 related to the japonica group	Thapa et al. (2020)
High temperature	20 linked markers	A 240 germplasm including elite breeding lines and landraces	Detected SSR marker RM547 associated with spikelet fertility under stress. Other markers, viz. RM228, RM205, RM247, RM242, INDEL3 and RM314, indirectly associated with high-temperature stress tolerance	Pradhan et al. (2016)
	13,160 SNPs	167 indica landraces and improved varieties	A total of 14 loci are significantly associated with spikelet sterility, capable of sensing abiotic stress and regulating cell division and gametophyte development during the stress	Lafarge et al. (2017)

(continued)

Table 2.1 (continued)

Stress	Genotyping platform/markers	Population/panel	Outcome	Reference
	45, 200 SNPs	209 rice genotypes of indica rice diversity panel	The study detected 38 loci for yield and quality traits during high night temperature, among which 18 and 20 loci were associated under control and high night temperature conditions, respectively	Bheemanahalli et al. (2021)
Metal toxicity	Genotyped with 44,000 SNPs	383 diverse rice accessions	48 loci associated with Al tolerance were identified. Four of them were co-localized with previous candidate genes	Famoso et al. (2011)
	The high-density rice array consisted of 7 lakh SNPs	211 accessions	60 putative QTLs and 5 toxicity tolerance QTLs were identified in the same genomic region under 2 stress conditions. Further, haplotype analysis revealed 22 candidate genes for 10 QTL regions	Zhang et al. (2017a)
	416,741 SNP	288 genotypes	Several highly significant SNP markers were detected. The significant marker SNP-2.22465867 caused an amino acid change in a gene (*LOC_Os02g37170*) with an unknown function. Candidate regions contained genes coding for a heavy metal transporter, peroxidase precursor and Mn^{2+} ion-binding proteins	Shrestha et al. (2018)
	700 K SNP assay	312 diverse rice accessions	The study revealed 14 QTLs associated with Cd accumulation. In addition, candidate gene *OsNRAMP2*was identified for high Cd accumulation	Zhao et al. (2018a)

	3.2 million SNPs	191 rice accessions belong to mini core collections of USDA	106 QTLs for ionomics flooded, 47 QTLs to ionomics unflooded, and 97 QTLs responsible for agronomic traits were identified. Three candidate genes, viz. *OsWRKY102 OsWRKY75*, and *OsMan07*, involved in Cd regulation and one gene *PIP2*;6 involved in control was also identified	Liu et al. (2020b)
	3,960,919 SNPs	338 diverse rice accessions	Identified35 QTLs are regulating Cd accumulation, in addition to a novel QTL, *qCd1*–3, for Cd stress tolerance	Pan et al. (2020)
Submergence	GBS	281 rice varieties belonging to the indica group	A total of 160 MTAs were identified under submergence and flooding conditions. In addition, candidate genes, viz. *OsOFP2, Dlf1, OsMADS56, SUI1, Sd1, OsCOL4, Nal1, OsphyB, GW5* and Ehd1, were also identified	Volante et al. (2017b)
	5291 SNP markers	432 indica varieties	The study detected 22 SNPs for traits associated with flooding tolerance. In addition, the candidate gene *LOC_Os06g03520* associated with haplotype showed high expression in anoxia	Zhang et al. (2017c)
	GBS	166 rice varieties	The study identified several QTLs and 50 candidate genes associated with submergence stress-associated traits	Gao et al. 92,020)
	50 K SNP chip	94 genotypes from Assam belongs to deep-water rice	GWAS detected 20 significant genes for anaerobic germination related traits. Genes *OsXDH1*and *SSXT* identified associated with anaerobic response index	Rohilla et al. (2020)

stresses. The available reports indicate the role of MYB family genes in stomatal movement regulated by ABA signalling and drought stress in *Arabidopsis* (Cominelli et al. 2005; Seo and Park 2010). The second gene, *LOC_Os02g40510,* is responsible for the regulator receiver domain-containing protein, which regulates ABA-mediated resistance linked to drought stress (Castells et al. 2010). The third gene, *LOC_Os02g40700,* encodes for cupin super family protein involved in protecting plants from environmental stress. Pantalião et al. (2016) employed GWAS in a panel of 175 rice accessions under with and without water conditions. They observed 13 SNP markers that were statistically associated with complex traits like grain yield under no-water conditions. They also report 50 genes, among which 30 were related to abiotic stress tolerance genes such as transcription factors WRKY, (*LOC_Os08g13840*), Apetala2 (*LOC_Os02g09650*), protein kinases and mitogen-activated protein kinase (MAPK) (Licausi et al. 2013; Chen et al. 2012; Kumar et al. 2008). Ma et al. (2016) identified six associated loci and two functional genes, namely, *OsPYL2,* an ABA receptor that regulates the stomatal behaviour, and *OsGA2ox9,* which is involved in the GA metabolic pathways for drought-resistant coefficient (Tian et al. 2015; Lo et al. 2008). Ma et al. (2016) revealed the new drought-tolerant candidate gene *OsRLK5* controlling the rate of water loss in leaves. GWAS was carried out by Li et al. (2017) in 529 accessions of natural population and depicted 143 prominent SNPs related to root traits under normal and drought stress conditions. Additionally, the study also identified SNPs associated with *DEEPER ROOTING 1* (*DRO1*), *WUSCHEL-related homeobox* gene *WOX11, OsJAZ1 (EG2), OsPID* and *Nal1,* which controls the root growth angle, root hair formation, spikelet development, adventitious root development, auxin transport, leaf width, spikelet number, photosynthesis rate and yield under drought conditions (Morita and Kyozuka 2007; Qi et al. 2008; Zhao et al. 2009; Fujita et al. 2013; Takai et al. 2013; Uga et al. 2013; Zhang et al. 2014; Zhao et al. 2015; Chen et al. 2015). Guo et al. (2018) conducted GWAS with automated non-destructive image-based phenotyping in a panel of 507 rice accessions and showed 437loci co-localized with previously reported drought-tolerant QTLs. Further, the study also showed a QTL *qCT1* for mean canopy temperature under drought. The LD block consists of seven genes, including mitochondrial fumarase and plasma membrane HAK transporter, which was involved in the accumulation of K^+, Cl^- and malate in guard cells which helps in the opening and closure of stomata during drought conditions. Hoang et al. (2019) discovered 14, 12, 9, 3 and 1 quantitative trait loci for leaf relative water content, drought sensitivity score, slope of relative water content, recovery ability and relative crop growth rate. Most of these QTLs were previously reported and related to drought tolerance.

2.4.2 Salt Tolerance

Salt stress is the second most important abiotic stress in rice production. Salt tolerance is a complex trait, and QTLs associated with tolerance include Na^{+}, and K^+ content of roots and shoots have been reported using bi-parental mapping populations by many researchers (Sabouri and Sabouri 2008; Ammar et al. 2009; Pandit et al. 2010; Islam et al. 2011; Mohammadi et al. 2013; Ghomi et al. 2013; Hossain et al. 2015). *SKC1* (*OsHKT1*;5) gene identified in Nona Bokra regulates the K^+/Na^+ homeostasis, which has a significant role in salt stress (Ren et al. 2005). EcoTILLING approach was employed in 392 rice accessions by Negrão et al. (2013) to detect genetic mechanisms responsible for Na^+/K^+ ratio, stress protection and signalling cascade. They found new allelic variants in coding sequences of five key salt-related genes. They recorded 11 SNPs in 4 candidate genes (*SalT, OsHKT1;5, OsNHX1 and OsCPK17*) significantly associated with salt-related traits. Kumar et al. (2015) identified SNPs associated with Na^+/K^+ ratio at reproductive stage and *SALTOL*, a major quantitative trait locus (QTL) for salinity tolerance at the seedling stage. *SALTOL* QTL region has three genes *SKC1*, *SalT* and *pectinesterase,* which help to control K^+ homeostasis under salinity (Claes et al. 1990; Bonilla et al. 2002; Ren et al. 2005). They also reported the expression level of *LOC_Os04g23580, LOC_Os04g23550, LOC_Os04g24110*, LOC_*Os04g57800,* LOC_*Os04g57760,* LOC_*Os04g57810* and LOC_*Os04g57850* genes during salt stress. Three genes on chromosome 6, namely, *LOC_ Os6g03700, LOC_Os6g03670* and *LOC_Os6g03750,* were associated with the Na^+/K^+ ratio, and these are encoding the proteins called calcium-dependent protein kinases (CDPK). Out of this, the locus *LOC_Os6g03670* characterized as *OsDREB1C* was responsible for salinity stress tolerance in rice and *Arabidopsis* (Yamaguchi-Shinozaki and Shinozaki 1994; Dubouzet et al. 2003).

An effort of GWAS using image-based phenotyping was carried out by Campbell et al. (2017). Four genomic regions associated with early growth salinity tolerance were detected on chromosome 3. Their study depicted that the genes present on chromosome 1 regulate the ionic stress and decline in early growth rate under salt stress. In one of the GWAS conducted, 22 significant SNPs were correlated with stress susceptibility indices for vigour index (SSI-VI) and mean germination time (SSI-MGT) under salt stress (Shi et al. 2017). The position of SSI-MGT was located on chromosome 1, which controls the Na^+, K^+ and Na^+/K^+ ratio. The region of SSI-VI contains two genes, viz. *OsNRT2.1* and *OsNRT2.2,* located on chromosome 2, which are responsible for nitrate transporter. In another study by Frouin et al. (2018), 50 QTL regions containing 300 genes were responsible for salt tolerance, among which 27 were validated. Some of the genes like *OsCBL7, OsCBL8, OsSAPK1, Os07g44330, OsCAX1, OsCAX2, OsMXH2* and *OsACA6* were responsible for the calcium-dependent ionic stress signalling pathway, kinase activity, antiporter of cation/H^+ or Mg^{2+}/H^+ exchange proteins and Ca^{2+}-ATPase, which plays a role in intracellular sodium ion homeostasis. Patishtan et al. (2018) detected 1200 candidate genes of different transcription factors and cation transporters having a significant role in salinity tolerance. They detected an association of QTL with gene *HKT* (*HKT1;3*), mediating the leaf rolling under stress (Véry et al. 2014).

Patishtan et al. (2018) identified nine nsSNPs in FBX289 that were highly relevant in salt tolerance. Two genes such as *LOC_Os01g45760* and *LOC_Os10g0486* having a role in auxin biosynthesis were detected. The identified SNPs were significantly associated with 903 genes, including two new genes, i.e. *LOC_Os02g49700* and *LOC_Os03g28300,* related to saline tolerance (Liu et al. 2019). They also reported class I transporters ofK^+ (HKT) mediating the leaf-blade Na^+ exclusion (Suzuki et al. 2016). The locus *LOC_Os05g31730* having known genes like *OsNHX5* and*OsNHX1,* was associated with SSI, which helps in Na^+ and K^+ compartmentalization between the cytoplasm and vacuole improves salt tolerance in rice (Fukuda et al. 2011). Two hundred loci associated with 448 SNPs were linked to the traits such as salt susceptibility index, filled grains, number of panicles and unfilled grains per plant (Lekklar et al. 2019). SNPs associated with the genes like *LOC_ Os10g03660, LOC_Os10g03620, LOC_Os10g03730, LOC_ Os10g03780, LOC_Os10g03740, LOC_Os10g05500* and *LOC_Os10g03930* regulate the abiotic stress responses in *Arabidopsis*, wheat and rice (Jain et al. 2007; Yan et al. 2011; Zhou et al. 2014; Jia et al. 2015; Gonzalez et al. 2017). Many of the detected genes, i.e. *LOC_Os01g66760, LOC_Os01g66740, LOC_Os02g02120* and *LOC_Os02g56630,* belong to the kinase family that encodes signalling factors under abiotic stresses (Sinha et al. 2011; Kovtun et al. 2000). Out of 54 QTLs detected, 17 loci were associated with dry weight ratio (DWR) during salt stress reported by An et al. (2020). An SNP at 22,580,051 on chromosome 12 has rapidly induced genes for PR10 protein associated with shoot length under salt and drought stresses (Hashimoto et al. 2004).

Another significant SNP at 7,088,028 close to the *OsPYL,* ABA receptors regulates the gene associated with drought and salt stress tolerance at the vegetative stage (Kim et al. 2014). MATE protein coding genes were identified in novel QTL qST7 and were located on chromosome 7. MATE proteins (*GrMATE18, GaMATE41, GrMATE34* and *GaMATE51)* were involved in the extrusion of citric acid and flavonoids or Al and other toxic compounds (Lu et al. 2018). Yuan et al. (2020) identified two genes *OsSTL1* and *OsSTL2* along with 21 QTLs. A total of 97 QTLs were associated with characters in the hydroponic system, whereas 74 QTLs with soil system and 11 QTLs were identified in both soil and hydroponic systems (Chen et al. 2020). Further, the study revealed 65 candidate genes consisting of *OsHKT1;5* and two post-translational modifications genes *OsSUMO1* and *OsSUMO2*. The candidate genes, namely, *OsNTL2, OsNAC4, OsNAC5, OsNAC3, OsbZIP23, OsPCF2, OsABF2, RSS1, DREB1C, OsBIHD1* and *OsGTg-1,* were identified. The *OsGTg-1* gene that was upregulated by the salt stress was located on chromosome 4. This was associated with the Na content and Na/K ratio in hydroponics and soil system, respectively (Fang et al. 2010). A total of 151 marker-trait associations were detected on chromosome 10 under salt stress condition by Nayyeripasand et al. (2021). QTL region consisted of candidate genes like *SalTol1* (cation chloride co-transporter), Os01g0624700 (WRKY transcription factor, WRKY 12), *Os01g0812000* (gibberellin-dependent alpha-amylase, GAMyb), *Os01g0966000* (plasma membrane H^+-ATPase), *Os01g0963000* (peroxidase BP1 protein) and *Os02g0730300*(K^+ transporter, HAK). Some other genes encoding retinoblastoma-related protein and pseudouridine synthase gene transcription factor were also identified.

2.4.3 Low- and High-Temperature Tolerance

Low temperature is another abiotic stress that threatens the adaptability of rice and its production all over the world. Cold stress vastly influences the grain yield and quality, especially when it coincided with the blooming/anthesis and grain filling stage (Kodra et al. 2011; Guirguis et al. 2011). Chilling tolerance in rice is controlled by many genes and influenced by the environment. More than 30 low-temperature tolerance QTLs were recognized at the germination phase using bi-parental mapping population (Zhou et al. 2010; Xu et al. 2008). Several investigators employed the GWAS approach for dissecting genomic regions controlling cold stress tolerance. Association study in 174 Chinese rice accessions mapped 51 QTLs for cold tolerance on rice genome, of which 22 and 33 QTLs were detected for germination and booting phases, respectively. The common QTL, namely, qLTSSvR6–2, for the survival rate of seedlings was identified for both japonica and indica panels (Pan et al. 2015). GWAS using SNP recognized about 132 QTLs to low-temperature stress, 57 QTLs for chilling and 63 QTLs for cold shock, and 12 common QTLs for chilling and cold stress were detected. Cold-tolerant genes *COLD1, Ctb1 (Os04g52830), OsRAN2 (Os05g49890), OsiSAP8 (Os06g41010), OsLti6a (Os07g44180)* and *OsMYB2 (Os03g20090)* were associated with QTL for cold stress (Ma et al. 2015; Lv et al. 2016). Wang et al. (2016) identified 67 QTLs on 11 chromosomes of rice with reference to low-temperature tolerance. They reported that qCTS3–9 having the candidate gene *Osryh1* encodes a GTP-binding protein (Bednarek et al. 1994). GWAS mapping using 400 rice accessions of Panel 1 of Rice Diversity (RDP1) and 700 K SNP markers for germination index revealed 42 QTLs linked with chilling tolerance at early developmental stages of seedlings (Shakiba et al. 2017). They detected 29 quantitative trait loci for low-temperature tolerance at the reproductive phase, among which 7 QTLs were linked to sterility (%), 10 QTLs were associated with seed weight panicle^{-1}, 14 QTLs were co-segregated with seed weight plant^{-1}, and a common QTL was linked with two traits. These QTLs were significantly linked with the enhancement of lipid metabolism, oxygen binding and response to biotic and abiotic stimuli. Sales et al. (2017) used a group of 200 landraces for GWAS to map low-temperature germination (LTG) regulation with the help of 1672 single nucleotide polymorphic markers. They found 31 SNP markers showing significant association with low-temperature tolerance, out of which 7 QTLs and 24 QTLs were linked to growth rate 25 °C and 15 °C, respectively. They recorded genes located in the QTL regions that have a role in the stress tolerance mechanisms like disease tolerance (*Os12g37280*) and oxidative stress (*Os01g07376*). They also found that two genes such as *Os06g06400* and *Os03g12820* were expressed differentially to various abiotic stresses. Schlappi et al. (2017) determined 48 QTLs throughout the rice genome at 39 regions. Two novel QTLs were identified, namely, qLTSS4–1 and qLTSS3–4 for LTSS (low-temperature seedling survivability). Two QTLs qPCGC9–2 closer to *OsWRKY76* and qCTS-9 near to gene *Os09g24440* having the ability to enhance chilling tolerance were detected (Yokotani et al. 2013; Zhao et al. 2017). Chawade et al. (2013) reported that qPGC6–1 having the gene *OsDREB1C* plays an important role in regulating low-temperature stress. Xiao et al. (2018) detected five cold tolerance-associated genetic loci for the booting phase and eight loci for the seedling phase. Identified locus *qPSR10* having *LOC_Os10g34840* candidate gene encodes a pectin lyase family protein responsible for seedling stage cold tolerance. Wang et al. (2018a) detected 53 QTLs associated with LTG; a major QTL having a causative gene encodes zinc-finger domain protein called *Stress-Associated Protein 16*. Further, they observed reduced germination due to loss of function of this gene. Zhang et al. (2018) recorded 47 prominent SNP loci correlated with seed survival rate (SR) and severity of damage (SD). They identified three genes like *LOC_Os01g55350, LOC_Os01g55560* and *LOC_Os01g55510,* which showed differential expression among cold-sensitive and tolerant varieties. The genes *LOC_Os01g55510* which codes dyne in light chain type 1 domain containing protein, *LOC_Os01g55560* which encodes ABIL3 protein, and *LOC_Os01g55350* which produces enzyme phosphoenolpyruvate carboxylase were reported to be responsible for abiotic stress tolerance (Sánchez et al. 2006; Jörgens et al. 2010). Twenty-one potentially novel QTL regions were identified for cold tolerance; among them, the indica subset shared a QTL, and the japonica subset shared 10 QTLs (Thapa et al. 2020). *OSWRKY76* and *OsDREB1C* genes were found near the identified QTLs, which were responsible for chilling stress tolerance.

Heat stress has become more hazardous due to greenhouse gas emissions from urbanization and industrialization. For the last 100 years, the average temperature rose by 0.6 °C globally; it has been predicted that, at the end of twenty-first century, the average temperature is expected to increase by 0.5–2.8 °C (Meehl et al. 2005; Root et al. 2003; Vuuren et al. 2008). High-temperature stress is considered a major threat to sustainable rice production around the world, particularly in India, Bangladesh, Pakistan, Thailand, China, Sudan and African countries. The temperature goes beyond 35 °C during the flowering period and adversely affects spikelets and pollens' fertility. Rice spikelet is exposed to 33.7 °C for <1 h at anthesis period which leads to spikelet sterility (Yoshida et al. 1981; Satake and Yoshida 1978; Jagadish et al. 2007). Using bi-parental populations, a number of QTLs were detected for high-temperature tolerance at the flowering phase (Cao et al. 2003; Zhang et al. 2008; Chen et al. 2008; Zhang et al. 2009; Jagadish et al. 2010; Xiao et al. 2011). Pradhan et al. (2016) revealed that spikelet fertility is linked with SSR marker RM547 under stress conditions and reported other microsatellite markers such as RM205, RM228, RM242, RM247, RM314 and INDEL3 indirectly linked to the QTL for heat tolerance. Lafarge et al. (2017) conducted genome-wide association study by employing 13,160 SNPs and reported that 14 loci were prominently linked with spikelet sterility. These loci have the ability to respond to abiotic stresses and regulate cell division and gametophyte development during stress. Associated regions have genes responsible for transcription regulation, namely, WAK (wall-associated kinase), HSP (heat shock protein), WRKY (wall receptor-like protein kinase) and serine carboxypeptidase, and proteins containing F-box domain along with abiotic stresses. They also reported that the genes DEFL peptides, TBC domain-containing protein, SNF2 family protein, *OsCHX15* and seed maturation protein PM23 influence cell division, plant reproduction system, gametophyte development, osmotic adjustment and senescence. A total of 38 loci for rice grain quality and yield characters were detected through GWAS. Out of which, 20 loci have a significant association with high night temperature (Bheemanahalli et al. 2021). These loci are associated with *GW5* candidate gene codes for calmodulin-binding motif family protein that controls rice grain width, and *GS3* regulates grain size and organ size (Weng et al. 2008). Yield QTL *Q3* and *Q8* were found near the *TPP7* gene responsible for abiotic stress tolerance (Li et al. 2011). Bheemanahalli et al. (2021) concluded that rice grain yield along with quality was significantly influenced by high temperature.

2.4.4 Metal Toxicity Tolerance

Metal toxicity is also major abiotic stress in rice cultivation. Aluminum (Al) is one of the main toxic ions that affect crop productivity, particularly under acidic soils (pH < 5.0). According to Uexkull and Mutert (1995), about 50% of the world agricultural land are suffering from this problem. In acidic condition, Al is solubilized and became a phytotoxic ion (Al^{3+}) and inhibits the root development. Famoso et al. (2011) used genome-wide association mapping and identified 48 QTLs associated with Al tolerance. Three rice mutants, namely, ART1, STAR2 and Nrat1, were identified as sensitive to Al toxicity. The haplotypes containing *Nrat1* contributed 40% variation for Al tolerance. The gene *LOC_Os02g0390* encoding *Nramp6* metal transporter was alternatively expressed in the roots of sensitive art1 mutant (Yamaji et al. 2009). This Nramp6 is a plasma membrane located transporter for Al, and it is designated as Nrat1 (Xia et al. 2010). *STAR2* encodes ATP-binding cassette (ABC), and it is an orthologue of Al-sensitive mutant als3 of *Arabidopsis* (Larsen et al. 2007; Huang et al. 2009).

Iron (Fe) and zinc (Zn) act as cofactors for many enzymes involved in the physiological and biochemical processes. Fe and Zn cause metal toxicity when present in larger quantities. In flooded acidic soils, the higher amount of Fe and Zn leads to nutrient imbalance by limiting the absorption of other nutrients. Many scientists investigated the toxic effects of Fe and Zn in rice (Borkert et al. 1998; De Dorlodot et al. 2005; Song et al. 2011; Vromman et al. 2013). Several genes, i.e. *OsFROs*, *OsNRAMPs*, *OsFERs*, *OsZIPs* and *OsYSLs*, were involved in Fe and Zn uptake, transport and accumulation in rice (Chandel et al. 2010). Many researchers identified QTLs for Fe or Zn toxicity tolerance and mapped using bi-parental population (Wu et al. 1998; Wan et al. 2003; Dong et al. 2006; Dufey et al. 2009; Zhang et al. 2013; Dufey et al. 2015; Liu et al. 2016). Zhang et al. (2017a) used the GWAS tool to discover the QTLs for Fe and Zn toxicity tolerance by utilizing 211 diverse rice genotypes. They observed 29 and 31 putative QTLs for various ionic concentrations in shoot at the seedling stage. Five QTLs for ferrous and zinc toxicity tolerance such as *qSdw3a*, *qSdw3b*, *qSFe5*, *qSZn5* and *qSdw12* were documented. The QTL region *qSFe2* was associated with the ferrous content, and it consists of candidate gene *LOC_Os02g48950* (ubiquitin-conjugating enzyme) responsible for abiotic stress tolerance (Zhou et al. 2010).

Toxic compounds released from smelting, mining, energy and allied industries and indiscriminate use of chemicals in agriculture contaminate the land, soil and water with heavy metals like Cd, Ni, Mn and As (Fasani et al. 2018). Manganese (Mn) is an important plant micronutrient required for several metabolic processes and photosynthesis and serves as a cofactor for few enzymes (Goussias et al. 2002; Hebbern et al. 2009). However, Mn can also has phytotoxic effects when it accumulates in a larger quantity in plant tissues (Millaleo et al. 2010). Several SNP markers and few QTLs were recorded for Mn toxicity tolerance using bi-parental rice populations (Wang et al. 2002; Shrestha et al. 2018). Several SNP markers linked with the tolerance to Mn toxicity were detected through GWAS in rice by Shrestha et al. (2018). These QTLs connected with the several candidate genes encoding Mn^{2+} ion binding proteins, peroxidase precursor and heavy metal transporter. An amino acid change was observed in the gene *LOC_Os02g37170* at SNP-2.22465867, but its function was unknown. The gene *OsNRAMP5* (*LOC_Os07g15370*) was linked with shoot Mn content. It is an orthologue of *AtNRAMP1 Arabidopsis* gene, which regulates the uptake and translocation of Mn from root to shoot (Cailliatte et al. 2010; Ishimaru et al. 2012; Yang et al. 2014).

Cadmium is the foremost toxic metal causing major problems in paddy fields. Usually, crops have a high affinity to assimilate Cd. Therefore, it enters the food chain easily (Liu et al. 2003; Meharg et al. 2013; Hu et al. 2016). Several quantitative trait loci linked to Cd accumulation have been identified and mapped to the rice genome (Xue et al. 2009; Ueno et al. 2009; Ishikawa et al. 2010; Norton et al. 2010; Abe et al. 2013). Many genes responsible for Cd intake and transport have been confirmed through cloning by several investigators in rice (Uraguchi and Fujiwara 2013; Clemens and Ma 2016). Zhao et al. (2018a) identified seven QTLs linked to the Cd accumulation in each indica and japonica group. Some of the identified quantitative trait loci were confined to a region containing candidate genes such as *OsNRAMP1, OsNRAMP5* and *OsHMA3*. Out of seven *NRAMP* genes identified, five genes, namely, *OsNRAMP1*, *OsNRAMP* 3, *OsNRAMP* 4, *OsNRAMP* 5 and *OsNRAMP* 6, have been characterized. Among which two genes, *OsNRAMP1* and *OsNRAMP5,* were associated with Cd transporters in rice (Takahashi et al. 2011; Clemens and Ma 2016; Mani and Sankaranarayanan 2018). GWAS conducted by Zhao et al. (2018c) revealed that the novel QTL called *qCd3*–2 contains candidate gene *OsNRAMP2* (functional Cd transporter). They observed four amino acid changes in the open reading frame of the *OsNRAMP2* gene between the contrasting Cd accumulating genotypes. GWAS done by Liu et al. (2020b) depicted 106 signifi cant QTLs associated with the concentration of Cd, Ni, Mo, Co, Sr, Cu, Rb, Zn and K in rice grain. Among which 40, 28, 11, 10, 4, 4, 3, 3 and 3 significant QTLs were associated with Mo, Cd, Co, Zn, Cu, Rb, K, Ni and Sr, respectively. A total of 47 quantitative loci were associated with Cd, Fe, Mo, Ni and Zn concentrations. Among those, 23, 7, 7, 7 and 3 QTLs were related with cadmium, iron, molybdenum, nickel and zinc, respectively. Three genes, namely, *CAL1*, *OsHMA2* and *rgMT,* were reported as cadmium accumulation tolerant genes (Jin et al. 2006; Satoh-Nagasawa et al. 2012; Luo et al. 2018). Pan et al. (2020) mapped 35 QTLs associated with cadmium accumulation using the GWAS approach. A novel QTL *qCd1*–3 consists of candidate gene *OsABCB24* associated with low Cd accumulation. QTLqCd6–2is located close to *OsLCT1,* and *qCd7*-1is located in the interval region of Cd transport gene *HMA3* (Uraguchi et al. 2011).

2.4.5 Submergence Tolerance

The waterlogging condition created by flooding is another major abiotic stress and affects rice production under lowland ecosystems. Flooding negatively affects rice cultivation in rainfed lowland regions of south and south-eastern Asia (Septiningsih et al. 2009). Flooding creates oxygen stress to seedlings; to overcome this problem, the rice coleoptiles need to increase their height as early as possible to attain the water surface and get sufficient oxygen (Magneschi et al. 2009; Alpi and Beevers 1983). During flooding, different metabolic changes occur physiologically, such as glycolysis, starch degradation and ethanol fermentation. Various enzymes related to metabolic processes such as alcohol dehydrogenase, alpha-amylases, fructose-6-phosphate-1, phosphofructokinase, pyruvate dehydrogenase and phosphotransferase were highly active under flooded conditions (Magneschi and Perata 2009; Lasanthi-Kudahettige et al. 2007; Gibbs et al. 2000).

Different mapping populations were used to detect various QTLs by Angaji et al. (2010). During a complete submergence situation, ERF (ethylene-responsive factor) genes help the survivability of rice seedlings up to 10–14 days (Xu et al. 2006; Fukao et al. 2006; Septiningsih et al. 2013). Volante et al. (2017b) observed 160 MTAs (marker-trait associations) through GWAS, and identified QTL regions correlated with *Dlf1, Ehd1, GW5, Nal1, OsOFP2, OsMADS56, OsphyB, OsCOL4, SUI1* and *Sd1* candidate genes. In low water conditions, complex mechanisms were involved in adaptability to decreased oxygen stress during flooding. This leads to increased ACC (the ethylene precursor, 1-aminocyclopropane-1-carboxylic acid) with decreased cytokinin and ABA (abscisic acid) in shoots (Price et al. 2013). Zhang et al. (2017c) recognized 9, 2 and 11 prominent SNPs for flooding tolerance index (FTI), normal coleoptile length (NCL) and flooded coleoptile length (FCL), respectively. They reported that the *LOC_Os06g03520* candidate gene is functionally characterized as protein containing DUF domain induced greatly by anoxia condition. The genomic region of qAG-7-2 at SNP seqrs3583 and one SNP seq-rs3970 was found near the *OsRAMY3D*-oxygen-deficiency-related gene (Nagai et al. 2010; Magneschi and Perata 2009). Gao et al. (2020) used transcriptomic, gene functional characterization and GWAS to identify 50 candidate genes and QTL for waterlogging stress. The candidate genes *LOC_Os11g47610, LOC_Os11g47590, LOC_Os11g47570* and *LOC_Os11g47550* were found adjacent to qAG11. About 20 important genes linked to AG-related characters were detected by Rohilla et al. (2020). The gene *OsXDH1* codes for enzyme *xanthine dehydrogenase 1* which acts as a ROS scavenger. The *SSXT* family protein is GRF1-interacting factor 3, the other potential gene, is involved in anaerobic germination. They reported that candidate gene *LOC_Os01g53920* (HXK6) coding for hexokinase was correlated to SNP

(S1_31006962). It serves as a glucose sensor and regulates the mechanisms related to sugar starvation and hypoxia (Yim et al. 2012; Lim et al. 2013).

2.4.6 Genomic Selection for Abiotic Stress Resilience Breeding

Even though the precision and advantage of genomic selection is accepted unanimously by several molecular rice breeders, the literature available on genomic selection for abiotic stress tolerance breeding in rice is very limited. Abiotic stress tolerance is influenced by the environment; therefore, prediction for G × E interaction at a molecular level is an important step in genomic selection. It measures genotypic response to stress based on yield loss under stress compared with under normal conditions. Several indexes have been proposed to evaluate the stress intensity and genotypic response in the G × E experiments (Fischer 2003). New G × E analysis methods are developed based on linear mixed models that connect the differential sensitivity of genotypes to environments and particular regions of the plant genome (Van Eeuwijk et al. 2010). Ben Hassen et al. (2018b) evaluated the effect of alternate wetting and drying (AWD) system on the performance of two rice breeding populations: a reference panel of 284 accessions and 97 advanced lines using 32 K SNP markers. They considered three traits – days to flowering, panicle weight and nitrogen balance index. The predicted unobserved phenotypes of untested entries were similar to the performance of single environment models with differences in predictive ability ranging from 6 to 4%. Hence, it is suggested to employ a multi-environment model for genomic prediction for abiotic stress tolerance in the genomic selection approach. The application of genomic selection for the prediction and advancement of superior stress-tolerant lines for abiotic stress is minimal, even though its advantage is well known. It is still an open avenue for researchers who wish to make a significant contribution to stress resistance rice breeding using genomic selection.

2.5 Application of GWAS and GS for Improvement of Biotic Stress Resistance in Rice

The consequences of changing climate impact rice development programs due to new pests and diseases (Hasan et al. 2015). Conventional breeding approaches over the years made a significant contribution to the development of sustainable rice cultivars against several biotic stresses. Recent advances in DNA marker technology paved the path to identifying genomic regions responsible for tolerance against several diseases. However, the occurrence of new stresses due to climatic factors or the evolution of new biotypes of pathogens demanded converging several resistance genes into a single high-yielding cultivar to provide broad-spectrum, durable resistance. Apart from this, molecular markers also provided the opportunity to track the resistance genes by following surrogate markers linked to each resistance gene, thus identifying plants carrying two or more resistance genes against targeted traits.

Among several biotic constraints restricting rice production, bacterial blight is one of the major constraints causing a drastic yield reduction through partial grain filling (Pradhan et al. 2015). Thus, developing resistant cultivars is the most efficient way since chemical control measures add extra cost to the production and create environmental hazards (Khush et al. 1989). Several resistance genes for bacterial blight have been identified and incorporated into high-yielding rice varieties (Kumar et al. 2014; Pradhan et al. 2015). Similar efforts have been made in controlling rice blast disease caused by *Magnaporthe oryzae* Barr. Extensive efforts lead to the identification of more than 100 blast resistant *R* genes through DNA molecular markers (Singh et al. 2015). Efforts are being made to investigate the causative genomic regions associated with resistance to other major diseases such as sheath blight and bakanae in rice breeding programs with conventional and molecular markers. Apart from the diseases, some pests like Asian rice gall midge and root-knot nematodes cause drastic yield reduction. Resistance sources to these pests have been investigated and utilized in developing high-yielding varietal development.

2.5.1 GWAS for Biotic Stress Resistance

The genetic architecture of these traits is complex and demands keen interest along with sophisticated and efficient screening methodologies to dissect resistance mechanisms. Several tolerant genomic regions have been mapped using bi-parental populations due to the abundance of molecular markers in rice. However, considering the limitations of bi-parental mapping with respect to allelic diversity and number of recombination, GWAS, a new way of investigating complex traits from a diverse panel of genotypes, dramatically improved the map resolution (Mitchell-Olds 2010). With research improvement and development of statistical algorithms with mixed model approaches (Zhou and Stephens 2012; Wang et al. 2014a, b, c), GWAS platform has been implemented in rice research to dissect complex traits (McCouch et al. 2016) and identified significant marker-trait association for major diseases in rice (Table 2.2). The genetic diversity that existed in rice across the globe favoured the implantation of GWAS for dissecting genomic regions controlling tolerance to many diseases. The genus *Oryza* being self-pollinated and contains many sub-populations exhibits differential population structure, promoting separate GWAS analysis for each species.

Among several diseases in rice, very few major diseases have been focused on dissecting the genomic regions using GWAS approach (Table 2.2); among them, investigations to dissect blast resistance take a major share. More than 500 resistance QTLs have been identified for blast resistance in bi-parental populations (Zheng et al. 2016; Wang et al. 2017a) and 102 *Pi* genes identified till date. However, the GWAS approach gained popularity in mining *R* genes using historical linkage disequilibrium by exploiting genome-wide marker polymorphisms. A study comprising 517 Chinese landraces with genome-wide coverage of SNP markers resulted in the identification of 30 marker-trait associations along with the identification of candidate genes (Wang et al. 2014a). Moreover, the use of SSR markers with low

Table 2.2 Some of the important genome-wide association studies (GWAS) conducted to dissect the marker-trait association for biotic tolerance in rice breeding

Stress	Genotyping platform/markers	Population used	Outcome	Reference
Sheath blight resistance	155 genome-wide markers	217 accessions of USDA core collection	The study showed 10 marker regions associated with sheath blight resistance. Line GSOR 310389 harboured the putative resistant alleles	Jia et al. (2012)
Blast resistance	Sequencing data from the Rice Haplotype Map Project Database	517 Chinese rice landraces	30 associated loci were identified. A candidate gene *Pif* *(Os11g0704100)* was identified. Besides, this study identified novel functional candidate genes	Wang et al. (2014a)
Blast resistance	118 SSR markers	226 japonica rice cultivars	The study identified 31 significant marker-trait associations with 17 SSR loci and 18 favourable alleles were identified	Guo et al. (2015)
Blast disease	44 K SNP chip	161 rice cultivars from rice diversity panel 1 (RDP1)	31 loci associated with blast resistance identified	Mgonja et al. (2016)
Nematode resistance (RKN)	Genotyped with 44,100 SNPs	332 accessions of RDP1	GWAS detected 11 QTLs. Further, candidate genes on chromosome 11 having homology with *HordeumMla* locus were identified	Dimkpa et al. (2016)
Blast resistance	Genotyped using 700,000 SNP array	RDP1	GWAS detected 97 loci for blast resistance. Among them, 82 were novel loci and 15 co-localized with known blast resistance loci	Kang et al. (2016)
Blast resistance	The 3835 high-quality SNP markers were selected from the 44-K SNP markers	RDP1	GWAS identified 16 LAFBRs. Among them, 13 are novel and the other 3 are co-localized with known blast resistance regions	Zhu et al. (2016)

(continued)

Table 2.2 (continued)

Stress	Genotyping platform/markers	Population used	Outcome	Reference
Blast resistance	160 SSR markers	276 *indica* landraces	The study identified 26 SSR markers significantly associated with blast resistance. Nineteen of the markers were associated with previously reported genes/QTL, and 7 were newly identified	Wu et al. (2016)
Blast resistance	Genotyping of 150 accessions with 10,937 SNPs. Indica panel 190 accessions were genotyped with 14,187 SNPs	150 accessions in a set from japonica group and another set consists of 190 accessions from indica group	A total of 7 SNPs co-localized with 4 NBS-LRR genes such as *Pi37* and *Pish*	Raboin et al. (2016)
Bacterial leaf blight resistance	GBS	285 rice accessions	GWAS revealed novel SNPs identified are linked with known bacterial blight resistance *Xa* genes	Dilla-Ermita et al. (2017)
Bakanae disease resistance	166,418 SNP markers by GBS	138 *japonica* rice collections	GWAS revealed two genomic regions associated with resistance to *bakanae* on chromosomes 1 (*qBK1_628091*) and 4 (as*qBK4_31750955*)	Volante et al. (2017a)
BLB resistance	317,894 SNPs	172 diverse accessions belonging to *Oryza sativa* ssp. *indica*	Twelve resistance loci were identified. Two hotspot regions (L11 and L12) were identified which were positioned within cloned *R* genes *xa25* and *Xa26* and one fine-mapped *R* gene *Xa4*	Zhang et al. (2017b)
Blast resistance	5291 SNPs from the custom-designed array	355 *indica* group accessions	In total, 127 associations were identified. Besides, 25 pleiotropic associations with more than 2 strains were identified. In addition,	Lu et al. (2019)

(continued)

Table 2.2 (continued)

Stress	Genotyping platform/markers	Population used	Outcome	Reference
			2341 non-redundant candidate genes, including 45 disease resistance-related genes, were predicted in a 200-kb genomic region for these associations	
Blast resistance	The 277,524 SNPs were utilized from 700 K SNPs	234 rice cultivars belong to RDP1	The study identified 56 QTLs for blast resistance. However, only one QTL was associated with resistance to all three isolates, and it was localized with the known R gene *Pik* locus	Li et al. (2019b)
BLB resistance	6 K SNP chip	120 JMAGIC lines	The detected quantitative trait nucleotides (QTNs)were delimited within two SNPs, 1,192,907 and 11,943,779	Kim and Reinke (2019)
Sheath blight resistance	2,977,750 single nucleotide	563 rice accessions	The study detected 134, 562 and 75 suggestive associations with culm length, lesion height and relative lesion height. More than 44% of detected relative lesion height-suggestive associated loci (RLH-SALs) harboured multiple QTLs/genes associated with sheath blight resistance. At the same time, the other RLH-SALs were putative novel sheath blight resistance loci. A total of 261 sheath blight resistance putative functional genes were screened from 23 RLH-SALs	Zhang et al. (2019)

(continued)

Table 2.2 (continued)

Stress	Genotyping platform/markers	Population used	Outcome	Reference
Sheath blight resistance	A total of 44,000 high-density SNPs	299 rice cultivars belong to RDP1	GWAS identified 11 significant SNPs for sheath blight resistance from the 3 independent trials. *qSB-3* and *qSB-6* on chromosomes 3 and 6 were identified were stable across trials	Chen et al. (2019)
Rice blast	37,423 SNP markers	A total of 311 accessions from temperate/ tropical japonica and indica group	GWAS revealed 14 MTA, of which 8 were identified under field conditions and 6 under controlled screening. Three stable marker-trait associations were identified under both conditions	Volante et al. (2020)
Root-knot nematode resistance	SNP genotyping with 50 K "OsSNPnks" genic Affymetrix chip	A total of 272 accessions belonging to *O. nivara, O. rufipogon, O. sativa* f. *spontanea* of wild rice species	The study identified 40 resistant accessions. Further, 17 new SNPs associated with resistant traits were identified. SNPs on chromosomes 1, 2, 3, 4, 6, 10 and 11 associated with the candidate genes such as NBS-LRR, Cf2/Cf5 resistance protein, *MYB, bZIP, ARF, SCARECROW* and *WRKY* TFs	Hada et al. (2020)
Bacterial streak resistance	176,820 SNPs	236 diverse rice accession	The study showed 12 QTLs conferring resistance to 5 Thai *Xoc*. Isolates, 5 of them conferred resistance to more than 1 isolate. QTLs, *qBLS5.1*and *qBLS2.3* were found promising and durable. QTL *qBLS5.1* harbours *xa5* as a potential candidate gene, while *qBLS2*.3 harboured	Sattayachiti et al. (2020)

(continued)

Table 2.2 (continued)

Stress	Genotyping platform/markers	Population used	Outcome	Reference
			putative candidate genes associated with pectinesterase inhibitor *(OsPEI)*, eukaryotic zinc-binding protein (*OsRAR1)* and NDP epimerase function	
Blast resistance	700, 000 SNPs	584 rice accessions of RDP2	GWAS detected 27 loci for rice blast resistance. 22 of them were not associated with any previously known *R* gene or QTLs	Liu et al. (2020b)
Blast resistance	9997 SNPs	139 temperate japonica rice strains	The study identified 3 novel QTLs other than *NIS1*. A rare allele of the *RRobN1*locus on chromosome 6 confers broad-spectrum resistance	Frontini et al. (2021)

genome coverage also allowed the identification of 13 MTAs, suggesting that lower marker density is sufficient to identify MTA in rice due to high LD owing to its self-pollination property (Guo et al. 2015). Frontini et al. (2021) made an effort to identify QTL in low and high nitrogen conditions and found that higher nitrogen levels increase the buffering capacity of susceptibility genes. Similar efforts have been made in other diseases like sheath blight (Jia et al. 2012; Zhang et al. 2019; Chen et al. 2019), bacterial leaf blight (Dilla-Ermita et al. 2017; Zhang et al. 2017b; Kim and Reinke 2019), nematode resistance (Dimkpa et al. 2016; Hada et al. 2020) and bakanae disease resistance (Volante et al. 2017a). Advances made in the sequencing approaches with the invention of second-generation sequencing tools provided strength to high-throughput genotyping of rice GWAS panels (Wang et al. 2018b). Using these modern sequencing tools, GWAS becomes handy and cost-effective. Further advanced statistical methods fill the gaps created by missing data generated during the genotyping process (Wang et al. 2018a). Moreover, genotyping of wild cultivars of rice is also improved with sequencing strategies which benefitted more to GWAS analysis in finding disease resistance QTL in wild relatives of rice (Huang et al. 2015).

2.5.2 Genomic Selection for Biotic Stress Resistance Breeding

Genomic selection is yet another modern breeding tool called a modified marker-assisted selection method, since it uses genome-wide molecular marker information to select future best individuals. The approach is applicable to complex traits under the control of minor genes, and its prediction/selection accuracy is more accurate than phenotype-based selection (Spindel and Iwata 2018). In rice, genomic selection is practiced for yield (Wang et al. 2017b) and related traits such as plant height (Spindel et al. 2016), panicle weight (Grenier et al. 2015), tiller number (Xu et al. 2014) and productive tiller number (Wang et al. 2017b). The GS accuracies for different traits revealed the best performance of rrBLUP model among the other GS models (Spindel et al. 2016). The GS approach has been employed for blast resistance in disease resistance breeding (Huang et al. 2019). They have implemented GS on two populations, one with 161 accessions of the African population and the other with 162 accessions from the USA. The African and USA panels were evaluated for six and eight different isolates of rice blast, respectively. The accuracy of different models was also tested by cross-validation and showed fixed effect BLUP (fgBLUP) model was more accurate than other methods. They also highlighted the importance and accuracy of GS in blast resistance rice breeding. The availability of literature in rice disease resistance breeding with an application of the GS is very limited; however, adapting GS will increase the effectiveness of selection and increase the success rate in biotic stress resistance breeding in rice in the near future. Application of deep learning and machine learning tools in predicting best performing genotype under disease pressure environments considering weather data along with genotype data will increase the selection accuracy (Gillberg et al. 2019). Integration of G $\times$ E component into GS models while selecting for disease resistance is still a challenging and required improvement of disease resistance in rice breeding.

2.6 GWAS and GS Perspectives in Stress Resilience Breeding of Rice

In rice, the application of GWAS for dissecting the genetic architecture of yield and yield-related traits has been started recently, and appreciable progress has been made in identifying genomic regions responsible for various target traits. However, GS application in rice breeding programs is still in the adaptation stage, and it's important that every rice breeder who wish to adopt GS in existing breeding activity should develop own standard operating procedure indicating exact steps for developing model and selecting genotypes based on GEBVs. From the different marker platforms, dense sets of markers now available have significantly modified the genetic toolkit for rice. Availability of sequencing platforms simplified the generation of genomic resources in the form of SNP markers covering the entire genome of rice, making it easy to adapt new breeding strategies in the form of GWAS and GS to accelerate the breeding process. It has been proved that GWAS and GS have significant contributions in achieving greater success in target-oriented breeding programs in cereal crops (Alqudah et al. 2020). Integrating GS models along with GWAS helps improve accuracy of genomic selection for target trait (Fig. 2.3). It is implausibly proved that GWAS explains the complete heritable variation of a complex trait, which is practically impossible in bi-parental QTL mapping. It's become easy to untangle the genomic architecture of complex traits by GWAS and identify novel allelic variation for breeding purposes (Sun et al. 2017). GWAS is also powerful in identifying candidate genes responsible for target traits, as demonstrated by many studies in different cereal crops; this feature of GWAS has been exploited in rice to identify target candidate genes for blast resistance (Volante et al. 2020). In the near future, the output of GWAS can be extensively used in candidate gene mapping and gene editing for stress resistance in rice. Deep analysis of GWAS by utilizing genotype by sequencing strategy and haplotype-based analysis is a key for success in detecting novel variation for stress tolerance in rice. Many studies used both bi-parental mapping and association mapping to detect and validate the same QTL for target traits in other crops like maize (Zhao et al. 2018c) and brassica (He et al. 2017); however, such studies for stress resistance breeding in rice are rather limited. A diverse panel including wild rice relative of association mapping population represents a source of diverse allelic variation for complex traits including biotic and abiotic stress tolerance. GWAS analysis allows identifying candidate genes, which can be further subjected to validation through gene editing and expression studies. Availability of different statistical tools enables integrating GWAS with omics tools which will help in the dissecting and improvement of stress resistance in rice.

As a modified version of marker-assisted selection, genomic selection is gaining popularity in plant breeding due to its effectiveness in selection. It estimates the breeding values of individuals in the population based on genome-wide marker information on which best individual will be selected. Even though the success of GS is evidenced in several other crops, implantation in rice is very limited, particularly in stress resistance breeding. Therefore, genomic selection may be considered as a potential breeding approach in stress resistance breeding of rice. The accurate identification of the best phenotype based on genotypic information in GS is much higher than marker-assisted selection. The genetic gain per unit cost and per breeding cycle is significantly higher in genomic selection, evidencing the importance of adapting in precision breeding programs. Integrating environmental variables with a genomic selection model helps to reduce the error rate in identifying appropriate genotypes and increases the precision of selection. Several genomic selection models have been developed to account for the interaction of genotypes with the environment and estimate the pure genetic or breeding value for selection. Utilizing these models in rice stress breeding programs helps to achieve fruitful results of stress-resilient rice cultivars.

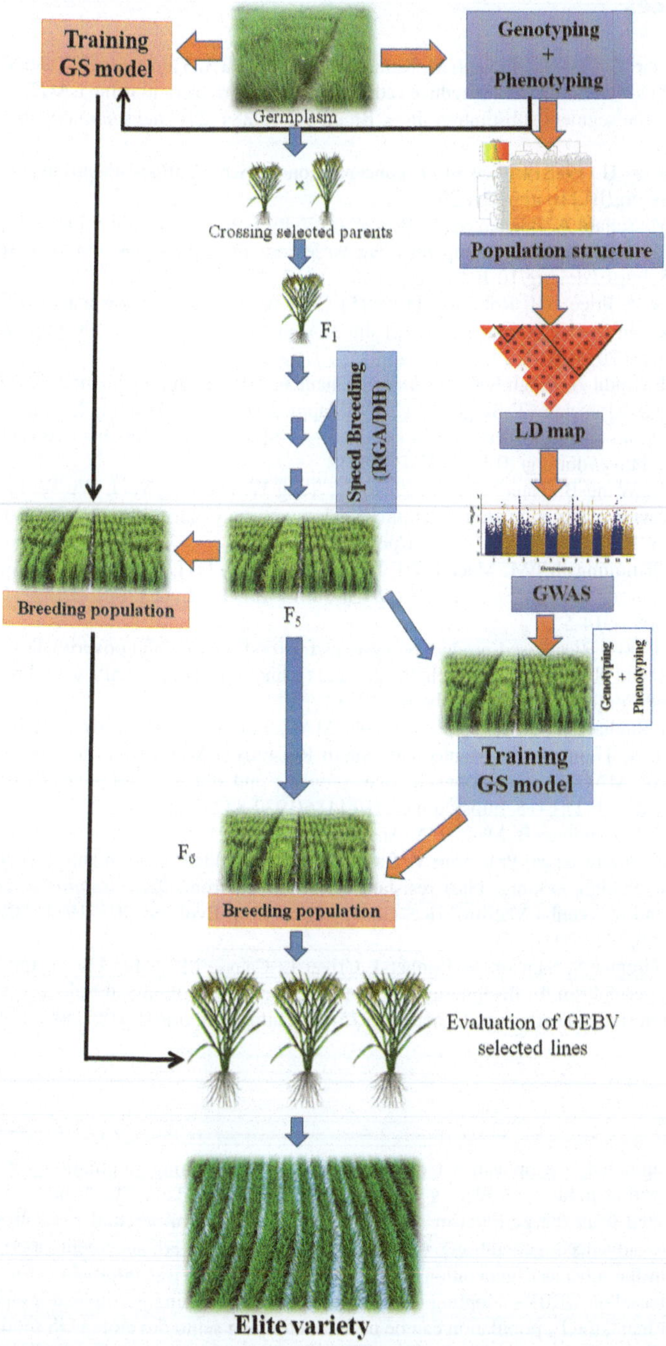

Fig. 2.3 Schematic representation of integrating genomic selection and GWAS techniques for improved genetic gain. On the right in the figure, GWAS can be started by genotyping and phenotyping a set of germplasm accession followed by performing population structure analysis and LD mapping to identify marker-trait associations for target trait. Simultaneously, the

References

Abe T, Nonoue Y, Ono N, Omoteno M, Kuramata M, Fukuoka S, Yamamoto T, Yano M, Ishikawa S (2013) Detection of QTLs to reduce cadmium content in rice grains using LAC23/Koshihikari chromosome segment substitution lines. Breed Sci 63:284–291. https://doi.org/10.1270/jsbbs.63.284

Alpi A, Beevers H (1983) Effects of O_2 concentration on rice seedlings. Plant Physiol 71:30–34. https://doi.org/10.1104/pp.71.1.30

Alqudah AM, Sallam A, Baenziger PS, Börner A (2020) GWAS: fast-forwarding gene identification and characterization in temperate cereals: lessons from Barley–A review. J Adv Res 22: 119–135. https://doi.org/10.1016/j.jare.2019.10.013

Al-Shugeairy Z, Price AH, Robinson D (2015) Genome wide association mapping for drought recovery trait in rice (*Oryza sativa* L.). Int J Appl Agric Sci 1(1):11–18. https://doi.org/10.11648/j.ijaas.20150101.12

Ammar MH, Pandit A, Singh RK, Sameena S, Chauhan MS, Singh AK, Sharma PC, Gaikwad K, Sharma TR, Mohapatra T, Singh NK (2009) Mapping of QTLs controlling Na^+, K^+ and Cl^- ion concentrations in salt tolerant indica rice variety CSR27. J Plant Biochem Biotechnol 18(2): 139–150. https://doi.org/10.1007/BF03263312

An H, Liu K, Wang B, Tian Y, Ge Y, Zhang Y, Tang W, Chen G, Yu J, Wu W, Liu X (2020) Genome-wide association study identifies QTLs conferring salt tolerance in rice. Plant Breed 139(1):73–82. https://doi.org/10.1111/pbr.12750

Angaji SA, Septiningsih EM, Mackill DJ, Ismail AM (2010) QTLs associated with tolerance of flooding during germination in rice (*Oryza sativa* L.). Euphytica 172(2):159–168.10.1007/s10681-009-0014-5

Badawi TA (2004) Rice-based production systems for food security and poverty alleviation in the Near East and North Africa: new challenges and technological opportunities. In: Proceedings of FAO Rice Conference, Rome, Italy, pp. 12–13

Bandillo N, Raghavan C, Muyco PA, Sevilla MAL, Lobina IT, Dilla-Ermita CJ, Tung CW, McCouch S, Thomson M, Mauleon R, Singh RK (2013) Multi-parent advanced generation inter-cross (MAGIC) populations in rice: progress and potential for genetics research and breeding. Rice 6(1):1–15. https://doi.org/10.1186/1939-8433-6-11

Basavaraj PS, Muralidhara B, Manoj CA, Anantha MS, Rathod S, Raju CD, Senguttuvel P, Madhav MS, Srinivasaprasad M, Prakasam V, Basavaraj K (2021) Identification and molecular characterization of high-yielding, blast resistant lines derived from *Oryza rufipogon* Griff. in the background of 'Samba Mahsuri' rice. Genet Resour Crop Evol 68:1905–1921. https://doi.org/10.1007/s10722-020-01104-1

Bauchet G, Grenier S, Samson N, Bonnet J, Grivet L, Causse M (2017) Use of modern tomato breeding germplasm for deciphering the genetic control of agronomical traits by genome wide association study. Theor Appl Genet 130:875–889. https://doi.org/10.1007/s00122-017-2857-9

Fig. 2.3 (continued) germplasm set can be considered as training population for developing genomic selection model considering genotype and phenotype data. The best performing elite parents selected from the germplasm set can be used to derive bi-parental progenies which are forwarded to advanced generations via accelerated generation advancement approaches. The population in the advanced generations like F_5 can be used as a test population for selection of individuals based on GEBVs adopting the GS model developed using germplasm as a training set. On the other hand, the F_5 population can be used as a training set to develop a GS model including the genotypic and phenotypic information of F_5 and also integrating results of GWAS for better accuracy. This model can be tested on the F_6 population for selection of elite lines which may be evaluated and released as a new elite variety

Bednarek SY, Reynolds TL, Schroeder M, Grabowski R, Hengst L, Gallwitz D, Raikhel NV (1994) A small GTP-binding protein from *Arabidopsis thaliana* functionally complements the yeast YPT6 null mutant. Plant Physiol 104(2):591–596. https://doi.org/10.1104/pp.104.2.591

Ben Hassen B, Cao TV, Bartholome J, Orasen G, Colombi C, Rakotomalala J, Razafinimpiasa L, Bertone C, Biselli C, Volante A, Desiderio F (2018a) Rice diversity panel provides accurate genomic predictions for complex traits in the progenies of biparental crosses involving members of the panel. Theor Appl Genet 131(2):417–435. https://doi.org/10.1007/s00122-017-3011-4

Ben Hassen M, Bartholomé J, Valè G, Cao T, Ahmadi N (2018b) Genomic prediction accounting for genotype by environment interaction offers an effective framework for breeding simultaneously for adaptation to an abiotic stress and performance under normal cropping conditions in rice. G3 8:2319–2332. https://doi.org/10.1534/g3.118.200098

Bernardo R (2010) Genomewide selection with minimal crossing in self-pollinated crops. Crop Sci 50:624–627. https://doi.org/10.2135/cropsci2009.05.0250

Bhandari A, Bartholomé J, Cao-Hamadoun TV, Kumari N, Frouin J, Kumar A, Ahmadi N (2019) Selection of trait-specific markers and multi-environment models improve genomic predictive ability in rice. PLoS One 14(5):e0208871. https://doi.org/10.1371/journal.pone.0208871

Bheemanahalli R, Knight M, Quinones C, Doherty CJ, Jagadish SVK (2021) Genome-wide association study and gene network analyses reveal potential candidate genes for high night temperature tolerance in rice. Sci Rep 11(6747):1–17. https://doi.org/10.1038/s41598-021-85921-z

Bonilla P, Dvorak J, Mackell D, Deal K, Gregorio G (2002) RFLP and SSLP mapping of salinity tolerance genes in chromosome 1 of rice (*Oryza sativa* L.) using recombinant inbred lines. Philippine Agricultural Scientist (Philippines)

Borkert CM, Cox FR, Tucker M (1998) Zinc and copper toxicity in peanut, soybean, rice, and corn in soil mixtures. Commun Soil Sci Plant Anal 29(19–20):2991–3005. https://doi.org/10.1080/00103629809370171

Bush WS, Moore JH (2012) Genome-wide association studies. PLoS Comput Biol 8(12):e1002822. https://doi.org/10.1371/journal.pcbi.1002822

Cailliatte R, Schikora A, Briat JF, Mari S, Curie C (2010) High-affinity manganese uptake by the metal transporter NRAMP1 is essential for Arabidopsis growth in low manganese conditions. Plant Cell 22(3):904–917. https://doi.org/10.1105/tpc.109.073023

Campbell MT, Knecht AC, Berger B, Brien CJ, Wang D, Walia H (2015) Integrating image-based phenomics and association analysis to dissect the genetic architecture of temporal salinity responses in rice. Plant Physiol 168(4):1476–1489. https://doi.org/10.1104/pp.15.00450

Campbell MT, Bandillo N, Al Shiblawi FR, Sharma S, Liu K, Du Q, Schmitz AJ, Zhang C, Véry AA, Lorenz AJ, Walia H (2017) Allelic variants of OsHKT1; 1 underlie the divergence between indica and japonica subspecies of rice (*Oryza sativa*) for root sodium content. PLoS Genet 13(6):e1006823. https://doi.org/10.1371/journal.pgen.1006823

Cao L, Zhao J, Zhan X, Li D, He L, Cheng S (2003) Mapping QTLs for heat tolerance and correlation between heat tolerance and photosynthetic rate in rice. Chin J Rice Sci 17(3): 223–227

Castells E, Portolés S, Huang W, Mas P (2010) A functional connection between the clock component TOC1 and abscisic acid signaling pathways. Plant Signal Behav 5(4):409–411. https://doi.org/10.4161/psb.5.4.11213

Chakraborti M, Anilkumar C, Verma RL, Fiyaz AR, Reshmi Raj KR, Patra BC, Balakrishnan D, Sarkar S, Mondal NP, Kar MK, Meher J, Sundaram RM, Subba Rao LV (2021) Rice breeding in India: eight decades of journey towards enhancing the genetic gain for yield, nutritional quality, and commodity value. ORYZA-An Int J Rice 58(Special Issue):69–88. https://doi.org/10.35709/ory.2021.58.spl.2

Chandel G, Banerjee S, Verulkar SB (2010) Expression profiling of metal homeostasis related candidate genes in rice (*Oryza spp.*) using semi quantitative RT-PCR analysis. Rice Genetics Newsletter 2010

Chawade A, Lindlöf A, Olsson B, Olsson O (2013) Global expression profiling of low temperature induced genes in the chilling tolerant japonica rice Jumli Marshi. PLoS One 8(12):e81729. https://doi.org/10.1371/journal.pone.0081729

Chen Q, Yu S, Li C, Mou T (2008) Identification of QTLs for heat tolerance at flowering stage in rice. Sci Agric Sin 41:315–321

Chen L, Song Y, Li S, Zhang L, Zou C, Yu D (2012) The role of WRKY transcription factors in plant abiotic stresses. Biochim Biophys Acta 1819:120–128. https://doi.org/10.1016/j.bbagrm.2011.09.002

Chen G, Feng H, Hu Q, Qu H, Chen A, Yu L, Xu G (2015) Improving rice tolerance to potassium deficiency by enhancing Os HAK 16p: WOX 11-controlled root development. Plant Biotechnol J 13(6):833–848. https://doi.org/10.1111/pbi.12320

Chen Z, Feng Z, Kang H, Zhao J, Chen T, Li Q, Gong H, Zhang Y, Chen X, Pan X, Liu W (2019) Identification of new resistance loci against sheath blight disease in rice through genome-wide association study. Rice Sci 26(1):21–31. https://doi.org/10.1016/j.rsci.2018.12.002

Chen C, Norton GJ, Price AH (2020) Genome-wide association mapping for salt tolerance of rice seedlings grown in hydroponic and soil systems using the Bengal and Assam Aus panel. Front Plant Sci 11:576479. https://doi.org/10.3389/fpls.2020.576479

Claes B, Dekeyser R, Villarroel R, Van den Bulcke M, Bauw G, Van Montagu M, Caplan A (1990) Characterization of a rice gene showing organ-specific expression in response to salt stress and drought. Plant Cell 2(1):19–27. https://doi.org/10.1105/tpc.2.1.19

Clemens S, Ma JF (2016) Toxic heavy metal and metalloid accumulation in crop plants and foods. Annu Rev Plant Biol 67(1):489–512. https://doi.org/10.1146/annurev-arplant-043015-112301

Cominelli E, Galbiati M, Vavasseur A, Conti L, Sala T, Vuylsteke M, Leonhardt N, Dellaporta SL, Tonelli C (2005) A guard-cell-specific MYB transcription factor regulates stomatal movements and plant drought tolerance. Curr Biol 15(13):1196–1200. https://doi.org/10.1016/j.cub.2005.05.048

Cooper M, Messina CD, Podlich D, Totir LR, Baumgarten A, Hausmann NJ, Wright D, Graham G (2014) Predicting the future of plant breeding: complementing empirical evaluation with genetic prediction. Crop Pasture Sci 65(4):311–336. https://doi.org/10.1071/CP14007

Courtois B, McLaren G, Sinha PK, Prasad K, Yadav R, Shen L (2000) Mapping QTL associated with drought avoidance in upland rice. Mol Breed 6:55–66. https://doi.org/10.1023/A:1009652326121

Das G, Patra JK, Baek KH (2017) Insight into MAS: a molecular tool for development of stress resistant and quality of rice through gene stacking. Front Plant Sci 8:985. https://doi.org/10.3389/fpls.2017.00985

De Dorlodot S, Lutts S, Bertin P (2005) Effects of ferrous iron toxicity on the growth and mineral composition of an interspecific rice. J Plant Nutrit 28(1):1–20. https://doi.org/10.1081/PLN-200042144

Desta ZA, Ortiz R (2014) Genomic selection: genome-wide prediction in plant improvement. Trends Plant Sci 19(9):592–601. https://doi.org/10.1016/j.tplants.2014.05.006

Dilla-Ermita CJ, Tandayu E, Juanillas VM, Detras J, Lozada DN, Dwiyanti MS, Vera Cruz C, Mbanjo EGN, Ardales E, Diaz MG, Mendioro M, Thomson MJ, Kretzschmar T (2017) Genome-wide association analysis tracks bacterial leaf blight resistance loci in rice diverse germplasm. Rice 10(1):1–17. https://doi.org/10.1186/s12284-017-0147-4

Dimkpa SON, Lahari Z, Shrestha R, Douglas A, Gheysen G, Price AH (2016) A genome-wide association study of a global rice panel reveals resistance in *Oryza sativa* to root-knot nematodes. J Exp Bot 67(4):1191–1200. https://doi.org/10.1093/jxb/erv470

Dong Y, Ogawa T, Lin D, Koh HJ, Kamiunten H, Matsuo M, Cheng S (2006) Molecular mapping of quantitative trait loci for zinc toxicity tolerance in rice seedling (*Oryza sativa* L.). Field Crops Res 95(2–3):420–425. https://doi.org/10.1016/j.fcr.2005.03.005

Dubouzet JG, Sakuma Y, Ito Y, Kasuga M, Dubouzet EG, Miura S, Seki M, Shinozaki K, Yamaguchi-Shinozaki K (2003) OsDREB genes in rice, *Oryza sativa* L., encode transcription activators that function in drought, high salt and cold responsive gene expression. Plant J 33(4): 751–763. https://doi.org/10.1046/j.1365-313X.2003.01661.x

Dufey I, Hakizimana P, Draye X, Lutts S, Bertin P (2009) QTL mapping for biomass and physiological parameters linked to resistance mechanisms to ferrous iron toxicity in rice. Euphytica 167(2):143–160. https://doi.org/10.1007/s10681-008-9870-7

Dufey I, Draye X, Lutts S, Lorieux M, Martinez C, Bertin P (2015) Novel QTLs in an interspecific backcross *Oryza sativa* × *Oryza glaberrima* for resistance to iron toxicity in rice. Euphytica 204(3):609–625. https://doi.org/10.1007/s10681-014-1342-7

Famoso AN, Zhao K, Clark RT, Tung CW, Wright MH, Bustamante C, Kochian LV, McCouch SR (2011) Genetic architecture of aluminum tolerance in rice (*Oryza sativa*) determined through genome-wide association analysis and QTL mapping. PLoS Genet 7(8). https://doi.org/10.1371/journal.pgen.1002221

Fang Y, Xie K, Hou X, Hu H, Xiong L (2010) Systematic analysis of GT factor family of rice reveals a novel subfamily involved in stress responses. Mol Gen Genomics 283(2):157–169. https://doi.org/10.1007/s00438-009-0507-x

Fang N, Wang R, He W, Yin C, Guan C, Chen H, Huang J, Wang J, Bao Y, Zhang H (2016) QTL mapping of panicle blast resistance in japonica landrace heikezijing and its application in rice breeding. Mol Breed 36(12):1–8. https://doi.org/10.1007/s11032-016-0603-7

Fasani E, Manara A, Martini F, Furini A, Dal Corso G (2018) The potential of genetic engineering of plants for the remediation of soils contaminated with heavy metals. Plant Cell Environ 41: 1201–1232. https://doi.org/10.1111/pce.12963

Fischer KS (2003) Breeding rice for drought-prone environments. Int Rice Res Inst 2003

Flint-Garcia SA, Thornsberry JM, Buckler ES (2003) Structure of linkage disequilibrium in plants. Ann Rev Plant Biol 54(1):357–374. https://doi.org/10.1146/annurev.arplant.54.031902.134907

Flowers TJ (2004) Improving crop salt tolerance. J Exp Bot 55(396):307–319. https://doi.org/10.1093/jxb/erh003

Frontini M, Boisnard A, Frouin J, Ouikene M, Morel JB, Ballini E (2021) Genome-wide association of rice response to blast fungus identifies loci for robust resistance under high nitrogen. BMC Plant Biol 21(1):99. https://doi.org/10.1186/s12870-021-02864-3

Frouin J, Languillaume A, Mas J, Mieulet D, Boisnard A, Labeyrie A, Bettembourg M, Bureau C, Lorenzini E, Portefaix M, Turquay P (2018) Tolerance to mild salinity stress in japonica rice: a genome-wide association mapping study highlights calcium signalling and metabolism genes. PLoS One 13(1):e0190964. https://doi.org/10.1371/journal.pone.0190964

Fujita D, Trijatmiko KR, Tagle AG, Sapasap MV, Koide Y, Sasaki K, Tsakirpaloglou N, Gannaban RB, Nishimura T, Yanagihara S, Fukuta Y (2013) NAL1 allele from a rice landrace greatly increases yield in modern indica cultivars. Proc Natl Acad Sci 110(51):20431–22036. https://doi.org/10.1073/pnas.1310790110

Fukao T, Xu K, Ronald PC, Bailey-Serres J (2006) A variable cluster of ethylene response factor–like genes regulates metabolic and developmental acclimation responses to submergence in rice. Plant Cell 18(8):2021–2034. https://doi.org/10.1105/tpc.106.043000

Fukuda A, Nakamura A, Hara N, Toki S, Tanaka Y (2011) Molecular and functional analyses of rice NHX-type Na+ /H+ antiporter genes. Planta 233:175–188. https://doi.org/10.1007/s00425-010-1289-4

Gao H, Zhang C, He H, Liu T, Zhang B, Lin H, Li X, Wei Z, Yuan Q, Wang Q, Yu C (2020) Loci and alleles for submergence responses revealed by GWAS and transcriptional analysis in rice. Mol Breed 40(8):1–6. https://doi.org/10.1007/s11032-020-01160-6

Ghomi K, Rabiei B, Sabouri H, Sabouri A (2013) Mapping QTLs for traits related to salinity tolerance at seedling stage of rice (*Oryza sativa* L.): an agri-genomics study of an Iranian rice population. OMICS 17(5):242–251. https://doi.org/10.1089/omi.2012.0097

Gibbs J, Morrell S, Valdez A, Setter TL, Greenway H (2000) Regulation of alcoholic fermentation in coleoptiles of two rice cultivars differing in tolerance to anoxia. J Exp Bot 51(345):785–796. https://doi.org/10.1093/jexbot/51.345.785

Gillberg J, Marttinen P, Mamitsuka H, Kaski S (2019) Modelling G× E with historical weather information improves genomic prediction in new environments. Bioinformatics 35(20): 4045–4052. https://doi.org/10.1093/bioinformatics/btz197

Gnanamanickam SS (2009) Major diseases of rice. In: Biological control of rice diseases. Progress in Biological Control, vol 8. Springer, Dordrecht. https://doi.org/10.1007/978-90-481-2465-7_2

Gonzalez LE, Keller K, Chan KX, Gessel MM, Thines BC (2017) Transcriptome analysis uncovers Arabidopsis F-BOX STRESS INDUCED 1 as a regulator of jasmonic acid and abscisic acid stress gene expression. BMC Genomics 18:533. https://doi.org/10.1186/s12864-017-3864-6

Goussias C, Boussac A, Rutherford AW (2002) Photosystem II and photosynthetic oxidation of water: an overview. Philos Trans R Soc Lond Ser B Biol Sci 357:1369–1381. https://doi.org/10.1098/rstb.2002

Gregorio GB, Islam MR, Vergara GV, Thirumeni S (2013) Recent advances in rice science to design salinity and other abiotic stress tolerant rice varieties. SABRAO J Breed Genet 45(1): 31–41

Grenier C, Cao T-V, Ospina Y, Quintero C, Châtel MH, Tohme J, Courtois B, Ahmadi N (2015) Accuracy of genomic selection in a rice synthetic population developed for recurrent selection breeding. PLoS One 10(8):e0136594. https://doi.org/10.1371/journal.pone.0136594

Guirguis K, Gershunov A, Schwartz R, Bennet S (2011) Recent warm and cold daily winter temperature extremes in the Northern Hemisphere. Geophys Res Lett S0094–8276. https://doi.org/10.1029/2011GL0487622

Guo Z, Tucker DM, Lu J, Kishore V, Gay G (2012) Evaluation of genome-wide selection efficiency in maize nested association mapping populations. Theor Appl Genet 124(2):261–275. https://doi.org/10.1007/s00122-011-1702-9

Guo Z, Tucker DM, Basten CJ, Gandhi H, Ersoz E, Guo B, Xu Z, Wang D, Gay G (2014) The impact of population structure on genomic prediction in stratified populations. Theor Appl Genet 127(3):749–762. https://doi.org/10.1007/s00122-013-2255-x

Guo L, Guo W, Zhao H, Wang J, Liu H, Sun J, Zheng H, Sha H, Zou D (2015) Association mapping and resistant alleles' analysis for japonica rice blast resistance. Plant Breed 134(6):646–652. https://doi.org/10.1111/pbr.12310

Guo Z, Yang W, Chang Y, Ma X, Tu H, Xiong F, Jiang N, Feng H, Huang C, Yang P, Zhao H, Chen G, Liu H, Luo L, Hu H, Liu Q, Xiong L (2018) Genome-wide association studies of image traits reveal genetic architecture of drought resistance in rice. Mol Plant 11(6):789–805. https://doi.org/10.1016/j.molp.2018.03.018

Gur A, Zamir D (2004) Unused natural variation can lift yield barriers in plant breeding. PLoS Biol 2(10):e245. https://doi.org/10.1371/journal.pbio.0020245

Hada A, Dutta TK, Singh N, Singh B, Rai V, Singh NK, Rao U (2020) A genome-wide association study in Indian wild rice accessions for resistance to the root-knot nematode *Meloidogyne graminicola*. PLoS One 15(9):1–24. https://doi.org/10.1371/journal.pone.0239085

Han B, Huang X (2013) Sequencing-based genome-wide association study in rice. Curr Opin Plant Biol 16(2):133–138. https://doi.org/10.1016/j.pbi.2013.03.006

Hasan MM, Rafii MY, Ismail MR, Mahmood M, Rahim HA, Alam MA, Ashkani S, Malek MA, Latif MA (2015) Marker-assisted backcrossing: a useful method for rice improvement. Biotechnol Biotechnol Equip 29(2):237–254. https://doi.org/10.1080/13102818.2014.995920

Hashimoto M, Kisseleva L, Sawa S, Furukawa T, Komatsu S, Koshiba T (2004) A novel rice PR10 protein, RSOsPR10, specifically induced in roots by biotic and abiotic stresses, possibly via the jasmonic acid signaling pathway. Plant Cell Physiol 45(5):550–559. https://doi.org/10.1093/pcp/pch063

He Y, Wu D, Wei D, Fu Y, Cui Y, Dong H, Tan C, Qian W (2017) GWAS, QTL mapping and gene expression analyses in *Brassica napus* reveal genetic control of branching morphogenesis. Sci Rep 7(1):1–9. https://doi.org/10.1038/s41598-017-15976-4

Hebbern CA, Laursen KH, Ladegaard AH, Schmidt SB, Pedas P, Bruhn D, Schjoerring JK, Wulfsohn D, Husted S (2009) Latent manganese deficiency increases transpiration in barley

(*Hordeum vulgare*). Physiol Plant 135(3):307–316. https://doi.org/10.1111/j.1399-3054.2008.01188.x

Hoang GH, Dinh LV, Nguyen TT, Ta NK, Gathignol F, Mai CD, Jouannic S, Tran KD, Khuat TH, Do VN, Lebrun M, Courtois B, Gantet P (2019) Genome-wide association study of a panel of vietnamese rice landraces reveals new QTLs for tolerance to water deficit during the vegetative phase. Rice 12(4):1–20. https://doi.org/10.1186/s12284-018-0258-6

Hoffstetter A, Cabrera A, Huang M, Sneller C (2016) Optimizing training population data and validation of genomic selection for economic traits in soft winter wheat. G3 6(9): 2919–2928.10.1534/g3.116.032532

Holland JB (2007) Genetic architecture of complex traits in plants. Curr Opin Plant Biol 10(2): 156–161. https://doi.org/10.1016/j.pbi.2007.01.003

Hossain H, Rahman MA, Alam MS, Singh RK (2015) Mapping of quantitative trait loci associated with reproductive-stage salt tolerance in rice. J Agron Crop Sci 201(1):17–31. https://doi.org/10.1111/jac.12086

Hu Y, Cheng H, Tao S (2016) The challenges and solutions for cadmium contaminated rice in China: a critical review. Environ Int 92:515–553. https://doi.org/10.1016/j.envint.2016.04.042

Huang CF, Yamaji N, Mitani N, Yano M, Nagamura Y, Ma JF (2009) A bacterial-type ABC transporter is involved in aluminum tolerance in rice. Plant Cell 21(2):655–667. https://doi.org/10.1105/tpc.108.064543

Huang X, Sang T, Zhao Q, Feng Q, Zhao Y, Li C, Zhu C, Lu T, Zhang Z, Li M, Fan D (2010) Genome-wide association studies of 14 agronomic traits in rice landraces. Nat Genet 42(11): 961. https://doi.org/10.1038/ng.695

Huang X, Zhao Y, Li C, Wang A, Zhao Q, Li W, Guo Y, Deng L, Zhu C, Fan D, Lu Y (2012) Genome-wide association study of flowering time and grain yield traits in a worldwide collection of rice germplasm. Nat Genet 44(1):32–39. https://doi.org/10.1038/ng.1018

Huang Y, Sun C, Min J, Chen Y, Tong C, Bao J (2015) Association mapping of quantitative trait loci for mineral element contents in whole grain rice (*Oryza sativa* L.). J Agric Food Chem 63: 10885–10892. https://doi.org/10.1021/acs.jafc.5b04932

Huang M, Balimponya EG, Mgonja EM, McHale LK, Luzi-Kihupi A, Wang GL, Sneller CH (2019) Use of genomic selection in breeding rice (*Oryza sativa* L.) for resistance to rice blast (*Magnaporthe oryzae*). Mol Breed 39(8):1-16.10.1007/s11032-019-1023-2

Ishikawa S, Abe T, Kuramata M, Yamaguchi M, Ando T, Yamamoto T, Yano M (2010) A major quantitative trait locus for increasing cadmium-specific concentration in rice grain is located on the short arm of chromosome 7. J Exp Bot 61:923–934. https://doi.org/10.1093/jxb/erp360

Ishimaru Y, Takahashi R, Bashir K, Shimo H, Senoura T, Sugimoto K, Ono K, Yano M, Ishikawa S, Arao T, Nakanishi H (2012) Characterizing the role of rice NRAMP5 in manganese, iron and cadmium transport. Sci Rep 2(1):1–8. https://doi.org/10.1038/srep00286

Islam MR, Hassan L, Salam MA, Collard BC, Singh RK, Gregorio GB (2011) QTL mapping for salinity tolerance at seedling stage in rice. Emir J Food Agric 15:137–146. https://doi.org/10.9755/ejfa.v23i2.6348

Jagadish SK, Craufurd PQ, Wheeler TR (2007) High temperature stress and spikelet fertility in rice (*Oryza sativa* L.). J Exp Bot 58(7):1627–1635. https://doi.org/10.1093/jxb/erm003

Jagadish SV, Muthurajan R, Oane R, Wheeler TR, Heuer S, Bennett J, Craufurd PQ (2010) Physiological and proteomic approaches to address heat tolerance during anthesis in rice (*Oryza sativa* L.). J Exp Bot 61(1):143–156. https://doi.org/10.1093/jxb/erp289

Jain M, Nijhawan A, Arora R, Agarwal P, Ray S, Sharma P, Kapoor S, Tyagi AK, Khurana JP (2007) F-box proteins in rice. Genome-wide analysis, classification, temporal and spatial gene expression during panicle and seed development, and regulation by light and abiotic stress. Plant Physiol 143:1467. https://doi.org/10.1104/pp.106.091900

Jia L, Yan W, Zhu C, Agrama HA, Jackson A, Yeater K, Li X, Huang B, Hu B, McClung A, Wu D (2012) Allelic analysis of sheath blight resistance with association mapping in rice. PLoS One 7(3). https://doi.org/10.1371/journal.pone.0032703

Jia FJ, Wang CY, Huang JG, Yang GD, Wu CG, Zheng CC (2015) SCF E3 ligase PP2-B11 plays a positive role in response to salt stress in Arabidopsis. J Exp Bot 66:4683–4697. https://doi.org/10.1093/jxb/erv245

Jin S, Cheng Y, Guan Q, Liu D, Takano T, Liu S (2006) A metallothionein-like protein of rice (rgMT) functions in E. coli and its gene expression is induced by abiotic stresses. Biotechnol Lett 28:1749–1753. https://doi.org/10.1007/s10529-006-9152-1

Jörgens CI, Grünewald N, Hülskamp M, Uhrig JF (2010) A role for ABIL3 in plant cell morphogenesis. Plant J 62:925–935. https://doi.org/10.1111/j.1365-313X.2010.04210.x

Kang H, Wang Y, Peng S, Zhang Y, Xiao Y, Wang D, Qu S, Li Z, Yan S, Wang Z, Liu W, Ning Y, Korniliev P, Leung H, Mezey J, McCouch SR, Wang GL (2016) Dissection of the genetic architecture of rice resistance to the blast fungus *Magnaporthe oryzae*. Mol Plant Pathol 17(6): 959–972. https://doi.org/10.1111/mpp.12340

Katara JL, Parameswaran C, Devanna BN, Verma RL, Anilkumar C, Patra BC, Samantaray S (2021) Genomics assisted breeding: the need and current perspective for rice improvement in India. ORYZA-An Int J Rice 58(1 Suppl):61–68. https://doi.org/10.35709/ory.2021.58.spl.1

Khan JA, Arshad MI, Jamil FF, Hasnain S (2009) Evaluation of rice genotypes against bacterial leaf blight (BLB) disease. Pak J Phytopathol 21(1):26–30

Khush GS, Jena KK (2009) Current status and future prospects for research on blast resistance in rice (*Oryza sativa* L.). In: Advances in genetics, genomics and control of rice blast disease. Springer, Dordrecht, pp 1–10. https://doi.org/10.1007/978-1-4020-9500-9_1

Khush GS, Mackill DJ, Sidhu GS (1989) Breeding rice for resistance to bacterial blight. Bacterial blight of rice, pp. 207–217

Kim H, Lee K, Hwang H, Bhatnagar N, Kim DY, Yoon IS, Byun MO, Kim ST, Jung KH, Kim BG (2014) Overexpression of PYL5 in rice enhances drought tolerance, inhibits growth, and modulates gene expression. J Exp Bot 65(2):453–464. https://doi.org/10.1093/jxb/ert397

Kim SM, Reinke RF (2019) A novel resistance gene for bacterial blight in rice, *Xa43(t)* identified by GWAS, confirmed by QTL mapping using a bi-parental population. PLoS One 14(2):e0211775. https://doi.org/10.1371/journal.pone.0211775

Kodra E, Steinhaeuser K, Auroop R (2011) Persisting cold extremes under 21st-century warming scenarios. Geophys Res Lett:S0094–S8276. https://doi.org/10.1029/2011GL047103

Koohafkan P, Furtado J (2004) Traditional rice–fish systems as globally important ingenious agricultural heritage systems. Int Rice Comm News Lett 53:66–73

Kovach MJ, McCouch SR (2008) Leveraging natural diversity: back through the bottleneck. Curr Opin Plant Biol 11(2):193–200. https://doi.org/10.1016/j.pbi.2007.12.006

Kovtun Y, Chiu WL, Tena G, Sheen J (2000) Functional analysis of oxidative stress activated mitogen-activated protein kinase cascade in plants. PNAS 97:2940–2945. https://doi.org/10.1073/pnas.97.6.2940

Kumar K, Rao KP, Sharma P, Sinha AK (2008) Differential regulation of rice mitogen activated protein kinase kinase (MKK) by abiotic stress. Plant Physiol Biochem 46(10):891–897. https://doi.org/10.1016/j.plaphy.2008.05.014

Kumar A, Dixit S, Ram T, Yadaw RB, Mishra KK, Mandal NP (2014) Breeding high-yielding drought-tolerant rice: genetic variations and conventional and molecular approaches. J Exp Bot 65(21):6265–6278. https://doi.org/10.1093/jxb/eru363

Kumar V, Singh A, Mithra SVA, Krishnamurthy SL, Parida SK, Jain S, Tiwari KK, Kumar P, Rao AR, Sharma SK, Khurana JP, Singh NK, Mohapatra T (2015) Genome-wide association mapping of salinity tolerance in rice (*Oryza sativa*). DNA Res 22(2):133–145. https://doi.org/10.1093/dnares/dsu046

Lafarge T, Bueno C, Frouin J, Jacquin L, Courtois B, Ahmadi N (2017) Genome-wide association analysis for heat tolerance at flowering detected a large set of genes involved in adaptation to thermal and other stresses. PLoS One 12(2):1–27. https://doi.org/10.1371/journal.pone.0171254

Lafitte HR, Ismail A, Bennett J (2004) Abiotic stress tolerance in rice for Asia: progress and the future. In: Fischer T, Turner N, Angus J, McIntyre L, Robertson M, Borrell A et al (eds) New

Directions for a Diverse Planet: Proceedings for the 4th International Crop Science Congress. The Regional Institute Ltd.. www.cropscience.org.au/icsc2004
Lafitte HR, Ismail A, Bennett J (2006) Abiotic stress tolerance in tropical rice: progress and future prospects. Oryza 43(3):171
Larsen P, Cancel J, Rounds M, Ochoa V (2007) Arabidopsis ALS1 encodes a root tip and stele localized half type ABC transporter required for root growth in an aluminum toxic environment. Planta 225:1447–1458. https://doi.org/10.1007/s00425-006-0452-4
Lasanthi-Kudahettige R, Magneschi L, Loreti E, Gonzali S, Licausi F, Novi G, Beretta O, Vitulli F, Alpi A, Perata P (2007) Transcript profiling of the anoxic rice coleoptile. Plant Physiol 144(1): 218–231. https://doi.org/10.1104/pp.106.093997
Lekklar C, Pongpanich M, Suriya-Arunroj D, Chinpongpanich A, Tsai H, Comai L, Chadchawan S, Buaboocha T (2019) Genome-wide association study for salinity tolerance at the flowering stage in a panel of rice accessions from Thailand. BMC Genomics 20(1):1–18. https://doi.org/10.1186/s12864-018-5317-2
Li HW, Zang BS, Deng XW, Wang XP (2011) Overexpression of the trehalose-6-phosphate synthase gene OsTPS1 enhances abiotic stress tolerance in rice. Planta 234(5):1007–1018. https://doi.org/10.1007/s00425-011-1458-0
Li X, Guo Z, Lv Y, Cen X, Ding X, Wu H, Li X, Huang J, Xiong L (2017) Genetic control of the root system in rice under normal and drought stress conditions by genome-wide association study. PLoS Genet 13(7):e1006889. https://doi.org/10.1371/journal.pgen.1006889
Li C, Wang D, Peng S, Chen Y, Su P, Chen J, Zheng L, Tan X, Liu J, Xiao Y, Kang H (2019a) Genome-wide association mapping of resistance against rice blast strains in South China and identification of a new *Pik* allele. Rice 12(1):1–9. https://doi.org/10.1186/s12284-019-0309-7
Li N, Zheng H, Cui J, Wang J, Liu H, Sun J, Liu T, Zhao H, Lai Y, Zou D (2019b) Genome-wide association study and candidate gene analysis of alkalinity tolerance in japonica rice germplasm at the seedling stage. Rice 12(1):11. https://doi.org/10.1186/s12284-019-0285-y
Licausi F, Ohme-Takagi M, Perata P (2013) Apetala2/ethylene responsive factor (AP2/ERF) transcription factors: mediators of stress responses and developmental programs. New Phytol 199:639–649. https://doi.org/10.1111/nph.12291
Lim MN, Lee SE, Yim HK, Kim JH, Yoon IS, Hwang YS (2013) Differential anoxic expression of sugar-regulated genes reveals diverse interactions between sugar and anaerobic signaling systems in rice. Mol Cells 36:169–176. https://doi.org/10.1007/s10059-013-0152-4
Lipka AE, Kandianis CB, Hudson ME, Yu J, Drnevich J, Bradbury PJ, Gore MA (2015) From association to prediction: Statistical methods for the dissection and selection of complex traits in plants. Curr Opin Plant Biol 24:110–118. https://doi.org/10.1016/j.pbi.2015.02.010
Liu J, Li K, Xu J, Liang J, Lu X, Yang J, Zhu Q (2003) Interaction of Cd and five mineral nutrients for uptake and accumulation in different rice cultivars and genotypes. Field Crops Res 83(3): 271–281. https://doi.org/10.1016/S0378-4290(03)00077-7
Liu H, Soomro A, Zhu Y, Qiu X, Chen K, Zheng T, Yang L, Xing D, Xu J (2016) QTL underlying iron and zinc toxicity tolerances at seedling stage revealed by two sets of reciprocal introgression populations of rice (*Oryza sativa* L.). Crop J 4(4):280–289. https://doi.org/10.1016/j.cj.2016.05.007
Liu C, Chen K, Zhao X, Wang X, Shen C, Zhu Y, Dai M, Qiu X, Yang R, Xing D, Pang Y, Xu J (2019) Identification of genes for salt tolerance and yield-related traits in rice plants grown hydroponically and under saline field conditions by genome-wide association study. Rice 12(1). https://doi.org/10.1186/s12284-019-0349-z
Liu MH, Kang H, Xu Y, Peng Y, Wang D, Gao L, Wang X, Ning Y, Wu J, Liu W, Li C, Liu B, Wang GL (2020a) Genome-wide association study identifies an NLR gene that confers partial resistance to *Magnaporthe oryzae* in rice. Plant Biotechnol J 18(6):1376–1383. https://doi.org/10.1111/pbi.13300
Liu S, Zhong H, Meng X, Sun T, Li Y, Pinson SR, Chang SK, Peng Z (2020b) Genome-wide association studies of ionomic and agronomic traits in USDA mini core collection of rice and

comparative analyses of different mapping methods. BMC Plant Biol 20(1):1–8. https://doi.org/10.1186/s12870-020-02603-0

Lo SF, Yang SY, Chen KT, Hsing YI, Zeevaart JA, Chen LJ, Yu SM (2008) A novel class of gibberellin 2-oxidases control semidwarfism, tillering, and root development in rice. Plant Cell 20(10):2603–2618. https://doi.org/10.1105/tpc.108.060913

Lu P, Magwanga RO, Guo X, Kirungu JN, Lu H, Cai X, Zhou Z, Wei Y, Wang X, Zhang Z, Peng R (2018) Genome-wide analysis of multidrug and toxic compound extrusion (MATE) family in Gossypium raimondii and Gossypium arboreum and its expression analysis under salt, cadmium, and drought stress. G3: Genes, Genomes, Genet 8(7):2483–2500. https://doi.org/10.1534/g3.118.200232

Lu Q, Wang C, Niu X, Zhang M, Xu Q, Feng Y, Yang Y, Wang S, Yuan X, Yu H, Wang Y (2019) Detecting novel loci underlying rice blast resistance by integrating a genome-wide association study and RNA sequencing. Mol Breed 39(6):1–10. https://doi.org/10.1007/s11032-019-0989-0

Luo JS, Huang J, Zeng DL, Peng JS, Zhang GB, Ma HL, Guan Y, Yi HY, Fu YL, Han B, Lin HX (2018) A defensin-like protein drives cadmium efflux and allocation in rice. Nat Commun 9(1): 1–9. https://doi.org/10.1038/s41467-018-03088-0

Lv Y, Guo Z, Li X, Ye H, Li X, Xiong L (2016) New insights into the genetic basis of natural chilling and cold shock tolerance in rice by genome-wide association analysis. Plant Cell Environ 39(3):556–570. https://doi.org/10.1111/pce.12635

Ma Y, Dai X, Xu Y, Luo W, Zheng X, Zeng D, Pan Y, Lin X, Liu H, Zhang D, Xiao J (2015) COLD1 confers chilling tolerance in rice. Cell 160(6):1209–1221. https://doi.org/10.1016/j.cell.2015.01.046

Ma X, Feng F, Wei H, Mei H, Xu K, Chen S, Li T, Liang X, Liu H, Luo L (2016) Genome-wide association study for plant height and grain yield in rice under contrasting moisture regimes. Front Plant Sci 7(1801):1–13. https://doi.org/10.3389/fpls.2016.01801

Magneschi L, Perata P (2009) Rice germination and seedling growth in the absence of oxygen. Ann Bot 103(2):181–196. https://doi.org/10.1093/aob/mcn121

Magneschi L, Kudahettige RL, Alpi A, Perata P (2009) Comparative analysis of anoxic coleoptile elongation in rice varieties: relationship between coleoptile length and carbohydrate levels, fermentative metabolism and anaerobic gene expression. Plant Biol 4:561–573. https://doi.org/10.1111/j.1438-8677.2008.00150.x

Mani A, Sankaranarayanan K (2018) In silico analysis of natural resistance associated macrophage protein (NRAMP) family of transporters in rice. Protein J 37:237–247. https://doi.org/10.1007/s10930-018-9773-y

Manolio TA (2010) Genome wide association studies and assessment of the risk of disease. N Engl J Med 363(2):166–176. https://doi.org/10.1056/NEJMra0905980

Marouli E, Graff M, Medina-Gomez C, Lo KS, Wood AR, Kjaer TR, Fine RS, Lu Y, Schurmann C, Highland HM, Rüeger S (2017) Rare and low-frequency coding variants alter human adult height. Nature 542(7640):186–190. https://doi.org/10.1038/nature21039

McCouch SR, Wright MH, Tung CW, Maron LG, McNally KL, Fitzgerald M, Singh N, DeClerck G, Agosto-Perez F, Korniliev P, Greenberg AJ (2016) Open access resources for genome-wide association mapping in rice. Nat Commun 7(1):1–14. https://doi.org/10.1038/ncomms10532

Meehl GA, Washington WM, Collins WD, Arblaster JM, Hu A, Buja LE, Strand WG, Teng H (2005) How much more global warming and sea level rise? Science 307(5716):1769–1772. https://doi.org/10.1126/science.1106663

Meharg AA, Norton G, Deacon C, Williams P, Adomako EE, Price A, Zhu Y, Li G, Zhao FJ, McGrath S, Villada A (2013) Variation in rice cadmium related to human exposure. Environ Sci Technol 47(11):5613–5618. https://doi.org/10.1021/es400521h

Mgonja EM, Balimponya EG, Kang H, Bellizzi M, Park CH, Li Y, Mabagala R, Sneller C, Correll J, Opiyo S, Talbot NJ, Mitchell T, Wang GL (2016) Genome-wide association mapping of rice resistance genes against *Magnaporthe oryzae* isolates from four African countries. Phytopathology 106(11):1359–1365. https://doi.org/10.1094/PHYTO-01-16-0028-R

Millaleo R, Reyes- Diaz M, Ivanov AG, Mora ML, Alberdi M (2010) Manganese as essential and toxic element for plants: transport, accumulation and resistance mechanisms. J Soil Sci Plant Nutr 10:470–481. https://doi.org/10.4067/S0718-95162010000200008

Mitchell-Olds T (2010) Complex-trait analysis in plants. Genome Biol 11(4):113. https://doi.org/10.1186/gb-2010-11-4-113

Mohammadi R, Mendioro MS, Diaz GQ, Gregorio GB, Singh RK (2013) Mapping quantitative trait loci associated with yield and yield components under reproductive stage salinity stress in rice (*Oryza sativa* L.). J Genet 92(3):433–443.10.1007/s12041-013-0285-4

Morita Y, Kyozuka J (2007) Characterization of OsPID, the rice ortholog of PINOID, and its possible involvement in the control of polar auxin transport. Plant Cell Physiol 48(3):540–549. https://doi.org/10.1093/pcp/pcm024

Mueller ND, Gerber JS, Johnston M, Ray DK, Ramankutty N, Foley JA (2012) Closing yield gaps through nutrient and water management. Nature 490:254–257. https://doi.org/10.1038/nature11420

Nagai K, Hattori Y, Ashikari M (2010) Stunt or elongate? Two opposite strategies by which rice adapts to floods. J Plant Res 123(3):303–309. https://doi.org/10.1007/s10265-010-0332-7

Nayyeripasand L, Garoosi GA, Ahmadikhah A (2021) Genome-wide association study (GWAS) to identify salt-tolerance QTLs carrying novel candidate genes in rice during early vegetative stage. Rice 14(1):1–21. https://doi.org/10.1186/s12284-020-00433-0

Negrão S, CecíliaAlmadanim M, Pires IS, Abreu IA, Maroco J, Courtois B, Gregorio GB, McNally KL, Margarida Oliveira M (2013) New allelic variants found in key rice salt-tolerance genes: an association study. Plant Biotechnol J 11(1):87–100. https://doi.org/10.1111/pbi.12010

Norton GJ, Deacon CM, Xiong L, Huang S, Meharg AA, Price AH (2010) Genetic mapping of the rice ionome in leaves and grain: identification of QTLs for 17 elements including arsenic, cadmium, iron and selenium. Plant Soil 329:139–153. https://doi.org/10.1007/s11104-009-0141-8

Onogi A, Ideta O, Inoshita Y, Ebana K, Yoshioka T, Yamasaki M, Iwata H (2015) Exploring the areas of applicability of whole-genome prediction methods for Asian rice (*Oryza sativa* L.). Theor Appl Genet 128(1):41–53. https://doi.org/10.1007/s00122-014-2411-y

Ozga JA, Kaur H, Savada RP, Reinecke DM (2017) Hormonal regulation of reproductive growth under normal and heat-stress conditions in legume and other model crop species. J Exp Bot 68(8):1885–1894. https://doi.org/10.1093/jxb/erw464

Pan Y, Zhang H, Zhang D, Li J, Xiong H, Yu J, Li J, Rashid MAR, Li G, Ma X, Cao G, Han L, Li Z (2015) Genetic analysis of cold tolerance at the germination and booting stages in rice by association mapping. PLoS One 10(3). https://doi.org/10.1371/journal.pone.0120590

Pan X, Li Y, Liu W, Liu S, Min J, Xiong H, Dong Z, Duan Y, Yu Y, Li X (2020) QTL mapping and candidate gene analysis of cadmium accumulation in polished rice by genome-wide association study. Sci Rep 10(1):1–11. https://doi.org/10.1038/s41598-020-68742-4

Pandit A, Rai V, Bal S, Sinha S, Kumar V, Chauhan M et al (2010) Combining QTL mapping and transcriptome profiling of bulked RILs for identification of functional polymorphism for salt tolerance genes in rice (*Oryza sativa* L.). Mol Gen Genomics 284:121–136. https://doi.org/10.1007/s00438-010-0551-6

Pantalião GF, Narciso M, Guimarães C, Castro A, Colombari JM, Breseghello F, Rodrigues L, Vianello RP, BorbaTO BC (2016) Genome wide association study (GWAS) for grain yield in rice cultivated under water deficit. Genetica 144(6):651–664. https://doi.org/10.1007/s10709-016-9932-z

Pathak MD, Khan ZR (1994) Insect pests of rice. Int Rice Res, Inst

Patishtan J, Hartley TN, Fonseca de Carvalho R, Maathuis FJ (2018) Genome-wide association studies to identify rice salt tolerance markers. Plant Cell Environ 41(5):970–982. https://doi.org/10.1111/pce.12975

Patra BC, Anilkumar C, Chakraborti M (2020) Rice breeding in India: a journey from phenotype based pure-line selection to genomics assisted breeding. Agric Res J57(6):816–825. https://doi.org/10.5958/2395-146X.2020.00120.9

Portwood JL, Woodhouse MR, Cannon EK, Gardiner JM, Harper LC, Schaeffer ML, Walsh JR, Sen TZ, Cho KT, Schott DA, Braun BL (2019) MaizeGDB 2018: the maize multi-genome genetics and genomics database. Nucleic Acids Res 47(D1):D1146–D1154. https://doi.org/10.1093/nar/gky1046

Pradhan SK, Nayak DK, Mohanty S, Behera L, Barik SR, Pandit E, Lenka S, Anandan A (2015) Pyramiding of three bacterial blight resistance genes for broad-spectrum resistance in deepwater rice variety, Jalmagna. Rice 8(1):1–14. https://doi.org/10.1186/s12284-015-0051-8

Pradhan SK, Barik SR, Sahoo A, Mohapatra S, Nayak DK, Mahender A, Meher J, Anandan A, Pandit E (2016) Population structure, genetic diversity and molecular marker-trait association analysis for high temperature stress tolerance in rice. PLoS One 11(8):1–23. https://doi.org/10.1371/journal.pone.0160027

Prakash A, Bentur JS, Prasad MS, Tanwar RK, Sharma OP, Bhagat S, Sehgal M, Singh SP, Singh M, Chattopadhyay C, Sushil SN (2014) Integrated pest management for rice. National Centre for Integrated Pest Management, LBS Building, IARI Campus, New Delhi, India, p 43

Price AH, Tomos AD (1997) Genetic dissection of root growth in rice (*Oryza sativa*L.). II: mapping quantitative trait loci using molecular markers. Theor Appl Genet 95:143–152. https://doi.org/10.1007/s001220050541

Price AH, Norton GJ, Salt DE, Ebenhoeh O, Meharg AA, Meharg C, Islam MR, Sarma RN, Dasgupta T, Ismail AM, McNally KL (2013) Alternate wetting and drying irrigation for rice in Bangladesh: Is it sustainable and has plant breeding something to offer? Food Energy Secur 2(2):120–129. https://doi.org/10.1002/fes3.29

Qi J, Qian Q, Bu Q, Li S, Chen Q, Sun J, Liang W, Zhou Y, Chu C, Li X, Ren F (2008) Mutation of the rice Narrow leaf1 gene, which encodes a novel protein, affects vein patterning and polar auxin transport. Plant Physiol 147(4):1947–1959. https://doi.org/10.1104/pp.108.118778

Raboin LM, Ballini E, Tharreau D, Ramanantsoanirina A, Frouin J, Courtois B, Ahmadi N (2016) Association mapping of resistance to rice blast in upland field conditions. Rice 9(1):1–2. https://doi.org/10.1186/s12284-016-0131-4

Ramegowda V, Senthil Kumar M (2015) The interactive effects of simultaneous biotic and abiotic stresses on plants: mechanistic understanding from drought and pathogen combination. J Plant Physiol 176:47–54. https://doi.org/10.1016/j.jplph.2014.11.008

Rebolledo MC, Dingkuhn M, Courtois B, Gibon Y, Clément-Vidal A, Cruz DF, Duitama J, Lorieux M, Luquet D (2015) Phenotypic and genetic dissection of component traits for early vigour in rice using plant growth modelling, sugar content analyses and association mapping. J Exp Bot 66(18):5555–5566. https://doi.org/10.1093/jxb/erv258

Ren ZH, Gao JP, Li LG, Cai XL, Huang W, Chao DY, Lin H-X (2005) A rice quantitative trait locus for salt tolerance encodes a sodium transporter. Nat Genet 37:1141–1146. https://doi.org/10.1038/ng1643

Riedelsheimer C, Endelman JB, Stange M, Sorrells ME, Jannink JL, Melchinger AE (2013) Genomic predictability of interconnected biparental maize populations. Genetics 194(2): 493–503. https://doi.org/10.1534/genetics.113.150227

Rohilla M, Singh N, Mazumder A, Sen P, Roy P, Chowdhury D, Singh NK, Mondal TK (2020) Genome-wide association studies using 50 K rice genic SNP chip unveil genetic architecture for anaerobic germination of deep-water rice population of Assam. India Mol Genet Genom 295(5): 1211–1226. https://doi.org/10.1007/s00438-020-01690-w

Root TL, Price JT, Hall KR, Schneider SH, Rosenzweigk C, Pounds JA (2003) Fingerprints of global warming on wild animals and plants. Nature 421(2):57–60. https://doi.org/10.1038/nature01333

Sabouri H, Sabouri A (2008) New evidence of QTLs attributed to salinity tolerance in rice. Afr J Biotechnol 7(24)

Sahebi M, Hanafi MM, Rafii MY, Mahmud TM, Azizi P, Osman M, Abiri R, Taheri S, Kalhori N, Shabanimofrad M, Miah G (2018) Improvement of drought tolerance in rice (*Oryza sativa* L.): genetics, genomic tools, and the WRKY gene family. BioMed Res Int 2018. https://doi.org/10.1155/2018/3158474

Sales E, Viruel J, Domingo C, Marqués L (2017) Genome wide association analysis of cold tolerance at germination in temperate japonica rice (*Oryza sativa* L.) varieties. PLoS One 12: e0183416. https://doi.org/10.1371/journal.pone.0183416

Sallam A, Martsch R (2015) Association mapping for frost tolerance using multiparent advanced generation inter-cross (MAGIC) population in faba bean (*Vicia faba* L.). Genetica 143(4): 501–514. https://doi.org/10.1007/s10709-015-9848-z

Sánchez R, Flores A, Cejudo FJ (2006) Arabidopsis phosphoenolpyruvate carboxylase genes encode immunologically unrelated polypeptides and are differentially expressed in response to drought and salt stress. Planta 223:901–909. https://doi.org/10.1007/s00425-005-0144-5

Satake T, Yoshida S (1978) High temperature induced sterility in indica rice at flowering. Jpn J Crop Sci 47:6–17. https://doi.org/10.1626/jcs.47.6

Satoh-Nagasawa N, Mori M, Nakazawa N, Kawamoto T, Nagato Y, Sakurai K, Takahashi H, Watanabe A, Akagi H (2012) Mutations in rice (*Oryza sativa*) heavy metal ATPase 2 (OsHMA2) restrict the translocation of zinc and cadmium. Plant Cell Physiol 53(1): 213–224. https://doi.org/10.1093/pcp/pcr166

Sattayachiti W, Wanchana S, Arikit S, Nubankoh P, Patarapuwadol S, Vanavichit A, Darwell CT, Toojinda T (2020) Genome-wide association analysis identifies resistance loci for bacterial leaf streak resistance in rice (*Oryza sativa* L.). Plan Theory 9(12):1673. https://doi.org/10.3390/plants9121673

Schlappi MR, Jackson AK, Eizenga GC, Wang A, Chu C, Shi Y, Shimoyama N, Boykin DL (2017) Assessment of five chilling tolerance traits and GWAS mapping in rice using the USDA mini-core collection. Front Plant Sci 8(957):1–13. https://doi.org/10.3389/fpls.2017.00957

Semagn K, Bjørnstad Å, Xu Y (2010) The genetic dissection of quantitative traits in crops. Electron J Biotechnol 13(5):16–17. https://doi.org/10.2225/vol13-issue5-fulltext-14

Seo P, Park C (2010) MYB96-mediated abscisic acid signals induce pathogen resistance response by promoting salicylic acid biosynthesis in Arabidopsis. New Phytol 186:471–483. https://doi.org/10.1111/j.1469-8137.2010.03183.x

Septiningsih EM, Pamplona AM, Sanchez DL, Neeraja CN, Vergara GV, Heuer S, Ismail AM, Mackill DJ (2009) Development of submergence-tolerant rice cultivars: the Sub1 locus and beyond. Ann Bot 103(2):151–160. https://doi.org/10.1093/aob/mcn206

Septiningsih EM, Ignacio JC, Sendon PM, Sanchez DL, Ismail AM, Mackill DJ (2013) QTL mapping and confirmation for tolerance of anaerobic conditions during germination derived from the rice landrace Ma-Zhan Red. Theor Appl Genet 126(5):1357–1366. https://doi.org/10.1007/s00122-013-2057-1

Shakiba E, Edwards JD, Jodari F, Duke SE, Baldo AM, Korniliev P, McCouch SR, Eizenga GC (2017) Genetic architecture of cold tolerance in rice (*Oryza sativa*) determined through high resolution genome-wide analysis. PLoS One 12(3):1–22. https://doi.org/10.1371/journal.pone.0172133

Shi Y, Gao L, Wu Z, Zhang X, Wang M, Zhang C, Zhang F, Zhou Y, Li Z (2017) Genome-wide association study of salt tolerance at the seed germination stage in rice. BMC Plant Biol 17(1): 1–11. https://doi.org/10.1186/s12870-017-1044-0

Shrestha A, Dziwornu AK, Ueda Y, Wu L-B, Mathew B, Frei M (2018) Genome-wide association study to identify candidate loci and genes for Mn toxicity tolerance in rice. PLoS One 13(2): 1–15. https://doi.org/10.1371/journal.pone.0192116

Singh AK, Singh PK, Arya M, Singh NK, Singh US (2015) Molecular screening of blast resistance genes in rice using SSR markers. Plant Pathol J 31(1):12. https://doi.org/10.5423/PPJ.OA.06.2014.0054

Singh R, Singh Y, Xalaxo S, Verulkar S, Yadav N, Singh S, Singh N, Prasad KSN, Kondayya K, Rao PR, Rani MG (2016) From QTL to variety-harnessing the benefits of QTLs for drought, flood and salt tolerance in mega rice varieties of India through a multi-institutional network. Plant Sci 242:278–287. https://doi.org/10.1016/j.plantsci.2015.08.008

Sinha AK, Jaggi M, Raghuram B, Tuteja N (2011) Mitogen-activated protein kinase signaling in plants under abiotic stress. Plant Signal Behav 6:196–203. https://doi.org/10.4161/psb.6.2.14701

Slatkin M (2008) Linkage disequilibrium - Understanding the evolutionary past and mapping the medical future. Nat Rev Genet 9(6):477–485. https://doi.org/10.1038/nrg2361

Song A, Li P, Li Z, Fan F, Nikolic M, Liang Y (2011) The alleviation of zinc toxicity by silicon is related to zinc transport and antioxidative reactions in rice. Plant Soil 344(1–2):319–333. https://doi.org/10.1007/s11104-011-0749-3

Song Q, Hyten DL, Jia G, Quigley CV, Fickus EW, Nelson RL, Cregan PB (2013) Development and evaluation of SoySNP50K, a high-density genotyping array for soybean. PLoS One 8(1): e54985. https://doi.org/10.1371/journal.pone.0054985

Sparks A, Nelson A, Castilla N (2012) Where rice pests and diseases do the most damage. Rice Today 11(4):26–27

Spindel J, Iwata H (2018) Genomic selection in rice breeding. In: Rice genomics, genetics and breeding. Springer, Singapore, pp 473–496. https://doi.org/10.1007/978-981-10-7461-5_24

Spindel J, Begum H, Akdemir D, Virk P, Collard B, Redona E, Atlin G, Jannink J-L, McCouch SR (2015) Genomic selection and association mapping in rice (*Oryza sativa*): effect of trait genetic architecture, training population composition, marker number and statistical model on accuracy of rice genomic selection in elite, tropical rice breeding lines. PLoS Genet 11(2):e1004982. https://doi.org/10.1371/journal.pgen.1004982

Spindel J, Begum H, Akdemir D, Collard B, Redoña E, Jannink J, McCouch S (2016) Genome-wide prediction models that incorporate de novo GWAS are a powerful new tool fortropical rice improvement. Heredity 116(4):395–408. https://doi.org/10.1038/hdy.2015.113

Sun C, Zhang F, Yan X, Zhang X, Dong Z, Cui D, Chen F (2017) Genome-wide association study for 13 agronomic traits reveals distribution of superior alleles in bread wheat from the Yellow and Huai Valley of China. Plant Biotechnol J 15(8):953–969. https://doi.org/10.1111/pbi.12690

Suzuki K, Yamaji N, Costa A, Okuma EF, Kobayashi NI, Kashiwagi T, Katsuhara M, Wang C, Tanoi K, Murata Y, Schroeder JI (2016) OsHKT1;4-mediated Na+ transport in stems contributes to Na+ exclusion from leaf blades of rice at the reproductive growth stage upon salt stress. BMC Plant Biol 16:22. https://doi.org/10.1186/s12870-016-0709-4

Swamy BPM, Shamsudin NAA, Rahman SNA, Mauleon R, Ratnam W, Cruz MT, Kumar A (2017) Association mapping of yield and yield-related traits under reproductive stage drought stress in rice (*Oryza sativa* L.). Rice 10(1). https://doi.org/10.1186/s12284-017-0161-6

Takahashi R, Ishimaru Y, Senoura T, Shimo H, Ishikawa S, Arao T, Nakanishi H, Nishizawa N (2011) The OsNRAMP1 iron transporter is involved in cd accumulation in rice. J Exp Bot 62: 4843–4850. https://doi.org/10.1093/jxb/err136

Takai T, Adachi S, Taguchi-Shiobara F, Sanoh-Arai Y, Iwasawa N, Yoshinaga S, Hirose S, Taniguchi Y, Yamanouchi U, Wu J, Matsumoto T (2013) A natural variant of NAL1, selected in high-yield rice breeding programs, pleiotropically increases photosynthesis rate. Sci Rep 3(1):1–11. https://doi.org/10.1038/srep02149

Thapa R, Tabien RE, Thomson MJ, Septiningsih EM (2020) Genome-wide association mapping to identify genetic loci for cold tolerance and cold recovery during germination in rice. Front Genet 11(22):1–11. https://doi.org/10.3389/fgene.2020.00022

Tian X, Wang Z, Li X, Lv T, Liu H, Wang L, Niu H, Bu Q (2015) Characterization and functional analysis of pyrabactin resistance-like abscisic acid receptor family in rice. Rice 8(1):1–3. https://doi.org/10.1186/s12284-015-0061-6

Ueno D, Kono I, Yokosho K, Ando T, Yano M, Ma JF (2009) A major quantitative trait locus controlling cadmium translocation in rice (*Oryza sativa*). New Phytol 182:644–653. https://doi.org/10.1111/j.1469-8137.2009.02784.x

Uexkull HRV, Mutert E (1995) Global extent, development and economic impact of acid soils. In: Date RA, Grundon NJ, Raymet GE, Probert ME (eds) Plant-soil interactions at low pH: principles and management. Kluwer Academic Publishers, Dordrecht, pp 5–19

Uga Y, Sugimoto K, Ogawa S, Rane J, Ishitani M, Hara N, Kitomi Y, Inukai Y, Ono K, Kanno N, Inoue H (2013) Control of root system architecture by DEEPER ROOTING 1 increases rice yield under drought conditions. Nat Genet 45(9):1097–1102. https://doi.org/10.1038/ng.2725

Uraguchi S, Fujiwara T (2013) Rice breaks ground for cadmium-free cereals. Curr Opin Plant Biol 16:328–334. https://doi.org/10.1016/j.pbi.2013.03.012

Uraguchi S, Kamiya T, Sakamoto T, Kasai K, Sato Y, Nagamura Y, Yoshida A, Kyozuka J, Ishikawa S, Fujiwara T (2011) Low-affinity cation transporter (*OsLCT1*) regulates cadmium transport into rice grains. Proc Natl Acad Sci 108(52):20959–20964. https://doi.org/10.1073/pnas.1116531109

Van Eeuwijk FA, Bink MC, Chenu K, Chapman SC (2010) Detection and use of QTL for complex traits in multiple environments. Curr Opin Plant Biol 13(2):193–205. https://doi.org/10.1016/j.pbi.2010.01.001

Verdeprado H, Kretzschmar T, Begum H, Raghavan C, Joyce P, Lakshmanan P, Cobb JN, Collard BC (2018) Association mapping in rice: basic concepts and perspectives for molecular breeding. Plant Prod Sci 21(3):159–176. https://doi.org/10.1080/1343943X.2018.1483205

Verulkar SB, Verma SK (2014) Screening protocols in breeding for drought tolerance in rice. Agric Res 3:32–40. https://doi.org/10.1007/s40003-014-0094-x

Véry AA, Nieves-Cordones M, Daly M, Khan I, Fizames C, Sentenac H (2014) Molecular biology of K+ transport across the plant cell membrane: what do we learn from comparison between plant species? J Plant Physiol 171(9):748–769. https://doi.org/10.1016/j.jplph.2014.01.011

Volante A, Tondelli A, Aragona M, Valente MT, Biselli C, Desiderio F, Bagnaresi P, Matic S, Gullino ML, Infantino A, Spadaro D (2017a) Identification of bakanae disease resistance loci in japonica rice through genome wide association study. Rice 10(1):1–16. https://doi.org/10.1186/s12284-017-0168-z

Volante A, Desiderio F, Tondelli A, Perrini R, Orasen G, Biselli C, Riccardi P, Vattari A, Cavalluzzo D, Urso S, Ben Hassen M, Fricano A, Piffanelli P, Cozzi P, Biscarini F, Sacchi GA, Cattivelli L, Valè G (2017b) Genome-wide analysis of japonica rice performance under limited water and permanent flooding conditions. Front Plant Sci 8:1862. https://doi.org/10.3389/fpls.2017.01862

Volante A, Tondelli A, Desiderio F, Abbruscato P, Menin B, Biselli C, Casella L, Singh N, McCouch SR, Tharreau D, Zampieri E, Cattivelli L, Valè G (2020) Genome wide association studies for japonica rice resistance to blast in field and controlled conditions. Rice 13(1). https://doi.org/10.1186/s12284-020-00431-2

Vromman D, Lutts S, Lefèvre I, Somer L, De Vreese O, Šlejkovec Z, Quinet M (2013) Effects of simultaneous arsenic and iron toxicities on rice (*Oryza sativa* L.) development, yield-related parameters and as and Fe accumulation in relation to as speciation in the grains. Plant Soil 371(1–2):199–217. https://doi.org/10.1007/s11104-013-1676-2

Vuuren DPV, Meinshausen M, Plattner GK, Joos F, Strassmann KM, Smith SJ, Wigley TM, Raper SC, Riahi K, De La Chesnaye F, Den Elzen MG (2008) Temperature increase of 21st century mitigation scenarios. Proc Natl Acad Sci 105(40):15258–15262. https://doi.org/10.1073/pnas.0711129105

Wambugu PW, Ndjiondjop HRJ (2018) Role of genomics in promoting the utilization of plant genetic resources in gene banks. Brief Funct Genomics 17(3):198–206. https://doi.org/10.1093/bfgp/ely014

Wan J-l, H-q Z, Wan J-m, Ikehashi H (2003) Detection and analysis of QTLs for ferrous iron toxicity tolerance in rice, *Oryza sativa* L. Euphytica 131(2):201–206. https://doi.org/10.1023/A:1023915710103

Wang YX, Wu P, Wu YR, Yan XL (2002) Molecular marker analysis of manganese toxicity tolerance in rice under greenhouse conditions. Plant Soil 238(2):227–233. https://doi.org/10.1023/A:1014487428033

Wang C, Yang Y, Yuan X, Xu Q, Feng Y, Yu H, Wang Y (2014a) Genome-wide association study of blast resistance in indica rice. BMC Plant Biol 14(1):1–11. https://doi.org/10.1186/s12870-014-0311-6

Wang Q, Tian F, Pan Y, Buckler ES, Zhang Z (2014b) A SUPER powerful method for genome wide association study. PLoS One 9(9):e107684. https://doi.org/10.1371/journal.pone.0107684
Wang X, Lee S, Wang J, Ma J, Bianco T, Jia Y, Bao J (2014c) Current advances on genetic resistance to rice blast disease. In: Rice – germplasm, genetics and improvement, pp 195–217. https://doi.org/10.5772/56824
Wang D, Liu J, Li C, Kang H, Wang Y, Tan X, Liu M, Deng Y, Wang Z, Liu Y, Zhang D, Xiao Y, Wang GL (2016) Genome-wide association mapping of cold tolerance genes at the seedling stage in rice. Rice 9(1):61. https://doi.org/10.1186/s12284-016-0133-2
Wang B, Ebbole DJ, Wang Z (2017a) The arms race between *Magnaporthe oryzae* and rice: diversity and interaction of Avr and R genes. J Integr Agric 16:2746–2760. https://doi.org/10.1016/S2095-3119(17)61746-5
Wang X, Li L, Yang Z, Zheng X, Yu S, Xu C, Hu Z (2017b) Predicting rice hybrid performance using univariate and multivariate GBLUP models based on North Carolina mating design II. Heredity 118(3):302–310. https://doi.org/10.1038/hdy.2016.87
Wang X, Zou B, Shao Q, Cui Y, Lu S, Zhang Y, Huang Q, Huang J, Hua J (2018a) Natural variation reveals that OsSAP16 controls low-temperature germination in rice. J Exp Bot 69(3):413–421. https://doi.org/10.1093/jxb/erx413
Wang W, Mauleon R, Hu Z, Chebotarov D, Tai S, Wu Z, Li M, Zheng T, Fuentes RR, Zhang F, Mansueto L (2018b) Genomic variation in 3,010 diverse accessions of Asian cultivated rice. Nature 557(7703):43–49. https://doi.org/10.1038/s41586-018-0063-9
Wang Q, Tang J, Han B, Huang X (2020) Advances in genome-wide association studies of complex traits in rice. Theor Appl Genet 133(5):1415–1425. https://doi.org/10.1007/s00122-019-03473-3
Weng J, Gu S, Wan X, Gao H, Guo T, Su N, Lei C, Zhang X, Cheng Z, Guo X, Wang J (2008) Isolation and initial characterization of GW5, a major QTL associated with rice grain width and weight. Cell Res 18(12):1199–1209. https://doi.org/10.1038/cr.2008.307
Werner K, Friedt W, Ordon F (2005) Strategies for pyramiding resistance genes against the barley yellow mosaic virus complex (BaMMV, BaYMV, BaYMV-2). Mol Breed 16:45–55. https://doi.org/10.1007/s11032-005-3445-2
Wu P, Hu B, Liao CY, Zhu JM, Wu YR, Senadhira D, Paterson AH (1998) Characterization of tissue tolerance to iron by molecular markers in different lines of rice. Plant Soil 203(2): 217–226. https://doi.org/10.1023/A:1004321218387
Wu YY, He JB, Li AH, Fang NY, He WW, Dang LL, Zeng GY, Huang J, Bao YM, Zhang HS (2016) Population structure analysis and association mapping of blast resistance in indica rice (*Oryza sativa* L.) landraces. Genet Mol Res 15:1–11. https://doi.org/10.4238/gmr.15038254
Xia J, Yamaji N, Kasai T, Ma J (2010) Plasma membrane-localized transporter for aluminium in rice. Proc Natl Acad Sci 107:18381. https://doi.org/10.1073/pnas.1004949107
Xiao Y, Pan Y, Luo L, Zhang G, Deng H, Dai L, Liu X, Tang W, Chen L, Wang GL (2011) Quantitative trait loci associated with seed set under high temperature stress at the flowering stage in rice (Oryza sativa L.). Euphytica 178(3):331–338. https://doi.org/10.1007/s10681-010-0300-2
Xiao N, Gao Y, Qian H, Gao Q, Wu Y, Zhang D, Zhang X, Yu L, Li Y, Pan C, Liu G (2018) Identification of genes related to cold tolerance and a functional allele that confers cold tolerance. Plant Physiol 177(3):1108–1123. https://doi.org/10.1104/pp.18.00209
Xu K, Mackill DJ (1996) A major locus for submergence tolerance mapped on rice chromosome 9. Mol Breed 2(3):219–224. https://doi.org/10.1007/BF00564199
Xu K, Xu X, Fukao T, Canlas P, Maghirang-Rodriguez R, Heuer S, Ismail AM, Bailey-Serres J, Ronald PC, Mackill DJ (2006) Sub1A is an ethylene-response-factor-like gene that confers submergence tolerance to rice. Nature 442(7103):705–708. https://doi.org/10.1038/nature04920
Xu LM, Zhou L, Zeng YW, Wang FM, Zhang HL, Shen SQ, Li ZC (2008) Identification and mapping of quantitative trait loci for cold tolerance at the booting stage in a japonica rice near-isogenic line. Plant Sci 174(3):340–347. https://doi.org/10.1016/j.plantsci.2007.12.003
Xu S, Zhu D, Zhang Q (2014) Predicting hybrid performance in rice using genomic best linear unbiased prediction. Proc Natl Acad Sci 111(34):12456–12461. https://doi.org/10.1073/pnas.1413750111

Xu Y, Liu X, Fu J, Wang H, Wang J, Huang C, Prasanna BM, Olsen MS, Wang G, Zhang A (2020) Enhancing genetic gain through genomic selection: from livestock to plants. Plant Commun 1(1):100005. https://doi.org/10.1016/j.xplc.2019.100005

Xue D, Chen M, Zhang G (2009) Mapping of QTLs associated with cadmium tolerance and accumulation during seedling stage in rice (*Oryza sativa*, L.). Euphytica 165:587–596. https://doi.org/10.1007/s10681-008-9785-3

Yamaguchi-Shinozaki K, Shinozaki K (1994) A novel cis-acting element in an Arabidopsis gene is involved in responsiveness to drought, low temperature or high-salt stress. Plant Cell 6:251–264. https://doi.org/10.1105/tpc.6.2.251

Yamaji N, Huang CF, Nagao S, Yano M, Sato Y, Nagamura Y, Ma JF (2009) A zinc finger transcription factor ART1 regulates multiple genes implicated in aluminium tolerance in rice. Plant Cell 21(10):3339–3349. https://doi.org/10.1105/tpc.109.070771

Yan YS, Chen XY, Yang K, Sun ZX, Fu YP, Zhang YM, Fang RX (2011) Overexpression of an F-box protein gene reduces abiotic stress tolerance and promotes root growth in rice. Mol Plant 4:190–197. https://doi.org/10.1093/mp/ssq066

Yang M, Zhang Y, Zhang L, Hu J, Zhang X, Lu K, Dong H, Wang D, Zhao FJ, Huang CF, Lian X (2014) OsNRAMP5 contributes to manganese translocation and distribution in rice shoots. J Exp Bot 65(17):4849–4861. https://doi.org/10.1093/jxb/eru259

Yim HK, Lim MN, Lee SE, Lim J, Lee Y, Hwang YS (2012) Hexokinase-mediated sugar signaling controls expression of the calcineurin B-like interacting protein kinase 15 gene and is perturbed by oxidative phosphorylation inhibition. J Plant Physiol 169:1551–1558. https://doi.org/10.1016/j.jplph.2012.06.003

Yokotani N, Sato Y, Tanabe S, Chujo T, Shimizu T, Okada K, Yamane H, Shimono M, Sugano S, Takatsuji H, Kaku H (2013) WRKY76 is a rice transcriptional repressor playing opposite roles in blast disease resistance and cold stress tolerance. J Exp Bot 64(16):5085–5097. https://doi.org/10.1093/jxb/ert298

Yoshida S, Satake T, Mackill DJ (1981) High temperature stress in rice (review). IRRI Res Paper Ser 67:5

Yuan J, Wang X, Zhao Y, Khan NU, Zhao Z, Zhang Y, Wen X, Tang F, Wang F, Li Z (2020) Genetic basis and identification of candidate genes for salt tolerance in rice by GWAS. Sci Rep 10(1):1–9. https://doi.org/10.1038/s41598-020-66604-7

Yue B, Xue WY, Xiong LZ, Yu XQ, Luo LJ, Cui KH, Jin DM, Xing YZ, Zhang QF (2006) Genetic basis of drought resistance at reproductive stage in rice: separation of drought tolerance from drought avoidance. Genet 172(2):1213–1228. https://doi.org/10.1534/genetics.105.045062

Zhang T, Yang L, Jiang K, Huang M, Sun Q, Chen W (2008) QTL mapping for heat tolerance of the tassel period of rice. Mol Plant Breed 6:867–873

Zhang G, Chen L, Xiao G, Xiao Y, Chen X, Zhang S (2009) Bulked segregant analysis to detect QTL related to heat tolerance in rice (*Oryza sativa* L.) using SSR markers. Agric Sci China 8(4): 482–487. https://doi.org/10.1016/S1671-2927(08)60235-7

Zhang J, Aijaz AS, Chai L, Cui Y, Wang X, Zheng T, Jianlong XU, Zhikang LI (2013) Mapping of QTL for iron and zinc toxicity tolerance at seedling stage using a set of reciprocal introgression lines of Rice. Acta Agron Sin 39(10):1754. https://doi.org/10.3724/SP.J.1006.2013.01754

Zhang GH, Li SY, Wang L, Ye WJ, Zeng DL, Rao YC, Peng YL, Hu J, Yang YL, Xu J, Ren DY (2014) LSCHL4 from japonica cultivar, which is allelic to NAL1, increases yield of indica super rice 93-11. Mol Plant 7(8):1350–1364. https://doi.org/10.1093/mp/ssu055

Zhang X, Lourenco D, Aguilar I, Legarra A, Misztal I (2016) Weighting strategies for single-step genomic BLUP: an iterative approach for accurate calculation of GEBV and GWAS. Front Genet 7:151. https://doi.org/10.3389/fgene.2016.00151

Zhang J, Chen K, Pang Y, Naveed SA, Zhao X, Wang X, Wang Y, Dingkuhn M, Pasuquin J, Li Z, Xu J (2017a) QTL mapping and candidate gene analysis of ferrous iron and zinc toxicity tolerance at seedling stage in rice by genome-wide association study. BMC Genom 18(1): 828. https://doi.org/10.1186/s12864-017-4221-5

Zhang F, Wu ZC, Wang MM, Zhang F, Dingkuhn M, Xu JL, Zhou YL, Li ZK (2017b) Genome-wide association analysis identifies resistance loci for bacterial blight in a diverse collection of indica rice germplasm. PLoS One 12(3):e0174598. https://doi.org/10.1371/journal.pone.0174598

Zhang M, Lu Q, Wu W, Niu X, Wang C, Feng Y, Xu Q, Wang S, Yuan X, Yu H, Wang Y, Wei X (2017c) Association mapping reveals novel genetic loci contributing to flooding tolerance during germination in indica rice. Front Plant Sci 8:678. https://doi.org/10.3389/fpls.2017.00678

Zhang M, Ye J, Xu Q, Feng Y, Yuan X, Yu H, Wang Y, Yang Y (2018) Genome-wide association study of cold tolerance of Chinese indica rice varieties at the bud burst stage. Plant Cell Rep 37(3):529–539. https://doi.org/10.1007/s00299-017-2247-4

Zhang F, Zeng D, Zhang CS, Lu JL, Chen TJ, Xie JP, Zhou YL (2019) Genome-wide association analysis of the genetic basis for sheath blight resistance in Rice. Rice 12(1):93. https://doi.org/10.1186/s12284-019-0351-5

Zhao Y, Hu Y, Dai M, Huang L, Zhou D-X (2009) The WUSCHEL-related homeobox gene WOX11 is required to activate shoot-borne crown root development in rice. Plant Cell 21(3):736–748. https://doi.org/10.1105/tpc.108.061655

Zhao K, Tung CW, Eizenga GC, Wright MH, Ali ML, Price AH, Norton GJ, Islam MR, Reynolds A, Mezey J, McClung AM (2011) Genome-wide association mapping reveals a rich genetic architecture of complex traits in *Oryza sativa*. Nat Commun 2(1):1–10. https://doi.org/10.1038/ncomms1467

Zhao Y, Cheng S, Song Y, Huang Y, Zhou S, Liu X, Zhou DX (2015) The interaction between rice ERF3 and WOX11 promotes crown root development by regulating gene expression involved in cytokinin signaling. Plant Cell 27(9):2469–2483. https://doi.org/10.1105/tpc.15.00227

Zhao J, Zhang S, Dong J, Yang T, Mao X, Liu Q, Wang X, Liu B (2017) A novel functional gene associated with cold tolerance at the seedling stage in rice. Plant Biotechnol J 15(9):1141–1148. https://doi.org/10.1111/pbi.12704

Zhao J, Yang W, Zhang S, Yang T, Liu Q, Dong J, Fu H, Mao X, Liu B (2018a) Genome-wide association study and candidate gene analysis of rice cadmium accumulation in grain in a diverse rice collection. Rice 11(61):1–15. https://doi.org/10.1186/s12284-018-0254-x

Zhao Q, Feng Q, Lu H, Li Y, Wang A, Tian Q, Zhan Q, Lu Y, Zhang L, Huang T, Wang Y (2018b) Pan-genome analysis highlights the extent of genomic variation in cultivated and wild rice. Nat Genet 50(2):278–284. https://doi.org/10.1038/s41588-018-0041-z

Zhao X, Luo L, Cao Y, Liu Y, Li Y, Wu W, Lan Y, Jiang Y, Gao S, Zhang Z, Shen Y (2018c) Genome-wide association analysis and QTL mapping reveal the genetic control of cadmium accumulation in maize leaf. BMC Genomics 19(1):1–13. https://doi.org/10.1186/s12864-017-4395-x

Zheng W, Wang Y, Wang L, Ma Z, Zhao J, Wang P, Zhang L, Liu Z, Lu X (2016) Genetic mapping and molecular marker development for Pi65(t), a novel broad-spectrum resistance gene to rice blast using next-generation sequencing. Theor Appl Genet 129:1035–1044. https://doi.org/10.1007/s00122-016-2681-7

Zhou X, Huang X (2019) Genome-wide association studies in rice: how to solve the low power problems? Mol Plant 12(1):10–12. https://doi.org/10.1016/j.molp.2018.11.010

Zhou X, Stephens M (2012) Genome-wide efficient mixed-model analysis for association studies. Nat Genet 44:821–824. https://doi.org/10.1038/ng.2310

Zhou GA, Chang RZ, Qiu LJ (2010) Over expression of soybean ubiquitin conjugating enzyme gene GmUBC2 confers enhanced drought and salt tolerance through modulating abiotic stress-responsive gene expression in Arabidopsis. Plant Mol Biol 72(4):357. https://doi.org/10.1007/s11103-009-9575-x

Zhou S, Sun X, Yin S, Kong X, Zhou S, Xu Y, Luo Y, Wang W (2014) The role of the F-box gene TaFBA1 from wheat (*Triticum aestivum* L.) in drought tolerance. Plant Physiol Biochem 84:213–223. https://doi.org/10.1016/j.plaphy.2014.09.017

Zhu D, Kang H, Li Z, Liu M, Zhu X, Wang Y, Wang D, Wang Z, Liu W, Wang GL (2016) A genome-wide association study of field resistance to *Magnaporthe oryzae* in rice. Rice 9(1):1–9. https://doi.org/10.1186/s12284-016-0116-3

Genome-Wide Association Studies and Genomic Predictions for Climate Change Resilience in Wheat

3

Rahul M. Phuke, Pratik I. Bhoyar, Divya Ambati, Sripada Udupa, Anil Gaddameedi, Vikas Gupta, Swapnil D. Tayde, and Gaurav N. Chaudhari

Abstract

Wheat is the most widely grown staple crop as compared to other food crops, covering 250 million hectares area around the world. It is predicted that many wheat-growing regions around the world are likely to face severe water and heat stress due to global climate change in the coming decades. Global climate change is largely assumed to drive the emergence of new abiotic and biotic stresses in wheat. Under such circumstances, understanding the underlying natural genetic variation and identifying novel tolerance alleles is a prerequisite for developing climate-resilient wheat cultivars. The most prevalent abiotic stresses, viz., heat and drought, are complex in genetic regulation and difficult to dissect in terms of high genotype x environment interaction and low heritability. However, the recent advances in wheat genomic approaches like genome-wide association mapping (GWAS) and genomic predictions (GP) can facilitate in understanding the genetic architecture of complex traits and identifying the novel tolerance alleles precisely on a wheat chromosome that are otherwise difficult through biparental mapping. GWAS utilizes the ancestral recombination events through a large population and creates an opportunity to identify closely linked markers,

R. M. Phuke (✉) · P. I. Bhoyar · D. Ambati
ICAR-Indian Agricultural Research Institute, Regional Station, Indore, India

S. Udupa
International Centre for Agricultural Research in the Dry Areas, Rabat, Morocco

A. Gaddameedi
International Crops Research Institute for the Semi Arid Tropics, Pantancheru, Hyderabad, India

V. Gupta
ICAR-Indian Wheat and Barley Research Institute, Karnal, India

S. D. Tayde · G. N. Chaudhari
Dr. Panjabrao Deshmukh Krishi Vidyapeeth, Akola, India

M. G. Mallikarjuna et al. (eds.), *Next-Generation Plant Breeding Approaches for Stress Resilience in Cereal Crops*, https://doi.org/10.1007/978-981-19-1445-4_3

where genomic prediction incorporates available linked markers information into the prediction model to predict the breeding value of selected genotypes, which is essential for efficient marker-assisted breeding programmes to develop stress-tolerant wheat cultivars. The application of GWAS and GS is gaining importance in stress-resilient wheat breeding. Here, we summarized the recent application of GWAS and GP in wheat breeding to develop climate-resilient cultivars.

Keywords

Wheat · Climate change · GWAS · Genomic prediction · Mapping population · G × E interaction

3.1 Introduction

Wheat (*Triticum* spp.) is the second most important cereal crop and a major food source for more than 35% of the world population, providing 55% of the carbohydrates and 20% of the food calories consumed globally (Shewry and Hey 2015). Wheat is grown in all the regions of the world, occupying nearly 250 million hectares of area, and is the main source of income to millions of smallholding farmers (https://wheat.org/). It supersedes maize and rice as a source of protein in developing nations and is consumed by more than 2.5 billion people (http://www.fao.org). Urbanization, rising incomes and an increase in per capita wheat consumption are driving a rapid rise in global wheat demand. By 2050, wheat demand is expected to rise by at least 50%; hence the global economies are concentrating on expanding wheat output and production (Enghiad et al. 2017). However, wheat yields continued to increase, and through much of the last century, wheat was the most produced crop in the world. Global wheat-breeding programme with long-term approaches to enhance the grain yield has solved various challenges faced by wheat farmers to ensure the better wheat crop (Ramadas et al. 2019). However, global wheat production faces serious challenges posed by global climate change in terms of many abiotic (environmental factors) and biotic factors (diseases and pests). Wheat breeding programmes all over the world are focusing on developing climate change-resilient varieties that can be grown in harsh climates and have a higher survival rate.

Global climate change has a substantial effect on agricultural productivity. The effect of climate change in the form of enhanced incidence of stresses like high temperature, drought, salinity, waterlogging and mineral toxicity and biotic stresses are the major concerns to wheat scientists. Drought stress can be simply defined as a scarcity of water, leading to dramatic changes in the plants morphological, biochemical, physiological and molecular features (Sallam et al. 2019). Heat stress induced by high temperature is expressed as an increase in air temperature beyond a particular threshold level and period, resulting in irreversible damage to the plant (Farooq et al. 2011). A meta-analysis of 1700 published simulations predicted a rise of 2 °C mean temperature in temperate and tropical regions which results in a significant yield loss in wheat (Challinor et al. 2014). Similarly, modelling studies predicted around 6% decrease in wheat production, which is equivalent to a possible reduction of 42 Mt. per degree rise in temperature (Asseng et al. 2015). In India, climate change is predicted to reduce the wheat yield from 6 to 23% by 2050 and 15 to 25% by 2080 (Kumar et al. 2014). Globally, over 20% of the cultivable land is affected by salinity and which is further expected to increase day by day owing to environmental changes and anthropogenic exercises (Munns and Tester 2008). According to estimates, high salinity affects 20% of cultivated agricultural lands and 33% of irrigated agricultural lands. Additionally, abiotic stress factors such as drought, salinity, extreme temperatures and acidity, account for 60 to 82% of yield loss. The increase or decrease in wheat yield losses due to changing climate will depend on climatic effects on pathogens and the host plant itself (Juroszek and von Tiedemann 2013). The potential risk of climate change may lead to increased losses, decreased resistance effectiveness and evolution of newer pathotypes/pathogens (Chakraborty and Newton 2011). The efficacy of many of the rust resistance genes is driven by temperature. Any change in temperature regimes due to climate change may alter the resistant status of the wheat genotypes carrying these temperature-sensitive resistance genes. Increased CO_2 concentration and elevated temperatures due to climate change may increase wheat biomass which in turn increase the total leaf area available for pathogen/pest attack leading to the build up of more inoculum, which may lead to severe disease epidemics problem in wheat. The conducive environment for rust pathogen may also lead to higher rates of new pathotype evolution in nature, leading to the breakdown of many deployed resistance genes (Chakraborty and Newton 2011). The evolution of newer rust and other pathogen races occurs due to changes in climate, monoculture, cultivation practices, etc. Many newer pathotypes are being continuously evolved in nature. Therefore, sustaining wheat productivity levels is a major challenge in present agriculture, and the mitigation strategies must be streamlined towards boosting grain yield under limited resources environment.

Being a complex trait, the grain yield shows low heritability and is greatly influenced by the G × E and G × E × M interactions.

In some parts of the world, there have been concerns about the stagnation or decline of staple crop production. Wheat yield stagnation has been reported in 37% of wheat-growing regions. Further, in many areas the wheat yields are sustained owing to the availability of genetic resources for crop improvement for introgression of desired target traits through conventional breeding. However, improving stress tolerance through conventional breeding is labour intensive and time consuming as it involves complex genetics owing to multigene families/QTLs mediated molecular and physiological stress responsive mechanisms. Thus, traditional approaches and methods are not enough to resolve global food security issues in changing climatic scenario. Available genetic variability in the primary gene pool has been integrating with long-term traditional breeding methodologies; thus, there is need to create the new variability to improve desired target traits by speed breeding, mutation breeding, rapid generation advancement techniques and genome-wide selection approaches.

Old-generation markers like RFLP, RAPD, AFLP and SSR have served for more than three decades as a tool for marker-assisted selection and QTL identification. Traditional markers have played a key role in the identification and mapping of different QTLs, and marker-assisted selection was accelerated with the identification of different markers like SSR, STS, SNP, DArT, etc. However, the old-generation markers have less genome coverage and less abundance throughout the genome, and their analysis is laborious and requires more time (Desta and Ortiz 2014). The advances in sequencing technology and reduction in cost and time requirement have revolutionized the marker system. Although the bi-parental populations were more popular for the mapping study, but it has weakness in identifying QTL with small effect and methods applied for identifying QTL may also hinder crop improvement. The bi-parental mapping in addition to these also has few limitations: (i) bi-parental population do not possess the same level of allelic diversity throughout the breeding programme which makes them unsuitable as representative of the populations; (ii) developing population and its maintenance become costly affairs; (iii) identified QTLs are needed to be validated which require further efforts; and (iv) QTLs with small effects are entirely missed due to stringent significant threshold. With the availability of the NGS, the bi-parental mapping is being slowly replaced by association mapping, which is more cost-effective as well as precise, for QTL mapping and trait investigations.

Breeding and trait development have been accelerating with recent genomic technologies and resulting cultivars with enhanced environmental resistance and productivity. The quick selection and breeding of elite varieties with new genetic combinations is enabled by the identification of loci that contribute to characteristics and together with genomic-assisted breeding. Further, the availability of NGS approaches made the exploration of genetic diversity at nucleotide-scale precision through genome-wide association and improved phenomics platforms. In addition to genome-wide breeding tools, advances in engineering the spatial and temporal regulation of genes and pathways are increasingly accelerated by the targeted editing of genomes for stress tolerance traits in wheat (Juliana et al. 2019). Wheat breeding, combined with genome-wide studies, enhances the accuracy of breeding practices and saves time to deliver new wheat cultivars to farming community (Ahmar et al. 2020).

3.2 Vulnerabilities of Global Climate Change in Wheat Production

Global climate change in terms of drought and heat stress, particularly at the reproductive stage of the crop, can be a great threat to food security. The yield losses of wheat due to these abiotic stresses vary substantially among the wheat-producing nations. Furthermore, the frequency and magnitude of stress-induced crop losses may increase in the future owing to projected global temperature rise by 0.6–2.5 °C by 2050 and 1.4–5.8 °C by 2100 which are accompanied with increased incidence and severity of drought conditions (IPCC 2007).

According to the global climate model predictions, cereal crops were found to be most affected by drought and heat stress, and drought conditions were found to be more extensive and persistent in the coming future (Seneviratne et al. 2012; Trenberth et al. 2014). However, the effects of these stresses on yield are complex, and stress at any growth stage can affect crop yield. Drought and heat stress can affect wheat germination, vegetative growth, tiller production, dry matter partition, reproductive organ development, grain filling and grain quality (Gooding et al. 2003; Prasad et al. 2008; Sehgal et al. 2017). Studies showed that wheat crop yields are reduced when exposed to heat stress during the growing season due to accelerated crop phenological stages, which affect photosynthesis and respiration (Lobell and Gourdji 2012; Rezaei et al. 2015). A more pronounced effect of these abiotic stresses is observed during the reproductive phase of wheat, i.e. grain filling stages affecting yield in both qualitative and quantitative terms (Sehgal et al. 2017; Kumar et al. 2020). Exposure of wheat crop to heat stress (>25 °C) for 2–5 days at the reproductive stage has resulted in substantial damage to florets' fertility. These stresses will limit the grain filling duration resulting in the reduction of grain weight, grain number and quality of grains (Wardlaw 2002; Farooq et al. 2011). The linear association was observed with increased high-temperature duration at this grain filling stage and the grain weight loss (Prasad and Djanaguiraman 2014). Protein quality of wheat grain under drought stress and heat stress indicates differences in the concentration of total nitrogen, protein and glutenin, gliadin and albumin concentrations compared to grain quality under optimum conditions (Barnabás et al. 2008). Dough quality was found to deteriorate due to a rise in gliadin content compared to glutenin, and the ratio of large polymers was found to decline when exposed to heat stress (Panozzo and Eagles 1998; DuPont and Altenbach 2003). During grain filling stage, the heat stress reduces the non-structural carbohydrates accumulation in the endosperm of wheat grain (Hurkman et al. 2003; Plaut et al. 2004). Along with losses associated with the quality and yield, plant diseases are mostly considered as one of the most formidable obstacles. Climate change is leading to an increase in the CO_2 concentration, which has surpassed 400 ppm, and it can increase the crop yields of C_3 crops and can surge disease severity in wheat (Vary et al. 2015). Few rust resistance genes in wheat are temperature sensitive, and variation in temperatures during the growing season can alter their resistance pattern in the future. Since 2000, the new races of rust fungus *Puccinia striiformis*, which causes yellow rust, have been more aggressive at higher temperatures and are becoming prevalent worldwide (Milus et al. 2009). Elevated CO_2 level has increased the susceptibility of wheat varieties to the fungal pathogen *Fusarium graminearum due to the increase in the virulence of fungus* (Vary et al. 2015). Changes in environmental temperature can modify insect/pest physiology, behaviour, voltinism and distribution (Sandra et al. 2021). With an increase in temperature during wheat crop season, the aphid population and their distribution can be increased, incurring more yield losses (Alford et al. 2014).

3.3 Genetic Behaviour of Complex Traits

Developing abiotic stress resilience especially drought and heat tolerance in wheat is of extreme importance, as wheat is the main contributor to the world food supply. Responses to drought, heat and other abiotic stresses are complex and governed by the up- and downregulation of several genes and pathways, and each may have a minor to major effect on traits (Bernardo 2008). Stress responses are composed of a network of the regulatory process comprised of upstream (stress hormones, reactive oxygen species, gaso-transmitters, polyamines, phytochromes and calcium) and downstream (transcription factors) signalling as well as structural modification (cuticle outside plants, electrolyte leakage) in response to environmental factors (He et al. 2018). Some genes, viz. quantitative trait loci (QTL), show additive and non-additive gene effects. As the responses having polygenic inheritance and genotype adopt by its interaction with the environment, abiotic stress resilience characteristically has little heritability (Mwadzingeni et al. 2016). The genomic-assisted selection has still to contribute for the improvement in the genotypes of wheat for such abiotic stresses due to the polygenic nature of these traits, complexity and large size of the genome (Berkman et al. 2012).

Phenotyping for tolerance against abiotic stresses is also a major challenge due to the complexity of regulatory networks behind tolerance against these stresses (Vandenbroucke and Metzlaff 2013). The phenotype represents the effect of either a single gene or multiple genes, which may express different phenotypic outcomes depending on its interaction with each other and environment. Phenotyping is done at each growth stage under stress, which may also show variation in the crop's tolerance and susceptibility for the stress. The phenotyping techniques also limit the application of genomic tools in these stresses as the phenotyping has not been standardized. However, with the advent of effective high-throughput phenotyping platforms and phenomics tools, it has become possible to screen larger populations for multiple characteristics non-destructively under stress conditions. Further, high-throughput phenomics also enables the genetic dissection of complex stress-tolerant adaptive traits through providing reliable and accurate phenotypic data (Yang et al. 2013).

The precise estimation of the phenotypic response requires sophisticated tools and techniques to analyse the key parameters unique to stress tolerance. Therefore, the degree of stress and extent of resistance or vulnerability of a cultivar have been assessed using several parameters (Collins et al. 2008). The development of a high-throughput phenotyping platform for precise phenotyping is of utmost importance for dissecting these complex traits in developing climate-resilient wheat. The high-throughput phenotyping can aid the application of genomics tools for the improvement of these traits. Consequently, advanced phenotyping and genotyping stand as a tool in precision genomic breeding through genomic-wide characterization, selection and marker discovery, gene/QTL mapping and candidate identification.

3.4 Genomics Opportunities for Climate-Resilient Breeding

Plant breeders have continuously improved the genetic architecture of crops through conventional breeding technologies from many decades; now, breeders need to focus on global climate change and its effects on crop production. The advances in current breeding techniques showed it has the capacity to drastically decrease time to deliver improved crop varieties resilient to recent climate change. These techniques include advanced genomic approaches, which bypass some traditional approaches of the long selection process and indirect selection of beneficial genes and alleles in elite wheat cultivars. Genomics approaches provide an understanding of many phenomena such as genotype × environment interaction, identification and mapping of genes related to environmental stress tolerance, indirect selection of complex abiotic stress-tolerant genes and introduction of valuable alleles from wild relatives to wheat cultivars through marker-assisted selection. Hence, we highlighted the two recent genomics approaches of viz. genome-wide association studies (GWAS) and genomic prediction (GP) in wheat crop for the development of climate-resilient wheat cultivars.

3.4.1 Genome-Wide Association Studies (GWAS)

The genome-wide association studies (GWAS) are used to compute the correlation between single genome positions, primarily SNP (single nucleotide polymorphism) and the phenotype or trait of interest. GWAS is proven to be a very efficient approach for locating marker trait associations (MTAs) in wheat from the last decade due to the decreasing cost of high-throughput genotyping. Thus, GWAS is becoming a powerful tool for detecting QTLs associated with important traits of wheat (Cericola et al. 2017; Lopes et al. 2015), which is also supported by high-density SNP- genotyping platforms developed by Illumina (Wang et al. 2014) and Affymetrix (Allen et al. 2017). Till now, the identification of genetic basis for underlying phenotypic variation in plants has been achieved through traditional linkage mapping based on genetic maps. The diversity of experimental population in traditional linkage mapping ranges from F_2 population to MAGIC (multiparent advanced generation intercross) population (Kover et al. 2009). However, in traditional linkage mapping, the RILs (recombinant inbred lines) are the most widely used population owing to immortal and completely homozygous nature, enabling the replication of each line throughout the locations and environmental seasons (Bergelson and Roux 2010). However, the traditional linkage mapping leads to two drawbacks: the limited genetic diversity, which is in the range of parental lines used to develop segregating populations, and the limited recombination events. These drawbacks can be overcome through GWAS, which takes advantage of ancestral recombination events accumulated over thousands of generations and uses natural linkage disequilibrium (LD) to identify polymorphism ultimately associated with phenotypic variation (Nordborg and Weigel 2010). For successful GWAS studies, it is a prerequisite to have the large diversity panel of germplasm having less similarity in the pedigree as well as with respect to adaptation and photoperiod requirement (Yu and Buckler 2006). GWAS makes use of natural linkage disequilibrium (LD) to identify polymorphism ultimately associated with phenotypic variations.

For successful GWAS studies, there are prerequisites of few important things, which include (1) diverse association mapping panel, (2) high-throughput genotyping platform, and (3) good phenotyping site, especially in the case of abiotic stresses like drought and heat and artificial epiphytotic platform (control environment) in the case of disease studies. The association mapping panel used in GWA mapping should be of appropriate size. The panel size is a very crucial factor in getting meaningful results; increasing population size will improve the power of association as it can define a sufficient portion of the phenotypic and genotypic variation. The individual genotypes of association mapping should represent sufficient diversity in terms of geographic origin, growth habit, etc. The known association mapping panels of wheat used globally includes, the global spring wheat association mapping panel consisting of 882 landraces and 912 improved accessions (493 experimental lines and 419 cultivars) originated from 107 countries which include old and new wheat accession from the year 1920 to 2012. In case of winter wheat the known GWAS panel are, NSGC core panel consist of 4007 accessions and other panel includes hard winter wheat association mapping panel 1 and 2 (HWWAMP 1 and HWWAMP 2). Recently in the case of durum wheat, a global durum wheat panel (GDP) of 1011 genotypes was developed that captures 94–97% of original diversity, and it consists of a wider representation of durum germplasms, landraces along with the selection of primitive tetraploid and emmer wheat (Mazzucotelli et al. 2020).

The second important requirement of GWAS is the high-throughput genotyping platform that can provide complete wheat genome coverage and fast genotyping for a large set of germplasm. The present next-generation sequencing (NGS) provides thousands of SNPs covering most of the genomic region of wheat. NGS techniques provide huge numbers of markers within a short time frame and can genotype a large number of genotypes simultaneously using genotyping arrays or chips. The first high-throughput genotyping array developed in wheat is popularly known as Illumina iSelect 9 K bead chip assay (Cavanagh et al. 2013). Following this high-throughput genotyping array, a 90 K iSelect Assay was developed consisting of allelic ratio deviating between hexaploid and tetraploid wheat, which includes approximately 90,000 gene-associated SNPs covering all 7 groups of wheat chromosomes (Wang et al. 2014). Recently the Breeders' 35 K Axiom® array, a 35,143 SNP-based genotyping assay, was derived from 8,19,517 previously characterized wheat markers and was developed in 384 samples format array. This assay is highly suited for genotyping of elite hexaploid wheat accessions and is most useful to characterize diverse global collections of wheat, including landraces and elite genotypes derived from the commercial breeding programmes. Additionally, Breeders' 35 K Axiom® array is found to be a cost-effective and efficient platform for screening a large number of wheat genotypes (Elbasyoni et al. 2019).

Even though genomics techniques are fast forward and gaining wide importance, the importance of phenotype is still evident from the fact that almost all genomic techniques, including QTL mapping, fine mapping, GWAS and genomic prediction, depend immoderately on precise and accurate phenotyping. The climate change imparts abiotic stress like drought and heat on wheat crop, and this stress creates huge threats to wheat production as its leads to change in plant's basic metabolism which ultimately represents the phenotype of crops; hence precise phenotyping is very much important to get success in both GWAS and genomic prediction. The screening of genotypes under ideal stress conditions will provide a true potential of given genotypes. A recent study by Mamrutha et al. (2020) prioritized hotspot locations in India for drought and heat screening of wheat, which reveals that Indore location in the state of Madhya Pradesh of India (ICAR- IARI, regional station, Indore) has the highest drought stress intensity index of 0.89 among 15 studied locations, and also it is observed that India can be a hub for wheat research across the globe for screening what germplasm for changing climatic conditions like heat and drought stress.

3.4.1.1 GWAS for Heat and Drought Stress in Wheat

Several GWAS studies have been conducted in wheat for climate change-associated stresses, which mostly include global stresses, viz. drought and high temperature. The summary of GWAS studies of wheat, including cultivated hexaploid and tetraploid wheat for both spring and winter wheat, along with synthetic wheat is discussed in this section (Table 3.1). The GWAS panels used in these studies were comprised of diverse genotypes, which include historic genotypes along with advanced cultivars; the diversity was maintained through core collection and pre-breeding lines developed from three-way crosses. The size of GWAS panels used in these studies was in the range of 91–2111 wheat accessions (Ayalew et al. 2018; Elbasyoni et al. 2017). The phenotypic traits or class of traits used in these GWAS studies are related to (i) agronomic traits, viz. grain yield, 1000 grain weight, spike length, tillers number, plant height, days flowering and days to maturity; (ii) physiological traits like normalized difference vegetation index (NDVI), canopy temperature (CT), SPAD, leaf rolling and biomass; (iii) biochemical traits such as cell membrane stability, proline content, etc.; and (iv) seedling traits, viz. seminal root angle, root length, seedling length, root:shoot ratio, days to wilting, number of leaves, water content and chlorophyll content. Most of these studies are conducted under field conditions by creating and maintaining stress conditions through different water regimes for drought stress phenotyping and late sowing for heat stress phenotyping. However, few studies are conducted under controlled environmental conditions by creating artificial drought stress through PEG 6000 solution treatment at the seedling stage (Elbasyoni et al. 2017; Lin et al. 2019) and for heat stress through control environment treatments of 45 °C (Elbasyoni et al. 2017) and 40/35 °C day/night temperature regimes (Maulana et al. 2018). The speed of these GWAS studies was truly accelerated by available NGS and high-throughput genotyping techniques. Millions of SNP markers spread throughout the wheat genome are now available through new genotyping techniques. The wheat GWAS

Table 3.1 List of MTAs identified for drought- and heat-related stress in wheat and its cultivated and wild tetraploid relatives (*T. turgidum* and *T. turgidum* ssp. *dicoccoides*)

S. No.	Trait class/ trait	GWAS panel and markers	Phenotyping condition	No of MTAs	Chromosome	Reference
1	Seedling length, days to wilting, leaf wilting	138 diverse spring wheat genotypes with 407 DArT markers		104	–	Ahmed et al. (2021)
2	Seminal root angle and nodal root angle	393 durum RILs form NAM genotype with 2541 genome-wide DArT markers	"Clear pot" method for root phenotyping	7	6A	Alahmad et al. (2019)
3	Four drought tolerance indices	382 advanced lines of spring wheat genotype with GBS markers	Rainfed condition for two crop seasons	175	4A is most prominent	Ballesta et al. (2020)
4	Cell membrane relative injury (RI%)	2111 spring wheat accession, 9 K SNP wheat iSelect assay	Drought—(60% PEG600) solution for 24 h at 10 °C in the dark Heat water bath at 45 °C for 1 h	Drought (RI %)—15 Heat (RI%) – 16	1A, 1B, 2A, 4A, 6B, 7B 1B, 2A, 4B, 6B, 7B	Elbasyoni et al. (2017)
5	Normalized difference vegetation index (NDVI)	248 elite durum wheat, Illumina iSelect 90 K wheat SNP assay	NDVI—Arial NDVI—Ground based At different growth stages and water regime	NDVI—Arial, 55 MTAs NDVI –ground based 41 MTAs SPAD—39 MTAs Leaf rolling—9 MTAs	1A, 1B, 2B, 4A, 4B, 5A, 6A, 6B, 7A 1A, 3B, 5A, 5B, 7A 3A	Condorelli et al. (2018)

6	Tillers number	92 Iranian wheat genotypes, 6349 SNP from 15 K wheat Infinium array	Two water regimes—Full irrigation and drought	Total tillers number—13 MTAs Fertile tillers number—11 MTAs	1A,2A,6A,6B,6D,7A 1A,2A,2B,5B,7A,7D	Bilgrami et al. (2020)
8	Stress tolerance indices	208 durum lines, 6211 DArTseq SNPs	Three treatments 1. Yield potential (YP) 2. Drought stress (DT) 3. Heat stress (HT)	YP—121 MTAs (8 traits) DT—159 MTAs (8 traits) HT—112 MTAs (7 traits)	All 14 chr. Mostly on 2A & 2B All 14 chr. Maximum on 2B & 5B All 14 chr. Except 3B	Sukumaran et al. (2018)
9	Agronomic traits 1. GY/plant 2. Effective spike no/plant 3. Spikelet no/spike 4. TGW, 5. Plant height 6. Spike length	277 winter wheat accessions, 395681 SNPs derived from transcriptome and genome sequencing	4 treatments (no stress + drought stress + heat stress + combine drought & heat stress) in 30 environments	Drought stress—65 MTAs Heat stress—27 MTAs Combined stress-30 MTAs	–	Li et al. (2019)
10	Stress tolerance indices	290 lines of WAMI population, 15,737 SNP markers	Under drought and irrigated condition	Irrigated—114 MTAs Drought—85 MTAs For 6 traits	1A, 1B, 2A, 2B, 3D, 5A, 6A, 6B, 7B, 7D 1B, 2A, 2B, 3A, 3B, 4B, 5A, 5B, 6A, 6B, 7A, 7B, 7D	Abou-Elwafa and Shehzad (2020)

(continued)

Table 3.1 (continued)

S. No.	Trait class/ trait	GWAS panel and markers	Phenotyping condition	No of MTAs	Chromosome	Reference
11	Agronomic traits (14 traits)	92 diverse bread wheat lines, 9236 polymorphic markers from 15 k Illumina chip	Four different stress treatments	Non stress—148 MTAs, Drought—95, Heat—151 and combined heat and drought—93	All chromosomes except 2A & 4D 1B, 2B, 3D, 4B, 6A, 6B, 6D, 7A, 7B, 7D 2A, 3A, 3D, 5B, 6A, 6D, 7A,7B, 7D	Qaseem et al. (2019)
12	Root length	91 winter wheat collections	Water stress (PEG 6000) and non-stress conditions	Stress—2 MTAs Non-stress—3 MTAs	2B & 3B 3B, 4A, 5B	Ayalew et al. (2018)
13	Seedling data 1. Leaf chlorophyll 2. Shoot length 3. No. of leaves/ seedling 4. Seedling recovery	200 red winter wheat genotypes	Heat stress (40/35 °C) dd Optimal temp (25/20 °C)	Leaf chlorophyll—6QTLs No of leaves/ seedling—3 QTLs Seedling recovery—6 QTLs	2B, 2D, 4A, 4B 3A, 3B, 5A 2A, 2B, 2D, 3A, 7A, 7B	Maulana et al. (2018)
14	Grain yield- and yield-related traits	123 synthetic hexaploidy wheat, 35648 GBS markers	Drought stress condition for 2 years	194 MTAs for GY and yield-related traits	All 21 chromosomes	Bhatta et al. (2018)
15	Drought tolerance indices	290 genotypes of WAMI population, 15735 SNP markers	Drought and irrigated condition	205 MTAs for six drought indices	Well water—1A, 1B, 2A, 2B, 2D, 3A, 3B, 3D, 4D, 5A, 5B, 6A, 6B, 7B, 7D Drought- 1B, 2A, 2B, 3A, 3B, 4B, 5A, 5B, 6A, 6B, 6D, 7A, 7B, 7D	Elwafa and Shehzad (2020)

16	Agronomic and physiological traits	339 pre-breeding lines derived from the three-way top-crosses, 7180 SNP markers	Well water and drought condition for 3 year	Well water—10 MTAs Drought—11 MTAs	NDVI—3A CT– 7B Plant ht.—7A DTH—3A, 6B DTM—1B G Y—4A, 6B Spike Len—2D, 3B No of grains/spike—6B, 3A Kernel abortion—3A, 3B TGW—3B	Shokat et al. (2020)
17	Biomass and agronomic traits	100 wheat genotypes, 15600 DArTseq-derived SNP markers	Drought stress and non-stress	Drought stress—46 MTAs Non-stress—41 MTAs	1B, 1D, 2A,2B,2D, 3B,4A, 4D, 5A,5B, 6A, 6B,7A, 7B, 7D 1A, 1B, 2A, 2B, 2D, 3A, 3B, 4A, 4D, 5A, 5B, 5D	Mathew et al. (2019)
18	Agronomic traits (10 traits) for heat stress	205 wheat accessions, 11911 SNP markers	Heat stress (late sown), 3 locations, 3 years	69 QTLS (10 significant +59 suggestive)	1A,1B,1D, 2A,2B,2D,3A,3B,3D,4D,5B,5D,6A,6B,7A,7B,7D	Kumar et al. (2020)
19	Yield and quality traits	96 bread wheat accessions, Illumina 90 K Infinium SNP array	Normal and water stress condition	28 MTAs—Normal condition 44 MTAs -drought stress condition	1A, 1B, 2A, 2B, 3B, 3D, 4D, 5A, 6B, 7A, 7B 1A,1B,2B,2D,3A,3B,3D,4A,4B,5A,5B5D,6B, 7B,7D	Ahmed et al. (2020)
20	Agronomic traits	320 spring wheat accessions, 9626 SNPs	Irrigated and rainfed	20 MTAs in irrigated, 19 MTAs in rainfed	1A, 1B, 2A, 2B, 2D, 3A, 3B, 4A, 4B, 5A, 5B, 5D, 6A, 6B, 6D, 7A, 7B (for five traits in both condition)	Gahlaut et al. (2019)

(continued)

Table 3.1 (continued)

S. No.	Trait class/ trait	GWAS panel and markers	Phenotyping condition	No of MTAs	Chromosome	Reference
21	Agronomic traits (7 drought indices)	287 wheat lines (WAMI II), 1863 DArT markers	3 contrasting irrigation condition with 2 locations	11 MTAs	1B, 2B, 6A (5 are unmapped)	Edae et al. (2013)
22	Yield and yield-related traits	123 Pakistan historic wheat cultivars, 14960 SNPs	Rainfed condition for 3 years	44MTAs	1AL, 1BS, 2AL, 2BS, 2BL, 4BL, 5BL, 6AL, 6BL	Ain et al. (2015)

studies for drought and heat stress discussed here also strongly depend on available fast genotyping techniques. The initial GWAS studies in wheat utilised DArTseq SNP markers (Edae et al. 2013), followed by newly developed wheat-specific genotyping assay or chips like 9 K SNP wheat iSelect Assay, Illumina iSelect 90 K, 15 K wheat Infinium array and Breeders' 35 K Axiom® array. All these genotyping platforms provide large number of SNP markers which can cover broad genotypic variations of the GWAS panel. The large SNP data generated through these platforms then need to be filtered based on missing data points, heterozygosity and minor allele frequencies before carrying out GWAS analysis. The different high-throughput techniques available in wheat have specific advantages and limitations (Chawade et al. 2019). Hence, the choice of genotyping platform should be objective specific and should also consider the knowledge about computational techniques for large data analysis. The GWAS studies using a large association panel with millions of SNPs can also have few limitations such as missing data, rare alleles, false discovery rates, etc. These GWAS limitations can be overcome through newer computational methods having improved statistics. The discussed wheat GWAS studies here applied different computational models, and the most used model in these studies was MLM (mixed linear model). The MLM model takes care of multiple levels of relatedness, effectively controls population structure and Type I and Type II error rates (Yu et al. 2006); however, the MLM model can be computational challenging for large datasets. The other analysis models used in these studies were GLM (General Linear Model), CMLM (compressed mixed linear model) and FarmCPU (Fixed and random model Circulating Probability Unification). The FarmCPU is a recent model, which combines both fixed effect and random effects in analysis and improves statistical power with reduced computational times (Liu et al. 2016). In the discussed studies for GWAS in wheat for drought and heat tolerance (Table 3.1), the FarmCPU model was used to identify significant MTAs for tillers number using 92 Iranian wheat genotypes in drought condition which identified 13 significant MTAs for tillers number and 11 MTAs for fertile tiller number on chromosome 1A, 2A, 2B, 5B, 6A, 6B, 6D, 7A, 7B and 7D. (Bilgrami et al. 2020). Another GWA study by Bhatta et al. (2018) using 123 synthetic hexaploid wheat accessions for grain yield and yield-related traits under drought condition using FarmCPU model identified 194 MTAs covering all 21 chromosomes of wheat.

GWAS mapping approaches also have few limitations, of which false positives are considered as the major limitation due to the large genetic diversity of the association panel, and to overcome this limitation, multiple correction methods are used. The significance of MTAs passing the threshold p-value (0.001) is usually determined by using Bonferroni correction (BC) and false discovery rate (FDR). These multiple correction methods were used to test the significance of millions of markers in GWAS mapping. The BC defines the threshold level of significance for several traits at once, while FDR calculates significance for each trait independently. It is suggested that the studies which focused on identifying candidate loci/genes for further genetic and molecular studies should use low FDR values (Alqudah et al. 2018).

Several GWAS studies in wheat have demonstrated the power of association mapping in identifying candidate genes for tolerance to climate change-related drought and heat stress. Alahmad et al. (2019) identified a major QTL region on the distal end of chromosome 6A for seminal root angle in drought condition, which overlaps with the gene model representing a *NAC* transcription factor, a fatty acid hydroxylase family protein and SAWADEE homeodomain protein 2. In the case of durum wheat, a GWAS study using 208 durum lines phenotyped with three different treatments, viz. yield potential, drought and heat stress, identified a QTL hotspot for stress tolerance indices on chromosomes 2A and 2B, and one SNP (100035706) in these regions was related to gene *DMAS1-A,* and the protein is characterised as deoxymgiretic acid syntheses 1 (Sukumaran et al. 2018). The GWAS study of 277 winter wheat accessions under drought and heat stress condition for six agronomic traits identifies haplotype blocks containing candidate gene for stress tolerance, which include a dwarfing gene *Rht-D1* located on haplotype block on chromosome 4D and another three *WRKY* genes (*TaWRKY 8, TaWRKY 45 and TaWRKY 70*) which confirm the genomic region with multiple abiotic stress tolerance on haplotype block on chromosomes 6A and 6D (Li et al. 2019). The GWA study of physiological traits like NDVI and CT under drought condition using 339 pre-breeding lines derived from three-way crosses observed a candidate gene *TauE/SaFE* responsible for taurine metabolism and anion export across cell membrane in stress condition on chromosome 4A; another candidate gene "*Loci09764454*" was observed coding for heat stress protein on chromosome 2D, and on chromosome 7B, two SNPs were associated with the candidate genes coding for Omega *glidin-D1* and asa-like protein of bread wheat (Shokat et al. 2020).

3.4.1.2 Genome-Wide Association Studies for Salinity Stress Tolerance in Wheat

Wheat production is globally hampered by soil salinity and sodicity. Moreover, there is less attention given to salinity stress due to lack of suitable phenotypic methods and lack of diversity with a narrow gene pool for salinity tolerance in wheat, which impaired the progress of salinity tolerance; hence the importance of GWAS comes into the picture to identify novel salinity tolerance genes in wheat. Unlike major abiotic stresses, drought and heat, very few GWA studies for salinity stress in wheat were conducted (Table 3.2). These studies are mainly based on phenotyping of Na^+ and K^+ ions accumulation at germination and seedling stage under salt stress conditions (Oyiga et al. 2018; Genc et al. 2019; Li et al. 2020; Chaurasia et al. 2020), whereas the physiological traits like root and shoot length were used by Liu et al. (2018) and Li et al. (2020) and grain yield in field condition by Hu et al. (2021). These GWA studies utilised diverse association mapping panels, which include exotic wheat cultivars, landraces, double haploids and synthetic wheat lines genotyped with SNP markers in all studies, expect a study by Liu et al. (2018), which used 546 SSR markers for genotyping of 277 wheat accessions for association mapping of salt tolerance indices at germination and seedling stage. The association mapping study by Genc et al. (2019) in 100 bread wheat accessions for leaf Na^+ accumulation in artificial pot treatment with Na^+ humate solution leads to

Table 3.2 List of MTAs identified for salinity stress tolerance in wheat

Sr.	Trait class/trait	GWAS panel and markers	Phenotyping condition	No. of MTAs	Chromosome	Reference
1	Ion traits (Na^+ and K^+ leaf content)	150 winter and facultative wheat, 18085 SNPs	Salt stress (150 mM NaCl) and non salt stress condition (3 location)	37 QTLs	1BS, 1DL, 2AL, 2BS, 3AL	Oyiga et al. (2018)
2	Leaf Na + accumulation	100 bread wheat entries, 41035 SNPs	Pot treatment (8 g/kg, Na^+ humate)	7 MTAs	2A, 2B, 2D, 4B, 4D, 5B, 7A	Genc et al. (2019)
3	Salt tolerance index (germination and seedling)	227 wheat accessions, 546 SSR markers	Artificial seawater solution treatment	24 MTAs (8 MTAs germination +16 MTAs seedling)	1A, 5A, 1B, 2B, 3B, 6B, 7B, 1D	Liu et al. (2018)
4	Yield and yield-related traits	191 wheat accessions, 387657 SNPs	Salt stress condition	48 SNPs for yield	2D, 3A, 5A, 5B, 7A	Hu et al. (2021)
5	Salt tolerance index (germination)	307 wheat accessions, 402176 SNPs	Different salt solutions	117 MTA	Mostly clustered on 1A (72MTAs), 3B (10 MTAs), 6B (20 MTAs)	Yu et al. (2020)
6	Leaf chlorophyll, membrane stability, ion traits (Na^+, K^+)	135 diverse wheat genotypes, 17927 SNPs	Vegetative salt tolerance (hydroponics) 100 mM NaCl	Na^+ (7MTAs) K^+ (5 MTAs) Na/K (4 MTAs)	(Na^+) 1BL, 2BL, 3AL, 3BL, 5Al, 5DL, 7BL (K^+) 2AL, 5AL, 5DL, 7BL	Chaurasia et al. (2020)

identification of 7 MTAs distributed on chromosomes 2A, 2B, 2D, 4B, 4D, 5B and 7A, and further four candidates genes, viz. calcium-transporting ATPase, $Na^{(+)}/H^{(+)}$ antiporter NhaB, AquaporinTIF1–4 and Aquaporin PIP2 having the potential function in Na^{+} accumulation, were identified.

3.4.2 Genomic Prediction

Genomic prediction (GP) is widely used in crops nowadays. GP explores available molecular markers to predict genomic estimated breeding values based on new marker-based models (Bhat et al. 2016; Xu et al. 2020). GP approach comprises two populations, viz. training population (reference population) and breeding population (testing population). Training set/population is employed to predict the genomic estimated breeding values for testing set/population based on a marker-based statistical model developed using phenotypic and genotypic information of the training population (Xu et al. 2020). GP for self- and cross-pollinated crops follows the different skim, as the training population and breeding population varies. The GP has two advantages over traditional MAS as there is no need to unearth the QTL related to target traits, and phenotyping for the breeding population can be exempted which reduce the time for GP. Thus, GP provides the opportunities to enhance the genetic gain of multigenic traits per unit time and cost. The high-throughput techniques of the genome-wide association have become cheaper, and several new markers have been developed in a large population with or without the reference genome sequence (Bhat et al. 2016). The next-generation sequencing has provided an SNP genotyping platform through genotyping by sequencing; hence the availability of the SNP markers for genome-wide studies has increased, so the precision in the marker-trait relation has also increased. The availability of such high-precision molecular marker and its platform made the GP routine work for crop improvement in both model and non-model crop species. Genotyping by sequencing using NGS has increased the precision in predicting the genomic-estimated breeding values (Xu et al. 2020). The GP must combine with high-throughput phenotyping to acquire maximum genetic gain from complex traits. The gradual decrease in sequencing cost has made sequencing of complete genome possible for all important crops, which will accelerate the genomic selection in present and future also.

Presently, there are plenty of models available for genomic predictions (Fig. 3.1) depending on the prediction accuracy and genetic gain from the selection. Every model of prediction has different responses due to the variety in assumption(s) for the variance of complex traits (Desta and Ortiz 2014). Several models were already used for prediction in wheat for different complex traits. Saint Pierre et al. (2016) evaluated 803 spring wheat lines at 5 locations with several traits characterized for grain yield and agronomic traits with the best linear unbiased prediction (BLUP) model, which suggested that the best prediction was observed when the genotypic and pedigree data combined in the model and their interaction with the environment. Heffner et al. (2011) used multifamily prediction models to enhance genomic selection accuracy by 28% compared to MAS for 374 winter wheat by comparing

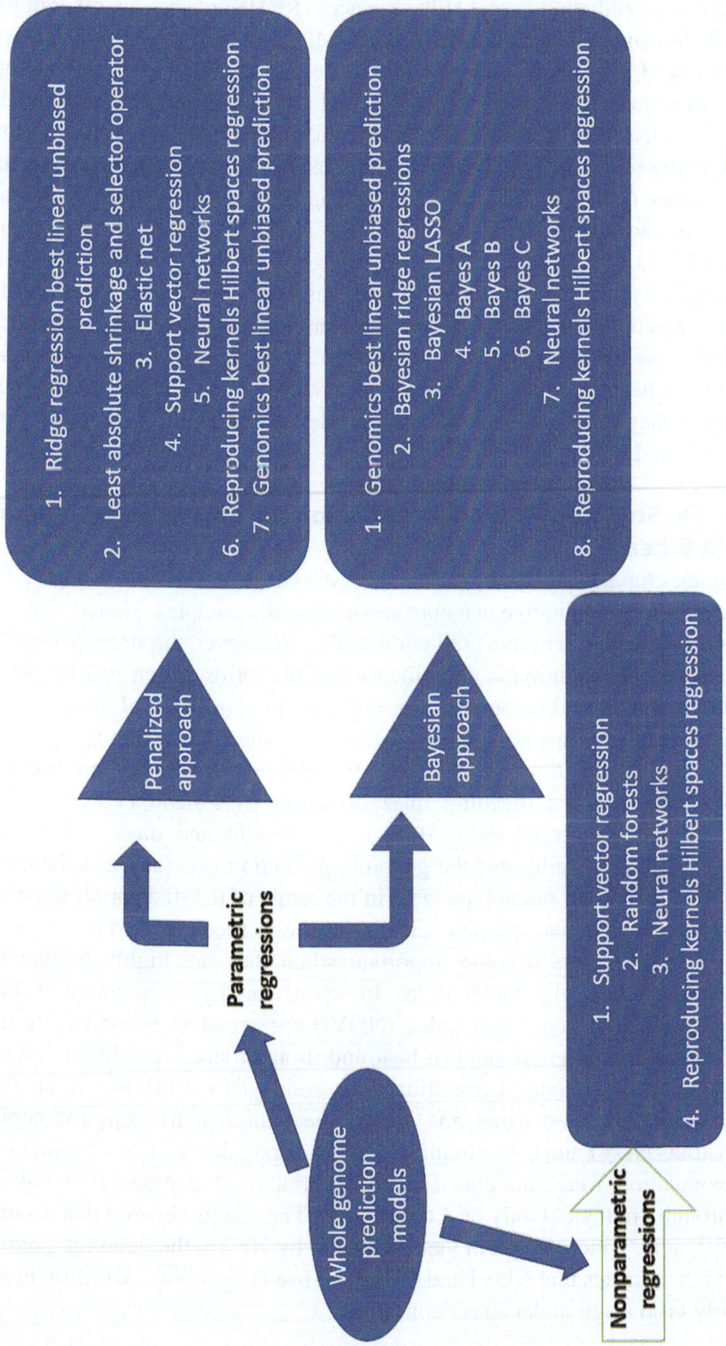

Fig. 3.1 Classification of whole-genome prediction models

13 agronomic traits. Rutkoski et al. (2012) used ridge regression (RR), Bayesian LASSO (BL), reproducing kernel Hilbert spaces (RKHS) regression, random forest (RF) regression and multiple linear regression (MLR) models for genomic prediction of fusarium head blight resistance in wheat and suggested that use of genome-wide marker apart from QTL-targeted marker has higher accuracy for genomic prediction with these models. The genomic best linear unbiased prediction (GBLUP) and a Bayesian regression method (BayesR) were used to predict genomic estimated breeding values (GEBVs) for rust resistance in 206 hexaploid wheat landraces, and the study showed that GBLUP has higher prediction accuracy when training population has a close relationship with reference population (Daetwyler et al. 2014). Sehgal et al. (2020) conducted a study on 4302 advanced bread wheat lines to integrate genetic architecture of grain yield and yield stability into the prediction model to increase the accuracy of the prediction. Though the different models have been evaluated for their prediction accuracy in wheat for different traits, the prediction accuracy may change for the different models based on the assumptions made and markers used.

3.4.2.1 Case Studies of Genomic Prediction for Abiotic Stress Tolerance in Wheat

The changing climate mostly triggers abiotic stresses like heat and drought. These abiotic stresses are quantitative in nature and genetically complex. Hence, these traits are ideal candidates for genomic prediction studies. However, the stress phenotyping requires specialised phenotypic equipments and platforms which restrict breeders with minimal budget and resources. The high phenotyping cost of abiotic stresses and availability of low-cost genotyping platforms in wheat make genomic prediction a more economical and attractive alternative for selection. A genomic prediction was applied in 254 advanced breeding lines of wheat by Poland et al. (2012) for agronomic traits, viz. grain yield, 1000 kernel weight and days to heading in contrasting irrigation conditions; the genomic prediction accuracy and the correlation between GEBV and phenotype were in the range of 0.3–0.5 for all three traits. For complex abiotic traits, indirect selection through secondary traits is common practice; secondary traits become important when they are highly heritable and genetically correlated with target traits. In wheat canopy temperature (CT) and normalized difference vegetation index (NDVI) are excellent secondary traits for genomic prediction for grain yield in heat and drought stress conditions owing to high heritability and genetic correlation with grain yield. Rutkoski et al. (2016) applied genomic selection using 555 bread wheat lines in five environments for secondary traits NDVI and CT to training and test population and grain yield only on training population were modulated as multivariate and compared to univariate models through grain yield only on a training set. The results showed that secondary traits NDVI and CT increase grain yield accuracy by 70% in the genomic prediction model, which indicates that NDVI and CT can be used in genomic selection in wheat during early crop stage under stress condition.

3.5 Prospects

It is long known that the abiotic stress in field conditions often occurs simultaneously rather than individual stress, which may create errors in phenotyping in field conditions. Hence, automated high throughput with high-precision phenotyping should be focused rather than normal field phenotyping with different irrigation and sowing dates. The traits used for phenotyping of stress tolerance should be prioritised based on their correlation with grain yield. Several agronomic, biochemical, physiological traits were used at different growth stages, and their relationship with grain yield is stage-specific, and each trait shows a differential association with grain yield. Hence, the target phenotyping stress indicator traits need to prioritise based on their relationship with grain yield and heritability of the trait(s). Major emphasis is needed to develop user-friendly and economic phenotypic platforms for continuous screening of elite wheat genotypes in breeding programmes and genomics studies. Likewise, the use of possible thermal infrared imaging and multispectral imaging in both ground- and aerial-based phenotypings should be considered.

GWAS has been attempted in wheat from more than a decade, although as stated above, GWAS faces new challenges with complex quantitative traits due to their genetic interaction (epistasis) and G $\times$ E interaction; the adoption of new statistical models and experimental design for these interactions should be considered in future. Now there is sufficient availability of stable and major genomic regions identified through GWAS studies for climate stress tolerance. Therefore, it demands the cloning of these genes identified in the genomic regions associated with target traits. The characterization of genes underlying these identified genomic will speed up the breeding programme for climate-resilient breeding.

References

Abou-Elwafa SF, Shehzad T (2020) Genetic diversity, GWAS and prediction for drought and terminal heat stress tolerance in bread wheat (*Triticum aestivum* L.). Genet Resour Crop Evol 68:711–728. https://doi.org/10.1007/s10722-020-01018-y

Ahmar S, Gill RA, Jung KH, Faheem A, Qasim MU, Mubeen M, Zhou W (2020) Conventional and molecular techniques from simple breeding to speed breeding in crop plants: recent advances and future outlook. Int J Mol Sci 21(7):2590

Ahmed HGM, Sajjad M, Zeng Y, Iqbal M, Khan SH, Ullah A, Akhtar MN (2020) Genome-wide association mapping through 90K SNP Array for quality and yield attributes in bread wheat against water-deficit conditions. Agriculture 10:392. https://doi.org/10.3390/agriculture10090392

Ahmed AA, Mohamed EA, Hussein MY, Sallam A (2021) Genomic regions associated with leaf wilting traits under drought stress in spring wheat at the seedling stage revealed by GWAS. Environ Exp Bot 184:104393

Ain Q, Rasheed A, Anwar A, Mahmood T, Imtiaz M, Mahmood T, Xia X, HeZand Quraishi UM (2015) Genome-wide association for grain yield under rainfed conditions in historical wheat cultivars from Pakistan. Front Plant Sci 6:743. https://doi.org/10.3389/fpls.2015.00743

Alahmad S, El Hassouni K, Bassi FM, Dinglasan E, Youssef C, Quarry G, Aksoy A, Mazzucotelli E, Juhász A, Able JA, Christopher J, Voss-Fels KP, Hickey LT (2019) A major root architecture QTL responding to water limitation in durum wheat. Front Plant Sci 10:436. https://doi.org/10.3389/fpls.2019.00436

Alford L, Andrade TO, Georges R, Burel F, Van Baaren J (2014) Could behavior and not physiological thermal tolerance determine winter survival of aphids in cereal fields? PLoS One 9:e114982. https://doi.org/10.1371/journal.pone.0114982

Allen AM, Winfield MO, Burridge AJ, Downie RC, Benbow HR, Barker GL, Wilkinson PA, Coghill J, Waterfall C, Davassi A, Scopes G, Pirani A, Webster T, Brew F, Bloor C, Griffiths S, Bentley AR, Alda M, Jack P, Phillips AL, Edwards KJ (2017) Characterization of a wheat Breeders' Array suitable for high-throughput SNP genotyping of global accessions of hexaploid bread wheat (*Triticum aestivum*). Plant Biotechnol J 15(3):390–401. https://doi.org/10.1111/pbi.12635

Alqudah AM, Youssef HM, Graner A, Schnurbusch T (2018) Natural variation and genetic make-up of leaf blade area in spring barley. Theor Appl Genet 131(4):873–886

Asseng S, Ewert F, Martre P, Rötter RP, Lobell DB, Cammarano D, Kimball BA, Ottman MJ, Wall GW, White JW, Reynolds MP (2015) Rising temperatures reduce global wheat production. Nat Clim Chang 5(2):143–147

Ayalew H, Liu H, Börner A, Kobiljski B, Liu C, Yan G (2018) Genome-wide association mapping of major root length QTLs under PEG induced water stress in wheat. Front Plant Sci 9:1759. https://doi.org/10.3389/fpls.2018.01759

Ballesta P, Mora F, Del Pozo A (2020) Association mapping of drought tolerance indices in wheat: QTL-rich regions on chromosome 4A. Sci Agric 77(2)

Barnabás B, Jäger K, Fehér A (2008) The effect of drought and heat stress on reproductive processes in cereals. Plant Cell Environ 31(1):11–38

Bergelson J, Roux F (2010) Towards identifying genes underlying ecologically relevant traits in Arabidopsis thaliana. Nat Rev Genet 11(12):867879

Berkman PJ, Lai K, Lorenc MT, Edwards D (2012) Next generation sequencing applications for wheat crop improvement. Am J Bot 99:365–371

Bernardo R (2008) Molecular markers and selection for complex traits in plants: learning from the last 20 years. Crop Sci 48:1649–1664

Bhat JA, Ali S, Salgotra RK, Mir ZA, Dutta S, Jadon V, Tyagi A, Mushtaq M, Jain N, Singh PK, Singh GP (2016) Genomic selection in the era of next generation sequencing for complex traits in plant breeding. Front Genet 7:221

Bhatta M, Morgounov A, Belamkar V, Baenziger PS (2018) Genome-wide association study reveals novel genomic regions for grain yield and yield-related traits in drought-stressed synthetic hexaploid wheat. Int J Mol Sci 19(10):3011

Bilgrami SS, Ramandi HD, Shariati V, Razavi K, Tavakol E, Fakheri BA, Nezhad NM, Ghaderian M (2020) Detection of genomic regions associated with tiller number in Iranian bread wheat under different water regimes using genome-wide association study. Sci Rep 10(1):1–17

Cavanagh CR, Chao S, Wang S, Huang BE, Stephen S, Kiani S, Forrest K, Saintenac C, Brown-Guedira GL, Akhunova A, See D (2013) Genome-wide comparative diversity uncovers multiple targets of selection for improvement in hexaploid wheat landraces and cultivars. Proc Natl Acad Sci 110(20):8057–8062

Cericola F, Jahoor A, Orabi J, Andersen JR, Janss LL, Jensen J (2017) Optimizing training population size and genotyping strategy for genomic prediction using association study results and pedigree information. A case of study in advanced wheat breeding lines. PLoS One 12(1): e0169606. https://doi.org/10.1371/journal.pone.0169606

Chakraborty S, Newton AC (2011) Climate change, plant diseases and food security: an overview. Plant Pathol 60(1):2–14

Challinor AJ, Watson J, Lobell DB, Howden SM, Smith DR, Chhetri N (2014) A meta-analysis of crop yield under climate change and adaptation. Nat Clim Chang 4(4):287–291

Chaurasia S, Singh AK, Songachan LS, Sharma AD, Bhardwaj R, Singh K (2020) Multi-locus genome-wide association studies reveal novel genomic regions associated with vegetative stage salt tolerance in bread wheat (*Triticum aestivum* L.). Genomics 112(6):4608–4621

Chawade A, van Ham J, Blomquist H, Bagge O, Alexandersson E, Ortiz R (2019) High-throughput field-phenotyping tools for plant breeding and precision agriculture. Agronomy 9(5):258

Collins NC, Tardieu F, Tuberosa R (2008) Quantitative trait loci and crop performance under abiotic stress: where do we stand? Plant Physiol 147:469–486

Condorelli GE, Maccaferri M, Newcomb M, Andrade-Sanchez P, White JW, French AN, Sciara G, Ward R, Tuberosa R (2018) Comparative aerial and ground based high throughput phenotyping for the genetic dissection of NDVI as a proxy for drought adaptive traits in durum wheat. Front Plant Sci 9:893. https://doi.org/10.3389/fpls.2018.00893

Daetwyler HD, Bansal UK, Bariana HS, Hayden MJ, Hayes BJ (2014) Genomic prediction for rust resistance in diverse wheat landraces. Theor Appl Genet 127(8):1795–1803

Desta ZA, Ortiz R (2014) Genomic selection: genome-wide prediction in plant improvement. Trends Plant Sci 19(9):592–601

DuPont FM, Altenbach SB (2003) Molecular and biochemical impacts of environmental factors on wheat grain development and protein synthesis. J Cereal Sci 38(2):133–146

Edae EA, Byrne PF, Haley SD, Lopes M, Reynolds MP (2013) Genome-wide association mapping of yield and yield components of spring wheat under contrasting moisture regimes. Theor Appl Genet 127(4):791–807. https://doi.org/10.1007/s00122-013-2257-8

Elbasyoni I, Saadalla M, Baenziger S, Bockelman H, Morsy S (2017) Cell membrane stability and association mapping for drought and heat tolerance in a worldwide wheat collection. Sustainability 9:1606. https://doi.org/10.3390/su9091606

Elbasyoni IS, El-Orabey WM, Morsy S, Baenziger PS, Al Ajlouni Z, Dowikat I (2019) Evaluation of a global spring wheat panel for stripe rust: resistance loci validation and novel resources identification. PLoS One 14(11):e0222755. https://doi.org/10.1371/journal.pone.0222755

Elwafa SFA, Shehzad Y (2020) Genetic diversity, GWAS and prediction for drought and terminal heat stress tolerance in bread wheat (*Triticum aestivum* L.). Genet Resour Crop Evol 68(2): 711–728. https://doi.org/10.1007/s10722-020-01018-y

Enghiad A, Ufer D, Countryman AM, Thilmany DD (2017) An overview of global wheat market fundamentals in an era of climate concerns. Int J Agron. https://doi.org/10.1155/2017/3931897

Farooq M, Bramley H, Palta JA, Siddique KH (2011) Heat stress in wheat during reproductive and grain-filling phases. CRC Crit Rev Plant Sci 30(6):491–507

Gahlaut V, Jaiswal V, Singh S, Balyan HS, Gupta PK (2019) Multi-locus genome wide association mapping for yield and its contributing traits in hexaploid wheat under different water regimes. Sci Rep 9(1):1–15

Genc Y, Taylor J, Lyons G, Li Y, Cheong J, Appelbee M, Oldach K, Sutton T (2019) Bread wheat with high salinity and sodicity tolerance. Front Plant Sci 10:1280

Gooding MJ, Ellis RH, Shewry PR, Schofield JD (2003) Effects of restricted water availability and increased temperature on the grain filling, drying and quality of winter wheat. J Cereal Sci 37(3): 295–309

He M, He CQ, Ding NZ (2018) Abiotic stresses: general defenses of land plants and chances for engineering multistress tolerance. Front Plant Sci 9:1771

Heffner EL, Jannink JL, Sorrells ME (2011) Genomic selection accuracy using multifamily prediction models in a wheat breeding program. Plant Genome 4(1)

Hu P, Zheng Q, Luo Q, Teng W, Li H, Li B, Li Z (2021) Genome-wide association study of yield and related traits in common wheat under salt-stress conditions. BMC Plant Biol 21(1):1–20

Hurkman WJ, McCue KF, Altenbach SB, Korn A, Tanaka CK, Kothari KM, Johnson EL, Bechtel DB, Wilson JD, Anderson OD, DuPont FM (2003) Effect of temperature on expression of genes encoding enzymes for starch biosynthesis in developing wheat endosperm. Plant Sci 164(5): 873–881

IPCC (2007) Intergovernmental panel on climate change. Fourth assessment report: Climate Change. Geneva

Juliana P, Poland J, Huerta-Espino J et al (2019) Improving grain yield, stress resilience and quality of bread wheat using large-scale genomics. Nat Genet 51:1530–1539. https://doi.org/10.1038/s41588-019-0496-6

Juroszek P, von Tiedemann A (2013) Climate change and potential future risks through wheat diseases: a review. Eur J Plant Pathol 136:21–33. https://doi.org/10.1007/s10658-012-0144-9

Kover PX, Valdar W, Trakalo J, Scarcelli N, Ehrenreich IM, Michael D (2009) A multiparent advanced generation inter-cross to fine- map quantitative traits in *Arabidopsis thaliana*. PLoS Genet 5:e1000551. https://doi.org/10.1371/journal.pgen.1000551

Kumar SN, Aggarwal PK, Swarooparani DN, Saxena R, Chauhan N, Jain S (2014) Vulnerability of wheat production to climate change in India. Clim Res 59(3):173–187

Kumar P, Gupta V, Singh G, Singh C, Tyagi BS, Singh GP (2020) Assessment of terminal heat tolerance based on agro-morphological and stress selection indices in wheat. Cereal Res Commun 49(2):217–226

Li L, Mao X, Wang J, Chang X, Reynolds M, Jing R (2019) Genetic dissection of drought and heat-responsive agronomic traits in wheat. Plant Cell Environ 42:2540–2553. https://doi.org/10.1111/pce.13577

Li L, Peng Z, Mao X, Wang J, Li C, Chang X, Ruilian Z (2020) Genetic insights into natural variation underlying salt tolerance in wheat. J Exp Bot 72(4):1135–1150. https://doi.org/10.1093/jxb/eraa500

Lin Y, Yi X, Tang S, Chen W, Wu F, Yang X, Jiang X, Shi H, Ma J, Chen G, Chen G, Zheng Y, Wei Y, Liu Y (2019) Dissection of phenotypic and genetic variation of drought-related traits in diverse Chinese wheat landraces. Plant Genome 12:190025. https://doi.org/10.3835/plantgenome2019.03.0025

Liu X, Huang M, Fan B, Buckler ES, Zhang Z (2016) Iterative usage of fixed and random effect models for powerful and efficient genome-wide association studies. PLoS Genet 12(2):1005767

Liu Y, Liu Y, Zhang Q, Fu B, Cai J, Wu J, Chen Y (2018) Genome-wide association analysis of quantitative trait loci for salinity-tolerance related morphological indices in bread wheat. Euphytica 214(10):1–11

Lobell DB, Gourdji SM (2012) The influence of climate change on global crop productivity. Plant Physiol 160(4):1686–1697

Lopes MS, Dreisigacker S, Peña RJ, Sukumaran S, Reynolds MP (2015) Genetic characterization of the wheat association mapping initiative (WAMI) panel for dissection of complex traits in spring wheat. Theor Appl Genet 128:453–464. https://doi.org/10.1007/s00122-014-2444-2

Mamrutha HM, Rinki K, Venkatesh K, Gopalareddy K, Khan H, Mishra CN, Kumar S, Kumar Y, Singh G, Singh GP (2020) Impact of high night temperature stress on different growth stages of wheat. Plant Physiol Rep 25(4):707–715

Mathew I, Shimelis H, Shayanowako AIT, Laing M, Chaplot V (2019) Genome-wide association study of drought tolerance and biomass allocation in wheat. PLoS One 14(12):e0225383. https://doi.org/10.1371/journal.pone.0225383

Maulana F, Ayalew H, Anderson JD, Kumssa TT, Huang W, Ma XF (2018) Genome-wide association mapping of seedling heat tolerance in winter wheat. Front Plant Sci 9:1272. https://doi.org/10.3389/fpls.2018.01272

Mazzucotelli E, Sciara G, Mastrangelo AM, Desiderio F, Xu SS, Faris J, Hayden MJ, Tricker PJ, Ozkan H, Echenique V, Steffenson BJ (2020) The global durum wheat panel (GDP): an international platform to identify and exchange beneficial alleles. Front Plant Sci 11:569905. https://doi.org/10.3389/fpls.2020.569905

Milus EA, Kristensen K, Hovmøller MS (2009) Evidence for increased aggressiveness in a recent widespread strain of *Puccinia striiformis* f. *sptritici* causing stripe rust of wheat. Phytopathology 99:89–94

Munns R, Tester M (2008) Mechanisms of salinity tolerance. Annu Rev Plant Biol 59:651–681

Mwadzingeni L, Shimelis H, Dube E, Laing MD, Tsilo TJ (2016) Breeding wheat for drought tolerance: progress and technologies. J Integr Agric 15(5):935–943

Nordborg M, Weigel D (2010) Next-generation genetics in plants. Nature 456:10–13. https://doi.org/10.1038/nature07629

Oyiga BC, Sharma RC, Baum M, Ogbonnaya FC, Leon J, Ballvora A (2018) Allelic variations and differential expressions detected at quantitative trait loci for salt stress tolerance in wheat. Plant Cell Environ 41(5):919–935

Panozzo JF, Eagles HA (1998) Cultivar and environmental effects on quality characters in wheat. I Starch Aust J Agric Res 49(5):757–766

Plaut Z, Butow BJ, Blumenthal CS, Wrigley CW (2004) Transport of dry matter into developing wheat kernels and its contribution to grain yield under post-anthesis water deficit and elevated temperature. Field Crops Res 86(2–3):185–198

Poland J, Endelman J, Dawson J, Rutkoski J, Wu S, Manes Y, Dreisigacker S, Crossa J, Sanchez-Villeda H, Sorrells M, Jannink JL (2012) Genomic selection in wheat breeding using genotyping-by-sequencing. Plant Genome 5(3):103–113

Prasad PV, Djanaguiraman M (2014) Response of floret fertility and individual grain weight of wheat to high temperature stress: sensitive stages and thresholds for temperature and duration. Funct Plant Biol 41(12):1261–1269

Prasad PVV, Staggenborg SA, Ristic Z (2008) Impacts of drought and/or heat stress on physiological, developmental, growth, and yield processes of crop plants. Response of crops to limited water: Understanding and modeling water stress effects on plant growth processes 1:301–355

Qaseem MF, Qureshi R, Shaheen H, Shafqat N (2019) Genome-wide association analyses for yield and yield-related traits in bread wheat (Triticum aestivum L.) under pre-anthesis combined heat and drought stress in field conditions. PLoS One 14(3):0213407. https://doi.org/10.1371/journal.pone.0213407

Ramadas S, Kiran Kumar TM, Singh GP (2019) Wheat production in India: trends and prospects recent advances in grain crops research. Recent advances in grain crops research publisher. Intech Open: https://doi.org/10.5772/intechopen.86341

Rezaei EE, Siebert S, Ewert F (2015) Intensity of heat stress in winter wheat—phenology compensates for the adverse effect of global warming. Environ Res Lett 10(2):024012

Rutkoski J, Benson J, Jia Y, Brown-Guedira G, Jannink JL, Sorrells M (2012) Evaluation of genomic prediction methods for fusarium head blight resistance in wheat. Plant Genome 5(2)

Rutkoski J, Poland J, Mondal S, Autrique E, Pérez LG, Crossa J, Reynolds M, Singh R (2016) Canopy temperature and vegetation indices from high-throughput phenotyping improve accuracy of pedigree and genomic selection for grain yield in wheat. G3 6(9):2799–2808

Saint Pierre C, Burgueño J, Crossa J, Dávila GF, López PF, Moya ES, Moreno JI, Muela VH, Villa VZ, Vikram P, Mathews K (2016) Genomic prediction models for grain yield of spring bread wheat in diverse agro-ecological zones. Sci Rep 6(1):1–11

Sallam A, Alqudah AM, Dawood MF, Baenziger PS, Borner A (2019) Drought stress tolerance in wheat and barley: advances in physiology, breeding and genetics research. Int J Mol Sci 20(13): 3137

Sandra MP, Purushothaman A, Padmakumar KB (2021) Prevalence of epibiosis in plankton community of the Indian EEZ: a review. Symbiosis 85:259–271. https://doi.org/10.1007/s13199-021-00816-x

Sehgal A, Sita K, Kumar J, Kumar S, Singh S, Siddique KH, Nayyar H (2017) Effects of drought, heat and their interaction on the growth, yield and photosynthetic function of lentil (Lens culinaris Medikus) genotypes varying in heat and drought sensitivity. Front Plant Sci 8:1776. https://doi.org/10.3389/fpls.2017.01776

Sehgal D, Rosyara U, Mondal S, Singh R, Poland J, Dreisigacker S (2020) Incorporating genome-wide association mapping results into genomic prediction models for grain yield and yield stability in CIMMYT spring bread wheat. Front Plant Sci 11:197. https://doi.org/10.3389/fpls.2020.00197

Seneviratne S, Nicholls N, Easterling D, Goodess C, Kanae S, Kossin J, Luo Y, Marengo J, McInnes K, Rahimi M, Reichstein M (2012) Changes in climate extremes and their impacts on the natural physical environment. EGU General Assembly 2012, held 22-27 April, 2012 in Vienna, Austria., p. 12566

Shewry PR, Hey SJ (2015) The contribution of wheat to human diet and health. Food Energy Secur 4(3):178–202

Shokat S, Sehgal D, Vikram P, Liu F, Singh S (2020) Molecular markers associated with agro-physiological traits under terminal drought conditions in bread wheat. Int J Mol Sci 21:3156. https://doi.org/10.3390/ijms21093156

Sukumaran S, Reynolds MP, Sansaloni C (2018) Genome-wide association analyses identify QTL hotspots for yield and component traits in durum wheat grown under yield potential, drought, and heat stress environments. Front Plant Sci 9:81. https://doi.org/10.3389/fpls.2018.00081

Trenberth KE, Dai A, Van Der Schrier G, Jones PD, Barichivich J, Briffa KR, Sheffield J (2014) Global warming and changes in drought. Nat Clim Chang 4(1):17–22

Vandenbroucke K, Metzlaff M (2013) Abiotic stress tolerant crops: genes, pathways and bottlenecks. Encycloped Sustainabil Sci Technol 10:1–3

Vary Z, Mullins E, McElwain JC, Doohan FM (2015) The severity of wheat diseases increases when plants and pathogens are acclimatized to elevated carbon dioxide. Glob Chang Biol 21: 2661–2669

Wang S, Wong D, Forrest K, Allen A, Chao S, Huang BE, Maccaferri M, Salvi S, Milner SG, Cattivelli L, Mastrangelo AM, Whan A, Stephen S, Barker G, Wieseke R, Plieske J, International Wheat Genome Sequencing Consortium, Lillemo M, Mather D, Appels R, Dolferus R, Brown-Guedira G, Korol A, Akhunova AR, Feuillet C, Salse J, Morgante M, Pozniak C, Luo MC, Dvorak J, Morell M, Dubcovsky J, Ganal M, Tuberosa R, Lawley C, Mikoulitch I, Cavanagh C, Edwards KJ, Hayden M, Akhunov E (2014) Characterization of polyploid wheat genomic diversity using a high-density 90 000 single nucleotide polymorphism array. Plant Biotechnol J 12:787–796. https://doi.org/10.1111/pbi.12183

Wardlaw IF (2002) Interaction between drought and chronic high temperature during kernel filling in wheat in a controlled environment. Ann Bot 90(4):469–476

Xu Y, Liu X, Fu J, Wang H, Wang J, Huang C, Prasanna BM, Olsen MS, Wang G, Zhang A (2020) Enhancing genetic gain through genomic selection: from livestock to plants. Plant Communi 1(1):100005

Yang W, Duan L, Chen G, Xiong L, Liu Q (2013) Plant phenomics and high throughput phenotyping: accelerating rice functional genomics using multidisciplinary technologies. Curr Opin Plant Biol 16:1–8

Yu J, Buckler ES (2006) Genetic association mapping and genome organization of maize. Curr Opin Plant Biol 17(2):155–160

Yu J, Pressoir G, Briggs WH, Vroh Bi I, Yamasaki M, Doebley JF, McMullen MD, Gaut BS, Nielsen DM, Holland JB, Kresovich S, Buckler ES (2006) A unified mixed-model method for association mapping that accounts for multiple levels of relatedness. Nat Genet 38(2):203–208. https://doi.org/10.1038/ng1702

Yu S, Wu J, Wang M, Shi W, Xia G, Jia J, Kang Z, Han D (2020) Haplotype variations in QTL for salt tolerance in Chinese wheat accessions identified by marker-based and pedigree-based kinship analyses. Crop J 8(6):1011–1024

Genomic Selection for Enhanced Stress Tolerance in Maize

4

Hirenallur Chandappa Lohithaswa,
Sowmya Muntagodu Shreekanth, Santhosh Kumari Banakara,
K. V. Sripathy, and Mallana Gowdra Mallikarjuna

Abstract

Maize is the fastest growing cereal in the world and serves as the most significant component of the global coarse grain trade. Interestingly, in addition to being a prime nutritional source, maize also has a variety of industrial applications. However, the crop is highly sensitive to various biotic and abiotic stresses, negatively affecting maize production worldwide. Thus, enhancing maize productivity is the central thrust area in maize breeding in the era of changing climate. The development and deployment of hybrids resistant or tolerant to biotic and abiotic stresses through genetic options is the most economical, sustainable and eco-friendly way to mitigate stress-mediated yield losses. Several breeding strategies are being employed to bring about the desired improvement in stress tolerance levels. With the advent of DNA markers, marker-assisted selection supplemented conventional breeding. However, marker-assisted introgression breeding has failed to significantly contribute to the improvement of quantitative traits in maize. Recently, genomic selection emerged as a potential breeding approach to deal with complex stress tolerance traits. Genomic selection (GS) or genomic prediction, which merges all the genome-wide marker information into a model to estimate the genetic worth of candidates for selection, appears to be very practical in biotic and abiotic stress tolerance maize breeding. Here, we

H. C. Lohithaswa (✉) · S. M. Shreekanth · S. K. Banakara
Department of Genetics and Plant Breeding, University of Agricultural Sciences, GKVK, Bangalore, India

K. V. Sripathy
Regional Centre, ICAR-Indian Institute of Seed Science, Bangalore, India

M. G. Mallikarjuna (✉)
Division of Genetics, ICAR-Indian Agricultural Research Institute, New Delhi, India
e-mail: MG.Mallikarjuna@icar.gov.in

M. G. Mallikarjuna et al. (eds.), *Next-Generation Plant Breeding Approaches for Stress Resilience in Cereal Crops*, https://doi.org/10.1007/978-981-19-1445-4_4

summarize the genomic selection efforts in maize breeding to deal with various abiotic and biotic stresses.

Keywords

Abiotic stresses · Biotic stresses · Climate change · Genomic selection · Genomic prediction · Maize

4.1 Introduction

Maize is entitled as 'Drosophila of Plant Breeding' owing to the vast amount of genetic diversity and amenable to undertake basic and applied studies. It is a member of Poaceae family with a chromosome number of $2n = 20$ and genome size of 2.5–2.7 Gb (Haberer et al. 2005). The existence of both dicliny and dichogamy made the crop highly cross-pollinated. Additionally, maize is one of the three major staple food crops. Maize has offered countless benefits to mankind since the prehistoric era. Maize as a crop showed its ability to support and uplift the farmer's standard of living, serve as a soil fertility indicator, generate income and feed the growing population. In addition to sustaining food and nutritional security, maize also serves as raw material for versatile industries such as starch and glucose production, biofuel processing, ethanol production and other sub-by-products.

Photosynthetically efficient (C_4), day-neutral and highly adaptive nature of the maize makes it suitable for most agro-climatic regions. Presently, maize is being grown in 169 countries across the globe (Anonymous 2019). Globally, maize is cultivated in 197.20 million hectares with a production of 1.15 billion tonnes and has a productivity of 5.82 tonnes/ha (Anonymous 2019). The major maize-growing countries in the world are the USA, China, Brazil and India. The USA is the major global corn-growing country with 30.21% of global production and accounting for 32.95 million hectares of the area under cultivation (Anonymous 2019). The low productivity in the developing world can be attributed to various biotic and abiotic stresses. Maize is highly sensitive to many pests, diseases and abiotic stresses like drought, salinity, nutrient deficiency and temperature stresses (Fig. 4.1). In addition, climate change induced abiotic constraints to have a wide range of yield-reducing effects on all field crops, including maize. Thus, they should be given high priority in maize improvement programmes (Gazal et al. 2018). Further, intensive cultivation of potential hybrids and varieties resistant to major diseases and pests leads to resurgence and increased crop vulnerability to minor pests and diseases.

Presently, more than 60 diseases are reported in maize. The major diseases which are severely affecting the production and productivity of maize are Northern and Southern corn leaf blights (NCLB, SCLB), sorghum downy mildew (SDM), brown spot (BS), polysora rust (PR), brown stripe downy mildew (BSDM), pre- and post-flowering stalk rots (PSR) and ear rots (ER) (Hooda et al. 2018). NCLB is the major foliar fungal disease affecting global maize production (Technow et al. 2013) and is reported to cause a yield loss of >50% (Raymundo and Hooker 1981; Perkins and

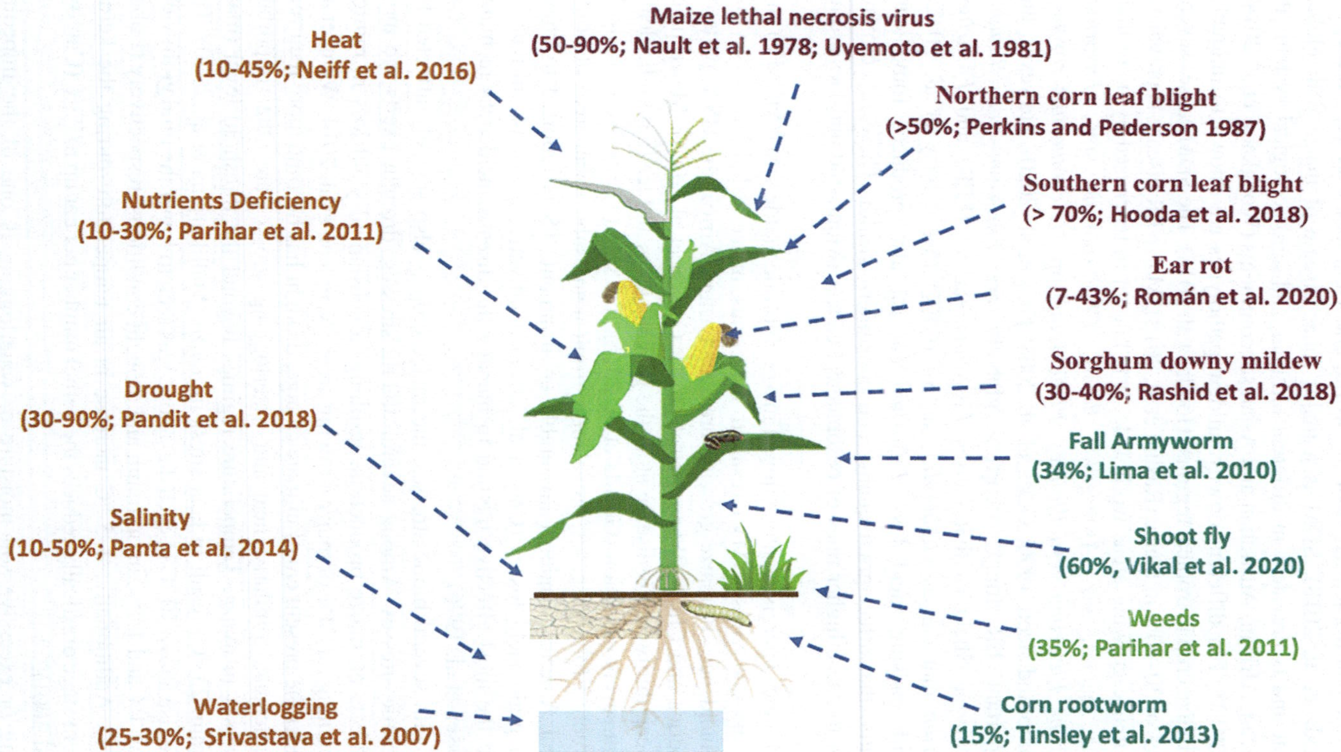

Fig. 4.1 The potential per cent grain yield losses due to various abiotic and biotic stress factors in maize. The abiotic stress factors, diseases, insect pests and weeds are represented with orange, magenta, olive green and light green colour texts

Pedersen 1987). Maydis leaf blight or SCLB is the most severe disease in warm and wet temperate and tropical areas of the world, and yield losses are reported up to 70% (Hooda et al. 2018). SDM is a major foliar disease of maize with global distribution and is prevalent in different altitudes and agro-ecological systems in the American, African, Australian and Asian subcontinents (Wongkaew et al. 2014; Lukman 2012). Ten different downy mildew pathotypes are known to uniquely infect the maize crop in tropical regions (Hooda et al. 2018). The SDM causes severe yield losses of 30–40% in maize (Rashid et al. 2018). Maize ER disease is prevalent in all the maize-growing areas worldwide. It is known to be caused by more than 20 fungal species, viz. *Aspergillus flavus*, *Cladosporium* spp., *Fusarium graminearum*, *Fusarium verticillioides*, *Penicillium* spp., *Trichothecium roseum*, etc. (Zummo and Scott 1990; Görtz et al. 2008; Guo et al. 2020). However, the most important ER fungi occurring globally are *Fusarium verticillioides* (*F. moniliforme* Sheldon) that causes *Fusarium* ear rot (FER) and *Fusarium graminearum* that causes *Gibberella* ear rot (GER) (Mesterhazy et al. 2012). Along with causing yield loss, *Fusarium* ear rot also produces fumonisins (mycotoxins) that affect grain quality and consumers' health. These mycotoxins are known to cause higher rates of oesophageal cancer (Munkvold and Desjardins 1997).

Recently, the synergistic interaction of maize chlorotic mottle virus (MCMV) and sugarcane mosaic virus (SMV) resulted in serious maize lethal necrosis virus (MLND) incidences in maize-growing areas of Eastern Africa, the USA, parts of Latin America and China (Wangai et al. 2012; Gowda et al. 2015). Maize plant shows susceptibility to this disease at all the growth stages and leading to the death of plants in severe condition (Gowda et al. 2015). Tar spot (TS) is one of maize's most destructive foliar diseases, prevalent in tropical and subtropical areas of South and Central America. Under a favourable environment, TS is reported to cause significant grain yield losses (Cao et al. 2017). Cultivating resistant cultivars is considered the most effective method to manage the disease incidences in maize (Dingerdissen et al. 1996).

Along with several biotic stresses, the maize crop is also severely affected by various abiotic stresses. Among several abiotic stresses, drought exerts the most harmful effects on maize production, resulting in substantial yield loss in rainfed areas accounting for 74% of maize-growing areas (Nepolean et al. 2014; Wang et al. 2019). Under the present era of climate change, drought has become more recurrent and unpredictable. Similarly, heat and waterlogging stresses are other important abiotic stresses in maize. Temperature regimes beyond the threshold level (max. 35 °C to min. 23 °C) result in heat stress in maize (Mallikarjuna et al. 2020). The one-degree rise above 30 °C in each day above was seen to lower the final grain yield of maize by 1% and 1.7%, in optimum and drought conditions, respectively (Lobell et al. 2011). Additionally, 4–5 °C increase in air temperature during the kernel development stage drastically reduces the kernel number per ear up to 73% (Cárcova and Otegui 2001).

Flooding or excessive soil moisture or waterlogging is one of the impeding abiotic stresses for maize production in South and South-East Asia owing to erratic rainfall patterns (Zaidi et al. 2004). Maize is a highly resource-demanding crop. Further, in maize production systems of developing world like sub-Saharan Africa, low nitrogen stress is one of the widespread problems, especially among marginal farming community (Ertiro et al. 2020). Phosphorus is a vital nutrient that mostly gets fixed in the soil largely by aluminium and calcium ions and becomes unavailable to the plants. Phosphorus starvation will severely affect the growth and development of maize, thereby decreasing the biomass and yield (Yu et al. 2018). The management of these stresses necessitates the development of stress-resilient maize cultivars in addition to system-specific management practices.

4.2 Genetic Improvement Strategies for Stress Tolerance in Maize

Genetic improvement of crops for enhanced stress tolerance is the most effective approach in managing unwarranted stress occurrence. The breeders employ various conventional and improved molecular tools in maize to improve the genetic tolerance for several abiotic and biotic stresses. The genetic improvement of maize through conventional breeding methods was mainly based on (i) selection and backcrossing, (ii) extensive screening for stress tolerance to derive improved stress-resilient germplasm via recurrent selection, (iii) utilization of alien genetic variation (pre-breeding activities) and (iv) breeding for early maturity and varieties with adaptation to specific ecologies (Gazal et al. 2018). The conventional methods of plant improvement are successful in increasing maize production by exploiting hybrid vigour, using male sterility systems through backcross methods, population improvement schemes for deriving the good inbred lines, synthetics and composites. As in conventional breeding, selection depends mainly on the phenotype that is highly sensitive to the environment; thus, selection efficiency is low. The advancement in molecular biology, i.e. marker technologies, supplemented the conventional breeding approaches by increasing the selection efficiency and contributed substantially to crop improvement. Molecular marker technology helps in reshaping the breeding activities and facilitates rapid gains from selection (Jannink et al. 2010; Liu et al. 2020).

Currently, the role of marker-assisted selection (MAS) in improving the polygenic traits is limited. However, it has been effectively utilized to improve traits with large effect alleles linked to markers (Zhong et al. 2007). The major limitation in improving quantitative traits is having the same QTL or genomic region expressing target traits across environments owing to QTL × environment interactions. Similarly, the genetic background of inbred lines limits the QTL expression across the germplasm set in a crop (Bernardo 2016). MAS and marker-assisted recurrent selection (MARS) depend mainly on significantly linked markers, tagged gene (s) or mapped quantitative trait loci (QTL). Furthermore, the MAS or MARS has main disadvantage in capturing the significant marker-QTL associations with minor effects (Heffner et al. 2009; Xu et al. 2012). Thus, marker-assisted selection has two components: first is to identify the QTL, and second is to estimate the effects

(Jannink et al. 2010). QTL identification using linkage mapping is carried out using biparental populations, but the power of detecting marker-trait association is poor because of the presence of chromosomes with low recombination rates and tedious and time-consuming nature (Guo et al. 2020).

To resolve the issue associated with linkage mapping for QTL detection, the concept of association mapping started during the early twentieth century to facilitate the identification of marker-trait association in non-biparental populations and for fine mapping of genomic regions with higher recombination rates. Nevertheless, even though it is advantageous, it has a drawback in identifying rare QTLs with minor genetic effects governing the economically important characteristics and is greatly influenced by the environment (Jannink et al. 2010). For instance, resistance for NCLB disease in maize showed many QTL dispersed throughout the genome (Van Inghelandt et al. 2012; Poland et al. 2011; Wisser et al. 2006; Ranganatha et al. 2021). Furthermore, high cross-pollination in maize resulted in the rapid decay of linkage disequilibrium (LD). Hence, maize demands a large number of polymorphic SNPs distributed throughout the genome (Gowda et al. 2015). To overcome the disadvantages associated with the above breeding approaches, Meuwissen et al. (2001) proposed the genomic selection (GS) to capture the total additive genetic variance using genome-wide molecular markers and to enhance the genetic gain for quantitative traits (Poland and Rutkoski 2016).

4.3 Genomic Selection in Maize: Need and Importance

With the advent of third-generation sequencing, longer sequence reads can be generated in a short period and at a significantly lower cost per run, which are subsequently helpful in the creation of fixed SNP-genotyping arrays that encompasses set of genome-wide dispersed genic and non-genic SNPs (Varshney et al. 2014). The cost of genotyping has reduced significantly relative to phenotyping costs; thus, GS becomes an attractive selection decision tool in breeding activities (Atanda et al. 2021). GS arose with an intention to utilize the available high-density parallel NGS technologies. Unlike other methods, GS capitalizes on all marker loci with and without significant trait association, thereby giving unbiased estimates of marker-trait association, and it is assumed that casual polymorphism would be coherent across the families, so the marker effects based on population-wide estimates would be meaningful (Jannink et al. 2010; Meuwissen et al. 2001; Guo et al. 2020). GS enhances the genetic gain through improved prediction accuracy of genomic estimated breeding values, shortening generation intervals and effective utilization of existing germplasm via genome-guided selection (Sonesson et al. 2010; Schierenbeck et al. 2011; Pryce et al. 2012). In different sets of maize, *Arabidopsis* and barley germplasm, the GS reduces the selection time by almost half per cycle compared to the phenotypic selection for most of the traits (Lorenzana and Bernardo 2009). The effectiveness of GS in predicting complex traits has been proven in various crops, including maize (Zhao et al. 2012; Rutkoski et al. 2014; Zhang et al. 2015). Further, GS with linkage and association mapping has improved breeding efficiency (Cao et al. 2017). However, GS is mainly used to predict the additive genetic value of the line, and non-additive genetics are often disregarded (Robertson et al. 2019).

GS can be employed to predict breeding values of the individuals with or without phenotypic information of their own. GS with phenotypic information improve the accuracy of selection, and without phenotypic information, it shortens the breeding cycle length by eliminating the need for phenotyping of the candidates before selection. Further, it is also possible to predict breeding values for a very large number of individuals, which cannot be phenotypically assessed, resulting in increased selection intensity. Additionally, GS can also be applied at several stages in the breeding process to enhance the genetic gain from selection (R total) (Poland and Rutkoski 2016).

In the genomic selection, marker effects are estimated based on the training set of genotypes, which are both phenotyped and fingerprinted with dense marker data. Based on these estimated marker effects, the individuals related to the training population that is only genotyped but not phenotyped are selected (Zhao et al. 2012). The estimated GEBVs are not the function of underlying genes; instead, they are the ideal selection criterion (Jannink et al. 2010). The major problem in the development of the prediction model is over-fitting, and such models can exaggerate minor variations in the data, and the prediction ability decreases (Jannink et al. 2010). Hence, the application of GS in breeding pipelines is influenced by several factors when the trait of interest is affected by a large number of loci. The training population size, genetic diversity and genetic relationship with the breeding or test population, i.e. the individuals of the training population or a close relative or distant relatives of the individuals of the breeding/test population, are the most important among other factors (Pszczola 2012). The heritability of the trait under selection, i.e. complex traits with low heritability and small marker effects, is suitable for genomic prediction/selection, whereas the oligogenic traits can be predicted accurately with few markers with relatively large effects (Daetwyler et al. 2010). The prediction accuracy is low for a complex trait(s) with a large number of markers when these markers are not in linkage disequilibrium with the QTL/genomic regions. However, the accuracy increases when the heritability of the trait and training population size increases (Isidro et al. 2015).

4.4 Genetic Resources for Genomic Selection in Maize for Stress Tolerance

Germplasm in crop plants serves as a valuable resource for crop improvement activities as they exhibit a high level of genetic diversity in many important agronomic traits. The usefulness of genetic resources or germplasm collections in achieving the improvement in grain yield and agronomic performance was unequivocally established by numerous reports. Presently, more than seven million crop accessions are presently preserved in the global gene banks worldwide, which represent the paramount but largely untapped opportunities for breaking productivity bottlenecks to accelerate genetic gain for yield and other traits (Wang et al. 2017). The major hindrances in utilizing the crops' genetic resources comprise the availability of larger germplasm collections and the lack of an integrated method to exploit the available germplasm resources. Recent advancements in high-throughput genotyping and phenotyping tools, along with evolving biotechnological tools, create opportunities to employ exotic germplasm in plant improvement programmes.

Phenotyping is the current bottleneck in plant breeding compared to genotyping, especially with the decline in genotyping cost by more than 100-fold in the last two decades. Therefore, the phenotyping cost needs to be optimized within the breeding programme. While designing the implementation of the GS scheme into the breeding cycle, the breeders need to select the optimal method for the selection of the training population so that the prediction accuracy increases and reduce the phenotyping cost with improvement in precision (Akdemir and Isidro-Sánchez 2019).

Various panels, training populations and biparental populations have been used to predict GEBVs for various stress-resilient traits in maize. Cao et al. (2017) have used the Drought Tolerant Maize for Africa (DTMA) association mapping panel to implement GS and GWAS analysis for tar spot complex in maize. The DTMA panel carries 282 tropical and subtropical maize inbreds developed at CIMMYT and comprised of lines with resistance or tolerance to an array of biotic and abiotic stresses, which affects the maize production, improved nitrogen use efficiency and grain nutritional quality. At the University of Hohenheim, 2 elite mapping panels comprising 130 dent and 114 flint lines of European origin were used to investigate the GWAS and genomic predictions for *Gibberella* ear rot (Han et al. 2018) and NCLB resistance in maize (Technow et al. 2013). Similarly, for *Fusarium* ear rot resistance, genomic predictions were performed in a panel of 874 lines encompassing the previous DTMA panel, CML lines and SYN_DH population (Liu et al. 2021) and tropical maize core collection (Ertiro et al. 2020). Three DH populations (CML550 × CML504, $N = 219$; CML550 × CML511, $N = 110$; CML550 × CML494, $N = 229$) and IMAS (Improved Maize for African Soil) panel were employed in genomic prediction for maize chlorotic mottle virus and maize lethal necrosis resistance (Sitonik et al. 2019) and nitrogen use efficiency (Ertiro et al. 2020). For insect pest resistance, Badji et al. (2021) employed a diverse tropical maize panel composed of 341 DH and inbred lines.

For drought and heat stresses, genomic predictions were carried in the diverse maize gene pools. A recent study employed 3068 DH lines derived from 54 biparental and test crosses generated by crossing an agronomically elite line with lines of drought-tolerant and farmer-preferred traits (Beyene et al. 2021). Further, a multiparent yellow synthetic maize population and rapid cycle genomic selection were employed to simultaneously improve drought and waterlogging stress tolerance in maize (Das et al. 2020). Many of the previous GWAS panel results can be employed to constitute the testing population to predict genomic breeding values for various abiotic stress tolerances like drought and heat (Shikha et al. 2017; Seetharam et al. 2021). Further, maize wild relatives and landraces harbour various abiotic and biotic stress tolerance genes (Table 4.1). Applying appropriate breeding tools like DH technology coupled with genomic selections could bring these valuable genes into

Table 4.1 Genetic resources for various stress tolerance in maize to develop training sets in genomic selection (Modified from Mammadov et al. 2018)

Stress	Wild maize relative	Reason of tolerance/ resistance	Reference
Biotic stress			
Insect resistance			
Fall armyworm (FAW) resistance	*Z. mays* subsp. *parviglumis*	Leaf trichomes and leaf toughness	Moya-Raygoza (2016)
	Z. diploperennis	Leaf chemical composition	Farias-Rivera et al. (2003)
	Z. mays spp. *parviglumis*	Higher expression of *wound inducible protein-1 (wip1), maize protease inhibitor (mpi)* and *pathogenesis-related protein (PR1)* genes	Szczepaniec et al. (2013)
	Teosinte; insect tolerant synthetic (ITS) G1	Release of herbivore-induced volatile compounds, viz. indole and various mono- and sesquiterpenes, resulting from FAW attracts FAW larval parasitoids, viz. *Cotesia marginiventris, Campoletis sonorensis* and *Meteorus laphygmae*	de Lange et al. (2014), de Lange et al. (2016), de Lange et al. (2018), Mammadov et al. (2018)
Maize spotted stalk borer resistance	*Z. mays* spp. *mexicana*	Possess high benzoxazinoid (BX) content	Frey et al. (2009), Glauser et al. (2011)
	Z. mays spp. *mexicana*	Oviposition of *Chilo partellus* produce the (E)-4,8-dimethyl-1,3,7-nonatriene and which attracts the egg (*Trichogramma bournieri*) and larval (*Cotesia sesamiae*) parasitoids of *Chilo partellus*	Mutyambai et al. (2015)
	Z. mays spp. *parviglumis*		
	Z. mays spp. *parviglumis*		
Western corn rootworm resistance	Teosinte (no information on specific species)	(E)-β-caryophyllene released from root herbivory by cutworm invites entomopathogenic nematode *Heterorhabditis megidis*	Rasmann et al. (2005)
Disease resistance			
SCLB resistance	*Z. diploperennis*	–	Wei et al. (2001)
NCLB resistance	*Z. diploperennis*	–	Wei et al. (2001)
	Tripsacum floridanum	*Ht3* gene	Hooker (1981)

(continued)

Table 4.1 (continued)

Stress	Wild maize relative	Reason of tolerance/ resistance	Reference
Grey leaf spot resistance	*Z. mays* subsp. *parviglumis*	–	Zhang et al. (2017)
Rust resistance	Eastern gamagrass	*Rp1* gene	Smith et al. (2004)
Corn smut disease resistance	Teosinte	–	Chavan and Smith (2014)
Maize chlorotic dwarf virus resistance	*Z. diploperennis*	–	Nault and Findley (1981), Nault et al. (1982)
Maize chlorotic mottle virus resistance	*Z. diploperennis*	–	Nault and Findley (1981), Nault et al. (1982)
Maize streak virus resistance	*Z. diploperennis*	–	Nault and Findley (1981), Nault et al. (1982)
Maize bushy stunt (mycoplasma) resistance	*Z. diploperennis*	–	Nault and Findley 91,981)
Maize stripe virus resistance	*Z. diploperennis*	–	Nault and Findley (1981)
Maize rayado fino virus resistance	*Z. diploperennis*	–	Nault and Findley (1981)
Weed resistance			
Striga hermonthica resistance	*Z. diploperennis*	Resisting to attachment germinating striga to roots and restricting the subsequent penetration into the vascular system via signalling that prevents the haustoria growth	Lane et al. (1997), Rich and Ejeta (2008), Gurney et al. (2003)
	Eastern gamagrass		Amusan et al. (2008)
	KSTP 94 (open-pollinated maize variety)	Post-attachment resistance to *S. hermonthica*	Mutinda et al. (2018)
Abiotic stress tolerance			
Drought tolerance	Eastern gamagrass	Deeply penetrating root system	Clark et al. (1998)
Acid soil and aluminium tolerance	Eastern gamagrass	–	Foy (1997)

(continued)

Table 4.1 (continued)

Stress	Wild maize relative	Reason of tolerance/ resistance	Reference
Waterlogging tolerance	*Z. nicaraguensis*	To prevent oxygen loss in stagnant deoxygenated conditions, adventitious roots develop radial oxygen barriers	Abiko et al. (2012)
	Z. luxurians	Developing root aerenchyma under anoxic conditions	Ray et al. (1999)
	Eastern gamagrass		

breeders' disposal for their rapid utility in the current breeding pipelines to deliver stress-resilient maize hybrids.

4.5 Statistical Models in Genomic Selection

In GS, various statistical methods have been employed to estimate marker effects which are classified into parametric, semi-parametric and nonparametric models. The accuracy of marker effects estimation using various statistical methods is a function of the target trait's genetic architecture (Daetwyler et al. 2010), the population structure (Habier et al. 2007; Zhong et al. 2009) and the marker's density (Meuwissen and Goddard 2010). The genomic selection models can be categorized into parametric, semi-parametric and nonparametric methods.

4.5.1 Classification of Statistical Models in Genomic Selection

4.5.1.1 Parametric Models in Genomic Selection

Linear least square regression model: Genomic selection is focused on predicting individuals' breeding value by modelling the association between individuals' genotype and phenotype. Linear least square regression (LLSR) is the simplest parametric model. The major problem associated with the LLSR model is that it is difficult to perform the estimation with much higher number of markers than the number of individuals with phenotypic information. Although an alternative approach, a subset of markers can be used; still the poor prediction accuracies are obtained if the ratio of the markers' number and the individuals' numbers is very large or has multicollinearity (Howard et al. 2014). Therefore, Meuwissen et al. (2001) suggested the modifications to the LLSR model to eliminate the problem of more independent variables (predictor) than dependent variables (regressands). However, it fails to fully take advantage of all the markers' information since the final model is based on markers with a significant effect only.

Ridge regression (RR): Multicollinearity between the marker data negatively affects the performance of variable selection methods. Ridge regression can be used when a large amount of marker information is available, so it can overcome the '$p > n$' problem of the least square regression model (Howard et al. 2014). RR in GS was implemented with an assumption of random marker effects ($m_{j's}\, j = 1 \ldots p$), and markers were drawn from a group with normal distribution and $\mathrm{Var}(m_j) = \sigma_m^2$, where $\sigma_m^2 = \frac{\sigma_a^2}{n_k\, \sigma_a^2}$ represents the additive component of genetic variance expressed among individuals and n_k is the number of marker loci (Meuwissen et al. 2001; Habier et al. 2007). The key property of RR is that it won't select a subset of predictors in contrast to other methods such as LASSO and elastic net (de Vlaming and Groenen 2015).

Best linear unbiased prediction (BLUP): The concept of BLUP theory and the mixed model formulation and their utility in animal and plant breeding were discussed by Henderson (1949) and Henderson et al. (1959). The BLUPs are useful to deal with unbalanced datasets, for instance, multilocational datasets, a discrepancy in the number of individuals, etc. (Bernardo 2010). Genomic BLUP (GBLUP) is based on a genomic relationship matrix that explains genetic relationships between individuals, which are calculated from genotypes at single-nucleotide polymorphisms (SNPs), whereas traditional pedigree BLUP (Henderson 1975) uses pedigree relationship matrix with a genomic relationship matrix (Habier et al. 2013).

Least absolute shrinkage and selection operator (LASSO): LASSO is a compelled form of ordinary least squares, which is developed to overcome the limitations of linear least squares by Tibshirani (1996) and in GS first implemented with cross-validation by Usai et al. (2009). LASSO is indifferent to closely correlated markers, i.e. LASSO picks one among the highly correlated markers and ignores the remaining (Wang et al. 2018). Being a penalized regression-based approach, LASSO gives better estimates when the number of markers is greater than the number of individuals ($p > n$) (Budhlakoti et al. 2020).

Bayesian alphabet models: The Bayesian alphabet models in the genomic selection were started with BayesA and BayesB models (Meuwissen et al. 2001). Later several models, viz. BayesCπ and BayesDπ (Habier et al. 2011), fast EM-BayesA (Sun et al. 2012), fast BayesB (Meuwissen et al. 2009), BRR (Bayesian ridge regression on markers) (VanRaden 2008), Bayesian LASSO (Park and Casella 2008), etc., were derived.

In BayesA and BayesB models, the data and the variances of the marker positions need to be modelled. The main difference between BayesA and BayesB is the prior for the variance components, i.e. in contrast to BayesA, BayesB assumes that not all the markers contribute to the genetic variation. The BayesCπ gives a more sensible formulation of the mixture. However, it poses the same spirit and limitations as BayesB (Gianola 2013). Park and Casella (2008) used the idea from Tibshirani (1996) to connect with Bayesian analysis to come up with Bayesian LASSO. Bayesian LASSO generates the models with non-null regression coefficients even if $p > n$ (Gianola and Fernando 2020). In other words, LASSO results in the sparse model, whereas Bayesian LASSO yields an effectively sparse specification like BayesB (Meuwissen et al. 2001). Yi and Xu (2008) first used the Bayesian LASSO model for QTL mapping followed by subsequent applications in genomic prediction by various researchers (de Los Campos et al. 2009; Legarra et al. 2011; Lehermeier et al. 2013).

4.5.1.2 Semi-Parametric Models

Reproducing kernel Hilbert space (RKHS): Gianola et al. (2006) proposed this semi-parametric model proposed by coalescing the best qualities of a nonparametric model with a mixed model framework (Howard et al. 2014). The RKHS model combines a genomic relationship matrix (G) and pedigree-based numerator relationship matrix (A) in a kernel matrix while making weaker assumptions on the compatibility of G and A (Rodríguez-Ramilo et al. 2014).

4.5.1.3 Nonparametric Models

Nadaraya-Watson estimator: Using Silverman's (1986) nonparametric kernel estimator, which is used in the estimation of *p* (x), Nadaraya (1964) and Watson (1964) estimated the conditional expectation function. The estimator is just a weighted sum of observations y_i, $i = 1, 2, 3\ldots.n$ and is called Nadaraya-Watson's equation. Nadaraya-Watson estimator is one of the most widely used nonparametric models for genomic selection. In the presence of additive effects, the prediction of Nadaraya-Watson estimator model is poor compared to other nonparametric models. However, in the presence of epistatic interactions, the performance of Nadaraya-Watson estimator was significantly better than the parametric methods (Howard et al. 2014).

Support vector machine (SVM) regression: Vapnik (1995) and Cortes and Vapnik (1995) proposed and discussed SVM approach. The SVM is originally employed in classification and regression analysis as supervised learning method. Here, the training dataset is used to create a maximum marginal classifier that results in the biggest possible separation between the comparing classes of observations. In plant breeding, the SVM regression explains the association between the marker's genotypes and the phenotypes which can be modelled with a linear or nonlinear mapping function that takes samples from a predictor space to an abstract, multi-dimensional feature space (Hastie et al. 2009; Long et al. 2011).

Neural networks (NNs): NNs are types of nonparametric GS models. NNs are originally developed to understand how neurons of the human brain interact, work and conduct computations (Bain 1873; James 1890; Hastie et al. 2009). The feed-forward model is a basic NN model, which is a two-stage network with three types of layers, i.e. an input layer, a hidden layer and an output layer. Nonparametric nature of NNs able to model both linear and complex nonlinear functions permits the quantification of additivity and epistasis interactions (Howard et al. 2014).

4.5.2 Genomic Selection Models: Predictive Abilities and Accuracies

Presently, there are various statistical models available to estimate genomic estimated breeding values. The selection of appropriate models is most crucial for effective genomic selection. Some models are fit better for extremely quantitative traits, while some are performing good for traits which fall between qualitative and quantitative nature (Poland and Rutkoski 2016).

An optimal model should give the highest possible prediction accuracy, limit overfitting on the training dataset and be based on maximum marker-QTL LD rather than on kinship (Habier et al. 2007). This makes models easy to implement as these are consistent across the broad range of phenotypes and datasets and computationally efficient (Heslot et al. 2012). Further, the prediction ability of GS models can be increased by correcting the field spatial variation, which includes use of blocking, with resolvable incomplete block designs such as the alpha-lattice being popular in early-generation testing (Patterson and Williams 1976; Ward et al. 2019).

Furthermore, genetic architecture and heritability have the utmost influence on estimates of prediction accuracy and mean squared error (MSE). Parametric methods give somewhat superior estimates than nonparametric methods for traits with additive genetic architecture. However, when the genetic architecture of the target trait is entirely under the interaction component, parametric methods fail to provide accurate estimates (Howard et al. 2014; Momen and Morota 2018). The parametric, semi-parametric and nonparametric models showed increased prediction accuracies with heritability and the number of markers and individuals. However, an inverse association was observed with the increase in the number of QTLs from 50 to 200 (Sahebalam et al. 2019).

4.6 Genomic Selection Strategies for Stress Tolerance in Maize

The genomic selection strategies can be grouped into three categories other than regular or basic GS in cereals. These strategies can be employed for the GS of desired traits depending on the germplasm relatedness, trait phenotyping and resources (Robertson et al. 2019).

4.6.1 Across-Breeding Cycle Genomic Selection

Across-breeding cycle GS necessitates the good association between training and test datasets. The relationship between training and the test data and high association between the datasets in subsequent years are most pre-requisite for across-breeding cycle GS. The association of training and test datasets can be achieved by including common parents in crossing plans for subsequent years, and/or the crossings must be based on the progeny of previous years that were used as parents (Robertson et al. 2019). In many cases, the varieties released by other breeders or germplasm of special interests enter the breeding pipelines as a source for germplasm diversification. At this juncture, to ensure the breeding materials with sufficient genetic relatedness to implement the across-breeding cycle GS is more challenging. Additionally, without any modifications of across-breeding cycle GS, 6 years is required to use the lines from the respective breeding programme as new parents (Robertson et al. 2019; Michel et al. 2016).

4.6.2 Within-Breeding Cycle Genomic Selection

In within-breeding cycle GS method, the lines from the same breeding cycle are used to constitute the training population for GS, for example, to predict GEBVs of the sister lines with missing phenotypic datapoints. GS within-breeding cycle is important when the aim is to reduce the phenotyping or environments or measuring the expensive and complex traits on selected portion of the progenies to predict for the rest. Generally, high GS prediction accuracies in the same generation are often associated with high genetic relatedness between lines, since multiple lines from each family are being tested. Therefore, prediction accuracy of GS selection is higher within the breeding cycle or generation (Robertson et al. 2019).

4.6.3 Genomic Selection Using Untested Parents for Breeding

In GS with untested parents' method, the untested parents refer to those lines which are started being used as parents without being tested in the field. It is a drastic way to use genomic selection wherein the phenotyping testing is skipped, at least for the portion of breeding programmes. Here, novel parental lines are selected solely on GEBVs. The use of untested parents can often significantly shorten the breeding cycle and allow faster genetic gain per year, especially when the breeding cycles are large owing to extensive phenotyping. In dairy cattle breeding, Schaeffer (2006) suggested the use of untested parents for the selection of bulls. Presently, use of untested parents in predicting the GS is more popular and revolutionizing dairy cattle breeding programme. In the case of agricultural crops and cereals where the extensive phenotyping is required, the use of untested parents could similarly revolutionize cereal's breeding approaches (Robertson et al. 2019).

4.7 Genomic Selection for Abiotic Stress Tolerance in Maize

Drought, heat, salinity, waterlogging and mineral nutrient stresses are the major abiotic constraints limiting maize production worldwide (Edmeades et al. 1989). The climate change effects resulted in increased frequency of moderate to severe drought, high air temperature and erratic rainfalls with high intensity. The major focus of maize research in the present scenario is to improve abiotic stress tolerance. However, identifying genetic components that provide abiotic stress tolerance is challenging and resource demanding.

The traits imparting abiotic stress tolerance are governed by several QTL with small individual effects on overall trait expression, which makes it difficult for its identification, modifications and introgression into elite cultivated varieties. Thus, marker-assisted selection and QTL mapping using linkage analysis fail significantly in bringing significant changes. Hence, with the advent of high-throughput genotyping, genomic selection is now being used in breeding for abiotic stress resistance in maize (Pace et al. 2015; Table 4.2).

4.7.1 Drought Tolerance

The genomic selection was attempted for drought tolerance in 240 maize subtropical lines employing 29,619 SNPs and assessed the genomic prediction accuracies with 7 GS models, i.e. BayesA, BayesB, elastic net, LASSO, random forest, reproducing kernel Hilbert space and ridge regression, for different agronomic target traits under drought stress environments. Of these seven genomic selection models, BayesB has been shown to have the highest prediction accuracy for the dataset. From the top 1053 SNPs, 77 SNPs were found to be associated with 10 drought-responsive transcription factors, which are associated with different physiological and molecular functions. Thus, these drought-related SNPs can be further employed for the development of drought-resilient maize cultivars (Shikha et al. 2017).

Rapid cycle genomic selection (RCGS) for drought resulted in a genetic gain of 110 and 135 kg ha^{-1} $year^{-1}$ in multi-parent yellow synthetic populations MSY-1 and MSY-2, respectively. The higher genetic gain for the trait of interest in biparental populations could be due to a change in the population structure of the base population. Further, the genetic diversity of MSY-1 and MSY-2 did not change significantly even after two cycles of GS, indicating that RCGS can be effectively used to achieve high genetic gains without loss of genetic diversity (Das et al. 2020).

Genomic prediction in 210 maize inbred lines under drought and well-watered conditions was conducted using all the SNPs, random SNPs and trait-associated SNPs. The investigation revealed the greater prediction accuracies with trait-associated SNPs across drought and well-watered conditions and all the traits such as grain yield, plant height, ear height, date of anthesis and silking and anthesis-silking interval (Wang et al. 2019). Recently, Beyene et al. (2021) employed genomic selection in 3068 DH lines derived from 54 biparental populations generated by crossing elite inbred line with lines showing tolerance to drought tolerance and other farmer-preferred traits. The study demonstrated that increasing the training set with genotyping and phenotyping data from the previous year along with combining 10–30% lines from the year of testing results in enhanced prediction accuracies. Additionally, Cerrudo et al. (2018) showed the superiority of GS over MAS for grain yield and physiological traits in the maize DH population across the water stress regimes.

4.7.2 Heat Tolerance

Along with drought stress, maize production is also constrained by damage caused by heat stress, which is more predominant in the present circumstances because of

Table 4.2 Genomic selection in maize for various abiotic stress tolerance traits using different training populations and GS models

S. no.	Stress	Trait studied	Population used	Genotyping platform	Model	Prediction accuracy/ genetic gain	Reference
1.	Drought	DTA, DTS, ASI and GY	DTMA panel with 300 lines	–	Bayesian LASSO and RKHS	0.41–0.79	Crossa et al. (2010)
2.	Drought	ASI, EG, EL, KR, KRN, GY and HKW	240 maize lines	Infinium MaizeSNP50 BeadChip	RR, LASSO, EN, RF RKHS, BayesA and BayesB	0.28–0.97	Shikha et al. (2017)
3.	Drought	GY, AD, PHT, ASI, NDVI, SEN, LR, GLAD, DSS	Doubled haploid	GBS	RR-BLUP	−0.07–0.49	Cerrudo et al. (2018)
4.	Drought	EL, EW, WPE, ERN, GNR, HGW, yield	Maize inbred lines	Illumina HiSeq™ 2000	RR-BLUP	a	Wang et al. (2019)
5.	Drought	PH, EH, RL, SL, SN, AD, SD, EPP, GY and MOI	Multi-parent yellow synthetic populations	GBS	RR-BLUP and GBLUP	0.11–0.13[a] t/ha/y	Das et al. (2020)
6.	Drought	GY, ASI, AD, PH	3068 DH lines	rAmpSeq	GBLUP	0.19–0.31	Beyene et al. (2021)
7.	Waterlogging	Crop yield susceptibility index	92 maize lines	GBS	RR-BLUP, Bayesian LASSO, Bayesian RR	0.06–0.44	Paril et al. (2017)
8.	Waterlogging	BR	Multi-parent synthetic populations	GBS	GBLUP	0.04–0.11[a] t/ha/y	Das et al. (2020)
9.	Heat	LGR, SD, SC, SN, LL, NL, DW, WC	Inbred lines	KASP SNP technology and RAD SNP	GBLUP	0.31–0.70	Inghelandt et al. (2019)

(continued)

Table 4.2 (continued)

S. no.	Stress	Trait studied	Population used	Genotyping platform	Model	Prediction accuracy/ genetic gain	Reference
10.	Heat	GY, FT, AD, ASI	Inbred lines	GBS	RR-BLUP	0.35–0.59	Yuan et al. (2019)
11.	Nitrogen stress	GY, AD, ASI, PH, EH, EPO, EPP, SEN	Mapping population	GBS	RR-BLUP, BLUE	0.20–0.71	Ertiro et al. (2020)
12.	Phosphorus stress	DTT, DTS, DTA, ASI, PH, EL, RN, GNPR, GN, HGW, GWPP	Inbred lines	FARM-CPU	RR-BLUP, GBLUP, BayesA, BayesB, BayesC	0.06–0.76	Xu et al. (2018)

Note: *AD* anthesis date, *ASI* anthesis-to-silking interval, *DSS* drought stress susceptibility, *DTA* days to anthesis, *DTS* days to silk, *DTT* days to tassel, *DW* shoot dry weight, *EG* the ear girth (centimetres), *EH* ear height, *EL* ear length (centimetres), *EN* elastic net, *EPO* ear position, *EPP* ear per plant, *ERN* ear row number, *EW* ear width, *GLAD* green leaf area duration, *GN* grain number, *GNPR* grain number per row, *GNR* grain number per row, *GWPP* grain weight per plant, *GY* grain yield (kilogrammes per plot), *HKW* 100-kernel weight (in grammes), *KR* kernel number per row in a cob, *KRN* kernel row number per cob, *LASSO* least absolute shrinkage and selection operator, *LL* leaf length, *LGR* leaf growth rate, *LR* leaf rolling, *NDVI* normalized difference vegetative index, *NL* leaf number, *PHT* plant height, *RF* random forest, *RKHS* reproducing kernel Hilbert space, *RR* ridge regression, *SC* leaf scorching, *SD* leaf greenness, *SEN* senescence, *SN* leaf senescence, *WC* shoot water content, *WPE* weight per ear

[a]Yield gain through genomic selection

[b]GP accuracies are presented in graphical form

global warming. Inghelandt et al. (2019) carried out an experiment to assess the diversity and QTL and predict the genomic values for heat tolerance-associated traits. The genome-wide prediction models' ability was tested by employing a large number of KASP (Kompetitive allele-specific PCR genotyping) and RAD (restriction site-associated DNA sequencing) SNPs. Additionally, the intra-, inter- and mixed pools prediction accuracies were also examined. Results from the study confirmed that the prediction ability of genome-wide prediction models was found high for within-population calibrations; hence, use of such approaches for selecting heat tolerance at the seedling stage is most preferred (Inghelandt et al. 2019). Genomic prediction for combined drought and heat stress in a panel of 300 maize lines of tropical and subtropical origin revealed that the genomic prediction accuracies obtained from marker trait-associated SNPs were comparatively greater (0.28 to 0.75) than those obtained from the genome-wide SNPs (0.13 to 0.64) for most of the targeted traits (Yuan et al. 2019).

4.7.3 Waterlogging Tolerance

Waterlogging tolerance in maize can be assessed easily by means of yield reduction. Three genomic selection models, viz. RR-BLUP, Bayesian RR and Bayesian LASSO, were employed in 92 sampled families from 390 S_1 families tested for waterlogging tolerance. The prediction accuracies from the three models were closer to 0 for crop yield susceptibility index and ranged from 0.16 to 0.44 for yield per se under normal and stressed conditions indicating the necessity of employing larger populations in genomic predictions (Paril et al. 2017). Under waterlogging stress, phenomic and genomic selection showed a genetic gain of 80 and 90 kg ha^{-1} in populations MSY-1 and MSY-2, respectively, whereas rapid cycle genomic selection resulted in a gain of 90 (MSY-1) and 43 kg ha^{-1} (MSY-2) (Das et al. 2020).

4.7.4 Nutrient Use Efficiency

Maize is a fertilizer-responsive crop and shows increased grain yield per unit fertilizer application associated with better nutrient use efficiencies. These nutrients are vital for plants to carry out many of the metabolic processes. Any deficiency of these nutrients inhibits plant growth and development, thereby affecting plant yield. Few studies were undertaken to predict genomic breeding values for major nutrients like nitrogen and phosphorus use efficiency traits in maize.

Under low phosphorus stress, the genomic prediction was undertaken in a maize panel with 410 maize inbred lines for 11 agronomic traits employing 5 classical models, viz. RR-BLUP, GBLUP and three Bayesian models (BayesA, BayesB and BayesC). The prediction accuracy was assessed by fivefold cross-validation. The predictive ability of all five models was comparable, although GBLUP outperformed the others. The prediction accuracies significantly varied between contrasting phosphorus environments. Under normal phosphorus conditions, the prediction accuracies were ranging from 0.40 (ASI in 2015) to 0.76 (days to tasselling in 2014), with a mean of 0.53, whereas, under low phosphorus stress, the predictions were 0.06 (ASI in 2015) to 0.73 (days to tasselling in 2015), with a mean of 0.45. Furthermore, traits with higher heritability mostly showed better prediction accuracies than those with relatively low heritability (Xu et al. 2018).

Genome breeding values were predicted in testcross progenies of 411 inbred lines selected from the IMAS panel and crossed with tester CML539 under both low and optimum nitrogen conditions. Moderate to high GEBVs were observed under both optimum and low nitrogen conditions. Under optimum nitrogen condition, the GEBVs of 0.42, 0.62, 0.59, 0.48, 0.60, 0.54, 0.29 and 0.52 were predicted for grain yield, anthesis date, ASI, plant height, ear height, ears per plant and senescence, respectively, whereas, under low nitrogen, the corresponding GEBVs were 0.45, 0.67, 0.64, 0.53, 0.64, 0.63, 0.42 and 0.24 (Ertiro et al. 2020).

4.8 Genomic Selection for Biotic Stress Tolerance in Maize

Maize production is limited by biotic stresses commonly induced by insect pests and/or diseases (Lodha et al. 2013). Maize is plagued by pests, including stem borers, pink borers, shoot fly, termites and various storage pests. In maize, resistance to various biotic stresses is controlled by various QTLs with small or minor effects (Gazal et al. 2018). Hence, the marker-assisted selection cannot serve the purpose. Thus, researchers are facing the genomic predictions and association mapping for the resistance breeding in maize. Presently, quite a good number of investigations were available on genomic perfections for biotic stresses, viz. insect pests and diseases, by fungal and viral pathogens (Table 4.3).

4.8.1 Fungal Diseases

Resistance to many fungal pathogens is complexly inherited. Thus, genomic prediction seems to be the viable option. Genomic predictions for NCLB resistance in the two heterotic groups ($N = 197$) of maize through the BLUP model showed greater prediction accuracies (~0.70) for both dent and flint heterotic groups (Technow et al. 2013). The application of the RR-BLUP model in $F_{2:3}$ populations derived from crosses CM212 × MAI 172 (population 1) and CM202 × SKV 50 (population 2) achieved the prediction accuracies of 0.83 (population 1) and 0.79 (population 2) for NCLB resistance, respectively (Balasundara et al. 2021).

A total of five biparental DH populations ($N = 635$) phenotyped for *Gibberella* ear rot incidence and three grain yield component traits were used to predict the GEBVs employing the RR-BLUP model. The prediction accuracies ranged from 0.20 to 0.39 among the DH populations. Within DH populations, the prediction accuracies were in agreement with theoretical expectations for the target traits showing moderate to high heritability. In contrast, the prediction accuracies are

Table 4.3 Genomic selection in maize using different training populations for biotic stress tolerance

S. no.	Stress	Trait studied	Population used	Genotyping platform	Model	Prediction accuracy	Reference
Fungal diseases							
1.	NCLB	NCLB resistance per se	300 inbred lines from DTMA panel	GBS	K, M-BL, KM-BL, M-RKHS, KM-RKHS	0.30–0.65	Crossa et al. (2011)
2	GLS	GLS resistance per se	300 inbred lines from DTMA panel	GBS	K, M-BL, KM-BL, M-RKHS, KM-RKHS	0.41–0.60	Crossa et al. (2011)
3	NCLB	NCLB resistance per se	Heterotic groups (100 dent and 97 flint lines)	Illumina SNP chip MaizeSNP50	Bayesian GBLUP	0.36–0.70	Technow et al. (2013)
4.	Ear rot	Rotten kernels % and ERIS	238 inbred lines	DArTseq	RR-BLUP; BSSV	0.19–0.93	dos Santos et al. (2016)
5.	TS complex	Resistance per se	DTMA panel and three DH populations: pop1 (CML 495 × La Posta Sequia C7 F64–2–6-2-2-B-B-B; $N = 174$); pop2 (CML451 × DTPYC9-F46–1–2-1-2-B-B-B; $N = 100$); pop3 (CML451× CML451 × DTPYC9-F46–1–2-1-2-B-B-B; $N = 111$)	GBS	RR-BLUP	0.55–0.74	Cao et al. (2017)
6.	*Fusarium* ear rot	DTA, DTS, EH, PH, erect plants, grain yield	494 $S_{0:1}$ family from a recurrent selection population	GBS	GBLUP, two-stage GBLUP, BayesCπ, Bayesian LASSO and extreme gradient boosting models	0.24–0.67	Holland et al. (2020)

(continued)

Table 4.3 (continued)

S. no.	Stress	Trait studied	Population used	Genotyping platform	Model	Prediction accuracy	Reference
7..	*Gibberella* ear rot (GER)	Disease severity, DS, PH, seed set	500 DH lines	High-density Affymetrix® axiom® maize genotyping array	RR-BLUP and wRR-BLUP	0.38–0.51	Gaikpa et al. (2020)
8.	Grey leaf spot	PH, GY, EH, ear position, anthesis date, grain moisture	IMAS maize panel (410 inbreds) and five DH populations	GBS	RR-BLUP	0.29–0.84	Kibe et al. (2020)
9.	NCLB	DFA, DFS, DSH, PH, EH, CL, EC, kernel rows per cob, KRC, TW, shelling % and plot GY	Two $F_{2:3}$ populations generated from CM212 × MAI172 CM202 × SKV 50 combinations	Illumina BeadXpress VeraCode reader	RR-BLUP	0.24–0.32	Balasundara et al. (2021)
Bacterial diseases							
10	Goss's wilt	Days to anthesis, AUPC	Goodman maize diversity panel, $N = 300$	Illumina MaizeSNP50 BeadChip	RR-BLUP	0.69	Cooper et al. (2019)
Viral diseases							
11.	MCMV and MLNV	Resistance per se	Three DH populations (CML550 × CML504, $N = 219$; CML550 × CML511, $N = 110$; CML550 × CML494, $N = 229$)	GBS	RR-BLUP	0.21–0.95	Sitonik et al. (2019)
12.	MLNV	Disease severity and AUPC	Doubled haploid populations ($N = 1400$)	GBS	RR-BLUP	0.36–0.72	Nyaga et al. (2020)

13.	MLNV	Disease resistance per se	615 (IMAS, 380, and DTMA, 235 panels)	GBS	RR-BLUP	a	Gowda et al. (2015)
Insect pests							
14.	Fall armyworm	DTA, stalk damage rating, tunnelling length, tunnel number	590 DH lines	Illumina MaizeSNP50 BeadChip	GBLUP	0.36–0.61	Foiada et al. (2015)
15.	Fall armyworm and maize weevil	Fall army worm resistance per se, GWL, adult progeny emergence, affected kernel number	Tropical maize panel with 341 DH and inbred lines	DArT	15 prediction models	−0.24–0.84	Badji et al. (2021)

Note: *ASI* anthesis-silking interval, *AUPC* area under disease progress curve, *BLUP* best linear unbiased predictions, *BSSV* Bayesian stochastic search variable, *CL* cob length, *DTA* date to anthesis, *DArT* diversity array technology, *DFA* days to 50% anthesis, *DFS* days to 50% silking, *DTS* days to silking, *DSH* days to 75% husk, *EC* ear circumference, *EH* ear height, *EN* elastic net, *ERIS* ear rot incidence score, *GBLUP* genomic best linear unbiased prediction, *GBS* genotyping by sequencing, *GWS* grain weight loss, *KRC* kernels per row per cob, *MCMV* maize chlorotic mottle virus, *MLNV* maize lethal necrosis virus, *LASSO* least absolute shrinkage and selection operator, *PH* plant height, *RF* random forest, *RKHS* reproducing kernel Hilbert space, *RR* ridge regression, *TS* tar spot, *TW* test weight, *wRR* weighted RR

[a]GP accuracies are presented in graphical form

declined by 42% when full-sib lines were replaced by half-sib lines (Riedelsheimer et al. 2013).

Two hundred and thirty-eight maize lines were clustered to identify the lines resistant to ear rot using 23,154 DArTseq markers. Bayesian stochastic search variable approach and RR-BLUP methods were employed to carry out genomic predictions, and both methods presented the equivalent predictive abilities (dos Santos et al. 2016). For FER resistance, the genomic predictions in a panel of 509 maize lines showed similar prediction ability of five GS models, viz. BayesA (0.355), BayesB (0.338), BayesC (0.357), GBLUP (0.367) and RR-BLUP (0.351) (Guo et al. 2020). Similar work on *Fusarium* ear rot and fumonisin contamination in maize using 449 $S_{0:1}$ lines derived from recurrent selection population was subjected to GBLUP, BayesCπ, Bayesian LASSO and extreme gradient boosting models.

The prediction accuracies showed a maximum value of 0.46 for FER and 0.67 for fumonisin (Holland et al. 2020). Further, the prediction accuracies for FER resistance estimated with genome-wide markers across the environments in the CML population, DTMA-AM panel and SYN_DH population and across the populations were 0.46, 0.53, 0.32 and 0.57, respectively. These prediction accuracies were improved (CML, 0.74; DATM, 0.62; SYN_DH, 0.63; and across populations, 0.65) when the models were framed with FER resistance-associated SNPs (Liu et al. 2021). However, quite low prediction accuracies were reported for FER (0.34) and starburst (0.4) in 320 tropical maize inbred lines using GBLUP, Bayesian LASSO and BayesC prediction models with 5000-fold cross-validations (Kuki et al. 2020).

In the case of tar spot disease, genomic predictions showed moderate to high prediction accuracy in different populations (DTMA, 0.55; pop1, 0.58; pop2, 0.74; and pop3, 0.69) employing several training populations and marker densities. When half of the population was included in the training set with 500 to 1000 SNPs, the prediction accuracy was more than 0.50 (Cao et al. 2017). There are no large effect resistant genes nor any practical control methods available to control Goss's wilt and leaf blight diseases. Additionally, the GWAS was not effective to identify variants that are significantly associated with Goss's wilt. However, genomic prediction with RR-BLUP showed prediction accuracy of 0.69, indicating the possible scope of GS in improving Goss's wilt and leaf blight diseases in maize (Cooper et al. 2019).

In maize, grey leaf spot (GLS) is one of the major diseases. GLS resistance is genetically controlled by multiple genes with cumulative effects. The genomic prediction was performed in biparental populations and association panel consisting of 410 maize lines employing RR-BLUP with fivefold cross-validation. The prediction accuracies within populations were low to moderate, i.e. 0.39, 0.37, 0.56, 0.30, 0.29 and 0.38 for IMAS association panel, DH pop1, DH pop2, DH pop3, F_3 pop4 and F_3 pop5, respectively. Further, across the populations, the prediction accuracy was greatly increased to 0.84. GP results further consolidated the resistant line development by incorporating both major and minor effect genes (Kibe et al. 2020).

4.8.2 Virus Diseases

Recently, the viral diseases are gaining importance owing to sudden outbreaks and devastating effects on maize production in the developing world. Among the viral diseases of maize, maize lethal necrosis (MLND) is the most prominent one. MLMD resulted through synergistic interaction of two viruses, viz. maize chlorotic mottle virus (MCMV) and sugarcane mosaic virus (SCMV). The fivefold cross-validation of ridge regression best linear unbiased prediction (RR-BLUP) model revealed higher prediction accuracy of 0.56 and 0.36 for IMAS-AM ($N = 380$) and DTMA-AM ($N = 235$) panels, respectively. Importantly, the addition of SNPs associated with MLND resistance in the prediction model improved the prediction accuracy from 0.41, which increased to 0.56 in the panels (Gowda et al. 2015). Genomic prediction for the MLND using 1400 diverse inbred lines showed an increase in prediction accuracy for disease severity and AUDPC with an increase in marker density and training population size from 500 to 6300 and 230 to 915, respectively (Nyaga et al. 2020). Similarly, genomic prediction for MLND which was carried out with RR-BLUP in three doubled haploid populations with fivefold cross-validation showed high prediction accuracy for the populations with high heritability and large population size compared to the others (Sitonik et al. 2019).

4.8.3 Bacterial Diseases

In maize, Goss's bacterial wilt and leaf blight are the major bacterial diseases. Under severe disease incidences, yield losses of >40 per cent have been seen in susceptible maize hybrids (Carson and Wicks 1991). Resistance to Goss's wilt is an intricate and polygenic trait with no large effect resistance genes or major QTL. Goodman maize diversity panel consisting of 223 diverse maize lines was evaluated to identify the genomic regions associated with Goss's wilt resistance by using the genomic prediction model RR-BLUP. The prediction accuracy of 0.69 was recorded (Cooper et al. 2019).

4.8.4 Insect Pests

Insect damage on maize plants in the field and stored grains severely affects food security in many countries across the globe (Demissie et al. 2008). Fall armyworm and stem borers are the major pests of maize that impede maize production in the field, and maize weevils are the major category of storage pests, causing the severe yield loss of up to 10 to 90 per cent. This, in turn, affects the grain marketability, and consumer health concerns may arise due to probable contamination of the grains with aflatoxins and mycotoxins (Tefera et al. 2019; Munyiri et al. 2013).

Individual and joint-population QTL analyses and genome-wide predictions with GBLUP for European corn borer stem damage resistance showed the superiority of the GBLUP model with the prediction accuracy of 0.70 over the QTL model despite the detection of QTL with large effects. The genomic trained model based on DH line per se performance was effective in predicting stalk breakage in test crosses (Foiada et al. 2015). Genomic prediction with 16 GP models on BLUPs and BLUEs for fall armyworm and maize weevil resistance was employed in 341 doubled haploid and inbred lines with ten- and fivefold cross-validation. The prediction accuracy realized with BLUPs was at least as twice as those with BLUEs. Additionally, genomic prediction models showed similar predictive abilities for all the studied traits, and a highly positive correlation (0.92) was witnessed between training population size and prediction accurecies in the random-based training set approach, and the reverse was seen in the pedigree-based training set approach (−0.44), owing to degree of kinship between the training and the breeding populations (Badji et al. 2021).

4.8.5 Weeds

Weeds cause devastating effects on maize yield potential. The striga (*Striga hermonthica*) parasitism is one of the major hurdles in the maize production system of sub-Saharan Africa. The genomic selection for striga resistance showed impressive gains in grain yield under striga-infested (498 kg ha^{-1} $cycle^{-1}$ or 16.9% $cycle^{-1}$) and optimal environments (522 kg ha^{-1} $cycle^{-1}$ or 12.6% $cycle^{-1}$), respectively. Additionally, the study revealed an enhanced genetic gain of grain yield per cycle in striga-infested condition was associated with enhanced plant and ear heights, resistance to root lodging, husk cover, ear parameters and striga tolerance level (Badu-Apraku et al. 2019).

4.9 Integrating Genomic Selection with Contemporary Maize Breeding Tools for Stress Tolerance

With improvement in biotechnological tools, the selection of plants has become more accurate and precise owing to the integration of both phenotypic and genotypic criteria in the selection process. Traditional marker-assisted selection methods with QTL or MAS served as a complementary tool to accelerate the selection in maize breeding programmes (Ribaut and Ragot 2007; Mayor and Bernardo 2009; Tuberosa and Salvi 2009; Beyene et al. 2016). But the identification of the QTLs that are showing expression constitutive across environments and populations with different genetic background is essential to use them in MAS (Bernier et al. 2008). G × E interactions reduce the correlation between the traits and QTL detected among the target environments (Bolanos and Edmeades 1996; Tuberosa et al. 2002). In practical breeding, the QTL identified for the target trait usually changes with different genetic backgrounds (Rong et al. 2007) and maize between the inbred lines per se and their testcross hybrids (Mei et al. 2005; Szalma et al. 2007). Numerous QTL mapping experiments conducted in the past have limited application in actual breeding because of the low marker densities in those studies which resulted in poor genetic resolution. Recent advances in genotyping techniques, such as genotyping by sequencing, have resulted in the availability of thousands of SNPs that are equally scattered throughout the genome (Elshire et al. 2011; Poland et al. 2012). The high-resolution genetic maps with high-density SNPs reduce the confidence interval of surrounding QTL, thereby allowing high-precision mapping.

Recently, MAS-based GS (GS-MAS) is considered as an upcoming strategy in maize breeding (Meuwissen et al. 2001). The GS-MAS allows the major benefit of capturing the minor effects in selection process. The traditional QTL-MAS demands the use of the flanking markers of target QTL or gene; however, the GS-MAS requires the large number of genome-wide distributed markers (Peng et al. 2014). For complex traits controlled by many QTLs with minor effects or low heritability, simulation and empirical analyses suggested the superiority of GS-MAS over QTL-MAS (Bernardo and Yu 2007; Mayor and Bernardo 2009; Heffner et al. 2010; Guo et al. 2013). Proper integration of GS-MAS in the breeding workflow can partially replace the field testing and reduce the line development time and cost of breeding activities (Heffner et al. 2010).

Phenotyping of the large-scale breeding material like doubled haploids is highly resource demanding and often exceeds the phenotyping capacity to evaluate all the lines in multi-environment trials. Therefore, partial use of genotypic data to select DH lines while improving the genetic gains for the key traits along with phenotypic selection can significantly save resources (Beyene et al. 2021). Hybrid breeding is also an evergreen area in maize research. Therefore, the application of genomic prediction in the pre-screening of hybrids could improve the efficiency and efficacy of maize hybrid breeding programmes. Among the various prediction models available, Bayesian models offer great flexibility for predicting and studying the hybrid performance (Alves et al. 2019). Additionally, bringing all the contemporary breeding tools on a platform with GS could enhance the genetic gain and efficiency of GS for stress resilience in maize. Especially, integrating GS with rapid generation advancement methods like doubled haploid (DH) technology, speed breeding coupled with precision phenotyping and high-throughput genotyping assisted by decision support tools could be useful in the rapid delivery of stress-resilient maize cultivars (Fig. 4.2).

4.10 Major Challenges for Genomic Selection in Maize Stress Tolerance Breeding

Genomic selection is yet to be popular among the plant breeding community, which necessitates more evidence for sensible and successful utility in ongoing breeding programmes. In fact, most of the studies on GS application rely on statistical models and simulations, which requires appreciable knowledge of both statistical genomics and quantitative genetics. Furthermore, many of the abiotic and biotic stress tolerance/resistances show complex inheritance and challenge the accuracy of GS as much as phenotypic selection. Since the statistical models in GS are trained with

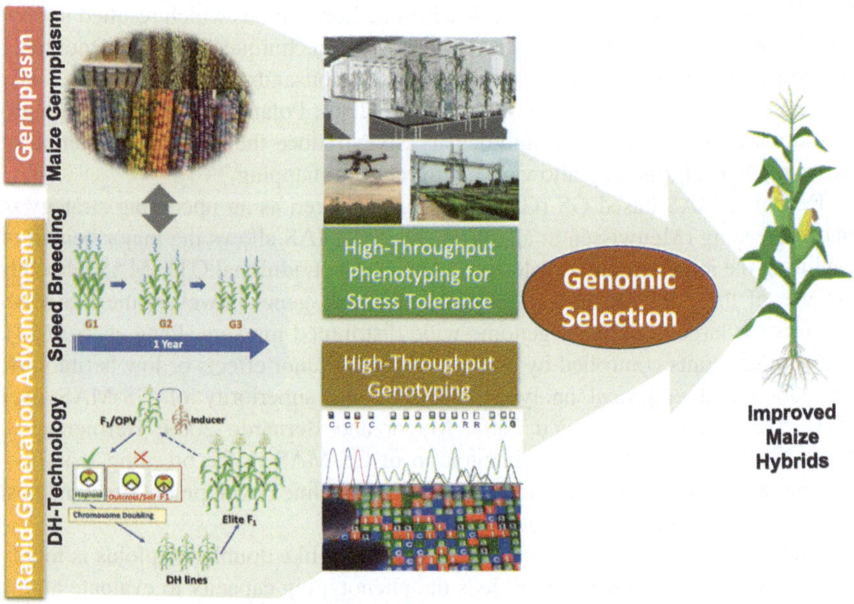

Fig. 4.2 Schematic representation of integrating genomic selection with rapid generation advancement tools like DH technology, speed breeding, high-throughput genotyping and novel high-throughput phenotyping for stress tolerance to enhance the efficiency and pace of stress-resilient maize breeding

phenotypic data, therefore, the reliability and successful utility of GS for stress resilience breeding depend on well-replicated phenotypic data (Juliana et al. 2018).

The applicability of GS is limited within its scope. The performance of GS is generally low when GS models are trained with completely unrelated germplasm or with lines evaluated in non-correlated environments (Juliana et al. 2018; Ertiro et al. 2020). The major goal of GS is to reduce the repeated phenotyping cost and accelerate the genetic gain. The GS requires high-throughput genotyping to capture the genomic contribution towards GEBVs for target traits. The necessity for genotyping with a large number of markers in every generation of selection adds considerably to the price of breeding programmes. Although NGS cost is reduced very significantly, still the prices are not affordable by many plant breeders of developing and underdeveloped worlds to incorporate the GS in their regular breeding programme.

Changes in the gene frequencies and interactions in the breeding generations influence the marker effects and subsequently GEBVs. Therefore, an amendment of the trained GS model in the breeding cycle with the addition/deletion of markers is required. Additionally, the accuracy of GEBVs has been evaluated with the additive component-based simulation models. However, these models ignore interaction components that do not seem to be realistic in practical plant breeding. Therefore, there is a need to develop statistical models which consider interaction effects in addition to additive genetic components.

The successful implementation of GS in stress-resilient breeding requires intensive infrastructure in terms of high-throughput genotyping and phenotyping, which are rapid, reliable and easy. Unfortunately, most of the moderate-sized public sector breeding programmes in the developing world don't possess high-end phenotypic platforms. Further, planning, integration and execution of GS with ongoing breeding programmes require breeders to reorient their strategies in their breeding programmes.

4.11 Prospects

Presently, GS is one of the most promising breeding methods for accelerating the development and release of new cultivars; as a consequence, the use of GS to shape the gene pools and breeding populations from gene bank accessions demands further focused investigation, especially given the vulnerability of elite inbred lines and hybrids to climate change-induced stresses. Furthermore, GS is mainly practised for a single trait; developing models to practise selection for multiple traits and including the component of G × E interaction would be more beneficial.

Genomic prediction requires the marker information that covers the entire genomic region. Thus, it becomes necessary to genotype the breeding material extensively. With the advancement in next-generation sequencing technologies, the genotyping has become easy and less resource driving. The most employed genotyping platforms like Ion Torrent, AmpSeq, GBTS and SNP-seq are reported to genotype thousands of SNPs at a time. A recent technique termed target SNP-seq conglomerates the benefits of high-throughput sequencing and multiplex PCR amplification. The genome-wide SNPs employed in the SNP-seq are poses the conserved flanking sequences, which facilitated capturing through PCR amplification. Furthermore, SNP-seq is suitable in developing countries owing to gain in several hundred SNPs while sequencing the SNP location with approximately a thousand times coverage within a short time and reduced cost.

Efficiency of genomic prediction is adversely affected by outliers, which may occur due to erroneous data imputation and outlying responses. Outlier detection in high-dimensional genomic data is difficult. Therefore, combining *p*-values based strategies to obtain a single *p*-value have been found to be very useful. The prediction accuracy of breeding values can be improved by considering the group means or group sums as a substitute to individual records for several traits which are difficult to phenotype but are economically important. For some of the economical and difficult-to-quantify traits, utilizing group means or group sums as an alternative to individual records can increase breeding value prediction accuracies. These prediction accuracies increase with increasing relationships between the group members.

Acknowledgements MGM is grateful to the 'Early Career Research Award scheme' (ECR/2017/000675), Science and Engineering Research Board (SERB), Government of India, and the network project on 'Computational Biology and Agricultural Bioinformatics (Agril.Edn.14(44)/2014-A&P)' for financial support to research work on stress genomics of maize. The funding agencies had no role in designing the study, data collection and analyses, the decision to publish or the preparation of the manuscript.

References

Abiko T, Kotula L, Shiono K, Malik AI, Colmer TD, Nakazono M (2012) Enhanced formation of aerenchyma and induction of a barrier to radial oxygen loss in adventitious roots of Zea nicaraguensis contribute to its waterlogging tolerance as compared with maize (*Zea mays* ssp. *mays*). Plant Cell Environ 35(9):1618–1630. https://doi.org/10.1111/j.1365-3040.2012.02513.x

Akdemir D, Isidro-Sánchez J (2019) Design of training populations for selective phenotyping in genomic prediction. Sci Rep 9:1446. https://doi.org/10.1038/s41598-018-38081-6

Alves FC, Granato IC, Galli G, Lyra DH, Fritsche-Neto R, de los Campos G (2019) Bayesian analysis and prediction of hybrid performance. Plant Methods 15:14. https://doi.org/10.1186/s13007-019-0388-x

Amusan IO, Rich PJ, Menkir A, Housley T, Ejeta G (2008) Resistance to *Striga hermonthica* in a maize inbred line derived from *Zea diploperennis*. New Phytol 178:157–166. https://doi.org/10.1111/j.1469-8137.2007.02355.x

Anonymous (2019) World food and agriculture – statistical pocketbook 2019. FAO, Rome

Atanda SA, Olsen M, Burgueño J, Crossa J, Dzidzienyo D, Beyene Y, Gowda M, Dreher K, Zhang X, Prasanna BM, Tongoona P, Danquah EY, Olaoye E, Robbins KE (2021) Maximizing efficiency of genomic selection in CIMMYT's tropical maize breeding program. Theor Appl Genet 134:279–294. https://doi.org/10.1007/s00122-020-03696-9

Badji A, Machida L, Kwemoi DB, Kumi F, Okii D, Mwila N, Agbahoungba S, Ibanda A, Bararyenya A, Nghituwamhata SN, Odong T, Wasswa P, Otim M, Ochwo-Ssemakula M, Talwana H, Asea G, Kyamanywa G, Rubaihayo P (2021) Factors influencing genomic prediction accuracies of tropical maize resistance to fall armyworm and weevils. Plan Theory 10(1):29. https://doi.org/10.3390/plants10010029

Badu-Apraku B, Talabi AO, Fakorede MAB, Fasanmade Y, Gedil M, Magorokosho C, Asiedu R (2019) Yield gains and associated changes in an early yellow bi-parental maize population following genomic selection for striga resistance and drought tolerance. BMC Plant Biol 19:129. https://doi.org/10.1186/s12870-019-1740-z

Bain A (1873) Mind and body: the theories of their relation. D. Appleton and Company, New York

Balasundara DC, Lohithaswa HC, Rahul M, Ravikumar RL, Pandravada A, Bhatia BS (2021) Genetic mapping and genomic prediction for northern corn leaf blight (*Exserohilum Turcicum* (pass.) Leonard and Suggs) resistance. Research-Square preprint server. https://doi.org/10.21203/rs.3.rs-618501/v1

Bernardo R (2010) Breeding for quantitative traits in plants. Stemma Press, Woodbury, MN 194:493–503

Bernardo R (2016) Bandwagons I, too, have known. Theor Appl Genet 129:2323–2332. https://doi.org/10.1007/s00122-016-2772-5

Bernardo R, Yu J (2007) Genome wide selection for quantitative traits in maize. Crop Sci 47:1082–1090. https://doi.org/10.2135/cropsci2006.11.0690

Bernier G, Atlin GN, Serraj R, Kumar A, Spaner D (2008) Breeding upland rice for drought resistance. J Sci Food Agric 88:927–939. https://doi.org/10.1002/jsfa.3153

Beyene Y, Gowda M, Pérez-Rodríguez P, Olsen M, Robbins KR, Burgueño J, Prasanna BM, Crossa J (2021) Application of genomic selection at the early stage of breeding pipeline in tropical maize. Front Plant Sci 12:685488. https://doi.org/10.3389/fpls.2021.685488

Beyene Y, Semagn K, Crossa J, Mugo S, Atlin GN, Tarekegne A (2016) Improving maize grain yield under drought stress and non-stress environments in sub-Saharan Africa using marker-assisted recurrent selection. Crop Sci 56:344–353. https://doi.org/10.2135/cropsci2015.02.0135

Bolanos J, Edmeades GO (1996) The importance of the anthesis-silking interval in breeding for drought tolerance in tropical maize. Field Crop Res 48:65–80. https://doi.org/10.1016/0378-4290(96)00036-6

Budhlakoti N, Rai A, Mishra DC (2020) Statistical approach for improving genomic prediction accuracy through efficient diagnostic measure of influential observation. Sci Rep 10:8408. https://doi.org/10.1038/s41598-020-65323-3

Cao S, Loladze A, Yuan Y, Wu Y, Zhang A, Chen J, Huestis G, Cao J, Chaikam V, Olsen M, Prasanna BM, Vicente FS, Zhang X (2017) Genome-wide analysis of tar spot complex resistance in maize using genotyping-by-sequencing SNPs and whole-genome prediction. Plant Genome 10(2):1–14. https://doi.org/10.3835/plantgenome2016.10.0099

Cárcova J, Otegui M (2001) Ear temperature and pollination timing effects on maize kernel set. Crop Sci 41(6):1809–1815. https://doi.org/10.2135/cropsci2001.1809

Carson M, Wicks Z (1991) Relationship between leaf freckles and wilt severity and yield losses in closely related maize hybrids. Phytopathology 81:95–98. https://doi.org/10.1094/Phyto-81-95

Cerrudo D, Cao S, Yuan Y, Martinez C, Suarez EA, Babu R, Zhang X, Trachsel S (2018) Genomic selection outperforms marker assisted selection for grain yield and physiological traits in a maize doubled haploid population across water treatments. Front Plant Sci 9:366. https://doi.org/10.3389/fpls.2018.00366

Chavan S, Smith SM (2014) A rapid and efficient method for assessing pathogenicity of *Ustilago maydis* on maize and teosinte lines. J Vis Exp 83:e50712. https://doi.org/10.3791/50712

Clark RB, Alberts EE, Zobel RW, Sinclair TR, Miller MS, Kemper WD, Foy CD (1998) Eastern gamagrass (*Tripsacum dactyloides*) root penetration into and chemical properties of claypan soils. Plant Soil 200(1):33–45. https://doi.org/10.1023/A:1004256100631

Cooper JS, Rice BR, Shenstone EM, Lipka AE, Jamann TM (2019) Genome-wide analysis and prediction of resistance to Goss's wilt in maize. Plant Genome 12:180045. https://doi.org/10.3835/plantgenome2018.06.0045

Cortes C, Vapnik V (1995) Support-vector networks. Mach learn 20(3):273–297. https://doi.org/10.1007/BF00994018

Crossa J, Campos GD, Pérez P, Gianola D, Burgueño J, Araus JL, Makumbi D, Singh RP, Dreisigacker S, Yan J, Arief V (2010) Prediction of genetic values of quantitative traits in plant breeding using pedigree and molecular markers. Genetics 186(2):713–724. https://doi.org/10.1002/tpg2.20035

Crossa J, Pérez P, de los Campos G, Mahuku G, Dreisigacker S, Magorokosho C (2011) Genomic selection and prediction in plant breeding. J Crop Improv 25(3):239–261. https://doi.org/10.1080/15427528.2011.558767

Daetwyler HD, Pong-Wong R, Villanueva B, Woolliams JA (2010) The impact of genetic architecture on genome-wide evaluation methods. Genetics 185:1021–1031. https://doi.otg/10.1534/genetics.110.116855

Das RR, Vinayan MT, Patel MB, Phagna RK, Singh SB, Shahi JP, Sarma A, Barua NS, Babu R, Seetharam K, Burgueño JA, Zaidi PH (2020) Genetic gains with rapid-cycle genomic selection for combined drought and waterlogging tolerance in tropical maize (*Zea mays* L.). Plant Genome 13:e20035. https://doi.org/10.1002/tpg2.20035

de Lange ES, Balmer D, Mauch-Mani B, Turlings TC (2014) Insect and pathogen attack and resistance in maize and its wild ancestors, the teosintes. New Phytol 204(2):329–341. https://doi.org/10.1111/nph.13005

de Lange ES, Farnier K, Degen T, Gaudillat B, Aguilar-Romero R, Bahena-Juárez F, Oyama K, Turlings TC (2018) Parasitic wasps can reduce mortality of teosinte plants infested with fall armyworm: support for a defensive function of herbivore-induced plant volatiles. Front Ecol Evol 6:55. https://doi.org/10.3389/fevo.2018.00055

de Lange ES, Farnier K, Gaudillat B, Turlings TC (2016) Comparing the attraction of two parasitoids to herbivore-induced volatiles of maize and its wild ancestors, the teosintes. Chemoecology 26(1):33–44. https://doi.org/10.1007/s00049-015-0205-6

de los Campos G, Naya H, Gianola D, Crossa J, Legarra A (2009) Predicting quantitative traits with regression models for dense molecular markers and pedigrees. Genetics 182(1):375–385. https://doi.org/10.1534/genetics.109.101501

de Vlaming R, Groenen PJ (2015) The current and future use of ridge regression for prediction in quantitative genetics. Biomed Res Int 2015:143712. https://doi.org/10.1155/2015/143712

Demissie G, Tefera T, Tadesse A (2008) Importance of husk covering on field infestation of maize by *Sitophilus zeamais* Motsch (coleoptera: Curculionidea) at Bako, Western Ethiopia African. J Biotechnol 7(20):3777–3782

Dingerdissen L, Geiger HH, Lee M, Schechert A, Welz HG (1996) Interval mapping of genes for quantitative resistance of maize to *Setosphaeria turcica*, cause of northern leaf blight, in a tropical environment. Mol Breed 2:143–156. https://doi.org/10.1007/BF00441429

Dos Santos JPR, Pires LMP, de Castro Vasconcellos RC, Pereira GS, Pinho RGV, Balestre M (2016) Genomic selection to resistance to *Stenocarpella maydis* in maize lines using DArTseq markers. BMC Genet 17:86. https://doi.org/10.1186/s12863-016-0392-3

Edmeades GO, Bolanos HR, Lafitte S, Rajram S, Pfeiffer W, Fischer RA (1989) Traditional approaches to breeding for drought resistance in cereals. In: Baker FWG (ed) . Drought Resistance in Cereals, ICSU and CABI, Wallingford, pp 27–52

Elshire RJ, Glaubitz JC, Sun Q, Poland JA, Kawamoto K, Buckler ES (2011) A robust, simple genotyping-by-sequencing (GBS) approach for high diversity species. PLoS One 6:1–10. https://doi.org/10.1371/journal.pone.0019379

Ertiro ET, Labuschagne M, Olsen M, Das B, Prasanna BM, Gowda M (2020) Genetic dissection of nitrogen use efficiency in tropical maize through genome-wide association and genomic prediction. Front Plant Sci 11:474. https://doi.org/10.3389/fpls.2020.00474

Farias-Rivera LA, Hernandez-Mendoza JL, Molina-Ochoa J, Pescador-Rubio A (2003) Effect of leaf extracts of teosinte, *Zea diploperennis* L., and a Mexican maize variety, criollo'Uruapeno', on the growth and survival of the fall armyworm (lepidoptera: Noctuidae). Fla Entomol 80(3): 239–243

Foiada F, Westermeier P, Kessel B (2015) Improving resistance to the European corn borer: a comprehensive study in elite maize using QTL mapping and genome-wide prediction. Theor Appl Genet 128:875–891. https://doi.org/10.1007/s00122-015-2477-1

Foy CD (1997) Tolerance of eastern gamagrass to excess aluminum in acid soil and nutrient solution. J Plant Nutr 20(9):1119–1136. https://doi.org/10.1080/01904169709365322

Frey M, Schullehner K, Dick R, Fiesselmann A, Gierl A (2009) Benzoxazinoid biosynthesis, a model for evolution of secondary metabolic pathways in plants. Phytochemistry 70:1645–1651. https://doi.org/10.1016/j.phytochem.2009.05.012

Gaikpa DS, Kessel B, Presterl T, Ouzunova M, Galiano-Carneiro AL, Mayer M, Melchinger AE, Schön C, Miedaner T (2020) Exploiting genetic diversity in two European maize landraces for improving *Gibberella* ear rot resistance using genomic tools. Theor Appl Genet 134:793–805. https://doi.org/10.1007/s00122-020-03731-9

Gazal A, Dar ZA, Lone AA (2018) Molecular breeding for abiotic stresses in maize (*Zea mays* L.). In: Maize germplasm - characterization and genetic approaches for crop improvement. Intechopen, London, pp 26–38. https://doi.org/10.5772/intechopen.71081

Gianola D (2013) Priors in whole-genome regression: the Bayesian alphabet returns. Genetics 194(3):573–596. https://doi.org/10.1534/genetics.113.151753

Gianola D, Fernando RL (2020) A multiple-trait Bayesian lasso for genome-enabled analysis and prediction of complex traits. Genetics 214(2):305–331. https://doi.org/10.1534/genetics.119.302934

Gianola D, Fernando RL, Stella A (2006) Genomic-assisted prediction of genetic value with semiparametric procedures. Genetics 173:1761–1776. https://doi.org/10.1534/genetics.105.049510

Glauser G, Marti G, Villard N, Doyen GA, Wolfender J-L, Turlings TCJ, Erb M (2011) Induction and detoxification of maize 1,4-benzoxazin-3-ones by insect herbivores. Plant J 68:901–911. https://doi.org/10.1111/j.1365-313X.2011.04740.x

Görtz A, Oerke EC, Steiner U, Waalwijk C, Vries I, Dehne HW (2008) Biodiversity of fusarium species causing ear rot of maize in Germany, cereal. Res Commun 36:617–622. https://doi.org/10.1556/CRC.36.2008.Suppl.B.51

Gowda M, Das B, Makumbi D, Babu R, Semagn K, Mahuku G, Olsen MS, Bright JM, Beyene Y, Prasanna BM (2015) Genome-wide association and genomic prediction of resistance to maize lethal necrosis disease in tropical maize germplasm. Theor Appl Genet 128(10):1957–1968. https://doi.org/10.1007/s00122-015-2559-0

Guo R, Dhilwayo T, Mageto EK, Palacios-Rojas N, Lee M, Yu D, Ruan Y, Zhang A, Vicente FS, Olsen M, Crossa J, Prasanna BM, Zhang L, Zhang X (2020) Genomic prediction of kernel zinc concentration in multiple maize populations using genotyping-by-sequencing and repeat amplification sequencing markers. Front Plant Sci 11:534. https://doi.org/10.3389/fpls.2020.00534

Guo Z, Tucker DM, Wang D, Basten CJ, Ersoz E, Briggs WH (2013) Accuracy of a cross-environment genome-wide prediction in maize nested association mapping populations. G3 (Bethesda) 3:263–272. https://doi.org/10.1534/g3.112.005066

Guo Z, Zou C, Liu X, Wang S, Li W, Jeffers D, Fan X, Xu M, Xu Y (2020) Complex genetic system involved in fusarium ear rot resistance in maize as revealed by GWAS, bulked sample analysis, and genomic prediction. Plant Dis 104:6. https://doi.org/10.1094/PDIS-07-19-1552-RE

Gurney AL, Grimanelli D, Kanampiu F, Hoisington D, Scholes JD, Press MC (2003) Novel sources of resistance to *Striga hermonthica* in *Tripsacum dactyloides*, a wild relative of maize. New Phytol 160(3):557–568. https://doi.org/10.1046/j.1469-8137.2003.00904.x

Haberer G, Young S, Bharti AK, Gundlach H, Raymond C, Fuks G, Butler E, Wing RA, Rounsley S, Birren B, Nusbaum C, Mayer KFX, Messing J (2005) Structure and architecture of the maize genome. Plant Physiol 139:1612–1624. https://doi.org/10.1104/pp.105.068718

Habier D, Fernando RL, Dekkers JCM (2007) The impact of genetic relationship information on genome-assisted breeding values. Genetics 177:2389–2397. https://doi.org/10.1534/genetics.107.081190

Habier D, Fernando RL, Garrick DJ (2013) Genomic BLUP decoded: a look into the black box of genomic prediction. Genetics 194:597–607. https://doi.org/10.1534/genetics.113.152207

Habier D, Fernando RL, Kizilkaya K, Garrick DJ (2011) Extension of the Bayesian alphabet for genomic selection. BMC Bioinformatics 12:186. https://doi.org/10.1186/1471-2105-12-186

Han S, Miedaner T, Utz HF (2018) Genomic prediction and GWAS of Gibberella ear rot resistance traits in dent and flint lines of a public maize breeding program. Euphytica 214:1–20. https://doi.org/10.1007/s10681-017-2090-2

Hastie T, Tibshirani R, Friedman J (2009) The elements of statistical learning: data mining, inference, and prediction. Springer, New York, New York. https://doi.org/10.1007/978-0-387-84858-7

Heffner EL, Lorenz AJ, Jannink JL, Sorrells ME (2010) Plant breeding with genomic selection: gain per unit time and cost. Crop Sci 50:1681–1690. https://doi.org/10.2135/cropsci2009.11.0662

Heffner EL, Sorrells ME, Jannink J (2009) Genomic selection for crop improvement. Crop Sci 49: 1–12. https://doi.org/10.2135/cropsci2008.08.0512

Henderson CR (1949) Estimates of changes in herd environment. J Dairy Sci 32:706

Henderson CR (1975) Best linear unbiased estimation and prediction under a selection model. Biometrics 1:423–447. https://doi.org/10.2307/2529430

Henderson CR, Kempthorne O, Searle SR, von Krosigk CM (1959) The estimation of environmental and genetic trends from records subject to culling. Biometrics 15:192

Heslot N, Yang HP, Sorrells ME, Jannink JL (2012) Genomic selection in plant breeding: a comparison of models. Crop Sci 52(1):146–160. https://doi.org/10.2135/cropsci2011.06.0297

Holland JB, Marino TP, Manching HC, Wisser RJ (2020) Genomic prediction for resistance to fusarium ear rot and fumonisin contamination in maize. Crop Sci 1–13. https://doi.org/10.1002/csc2.20163

Hooda KS, Bagaria PK, Khokhar, Mukesh, Kaur, Harleen, Rakshit Sujay (2018) Mass screening techniques for resistance to maize diseases. ICAR-Indian Institute of Maize Research, PAU Campus, Ludhiana, pp. 93

Hooker AL (1981) Resistance to *Helminthosporium turcicum* from *Tripsacum floridanum* incorporated into corn. Maize Genet Coop Newslett 55:87–88

Howard R, Carriquiry AL, Beavis WD (2014) Parametric and nonparametric statistical methods for genomic selection of traits with additive and epistatic genetic architectures. G3-Genes Genom Genet 4:1027. https://doi.org/10.1534/g3.114.010298

Inghelandt D, Felix P, Frey RD, Stich B (2019) QTL mapping and genome-wide prediction of heat tolerance in multiple connected populations of temperate maize. Sci Rep 9:14418. https://doi.org/10.1038/s41598-019-50853-2

Isidro J, Jannink JL, Akdemir D, Poland J, Heslot N, Sorrells ME (2015) Training set optimization under population structure in genomic selection. Theor Appl Genet 128(1):145–158. https://doi.org/10.1007/s00122-014-2418-4

James W (1890) The principles of psychology. H. Holt and Company, New York

Jannink J, Aaron J, Lorenz IH (2010) Genomic selection in plant breeding: from theory to practice. Brief Funct Genomics 9(2):166–177. https://doi.org/10.1093/bfgp/elq001

Juliana P, Singh RP, Poland J, Mondal S, Crossa J, Montesinos-López OA, Dreisigacker S, Pérez-Rodríguez P, Huerta-Espino J, Crespo-Herrera L, Govindan V (2018) Prospects and challenges of applied genomic selection—a new paradigm in breeding for grain yield in bread wheat. Plant Genome 11:180017. https://doi.org/10.3835/plantgenome2018.03.0017

Kibe M, Nair SK, Das B, Bright JM, Makumbi D, Kinyua J, Suresh LM, Beyene Y, Olsen MS, Prasanna BM, Gowda M (2020) Genetic dissection of resistance to gray leaf spot by combining genome-wide association, linkage mapping, and genomic prediction in tropical maize germplasm. Front Plant Sci 11:572027. https://doi.org/10.3389/fpls.2020.572027

Kuki MC, Pinto RJB, Augusto F, Bertagna B, Tessmann DJ, Teixeira do Amaral A Jr, Scapim CA, Holland JB (2020) Association mapping and genomic prediction for ear rot disease caused by *fusarium verticillioides* in a tropical maize germplasm. Crop Sci 60(6):2867–2881. https://doi.org/10.1002/csc2.20272

Lane JA, Child DV, Moore TH, Arnold GM, Bailey JA (1997) Phenotypic characterisation of resistance in *Zea diploperennis* to *Striga hermonthica*. Maydica 42(1):45–51

Legarra A, Robert-Granié C, Croiseau P, Guillaume F, Fritz S (2011) Improved LASSO for genomic selection. Genet Res 93(1):77–87. https://doi.org/10.1017/S0016672310000534

Lehermeier C, Wimmer V, Albrecht T, Auinger HJ, Gianola D, Schmid VJ, Schön CC (2013) Sensitivity to prior specification in Bayesian genome-based prediction models. Stat Appl Genet Mol Biol 12(3):375–391. https://doi.org/10.1515/sagmb-2012-0042

Lima MS, Silva PS, Oliveira OF, Silva KM, Freitas FC (2010) Corn yield response to weed and fall armyworm controls. Planta Daninha 28:103–111. https://doi.org/10.1590/S0100-83582010000100013

Liu J, Alisdai R, Fernie YJ (2020) The past, present, and future of maize improvement: domestication, genomics, and functional genomic routes toward crop enhancement. Plant Comm 1:1–19. https://doi.org/10.1016/j.xplc.2019.100010

Liu Y, Hu G, Zhang A, Loladze A, Hu Y, Wang H, Qu J, Zhang X, Olsen M, Vicente FS, Crossa J, Lin F, Prasanna BM (2021) Genome-wide association study and genomic prediction of fusarium ear rot resistance in tropical maize germplasm. Crop J 9:325–341. https://doi.org/10.1016/j.cj.2020.08.008

Lobell DB, Bänziger M, Magorokosho C, Vivek BS (2011) Nonlinear heat effects on African maize as evidenced by historical yield trials. Nat Clim Chang 1:42–45. https://doi.org/10.1038/nclimate1043

Lodha T, Hembram P, Basak N (2013) Proteomics: a successful approach to understand the molecular mechanism of plant-pathogen interaction. Am J Plant Sci 4:1212–1226. https://doi.org/10.4236/ajps.2013.46149

Long N, Gianola D, Rosa GJM, Weigel KA (2011) Application of support vector regression to genome-assisted prediction of quantitative traits. Theor Appl Genet 123:1065–1074. https://doi.org/10.1007/s00122-011-1648-y

Lorenzana RE, Bernardo R (2009) Accuracy of genotypic value predictions for marker-based selection in biparental plant populations. Theor Appl Genet 120(1):151–161. https://doi.org/10.1007/s00122-009-1166-3

Lukman R (2012) Unraveling the genetic diversity of maize downy mildew in Indonesia. J Plant Pathol Microbiol 4:2. https://doi.org/10.4172/2157-7471.1000162

Mallikarjuna MG, Bhat JS, Hossain F, Veeraya P, Tyagi A, Karjagi CG, Lohithaswa HC (2020) Genetic enhancement of heat tolerance in maize through conventional and modern strategies. In: Heat stress in food grain crops: plant breeding and omics research, pp 28–66. https://doi.org/10.2174/9789811473982120010004

Mammadov J, Buyyarapu R, Guttikonda SK, Parliament K, Abdurakhmonov IY, Kumpatla SP (2018) Wild relatives of maize, rice, cotton, and soybean: treasure troves for tolerance to biotic and abiotic stresses. Front Plant Sci 9:886. https://doi.org/10.3389/fpls.2018.00886

Mayor PJ, Bernardo R (2009) Genome wide selection and marker-assisted recurrent selection in doubled haploid versus F_2 populations. Crop Sci 49:1719–1725. https://doi.org/10.2135/cropsci2008.10.0587

Mei HW, Li ZK, Shu QY, Guo LB, Wang YP, Yu XQ (2005) Gene actions of QTLs affecting several agronomic traits resolved in a recombinant inbred rice population and two backcross populations. Theor Appl Genet 110:649–659. http://doi.org/https://doi.org/10.1007/s00122-004-1890-7

Mesterhazy A, Lemmens M, Reid LM (2012) Breeding for resistance to ear rots caused by fusarium spp. in maize–a review. Plant Breed 131:1–9. https://doi.org/10.1111/j.1439-0523.2011.01936.x

Meuwissen THE, Goddard ME (2010) Accurate prediction of genetic values for complex traits by whole-genome resequencing. Genetics 185:623–631. https://doi.org/10.1534/genetics.110.116590

Meuwissen THE, Hayes BJ, Goddard ME (2001) Prediction of total genetic value using genome-wide dense marker maps. Genetics 157:1819–1829. https://doi.org/10.1093/genetics/157.4.1819

Meuwissen TH, Solberg TR, Shepherd R, Woolliams JA (2009) A fast algorithm for BayesB type of prediction of genome-wide estimates of genetic value. Gen Sel Evol 41:2. https://doi.org/10.1186/1297-9686-41-2

Michel S, Ametz C, Gungor H, Epure D, Grausgruber H, Loschenberger F, Buerstmayr H (2016) Genomic selection across multiple breeding cycles in applied bread wheat breeding. Theor Appl Genet 129:1179–1189. https://doi.org/10.1007/s00122-016-2694-2

Momen M, Morota G (2018) Quantifying genomic connectedness and prediction accuracy from additive and non-additive gene actions. Genet Sel Evol 50:45. https://doi.org/10.1186/s12711-018-0415-9

Moya-Raygoza G (2016) Early development of leaf trichomes is associated with decreased damage in teosinte, compared with maize, by *Spodoptera frugiperda* (lepidoptera: Noctuidae). Ann Entomol Soc Am 109(5):737–743. https://doi.org/10.1093/aesa/saw049

Munkvold GP, Desjardins AE (1997) Fumonisins in maize: can we reduce their occurrence? Plant Dis 81:556–565. https://doi.org/10.1094/PDIS.1997.81.6.556

Munyiri SW, Mugo SN, Otim M, Mwololo JK, Okori P (2013) Mechanisms and sources of resistance in tropical maize inbred lines to *Chilo partellus* stem borers. J Agric Sci 5(7): 51–60. http://www.ccsenet.org/journal/index.php/jas/article/view/26063

Mutinda SM, Masanga J, Mutuku JM, Runo S, Alakonya A (2018) KSTP 94, an open-pollinated maize variety has postattachment resistance to purple witchweed (Striga hermonthica). Weed Sci 66(4):525–529. https://doi.org/10.1017/wsc.2018.24

Mutyambai DM, Bruce TJ, Midega CA, Woodcock CM, Caulfield JC, Van Den Berg J, Pickett JA, Khan ZR (2015) Responses of parasitoids to volatiles induced by *Chilo partellus* oviposition on teosinte, a wild ancestor of maize. J Chem Ecol 41(4):323–329. https://doi.org/10.1007/s10886-015-0570-1

Nadaraya EA (1964) On estimating regression. Theory Prob Appl 9:141–142

Nault LR, Findley WR (1981) *Zea diploperennis*: a primitive relative offers new traits to improve corn. Ohio Rep Res Develop Agri Home Econ Nat Resour 66(6):90–92

Nault LR, Gordon DT, Damsteegt VD, Iltis HH (1982) Response of annual and perennial teosintes (*Zea*) to six maize viruses. Plant Dis 66(1):61–62

Nault LR, Styer WE, Coffey ME, Gordon DT, Negi LS, Niblett CL (1978) Transmission of maize chlorotic mottle virus by chrysomelid beetles. Phytopathology 68(7):1071–1074

Neiff N, Trachsel S, Valentinuz OR, Balbi CN, Andrade FH (2016) High temperatures around flowering in maize: effects on photosynthesis and grain yield in three genotypes. Crop Sci 56(5): 2702–2712. https://doi.org/10.2135/cropsci2015.12.0755

Nepolean T, Hossain F, Arora K, Sharma R, Shiriga K, Mittal S (2014) Functional mechanisms of drought tolerance in subtropical maize (*Zea mays* L.) identified using genome-wide association mapping. BMC Genomics 15(1182). https://doi.org/10.1186/1471-2164-15-1182

Nyaga C, Gowda M, Beyene Y, Muriithi WT, Makumbi D, Olsen MS, Suresh LM, Bright JM, Das B, Prasanna BM (2020) Genome-wide analyses and prediction of resistance to MLN in large tropical maize germplasm. Genes 11:16. https://doi.org/10.3390/genes11010016

Pace J, Gardner C, Romay C, Ganapathysubramanian B, Lubberstedt T (2015) Genome-wide association analysis of seedling root development in maize (*Zea mays* L.). BMC Genomics 16(1):47. https://doi.org/10.1186/s12864-015-1226-9

Pandit M, Sah RP, Chakraborty M, Prasad K, Chakraborty MK, Tudu V, Narayan SC, Kumar A, Manjunatha N, Kumar A, Rana M (2018) Gene action and combining ability for dual purpose traits in maize (Zea mays L.) under water deficit stress prevailing in eastern India. Range Mgmt Agroforestry 39(1):29–37

Panta S, Flowers T, Lane P, Doyle R, Haros G, Shabala S (2014) Halophyte agriculture: success stories. Environ Exp Bot 107:71–83. https://doi.org/10.1016/j.envexpbot.2014.05.006

Parihar CM, Jat SL, Singh AK, Kumar RS, Hooda KS, Chikkappa GK, Singh DK (2011) Maize production technologies in India. DMR technical bulletin. Directorate of Maize Research, New Delhi

Paril FJ, Sanchez MAB, Salazar AM, LAlsin AG, Cruz PS and Ocampo ETM (2017) Genomic selection in maize (*Zea mays* L.) population improvement for waterlogging tolerance. Philippine J Crop Sci 42 (1):15–26

Park T, Casella G (2008) The Bayesian lasso. J Am Stat Assoc 103(482):681–686. https://doi.org/10.1198/016214508000000337

Patterson HD, Williams ER (1976) A new class of resolvable incomplete block designs. Biometrika 63(1):83–92. https://doi.org/10.1093/biomet/63.1.83

Peng T, Sin X, Mumm RH (2014) Optimized breeding strategies for multiple trait integration: II. Process efficiency in event pyramiding and trait fixation. Mol Breed 33:105–115. https://doi.org/10.1007/s11032-013-9937-6

Perkins JM, Pedersen WL (1987) Disease development and yield losses associated with northern leaf blight on corn. Plant Dis 71:940–943. https://doi.org/10.1094/PD-71-0940

Poland JA, Bradbury PJ, Buckler ES, Nelson RJ (2011) Genome-wide nested association mapping of quantitative resistance to northern leaf blight in maize. Proc Natl Acad Sci 108:6893–6898. https://doi.org/10.1073/pnas.1010894108

Poland JA, Brown PJ, Sorrells ME, Jannink JL (2012) Development of high-density genetic maps for barley and wheat using a novel two enzyme genotyping-by-sequencing approach. PLoS One 7:e32253. https://doi.org/10.1371/journal.pone.0032253

Poland J, Rutkoski J (2016) Advances and challenges in genomic selection for disease resistance. Annu Rev Phytopathol 54:79–98. https://doi.org/10.1146/annurev-phyto-080615-100056

Pryce JE, Hayes BJ, Goddard ME (2012) Novel strategies to minimize progeny inbreeding while maximizing genetic gain using genomic information. J Dairy Sci 95:377–388. https://doi.org/10.3168/jds.2011-4254

Pszczola M (2012) Reliability of direct genomic values for animals with different relationships within and to the reference population. J Dairy Sci 95:389–400. https://doi.org/10.3168/jds.2011-4338

Ranganatha HM, Lohithaswa HC, Pandravada A (2021) Mapping and validation of major quantitative trait loci for resistance to northern corn leaf blight along with the determination of the relationship between resistances to multiple foliar pathogens of maize (*Zea mays* L.). Front Genet 11:548407. https://doi.org/10.3389/fgene.2020.548407

Rashid Z, Singh PK, Vemuri H, Zaidi PH, Prasanna BM, Nair SK (2018) Genome-wide association study in Asia-adapted tropical maize reveals novel and explored genomic regions for sorghum downy mildew resistance. Sci Rep 8(1):366. https://doi.org/10.1038/s41598-017-18690-3

Rasmann S, Köllner TG, Degenhardt J, Hiltpold I, Toepfer S, Kuhlmann U, Gershenzon J, Turlings TC (2005) Recruitment of entomopathogenic nematodes by insect-damaged maize roots. Nature 434(7034):732–737. https://doi.org/10.1038/nature03451

Ray JD, Kindiger B, Sinclair TR (1999) Introgressing root aerenchyma into maize. Maydica 44(2): 113–117

Raymundo AD, Hooker AL (1981) Measuring the relationship between northern corn leaf blight and yield losses. Plant Dis 65(4):325–327. https://doi.org/10.1094/PD-65-325

Ribaut JM, Ragot M (2007) Marker-assisted selection to improve drought adaptation in maize: the backcross approach, perspectives, limitations, and alternatives. J Exp Bot 58:351–360. https://doi.org/10.1093/jxb/erl214

Rich PJ, Ejeta G (2008) Towards effective resistance to striga in African maize. Plant Signal Behav 3(9):618–621. https://doi.org/10.4161/psb.3.9.5750

Riedelsheimer C, Endelman JB, Stange M, Sorrells ME, Jannink JL, Melchinger AE (2013) Genomic predictability of interconnected biparental maize populations. Genetics 194:493–503. https://doi.org/10.1534/genetics.113.150227

Robertson CD, Hjortshoj RL, Janss LL (2019) Genomic selection in cereal breeding. Agron 9(2): 95. https://doi.org/10.3390/agronomy9020095

Rodríguez-Ramilo ST, García-Cortés LA, González-Recio Ó (2014) Combining genomic and genealogical information in a reproducing kernel Hilbert spaces regression model for genome-enabled predictions in dairy cattle. PLoS One 9(3):e93424. https://doi.org/10.1371/journal.pone.0093424

Román SG, Quiroz-Chávez J, Villalobos M, Urías-Gutiérrez V, Nava-Pérez E, Ruíz-May E, Singh RK, Sharma L, Quiroz-Figueroa FR (2020) A global screening assay to select for maize phenotypes with a high tolerance or resistance to *fusarium verticillioides* (Sacc.) Nirenberg rots. Agronomy 10(12):1990. https://doi.org/10.3390/agronomy10121990

Rong J, Feltus FA, Waghmare VN, Pierce GJ, Chee PW, Draye X (2007) Meta-analysis of polyploid cotton QTL shows unequal contributions of subgenomes to a complex network of genes and gene clusters implicated in lint fiber development. Genet 176:2577–2588. https://doi.org/10.1534/genetics.107.074518

Rutkoski JE, Poland JA, Singh RP, Huerta-Espino J, Bhavani S, Barbier H, Rouse MN, Jannink JL, Sorrells ME (2014) Genomic selection for quantitative adult plant stem rust resistance in wheat. Plant. Genome 7(3). https://doi.org/10.3835/plantgenome2014.02.0006

Sahebalam H, Gholizadeh M, Hafezian H, Farhadi A (2019) Comparison of parametric, semiparametric and nonparametric methods in genomic evaluation. J Genet 98:102. https://doi.org/10.1007/s12041-019-1149-3

Schaeffer LR (2006) Strategy for applying genome-wide selection in dairy cattle. J Anim Breed Genet 123:218–223. https://doi.org/10.1111/j.1439-0388.2006.00595.x

Schierenbeck S, Pimentel ECG, Tietze M, Koerte J, Reents R (2011) Controlling inbreeding and maximizing genetic gain using semi-definite programming with pedigree-based and genomic relationships. J Dairy Sci 94:6143–6152. https://doi.org/10.3168/jds.2011-4574

Seetharam K, Kuchanur PH, Koirala KB, Tripathi MP, Patil A, Sudarsanam V, Das RR, Chaurasia R, Pandey VH, Vinayan MT, Nair SK, Babu R, Zaidi PH (2021) Genomic regions associated with heat stress tolerance in tropical maize (*Zea mays* L.). Sci Rep:11:13730. https://doi.org/10.1038/s41598-021-93061-7

Shikha M, Kanika A, Rao AR, Mallikarjuna MG, Gupta HS, Nepolean T (2017) Genomic selection for drought tolerance using genome wide SNPs in maize. Front Plant Sci 8:550. https://doi.org/10.3389/fpls.2017.00550

Silverman BW (1986) Density estimation for statistics and data analysis. Monographs on statistics and applied probability. Chapman and Hall, London

Sitonik C, Suresh LM, Beyene Y, Olsen MS, Makumbi D, Oliver K, Das B, Bright JM, Mugo S, Crossa J, Tarekegne A, Prasanna BM, Gowda M (2019) Genetic architecture of maize chlorotic mottle virus and maize lethal necrosis through GWAS, linkage analysis and genomic prediction in tropical maize germplasm. Theor Appl Genet 132(8):2381–2399. https://doi.org/10.1007/s00122-019-03360-x

Smith SM, Pryor AJ, Hulbert SH (2004) Allelic and haplotypic diversity at the *rp1* rust resistance locus of maize. Genetics 167(4):1939–1947

Sonesson AK, Meuwissen THE, Goddard ME (2010) The use of communal rearing of families and DNA pooling in aquaculture genomic selection schemes. Gen Sel Evol 42(1):41. https://doi.org/10.1186/1297-9686-42-41

Srivastava JP, Gangey SK, Shahi JP (2007) Waterlogging resistance in maize in relation to growth, mineral composition and some biochemical parameters. Ind J Plant Physiol 12(1):28–33

Sun X, Qu L, Garrick DJ, Dekkers JC, Fernando RL (2012) A fast EM algorithm for BayesA-like prediction of genomic breeding values. PLoS One 7(11):e49157. https://doi.org/10.1371/journal.pone.0049157

Szalma SJ, Hostert BM, LeDeaux JR, Stuber CW, Holland JB (2007) QTL mapping with near-isogenic lines in maize. Theor Appl Genet 114:1211–1228. https://doi.org/10.1007/s00122-007-0512-6

Szczepaniec A, Widney SE, Bernal JS, Eubanks MD (2013) Higher expression of induced defenses in teosintes (*Zea* spp.) is correlated with greater resistance to fall armyworm, *Spodoptera frugiperda*. Entomol Exp Appl 146(2):242–521. https://doi.org/10.1111/eea.12014

Technow F, Burger A, Melchinger AE (2013) Genomic prediction of northern corn leaf blight resistance in maize with combined or separated training sets for heterotic groups. Genes Genomes Genet 3:197–203. https://doi.org/10.1534/g3.112.004630

Tefera T, Goftishu M, Ba M, Rangaswamy MA (2019) Guide to biological control of fall armyworm in Africa using egg parasitoids, 1st edn. Nairobi, Kenya

Tibshirani R (1996) Regression shrinkage and selection via the LASSO. J R Statist Soc B 58(1):267–288

Tinsley NA, Estes RE, Gray ME (2013) Validation of a nested error component model to estimate damage caused by corn rootworm larvae. J Appl Entomol 137:161–169. https://doi.org/10.1111/j.1439-0418.2012.01736.x

Tuberosa R, Salvi S (2009) QTL for agronomic traits in maize production. In: Bennetzen JL, Hake SC (eds) Handbook of maize: its biology. Springer, New York, NY. https://doi.org/10.1007/978-0-387-79418-1_26

Tuberosa R, Salvi S, Sanguineti MC (2002) Mapping QTLs regulating morpho-physiological traits and yield: case studies, shortcomings and perspectives in drought-stressed maize. Ann Bot 89:941–963. https://doi.org/10.1093/aob/mcf134

Usai MG, Goddard ME, Hayes BJ (2009) LASSO with cross-validation for genomic selection. Genet Res (Camb) 91(6):427–436. https://doi.org/10.1017/S0016672309990334

Uyemoto JK, Claflin LE, Wilson DL, Raney RJ (1981) Maize chlorotic mottle and maize dwarf mosaic viruses; effect of single and double inoculations on symptomatology and yield. Plant Dis 65(1):39–41

Van Inghelandt D, Melchinger AE, Martinant JP, Stich B (2012) Genome-wide association mapping of flowering time and northern corn leaf blight (*Setosphaeria turcica*) resistance in a vast commercial maize germplasm set. BMC Plant Biol 12(1):1–5. https://doi.org/10.1186/1471-2229-12-56

VanRaden PM (2008) Efficient methods to compute genomic predictions. J Dairy Sci 91:4414–4423. https://doi.org/10.3168/jds.2007-0980

Vapnik V (1995) The nature of statistical learning theory, Ed. 2 edn. Springer, New York

Varshney RK, Terauchi R, McCouch SR (2014) Harvesting the promising fruits of genomics: applying genome sequencing technologies to crop breeding. PLoS Biol 12(6):e1001883. https://doi.org/10.1371/journal.pbio.1001883

Vikal Y, Kaur A, Jindal J, Kaur K, Pathak D, Garg T, Singh A, Singh P, Yadav I (2020) Identification of genomic regions associated with shoot fly resistance in maize and their syntenic relationships in the sorghum genome. PLoS One 15(6):e0234335. https://doi.org/10.1371/journal.pone.0234335

Wang C, Hu S, Gardner C, Lübberstedt T (2017) Emerging avenues for utilization of exotic germplasm. Trends Plant Sci 22(7):624–637. https://doi.org/10.1016/j.tplants.2017.04.002

Wang N, Liu B, Liang X, Zhou Y, Song J, Yang J, Yong H, Weng J, Zhang D, Li M, Nair S (2019) Genome-wide association study and genomic prediction analyses of drought stress tolerance in China in a collection of off-PVP maize inbred lines. Mol Breed 39(8):1–6. https://doi.org/10.1007/s11032-019-1013-4

Wang B, Liu C, Zhang D (2019) Effects of maize organ-specific drought stress response on yields from transcriptome analysis. BMC Plant Biol 19:335. https://doi.org/10.1186/s12870-019-1941-5

Wang X, Xu Y, Hu Z, Xu C (2018) Genomic selection methods for crop improvement: current status and prospects. Crop J 6(4):330–340. https://doi.org/10.1016/j.cj.2018.03.001

Wangai AW, Redinbaugh MG, Kinyua ZM, Miano DW, Leley PK, Kasina M, Mahuku G, Scheets K, Jeffers D (2012) First report of maize chlorotic mottle virus and maize lethal necrosis in Kenya. Plant Dis 96(10):1582. https://doi.org/10.1094/PDIS-06-12-0576-PDN

Ward BP, Brown-Guedira G, Tyagi P, Kolb FL, Van Sanford DA, Sneller CH, Griffey CA (2019) Multienvironment and multitrait genomic selection models in unbalanced early-generation wheat yield trials. Crop Sci 59:491–507. https://doi.org/10.2135/cropsci2018.03.0189

Watson GS (1964) Smooth regression analysis. Sankhya Ser A 26:359–372

Wei WH, Qin R, Song YC, Guo LQ, Gu MG (2001) Comparative analyses of disease resistant and nonresistant lines from maize x Zea diploperennis by GISH. Bot Bull Acad Sin 42:109–114. https://ejournal.sinica.edu.tw/bbas/content/2001/2/bot422-04.html

Wisser RJ, Balint-Kurti PJ, Nelson RJ (2006) The genetic architecture of disease resistance in maize: a synthesis of published studies. Phytopathology 96:120–129. https://doi.org/10.1094/PHYTO-96-0120

Wongkaew A, Phumichai C, Chunwongse J, Jampatong S, Grudloyma P, Pulam T, Doungchan W (2014) Detection of candidate R genes and single nucleotide polymorphisms for downy mildew resistance in maize inbred lines by association analysis. Euphytica 197:109–118. https://doi.org/10.1007/s10681-013-1056-2

Xu Y, Lu Y, Xie C, Gao S, Wan J, Prasanna BM (2012) Whole-genome strategies for marker-assisted plant breeding. Mol Breed 29:833–854. https://doi.org/10.1007/s11032-012-9699-6

Xu C, Zhang H, Sun J, Guo Z, Zou C, Li W, Xie C, Huang C, Xu R, Liao H, Wang J, Xu X, Wang S, Xu Y (2018) Genome-wide association study dissects yield components associated with low-phosphorus stress tolerance in maize. Theor Appl Genet. https://doi.org/10.1007/s00122-018-3108-4

Yi N, Xu S (2008) Bayesian LASSO for quantitative trait loci mapping. Genetics 179(2):1045–1055. https://doi.org/10.1534/genetics.107.085589

Yu P, Wang C, Baldauf JA, Tai H, Gutjahr C, Hochholdinger F (2018) Root type and soil phosphate determine the taxonomic landscape of colonizing fungi and the transcriptome of field-grown maize roots. New Phytol 217:1240–1253. https://doi.org/10.1111/nph.14893

Yuan Y, Cairns JE, Babu R, Gowda M, Makumbi D, Magorokosho C, Zhang A, Liu Y, Wang N, Hao Z, San Vicente F, Olsen MS, Prasanna BM, Lu Y, Zhang X (2019) Genome-wide association mapping and genomic prediction analyses reveal the genetic architecture of grain yield and flowering time under drought and heat stress conditions in maize. Front Plant Sci 9: 1919. https://doi.org/10.3389/fpls.2018.01919

Zaidi PH, Rafique S, Rai PK, Singh NN, Srinivasan G (2004) Tolerance to excess moisture in maize (*Zea mays* L.): susceptible crop stages and identification of tolerant genotypes. Field Crops Res 90:189–202. https://doi.org/10.1016/j.fcr.2004.03.002

Zhang X, Pérez-Rodríguez P, Semagn K (2015) Genomic prediction in biparental tropical maize populations in water-stressed and well-watered environments using low-density and GBS SNPs. Heredity 114:291–299. https://doi.org/10.1038/hdy.2014.99

Zhang X, Yang Q, Rucker E, Thomason W, Balint-Kurti P (2017) Fine mapping of a quantitative resistance gene for gray leaf spot of maize (*Zea mays* L.) derived from teosinte (Z *mays* ssp. *parviglumis*). Theor Appl Genet 130(6):1285–1295. https://doi.org/10.1007/s00122-017-2888-2

Zhao Y, Gowda M, Liu W, Würschum T, Maurer HP, Longin FH, Ranc N, Reif JC (2012) Accuracy of genomic selection in European maize elite breeding populations. Theor Appl Genet 124(4):769–776. https://doi.org/10.1007/s00122-011-1745-y

Zhong SQ, Dekkers JCM, Fernando RL, Jannink JL (2009) Factors affecting accuracy from genomic selection in populations derived from multiple inbred lines: a barley case study. Genetics 182:355–364. https://doi.org/10.1534/genetics.108.098277

Zhong S, Toubia-Rahme H, Steffenson BJ (2007) Molecular mapping and marker-assisted selection of genes for Septoria speckled leaf blotch resistance in barley. Phytopathology 96:993–997. https://doi.org/10.1094/PHYTO-96-0993

Zummo N, Scott GE (1990) Cob and kernel infection by *aspergillus flavus* and *fusarium moniliforme* in inoculated, field-grown maize ears. Plant Dis 74:627–631. https://doi.org/10.1094/PD-74-0627

Genome-Wide Association Studies and Genomic Selection for Nutrient Use Efficiency in Cereals

5

Bingi Pujari Mallikarjuna, Niveditha Shettigar, Dattatraya Hegde Radhika, E. Lamalakshmi Devi, Jayant S. Bhat, B. S. Patil, Hirenallur Chandappa Lohithaswa, and Mallana Gowdra Mallikarjuna

Abstract

Cereals are essential food crops ensuring global food and nutritional security by providing more than 60% of global calories requirement. However, cereal production is under threat owing to various climate change-mediated abiotic and biotic stresses. Additionally, the low and injudicious usage of nutrients is a major impediment to achieve nutrient use efficiency in cereals. Among essential nutrients, nitrogen (N), phosphorus (P) and potassium (K) are the major nutrients required in greater amounts for the proper growth and development of crops. Besides better agronomic practices, the development of cereal cultivars with genetically enhanced nutrient use efficiency is the most sustainable approach to improve NUE and reduce the cost of cultivation and environmental pollution. The availability of complete genome sequences in cereal crops has greatly contributed to enormous molecular markers and high-density linkage maps to implement the next-generation breeding approaches to enhance the genetic gain through nutrient

B. P. Mallikarjuna · J. S. Bhat · B. S. Patil
ICAR-IARI's Regional Research Centre, Dharwad, India

N. Shettigar
ICAR-Research Complex for NEH Region, Umiam, Meghalaya, India

D. H. Radhika
University of Horticultural Sciences, Bagalkote, Karnataka, India

E. L. Devi
ICAR-Research Complex for NEH Region, Sikkim Centre, Tadong, Sikkim, India

H. C. Lohithaswa
Department of Genetics and Plant Breeding, University of Agricultural Sciences, Bangalore, India

M. G. Mallikarjuna (✉)
Division of Genetics, ICAR-Indian Agricultural Research Institute, New Delhi, India
e-mail: MG.Mallikarjuna@icar.gov.in

M. G. Mallikarjuna et al. (eds.), *Next-Generation Plant Breeding Approaches for Stress Resilience in Cereal Crops*, https://doi.org/10.1007/978-981-19-1445-4_5

use efficiency. Among the various genetic tools in crops, genome-wide association studies (GWAS) and genomic selection (GS) can improve complex traits like nutrient use efficiency traits in cereals by altering functional adaptive traits. Further, the developments in phenotyping approaches coupled with GS and GWAS revealed various candidate genes for nutrient use efficient adaptive traits and their possible mechanisms in enhancing the major nutrient use efficiency in cereals. Here, we presented key updates on the application and utility of GS and GWAS in cereals to improve the N, P and K use efficiency in cereals.

Keywords

Nutrient use efficiency (NUE) · Genetic gain · Genomic selection · Genome-wide association analysis Cereals

5.1 Introduction

Cereals play a central role in providing major food calories to the human population. Thus, improving cereal yield to fulfil the demand of the increasing human population is becoming a challenging task. Global cereal production is determined by various genetic, edaphic and environmental factors. Among various crop growth determinants, nutrients are the important factors influencing grain yield and the nutritional profile of cereals. Nitrogen (N), phosphorus (P) and potassium (K) are the important key nutrients capable of producing major impacts on cereal production and the most prominent external inputs in modern agriculture. These major nutrients, viz. N, P and K, are the essential components of protein, nucleic acid, chlorophyll, cell wall, energy carriers, osmoregulation, photosynthesis and several secondary metabolites (Shrivastav et al. 2020; Sanchez-Bragado et al. 2017). The external application of NPK fertilizers has significantly increased the global cereals' yields and heralded the green revolution in Southeast Asia. The FAO reported the huge rise in global demand for NPK from 186,625 thousand tonnes (2016) to 199,006 thousand tonnes (2019) (FAO 2016). However, the improper and non-judicious application of major fertilizers results in lower NUE and environmental pollutions (Wuebbles 2009; Pingali 2012; Ng et al. 2016). Though the use of synthetic fertilizers appreciably improves crop performance in terms of grain yield, plants could be able to absorb only 30–40% of the externally applied fertilizers (Curci et al. 2017).

The nitrogen utilization efficiency (NtUtE) is only 40% out of 94 million tonnes of externally applied N fertilizers (Plett et al. 2018). Further, high application of N fertilizers coupled with improper agronomic practices, viz. wrong irrigation, ploughing, fertilizer application patterns, etc., is increasing the N losses through denitrification/volatilization, leaching and immobilization and creating a global hazard to the environment. Insufficient application of N reduces the crop yield, whereas excess application expands the vegetative growth phase and susceptibility to pests and diseases and creates several environmental pollutions (Dogan and Bilgili 2010; Liu and Shi 2013).

The P is the next important major nutrient after N, and any deficiency severely affects the crop performance (Hussain et al. 2008; Ziadi et al. 2008; Haileselassie et al. 2014; Jeong et al. 2017). Soils with the highest P fixation show high P deficiency and reduced crop production. The tropical regions are home to approximately 1018 million hectares (ha) of land with problems like P fixation and P deficiency (Sanchez and Logan 1992). Nearly 67% of world agricultural land is deficient (Batjes 1997; Hinsinger 2001). The manufacture of phosphorus mostly depends on rock phosphate, the common and primary non-renewable source of P. The P fertilizers' prices are significantly inflated due to the possibility of rock phosphate depletion in the preceding 50 to 100 years (Cordell et al. 2009). The P availability to plants is influenced by various soil factors, viz. pH, alkalinity, acidity, etc. (Lindsay et al. 1989; Marschner et al. 1986). Besides immobile nature in soil, the P losses occur in sandy soils, soils with high organic content and soils with overapplied P fertilizers. The lost P in the environment is detrimental to the aquatic ecosystems and results in aquatic blooms (Sims et al. 1998; Ashley et al. 2011). The third most major nutrient is K and plays a prominent role in osmoregulation, protein metabolism, enzyme activity, photosynthesis and photoregulation (Gattward et al. 2012; Hastings and Gutknecht 1978; Schachtman and Shin 2007; Safdar et al. 2020). Additionally, K is also known to influence NtUE, tolerance to pests and diseases and product quality (Brar et al. 2011; Shabala and Pottosin 2014). The neglected K management is one of the primary reasons for low productivity in the agricultural production systems of the developing world.

To maximize the genetic gain in cereals and to discourage the high-input agriculture system and minimize collateral damage to the environment and society, nutrient use efficiency (NUE) of N, P and K remains one of the crucial strategies. Several classical and modern breeding approaches have been devoted to improving the N, P and K use efficiency in various cereals. Here, we have briefly discussed how plants acquire N, P and K nutrients at the molecular level and how the use efficiency of nutrients can be maximized by deploying molecular tools like genome-wide association mapping and genomic selection with particular reference to major cereals. Several techniques have been proposed at the agronomic and physiological levels to improve high NUE. In cereals, exploring genetic variability across the various crop gene pools could be one of the efficient and sustainable approaches to improve the nutrient use efficiency in addition to minimizing the overusage of fertilizers that cause environmental pollution. Multiple genetic factors influence the NUE, and the large genotype by environment interaction makes genetic dissection quite challenging. Several genes and QTLs have been elucidated based on classical genetics and bi-parental mapping populations in various cereal crops. To understand NUE, the availability of high-throughput molecular markers, especially single nucleotide polymorphisms (SNPs), can further shed light on the underlying candidate gene (s)/QTLs controlling NUE across the whole genome level. Thus, these SNP markers could greatly facilitate practising genomic selection to select high N, P and K use efficient breeding lines in cereal crops.

Likewise, breakthroughs in functional genomics and the availability of complete genome sequences of cereal crops have greatly allowed us to pinpoint the candidate gene(s) and their possible function controlling NUE. However, phenotyping of this trait remains a major hurdle in improving high NUE in crops. Thus, emerging high-throughput phenotyping and machine learning approaches could increase our understanding of NUE at the phenotypic level. We have also discussed the scope of employing other powerful breeding techniques like genomic selection, genome-wide association mapping and CRISPR/Cas9-based genome editing technology to improve the NUE with suitable examples.

5.2 Functional Adaptive Traits for Nutrient Use Efficiency in Cereals

Understanding the traits and metabolic and physiological processes governing NUE, resulting in improvement in yield by any increment in nitrogen, phosphorus and potassium application or sustaining its productivity in low or moderate nutrient stress conditions, is vital to breed plants for nutrient use efficiency or nutrient stress-tolerant lines. Functional traits that include morphological, biochemical, physiological, structural, phenological or behavioural traits and their response to the environment and effects on the ecosystem properties should be given thrust (Violle et al. 2007). These are of two types, i.e. effect traits and response traits. Effect traits are those that have an impact on the ecosystem and the services or disservices that it provides to human societies. Response traits are the ones that impact the colonization, flourishment and spread of a species and its sustainment in the changing environment. Some of the characters can act as both effect and response traits. Functional adaptive traits are the traits that help in the survival of species in the target environment. Various functional adaptive traits are found in the cereals, such as relative growth rate, germination rate, leaf mass index, frost tolerance, potential photosynthetic rate, etc., and several are associated with the NUE (Youngquist et al. 1992; Maranville and Madhavan 2002; Jia et al. 2008; Ning et al. 2013; Wang et al. 2017a, b; Silva et al. 2016; Wang et al. 2019b; Sharma et al. 2021). Nutrient usage is divided into several stages, for instance, in the case of N, uptake phase, reduction of N into usable forms, absorption into different components of biomolecules and finally reallocation from different tissues to the reproductive part (Masclaux-Daubresse et al. 2010). Similar processes are involved in P and K use in plants. The various target phenotypes and physio-biochemical traits that are used to enhance the major nutrient use efficiency in cereals are summarized in Table 5.1.

Table 5.1 Various morphological, physiological and biochemical adaptive traits for major nutrient use efficiency in cereals

Stages	Nitrogen use efficiency	Shoot-specific target traits	Reference
Morphological traits	Shoot and leaf traits	• Days to germination • Green leaf number/plant • Yellow leaf number/plant • Total leaf number/plant • Number of senesced and green leaves • Number of purple leaves • Days to onset of flowering • Days to 50% flowering • Anthesis and silking interval • Stem thickness • Stalk diameter • Plant height • Shoot length • Biomass (dry and fresh) • Total seeds/panicle or ear • Per cent unfilled and filled ears/spikelet • Grain weight • Harvest index	Andresen et al. (2016) Bruen and Struik (2017) Ciampitti and Vyn (2012) Ehdaie et al. (2010) Sharma et al. (2021) Guttieri et al. (2017) Hirel et al. (2001) Maranville and Madhavan (2002) Tollenaar and Lee (2011) Wang et al. (2019a)
	Root traits	• Root length • Root biomass • Root density • Number of roots (seminal) • Fine hairs in root • Lateral root count	
Biochemical and physiological traits	Shoot and leaf traits	• Stay-green trait of leaf • Chlorophyll content • Leaf area index • Leaf photosynthetic rate • N/P/K uptake • Leaf area index (LAI) • Carbon exchange rate • Carbon isotope ratio • Assimilation efficiency indices • NADP malic enzyme activity • Total soluble protein • RuBisCO activity • Glutamine synthetase activity • Nitrate reductase activity • PEPCase activity • Nitrogen internal efficiency • Nitrogen response efficiency	

(continued)

Table 5.1 (continued)

Stages	Nitrogen use efficiency	Shoot-specific target traits	Reference
		• Nitrogen harvest index • Nitrogen remobilization ratio • Nitrogen contribution ratio • Photosynthetic nutrient use efficiency • Phosphorus concentration in stems and leaves • Phosphorus concentration in grain • Phosphorus harvest index (%) • Phosphorus acquisition efficiency (PAE) • Phosphorus internal efficiency (PIE) • Phosphorus biological ratio (PEBR) • K concentration in stems and leaves • K concentration in grain • K harvest index (%) • K uptake efficiency (KUpE) • K utilization efficiency (KUtE)	
	Root	• Hydroponic root exudate estimation	

5.2.1 Types of Traits Associated with Nitrogen Use Efficiency in Cereals

5.2.1.1 Morphological Traits Associated with Nitrogen Use Efficiency

NtUE is usually defined as the uptake, utilization and physiological efficiency of N by the plants. As far as agronomical efficiency is concerned, the yield increment per unit of N applied or the output-to-input ratio is the main criteria for NtUE (Raghuram and Sharma 2019). There are different N-responsive traits in the plants like germination percentage, green leaf number at the vegetative and flowering stages, yellow leaf number at the vegetative and flowering stages, total leaf number at the vegetative and flowering stage, leaf width, stem thickness, shoot length before and after harvest, specific leaf area, leaf life span, leaf senescence, fresh and dry biomass of root and shoot, root length, total plant height, days to flowering, unfilled grain weight, filled grain weight, panicle weight, filled grain percentage, harvest index, root absorption capacity, number of ears/plant, number of grains/ear, thousand grain weight, grain yield/plant, weight of panicle remains, etc. (Lammerts van Bueren and Struik 2017; Sharma et al. 2021). Decreased grain yield up to 37% in low nitrogen level in comparison to high nitrogen level was observed in experiments conducted by Presterl et al. (2003) in European maize. Reduced kernel abortion, anthesis-silking interval and ear number per plant were found to be stress indicators associated with N use efficiency (Gallais and Coque 2005; Geiger 2009). A strong genetic correlation between the plant height and flowering days with N use efficiency was found in hard winter wheat (Guttieri et al. 2017). Further, the studies showed that under limited N, the larger root system of efficient genotypes showed higher N uptake and did not necessarily decrease significant grain yield in winter wheat (Ehdaie et al. 2010; Andresen et al. 2016). Similarly, the wheat genotypes grown in deep tube rhizotrons under limited N showed significant differences in the spatial distribution of root architecture and root biomass, suggesting that the improved root growth in the initial growth phase adapts to the N starvation better (Andresen et al. 2016).

5.2.1.2 Physiological and Biochemical Traits Affecting Nitrogen Use Efficiency

Various N-responsive physiological traits were studied by various researchers, such as leaf chlorophyll concentration, carbon exchange rates (CER), PEPCase activity, NADP-malic enzyme activity, RuBisCO activity, photosynthesis rate, plant total N concentration, plant total protein content, leaf N content, etc. (Maranville and Madhavan 2002; Wang et al. 2019a). N uptake efficiency (NtUpE) or N recovery efficiency (NtRE) and N utilization efficiency (NtUtE) or nitrogen internal efficiency (NtIE) are the components contributing to N use efficiency. Agronomically, NtUE is the product of NtRE and NtIE, i.e. NtUE = NtRE × NtIE (Moll et al. 1982; Ciampitti and Vyn 2012). NtRE is important in high N supply environments, whereas NtIE is imperative in low N environments. NtUpE is the amount of N taken up from soil which usually depends on the root system architecture (RSA) and its capacity to mine (Eghball and Maranville 1993). According to Moll et al. (1982), NtUtE is the grain yield produced per unit of plant N. Maranville et al. (1980) believed both grain and forage produced per unit of plant N is important in NtIE. Ciampitti and Vyn (2011) and Ciampitti et al. (2013) explained other parameters important in NtUE estimations such as N harvest index (NtHI), N remobilization ratio (NtRR) and N contribution ratio (NtCR).

Wang et al. (2019a) conducted an experiment in two commercial hybrids and their parents in maize and concluded that 52% of the total variation was accounted by NtIE and said NtUE is ascribed by a pre-anthesis accumulation of N results in the faster appearance of the leaves with maximum leaf area index, PtNUE and faster remobilization of N from leaves and stalk. However, the stay-green trait and reduced grain N concentration were also reported by Ciampitti and Vyn (2012). From this, it could be understood that N utilization efficiency/NIE decreases the nitrogen in grains to maintain the yield in N stress conditions. Physiological adaptation in sorghum for NtUE was studied by Maranville and Madhavan (2002) by comparing two high NU efficient Chinese lines and two less NU efficient US lines and suggested that PEPCase and enzymes that are connected with phosphoenolpyruvate production play the roles in sustaining photosynthetic efficiency under N stress conditions. In maize, NtUE was increased by selecting genotypes with a higher

NO_3^- storing capacity in leaves, leaf longevity or stay-green trait and prolonged reproductive phase N accumulation (Hirel et al. 2001; Tollenaar and Lee 2011). Various biochemical traits such as assimilation efficiency indices (ACi), glutamine synthetase (GS), nitrate reductase (NR) and protein content in grain and leaf were found to be important in NtUE (Maranville and Madhavan 2002; Vijayalakshmi et al. 2015). Maranville and Madhavan (2002) confirmed with their experiment that high CO_2 assimilation is linked to higher biomass production in low leaf N conditions in sorghum. In aromatic rice genotypes, the GS activity was more in low N conditions and high NU efficient lines, whereas NR activity was more in low NtUE genotypes (Vijayalakshmi et al. 2015). In addition, protein content in grain was found to decrease in low N conditions. Osman et al. (2012) insisted on improving the N uptake efficiency to maintain the grain nitrogen in bread wheat as it is very crucial to the protein quality of bread. Increasing N application enhanced grain protein content and protein yield in six spring wheat genotypes (Gauer et al. 1992). Tiong et al. (2021) utilized the genetically modified rice lines overexpressing *alanine aminotransferase* for studying the changes in pathways for NtUE, and it was found that carbon metabolites, especially those associated with glycolysis and TCA (tricarboxylic acid) cycle, were significantly changed in roots suggesting high metabolic turnover and its upregulation in low N stress conditions. This could result in better energy production and higher N assimilation and, in turn, enhance the biomass. Phytohormonal and secondary metabolite changes are also potential mechanisms in the high NtUE phenotype.

5.2.2 Types of Traits Associated with Phosphorus Use Efficiency in Cereals

5.2.2.1 Morphological Traits Associated with Phosphorus Use Efficiency

Root traits are considered important to scavenge phosphorus from soil. Root characteristics such as more adventitious roots and lateral root spreads, smaller root diameter, shallower basal roots, good root biomass and longer and denser root hair are important to improve the P uptake in soil (Wang et al. 2004; Yan et al. 2004; Lynch 2007; Richardson et al. 2011; Silva et al. 2016). Increased axial root length without lateral root branching is seen in maize as exploratory behaviour (Richardson et al. 2011). Topsoil is richer in P availability; hence, shallower basal root and increased root density in upper layers are well-balanced adaptive characteristics for high PUE. Screening of wheat genotypes for PUE showed an increased biomass and root/shoot ratio in P efficient genotype compared to inefficient genotype (Yan et al. 2010). Traits which originate from the stem or from other tissues, such as crown root formation in maize, can also be helpful for phosphorus uptake (Ochoa et al. 2006). Along with the above-mentioned traits, root and shoot fresh and dry weight, tiller numbers per plant and root to shoot biomass are notable traits for PUE in cereals. Li et al. (2021) reported a decline in PUE and phosphorus acquisition efficiency (PAE) from founder to elite flints and confirmed the shorter root hair and smaller root system at low P as beneficial traits.

5.2.2.2 Physiological and Biochemical Traits Associated with Phosphorus Use Efficiency

Phosphorus use efficiency is divided into two components PUE or PAE and phosphorus internal efficiency (PIE). Root exudates play a role in improving the PAE. These exudates comprise protons and organic acids such as citrate, malate oxalate, etc. Acid phosphatases and ribonucleases upon exudation are known to release fixed P in soil (Vance et al. 2003). P transporters located in cell membranes are also important in P acquisition. PIE depends on the optimal allocation of P inside the plant system. In cells, P is present in two forms, i.e. free inorganic orthophosphate and organic phosphate esters. Inorganic phosphate (Pi) is influenced by P supply (White and Hammond 2008). Excess Pi is stored in vacuoles which will be utilized in P-deprived conditions (Mimura et al. 1990). Organic phosphate is present in nucleic acids, phospholipids, metabolites and proteins. Large P concentration in the seed is not suitable for monogastric animals as their intestine cannot absorb the phytate form of P present in seeds, and it goes to the environment and results in pollution. Seed P concentration is decreased gradually upon breeding for high-yielding varieties, but this will affect the seed vigour in order to compensate that seeds can be coated with P fertilizer (Veneklaas et al. 2012). Remobilizing the P from senescing plants to the growing plant parts and grains is also an important criterion to improve the internal P use efficiency. Phospholipids present in the cell membrane can be bred to be replaced by non-phosphorus compounds such as sulpholipids and galactolipids. It can be replaced either constitutively or in response to P deficiency (Lambers et al. 2012). Cell walls can be adapted by synthesizing P-free polysaccharides such as cellulose (Rao and Terry 1995). Wang et al. (2017a) concluded in their experiment in rice that low P in straw and better grain yield indicated improved P translocation and translocation efficiency of P. Gill et al. (2004) screened 30 spring wheat varieties for their P uptake and use efficiency and could identify high grain yield and high P uptake genotypes (WH711 and PBW343) and high grain yield and low P uptake varieties (Raj3765 and WH283).

5.2.3 Types of Traits Associated with Potassium Use Efficiency in Cereals

5.2.3.1 Morphological Traits Associated with Potassium Use Efficiency

Morphological traits, especially root traits, are important to acquire soil potassium at low K^+ concentration and proliferate into deeper layers in search of K^+ and its ability to extract non-exchangeable K^+ (White et al. 2013; Steingrobe and Claassen 2000; Wang et al. 2011). A larger root system and increased root density help in greater K^+ acquisition (Zörb et al. 2014). The uptake increases in roots with the larger specific surface area, which is achieved in roots with more branches and finer root hairs (White et al. 2013). The deeper root system helps in K^+ uptake from subsoils (Ehdaie et al. 2010). Samal et al. (2010) reported that wheat acquired more K^+ due to greater root length to shoot length. Grain yield is also known to increase in response to K^+ fertigation in maize (Ebelhar and Varsa 2000). Part of yield increase may be accounted for improved stalk strength (reduced lodging), particularly when high K^+ and N fertilizers are applied (Welch and Flannery 1985). Similar results were reported in the wheat crop (Beaton and Sekhon 1985; Haeder and Beringer 1981). Jan et al. (2018) reported a significant effect of potassium on crop phenology, growth and yield traits. In rice, Jia et al. (2008) observed that K^+ efficient lines had more fine roots and root surface compared to inefficient lines. Larger root/shoot biomass ratios in rapidly growing crops have greater K^+ demand, and they are often met by greater K uptake capacities. Roots of cereals possess larger uptake capacities (Pettersson and Jensén 1983).

5.2.3.2 Physiological and Biochemical Traits Associated with Potassium Use Efficiency

In agronomy terms, potassium use efficiency (KUE) is the grain yield produced per unit of available K^+, which is divided into components K uptake efficiency (KUpE) and K utilization efficiency (KUtE). KUE is also measured in terms of the response of grain yield to K availability, tissue K^+ concentration to available K^+ (White 2013) and the response of yield to plant K^+ content. Physiological K^+ requirement in plant accomplishes 90% of its growth and growth rate at critical tissue K^+ concentration (White 2013). Physiological KUtE can be improved by the replacement of vacuolar K^+ with other solutes and increasing remobilization of K from older leaves to other growing and younger parts. Physiological K^+ efficiency is also depending on K^+ transport channels. Many transport proteins are involved in various cellular membranes. These transporters are precisely regulated to modulate the K^+ homeostasis in cellular compartments (White and Karley 2010; Véry et al. 2014). Root exudates also play a role in the K^+ uptake capacities of species. Carboxylates such as citrate, malate and oxalate can dissolute feldspars and micas to release potassium (Marchi et al. 2012). Root-induced acidification of soil releases non-exchangeable K in soil (Giles et al. 2017). All these vary significantly between species and genotypes within species. Potassium utilization efficiency is significantly correlated with K^+ translocation ability, which in turn affects the grain yield, biomass production in seedling tillering stages and harvest index in rice (Yang et al. 2004). The malic acid exudate was increased upon K^+ supply as observed in maize by Kraffczyk et al. (1984). The experiment in maize comparing the accumulation and remobilization of nutrients (NPK) confirmed that new varieties took up more N, P and K during the post-silking stage and remobilized well to the grain in comparison to old varieties (Ning et al. 2013).

5.3 Strategies to Improve Nutrient Use Efficiency in Cereals

The demand for food is increasing every year because of the growing population and lower rate of crop yield per unit area. Repeated cropping of high-yielding varieties takes up excess nutrients from the soil, resulting in poor fertility, and will create environmental stress in soil. The major challenge of feeding the population can be achieved by increasing the production per unit area and maintaining soil health (Atiq et al. 2017; Hussain et al. 2002; Leghari et al. 2016). Many a time, the availability of the nutrient is the limiting factor for the yield; on the contrary, in the high input agriculture system, farmers apply a higher dose of chemical fertilizers. Excess chemical fertilizers created a significant environmental concern in several aspects (Vitousek et al. 2009). Compared to biotic and abiotic stresses, nutrient management is the least attended aspect in plant breeding, even though it was well established that nutrient management contributes to the higher productivity of a cultivar.

NUE refers to production of yield per unit of nutrient or fertilizers applied to field (Ortiz-Monasterio et al. 2001). It comprises two issues: 1) the ability of crop to uptake the nutrients from the soil through the roots and 2) the ability to mobilize these nutrients towards an increased yield (McDonald et al. 2013). The NUE is a complex phenomenon, where it is affected by several environmental factors, rhizosphere condition, plant root architecture, genetic makeup and physio-biochemical and biological condition of the plant. Across the world, many experiments were performed on the effects of fertilizers on yield and soil fertility (Berzsenyi et al. 2000; Zhang et al. 2009; Duncan et al. 2018; Gulser et al. 2019).

Despite significant investments in NUE research, very few crop varieties have been released with nutrient use efficiency. Because of the NUE-associated phenological and physiological trait complexity, there are no single or few traits for assessing NUE. Therefore, there is a need to select several NUE- related traits and assess the cultivars with respect to NUE. Though conventional breeding strategies to enhance NUE were considerably applied in the important crops, like rice, maize and wheat, very few efforts have been attempted to explore the candidate genes associated with NUE characterization and their association with NUE phenotypes in cereals.

5.3.1 Improving Root Architecture

The RSA contribute significantly to crop productivity, since roots extract essential nutrients from the soil. The importance of root morphology parameters in the uptake of a variety of nutrients was indicated by the mechanistic mathematical models based on ion uptake, soil nutrient supply and root morphology (Barber and Cushman 1981; Barber and Silverbush 1984). Therefore, better root growth is considered as prerequisite for healthy plant growth. Differential transcriptome expression analysis of roots in the low and high NUE crop gives an idea about the root architecture. Lateral growth of the root in cereals enhanced the NUE, where overexpression of *OsNPF8.20(OsPTR9)*, a lateral root formation promoting gene, resulted in higher lateral root formation and efficient N uptake and, as a result, increased tiller and effective panicle number and grain yield (Fang et al. 2013). Therefore, targeting various attributes of RSA is one of the major strategies in NUE breeding of cereals.

5.3.2 Genetics of Root-Microbe Interaction

Nutrient uptake is determined by root growth and the bioavailability of nutrients in the rhizosphere. When different NUE responding lines were selected for studying microbial communities in their rhizosphere, different microbial communities and metabolic pathways were observed. Different transcriptional activities like N mineralization, ammonification, nitrification and de-nitrification were evident along with differential expression of subunits of the same genes, denoting that the two plants with different NUE not only were chosen for particular microbial community in rhizosphere but also induced the gene expression (Pathan et al. 2018). Dual transcriptome analysis of the rhizosphere gives a clear picture of gene expression and pathways. A transcription profile will help to identify genes involved in nutrient mineralization, proper interaction, suction and assimilation of the nutrients. Many plant growth-promoting bacteria (PGPB) improve root growth; however, their effectiveness could be determined by the nutrient status in rhizosphere. The attraction of the microbial biome depends on the root exudates. Therefore, modifying the cereals' root exudates could change the nutrient uptake and is expected to enhance NUE. The cereal genotypes showing efficient root exudates to facilitate the colonization of NPK mobilizing microbes could be an added strategy to improve NUE in cereals.

5.3.3 Identification of Candidate Genes Related to Nutrient Use Efficiency

Breeding efforts are to be made to enhance the NUE of crops specifically to obtain higher yields under the low nutrient status of the soil, since there is no clear single phenotypic characteristic or any single gene/QTL for differentiating high or low NUE or that exclusively increases the grain yield. Nevertheless, previous QTL studies identified genomic regions for grain quality- and quantity-related traits, i.e. ear leaf area (ELA), plant height (PHT), grain yield (14% moisture) per plant (GYP), number of ears per plant (EPP) and number of kernels per ear (NKE) and kernel weight (KWT) (Agrama et al. 1999). These traits showed comparatively higher heritability correlation >0.5 under different N levels. With respect to P and K, there are no clearly defined phenotypic traits as of now. Since there are very few phenotypic markers, a tremendous opportunity is available to utilize genetic markers like SNPs, ISSRs, SSRs, etc. Once nutrient is taken up by the plant, there will be switching on of different pathways till it reaches the yield/grain formation stage. Recent innovations in the next-generation sequencing (NGS) platforms made them a highly reliable tool in understanding the functional genomics of the low and high NUE crops. Application of different 'omics' could hasten the current studies on NUE. Based on the previous studies, genes related to *glutamate-pyruvate transaminase* (*GPT*), *glutamate-glyoxylate aminotransferase* (*GGT*), *high-affinity nitrate transporters* (*NRT2*) and the associated partner protein (*NAR2*) families were considered as candidate genes for N use efficiency (Araki and Hasegawa 2006; Cai et al. 2008; Feng et al. 2011; Hu et al. 2015).

5.3.4 Genetic Engineering to Increase Nutrient Use Efficiency

Several NUE-related candidate genes identified can be exploited either through a transgenic approach or through gene editing to rebuild the metabolic pathway or increase the specific gene expression to increase the NUE. In rice, few members of NRT1/PTR, 4 NRT2 and 2 NAR2 signal transporter gene families have been functionally characterized. Signal transporter gene expression at the roots enhanced the yield by 30–40% compared to their mutant (Sánchez-Calderón et al. 2006). The comparative genomic study is helpful to explore more genes in the other cereals too. In wheat, *alanine aminotransferase* gene transferred from barley enhanced the N use efficiency in greenhouse conditions (Ahmed et al. 2020). Targeting primary assimilation was also found beneficial and proven that overexpression of *cytosolic glutamine synthetase* (*GS*) isoform in maize increased the kernel number and grain yield by nearly 30% against control type (Martin et al. 2006). In rice, various transporter gene families for the same nutrient were discovered, but allelic variation altered uptake kinetics of nitrate transporter, and differential uptake capability (Hu et al. 2015) between two subspecies was observed. In such cases, gene editing is the best tool to modify the targeted genes.

Only N is the most abundantly studied nutrient in model plants. Still, there is scope to understand and identify NUE candidate genes and trait selection for phenotyping for other nutrients. Along with genetic improvement, good agronomic practices can effectively aid in exploiting the full genetic potential of the cultivar. It is always advocated to conserve the optimum rhizosphere conditions such as pH, temperature, water level, healthy synergetic microbial load and soil aeration.

5.4 Genetic Resources for Genome-Wide Association Analysis and Genomic Selection in Cereals for Nutrient Use Efficiency

Cereals like rice, wheat and maize are the principal sources of food and nutrition to the human population. With the rising global population, there is a demand for adequate production of food grains. Hence, there is a need to improve crop yields through the efficient use of resources, including NPK fertilizers, to achieve the sustainability of food production. Crop yield can be improved through the breeding cultivars that high yields high with limited fertilizer inputs through utilization of cereal genetic resources. The genetic variability in the elite germplasm is essential to improve quantitative traits, including NUE. Decades of breeding cereals for high-yield and high-input agriculture have developed the cultivars poorly adapted to low nutrient availability. Interestingly, there are few reports on modern nutrient-responsive germplasm in crops like wheat and maize (Hirel et al. 2007; Moose and Below 2009). Therefore, it shows the presence of genetic variation for NPK use efficiency and component traits to explore (Garnett et al. 2015; van de Wiel et al. 2016; Maharajan et al. 2021). Several nutrient use efficient genotypes were identified in major cereals. Recently, Jia et al. (2020) reported four rice lines, viz. 99–28, Shennong 315, Teyou 2 and Xindao 41, for NtUE through screening at four levels of N supply 0, 104, 207 and 311 kg/ha. Similar, more than 100 rice landraces given relatively higher yield under treatment of no N application. This study suggests the importance of land races as source of breeding material for NUE in cereals (Rao et al. 2018).

As compared to N, quite few reports are available on the screening of cereal germplasm for P and K stress owing to difficulty in creation of P and K sick plots. In rice, several lines were reported for PUE, viz. Wazuhophek (Swamy et al. 2019), ULR026, ULR031, ULR124, ULR145, ULR180, ULR183, ULR185, ULR186, ULR213, ULR260 and ULR305 (Chankaew et al. 2019). Similarly, in the case of wheat, Nisar et al. (2016) reported NR-397, NR-379, NR-390, NR-403, NR-401, NR-378 and NR-404 as the most efficient lines. Additionally, Hari-Gowthem et al. (2019) reported wheat lines pau16059, pau16063, pau16065, pau16066 and pau16067 for enhanced PUE. Further, heritable variations for NUE in exotic germplasm and populations evolved under low input agricultural systems may also serve as treasures of NUE genes. Unfortunately, very limited efforts were directed in the utilization of germplasm of exotic and low input agricultural systems in the evaluation and improvement of NUE or component traits, owing to lack of knowledge base and difficulty in phenotyping of NUE phenologies (Ranjan and Yadav 2019).

The variation between genotypes can be used to select superior genotypes and/or genes that play an important role in NUE (Mohammed 2018). Nutrient uptake mainly depends on the genotype and the interaction between genotype and the environment resulting in significant differences in nutrient uptake and utilization efficiency and composition (Zhang et al. 2020). Using genotypes with more efficient nutrient absorption efficiency at low nutrient soil leads to result in increased crop yield (Baligar et al. 2001). Since NUE is a complex trait, the QTL mapping approach with a huge QTL and minimal overlap between studies is of limited use for improving NUE. Therefore, it is essential to undertake genetic dissection of NUE traits pertinent to the cropping region using suitable mapping panels or populations. Further, the precision can be further enhanced by the application of high-throughput phenotyping and modern biotechnological tools.

By utilizing the genetic variation for NUE-related traits, mapping approaches such as genome-wide association studies (GWAS) can be employed to dissect the QTL or genes associated with important NUE traits, particularly when merged with improved and precise phenotyping techniques (Poland et al. 2012; Cooper et al. 2014). The genomic regions and candidate genes identified through mapping approaches can be further analysed using forward and reverse genetics and transgenic approaches to improve crop yield (Wan et al. 2017). Further, genetic variation existing for agronomically important quantitative traits governed by small effect genes can be improved by novel breeding technique, i.e. genomic selection (GS), by predicting breeding values of individuals based on genome-wide marker data. The implementation of novel breeding tools will fasten the rate of progress in genetic enhancement of NUE in major cereal crops, including rice, maize and polyploid with large and complex genomes such as wheat.

Different types of populations were used as association panels to carry out GWAS analysis. The existing varieties as a source of genetic variation have been utilized for association mapping approaches. The study of Monostori et al. (2017) used an elite germplasm set of 93 wheat varieties adapted to the Central European region. Significant phenotypic differences were observed for 15 investigated traits, including grain yield under low and normal N conditions. In another study, Rao et al. (2018) used 472 rice genotypes comprising landraces and breeding lines in a GWAS study and identified over a hundred genotypes with relative higher yield under low N conditions. In maize, association panel consisting of inbred lines and elite introgression lines was used for GWAS analysis for dissecting N and P use efficiency-related traits under low and optimum nutrient conditions (Xu et al. 2018; Ertiro et al. 2020; Wang et al. 2019b; Ma et al. 2020; Sun et al. 2020). Morosini et al. (2017) used an association panel comprising 64 inbred lines contrasting for N use efficiency and evaluated for N use efficiency-related traits such as total root length (TRL) and low nitrogen tolerance index (LNTI). These genetic resources possess different nutrient uptake and utilization mechanisms which are highly useful in developing nutrient use efficient varieties with higher grain yield. Further, well-characterized genotypes showing nutrient-responsive component traits and harbouring important candidate genes for NUE are valuable genetic resources for modern NUE breeding.

The analysis of QTLs with minor effects using traditional linkage mapping often present several limitations for complex polygenic traits like NUE owing to imprecise estimation and discrepancy in the detection of most of QTLs across mapping populations and target environments (Xu 2010). The advances in molecular breeding technologies helped breeders gain access to innovative genomic tools to gain high-density markers with genome-wide distribution. The genome-wide markers facilitate the genomic selection where genomic breeding values are estimated based on cumulative effects of all these markers' models. Genomic prediction affected by the size and genetic diversity of the training population and its relationship with the testing population (Pszczola et al. 2012). In rice, Liu et al. (2016) reported donor parent for plant height ratio of low N/normal N (PHR) and tiller number ratio of low N/normal N (TNR) through both the association analysis and genomic prediction approaches. Also, this study suggested that through genomic prediction, germplasms which have both high and low breeding values, respectively, can be selected by combining both PHR and TNR traits. The study of Fritsche-Neto et al. (2012) used 41 single-cross maize hybrids and observed higher genome-wide selection accuracy for root traits under low N and P stresses compared to phenotypic selection accuracy. Further, Lyra et al. (2017) used 49 maize inbred lines contrasting for N use efficiency to develop 738 single-cross hybrids and applied multi-trait genomic prediction for nitrogen response indices using different selection indices. The use of historical datasets generated from multi-environment trials in GS for N use efficiency helps to achieve wide adaptation. A recent study by Mastrodomenico et al. (2019) evaluated 552 maize hybrids under low (0 kg Nha^{-1}) and high N (252 kg Nha^{-1}) conditions across 10 environments and observed best GS in the training population when both parents were present in the training and validation sets with larger training population size. Similarly, Ertiro et al. (2020) evaluated testcross hybrids of maize across 9 optimum and 13 managed low N-stressed sites and obtained moderate to high prediction accuracies for target traits under optimum and low N conditions. The above representative studies suggest that genomic selection for NUE-related traits in diverse cereal germplasm could benefit for NUE more than phenotypic selection and marker-assisted selection.

5.5 Different Genomic Approaches to Improve NUE in Major Cereals

Nitrogen, phosphorus and potassium (N, P, K) constitute the primary macronutrients required for optimum crop growth and yield. Plant ability to absorb and utilize nutrients largely depends on the genetic makeup and molecular and physiological mechanisms (Baligar et al. 2001). The in-depth knowledge on genetic basis of the molecular pathways underlying the nutrient use efficiency (NUE)-related traits is critical to optimize NUE and to improve crop yield. Genomic approaches such as genetic linkage mapping and quantitative trait locus (QTL) analysis are being performed to identify the loci governing the agronomically important traits, including NUE traits in crop plants (Ali et al. 2018; Hartley et al. 2020; Ranjan and Yadav 2019). The advent of decoded genomes and advances in genome sequencing technologies, along with the discovery of novel genome analysis computations, have led to the development of high-throughput, cost-effective single nucleotide polymorphisms (SNPs). SNP markers are widely employed for the construction of high-resolution genetic maps to dissect complex QTLs and the annotating function of underlying candidate genes of target traits (Alseekh et al. 2021). The GWAS provide for high-resolution mapping using a set of diverse genotypes and map-based cloning of complex trait genes. Genomic selection (GS) is another potential approach that uses markers covering the entire genome to predict genomic-estimated breeding values (GEBVs) of individuals. GS enhances the genetic gain and improves speed and efficiency of the breeding programmes (Spindel et al. 2015). A genome-assisted breeding approach for developing NUE efficient crop varieties is illustrated in Fig. 5.1. This section illustrates genomic approaches such as GWAS and GS that are being applied to genetically dissect various NUE traits in major cereal crops.

5.5.1 Genome-Wide Association Analysis

GWAS or linkage disequilibrium (LD) mapping is an approach for identifying the associations between traits and genetic markers in a large population (Mackay and Powell 2007). GWAS uses the diverse panel of genotypes (such as landraces, diverse germplasm, breeding populations, doubled haploid populations, etc.) to identify significant marker-trait associations (MTAs) with the power to identify multiple

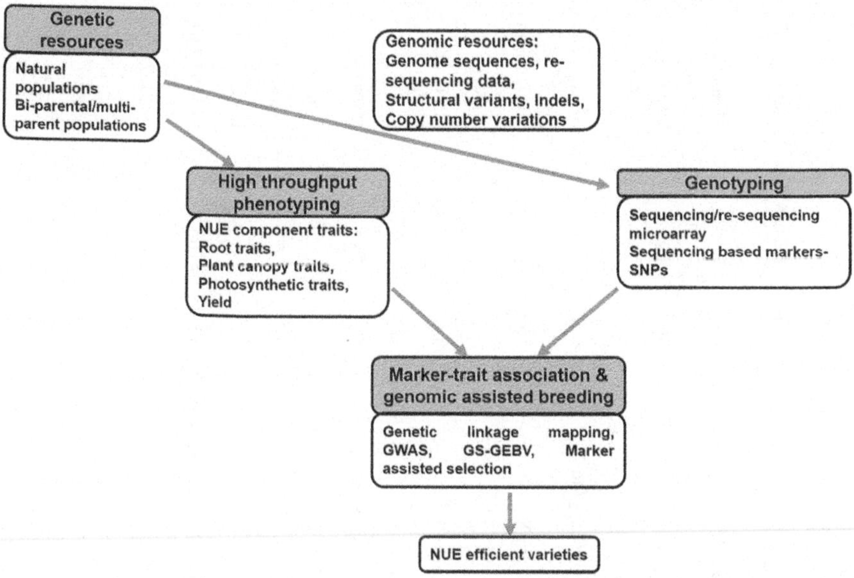

Fig. 5.1 Genome-assisted breeding approach for developing NUE efficient crop varieties: The genetic variation present in the crop germplasm pool can be identified using high-throughput phenotyping of NUE component traits for breeding higher NUE. The advances in genomic technologies in recent years have led to the development of large-scale genomic resources such as genome sequences and millions of genome-wide variations (such as SNPs, indels, SVs, CNVs). Using high-throughput genotyping and high-throughput precise phenotyping approaches, complex traits such as NUE component traits can be dissected at the genetic level by using genomic approaches such as QTL mapping, GWAS and whole genome prediction using genomic selection models and marker-assisted selection for developing NUE efficient crop varieties

loci with several alleles simultaneously (Wang et al. 2019b) and provides a very high genetic resolution based on historical as well as evolutionary recombination events (Chang et al. 2018). With the latest updates in NGS techniques coupled with computational tools, GWAS has become a potent technique for detecting the natural variations and QTLs governing the target phenotype. It has widely been used to understand the genetic basis of economically important complex traits in various crop plants including cereals (Zhao et al. 2011; Yang et al. 2014; Wang et al. 2017b; Liu and Yan 2018). Further, GWAS have been reported to delineate the nutrients' (N, P, K) use efficiency-related traits in different crop species, including major cereals such as rice, wheat and maize (Table 5.2).

5.5.2 Nitrogen Use Efficiency

Nitrogen is the most crucial nutrient element required for the growth of crop plants in natural ecosystems. Nitrogen use efficiency is reported to be a complex attribute governed by multiple genes (Yang et al. 2017), and its expression is regulated at

Table 5.2 List of association mapping studies related to nutrient use efficiency in major cereals

Crop	Association panel	Markers used	Environments	Year	Site	Treatments	Traits	QTL/MTA	Reference
Nitrogen use efficiency									
Rice	184 varieties	157 SSRs	1	1	1	Low and standard N	3	2	Liu et al. (2016)
	472 landraces and breeding lines	50 SSRs	1	2	1	LN=HN-100 kg N	9	12	Rao et al. (2018)
	461 accessions	1,531,224 SNPs	1	3	1	LN-0 N and HN-300 kg/ha N	6	7	Tang et al. (2019)
	190 japonica varieties	38,390 SNPs	1	2	1	LN and HN-120 kg N	16		Rakotoson et al. (2021)
Wheat	260 lines of core collection	3 TaGS2	1	1	1	LN and HN	5	34	Li et al. (2011)
	196 lines of core collection	899 (DArT, SSR, SNP)	12	2	3	LN=HN (35–120 kg N)	8	54	Bordes et al. (2013)
	214 commercial varieties	23,503 SNP	8	2	3	LN=HN-100 kg N	28	333	Cormier et al. (2014)
	93 cultivars	12,293 SNPs 13,160 silicoDArT markers	1	3	1	No N and N120 (120 kg N per ha)	16	183	Monostori et al. (2017)

Maize	64 inbred lines	616,201 SNPs	2	2	2	Low and ideal N	2	7	Morosini et al. (2017)
	411 inbred lines	955,120 SNPs	2	3	14	Low and optimum N	8	38—Optimum 45—Low	Ertiro et al. (2020)
	461 inbred lines	542,796 high-density SNPs	1	2	1	Low nitrate (LN, 0.05 mmol L^{-1}) and high nitrate (HN, 5 mmol L^{-1})	24	328	Sun et al. (2020)
	226 doubled haploid lines	61,634 SNPs	1	1	1	HN (15 mmol L^{-1} NO_3^-) and LN (1.5 mmol L^{-1} NO_3^-)	24	84	Ma et al. (2020)
	139 inbred lines	50,790 SNPs	1	1	1	Normal N supply and low N supply	6	50	He et al. (2020)
Phosphorus use efficiency									
Rice	292 diverse accessions	44 K rice SNPs	2	2	2	Low P and a + P control	9	6	Wissuwa et al. (2015)
Wheat	82 diverse accessions	35 K SNP	2	2	1	Low (2.67 mg P kg^{-1} soil) and sufficient P (42.2 mg P kg^{-1} soil)	15	78	Soumya et al. (2021)
Maize	Panel a—410 inbred lines Panel B—475 inbred lines	55 K SNPs	2	2	2	NP and LP	11	264	Xu et al. (2018)
Maize	356 diverse inbred lines	56,110 SNPs	1	2	3	P-deficient (nutrient solution without $NH_4H_2PO_4$) and P-sufficient (complete nutrient solution containing 1 mmol L^{-1} $NH_4H_2PO_4$)	13	1157 SNPs across 1238 genes	Wang et al. (2019b)

(continued)

Table 5.2 (continued)

Crop	Association panel	Markers used	Environments	Year	Site	Treatments	Traits	QTL/MTA	Reference
Potassium use efficiency									
Rice	324 cultivars	SNPs	1	1	1	K + −deficient (0.1 mM) and K + −replete (1 mM) nutrient solutions	7	10 SNPs	Hartley et al. (2020)
Wheat	150 spring wheat varieties	90 K Infinium iSelect SNP array	1	3	1	Low K (0.09 g/kg of soil) and normal K (0.18 g/kg of soil)	18	534 mta 11 loci	Safdar et al. (2020)

different levels by transcription factors, allosteric regulation and post-transcriptional modification (Ranjan and Yadav 2019). Using bi-parental mapping populations, QTLs underlying NtUE have been identified in major cereals. Several researchers have located the genes encoding glutamate synthase (GS, GOGAT) or nitrate reductase (NR) in identified QTLs for N uptake and remobilization in different crop plants. In rice, Liu et al. (2016) used a population of 184 varieties and studied NUE traits (plant height, tiller number and grain length) in low and optimum N conditions. Association mapping, using genotyping data of 157 genome-wide simple sequence repeat (SSR) markers, identified 8 markers showing significant association with NUE traits. Of these, the genomic regions of two loci at RM5639 and RM3628 contained key NUE-related genes GS1:2 and AspAt3, respectively. Grain yield is generally used as an indicator of NUE, and genotypes with higher NUE have the capacity to uptake N efficiently and divert it for grain yield production (Ali et al. 2018). A set of 472 landraces and breeding lines of rice were screened under low and recommended nitrogen (100 kg ha^{-1}) in field condition (Rao et al. 2018). The study revealed that traits such as grains on secondary branches, grain N concentration and yield are the likely target traits for selection. Further, GWAS analysis using a set of 50 SSR markers revealed about 12 genomic regions associated with yield and related traits under low nitrogen. Subsequent analysis of QTL regions detected three candidate genes (2-oxoglutarate/malate translocator, alanine aminotransferase and pyridoxal phosphate-dependent transferase) that showed enhanced expression in high-yielding genotypes under low N conditions. Tang et al. (2019) integrated GWAS with functional characterization of NtUE genes using a population consisting of rice landraces and identified an $OsNPF6.1^{HapB}$, a rare variant of nitrate transporter *OsNPF6.1* that enhances nitrogen use efficiency by increasing effective panicle number and yield per plant. Recently, Rakotoson et al. (2021) reported 369 significant SNPs belonging to 46 distinct haplotype groups associated with NtUE and yield-related component traits. Further, SNPs showing significant association with NtUE and yield traits co-localized with the genes are involved in N metabolism and transport. Also, the authors found that complex traits like grain yield and nitrogen use efficiency are governed by a several number of QTLs with minimal effects and such small effects can be captured through GWAS and genomic selection approaches.

In wheat, the function of the glutamine synthetase (GS) enzyme in controlling NtUE was proved through the correlation studies (Kichey et al. 2007). Further, haplotype studies of the genes encoding GS plastic isoforms and their association with N use and yield-associated traits revealed four favourable TaGS2 haplotypes (A1b, B1a, B1b, D1a) that may provide better growth, agronomic performance and N uptake for vegetative growth (Li et al. 2011). Previous studies had mapped QTLs related to N use and yield on the chromosomal location containing *GS2* in wheat (Yang et al. 2007; Laperche et al. 2007), indicating the importance of genomic regions surrounding *GS2* gene for breeding wheat cultivars with enhanced N use efficiency and yield. Further, Bordes et al. (2013) used an association panel of 196 accessions of a wheat core collection for GWAS analysis and identified 23 regions, spread over 16 chromosomes, for response to nitrogen level. Similarly,

Cormier et al. (2014) identified 333 genomic regions associated with 28 traits associated with NtUE in a panel of 214 European winter wheat varieties. These studies not only provided new insights on NUE genetic determinism but also assessed QTLs' co-localizations with known N uptake or assimilation enzymes. Recently, Monostori et al. (2017) used DArTseq markers in a GWAS study and identified 183 marker-trait associations (MTA) affecting N use-related complex agronomic traits. These significant genomic regions overlapped with the regions previously mapped for N uptake (Laperche et al. 2007; Xu et al. 2014) and N utilization efficiency (Guo et al. 2012) in wheat.

In maize, using 64 tropical maize inbred lines in a GWAS study, Morosini et al. (2017) showed 7 significant SNPs for low N tolerance index and total root length. Further, the candidate genes that were predicted within the mapped region were mostly engaged with transcriptional regulation and enzyme activity in the N cycle. Breeding maize for NUE is hampered by costly phenotypic screenings and trait complex nature of traits under low N. To circumvent this, Ertiro et al. (2020) identified 38 and 45 SNPs which showed significant association with grain yield (GY) and other traits under optimum and low N conditions, respectively, in a testcross progeny of 411 maize inbred lines. The significant SNPs were further analysed to predict 136 putative candidate genes. Sun et al. (2020) conducted GWAS and candidate gene mining for maize root traits under low N stress using a panel of 461 maize inbred lines. As a result, 328 significant SNPs associated with root and shoot traits were obtained. Upon mining of candidate genes, four genes within the 100-kb intervals flanking the SNPs were identified. Further, Ma et al. (2020) grew 226 DH population of maize under growth chamber with HN (15 mmol L^{-1} $NO3^{-}$) or LN (1.5 mmol L^{-1} $NO3^{-}$) and identified 51 and 33 SNPs, respectively, associated with RSA traits. Using these SNPs, candidate genes involved in seedling, seed and root system development or N metabolism were predicted. Recently, He et al. (2020) used a panel of 139 maize inbred lines to map 27 and 23 SNPs associated with complex NUE-related traits under normal and low N levels, respectively. Among the candidate genes identified, two genes, viz. *Zm00001d025831* and *Zm00001d004633*, encode ammonium transporter 1 and transmembrane amino acid transporter family protein, respectively.

5.5.2.1 Phosphorus Use Efficiency

Like any other nutrients, phosphorus use efficiency (PUE) has been described by two components, viz. P uptake and P utilization efficiency (Wang et al. 2010), and improving both components would be the appropriate approach for improved tolerance to P deficiency. In this direction, various studies have reported genes and QTLs controlling agronomic traits related to PUE in different crops (Bovill et al. 2013). However, relatively few studies have applied the GWAS approach to identify genes/QTLs for PUE in crop species. Wissuwa et al. (2015) characterized the genotypic variation for PUE using a rice panel comprising 292 diverse accessions by using a hydroponic system. GWAS analysis using 44 K rice SNPs identified several loci for PUE on chromosomes 1, 4, 11 and 12. Subsequent coding regions and expression analysis between genotypes of contrasting haplotypes revealed functional changes in two predicted nucleic acid-interacting proteins that are likely causative factors for the observed haplotype-associated variations in PUE. In wheat, Soumya et al. (2021) phenotyped 82 bread wheat genotypes in soil and hydroponics at low and optimum P and performed GWAS analysis with 35 K SNPs. The study showed 78 marker-trait associations (MTAs) and 297 candidate genes involved in key biological processes. Maize is an important cereal showing enormous genetic variation and rapid LD decay, which is quite appropriate for GWAS. For low P stress tolerance, Xu et al. (2018) performed a GWAS using 2 natural populations of maize and identified 259 candidate genes that are associated with transcriptional regulation, scavenging of reactive oxygen species, hormone regulation and cell wall remodelling. Similarly, using 356 diverse inbred lines of maize, Wang et al. (2019b) obtained significant SNPs for 13 traits under P-sufficient and P-deficient conditions. Also, natural variations and haplotypes within the low stress-responsive genes associated with low P stress were detected for root traits. Further, different expression levels of candidate genes in response to low P stress identified candidate genes such as *GRMZM2G466545, GRMZM2G024530, GRMZM2G398848, GRMZM2G143204, GRMZM2G100652, GRMZM2G117250* and *GRMZM2G301738* that are previously reported by Zhang et al. (2014) under low P stress.

5.5.2.2 Potassium Use Efficiency

Potassium use efficiency (KUE) is a complex trait and combines of K uptake efficiency (KUpE) and K utilization efficiency (KUtE). Therefore, the genetic improvement of crops for KUE is carried out by identifying key genomic regions containing QTLs/genes associated with these traits. Despite this, a few QTL studies for KUE-related traits have been reported in major cereal crops using bi-parental mapping populations (Hartley et al. 2020; Ali et al. 2018; Safdar et al. 2020). With the availability of genome-wide SNP markers for the genotypes that represent the diverse background, the information on QTLs governing KUE using the GWAS approach is beginning to accumulate. In rice, a GWAS study with diverse genotypes identified ten SNPs for physiological responses to low potassium stress, including a sodium transporter gene *OsHKT2;1*, a key factor that impacts KUE (Hartley et al. 2020). In this study, the RGR-K signal identified on chromosome 1 overlapped with the QTL identified previously by Fang et al. (2015). Also, the tissue sodium-associated signals found on chromosome 6 related to Na^+ uptake in this study were earlier described by Miyamoto et al. (2012). Similarly, in wheat, the study by Safdar et al. (2020) used a panel of 150 spring wheat varieties to identify 534 significant associations. Further analysis of these marker-trait associations led to the detection of 11 stable loci that are associated with potassium use efficiency and other important agronomic traits.

5.5.3 Genomic Selection

Most agriculturally important traits, including NUE-related traits, are reported to be polygenic in nature and governed by many genes with minor effects accounting for a small proportion of total genetic variances (Robertsen et al. 2019). These small effect genes/QTLs are difficult to map and use simultaneously in the breeding through traditional linkage and QTL mapping (Lande and Thompson 1990). As a consequence, marker-assisted selection (MAS) has a limited success in improving such traits (Heffner et al. 2009). Genomic selection (GS) is a potential tool that overcomes the limitations of MAS for quantitative traits. GS uses genome-wide markers to predict the individual's genetic potential instead of identifying the specific QTL. Advances in NGS technologies, including the availability of high-throughput, cost-effective, informative SNP arrays and improved statistical methods to accurately predict marker effects, have led to the application of GS in making selection decisions in crop plants. GS greatly improves the accuracy of selection, speed and efficiency of breeding programmes. GS has been widely applied to enhance grain yield and other agronomical traits in major crop plants (Robertsen et al. 2019; Srivastava et al. 2020).

Liu et al. (2016) in rice explored the potential of marker-based prediction as a novel approach for NtUE breeding. For this, they used 157 genome-wide SSR marker data for GS by ridge regression and best linear unbiased prediction mixed models (RR-BLUP) to assess the genomic prediction accuracy for plant height ratio and tiller number ratio under normal and low N conditions and found high prediction accuracies for plant height ratio. Root traits are crucial for the uptake of nutrients in maize. Fritsche-Neto et al. (2012) assessed the accurateness of the genome-wide selection (GWS) in maize for root traits under N and P stress using 41 single-cross hybrids. It was showed that, based on hybrid data, the genomic prediction Scheme (RR-BLUP) generated higher GWS accuracy than the phenotypic selection for all the traits. Evaluation and comparison of prediction accuracies by single- and multi-trait models were performed in 738 maize single-cross hybrids derived from 49 tropical inbred lines contrasting for N regimes. The study reported the suitability of multi-trait genomic prediction with a combination of different selection indices and showed the advantage of using single-trait RKHS and GK multi-trait than GBLUP (Lyra et al. 2017). Similarly, evaluation of 552 maize hybrids under low (0 kg Nha^{-1}) and optimum N (252 kg Nha^{-1}) situations across 10 environments showed improved prediction accuracies in larger training and test population when parental lines are included. However, the prediction accuracy on response to training population size and composition was found to be dependent on the N use trait (Mastrodomenico et al. 2019). Additionally, moderate to high prediction accuracies for grain yield and other traits under low N conditions were reported in maize (Ertiro et al. 2020). For low phosphorus stress tolerance in maize, Xu et al. (2018) validated 5 classical genomic selection models for 11 traits under low P (0 kg/ha P_2O_5) and normal P (120 kg/ha P_2O_5) conditions and found that traits with higher heritability had higher prediction accuracy and, with respect to marker density, a moderate density of SNP markers (8000 SNPs) would be appropriate to achieve precise predictions on low phosphorus tolerance traits. Several of these studies used GWAS in conjunction with GS and found that integrating the powerful GWAS results increased prediction accuracy of GS and will improve breeding efficiency for higher nutrient use efficiency.

5.6 Prospects and Conclusion

N, P and K are the vital macronutrients required for plant growth and development, including crop yield and quality. Injudicious use of inorganic fertilizers to meet the nutrient demand of crop plants for achieving higher crop yields is a major cause of environmental pollution and financial burden to the farmers. In light of scarce resources and increased cost of fertilizer production, the development of cultivars with higher nutrient use efficiency is the most feasible approach for sustainable crop growth and yield, especially under low nutrient soils (Baligar et al. 2001; Sarkar and Baishya 2017). To genetically improve nutrient uptake and utilization efficiency in crop plants, we need to understand the molecular genetic mechanisms underlying nutrient use efficiency in crops. Breeding for improved NUE relies on the identification of genetic variation in component traits within germplasm lines, high-throughput precise phenotyping of NUE-related traits in large number of germplasm lines, molecular tagging of NUE phenotypes and finally introgressing beneficial traits into elite cultivars or locally adapted germplasm (White 2013; White and Bell 2017). A large useful genetic variation for component traits related to N, P and K use efficiency has been reported in major cereals (rice, wheat, maize), which provides an opportunity to exploit diverse germplasm lines to identify efficiency alleles and breed for genotypes with higher NUE (White 2013). Such genetic material needs to be screened for NUE parameters using appropriate phenotyping techniques targeting canopy, photosynthetic traits using optical sensors (Erdle et al. 2011) and crop indices such as NDVI (normalized difference vegetation index) (Aparicio et al. 2000) to measure canopy development and canopy nutritional status with both ground-based and aerial imagery devices (Knyazikhin et al. 2013; Li et al. 2013). In addition, efficient uptake of nutrients by root systems is critical to improve the NUE in cereal crops; hence, there is a scope for genetic improvement of root traits.

At the molecular level, plant NPK use efficiency is highly complex involving the integration of many genes and regulatory elements for nutrient sensing, uptake, translocation, assimilation and remobilization which are under the strong influence of environmental variation (Yang et al. 2017; Wang et al. 2014; Gong et al. 2015). Therefore, identification of large-effect QTLs/genes and molecular regulators is challenging. Despite this, many researchers attempted to map complex NPK use efficiency and component traits in major cereals (rice, wheat, maize) with varying degrees of phenotypic variation using molecular markers and advanced biotechnological tools. Combined genomic and phenomic studies so far identified several QTLs and genes for NPK acquisition and transportation in variable genetic backgrounds under diverse doses of NPK. In recent years, better statistical tools for genetic mapping have been developed, and it has been recognized the necessity for more careful experimental design and replicate testing (Myles et al. 2009; Tong et al. 2014). Further, recent innovations in molecular marker tools and sequencing chemistries have led to the development of a cost-effective integrative SNP array for diverse breeding applications, including GWAS and GS. Several studies used the GWAS approach by integrating 'omics' data and identified a number of markers associated with NUE-related traits and key loci/genes governing plant yield along with NPK uptake and utilization in major cereals. However, functional validation of NUE-associated structural or regulatory genes was rarely successful. With the availability of genome-wide SNP markers and powerful computational methods to accurately predict marker effects, novel breeding approaches such as genomic selection (GS) with whole genome prediction models have become convincing strategy to select even for minor QTLs and accelerate the genetic gain. All these efforts require a collective holistic strategy integrating with novel omics tools for the effective implementation of NUE breeding programmes for developing well-adapted and more nutrient-efficient cultivars.

Acknowledgements MGM is grateful to the 'Early Career Research Award scheme' (ECR/2017/000675), Science and Engineering Research Board (SERB), Government of India, and the network project on 'Computational Biology and Agricultural Bioinformatics (Agril.Edn.14(44)/2014-A&P)' for financial support to research work on stress genomics in maize. The funding agencies had no role in designing the study, data collection and analyses, the decision to publish or the preparation of the manuscript.

References

Agrama HAS, Zakaria AG, Said FB, Tuinstra M (1999) Identification of quantitative trait loci for nitrogen use efficiency in maize. Mol Breed 5:187–195. https://doi.org/10.1023/A:1009669507144

Ahmed M, Saeed NA, Ashraf YM, Mukhtar Z, Mansoor S (2020) Improving nitrogen use efficiency in wheat (*Triticum aestivum* L.) through transformation of codon optimized *alanine aminotransferase* gene. Pak J Agri Sci 57(3):707–714. https://doi.org/10.21162/PAKJAS/20.9032

Ali J, Jewel ZA, Mahender A, Anandan A, Hernandez J, Li Z (2018) Molecular genetics and breeding for nutrient use efficiency in rice. Int J Mol Sci 19(6):1762. https://doi.org/10.3390/ijms19061762

Alseekh S, Kostova D, Bulut M et al (2021) Genome-wide association studies: assessing trait characteristics in model and crop plants. Cell Mol Life Sci 78:5743–5754. https://doi.org/10.1007/s00018-021-03868-w

Andresen M, Dresbøll DB, Stoumann Jensen L, Magid J, Thorup-Kristensen K (2016) Cultivar differences in spatial root distribution during early growth in soil, and its relation to nutrient uptake-a study of wheat, onion and lettuce. Plant Soil 408:255–270. https://doi.org/10.1007/s11104-016-2932-z

Aparicio N, Villegas D, Casadesus J, Araus JL, Royo C (2000) Spectral vegetation indices as non-destructive tools for determining durum wheat yield. Agron J 92:83–91. https://doi.org/10.2134/agronj2000.92183x

Araki R, Hasegawa H (2006) Expression of rice (*Oryza sativa* L.) genes involved in high-affinity nitrate transport during the period of nitrate induction. Breed Sci 56:295–302. https://doi.org/10.1270/jsbbs.56.295

Ashley K, Cordell D, Mavinic D (2011) A brief history of phosphorus: from the philosopher's stone to nutrient recovery and reuse. Chemosphere 84(6):737–746. https://doi.org/10.1016/j.chemosphere.2011.03.001

Atiq M, Javed N, Urooj S, Bukhari A, Ali Y, Zeeshan A, Jabbar A (2017) Management of leaf rust of wheat through different levels of NPK and sowing times. Ad Zool Bot 5:39–44. https://doi.org/10.13189/azb.2017.050401

Baligar VC, Fageria NK, Hea ZL (2001) Nutrient use efficiency in plants. Commun Soil Sci Plant Anal 32:7–8. https://doi.org/10.1081/CSS-100104098

Barber SA, Cushman JH (1981) Nitrogen uptake model for agronomic crops. In: Iskander IK (ed) Modeling wastewater renovation-land treatment. Wiley-Interscience, New York, pp 382–409

Barber SA, Silverbush M (1984) Plant root morphology and nutrient uptake. In: Barber SA, Bouldin DR, Kral DM, Hawkins SL (eds) Roots, nutrient and water influx and plant growth, ASA special publication number, vol 49. American Society of Agronomy, Madison, WI, pp 65–88

Batjes NH (1997) A world data set of derived soil properties by FAO UNESCO soil unit for global modelling. Soil Use Manag 13:9–16. https://doi.org/10.1111/j.1475-2743.1997.tb00550.x

Beaton JD, Sekhon GS (1985) Potassium nutrition of wheat and other small grains. In: Munson RD (ed) Potassium in agriculture. ASA, CSSA and SSSA, Madison, WI, pp 701–752

Berzsenyi Z, Gyorffy B, Lap D (2000) Effect of crop rotation and fertilisation on maize and wheat yields and yield stability in a long-term experiment. Eur J Agron 13(2–3):225–244. https://doi.org/10.1016/S1161-0301(00)00076-9

Bordes J, Ravel C, Jaubertie JP, Duperrier B, Gardet O et al (2013) Genomic regions associated with the nitrogen limitation response revealed in a global wheat core collection. Theor Appl Genet 126:805–822. https://doi.org/10.1007/s00122-012-2019-z

Bovill WD, Huang CY, McDonald GK (2013) Genetic approaches to enhancing phosphorus-use efficiency (PUE) in crops: challenges and directions. Crop Pasture Sci 64:179–198. https://doi.org/10.1071/CP13135

Brar MS, Bijay-Singh BSK, Srinivasarao C (2011) Role of potassium nutrition in nitrogen use efficiency in cereals. Res Find e-ifc 29:20–27

Bruen ETLV, Struik PC (2017) Diverse concept of breeding for nitrogen use efficiency. A review. Agron Sustain Dev 37:50. https://doi.org/10.1007/s13593-017-0457-3

Cai C, Wang JY, Zhu YG, Shen QR, Li B, Tong YP, Li ZS (2008) Gene structure and expression of the high-affinity nitrate transport system in rice roots. J Integr Plant Biol 50:443–451. https://doi.org/10.1111/j.1744-7909.2008.00642.x

Chang F, Guo C, Sun F, Zhang J, Wang Z, Kong J et al (2018) Genome-wide association studies for dynamic plant height and number of nodes on the main stem in summer sowing soybeans. Front Plant Sci 9:1184. https://doi.org/10.3389/fpls.2018.01184

Chankaew S, Monkham T, Pinta W, Sanitchon J, Kaewpradit W, Srinives P (2019) Screening tolerance to phosphorus deficiency and validation of phosphorus uptake 1 (Pup1) gene-linked markers in Thai indigenous upland Rice germplasm. Agronomy 9(2):81. https://doi.org/10.3390/agronomy9020081

Ciampitti IA, Murrell ST, Camberato JJ, Tuinstra M, Xia Y, Friedemann P, Vyn TJ (2013) Physiological dynamics of maize nitrogen uptake and partitioning in response to plant density and nitrogen stress factors: II. Reproductive phase. Crop Sci 53:2588–2602. https://doi.org/10.2135/cropsci2013.01.0041

Ciampitti IA, Vyn TJ (2011) A comprehensive study of plant density consequences on nitrogen uptake dynamics of maize plants from vegetative to reproductive stages. Field Crops Res 121:2–18. https://doi.org/10.1016/j.fcr.2010.10.009

Ciampitti IA, Vyn TJ (2012) Physiological perspectives of changes over time in maize yield dependency on nitrogen uptake and associated nitrogen efficiencies: a review. Field Crops Res 133:48–67. https://doi.org/10.1016/j.fcr.2012.03.008

Cooper M, Messina CD, Podlich D, Totir LR, Baumgarten A, Hausmann NJ, Wright D, Graham G (2014) Predicting the future of plant breeding: complementing empirical evaluation with genetic prediction. Crop Pasture Sci 65(4):311–336. https://doi.org/10.1071/CP14007

Cordell D, Drangert JO, White S (2009) The story of phosphorus: global food security and food for thought. Glob Environ Chang 19(2):292–305. https://doi.org/10.1016/j.gloenvcha.2008.10.009

Cormier F, Gouis JL, Dubreuil P, Lafarge S, Praud S (2014) A genome-wide identification of chromosomal regions determining nitrogen use efficiency components in wheat (*Triticum aestivum* L.). Theor Appl Genet 127:2679–2693. https://doi.org/10.1007/s00122-014-2407-7

Curci PL, Cigliano RA, Zuluaga DL, Janni M, Sanseverino W, Sonnante G (2017) Transcriptomic response of durum wheat to nitrogen starvation. Sci Rep 7:1176. https://doi.org/10.1038/s41598-017-01377-0

Dogan R, Bilgili U (2010) Effects of previous crop and N-fertilization on seed yield of winter wheat (*Triticum aestivum* L.) under rain-fed Mediterranean conditions. Bulgarian J Agric Sci 16:733–739

Duncan EG, O'Sullivan CA, Roper MM, Palta J, Whisson K, Peoples MB (2018) Yield and nitrogen use efficiency of wheat increased with root length and biomass due to nitrogen, phosphorus, and potassium interactions. J Plant Nutr Soil Sci 181(3):364–373. https://doi.org/10.1002/jpln.201700376

Ebelhar SA, Varsa EC (2000) Tillage and potassium placement effects on potassium utilization by corn and soybean. Commun Soil Sci Plant Anal 31:11–14. https://doi.org/10.1080/00103620009370591

Eghball B, Maranville JW (1993) Root development and nitrogen influx of corn genotypes grown under combined water and nitrogen stresses. Agron J 85:147–152. https://doi.org/10.2134/agronj1993.00021962008500010027x

Ehdaie B, Merhaut DJ, Ahmadian S, Hoops AC, Khuong T, Layne AP, Waines JG (2010) Root system size influences water-nutrient uptake and nitrate leaching potential in wheat. J Agron Crop Sci 196:455–466. https://doi.org/10.1111/j.1439-037X.2010.00433.x

Erdle K, Mistele B, Schmidhalter U (2011) Comparison of active and passive spectral sensors in discriminating biomass parameters and nitrogen status in wheat cultivars. Field Crops Res 124(1):74–84. https://doi.org/10.1016/j.fcr.2011.06.007

Ertiro BT, Labuschagne M, Olsen M, Das B, Prasanna BM, Gowda M (2020) Genetic dissection of nitrogen use efficiency in tropical maize through genome-wide association and genomic prediction. Front Plant Sci 11:474. https://doi.org/10.3389/fpls.2020.00474

Fang Y, Wu W, Zhang X, Jiang H, Lu W, Pan J, Hu J, Guo L, Zeng D, Xue D (2015) Identification of quantitative trait loci associated with tolerance to low potassium and related ions concentrations at seedling stage in rice (*Oryza sativa* L.). Plant Growth Regul 77:157–166. https://doi.org/10.1007/s10725-015-0047-9

Fang Z, Xia K, Yang X, Grotemeyer MS, Meier S, Rentsch D, Xu X, Zhang M (2013) Altered expression of the PTR/NRT1 homologue OsPTR9 affects nitrogen utilization efficiency, growth and grain yield in rice. Plant Biotechnol J 11:446–458. https://doi.org/10.1111/pbi.12031

FAO (2016) World fertilizer trends and outlook to 2019. Food and Agriculture Organization of the United Nations, Rome. http://www.fao.org/3/a-i5627e.pdf

Feng H, Yan M, Fan X, Li B, Shen Q, Miller AJ, Xu G (2011) Spatial expression and regulation of rice high-affinity nitrate transporters by nitrogen and carbon status. J Exp Bot 62:2319–2332. https://doi.org/10.1093/jxb/erq403

Fritsche-Neto R, DoVale JC, Lanes ECM, Resende MDV, Miranda GV (2012) Genome-wide selection for tropical maize root traits under conditions of nitrogen and phosphorus stress. Acta Scientiarum 34(4):389–395. https://doi.org/10.4025/actasciagron.v34i4.15884

Gallais A, Coque M (2005) Genetic variation and selection for nitrogen use efficiency in maize: a synthesis. Maydica 50:531–547

Garnett T, Plett D, Heuer S, Okamoto M (2015) Genetic approaches to enhancing nitrogen-use efficiency (NUE) in cereals: challenges and future directions. Funct Plant Biol 42:921–941. https://doi.org/10.1071/FP15025

Gattward JN, Almeida AA, Souza JO Jr, Gomes FP, Kronzucker HJ (2012) Sodium-potassium synergism in Theobroma cacao: stimulation of photosynthesis, water-use efficiency and mineral nutrition. Physiol Plant 146:350–362. https://doi.org/10.1111/j.1399-3054.2012.01621.x

Gauer LE, Grant CA, Gehl DT, Bailey LD (1992) Effects of nitrogen fertilization on grain protein content, nitrogen uptake, and nitrogen use efficiency of six spring wheat (*Triticum aestivum* L.) cultivars, in relation to estimated moisture supply. Can J Plant Sci 72:235–241. https://doi.org/10.4141/cjps92-026

Geiger HH (2009) Agronomic traits and maize modifications: nitrogen use efficiency. In: Bennetzen JL, Hake SC (eds) Handbook of maize: its biology. Springer Science, New York, pp 405–417

Giles CD, Brown LK, Adu MO, Mezeli MM, Sandral GA, Simpson RJ, Wendler R, Shand CA, Menezes-Blackburn D, Darch T, Stutter MI, Lumsdon DG, Zhang H, Blackwell MSA, Wearing C, Cooper P, Haygarth PM, George TS (2017) Response-based selection of barley cultivars and legume species for complementarity: root morphology and exudation in relation to nutrient source. Plant Sci 255:12–28. https://doi.org/10.1016/j.plantsci.2016.11.002

Gill HS, Anoop S, Sethi SK, Behl RK (2004) Phosphorous uptake and use efficiency in different varieties of bread wheat (*Triticum aestivum* L). Arch Agron Soil Sci 50:563–572. https://doi.org/10.1080/03650340410001729708

Gong X-P, Liang X, Guo Y, Wu C-H, Zhao Y, Li X-H, Li S-S, Kong F-M (2015) Quantitative trait locus mapping for potassium use efficiency traits at the seedling stage in wheat under different nitrogen and phosphorus treatments. Crop Sci 55:2690–2700. https://doi.org/10.2135/cropsci2014.10.0711

Gulser C, Zharlygasov Z, Kızılkaya R, Kalimov N, Akca I, Zharlygasov Z (2019) The effect of NPK foliar fertilization on yield and macronutrient content of grain in wheat under Kostanai-Kazakhstan conditions. Eurasian J Soil Sci 8:275–281. https://doi.org/10.18393/ejss.575026

Guo Y, Kong FM, Feng XY, Zhao Y, Liang X, Wang Y (2012) QTL mapping for seedling traits in wheat grown under varying concentrations of N, P and K nutrients. Theor Appl Genet 124:851–865. https://doi.org/10.1007/s00122-011-1749-7

Guttieri MJ, Frels K, Regassa T, Waters BM, Baenziger PS (2017) Variation for nitrogen use efficiency traits in current and historical great plains hard winter wheat. Euphytica 213:87. https://doi.org/10.1007/s10681-017-1869-5

Haeder HE, Beringer H (1981) Influence of potassium nutrition and water stress on the content of abscisic acid in grains and flag leaves of wheat during grain-development. J Sci Food Agric 32: 552–556. https://doi.org/10.1002/jsfa.2740320605

Haileselassie B, Habte D, Haileselassie M, Gebremeskel G (2014) Effects of mineral nitrogen and phosphorus fertilizers on yield and nutrient utilization of bread wheat (*Triticum aestivum*) on the sandy soils of Hawzen District, northern Ethiopia. Agric Fish 3:189–198. https://doi.org/10.11648/j.aff.20140303.18

Hari-Gowthem G, Kaur S, Sekhon BS, Sharma P, Chhuneja P (2019) Genetic variation for phosphorus-use efficiency in diverse wheat germplasm. J Crop Improv 33(4):536–550. https://doi.org/10.1080/15427528.2019.1627633

Hartley TN, Thomas AS, Maathuis FJM (2020) A role for the OsHKT 2;1 sodium transporter in potassium use efficiency in rice. J Exp Bot 71(2):699–706. https://doi.org/10.1093/jxb/erz113

Hastings DF, Gutknecht J (1978) Potassium and turgor pressure in plants. J Theor Biol 73:363–366. https://doi.org/10.1016/0022-5193(78)90197-2

He K, Xu S, Zhang X, Li Y, Chang L, Wang Y, Shi Y, Cui T, Dong Y, Lan T, Liu X, Du Y, Zhang R, Liu J, Xue J (2020) Mining of candidate genes for nitrogen use efficiency in maize based on genome-wide association study. Mol Breed 40:83. https://doi.org/10.1007/s11032-020-01163-3

Heffner EL, Sorrells ME, Jannink JL (2009) Genomic selection for crop improvement. Crop Sci 49: 1–12. https://doi.org/10.2135/cropsci2008.08.0512

Hinsinger P (2001) Bioavailability of soil inorganic P in the rhizosphere as affected by root-induced chemical changes: a review. Plant Soil 237:173–195. https://doi.org/10.1023/A:1013351617532

Hirel B, Berlin P, Quillere I, Bourdoncle W, Attagnant C, Dellay C, Gouy A, Cadiou S, Retailliau C, Falque M, Gallais A (2001) Towards a better understanding of the genetic and physiological basis for nitrogen use efficiency in maize. Plant Physiol 125:1258–1270. https://doi.org/10.1104/pp.125.3.1258

Hirel B, Le Gouis J, Ney B, Gallais A (2007) The challenge of improving nitrogen use efficiency in crop plants: towards a more central role for genetic variability and quantitative genetics within integrated approaches. J Exp Bot 58(9):2369–2387. https://doi.org/10.1093/jxb/erm097

Hu B, Wang W, Ou S, Tang J, Li H, Che R, Zhang Z, Chai X, Wang H, Wang Y, Liang C, Liu L, Piao Z, Deng Q, Deng K, Xu C, Liang Y, Zhang L, Li L, Chu C (2015) Variation in *NRT1.1B* contributes to nitrate-use divergence between rice subspecies. Nat Genet 47(7):834–838. https://doi.org/10.1038/ng.3337

Hussain N, Khan MB, Ahmad R (2008) Influence of phosphorus application and sowing time on performance of wheat in calcareous soils. Int J Agric Biol 10:399–404

Hussain MI, Shah SH, Hussain S, Iqbal K (2002) Growth, yield and quality response of three wheat (*Triticum aestivum* L.) varieties to different levels of N, P and K. Int J Agric Biol 4:362–364

Jan MF, Khan AA, Liaqat W, Ahmad H, Rehan W (2018) Phenology, growth, yield and yield components of maize (*Zea mays* L.) hybrids to different levels of mineral potassium under semiarid climate. Agri Res Tech 15:0027–0031. https://doi.org/10.19080/ARTOAJ.2018.15.555943

Jeong K, Julia CC, Waters DLE, Pantoja O, Wissuwa M, Heuer S, Liu L, Rose TJ (2017) Remobilisation of phosphorus fractions in rice flag leaves during grain filling: implications for photosynthesis and grain yields. PLoS One 12(11):e0187521. https://doi.org/10.1371/journal.pone.0187521

Jia C, Wang F, Yuan J, Zhang Y, Zhao Z, Abulizi B et al (2020) Screening and comprehensive evaluation of rice (Oryza sativa L. subsp. japonica Kato) germplasm resources for nitrogen efficiency in Xinjiang, China. Plant Gen Resources 18(3):179–189. https://doi.org/10.1017/S1479262120000118

Jia Y, Yang X, Feng Y, Jilani G (2008) Differential response of root morphology to potassium deficient stress among rice genotypes varying in potassium efficiency. J Zhejiang Univ Sci 9:427–434. https://doi.org/10.1631/jzus.B0710636

Kichey T, Hirel B, Heumez E, Dubois F, Le Gouis J (2007) In winter wheat (*Triticum aestivum* L.), post-anthesis nitrogen uptake and remobilisation to the grain correlates with agronomic traits and nitrogen physiological markers. Field Crop Res 102:22–32. https://doi.org/10.1016/j.fcr.2007.01.002

Knyazikhin Y, Schull MA, Stenberg P, Mottus M, Rautiainen M, Yang Y, Marshak A, Carmona PL, Kaufmann RK, Lewis P et al (2013) Hyperspectral remote sensing of foliar nitrogen content. Proc Natl Acad Sci U S A 110:E185–E192. https://doi.org/10.1073/pnas.1210196109

Kraffczyk I, Trolldenier G, Beringer H (1984) Soluble root exudates of maize: influence of potassium supply and rhizosphere microorganisms. Soil Biol Biochem 16:315–322. https://doi.org/10.1016/0038-0717(84)90025-7

Lambers H, Cawthray GR, Giavalisco P, Kuo J, Laliberté E, Pearse SJ, Scheible WR, Stitt M, Teste F, Turner BL (2012) Proteaceae from severely phosphorus-impoverished soils extensively replace phospholipids with galactolipids and sulfolipids during leaf development to achieve a high photosynthetic phosphorus-use-efficiency. New Phytol 196:1098–1108. https://doi.org/10.1111/j.1469-8137.2012.04285.x

Lammerts van Bueren ET, Struik PC (2017) Diverse concepts of breeding for nitrogen use efficiency: a review. Agron Sustain Dev 37:50. https://doi.org/10.1007/s13593-017-0457-3

Lande R, Thompson R (1990) Efficiency of marker-assisted selection in the improvement of quantitative traits. Genetics 124:743–756. https://doi.org/10.1093/genetics/124.3.743

Laperche A, Brancourt-Hulmel M, Heumez E, Gardet O, Hanocq E, Devienne-Barret F, Le Gouis J (2007) Using genotype × nitrogen interaction variables to evaluate the QTL involved in wheat tolerance to nitrogen constraints. Theor Appl Genet 115:399–415. https://doi.org/10.1007/s00122-007-0575-4

Leghari A, Laghari GM, Ansari MA, Mirjat MA, Laghari UA, Leghari SJ, Laghari AH, Abbasi ZA (2016) Effect of NPK and boron on growth and yield of wheat variety TJ-83 at Tandojam soil. Adv Environ Biol 10:209–216

Li X, Mang M, Piepho HP, Melchinger A, Ludewig U (2021) Decline of seedling phosphorous use efficiency in the heterotic pool of flint maize breeding lines since the onset of hybrid breeding. J Agro Crop Sci 207:857–872. https://doi.org/10.1111/jac.12514

Li XP, Zhao XQ, He X, Zhao GY, Li B, Liu DC, Zhang AM, Zhang XY, Tong YP, Li ZS (2011) Haplotype analysis of the genes encoding glutamine synthetase plastic isoforms and their association with nitrogen-use- and yield-related traits in bread wheat. New Phytol 189(2): 449–458. https://doi.org/10.1111/j.1469-8137.2010.03490.x

Li HL, Zhao CJ, Huang WJ, Yang GJ (2013) Non-uniform vertical nitrogen distribution within plant canopy and its estimation by remote sensing: a review. Field Crops Res 142:75–84. https://doi.org/10.1016/j.fcr.2012.11.017

Lindsay WL, Vlek PLG, Chien SH (1989) Phosphate minerals. In: Dixon JB, Weed SB (eds) Minerals in soil environments, 2nd edn. Soil Science Society of America, Madison, WI, pp 1089–1130

Liu D, Shi Y (2013) Effects of different nitrogen fertilizer on quality and yield in winter wheat. Adv J Food Sci Technol 5(5):646–649. https://doi.org/10.19026/ajfst.5.3141

Liu HJ, Yan JB (2018) Crop genome-wide association study: a harvest of biological relevance. Plant J 97:8–18. https://doi.org/10.1111/tpj.14139

Liu ZY, Zhu CS, Jiang Y, Tian YL, Yu J, An HZ, Tang WJ, Sun J, Tang JP, Chen GM, Zhai HQ, Wang CM, Wan JM (2016) Association mapping and genetic dissection of nitrogen use efficiency-related traits in rice (*Oryza sativa* L.). Funct Integr Genomics 16:323–333. https://doi.org/10.1007/s10142-016-0486-z

Lynch JP (2007) Roots of the second green revolution. Aust J Bot 55:1–20. https://doi.org/10.1071/BT06118

Lyra DH, de Freitas ML, Galli G, Alves FC et al (2017) Multi-trait genomic prediction for nitrogen response indices in tropical maize hybrids. Mol Breed 37:80. https://doi.org/10.1007/s11032-017-0681-1

Ma L, Qing C, Frei U, Shen Y, Lübberstedt T (2020) Association mapping for root system architecture traits under two nitrogen conditions in germplasm enhancement of maize doubled haploid lines. Crop J 8(2):213–226. https://doi.org/10.1016/j.cj.2019.11.004

Mackay I, Powell W (2007) Methods for linkage disequilibrium mapping in crops. Trends Plant Sci 12:57–63. https://doi.org/10.1016/j.tplants.2006.12.001

Maharajan T, Roch GV, Ceasar SA (2021) Recent advancements of molecular breeding and functional genomics for improving nitrogen-, phosphorus-, and potassium-use efficiencies in wheat. Molecular breeding in wheat, maize and sorghum: Strategies for improving abiotic stress tolerance and yield. pp: 170–96. doi: https://doi.org/10.1079/9781789245431.0009

Maranville JW, Clark RB, Ross WW (1980) Nitrogen efficiency in grain sorghum. J Plant Nutr 2: 577–589. https://doi.org/10.1080/01904168009362800

Maranville JW, Madhavan S (2002) Physiological adaptations for nitrogen use efficiency in sorghum. Plant Soil 245:25–34

Marchi G, Silva VA, Guilherme LRG, Lima JM, Nogueira FD, Guimaraes PTG (2012) Potassium extractability from soils of Brazilian coffee regions. Biosci J 28:913–919

Marschner H, Römheld V, Horst WJ, Martin P (1986) Root-induced changes in the rhizosphere: importance for the mineral nutrition of plants. J Plant Nutr Soil Sci 149(4):441–456. https://doi.org/10.1002/jpln.19861490408

Martin A, Lee J, Kichey T, Gerentes D, Zivy M, Tatout C, Dubois F, Balliau T, Valot B, Davanture M, Terc'e-Laforgue T, Quiller'e I, Coque M, Gallais A, Gonzalez Moro M B,

Bethencourt L, Habash DZ, Lea PJ, Charcosset A, Perez P, Murigneux A, Sakakibara H, Edwards KL, Hirel B (2006) Two cytosolic glutamine synthetase isoforms of maize are specifically involved in the control of grain production. Plant Cell 18:3252–3274. https://doi.org/10.1105/tpc.106.042689

Masclaux-Daubresse C, Daniel-Vedele F, Dechorgnat J, Chardon F, Gaufichon L, Suzuki A (2010) Nitrogen uptake, assimilation and remobilization in plants: challenges for sustainable and productive agriculture. Ann Bot 105:1141–1157. https://doi.org/10.1093/aob/mcq028

Mastrodomenico AT, Bohn MO, Lipka AE, Below FE (2019) Genomic selection using maize ex-plant variety protection germplasm for the prediction of nitrogen-use traits. Crop Sci 59:212–220. https://doi.org/10.2135/cropsci2018.06.0398

McDonald G, Bovill W, Huang C, Lightfoot D (2013) Nutrient use efficiency. In: Kole C (ed) Genomics and breeding for climate-resilient crops. Springer, Berlin, Heidelberg, pp 333–393

Mimura T, Dietz KJ, Kaiser W, Schramm MJ, Kaiser G, Heber U (1990) Phosphate transport across biomembranes and cytosolic phosphate homeostasis in barley leaves. Planta 180:139–146. https://doi.org/10.1007/BF00193988

Miyamoto T, Ochiai K, Takeshita S, Matoh T (2012) Identification of quantitative trait loci associated with shoot sodium accumulation under low potassium conditions in rice plants. Soil Sci Plant Nutr 58:728–736. https://doi.org/10.1080/00380768.2012.745797

Mohammed NAA (2018) Exploring Rice genetic resources to improve nutrient use efficiency. PhD Thesis, University of York

Moll RH, Kampreth EJ, Jackson WA (1982) Analysis and interpretation of factors which contribute to efficiency of nitrogen utilisation. Agron J 74:562–564. https://doi.org/10.2134/agronj1982.00021962007400030037x

Monostori I, Szira F, Tondelli A, Arendas T, Gierczik K, Cattivelli L et al (2017) Genome-wide association study and genetic diversity analysis on nitrogen use efficiency in a central European winter wheat (*Triticum aestivum* L.) collection. PLoS One 12(12):e0189265. https://doi.org/10.1371/journal.pone.0189265.g004

Moose SP, Below FE (2009) Biotechnology approaches to improving maize nitrogen use efficiency. In: Kriz AL, Larkins BA (eds) Molecular genetic approaches to maize improvement. Biotechnology in agriculture and forestry, vol 63. Springer, Berlin Heidelberg, pp 65–77

Morosini JS, Mendonça LDF, Lyra DH, Galli G, Vidotti MS, Fritsche-Neto R (2017) Association mapping for traits related to N use efficiency in tropical maize lines under field conditions. Plant Soil 421:453–463. https://doi.org/10.1007/s11104-017-3479-3

Myles S, Peiffer J, Brown PJ, ErsozES ZZ et al (2009) Association mapping: critical considerations shift from genotyping to experimental design. Plant Cell 21:2194–2202. https://doi.org/10.1105/tpc.109.068437

Ng JMS, Han M, Beatty PH, Good A (2016) 'Genes, meet gases': the role of plant nutrition and genomics in addressing greenhouse gas emissions. In: Edwards D, Batley J (eds) Plant genomics and climate change. Springer, New York, NY, pp 149–172. https://doi.org/10.1007/978-1-4939-3536-9_7

Ning P, Li S, Yu P, Zhang Y, Li C (2013) Post-silking accumulation and partitioning of dry matter, nitrogen, phosphorous and potassium in maize varieties differing in leaf longevity. Field Crops Res 144:19–27. https://doi.org/10.1016/j.fcr.2013.01.020

Nisar A, Khan SU, Shah AH (2016) Screening and evaluation of wheat germplasm for phosphorus use efficiency. Iran J Sci Technol Trans Sci 40:201–207. https://doi.org/10.1007/s40995-016-0085-9

Ochoa IE, Blair MW, Lynch JP (2006) QTL analysis of adventitious root formation in common bean (*Phaseolus vulgaris* L.) under contrasting phosphorus availability. Crop Sci 46:1609–1621. https://doi.org/10.2135/cropsci2005.12-0446

Ortiz-Monasterio J, Manske G, Van Ginkel M (2001) Nitrogen and phosphorus use efficiency. In: Application of physiology in wheat breeding. CIMMYT, Mexico, pp 200–207

Osman AM, Struik PC, Lammerts van Bueren ET (2012) Perspectives to breed for improved baking quality for wheat varieties adapted to organic growing conditions. J Sci Food Agri 92:207–215. https://doi.org/10.1002/jsfa.4710

Pathan SI, Větrovský T, Giagnoni L, Dutta R, Baldrian P, Nannipieri P, Renella G (2018) Microbial expression profiles in the rhizosphere of two maize lines differing in N use efficiency. Plant Soil 433:401–413. https://doi.org/10.1007/s11104-018-3852-x

Pettersson S, Jensén P (1983) Variation among species and varieties in uptake and utilization of potassium. Plant Soil 72:231–237. https://doi.org/10.1007/BF02181962

Pingali PL (2012) Green revolution: impacts, limits, and the path ahead. Proc Natl Acad Sci 109(31):12302–12308. https://doi.org/10.1073/pnas.0912953109

Plett DC, Holtham LR, Okamoto M, Garnett TP (2018) Nitrate uptake and its regulation in relation to improving nitrogen use efficiency in cereals. Semin Cell Dev Biol 74:97–104. https://doi.org/10.1016/j.semcdb.2017.08.027

Poland J, Endelman J, Dawson J, Rutkoski J, Wu S, Manes Y, Dreisigacker S, Crossa J, Sánchez-Villeda H, Sorrells M, Jannink J-L (2012) Genomic selection in wheat breeding using genotyping-by-sequencing. Plant Genome 5(3):103–113. https://doi.org/10.3835/plantgenome2012.06.0006

Presterl I, Seitz G, Landbeck M, Theimt EM, Schmidt W, Geiger HH (2003) Improving nitrogen-use efficiency in European maize; estimation of quantitative genetic parameters. Crop Sci 43: 1259–1265. https://doi.org/10.2135/cropsci2003.1259

Pszczola M, Strabel T, Mulder H, Calus M (2012) Reliability of direct genomic values for animals with different relationships within and to the reference population. J Dairy Sci 95:389–400. https://doi.org/10.3168/jds.2011-4338

Raghuram N, Sharma N (2019) Improving crop nitrogen use efficiency. In: Moo-Young M (ed) Comprehensive biotechnology, vol 4. Elsevier, Pergamon, pp 211–220. https://doi.org/10.1016/B978-0-444-64046-8.00222-6

Rakotoson T, Dusserre J, Letourmy J, Frouin J, Ratsimiala IR, Rakotoarisoa NV et al (2021) Genome-wide association study of nitrogen use efficiency and agronomic traits in upland rice. Rice Sci 28(4):379–390. https://doi.org/10.1016/j.rsci.2021.05.008

Ranjan R, Yadav R (2019) Targeting nitrogen use efficiency for sustained production of cereal crops. J Plant Nutr 42(9):1086–1113. https://doi.org/10.1080/01904167.2019.1589497

Rao IS, Neeraja CN, Srikanth B, Subrahmanyam D, Swamy N, Rajesh K et al (2018) Identification of rice landraces with promising yield and the associated genomic regions under low nitrogen. Sci Rep 8:9200. https://doi.org/10.1038/s41598-018-27484-0

Rao IM, Terry N (1995) Leaf phosphate status, photosynthesis, and carbon partitioning in sugar beet. IV. Changes with time following increased supply of phosphate to low phosphate plants. Plant Physiol 107:1313–1321. https://doi.org/10.1104/pp.107.4.1313

Richardson AE, Lynch JP, Peter RR, Emmanuel D, Smith FA, Smith SE, Harvey PR, Ryan MH, Veneklaas EJ, Lambers H, Oberson A, Culvenor RA, Simpson RJ (2011) Plant and microbial strategies to improve phosphorous efficiency in agriculture. Plant Soil 349:121–156. https://doi.org/10.1007/s11104-011-0950-4

Robertsen CD, Hjortshøj RL, Janss LL (2019) Genomic selection in cereal breeding. Agronomy 9(2):95. https://doi.org/10.3390/agronomy9020095

Safdar LB, Andleeb T, Latif S, Umer MJ, Tang M, Li X, Liu S, Quraishi UM (2020) Genome-wide association study and QTL meta-analysis identified novel genomic loci controlling potassium use efficiency and agronomic traits in bread wheat. Front Plant Sci 11:70. https://doi.org/10.3389/fpls.2020.00070

Samal D, Kovar JL, Steingrobe B, Sadana US, Bhadoria PS, Claassen N (2010) Potassium uptake efficiency and dynamics in the rhizosphere of maize (*Zea mays* L.), wheat (*Triticum aestivum* L.), and sugar beet (*Beta vulgaris* L.) evaluated with a mechanistic model. Plant Soil 332:105–121. https://doi.org/10.1007/s11104-009-0277-6

Sanchez PA, Logan TJ (1992) Myths and science about the chemistry and fertility of soils in the tropics. Myths Sci Soils Tropics 29:35–46

Sanchez-Bragado R, Serret MD, Araus JL (2017) The nitrogen contribution of different plant parts to wheat grains: exploring genotype, water, and nitrogen effects. Front Plant Sci 7:1986. https://doi.org/10.3389/fpls.2016.01986

Sánchez-Calderón L, López-Bucio J, Chacón-López A, Gutiérrez-Ortega A, Hernández-Abreu E, Herrera-Estrella L (2006) Characterization of low phosphorus insensitive mutants reveals a crosstalk between low phosphorus-induced determinate root development and the activation of genes involved in the adaptation of Arabidopsis to phosphorus deficiency. Plant Physiol 140(3): 879–889. https://doi.org/10.1104/pp.105.073825

Sarkar D, Baishya LK (2017) Nutrient use efficiency. In: Naeem M, Ansari AA, Gill SS (eds) Essential plant nutrients uptake, use efficiency, and management. Springer, Cham, pp 119–146. https://doi.org/10.1007/978-3-319-58841-4

Schachtman DP, Shin R (2007) Nutrient sensing and signaling: NPKS. Annu Rev Plant Biol 58:47–69. https://doi.org/10.1146/annurev.arplant.58.032806.103750

Shabala S, Pottosin I (2014) Regulation of potassium transport in plants under hostile conditions: implications for abiotic and biotic stress tolerance. Physiol Plant 151(3):257–279. https://doi.org/10.1111/ppl.12165

Sharma N, Sinha VB, Kumar NAP, Subrahmanyam D, Neeraja CN, Kuchi S, Jha A, Prasad R, Sitaramam V, Rghuram N (2021) Nitrogen use efficiency phenotype and associated genes: roles of germination, flowering, root/shoot length and biomass. Front Plant Sci 11:1–20. https://doi.org/10.3389/fpls.2020.587464

Shrivastav P, Prasad M, Singh TB, Yadav A, Goyal D, Ali A, Dantu PK (2020) In: Naeem M, Ansari A, Gill S (eds) Role of nutrients in plant growth and development. Contaminants in agriculture. Springer International Publishing AG, Cham, pp 43–59

Silva AD, Bruno IP, Franzini VI, Nericlenes CM, Leticia B, Muraoka T (2016) Phosphorous uptake efficiency, root morphology and architecture in Brazilian wheat cultivars. J Radioanal Nucl Chem 307:1055–1063. https://doi.org/10.1007/s10967-015-4282-3

Sims JT, Simard RR, Joern BC (1998) Phosphorus loss in agricultural drainage: historical perspective and current research. J Environ Qual 27(2):277–293. https://doi.org/10.2134/jeq1998.00472425002700020006x

Soumya PR, Burridge AJ, Singh N. Batra R, Pandey R et al. (2021) Population structure and genome-wide association studies in bread wheat for phosphorus efficiency traits using 35K wheat Breeder's Affymetrix array. Sci Rep 11:7601. doi: https://doi.org/10.1038/s41598-021-87182-2

Spindel J, Begum H, Akdemir D, Virk P, Collard B, Redona E et al (2015) Genomic selection and association mapping in rice (*Oryza sativa*): effect of trait genetic architecture, training population composition, marker number and statistical model on accuracy of rice genomic selection in elite, tropical rice breeding lines. PLoS Genet 11:e1004982. https://doi.org/10.1371/journal.pgen.1004982

Srivastava RK, Singh RB, Pujarula VL, Bollam S, Pusuluri M, Chellapilla TS, Yadav RS, Gupta R (2020) Genome-wide association studies and genomic selection in pearl millet: advances and prospects. Front Genet 10:1389. https://doi.org/10.3389/fgene.2019.01389

Steingrobe B, Claassen N (2000) Potassium dynamics in the rhizosphere and K efficiency of crops. J Plant Nutr Soil Sci 163:101–106. https://doi.org/10.1002/(SICI)1522-2624(200002)163:1<101::AID-JPLN101>3.0.CO;2-J

Sun X, Ren W, Wang P, Chen F, Yuan L, Pan Q, Mi G (2020) Evaluation of maize root growth and genome-wide association studies of root traits in response to low nitrogen supply at seedling emergence. Crop J 9(4):794–804. https://doi.org/10.1016/j.cj.2020.09.011

Swamy HKM, Anila M, Kale RR et al (2019) Phenotypic and molecular characterization of rice germplasm lines and identification of novel source for low soil phosphorus tolerance in rice. Euphytica 215:118. https://doi.org/10.1007/s10681-019-2443-0

Tang W, Ye J, Yao X, Zhao P, Xuan W, Tian Y, Zhang Y, Xu S, An H, Chen G et al (2019) Genome-wide associated study identifies NAC42-activated nitrate transporter conferring high nitrogen use efficiency in rice. Nat Commun 10(1):5279. https://doi.org/10.1038/s41467-019-13187-1

Tiong J, Sharma N, Sampath R, MacKenzie N, Watanabe S, Metot C, Lu Z, Skinner W, Lu Y, Kridl J, Baumann U, Heuer S, Kaiser B, Okamoto M (2021) Improving nitrogen use efficiency through over expression of alanine aminotransferase in rice, wheat and barley. Front Plant Sci 12:628521. https://doi.org/10.3389/fpls.2021.628521

Tollenaar M, Lee E (2011) 2 strategies for enhancing grain yield in maize. Plant Breeding Rev 34: 37–82. https://doi.org/10.1002/9780470880579.ch2

Tong C, Shen L, Lv Y, Wang Z, Wang X et al (2014) Structural mapping: how to study the genetic architecture of a phenotypic trait through its formation mechanism. Brief Bioinform 15:43–53. https://doi.org/10.1093/bib/bbs067

van de Wiel CCM, van der Linden CG, Scholten OE (2016) Improving phosphorus use efficiency in agriculture: opportunities for breeding. Euphytica 207:1–22. https://doi.org/10.1007/s10681-015-1572-3

Vance CP, Uhde-Stone C, Allan DL (2003) Phosphorus acquisition and use: critical adaptations by plants for securing a non-renewable resource. New Phytol 157:423–447. https://doi.org/10.1046/j.1469-8137.2003.00695.x

Veneklaas EJ, Lambers H, Bragg J, Finnegan PM, Lovelock CE, Plaxton WC, Price CA, Scheible W-R, Shane MW, White PJ, Raven JA (2012) Opportunities for improving phosphorous use efficiency in crop plants. New Phytol 195(2):306–320. https://doi.org/10.1111/j.1469-8137.2012.04190.x

Véry AA, Nieves-Cordones M, Daly M, Khan I, Fizames C, Sentenac H (2014) Molecular biology of K+ transport across the plant cell membrane: what do we learn from comparison between plant species? J Plant Physiol 171:748–769. https://doi.org/10.1016/j.jplph.2014.01.011

Vijayalakshmi P, Vishnukiran T, Kumari BR, Srikanth B, Rao IS, Swamy KN, Surekha K, Sailaja N, Subbarao LV, Rao PR, Subrahmanyam D, Neeraja CN, Voleti SR (2015) Biochemical and physiological characterisation for nitrogen use efficiency in aromatic rice genotypes. Field Crop Res 179:132–143. https://doi.org/10.1016/j.fcr.2015.04.012

Violle C, Navas M-L, Vile D, Kazakou E, Fortunel C, Hummel I, Garnier E (2007) Let the concept of trait be functional! Oikos 116(5):882–892. https://doi.org/10.1111/j.0030-1299.2007.15559.x

Vitousek PM, Naylor R, Crews T, David MB, Drinkwater LE, Holland E, Johnes PJ, Katzenberger J, Martinelli LA, Matson PA et al (2009) Nutrient imbalances in agricultural development. Science 324(5934):1519–1520. https://doi.org/10.1126/science.1170261

Wan TE, Xue HE, TONG YP (2017) Transgenic approaches for improving use efficiency of nitrogen, phosphorus and potassium in crops. J Integr Agri 16(12):60345–60347. doi: https://doi.org/10.1016/S2095-3119(17)61709-X

Wang K, Cui K, Liu G, Luo X, Huang J, Nie L, Wei D, Peng S (2017a) Low straw phosphorous concentration is beneficial for high phosphorous use efficiency for grain production in rice recombinant inbred lines. Field Crop Res 203:65–73. https://doi.org/10.1016/j.fcr.2016.12.017

Wang K, Cui K, Liu G, Xie W, Yu H, Pan J, Huang J, Nie L, Shah F, Peng S (2014) Identification of quantitative trait loci for phosphorus use efficiency traits in rice using a high density SNP map. BMC Genet 15:155. https://doi.org/10.1186/s12863-014-0155-y

Wang LD, Liao H, Yan XL, Zhuang BC, Dong YS (2004) Genetic variability for root hair traits as related to phosphorus status in soybean. Plant Soil 261:77–84. http://www.jstor.org/stable/24124282

Wang Z, Ma Bao-Luo YX, Gao J, Sun J, Su Z, Yu S (2019a) Physiological basis of heterosis for nitrogen use efficiency of maize. Sci Rep 9:18708. https://doi.org/10.1038/s41598-019-54864-x

Wang H-Y, Shen Q-H, Zhou J-M, Wang J, Du C-W, Chen X-Q (2011) Plants use alternative strategies to utilize nonexchangeable potassium in minerals. Plant Soil 343:209–220. https://doi.org/10.1007/s11104-011-0726-x

Wang X, Yan X, Liao H (2010) Genetic improvement for phosphorus efficiency in soybean: a radical approach. Ann Bot 106:215–222. https://doi.org/10.1093/aob/mcq029

Wang QJ, Yuan Y, Liao Z, Jiang Y, Wang Q, Zhang L, Gao S, Wu F et al (2019b) Genome-wide association study of 13 traits in maize seedlings under low phosphorus stress. Plant Genome 12: 190039. https://doi.org/10.3835/plantgenome2019.06.0039

Wang SX, Zhu YL, Zhang DX, Shao H, Liu P et al (2017b) Genome-wide association study for grain yield and related traits in elite wheat varieties and advanced lines using SNP markers. PLoS One 12(11):e0188662. https://doi.org/10.1371/journal.pone.0188662

Welch LF, Flannery RL (1985) Potassium nutrition of corn. In: Munson RD (ed) Potassium in agriculture, ASA, CSSA and SSSA, Madison, WI, pp 647–664

White PJ (2013) Improving potassium acquisition and utilisation by crop plants. J Plant Nutr Soil Sci 176:305–316. https://doi.org/10.1002/jpln.201200121

White PJ, Bell MJ (2017) The genetics of potassium uptake and utilization in plants. In: Murrell TS, Mikkelsen RL (eds) Proceedings for the frontiers of potassium science conference, 25–-27 January 2017. International Plant Nutrition Institute, Peachtree Corners, Rome, pp 46–65. https://www.apni.net/k-frontiers/. Accessed 29 May 2020

White PJ, George TS, Gregory PJ, Bengough AG, Hallett PD, McKenzie BM (2013) Matching roots to their environment. Ann Bot 112:207–222. https://doi.org/10.1093/aob/mct123

White PJ, Hammond JP (eds) (2008) The ecophysiology of plant–phosphorus interactions. Springer, Dordrecht

White PJ, Karley AJ (2010) Potassium. In: Hell R, Mendel R-R (eds) Cell biology of metals and nutrients. Springer, Berlin, pp 199–224. https://doi.org/10.1007/978-3-642-10613-2_9

Wissuwa M, Kondo K, Fukuda T, Mori A, Rose MT, Pariasca-Tanaka J, Kretzschmar T, Haefele SM, Rose TJ (2015) Unmasking novel loci for internal phosphorus utilization efficiency in Rice germplasm through genome-wide association analysis. PLoS One 10(4):e0124215. https://doi.org/10.1371/journal.pone.0124215

Wuebbles DJ (2009) Nitrous oxide: no laughing matter. Science 326:56–57. https://doi.org/10.1126/science.1179571

Xu Y (2010) Molecular plant breeding. CIMMYT, Mexico

Xu Y, Wang R, Tong Y, Zhao H, Xie Q, Liu D et al (2014) Mapping QTLs for yield and nitrogen-related traits in wheat: influence of nitrogen and phosphorus fertilization on QTL expression. Theor Appl Genet 127(1):59–72. https://doi.org/10.1007/s00122-013-2201-y

Xu C, Zhang H, Sun J, Guo Z, Zou C et al (2018) Genome-wide association study dissects yield components associated with low-phosphorus stress tolerance in maize. Theor Appl Genet 131: 1699–1714. https://doi.org/10.1007/s00122-018-3108-4

Yan XL, Liao H, Beebe SE, Blair MW, Lynch JM (2004) QTL mapping of root hair and acid exudation traits and their relationship to phosphorus uptake in common bean. Plant Soil 265:17–29. https://doi.org/10.1007/s11104-005-0693-1

Yan H, Wenjia L, Liu X, Li G, Zhang S (2010) Comparison of rhizosphere impacts of wheat genotypes differing in phosphorous utilisation efficiency. Canadian J Pl Sci 90:311–317. https://doi.org/10.4141/CJPS09005

Yang DL, Jing RL, Chang XP, Li W (2007) Identification of quantitative trait loci and environmental interactions for accumulation and remobilization of water-soluble carbohydrates in wheat (*Triticum aestivum* L.) stems. Genetics 176:571–584. https://doi.org/10.1534/genetics.106.068361

Yang XE, Liu JX, Wang WM, Ye ZQ, Luo AC (2004) Potassium internal use efficiency relative to growth vigor, potassium distribution, and carbohydrate allocation in Rice genotypes. J Plant Nutr 27:837–852. https://doi.org/10.1081/PLN-120030674

Yang N, Lu Y, Yang X, Huang J, Zhou Y, F. Ali, et al. (2014) Genome wide association studies using a new nonparametric model reveal the genetic architecture of 17 agronomic traits in an enlarged maize association panel. PLoS Genet 10:E1004573. doi: https://doi.org/10.1371/journal.pgen.1004573

Yang X, Xia X, Zhang Z, Nong B, Zeng Y, Xiong F, Wu Y, Gao J, Deng G, Li D (2017) QTL mapping by whole genome resequencing and analysis of candidate genes for nitrogen use efficiency in rice. Front Plant Sci 8:1–10. https://doi.org/10.3389/fpls.2017.01634

Youngquist JB, Bramel-Cox P, Maranville JW (1992) Evaluation of alternative screening criteria for selecting nitrogen use efficient genotypes in sorghum. Crop Sci 32:1310–1313. https://doi.org/10.2135/cropsci1992.0011183X003200060002x

Zhang Z, Gao S, Chu C (2020) Improvement of nutrient use efficiency in rice: current toolbox and future perspectives. Theor Appl Genet 133:1365–1384. https://doi.org/10.1007/s00122-019-03527-6

Zhang S, Lovdahl L, Grip H, Tong Y, Yang X, Wang Q (2009) Effects of mulching and catch cropping on soil temperature, soil moisture and wheat yield on the loess plateau of China. Soil Tillage Res 102(1):78–86. https://doi.org/10.1016/j.still.2008.07.019

Zhang H, Uddin M, Zou C, Xie C, Xu Y, Li W (2014) Meta-analysis and candidate gene mining of low-phosphorus tolerance in maize. J Integr Plant Biol 56:262–270. https://doi.org/10.1111/jipb.12168

Zhao K, Tung CW, Eizenga GC, Wright MH, Ali ML, Price AH, Norton GJ et al (2011) Genome-wide association mapping reveals a rich genetic architecture of complex traits in Oryza sativa. Nat Commun 2:467. https://doi.org/10.1038/ncomms1467

Ziadi N, Brassard M, Belanger G, Claessens A, Tremblay N, Cambouris AN, Nolin MC, Parent L-E (2008) Chlorophyll measurements and nitrogen nutrition index for the evaluation of corn nitrogen status. Agron J 100(5):1264–1273. https://doi.org/10.2134/agronj2008.0016

Zörb C, Senbayram M, Peiter E (2014) Potassium in agriculture – status and perspectives. J Plant Physiol 171:656–669. https://doi.org/10.1016/j.jplph.2013.08.008

Doubled Haploidy: An Accelerated Breeding Tool for Stress Resilience Breeding in Cereals

6

Devender Sharma, N. C. Gahtyari, Parul Sharma, R. K. Khulbe, R. S. Pal, and Lakshmi Kant

Abstract

Doubled haploid (DH) technology in cereals has emerged as a promising tool for accelerating the development of completely homozygous lines in a much shorter time than conventional breeding methods. The rapid doubled haploid line production method reduces the breeding cycle's length and increases the genetic gain. In cereals, mainly conventional approaches, such as in vitro and *in planta* methods, have been employed to generate haploids that are subsequently converted to doubled haploids by spontaneous or induced chromosome doubling. The use of in vitro methods is limited owing to genotypic specificity and tissue culture dependence. *In planta* methods prevent the need for difficult tissue culture procedures and enhance the haploid recovery rate. Further, haploid induction through inducer lines can generate the haploid embryos when crossed with other plants. In cereals, the availability of commercially usable maternal and paternal haploid inducers at present is limited to maize. Mutations in the genomic region coding for phospholipases, *MTL, NLD* and *PLA1*, have been established to be responsible for the haploid induction in maize and other cereals such as rice. With the technology advancement, improved and highly efficient genetic engineering methods lead to the development of haploid inducers. *MTL* gene is targeted with CRISPR/Cas9 for haploid induction in maize and rice. The present updates on the genetic and molecular basis of doubled haploidy have opened new cereal breeding prospects for undertaking targeted and precise genetic improvement programmes in a shorter time. Doubled haploidy can be integrated with

D. Sharma (✉) · N. C. Gahtyari · R. K. Khulbe · R. S. Pal · L. Kant
ICAR-Vivekananda Parvatiya Krishi Anusandhan Sansthan, Almora, India
e-mail: devender.kumar1@icar.gov.in

P. Sharma
Punjab Agricultural University, Ludhiana, India

M. G. Mallikarjuna et al. (eds.), *Next-Generation Plant Breeding Approaches for Stress Resilience in Cereal Crops*, https://doi.org/10.1007/978-981-19-1445-4_6

marker-assisted breeding to fix favourable alleles of multiple traits in a single DH line. The technology can also be used for unlocking the genetic variability present in the unexploited germplasm/landraces, CMS line production and reverse breeding. DH breeding's future is promising due to the availability of robust DH production protocols and closer integration with marker-assisted technologies.

Keywords

Accelerated breeding · Cereals · CRISPR/Cas9 · Doubled haploidy · Haploid inducers · Stress resilience

6.1 Introduction

Genetic gains in the grain yield and quality are an ever-going effort in crop improvement to be realized by humankind. Genetic gain directly depends on the genetic variance, an important component of the breeder's equation. Since genetic variance is much higher for a doubled haploid (DH) population than an F_2 intercross-derived population, chances of finding a superior progeny are much higher in a DH population (Strigens et al. 2013). Then, genetic gains are also inversely proportional to the length of a breeding cycle. Hence, doubled haploids are very efficient in realizing genetic gain as homozygous inbreds or pure lines are derived in a single generation against the seven to eighth generation in conventional breeding (Chaikam and Prasanna 2012). Moreover, the DH population has only the additive component of genetic variance that responds to selection; hence, genetic gains for the trait of interest can be maximized using doubled haploids. However, one major disadvantage with DH is that they undergo only one cycle of recombination. Hence, chances to break undesirable linkages are much lower than conventional self–/sib pollination methods undergoing six to seven generations of recombinations, and consequently, many novel recombinants may not appear (Boerman et al. 2020).

Due to the potential benefits, haploid technology is gaining widespread recognition in crop improvement programmes of several crops, viz. cereals, oilseeds, legumes and other crops. Understanding the mechanisms of haploid induction and improving its efficiency are essential to reap the potential benefits in the future. QTLs and genomic regions involved in the induction of haploids are studied in due course of time. Significant QTLs, their genomic location and the cloning of the important genes have helped characterize genes involved in the haploid induction, which are now being targeted through genetic engineering approaches (Prigge et al. 2012a). Haploid induction is found to be a polygenic trait. Many important characterization studies have been performed in maize crops with genes like *MTL*, *ZmPLA1* and *NLD* being cloned and studied (Kelliher et al. 2017). These genes are now targeted through genetic engineering tools like CRISPR/Cas9 for artificial haploid induction (Hooghvorst and Nogués 2020).

Histological and cytological studies have helped to understand the molecular basics of haploid production, which revolves mainly around two hypotheses: (i) a 'pre-zygotic' hypothesis where haploid induction occurs due to failure of regular double fertilization wherein sperm nuclei fail to fertilize the egg nuclei (Swapna and Sarkar 2011) and (ii) a 'post-zygotic' hypothesis where although transient zygotes form, gradual loss of genome of one parent leads to the haploid formation (Kelliher et al. 2019). Identifying molecular reasons behind the selective chromosomal elimination of one of the parents is a vital research area that has been researched much in the past. Nevertheless, it requires continued attention for determining the underlying mechanisms and processes leading to haploid formation. Synchronization in the cell cycle, malfunctions in spindle separation, dysfunctional centromere and kinetochore activity, among others, are some of the identified reasons for haploid induction (Laurie and Bennett 1989; Mochida et al. 2004; Komeda et al. 2007; Ishii et al. 2016). Also, genes and QTLs conferring microspore embryogenesis capabilities in a few crops like wheat and barley have been identified (Maraschin et al. 2006; Sánchez-Díaz et al. 2013). Identification of a few of such mechanisms is greatly helping in the opening of novel research frontiers in the form of genetically engineered *CENH3* gene, which can significantly enhance the production capability of DHs (Britt and Kuppu 2016). Also, identification and knock-down of genes like *MATL*, *PLA1* and *NLD* in monocots and *CENH3* in dicots through genome editing tools like CRISPR-Cas can not only help in haploid generation but also in widening the genetic base of the inducer lines (Jacquier et al. 2020).

Haploids can be generated mainly by in vitro or in vivo methods. The in vitro technique generates the haploid either through androgenesis or gynogenesis, with the former being more prevalent. Androgenic haploids can be generated through anther culture or pollen culture methods and have been successfully developed in many crops like wheat, rice, oat, rye, triticale and rapeseed. (Mujeeb-Kazi et al. 2006; Bernardo 2009; Basu et al. 2010; Germanà 2011; Tripathy et al. 2019). Unfertilized ovaries or ovules are the starting point for developing gynogenic haploids that have been successfully demonstrated in many cereals, vegetables and horticultural crops (Tang et al. 2006; Agnieszka and Adela 2010). Despite that, gynogenesis is not the preferred choice among researchers due to its lower efficiency than the androgenic mode of haploid production (Rakha et al. 2012). In vitro haploid production requires sophisticated tissue culture techniques and suffers from disadvantages like higher frequency of albinos, genotype specificity, recalcitrancy, minor epigenetic modifications and need of hardening before field evaluation associated with DHs generated through in vitro methods (Niu et al. 2014; Tefera 2017). The *in planta* haploid production methods are gaining popularity, especially among the more extensive breeding programmes. Wide hybridization or inducer crosses lead to selective chromosomal elimination of the unwanted parent leading to haploid induction in a material of choice. Wheat haploids can be generated by methods like wide hybridization of wheat with *H. bulbosum* (Bulbosum method) (Zenketler and Straub 1979), maize (wheat × maize system) (Laurie and Bennett 1989) and *Imperata cylindrica* (Chaudhary et al. 2005) among which wheat × maize system is the most popular across the world. Low haploid induction rates with the existing technologies indicate the future research requirement for improving the haploid induction rates.

Novel methods like wheat × *Imperata cylindrica* have been reported to have potential advantages over the existing methods in the form of higher haploid induction rates and applicability in durum and triticale breeding as well and thus signify progress in the direction of improving DH production efficiency. The method, however, needs more validation before it can be adopted for large-scale DH production (Chaudhary et al. 2019; Sharma et al. 2019a; Sharma et al. 2019b). The primary beneficiary of in vivo haploid production has been maize, where inducer stocks are extensively used for doubled haploid production and thus integrated into the breeding program. Inducers stocks can be used either as male or female parents according to the need of the breeding programme. For example, while maternal haploid inducers are used for rapid derivation of fixed lines from heterozygous source populations, paternal haploids are useful in producing novel CMS lines (when CMS is present in the paternal inducer) (Ren et al. 2017). Nowadays, genome editing methods like ZFNs, TALENs and CRISPR/cas9 are frequently used in conjugation with haploids (Kelliher et al. 2019; Hooghvorst and Nogués 2020). DHs have a role in transgenics as well. It is much easier to identify haploids with transgene integration, which can subsequently be doubled to obtain individuals homozygous for the transgene. For example, wheat transgenic with barley drought tolerance *HVA1* gene was doubled to obtain a homozygous copy with stable transgenerational expression (Chauhan and Khurana 2011).

Derivation of doubled haploids from any material of choice is a stepwise process involving several operational and genetic challenges. Differentiating putative haploids from the diploids requires screening methods such as the 'inverted light technique,' differences in karyotype and DNA content, pigmentation differences in seed and/or roots and several other morphological and biochemical differences (Arumuganathan and Earle 1991a, b; Bains et al. 1998; Chaikam et al. 2015). *R1-nj* pigmentation marker on seed, also called '*Navajo*' phenotype, is most commonly used to separate putative haploids in maize. However, methods amenable to larger scalability like the recent NMR-based differentiation method based on threshold seed oil content can improve the overall efficiency of the doubled haploid production (Qu et al. 2021). Haploids generated through in vitro or in vivo methods must be doubled using antimitotic chemicals, and colchicine is currently the most widely used doubling agent. However, research initiatives are much needed to find alternatives to colchicine due to its toxicity to humans and the environment (Boerman et al. 2020). Certain dinitroanilines chemicals and herbicides like APM and pronamide were found to be effective doubling agents (Hooghvorst et al. 2020). However, their ability to replace colchicine is still questionable, and hence more research initiatives are required in times to come. Identifying genes and QTLs conferring spontaneous haploid genome doubling (SHGD) capabilities in maize crop is another novel frontier that can be an alternative to colchicine use. Orthologous genes and QTLs can be discovered in other crops, making the doubled haploid production labour-friendly and *in planta* in the true sense (Ren et al. 2017; Molenaar et al. 2019a, b; Boerman et al. 2020).

From the accidental discovery of haploids from the anthers of *Datura innoxia* (Guha and Maheshwari 1964), doubled haploid research has come a long way to the point where it is routinely being utilized in crop improvement of many crop species (Chaikam et al. 2019a, b; Patial et al. 2019). The complete homozygosity associated with doubled haploids makes them quite valuable for several kinds of genetic studies. The complete homozygosity nullifies the background noise associated with residual heterozygosity in other bi- or multi-parental populations, making it ideal for identifying G × E interactions (Boerman et al. 2020). Thus, doubled haploid mapping populations are frequently used to find QTLs and candidates genes for various traits in different crop species (Gahlaut et al. 2017; Jiao et al. 2020; Kwon et al. 2021). During the process of domestication, many useful alleles present in the progenitors failed to funnel into the domesticates and, consequently, are absent in the modern-day cultivars. These alleles are increasingly gaining relevance in the present climate change regime that has engendered a host of hitherto unknown abiotic and biotic stresses. DHs can be useful in the swift recovery of the useful alleles into the modern-day breeding material (Tefera 2017; Patial et al. 2019; Samantaray et al. 2021). These alleles may come in the form of synthetics in wheat or wide-hybridized populations in maize and other crops, which can be further utilized to improve the germplasm and breeding material (Mujeeb-Kazi et al. 2008; Strigens et al. 2013). DH has revolutionized modern plant breeding approaches by accelerating genetic gains (Fig. 6.1). DHs are also very useful in mutation breeding. Thousands of haploid cell lines can be screened in vitro against various biotic and abiotic stresses. Also, in the field conditions, complete homozygosity of the DHs makes mutant identification relatively easy, along with shortening the mutant development time compared to conventional mutation breeding (Rahman et al. 1995; Szarejko and Forster 2007). In addition to this, the potential of reverse breeding can be harnessed through DHs, which will improve with the advancement of technology (Dirks et al. 2009). Hence, due to the benefits associated with DHs, their role in hybrid/varietal development is ever increasing. Successful crop varieties and hybrids are increasingly being developed in wheat, rice, maize, barley and other crops (Sugimoto and Arai 2002; Thomas et al. 2003; Chaudhary et al. 2015; Patial et al. 2019). Thus, in the present chapter, we will discuss the basics and advances of DH production, associated challenges and their potential application in crop improvement in detail.

6.2 Haploid Induction in Cereals: Conventional to Transgenics

Haploid induction is generally achieved through three systems, in vitro, in vivo/*in planta* and transgenic. The mechanisms involved in these systems are discussed below:

6.2.1 In Vitro Methods of Haploid Induction

6.2.1.1 Androgenesis

It involves the production of haploids through anther culture or microspore culture.

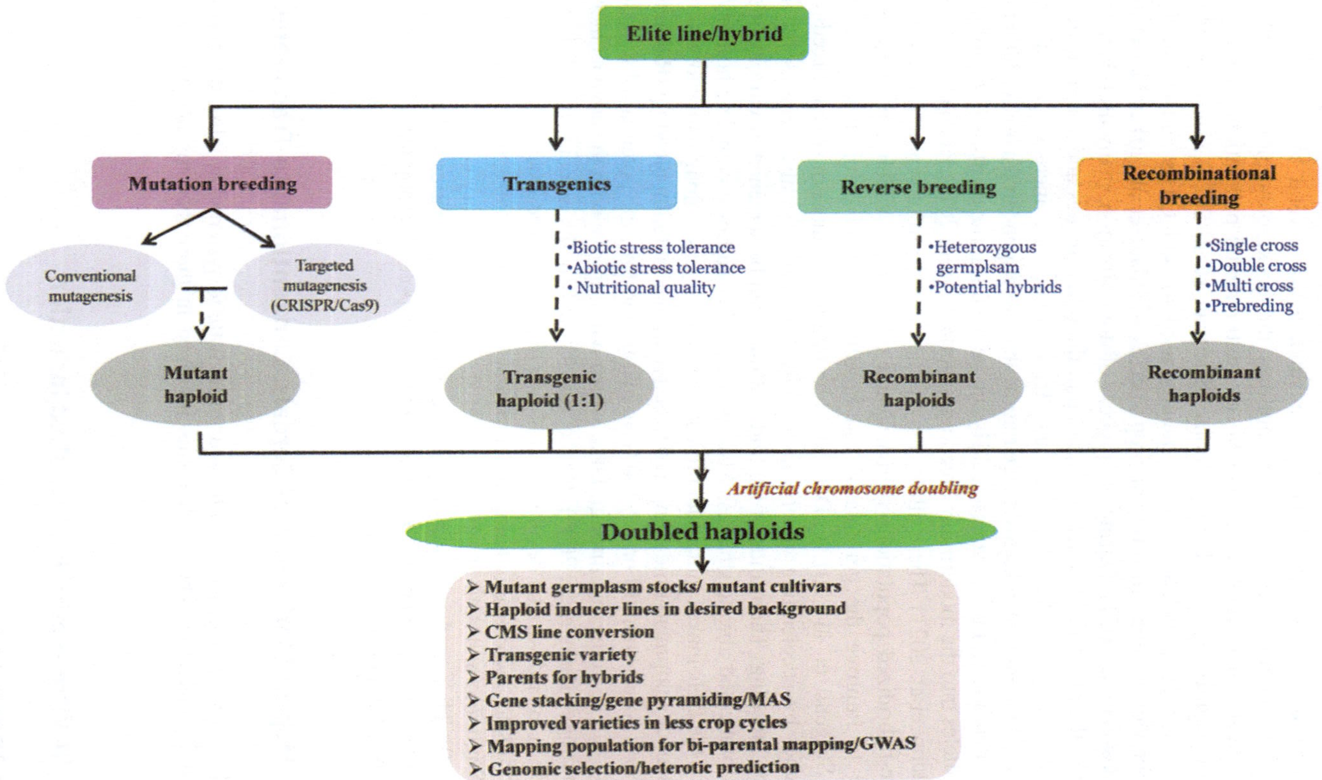

Fig. 6.1 Linking of doubled haploidy with modern plant breeding approaches

Anther Culture

The anther culture involves the isolation of haploid plants obtained by culturing immature anthers in artificial media under controlled conditions. The first spontaneous haploid was documented in *Datura stramonium* L. (Blakeslee et al. 1922), and the development of an efficient anther culture technique 40 years later by Guha and Maheshwari (1964, 1966) in *Datura innoxia* brought a revolution in haploid breeding. Since then, many researchers have adopted anther culture technique for haploid production in several crops, including those belonging to Solanaceae, Brassicaceae and Gramineae families. Anther culture is one of the important approaches of plant tissue culture used for shortening the breeding cycle through the induction of haploids (Kasha and Maluszynski 2003; El-Hennawy et al. 2011). Several factors, *viz.* genotype, growth conditions, stage of microspores, pretreatment, physiological status of the donor plants, the composition of culture media and cultural conditions affect the efficiency of haploid induction through anther culture (Lazar et al. 1985; Anderson et al. 1987; Zhou and Konzak 1989; Ekiz and Konzak 1991; Zheng and Konzak 1999; Kasha and Maluszynski 2003). Some crops like barley, rapeseed, tobacco and wheat are considered model plants to study the process of microspore embryogenesis because of their efficiency and high responsiveness to anther culture (Forster et al. 2007). This technique has been successfully adopted in many crops (Forster et al. 2007; Dunwell 2010; Germanà 2011), including wheat (Ouyang et al. 1973; Touraev et al. 1996; García-llamas et al. 2004; MujeebKazi et al. 2006), maize (Gaillard et al. 1991; Bernardo 2009), rice (Genovesi and Clint 1979; Zhahg-Yi et al. 2008; Tripathy et al. 2019), barley (Clapham 1973), oat (Kiviharju et al. 2005), rye and triticale (Basu et al. 2010) for production of doubled haploids. For large-scale production of doubled haploids, anther culture has been reported to be more economical than wide hybridization approaches (Snape et al. 1986). However, anther culture is associated with drawbacks such as species and genotype specificity, high frequency of albinism, low efficiency of DH production, segregation distortion and higher labour requirement and cost, among others (Redha and Talaat 2008; Dunwell 2010; Grauda et al. 2010), which make other techniques of haploid induction more attractive.

Microspore Culture

Microspore culture (Nitsch and Nitsch 1969), also known as pollen culture, is a technique where immature pollens at a specific stage (most preferably at the uninucleated stage) are removed from the anther under aseptic conditions and then cultured artificially on a nutrient medium. Isolated microspore culture is considered more advantageous than other commonly used techniques (Touraev et al. 2001). There are several advantages of microspore culture over anther culture: (i) haploid nature of microspores resulting in easy genetic manipulation, (ii) elimination of diploid tissues like anther wall and connective tissues of anthers in developing sporophyte and (iii) increase in the frequency of spontaneous chromosomes doubling (Castillo et al. 2009; Ferrie and Caswell 2011; Shariatpanahi and Ahmadi 2016). Microspore/pollen culture has been used in barley (Köhler and Wenzel 1985; Hoekstra et al. 1993), wheat (Hu et al. 1995; Scagliusi 2014) and oats (Sidhu and

Davies 2009). The major disadvantages of this technique are genotype dependence (Murovec and Bohanec 2012) and difficulty in identifying the appropriate stage of microspores for culturing.

6.2.1.2 Gynogenesis

Gynogenesis is the development of sporophytes by culturing female gametophytes, non-fertilized ovaries or ovules on nutritional media. Although haploids generated through gynogenesis are highly genetically stable and show a low frequency of albinos compared to androgenetic regenerants, gynogenesis is used only as a substitute when other in vitro techniques fail to produce haploids (Rakha et al. 2012). Gynogenesis offers an advantage over anther culture in the form of an increase in haploid green plant regeneration frequency. It is suitable for cultivars that show a low frequency of haploid induction (Zhou and Yang 1981). However, despite being more efficient than microspore and anther culture, gynogenesis is still a rare choice owing to the presence of only a few embryo sacs per ovary and is limited only to a few crop species. In cereals, gynogenesis-mediated haploid induction has been successfully used by the researchers in wheat, rice, barley, maize, onion, sugar beet, cucumber, cotton, potato, carrot, squash, gerbera, watermelon and sunflower (Zhou and Yang 1981; Zhu et al. 1981; Tang et al. 2006). However, its practical utility is mainly limited to onion and sugar beet.

6.2.2 In Planta Methods of Haploid Induction

Due to the shortcomings of in vitro methods of haploid induction, *in planta* or in vivo methods, including chromosome elimination-mediated doubled haploidy breeding and in vivo maternal haploid induction in maize, are the potential approaches that are being used by the researchers in various crops.

6.2.2.1 Chromosome Elimination-Mediated Doubled Haploidy Breeding

Genetic variation is a fundamental requirement of plant breeding, and wide hybridization is one of the proven tools for introducing variation through interspecific and intergeneric crossing programmes. However, pre-fertilization and post-fertilization barriers are major bottlenecks for wide hybridization that slow down the progress of crop improvement programmes. Pre-fertilization obstructions mainly include pollen stigma incompatibility or failure of fertilization due to short pollen tubes. Post-fertilization barriers include failure of zygote development and preferential chromosome elimination of one parent. With technological advancement, uniparental chromosome elimination has become a potential tool for haploid induction in many crops (Devaux and Pickering 2005). Different chromosome elimination approaches followed for haploid induction have been described under the following heads:

Bulbosum Method

Bulbosum method of haploid induction was first reported in *Hordeum vulgare* × *H. bulbosum* wide hybridization, where elimination of chromosomes of *H. bulbosum* ($2n = 2x = 14$) during earlier stages of embryogenesis resulted in the production of haploid embryos (Kasha and Kao 1970; Lange 1971). This chromosome elimination-mediated approach was utilized in breeding programmes to produce several haploids in different genotypes, keeping the technique's advantages over anther culture in barley. Likewise, Barclay (1975), for the first time, successfully utilized bulbosum method in wheat (Chinese Spring variety) to produce haploids (Barclay 1975; Zenketler and Straub 1979). However, in other wheat varieties, the effectiveness of bulbosum method was affected by the presence of dominant crossability inhibitor alleles, *viz.*, *Kr1, Kr2, Kr3* and *Kr4* located on 5B, 5A, 5D and 1A (Riley and Chapman 1967a, b; Krolow 1970; Zheng et al. 1992), which restricted its use in wheat and other breeding programmes.

Wheat × Maize System

For the induction of haploids, the wheat × maize system was first utilized and reported by Zenkteler and Nitzsche (1984), where haploid wheat embryos were found in crosses between hexaploid wheat and diploid maize. Subsequently, Laurie and Bennett (1986) confirmed these results through cytology and reported elimination of maize chromosomes after three to four mitotic cell divisions (Laurie and Bennett 1988a, b; Laurie and Bennett 1989) as the underlying mechanism. This method being genotype independent, as maize pollen showed unresponsiveness to dominant crossability inhibitors alleles (Sitch et al. 1985; Laurie and Bennett 1989), led to a revolution in wheat improvement programmes by allowing haploid induction in a wide range of wheat cultivars. Furthermore, several studies have shown higher haploid induction efficiency, broad genotypic specificity and lack of albinism in wheat × maize system compared to other grass species as pollen sources (Kisana et al. 1993; Inagaki and Mujeeb-Kazi 1995; Pratap et al. 2006; Wang et al. 1991). However, this system suffers from non-synchronization of wheat and maize flowering, higher cost and ineffectiveness in triticale × wheat and wheat × rye derivatives (Kishore et al. 2011).

Wheat × Imperata cylindrica System

Efforts to overcome the constraints of wheat × maize system led to the discovery of a superior and efficient alternative pollen source, *Imperata cylindrica*, a weedy perennial grass with chromosome number $2n = 20$ that has emerged as a competent source of haploid induction in wheat (Chaudhary et al. 2005; Pratap et al. 2005). *I. cylindrica* system has proven efficient over wheat × maize in hexaploid wheat (Chaudhary et al. 2005; Chaudhary et al. 2013; Chaudhary et al. 2019; Sharma et al. 2019a), durum wheat (Mahato and Chaudhary 2015; Mehta et al. 2020) and wheat, triticale and their derivatives (Kishore et al. 2011; Sharma et al. 2019b). The major advantages of wheat × *I. cylindrica* system over wheat × maize system include the non-requirement of greenhouse facilities, coincidence of flowering with that of wheat, insensitivity to crossability inhibitor genes, higher embryo formation frequency and effectiveness in triticale and rye hybridization (Chaudhary et al. 2005; Kishore et al. 2011; Chaudhary et al. 2013; Chaudhary et al. 2019; Sharma et al. 2019a; Sharma et al. 2019b). Further, in this approach, there is no endosperm formation due to the elimination of *I. cylindrica* chromosomes in the first zygotic division compared to maize, where elimination takes place in three or four mitotic divisions (Komeda et al. 2007). Apart from these systems, several other distantly related species have been used in wheat for developing haploids (Laurie 1989; Liu et al. 2014). Still, the wheat × maize system is being preferred and widely used over other haploid production systems for the commercial DH production in wheat.

6.2.2.2 In Vivo Haploid Induction in Maize

During the last three decades, in vivo haploid induction method has been extensively used in commercial maize breeding programmes. It all began with discovering and developing naturally occurring haploid lines in maize (Chase 1969). The major breakthrough in haploid breeding of maize was achieved with the study of Coe (1959), who used haploid inducer 'Stock 6' to produce haploids in maize. Haploid inducers are broadly categorized into paternal and maternal inducers. For deriving paternal haploids, haploid inducers are used as the female parent, while in maternal haploids, haploid inducers are used as pollen parents. The gene *ig1* (indeterminate gametophyte 1) was identified as a trigger for paternal haploid induction (Kermicle 1969; Evans 2007). However, in maize, the paternal haploid induction method is less preferred because of the low frequency of haploid induction (Kermicle 1994) and the inheritance of cytoplasm from the inducer line in haploids (Kermicle 1973).

In contrast, maternal haploids receive both the cytoplasm and nucleus from the same female parent, making maternal haploid induction the preferred method. The efficiency of this system improved with the development of temperate inducers (WS14, MHI, PHI, CAUHOI and RWS) with a higher haploid induction rate than Stock 6 (Wu et al. 2014), which have been widely used in maize breeding programmes. In addition, the drawbacks associated with temperature inducers have been overcome by developing tropically adapted haploid inducer lines (TAILs and CIM2GTAILs) with higher induction rates and superior agronomic performance (Prigge et al. 2012b; Chaikam et al. 2016). However, the main limitation of this method is the non-availability of good inducer lines to all the breeders due to proprietary terms.

6.2.2.3 Centromere-Mediated Genome Elimination Approach

Ravi and Chan (2010) defined an innovative technique of in vivo haploid induction through the modification of CENH3 that resulted in centromere-mediated genome elimination. They reported the induction of haploids through a cross between CENH3 mutant and WT *Arabidopsis thaliana*. This approach overcomes the main limitation of other in vivo haploid induction technologies, whether genotype or crop-specific. The easy generation of haploid seeds by crossing with an inducer, either male or female, is the key feature of this technology that makes it unique for haploid induction with a wide range of implications (Ravi et al. 2014). Researchers have utilized different approaches to modify CENH3 for haploid induction (Wang et al.

2019). In addition, various workers have carried out N-terminal tail editings in diverse species (Britt and Kuppu 2016). However, the practical utility of this approach is reported only in maize and rice (Kelliher et al. 2017; Kalinowska et al. 2019). In wheat, where two CENH3 genes (αCENH3 and βCENH3) are known (Yuan et al. 2015), no report of modification in CENH3 is available.

6.2.3 Transgenic and Genome Editing Methods

In the last decade, the introduction of genome editing technologies has modernized every facet of plant science. In genome editing technology, sequence-specific nucleases play a crucial role in generating double-stranded DNA breaks (DSBs) at targeted sites by identifying specific DNA sequences. At present, three classes of sequence-specific nucleases have been used in plants: zinc finger nucleases (ZFNs), transcription activator-like effector nucleases (TALENs) and the clustered regularly interspaced short palindromic repeats/CRISPR-associated protein (CRISPRs/Cas) system. The advantages of CRISPR over zinc finger nucleases (ZFNs) and transcription activator-like effector nucleases (TALENs) make it a tool of choice for gene editing in plants. Combining this technology with microspore technology, Bhowmik et al. (2018) developed an improved haploid mutagenesis system to alter the wheat genome genetically. It evidenced directed changes in an exogenous and two endogenous wheat genes, viz. *DsRed* gene, *TaLox2* and *TaUbiL1*, respectively. In another study, Liu et al. (2019) utilized CRISPR technology to produce double knockout mutants in wheat for *MTL/ZmPLA1/NLD* homolog in *qhir1*, a gene known for stimulating the induction of haploids in maize (Kelliher et al. 2017; Liu et al. 2017a, b). Knocking out of specific genes, *OsMATL* in rice (Yao et al. 2018) and *ZmDMP* in maize, in the presence of *MTL/ZmPLA/NLD* (Zhong et al. 2019) using CRISPR technology has been reported to increase haploid induction rates. Haploid induction editing technology (HI-Edit) given by Kelliher et al. (2019) showed the potential of CRISPR technology through standardization of one-step genome editing protocol for haploid induction, which opens new vistas of improvement of crops like maize (Kelliher et al. 2019).

6.3 Genetic Basis of Haploid Induction

The *in planta*/in vivo haploid induction in wheat is attributed to wide hybridization, whereas in maize, it is due to the paternal and maternal haploid inducers. In wide hybridization (wheat × *H. bulbosum*, wheat × maize and wheat × *I. cylindrica),* wheat is crossed as a female parent to *H. bulbosum*/maize/*I. cylindrica,* which acts as a pollen parent. Therefore *H. bulbosum*/maize/*I. cylindrica* acts as inducers for haploid induction system in wheat. However, the crossability of wheat to haploid inducers is under the control of crossability inhibitor genes *Kr1* and *Kr2* situated on 5B and 5A, respectively, which affects the pollen tube growth (Riley and Chapman 1967a, b; Snape et al. 1986). These cross-compatibility genes are most pronounced (dominant) in the western wheat lines. Further, the genetics of haploid induction in wheat is not well established and is based mainly on mitotic/meiotic irregularities. However, maize being a model crop to study various genetic principles, the genetics of haploid induction based on maternal haploid inducers is well established.

Maternal haploid induction in maize is under polygenic control and is governed by various minor genes. The haploid induction trait identified in 'Stock 6' was dominant and governed by few nuclear genes. The inducer × non-inducer crosses were attempted to generate the mapping populations, and several quantitative trait loci (QTLs) responsible for haploid induction in maize were identified. Deimling et al. (1997) showed that two QTLs for haploid induction were located on chromosomes 1 and 2, explaining 17.9% phenotypic variation. Barret et al. (2008) employed segregation-distortion-based QTL mapping and revealed that one major QTL for haploid induction trait was located on chromosome 1. Mapping results based on four mapping populations showed that two major QTLs, *qhir1* located on chromosome 1 (bin 1.04) and *qhir8* located on chromosome 9 (bin 9.01), explain about 66% and 20% phenotypic variation, respectively, for haploid induction (Prigge et al. 2012a). Based on the mapping results, bin 1.04, which harbours QTL *qhir1*, emerged as critical for conditioning the haploid induction trait. The fine-mapping efforts narrowed down the *qhir1* locus to 243 kb in length. Based on subsequent genome-wide association studies (GWAS), including inducers and non-inducers, *qhir1* was separated into two regions *qhir11* and *qhir11* + *qhir12*. However, further evaluation of these regions revealed that *qhir11* had a significant effect on haploid induction rate (HIR) (Nair et al. 2017). The gene in the *qhir11* region was cloned in three independent studies and named *MATRILINEAL* (*MTL*), *patatin-like phospholipase A (ZmPLA1)* and *NOT LIKE DAD* (*NLD*) genes (Gilles et al. 2017; Kelliher et al. 2017; Liu et al. 2017a, b). It was also reported that haploid induction is a post-zygotic activity due to a frame shift mutation in the *MTL/ZmPLA/NLD* gene. This gene encodes a phospholipase A specifically present in the sperm cell cytoplasm. The effect of the mutant allele has been identified and verified by various technologies such as backcrossing, fine mapping, genome sequencing, TALENs and CRISPR/cas9 and has revealed a wide variation for haploid induction rate (0.5–12.5%). The HIR variations revealed the quantitative nature of the maternal haploid induction, which is affected by various minor genes in the background. In a recent study, the second most important QTL, *qhir8,* has been cloned and identified as a *ZmDMP* gene encoding a membrane protein (Zhong et al. 2019). This mutant allele conditions very low HIR (0.1–0.3%) but dramatically increases HIR fivefold to sixfold in the presence of *MTL/ZmPLA/NLD.* The findings revealed that *MTL/ZmPLA/NLD* is the most critical gene for conditioning the HIR, and other minor genes could further enhance the effect of this allele on HIR.

Apart from the haploid inducer, the HIR is also influenced by the source germplasm/population from which haploids are derived (Kebede et al. 2011). Additionally, the haploid induction rate was found to vary among tropical germplasm as higher HIR was observed in single crosses and landraces than open-pollinated varieties and local cultivars (Prigge et al. 2011). Two QTLs, *qmhir1* (on chromosome 1) and *qmhir2* (on chromosome 3) that contribute to maternal genetics of haploid induction, were reported to account for phenotypic variations of 14.7% and 8.4%, respectively (Wu et al. 2014). In addition, the haploid induction rate is also affected by meteorological conditions, rainfall patterns and abiotic stresses (Geiger 2009; De La Fuente et al. 2018). Combining ability studies revealed significant general combining ability (Kebede et al. 2011) and specific combining ability effects (De La Fuente et al. 2018) on HIR in maize. De La Fuente et al. (2018) reported significant reciprocal effects on HIR in maize. The direction of crosses, that is, the choice of male and female in the parental cross for generating the source population, also affects the HIR in maize.

6.4 Molecular Basis/Mechanisms of Haploid Induction

The actual molecular mechanism underlying haploid induction still remains a mystery, and various researchers have proposed different hypotheses. The proposed hypothesis for chromosome elimination in wheat includes asynchronous cell cycling that leads to discrepancy in duration of essential mitotic processes (Gupta 1969), synthesis of multipolar spindles (Subrahmanyam and Kasha 1973), irregularity in the synthesis of nucleoprotein that results in the elimination of the most lagging chromosomes (Laurie and Bennett 1989), the disjunction of chromosomes at different stages of the cell cycle (Finch and Bennett 1982; Schwarzacher Robinson et al. 1987), loss of activity of centromeres (Finch 1983) and degeneration of foreign chromosomes by species- and genotype-specific nucleases (Davies 1974). Some additional concepts for uni-parental chromosome elimination involved lack or insufficiency of factors that are responsible for the movement of chromosomes during the cell cycle (Mochida et al. 2004), elimination dependent on mitosis (Gernand et al. 2005), development of additional nuclear extrusions that leads to genome elimination (Gernand et al. 2006), lack of functional kinetochores that results in abrupt segregation of chromosomes (Komeda et al. 2007) and many others. Nowadays, the most popular hypothesis for preferential elimination of chromosomes during wide hybridization is centromere-mediated genome elimination involving manipulation of *CENH3*, a variant of the centromere-specific histone H3 (Maruthachalam and Chan 2010; Ishii et al. 2016).

Various genes associated with microspore embryogenesis have been identified in different crops like wheat (Sánchez-Díaz et al. 2013), barley (Maraschin et al. 2006), rapeseed (Tsuwamoto et al. 2007; Joosen et al. 2007), hexaploid triticale (Żur et al. 2014), etc. For example, in wheat, Sánchez-Díaz et al. (2013) identified 14 genes, viz., *TaTPD1-like, TAA1b, GSTF2, GSTA2, TaNF-YA, TaAGL14, TaFLA26, CHI3, XIP-R, Tad1, WALI6, TaEXPB4, TaAGP31-LIKE* and *TaME1* related to different stages of microspore embryogenesis. However, genotypic specificity and differential responses of wheat cultivars for anther culture were explained based on shifting activated genes to earlier stages before rupturing of the exine (Sánchez-Díaz et al. 2013).

Additionally, the exact mechanism of maternal haploid induction in maize has yet to be conclusively elucidated. Two hypotheses, single fertilization and chromosomal elimination after regular double fertilization, have been propounded to explain the possible mechanism of maternal haploid induction in maize (Zhao et al. 2013). In regular fertilization, two sperm nuclei are formed in the single pollen tube of a normal pollen grain. One sperm nuclei fuse with the egg cell forming the diploid embryo, and the other sperm nuclei fuse with the two polar nuclei resulting in triploid endosperm formation. However, in the maternal haploid inducer, the pollen grains induce haploid induction in the maternal plant due to distorted fertilization events. One sperm nuclei of the inducer pollen grain fertilize the central cells forming the normal dividing cells and triploid endosperm. But the other sperm nuclei fail to fertilize the egg cells of the female gametophyte. As a result, the normal central dividing cells induce the unfertilized haploid egg cells to form the haploid embryo (Chase 1969; Swapna and Sarkar 2011). Such single fertilization events may be attributed to the defects in the pollen grain or single sperm cells in the pollen grain, resulting in the single fertilization event and ultimately a haploid embryo. In an in vitro study of germinating pollen grains from the inducer and non-inducers, the pollen grains of the inducers were observed to have two different pollen tubes growing at a variable rate, which was related to haploid inducibility (Pogna and Marzetti 1977).

Other reproductive abnormalities recorded in the haploid induction include high frequency (6.3%) of morphologically different sperm nuclei (Bylich and Chalyk 1996); microsporocyte with aneuploidy (Qiu et al. 2014; Li et al. 2017); and the poor competitive ability of the inducer pollen compared to non-inducer pollen attributed to delayed germination (Xu et al. 2013). Higher levels of heterofertilization (when the egg and central cell are fertilized by sperm cells of two different pollen grains) (Sarkar and Coe Jr 1971), embryo abortion and single fertilized ovule after fertilization upon using inducer pollen could be the result of the defects in the sperm nuclei or pollen grains. These reproductive anomalies point towards a single fertilization event instead of a double fertilization event. Using advanced microscopy, single fertilized ovules were identified after pollination with the haploid inducer (Swapna and Sarkar 2011; Tian et al. 2018), presenting strong evidence for a single fertilization event. Many studies report that phenotypic abnormalities, such as embryo abortion/endosperm abortion, are associated with haploid induction and haploid inducibility (Dong et al. 2013; Xu et al. 2013), which is attributed to abnormalities in the double fertilization process.

The second proposed hypothesis involves the loss of the inducer chromosomes after a normal double fertilization event. The substantial evidence that supported this hypothesis is the occurrence of inducer chromosomal segments in the maternal haploids and derived doubled haploid lines (Fischer 2004; Li et al. 2009; Qiu et al. 2014). This observation corresponds to haploid induction, a post-zygotic activity that eliminates the inducer chromosomes from the developing embryo. The evidence involves the transfer of physiological, phenological and genetic markers to the haploids from the inducers. Haploids with weak anthocyanin expression and high oil content were observed when inducers were equipped with seed anthocyanin and high oil markers, respectively (Li et al. 2019). B chromosomes were observed in low proportions when the inducer was equipped with a cytogenetic marker like the B chromosome (Zhao et al. 2013).

Further mosaic endosperms were observed in sweet corn when sweet corn with shrunken endosperm was pollinated with a haploid inducer with a normal endosperm, indicating the loss of inducer chromosomes (Zhang et al. 2008; Qiu et al. 2014). Other anomalies involve mixoploidy, aneuploidy, lagging chromosomes and micronuclei in the mitotic cells of developing embryo/endosperm (Wedzony et al. 2002; Zhang et al. 2008; Qiu et al. 2014; Li et al. 2017). Tian et al. (2018) reported that both single fertilization and genome elimination are responsible for the haploid induction in maize. The evidence available so far indicates that multiple mechanisms are responsible for maternal haploid induction.

6.5 Haploid Identification and Verification

Haploid identification is the second most crucial step after haploid induction in wheat and maize. Different techniques and methods are employed for the differentiation of haploid embryos/seeds from diploids. The haploid identification and verification techniques commonly used in wheat are discussed below:

6.5.1 Haploid Identification in Wheat

6.5.1.1 Inverted Light Technique

The 'inverted light technique' is the most common, simple and effective method to identify haploid embryo-carrying seeds in wheat × maize and wheat × *Imperata cylindrica* hybrids. The pollinated spikes of wheat are harvested after 15–18 days of pollination, and haploid embryos carrying immature seeds (known as pseudoseeds) are identified before dissection for embryo rescue. First of all, normal and haploid embryo-carrying pseudoseeds are differentiated by using morphological markers, i.e. the absence of endosperm in hybrid seeds as a morphological marker. Subsequently, after differentiating the selfed seeds from hybrid seeds, haploid embryo-carrying seeds are identified using the inverted light technique by placing a light source above the pseudoseeds that allows the visibility of embryos within them when observed underneath (Bains et al. 1998).

6.5.1.2 Botanical Features

Haploid identification based on morphological appearance is an indirect means of selection wherein plant height, length of guard cells, floral biology and fertility differentiate haploids from normal diploid plants. The haploids show poor plant vigour with short height, small guard cell length and degenerated anthers that are mostly sterile or show significantly reduced fertility. In contrast, diploid plants are invariably characterized by normal plant height, long guard cell length, normal flower and pollen development (Hua et al. 2008; Zhang et al. 2014).

6.5.1.3 Cytology

Chromosome counting at specific stages of mitotic or meiotic cell division is the most common and effective method of ploidy level identification. Between two cell divisions, mitotic cell division is considered the best for counting chromosomes that can be performed easily using root tips or other meristematic tissues (Maluszynska 2003).

6.5.1.4 Flow Cytometry

Flow cytometry is one of the most reliable methods that have been utilized for determining the nuclear DNA content of plants (Galbraith et al. 1983; Ochatt 2008). The main advantages of the system involve simplicity, determination during early developmental phases, permitting the identification of mixoploid regenerants, suitable for any nuclei-carrying tissue and quick method that make this method a suitable option (Arumuganathan and Earle 1991a, b).

6.5.1.5 Haploid Identification in Maize

In maize, the induction crosses generally result in 5–15% putative haploids using maternal inducers. The rest of the seeds are of no use in DH production. Therefore, identifying putative haploids from the induction crosses is a critical and labour-intensive step that needs significant time. Haploids can be separated from the diploids at the seed stage, seedling stage and adult plant stage. Various genetic markers are integrated into the inducers, and their expression assists in the haploid identification. Direct methods include cytogenetic techniques, chromosome counting (Couto et al. 2013) and flow cytometry (Bohenac 2003) for assessing the DNA content and molecular markers. Chromosome counting requires expertise and is a time-consuming process.

In contrast, flow cytometry requires costly initial setup and expertise. Various workers used the SSR molecular markers (Belicuas et al. 2007; Battistelli et al. 2013; Couto et al. 2013) to identify haploids from diploids and also distinguished homozygous, haploids and diploids for induction crosses (Ribeiro et al. 2018). However, molecular markers are an efficient method for haploid identification but require skills, cost and special preparation. Some of the potential phenotypic and other seedling trait-based marker systems are discussed below.

6.5.1.6 Phenotypic Markers

The dominant genetic markers that express typical phenotype are incorporated in the inducer line to assist haploid identification. Generally, both maternal and paternal chromosome complements are expressed in diploids, whereas haploids inherit only the maternal complement. But the seeds of both haploid and diploid look alike, and their ploidy status cannot be assessed visually. Consequently, any dominantly expressing genetic marker system integrated into the haploid inducer is very useful in separating the haploid seeds from diploid seeds. *R1-nj* is the dominantly expressed anthocyanin-based marker integrated into all the available inducers that helps identify putative haploids (Nanda and Chase 1966; Chaikam and Prasanna 2012; Melchinger et al. 2013). The other genes responsible for the anthocyanin pigmentation pathway (*A1, A2, C1, C2, Bz1* and *Bz2* and *C1*) are required for the typical phenotypic expression and must be present in the haploid inducers (Coe 1994). The typical phenotype produced through *R1-nj* expression is known as the '*Navajo*' phenotype. It is characterized by purple/red pigmentation on the aleurone layer of the endosperm and scutellum of the embryo. An induction cross of the inducer with the source population results in four seed phenotypes: (i) hybrid of inducer and source germplasm in which purple colour pigmentation is present on both embryo and endosperm, (ii) putative haploids in which purple pigmentation is present only on the endosperm, (iii) no pigmentation due to inhibition due to *C1* anthocyanin inhibitory locus present in some tropical germplasm (Chaikam et al. 2015) and (iv) outcrossed/self with other/own pollen. To ultimately obtain 1000 putative haploid seeds, it would generally require approximately 10,000–20,000 seeds to be inspected for the presence of *Navajo* phenotype, which is a time- and labour-consuming process. To deal with this issue, high-throughput methods were developed to optimize mechanical sorting through fluorescence imaging (Boote et al. 2016) and multispectral and hyperspectral technologies (Wang et al. 2018).

Nevertheless, the *R1-nj* marker system is widely used in maize breeding programmes. Still, it is associated with practical problems—the typical crown pigmentation varies from a small spot on the crown to the whole crown. Even the intensity of the colour pigmentation varies from pale to deep (Prasanna 2012; Khulbe et al. 2019), which is also attributed to the kernel moisture content at the harvesting stage (Rotarenco et al. 2010). Besides, seasonal differences also occur; anthocyanin pigmentation inhibition is the major problem that ranges from partial inhibition to complete inhibition. Chaikam et al. (2015) reported that ~30% of elite tropical germplasm is associated with the anthocyanin inhibition genes, limiting the widespread use of *R1-nj* expression. Molecular markers were designed based on the sequence variation in the *C1-I* gene to predict the *R1-nj* inhibition pattern (Chaikam et al. 2015). Complete inhibition leads to a high frequency of false positives and false negatives (loss of putative haploids in diploid fraction) (Röber et al. 2005; Melchinger et al. 2014; Chaikam et al. 2016). These limitations could be overcome by integrating the other marker system in the *R1-nj* gene to enhance the haploid recovery. Chaikam et al. (2015) showed that gene-specific markers and single nucleotide variation in the *C1-I* gene could easily predict the inhibition pattern in the tropical germplasm present in the CIMMYT. Moreover, haploid sorting can be performed by manual sorting through visual selection, optical sorting and stereoscopic methods. Manual sorting is prone to human errors, whereas up to 85% and 100% accuracy was reported by using near-infrared stereoscopic (NIR) for haploids and diploids, respectively (Davrieux et al. 2010; Jones et al. 2012; Fox et al. 2013). But this method requires multiple scans, so the time to sort a large number of seeds can take even much longer than the visual sorting method.

Another marker system includes the high oil xenia effect. High oil trait is introduced in various inducers such as CAUHOI, UH600 and UH601 (Li et al. 2009; Melchinger et al. 2013). The derived haploid lines possess lower oil content compared to the diploids, but the oil content is also dependent on the oil content of the inducer and source population. The automation system developed for oil content-based classification of haploids based on nuclear magnetic resonance (NMR) was developed for the automated sorting of haploid seeds from diploid seeds (Rotarenco et al. 2007; Wang et al. 2016; Melchinger et al. 2018). Moreover, the oil content marker is not genotype-dependent, enabling its use in the tropical germplasm, landraces and wild relatives more efficiently in contrast to *R1-nj*-based haploid classification.

In the seedling stage, other marker systems were proposed for the classification of haploids and diploid seeds. The red root colour in the diploids is expressed dominantly. The haploid with white roots could be easily separated (Chaikam et al. 2016). Several inducers with both *R1-nj* and red root markers have been developed to enhance the recovery of haploids. It was shown that the red root phenotype complements the *R1-nj*-based sorting of the haploid seeds. The only associated limitation with these systems is germinating all the seedlings to observe the root colour, enhancing labour cost and time. But these systems result in the timely removal of the false-positive seedlings before taking them to the field. The absence of ligules controlled by three recessive genes, *lg1, lg2* and *lg3,* also facilitates identifying haploids from diploids. Haploid plants are characterized by erect leaf architecture and the absence of ligules (Prigge et al. 2012b; Couto et al. 2013). Another marker system, i.e. the purple sheath marker system, was introduced in some inducers to reduce the false-positive rate. The manipulated *Purple1* (*Pl1*) gene imparts sunlight-independent anthocyanin purple pigmentation to the aboveground tissues. But the major disadvantage of the system was the expression of the marker system in the later vegetative stage (Röber et al. 2005); for that, one has to give the chromosome doubling treatment, which also enhances the cost and time. The *Booster1* (*B1*) gene was also investigated a decade ago to impart sunlight-dependent purple pigmentation to the aboveground tissues (Coe 1994). Both *Pl1* and *B1* genes were utilized to augment the *R1-nj* system (Rotarenco et al. 2010). But environmental factors, growing up to the vegetative stage, sunlight intensity and temperature limit their widespread use in haploid identification. So this system was significantly less in use in the maize DH breeding programmes. Some transgenic marker systems based on 35-S-derived engineered green fluorescent protein (*GFP*) (Zhu et al. 1999; Yu and Birchler 2016) and BAST herbicide resistance gene (Geiger et al. 1994) were integrated into some haploid inducer lines. Like the red root and purple sheath marker, the transgenic marker system also involves a higher cost and is time-consuming. Additionally, transgenics are regulated under the legal framework in many countries, which limits their widespread use.

6.5.1.7 Haploid Identification Based on the Vegetative Difference Among Haploid and Diploids

Haploid and diploids have ploidy differences which manifest in their vegetative performance. Such natural phenotypic variations could be exploited for the classification of haploid seedlings from diploids. The seed weight is less in the haploids than the diploids at the seed stage, but this also overlaps with the populations. In the seedling stage, radical length (Rotarenco et al. 2010), radical colour (Chaikam et al. 2016), coleoptile length, plant vigour (Battistelli et al. 2013), no. of seminal roots, plume length and other seedling traits (Chaikam et al. 2015) vary among haploid and diploids. The haploids possess lower trait values compared to the diploids (Chaikam et al. 2017). A significant difference in the stomatal sizes among haploid and diploids was also reported (Choe et al. 2012). The seedling traits also help in sorting the haploid seedlings from diploid seedlings, but this also necessitates the germination of all seedlings. These seedlings' markers may complement the *R1-nj*-based haploid identification and reduce the false positives. Recently, flow cytometry has also been used to classify haploids from diploids following the chromosome doubling (Molenaar et al. 2019a, b).

Moreover, the vegetative and reproductive differences among haploids and diploids exist, which also assist in identifying haploids and diploids in the fields. Haploids are generally characterized by poor vigour, erect and narrow leaf, pale/light green leaf colour, poor pollen production, lack of pollen production and less seed production (Chase 1969; Liu et al. 2017a, b; Wu et al. 2017). Although these are not reliable measures, they remove the false positives in the vegetative stage, leading to cost- and time-saving.

There are many methods for differentiating haploids from diploids that would enhance the accuracy of haploid identification. But mainly *R1-nj*, high oil content and red root marker in which automation in the high oil content marker and removal of diploids based on *R1-nj* and red root marker system enhances the accuracy by reducing the number of false positives. In CIMMYT, earlier sorting was based on *R1-nj* alone. Using a combination of *R1-nj* with red root marker and seedling traits for haploid identification has reduced false positives from 15–40% to less than 5%. The accuracy in the haploid identification saves resources significantly and accelerates the DH line production efficiency.

6.6 Development of Stress-Tolerant Genetic Stocks Through Doubled Haploidy

6.6.1 Doubled Haploids Against Biotic and Abiotic Stresses

Haploids have been successfully utilized in screening against biotic and abiotic stresses in in vitro and in vivo conditions in various crop species (Fig. 6.2). Screening thousands of haploid cell lines under artificially created abiotic or biotic stresses is an alternative or good exercise before evaluation under the glasshouse/field conditions. A large number of individuals can be screened in vitro, which is not possible otherwise (Rahman et al. 1995). Salt-tolerant DH lines have been derived in rice and barley through in vitro selection in anther culture under varying NaCl concentrations (Ye et al. 1987; Lee et al. 2003).

The second advantage of DHs is their ability to work in conjugation with recombination and mutation breeding (Fig. 6.1). Hybridization of desirable parents leads to recombinants, which take 6–7 generations of selfing to become stably homozygous. DHs come in handy in fixing the rare recombinant in a single generation. F_1 hybrids can be irradiated, and anthers can be used to produce DHs, which are

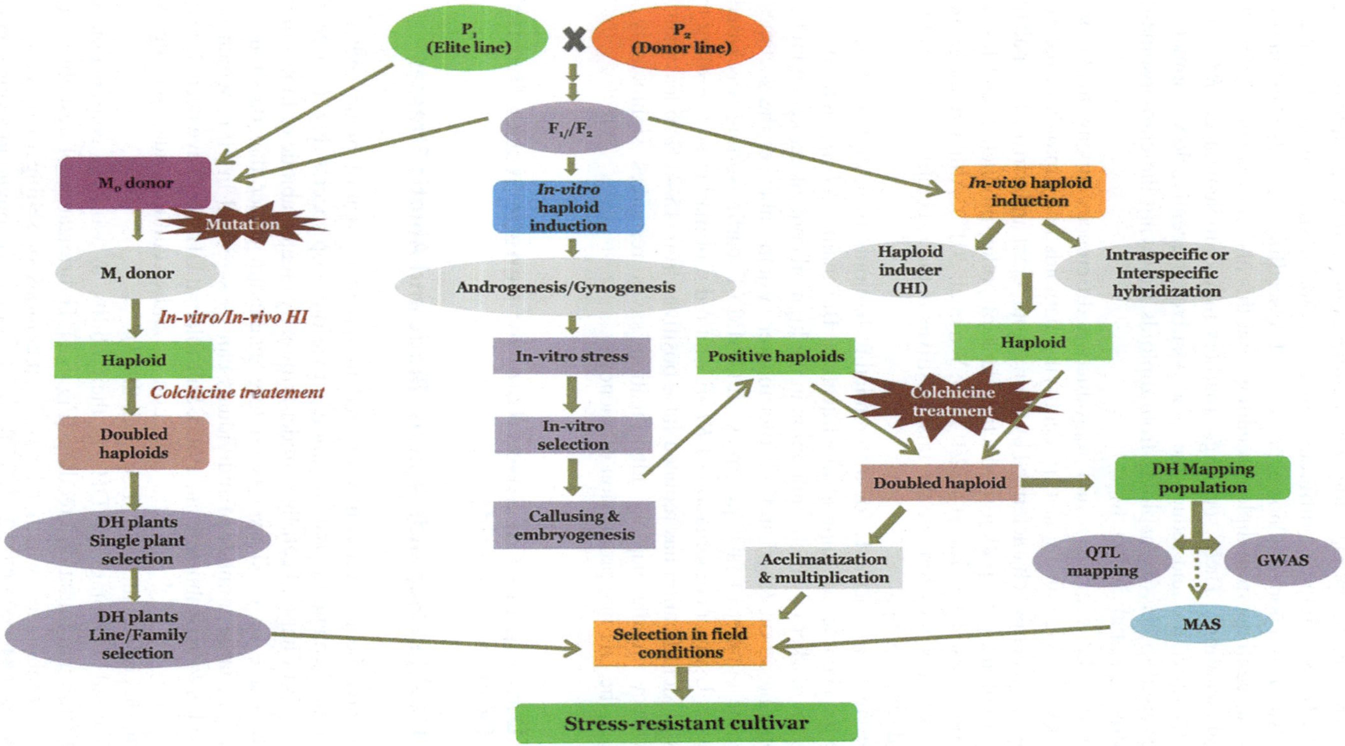

Fig. 6.2 Application of doubled haploidy in stress tolerance breeding of cereals

mutated and screened for desired traits like disease resistance, earliness and grain yield. In a study in wheat, 250–300 F_1 plants each of five potential crosses were irradiated with 150 Gy gamma irradiation to induce mutations followed by the production of DHs. Subsequent screening of DHs under moisture-deficit conditions led to identifying nine DH lines, two of which outperformed existing checks in grain yield and other agronomic traits in multilocation field trials (Khan et al. 2001). In maize crop, DHs have been used to generate hybrids, many of which perform well under drought conditions. The best hybrid recorded a yield advantage of 44% over the best available hybrid checked in eastern African regions of Uganda and Tanzania (Sserumaga et al. 2018). Moreover, such hybrids were found stable across different drought stress environments.

The third advantage of the haploid cells is that microspores, microspore-derived embryoids and haploid protoplast can be used to induce mutation, which can be further selected for the trait of interest. DHs are ideal for mutation detection as false-positive phenotypes are much less identified due to complete homozygosity than the lines developed by selfing with residual heterozygosity. As most desirable traits are quantitative, conventional mutation breeding needs M_3 generation; however, M_1 plants can be used as donors for DH production where selection can be exercised more efficiently due to their level of homozygosity (Szarejko and Forster 2007). Mutant lines surpassing the original cultivar for grain yield have been identified in barley using the technique where the anther and microspore culture of the M_1 plants was utilized to develop haploids. Thus, DH technology can be combined with mutation breeding to achieve much quicker genetic gains than conventional mutagenesis programmes. Microspores or their derived lines are extensively being utilized in *Brassica* spp. for generating useful mutant improved in quality (higher oleic acid, lower linoleic acid), lower in antinutritional factors (erucic acid, glucosinolates) and having enhanced disease and cold tolerance (Barro et al. 2001, 2002; McClinchey and Kott 2008). Likewise, the technique has great potential in cereal crops as well. Microspore-derived lines tolerant to herbicide (cyhalofop-butyl) have been derived in rice crop (Bae et al. 2002). Similarly, isolated microspore or derived lines have been mutated to obtain lines improved in morphological traits, e.g. in barley using NaN_3 (Castillo et al. 2001) and in rice using gamma rays (Kim et al. 2003).

6.6.2 Doubled Haploids in QTL Detection

Immortal or permanent mapping populations due to their near to complete homozygosity are best suited for associating a marker with a trait of interest. Conventional immortal populations like recombinant inbred lines (RILs), backcross inbred lines (BILs) or near-isogenic lines (NILs) require a higher number of generations (6–7) to become homozygous enough to be effectively utilized in QTL mapping. On the other hand, DHs can achieve homozygosity in one or two generations and have the added advantage of being completely homozygous, making them highly attractive for QTL and GWAS studies. The complete homozygosity reduces the variation caused by genetic background and thus doesn't confound with other studied G × E interactions. But one major disadvantage with DH is that they undergo only one cycle of recombination. Hence, many novel recombinations or chances to break undesirable linkages are much lower than conventional self–/sib-pollination methods undergoing six to seven generations of recombination. Several genetic studies in crop species like wheat, barley, rice, maize and other crops have utilized DHs to locate a QTL genomic position. DHs have helped in detecting QTLs for traits ranging from disease resistance (rusts in wheat), insect resistance (brown plant hopper resistance in rice), abiotic stresses (salt tolerance, cadmium tolerance and drought tolerance), quality traits (bread quality in wheat and malting in barley), yield and yield-attributing traits in many crop species. The QTLs identified for different traits in major cereal crops are summarized in Table 6.1.

6.6.3 DH in Varietal Development

DH is an important technology in the breeder's toolbox for developing a variety and improving the genetic base of a population. Its ability to significantly reduce the time for developing a completely homozygous inbred or pure line gives it a significant edge over the conventional breeding methods directed towards a similar goal. Additionally, it can be convenient to introduce wild genes from progenitors or wild relatives, as the technique involves rescuing the embryo and doubling the haploid plant (Patial et al. 2019). Apart from the time-saving, it saves on resources due to higher selection efficiency for a favourable homozygous progeny, which is $(1/2)^n$ compared to a similar bi-parental progeny segregating for $(1/4)^n$ for '*n*' segregating loci. Furthermore, paternal haploids are very useful in converting any inbred of choice into a male sterile background in just two generations, which can be utilized in maize in the presence of the *ig1* gene (*indeterminate gametophyte 1*) system (Evans 2007; Ravi et al. 2014). DHs are also an integral part of the reverse breeding technique, which aims to resynthesize inbreds from a successful hybrid by inhibiting the meiotic crossovers in F_1. The complementary DHs thus generated can be crossed to reconstitute the original hybrid (Dirks et al. 2009). Due to these advantages, DHs are increasingly being used in varietal development processes, especially in the private sector, which retains the parentage details as proprietary. Similarly, many private companies like Pioneer and others extensively utilize DH technology to develop newer corn hybrids (Rajcan et al. 2011). A detailed tabulation of successful varieties bred through DHs is presented in Table 6.2.

6.7 Genetic Challenges with Haploids

DH can be created using in vitro or in vivo techniques. However, limitations exist in both methods at each DH production step, limiting the number of obtained DH plants. In the in vitro process, a DH plantlet generation has to overcome several obstacles like low frequency of callus induction and plantlet regeneration, high

Table 6.1 The tabulation of various DH population-based QTL mapping studies employed to detect the genomic regions associated with stress resilience in cereals

Crop	Parents	Population size	Traits associated	Chromosomes with stable/major QTLs	PVE	Reference
Barley	Rec × Dom	87	Cd tolerance	2H and 5H	38.6–47.2%	Derakhshani et al. (2020)
	Clipper × Sahara 3771	146	Grain yield	2H and 6H	12–14%	Vafadar Shamasbi et al. (2017)
	Clipper × Sahara 3771	146	TKW	2H	69%	Vafadar Shamasbi et al. (2017)
	Sofiara × Victoriana	100	Malting quality in drought	3H	13.8–37.4%	Kochevenko et al. (2018)
	ND24260 × Flagship	100	Stay green in heat and water stress	5H and 6H	13.7–17.4%	Gous et al. (2016)
Wheat	Excalibur × Kukri	192	Drought tolerance	5A and 7A	16.6–20.4%	Gahlaut et al. (2017)
	Chinese spring (CS) × SQ1	96	Drought tolerance	4B, 5A, 5B, 6B, 7A	13–36.0%	Dashti et al. (2007)
	Opata × SH223	140	Chlorophyll content and kinetics in drought	1B	10.1%	Ilyas et al. (2014)
	Hanxuan 10 × Lumai 14	150	Root traits and grain yield in water stress	3B	9.1–13.8%	Liu et al. (2013)
	Excalibur × Kukri	212	Salt tolerance	2B	12.1%	Asif et al. (2018)
	Kariega × Avocet S	254	Stripe, leaf and stem rust	2B	17.7–45.7%	Prins et al. (2011)
	CA9613 × H1488	113	Stem strength	3A and 3B	10.6–16.6%	Hai et al. (2005)
	TA4152–60 × ND495	120	APR to leaf rust	3AL and 3BL	12.0–36.0%	Chu et al. (2009)
Rice	IR64 (indica) × Azucena (japonica)	135	BPH resistance	12	N.A.	Huang et al. (1997)
	93–11 (indica) × Milyang 352 (japonica)	117	Shoot branching	2 & 4	14.5–20.5%	Kwon et al. (2021)
		70		8	40.0–47.0%	Park et al. (2014)

(continued)

Table 6.1 (continued)

Crop	Parents	Population size	Traits associated	Chromosomes with stable/major QTLs	PVE	Reference
	Cheongcheong (indica) × Nagdong (japonica)		Panicle per plant and grain yield			
	R05F102 × IR69428	148	Iron and Zn	9 and 12	11.8–15.3%	Calayugan et al. (2020)
	Baiyeqiu (indica) × Maybelle (japonica)	251	Sheath blight resistance	1	8.9–13.2%	Xu et al. (2011)
	IR64 (indica) × Azucena (japonica)	131	Brown planthopper resistance	4	10.1–16.6%	Alam and Cohen (1998)
Maize	Qi319 × Chang 7–2	119	Haploid male fertility (anther emergence score)	3 and 5	10.6–11.3%	Jiao et al. (2020)
	Qi319 × Chang 7–2	119	Haploid male fertility (pollen production score)	1 and 5	14.3–16.1%	Jiao et al. (2020)

Table 6.2 The successful varieties released in cereals using DH approaches

Crop	Cultivar Name	Country	Method	Reference
Wheat	Florin	France	Anther culture and colchicine	Buyser et al. (1987)
	Hua Pei 1, Jinghua 1, Yunhua 1 and 2, Jing-Hua 1, 3 and 5	China	Anther culture	Han (1986), Singh (1998)
	McKenzie	Canada	Anther culture	Graf et al. (2013)
	Kharoba	Morocco	Anther culture	Dwivedi et al. (2015)
	Glosa, Gruia, Litera, Miranda	Romania	Wheat × maize	Depauw et al. (2005)
	BRS 328	Brazil	Wheat × maize	Scheeren et al. (2014)
	Him Pratham	India	Wheat × *Imperata cylindrica*	Chaudhary et al. (2015)
	AAC Elevate, AAC Connery, Emerson Sunrise, Snowstar, Lillian	Canada	Wheat × maize	Patial et al. (2019)
Rice	Parag 401, Risabell, Janka, Abel, CR Dhan10, CR Dhan 801	India	Anther culture	Patil et al. (1997), Pauk et al. (2009), Mishra and Rao (2016)
	Huayu I and II, Tunghua 1, 2 and 3, Zhong-Hua 8, 9, 10 and 11, Nanhua 5, Huahanzao, Guan 18, Huayu 15, Milyang 90	China	Anther culture	Mishra and Rao (2016) Yang and Fu (1989)
	Hwacheongbyeo, Joryeongbyeo, Hwajinbyeo	South Korea	Anther culture	Lee et al. (1989)
	Bicoll	Philippines	Anther culture	Senadhira et al. (2002)
	Joiku N 394, Hirohikari, Hirohonami, AC. No.1, Kibinohana	Japan	Anther culture	Singh (1998)
	Shirayukihime	Japan	Anther culture	Sugimoto and Arai (2002)
	Dama	Hungary	Somaclonal selection	Heszky and Simon-Kiss (1992)
Barley	Flag, Ladoga, Lyric, Naomie	France	Anther culture	Thomas et al. (2003)

frequency of albinos and lower number of doubled plants. Though a large number of cell lines can be screened for biotic and abiotic stresses, only selective genotypes are amenable to morphogenesis and ultimately plantlet development (Al-Ashkar et al 2019). In more than 4300 and 3100 anthers in liquid and solid media, respectively, the number of green plantlets was always fewer than albinos, ranging from 7.2 to 82.0%. This trait is heritable and hence needs to be transferred as well (Al-Ashkar et al. 2019). Even after obtaining a haploid, doubling its genome is a genetic challenge, and this ability is variable among the different crops and genotypes among the same species. Stress treatments are generally given to the gametophytic cells to stimulate them for embryogenesis. However, they induce unwanted oxidative stress, higher reactive oxygen species (ROS) and nuclear or extranuclear genome changes that result in a higher frequency of albinos and lower plantlet regeneration. Studies have even indicated epigenetic modifications in the plant tissue due to stresses in culture conditions. Some additives can be added to lower the induced unwanted changes, e.g. plant growth regulators, cycocel, polyamines, osmoprotectants, DNA demethylators, histone deacetylase inhibitors, antioxidants like L-ascorbic acid and additives like polyvinylpyrrolidone (PVP), activated charcoal, silver nitrate, etc. (Niazian and Shariatpanahi 2020).

In vivo haploid production is of more commercial interest as it can better integrate existing plant breeding programmes with public or private institutions. But at each step of the DH line production, the in vivo maternal haploid induction system faces several challenges. The haploid induction rate is a quantitatively governed trait and is affected by genetic backgrounds (Prigge et al. 2011). Genetic backgrounds also affect the proportion of identifiable haploid seeds. Diploidization frequency is also reported to exhibit genotype specificity, which may be attributed to variation in SHGD among source populations. The haploid male is particularly low, resulting in less seed recovery (Kleiber et al. 2012). The haploid induction rate is also affected by the germplasm used as the source population (Eder and Chalyk 2002; Khulbe et al. 2020). Maintaining HIR in the inducer is also a challenging task, as 58% of the HIR is observed in the single cycle of self-pollination (Ribeiro et al. 2018). Various climatic factors like high temperature and high humidity affect the HIR, especially during pollination (Kebede et al. 2011).

The mutated genes of haploid induction include *MATRILINEAL* (*MTL*) or *NOT LIKE DAD* (*NLD*) and *phospholipase A1* (*PLA1*), which can interfere with the processes like pollen tube growth or pollen germination, etc., essential for maintaining the normal diploid chromosome number. Although it had been utilized in monocots (rice, wheat) for haploid induction, *MTL* or *NLD* like genes had not been successful in dicots due to their inability to find relevant orthologues (Jacquier et al. 2020). In dicots, engineering the *CENH3* gene, which interferes with the normal segregation during mitosis due to a dysfunctional histone of the centromere, is being utilized, though with limited success in only a few crop species (Britt and Kuppu 2016). Though HI lines created by engineering the *CENH3* gene were produced in *Arabidopsis*, they are not as widespread as in maize, where the technology is routinely utilized to produce DHs. Also, the *CENH3* haploid induction system has a low haploid induction in the tested species. Apart from *CENH3* mutants, no other alternative mutations have been found to induce haploid production.

Advanced genomic tools like CRISPR/Cas9 capable of inducing targeted mutations have offered hope to produce HI lines in crops that lack naturally occurring genes inducing haploidy (Hooghvorst 2020). One such technique called the 'HI-edit system' can turn any material of choice into a haploid inducer line. It utilizes the knock-down capabilities of the CRISPR-Cas9 by targeting the *MTL* or *CENH3* gene. Though CRISPR is very effective in knocking down a targeted gene, its efficacy is limited by the specificity of the used genotype to respond to the delivery system (*Agrobacterium*/particle bombardment). The DH lines derived using CRISPR-/Cas9-generated HI can safely be kept out of the GMO purview as (i) the genome of HI does not integrate with the maternal genome, (ii) the haploids originate from the maternal genome only (Hooghvorst 2020) and (iii) the resultant haploid lacks CRISPR edit machinery used to induce HI line, thus making the derived product out of the transgenic debate. Therefore, breeding programmes can effectively utilize them without too much labour and maintenance (Kelliher et al. 2019).

The genome of haploid plants generated either through in vitro or in vivo techniques must be doubled to obtain the desired DH plants. Colchicine is the most widely used doubling agent due to its ability to arrest microtubule polymerization during the M phase of the cell cycle (Tefera 2017; Boerman et al. 2020). The replacement of colchicine with alternative anti-mitogens is highly desirable because of its human toxicity and potential environmental hazards. Some herbicides, either singly or in combination, have been found as potential substitutes. Herbicide combination of APM and pronamide was found effective and is much less toxic to humans and the environment. Researchers have compared colchicine with other potential antimitotic agents like dinitroanilines (trifluralin and oryzalin), but colchicine is still the most efficient in doubling the chromosome number (Hooghvorst et al. 2020). Apart from its hazardous nature, the use of colchicine requires technical skills and is also highly labour-intensive, making it unfriendly towards the larger breeding programmes. There is yet another alternative available to double the genome of haploid plants, which can make the process *in planta* in a true sense, i.e. without using an antimitotic agent like colchicine. The method known as spontaneous haploid genome doubling (SHGD) utilizes few genetic loci capable of spontaneously doubling the genome of a haploid plant.

SHGD has been reported in several crop species like wheat, rice and barley with the frequency of doubling ranging from 10 to 70%, and some genotypes in barley and rye exhibit doubling rate as high as 90% (Ren et al. 2017). There had been an attempt to identify genomic regions responsible for the SHGD phenomenon using bi-parental QTL mapping and GWAS. In maize, these studies have indicated QTLs for SHGD to be dispersed across almost all chromosomes; however, similar bin regions on chromosomes 3, 6, 7 and 10 in multiple studies reconfirm their potential use (Boerman et al. 2020). A large effect and stable QTL for SHGD in maize has been identified on chromosome 5 (Ren et al. 2019; Trampe et al. 2020). QTL region on chromosome 6 was identified with a potential candidate gene causing the absence of first division (*adf1*) (Trampe et al. 2020). A wide range of genotypic variation (less than 5% to as high as 50%) is available for SHGD in maize (Chalyk 1994; Ma et al. 2018; Chaikam et al. 2019a, b). Notably, this trait is reportedly intensified by recurrent selection and can be enhanced to the tune of 10–30% depending upon the genetic background of the resultant line (Molenaar et al. 2019a, b).

The genetic challenge with the SHGD technique is that apart from spontaneous doubling genetic capability, the resultant haploid DH needs to have high male fertility (HMF). Male fertility is an issue in haploid plants due to their overall weak establishment in the field conditions. Female fertility is generally not a major concern as 97–100% haploid plants produce seed with pollen from normal diploid plants (Chalyk 1994, Geiger et al. 2006). Generally, female fertility is much higher than male fertility, and only limited anthers in their tassel are reported to produce viable pollen. Male fertility in maize is also influenced by genotypic and environmental variables, like temperate vs. tropical germplasm, greenhouse vs. field conditions, etc. In comparison, the former factor has higher male fertility (Kleiber et al. 2012). Also, genetic variation is reported for the trait (Jiao et al. 2020). Thus, mapping HMF becomes important, and associated QTLs have been identified on all maize chromosomes except 6 and 8 in a study using DH mapping population (Jiao et al. 2020). However, large effect and stable QTLs were detected on chromosomes 1 and 5, which can be used in MAS (Jiao et al. 2020). However, converting the existing germplasm to have SHGD capabilities is a significant challenge (Trampe et al. 2020).

Identification of putative haploids is an important step in DH production. However, there exist several operational and genetic challenges in the identification process. In maize, *R1-nj*-based phenotypic marker on the kernel endosperm and embryo is presently the most widely used identification system. However, *R1-nj* phenotypic marker fails to work well in all genetic backgrounds, particularly flint corn and landraces. Another marker system, *Pl1* (purple 1), is anthocyanin pigmentation imparting red/purple root colouration, where putative haploids are colourless, whereas diploids have coloured roots (Vanous et al. 2017). Differentiating haploids based on the oil content in the kernel has been tried using NMR or near infrared spectroscopy (NIR), wherein haploids have less oil content than diploids (Tefera 2017; Boerman et al. 2020). However, in vivo haploid production also produces aborted embryo kernel (EmA), whose low oil content may lead it to be falsely identified as putative haploid. Hence, rather than a single threshold oil content value between the haploids and diploids, a double threshold limit is better suited to identify real haploids. In this NMR-based discrimination method, seeds with oil content below a threshold level are EmA kernels, whereas diploid kernels are above the threshold limit. Hence, kernels between the upper and lower boundary are the putative haploids (Qu et al. 2021).

6.8 Conclusion and Prospects

Doubled haploidy has emerged as an efficient tool in accelerating the line development process in various crops. Finished DH lines can be harvested in the two to three crop seasons compared to six to seven seasons in the conventional breeding programmes. DH lines are generated through both in vitro and in vivo methods. In vitro methods have proven their significance in the rice DH line development programmes. However, in vitro methods face challenges of protocol standardization, genotype dependency and recalcitrant nature of some crops. The in vivo/*in planta* methods have been developed in wheat, barley and maize. The *in planta* methods have been successfully utilized in *H. vulgare* × *H. bulbosum*, wheat × maize and wheat × *I. cylindrica* systems for DH line development in barley and wheat and are characterized by chromosome/genome elimination of the donor parent. In these intergeneric crosses, maize, *I. cylindrica* and *H. bulbosum* act as an inducer for the haploid induction in wheat. In maize, haploid inducers, particularly maternal haploid inducers, have been widely employed for DH line development. The genes *ig1* and *MTL/NLD/PLA1* are responsible for haploid induction in maize by paternal and maternal inducers, respectively, that act through mechanisms that interfere with single fertilization and chromosome/genome elimination.

DH lines are a vital genetic resource for unlocking the genetic variation present in the wild germplasm or landraces. Moreover, QTL mapping and various genetic and genomic studies have been carried out globally to identify and introgress the QTLs/loci/gene conferring tolerance to biotic-abiotic stresses, nutritional quality and yield-related traits. Integrating DH technology with MAS and genomic selection in cereals would reduce the breeding cycles and maximize the genetic gain per unit time. DH technology has its applications in CMS line development, reverse breeding and gene stacking. Moreover, several cultivars developed and released for commercial cultivation in many countries are derived from the DH lines. In future, for in vitro methods, more focus needs to be given on the direct somatic embryogenesis from the microspores, which is a cost-effective method. The molecular mechanism of haploid induction is still not clear and needs further elucidation. Further understanding of the mechanism of haploid induction will open new avenues of haploid induction in other crops also. The non-toxic chromosome doubling alternatives to colchicine need to be identified to avoid environmental hazards. The potential of CRISPR/cas9 needs to be utilized to develop new haploid inducer lines with high HIR. The genes responsible for haploid induction and their orthologues need to be explored and suitably manipulated to introduce haploid induction mechanisms (or establish haploid induction systems) in the novel species. More extensive research on spontaneous chromosome doubling needs to be done to avoid the hazardous chemicals and achieve maximum DH recovery. There is a need for substantial improvement in the DH programmes for their full exploitation in the cultivar/line development in a shorter time for accelerating the genetic gains.

References

Agnieszka K, Adela A (2010) In vitro culture of unfertilized ovules in carrot (*Daucus carota* L.). Plant Cell Tissue Organ Cult 102:309–319

Alam SN, Cohen MB (1998) Detection and analysis of QTLs for resistance to the brown planthopper, Nilaparvata lugens, in a doubled-haploid rice population. Theor Appl Genet 97: 1370–1379. https://doi.org/10.1007/s001220051031

Al-Ashkar A, El-Hendawy A-S, El-Kafafi S, Seleiman MF (2019) Detecting salt tolerance in doubled haploid wheat lines. Agronomy 9(4):211. https://doi.org/10.3390/agronomy9040211

Anderson SB, Due IK, Olsen A (1987) The response of anther culture in a genetically wide material of winter wheat (*Triticum aestivum* L.). Plant Breed 99:181–186. https://doi.org/10.1111/j.1439-0523.1987.tb01170.x

Arumuganathan K, Earle ED (1991a) Nuclear DNA content of some important plant species. Plant Mol Biol Rep 9:208–218. https://doi.org/10.1007/BF02672069

Arumuganathan K, Earle ED (1991b) Estimation of nuclear DNA contents of plants by flow cytometry. Plant Mol Biol Rep 9:229–241. https://doi.org/10.1038/nprot.2007.310

Asif MA, Schilling RK, Tilbrook J, Brien C, Dowling K, Rabie H et al (2018) Mapping of novel salt tolerance QTL in an Excalibur × Kukri doubled haploid wheat population. Theor Appl Genet 131:2179–2196. https://doi.org/10.1007/s00122-018-3146-y

Bae CH, Lee YI, Lim YP, Seo YW, Lee DJ, Yang DC, Lee HY (2002) Detection of herbicide tolerant cell lines from c-ray irradiated cell cultures in rice (Oryza sativa L. cv. Ilpumbyeo). J Plant Biotech 4:123–127

Bains NS, Mangat GS, Singh K, Nanda GS (1998) A simple technique for the identification of embryo carrying seeds from wheat × maize crosses prior to dissection. Plant Breed 117:191–192

Barclay IR (1975) High frequencies of haploid production in wheat (*Triticum aestivum*) by chromosome elimination. Nature 256:410–411. https://doi.org/10.1038/256410a0

Barret P, Brinkmann M, Beckert M (2008) A major locus expressed in the male gametophyte with incomplete penetrance is responsible for in situ gynogenesis in maize. Theor Appl Genet 117: 581–594. https://doi.org/10.1007/s00122-008-0803-6

Barro F, Fernandez-Escobar J, De la Vega M, Martin A (2001) Doubled haploid lines of *Brassica carinata* with modified erucic acid content through mutagenesis by EMS treatment of isolated microspores. Plant Breed 120:262–264. https://doi.org/10.1046/j.1439-0523.2001.00602.x

Barro F, Fernandez-Escobar J, De La Vega M, Martin A (2002) Modification of glucosinolate and erucic acid contents in doubled haploid lines of *Brassica carinata* by UV treatment of isolated microspores. Euphytica 129:1–6. https://doi.org/10.1023/A:1021578318098

Basu SK, Eudes F, Kovalchuk I (2010) Role of *recA/RAD51* gene family in homologous recombination repair and genetic engineering of transgenic plants. In: Kumar A, Sopory S (eds) Applications of plant biotechnology: In vitro propagation, plant transformation and secondary metabolite production. I.K. International Publishing House Pvt Ltd., New Delhi, pp 231–255

Battistelli GM, VonPinho RG, Justus A, Couto EG, Balestre M (2013) Production and identification of doubled haploids in tropical maize. Gen Mol Res 12:4230–4242. https://doi.org/10.4238/2013.October.7.9

Belicuas PR, Guimarães CT, Paiva LV, Duarte JM, Maluf WR, Paiva E (2007) Androgenetic haploids and SSR markers as tools for the development of tropical maize hybrids. Euphytica 156:95–102. https://doi.org/10.1007/s10681-007-9356-z

Bernardo R (2009) Should maize doubled haploids be induced among F1 or F2 plants? Theor Appl Genet 119:255–262. https://doi.org/10.1007/s00122-009-1034-1

Bhowmik P, Ellison E, Polley B, Bollina V, Kulkarni M, Ghanbarnia K, Kagale S (2018) Targeted mutagenesis in wheat microspores using CRISPR/Cas9. Sci Rep 8:6502. https://doi.org/10.1038/s41598-018-24690-8

Blakeslee AF, Belling J, Farnham ME, Bergner AD (1922) A haploid mutant in the Jimson weed, *Datura stramonium*. Science 55:646–647

Boerman NA, Frei UK, Lübberstedt T (2020) Impact of spontaneous haploid genome doubling in maize breeding. Plan Theory 9(3):369. https://doi.org/10.3390/plants9030369

Boote BW, Freppon DJ, De La Fuente GN et al (2016) Haploid differentiation in maize kernels based on fluorescence imaging. Plant Breed 135:439–445

Britt AB, Kuppu S (2016) Cenh3: an emerging player in haploid induction technology. Front Plant Sci 7:357. https://doi.org/10.3389/fpls.2016.00357

Buyser J, Henry Y, Lonnet P, Hertzog R, Hespel A (1987) 'Florin': a doubled haploid wheat variety developed by the anther culture method. Plant Breed 98:53–56. https://doi.org/10.1111/J.1439-0523.1987.TB01089.X

Bylich VG, Chalyk ST (1996) Existence of pollen grains with a pair of morphologically different sperm nuclei as a possible cause of the haploid-inducing capacity in ZMS line. Maize Genet Coop News Lett 70:33–33

Calayugan MIC, Formantes AK, Amparado A, Descalsota-Empleo GI, Nha CT, Inabangan-Asilo MA et al (2020) Genetic analysis of agronomic traits and grain iron and zinc concentrations in a doubled haploid population of rice (*Oryza sativa* L.). Sci Rep 10:1–14. https://doi.org/10.1038/s41598-020-59184-z

Castillo AM, Cistue L, Valles MP, Sanz L, Romagosa I, Molina-Cano JL (2001) Efficient production of androgenic doubled-haploid mutants in barley by the application of sodium azide to anther and microspore cultures. Plant Cell Rep 20:105–111

Castillo AM, Cistué L, Vallés MP et al (2009) Chromosome doubling in monocots. In: Touraev A, Forster BP, Jain SM (eds) Advances in haploid production in higher plants. Springer, Berlin, pp 329–338

Chaikam V, Prasanna BM (2012) Maternal haploid detection using anthocyanin markers. In: Prasanna BM, Chaikam V, Mahuku G (eds) Doubled haploid technology in maize breeding: theory and practice. CIMMYT, Mexico, pp 20–23

Chaikam V, Nair SK, Babu R et al (2015) Analysis of effectiveness of *R1-nj* anthocyanin marker for *in vivo* haploid identification in maize and molecular markers for predicting the inhibition of *R1-nj* expression. Theor Appl Genet 128:159–171. https://doi.org/10.1007/s00122-014-2419-3

Chaikam V, Martinez L, Melchinger AE et al (2016) Development and validation of red root marker-based haploid inducers in maize. Crop Sci 56:1678–1688. https://doi.org/10.2135/cropsci2015.10.0653

Chaikam V, Lopez LA, Martinez L et al (2017) Identification of in vivo induced maternal haploids in maize using seedling traits. Euphytica 213:177. https://doi.org/10.1007/s10681-017-1968-3

Chaikam V, Gowda M, Nair SK, Melchinger AE, Boddupalli PM (2019a) Genome-wide association study to identify genomic regions influencing spontaneous fertility in maize haploids. Euphytica 215:138. https://doi.org/10.1007/s10681-019-2459-5

Chaikam V, Molenaar W, Melchinger AE et al (2019b) Doubled haploid technology for line development in maize: technical advances and prospects. Theor Appl Genet 132:3227–3243. https://doi.org/10.1007/s00122-019-03433-x

Chalyk ST (1994) Properties of maternal haploid maize plants and potential application to maize breeding. Euphytica 79:13–18. https://doi.org/10.1007/BF00023571

Chase SS (1969) Monoploids and monoploid-derivatives of maize (*Zea mays* L.). Bot Rev 35:117–168. https://doi.org/10.1007/BF02858912

Chaudhary HK, Sethi GS, Singh S, Pratap A, Sharma S (2005) Efficient haploid induction in wheat by using pollen of *Imperata cylindrica*. Plant Breed 124:96–98. https://doi.org/10.1111/j.1439-0523.2004.01034.x

Chaudhary HK, Tayeng T, Kaila V, Rather SA (2013) Enhancing the efficiency of wide hybridization mediated chromosome engineering for high precision crop improvement with special reference to wheat × *Imperata cylindrica* system. Nucleus 56:7–14. https://doi.org/10.1007/s13237-013-0077-5

Chaudhary HK, Badiyala A, Jamwal NS (2015) New frontiers in doubled haploidy breeding in wheat. Agric Res J 52(4):1–12. https://doi.org/10.5958/2395-146X.2015.00053.8

Chaudhary HK, Sharma P, Manoj NV, Singh K (2019) New frontiers in chromosome elimination-mediated doubled haploidy breeding: Focus on speed breeding in bread and durum wheat. Indian J Genet 79:254–263. https://doi.org/10.5958/2395-146X.2015.00053.8

Chauhan H, Khurana P (2011) Use of doubled haploid technology for development of stable drought tolerant bread wheat (*Triticum aestivum* L.) transgenics. Plant Biotechnol J 9:408–417. https://doi.org/10.1111/j.1467-7652.2010.00561.x

Choe E, Carbonero CH, Mulvaney K, Rayburn AL, Mumm RH (2012) Improving *in vivo* maize doubled haploid production efficiency through early detection of false positives. Plant Breed 131:399–401. https://doi.org/10.1111/j.1439-0523.2012.01962.x

Chu CG, Friesen TL, Xu SS, Faris JD, Kolmer JA (2009) Identification of novel QTLs for seedling and adult plant leaf rust resistance in a wheat doubled haploid population. Theor Appl Genet 119:263–269. https://doi.org/10.1007/s00122-009-1035-0

Clapham D (1973) Haploid *Hordeum* plants from anthers in vitro. Z Pflanzenzüchtg 69:142–155

Coe EH (1959) A line of maize with high haploid frequency. Am Nat 93:381–382

Coe EH (1994) Anthocyanin genetics. In: Freeling M, Walbot V (eds) The maize handbook. Springer- Verlag, New York, pp 279–281

Couto EGO, Davide LMC, Bustamante FO, Von Pinho RG, Silva TN (2013) Identification of haploid maize by flow cytometry, morphological and molecular markers. Ciên Agrotecnol 37: 25–31. https://doi.org/10.1590/S1413-70542013000.100003

Dashti H, Yazdisamadi B, Reza Bihamta M, Reza Naghavi M, Yazdi-samadi B, Ghannadha M et al (2007) QTL analysis for drought resistance in wheat using doubled haploid lines. Int J Agric Biol 9:98–102. http://www.fspublishers.org/published_papers/71708_..pdf. Accessed 22 April 2021

Davies DR (1974) Chromosome elimination in inter-specific hybrids. Heredity 32:267–270

Davrieux F, Allal F, Piombo G, Kelly B, Okulo JB, Thiam M, Bouvet JM (2010) Near infrared spectroscopy for high- throughput characterization of shea tree (*Vitellaria paradoxa*) nut fat profiles. J Agric Food Chem 58:7811–7819. https://doi.org/10.1021/jf100409v

De La Fuente GN, Frei UK, Trampe B, Nettleton D, Zhang W, Lubberstedt T (2018) A diallel analysis of a maize donor population response to *in vivo* maternal haploid induction I: inducibility. Crop Sci 58:1830–1837. https://doi.org/10.2135/cropsci2017.05.0285

Deimling S, Röber F, Geiger HH (1997) Methodik und genetik der in vivo-haploideninduktion bei mais. Vor Pflanzenzüchtg 38:203–224

DePauw RM, Townley-Smith TF, Humphreys G (2005) Lillian hard red spring wheat. Can J Plant Sci 85:397–401. https://doi.org/10.4141/P04-137

Derakhshani B, Jafary H, Zanjani BM, Hasanpur K, Mishina K, Tanaka T et al (2020) Combined QTL mapping and RNA-Seq profiling reveals candidate genes associated with cadmium tolerance in barley. PLoS One 15:e0230820. https://doi.org/10.1371/journal.pone.0230820

Devaux P, Pickering R (2005) Haploids in the improvement of Poaceae. In: Haploids in crop improvement II. Springer, Berlin, pp 215–242

Dirks R, Dun K, Snoo CB, de Berg M, van den Lelivelt CLC, Voermans W et al (2009) Reverse breeding: a novel breeding approach based on engineered meiosis. Plant Biotechnol J 7:837. https://doi.org/10.1111/J.1467-7652.2009.00450.X

Dong X, Xu X, Miao J et al (2013) Fine mapping of qhir1 influencing in vivo haploid induction in maize. Theor Appl Genet 126:1713–1720. https://doi.org/10.1007/s00122-013-2086-9

Dunwell JM (2010) Haploids in flowering plants: origins and exploitation. Plant Biotechnol J 8: 377–424. https://doi.org/10.1111/j.1467-7652.2009.00498.x

Dwivedi SL, Britt AB, Tripathi L, Sharma S, Upadhyaya HD, Ortiz R (2015) Haploids: constraints and opportunities in plant breeding. Biotechnol Adv 33:812–829

Eder J, Chalyk ST (2002) *In vivo* haploid induction in maize. Theor Appl Genet 104:703–708. https://doi.org/10.1007/s00122-001-0773-4

Ekiz H, Konzak CF (1991) Nuclear and cytoplasmic control of anther culture response in wheat. III Common wheat crosses Crop Sci 31:1432–1436

El-Hennawy MA, Abdalla AF, Shafey SA, Al-Ashkar IM (2011) Production of doubled haploid wheat lines (*Triticum aestivum* L.) using anther culture technique. Ann Agri Sci 56:63–72. https://doi.org/10.1016/j.aoas.2011.05.008

Evans MMS (2007) The indeterminate gametophyte1 gene of maize encodes a LOB domain protein required for embryo sac and leaf development. Plant Cell 19:46–62. https://doi.org/10.1105/tpc.106.047506

Ferrie AMR, Caswell KL (2011) Isolated microspore culture techniques and recent progress for haploid and doubled haploid plant production. Plant Cell Tissue Organ Cult 104:301–309. https://doi.org/10.1007/s11240-010-9800-y

Finch RA (1983) Tissue-specific elimination of alternative whole parental genomes in one barley hybrid. Chromosoma 88:386–393. https://doi.org/10.1007/BF00285861

Finch RA, Bennett MD (1982) The mechanism of somatic chromosome elimination in *Hordeum*. In: Brandham PE, Bennett MD (eds) Kew Chromosome Conference II. Allen & Unwin, London, pp 146–153

Fischer E (2004) Molecular genetic studies on the occurrence of paternal DNA transmission during in vivo haploid induction in maize (*Zea mays*). Doctoral dissertation. University of Hohenheim, Germany

Forster BP, Heberle-Bors E, Kasha KJ, Touraev A (2007) The resurgence of haploids in higher plants. Trends Plant Sci 12:368–375. https://doi.org/10.1016/j.tplants.2007.06.007

Fox G, Wu A, Yiran L, Force L (2013) Variation in caffeine concentration in single coffee beans. J Agric Food Chem 61(45):10772–10778. https://doi.org/10.1021/jf4011388

Gahlaut V, Jaiswal V, Tyagi BS, Singh G, Sareen S, Balyan HS et al (2017) QTL mapping for nine drought-responsive agronomic traits in bread wheat under irrigated and rain-fed environments. PLoS One 12:e0182857. https://doi.org/10.1371/journal.pone.0182857

Gaillard A, Vergne P, Beckert M (1991) Optimization of maize microspore isolation and culture conditions for reliable plant regeneration. Plant Cell Rep 10:55–58. https://doi.org/10.1007/BF00236456

Galbraith DW, Harkins KR, Maddox JM, Ayres NM, Sharma DP, Firoozbady E (1983) Rapid flow cytometric analysis of cell cycle in intact plant tissue. Science 220:1049–1051. https://doi.org/10.1126/science.220.4601.1049

García-llamas C, Martín A, Ballesteros J (2004) Differences among auxin treatments on haploid production in durum wheat × maize crosses. Plant Cell Rep 23:46–49. https://doi.org/10.1007/s00299-004-0786-y

Geiger HH (2009) Doubled haploids. In: Bennetzen JL, Hake S (eds) Handbook of maize. Springer, New York, pp 641–657. https://doi.org/10.1007/978-0-387-77863-1_32

Geiger HH, Roux SR, Deimling S (1994) Herbicide resistance as a marker in screening for maternal haploids. Maize Genet Coop News Lett 68:99

Genovesi A, Clint W (1979) Improved rate of callus and green plant production from rice anther culture following cold shock. Crop Sci 19:662–664

Germanà MA (2011) Gametic embryogenesis and haploid technology as valuable support to plant breeding. Plant Cell Rep 30:839–857. https://doi.org/10.1007/s00299-011-1061-7

Gernand D, Rutten T, Varshney A, Rubtsova M, Prodanovic S et al (2005) Uniparental chromosome elimination at mitosis and interphase in wheat and pearl millet crosses involves micronucleus formation, progressive heterochromatinization, and DNA fragmentation. Plant Cell 17: 2431–2438. https://doi.org/10.1105/tpc.105.034249

Gernand D, Rutten T, Pickering R, Houben A (2006) Elimination of chromosomes in *Hordeum vulgare* × *H. bulbosum* crosses at mitosis and interphase involves micronucleus formation and progressive heterochromatinization. Cytogenet Genome Res 114:169–174. https://doi.org/10.1159/000093334

Gilles LM, Khaled A, Laffaire J et al (2017) Loss of pollen-specific phospholipase NOT LIKE DAD triggers gynogenesis in maize. EMBO J 36:707–717. https://doi.org/10.15252/embj.201796603

Gous PW, Christopher LH, Franckowiak J, Fox GP (2016) Discovery of QTL for stay-green and heat-stress in barley (*Hordeum vulgare*) grown under simulated abiotic stress conditions. Euphytica 207:305–317. https://doi.org/10.1007/s10681-015-1542-9

Graf RJ, Beres BL, Laroche A (2013) Emerson hard red winter wheat. Can J Plant Sci 93:741–748. https://doi.org/10.4141/cjps2012-262

Grauda D, Lepse N, Strazdina V et al (2010) Obtaining of doubled haploid lines by anther culture method for the Latvian wheat breeding. Agron Res 8:545–552

Guha S, Maheshwari SC (1964) In vitro production of embryos from anthers of *Datura*. Nature 204: 497–498. https://doi.org/10.1038/204497a0

Guha S, Maheshwari SC (1966) Cell division and differentiation of embryos in the pollen grains of Datura in vitro. Nature 212:97–98. https://doi.org/10.1038/212097a0

Gupta SB (1969) Duration of mitotic cycle and regulation of DNA replication in *Nicotiana plumbaginifolia* and a hybrid derivative of *N. tabacum* showing chromosome instability. Can J Genet Cytol 11:133–142. https://doi.org/10.1139/g69-017

Hai L, Guo H, Xiao S, Jiang G, Zhang X, Yan C et al (2005) Quantitative trait loci (QTL) of stem strength and related traits in a doubled-haploid population of wheat (*Triticum aestivum* L.). Euphytica 141:1–9. https://doi.org/10.1007/s10681-005-4713-2

Han H (1986) Wheat: improvement through anther culture Biotechnology in agriculture and forestry 2 by Bajaj YPS, vol 2. Springer, Berlin, pp 55–722. https://doi.org/10.1007/978-3-642-61625-9_3

Heszky LE, Simon-Kiss I (1992) The first plant variety of biotechnology origin in Hungary, registered in 1992. Hungar Agric Res 1:30–32

Hoekstra S, Van Zijderveld MH, Heidekamp E, Van der Mark E (1993) Microspore culture of *Hordeum vulgare* L.: the influence of density and osmolality. Plant Cell Rep 12:661–665. https://doi.org/10.1007/BF00233415

Hooghvorst I (2020) Opportunities and challenges in doubled haploids systems in cucurbits. Agron 10:1441. https://doi.org/10.3390/agronomy10091441

Hooghvorst I, Ribas P, Nogués S (2020) Chromosome doubling of androgenic haploid plantlets of rice (Oryza sativa) using antimitotic compounds. Plant Breed 139:754–761. https://doi.org/10.1111/pbr.12824

Hu TC, Ziauddin A, Simion E, Kasha KJ (1995) Isolated microspore culture of wheat *(Triticum aestivum* L.) in a defined media: I. Effects of pretreatment, isolation methods, and hormones. In Vitro Cell Dev Biol 31:79–83. https://doi.org/10.1007/BF02632241

Hua C, Chuan-xi MA, Yu-qiang Q (2008) Identification of morphology and cytology in wheat haploid through wheat × maize. J Nucl Agric Sci 22:127–130

Huang N, Parco A, Mew T, Magpantay G, McCouch S, Guiderdoni E et al (1997) RFLP mapping of isozymes, RAPD and QTLs for grain shape, brown planthopper resistance in a doubled haploid rice population. Mol Breed 3:105–113. https://doi.org/10.1023/A:1009683603862

Ilyas M, Ilyas N, Arshad M, Kazi AG, Kazi AM, Waheed A (2014) QTL mapping of wheat doubled haploids for chlorophyll content and chlorophyll fluorescence kinetics under drought stress imposed at anthesis stage. Pak J Bot 46:1889–1897

Inagaki MN, Mujeeb-Kazi A (1995) Comparison of polyhaploid production frequencies in crosses of hexaploid wheat with maize, pearl millet and sorghum. Breed Sci 45:157–161. https://doi.org/10.1270/jsbbs1951.45.157

Ishii T, Karimi-Ashtiyani R, Houben A (2016) Haploidization *via* chromosome elimination: means and mechanisms. Annu Rev Plant Biol 67:421–438. https://doi.org/10.1146/annurev-arplant-043014-114714

Jacquier NMA, Gilles LM, Pyott DE, Martinant JP, Rogowsky PM, Widiez T (2020) Puzzling out plant reproduction by haploid induction for innovations in plant breeding. Nat Plants 6:610–619. https://doi.org/10.1038/s41477-020-0664-9

Jiao Y, Li J, Li W, Chen M, Li M, Liu W et al (2020) QTL mapping and prediction of haploid male fertility traits in maize (*Zea mays* L.). Plan Theory 9:1–12. https://doi.org/10.3390/plants9070836

Jones RW, Reinot T, Frei UK, Tseng Y, Lübberstedt T, McClelland JF (2012) Selection of haploid maize kernels from hybrid kernels for plant breeding using near infrared spectroscopy and SIMCA analysis. Appl Spectrosc 66:447–450. https://doi.org/10.1366/2F11-06426

Joosen R, Cordewener J, Supena EDJ, Vorst O, Lammers M, Maliepaard C et al (2007) Combined transcriptome and proteome analysis identifies pathways and markers associated with the establishment of rapeseed microspore-derived embryo development. Plant Physiol 144:155–172. https://doi.org/10.1104/pp.107.098723

Kalinowska K, Chamas S, Unkel K, Demidov D, Lermontova I, Dresselhaus T, Kumlehn J, Dunemann F, Houben A (2019) State-of-the-art and novel developments of in vivo haploid technologies. Theor Appl Genet 132:593–605. https://doi.org/10.1007/s00122-018-3261-9

Kasha KJ, Kao KN (1970) High frequency haploid production in barley (*Hordeum vulgare* L.). Nature 225:874–876. https://doi.org/10.1038/225874a0

Kasha KJ, Maluszynski M (2003) Production of doubled haploids in crop plants. In: Maluszynski M, Kasha KJ, Forster BP, Szarejko I (eds) Doubled haploid production in crop plant. Kluwer Academic Publishers, Dordrecht, pp 1–4

Kebede AZ, Dhillon BS, Schipprack W et al (2011) Effect of source germplasm and season on the *in vivo* haploid induction rate in tropical maize. Euphytica 180:219–226. https://doi.org/10.1007/s10681-011-0376-3

Kelliher T, Starr D, Richbourg L et al (2017) MATRILINEAL, a sperm-specific phospholipase, triggers maize haploid induction. Nature 542:105–109. https://doi.org/10.1038/nature20827

Kelliher T, Starr D, Su X, Tang G, Chen Z, Carter J et al (2019) One-step genome editing of elite crop germplasm during haploid induction. Nat Biotechnol 37:287–292. https://doi.org/10.1038/s41587-019-0038-x

Kermicle JL (1969) Androgenesis conditioned by a mutation in maize. Science 166:1422–1424. https://doi.org/10.1126/science.166.3911.1422

Kermicle J (1973) Androgenesis and the indeterminate gametophyte mutation: source of the cytoplasm. Maize Genet Coop News Lett 47:208–209

Kermicle JL (1994) Indeterminate gametophyte (ig): biology and use. In: Freeling M, Walbot V (eds) The maize handbook. Springer, New York, pp 388–393. https://doi.org/10.1007/978-1-4612-2694-9_58

Khan AJ, Hassan S, Tariq M, Khan T (2001) Haploidy breeding and mutagenesis for drought tolerance in wheat. Euphytica 120:409–414. https://doi.org/10.1023/A:1017598202368

Khulbe RK, Pattanayak A, Panday V (2019) *R1-nj* expression in parental inbreds as a predictor of amenability of maize hybrids to *R1-nj*-based doubled haploid development. Indian J Genet 79: 678–684. https://doi.org/10.31742/IJGPB.79.4.5

Khulbe RK, Pattanayak A, Lakshmi K, Bisht GS, Pant MC, Pandey V, Rohit K, Mishra NC (2020) Doubled haploid production in maize under submontane Himalayan conditions using *R1-nj*-based haploid inducer TAILP1. Indian J Genet 80:261–266. https://doi.org/10.31742/IJGPB.80.3.4

Kim DS, Lee IS, Jang CS, Hyun DY, Lee SJ, Seo YW, Lee YI (2003) Selection of 5-methyltryptophan and S-(2-aminoethyl)-L-cysteine resistant microspore-derived rice cell lines irradiated with gamma rays. J Plant Biotech 5:33–41

Kisana NS, Nkongolo KK, Quick JS, Johnson DL (1993) Production of doubled haploids by anther culture and wheat × maize method in a wheat breeding programme. Plant Breed 110:96–102

Kishore N, Chaudhary HK, Chahota RK, Kumar V, Sood SP, Jeberson S, Tayeng T (2011) Relative efficiency of the maize and *Imperata cylindrica*-mediated chromosome elimination approaches for induction of haploids of wheat-rye derivatives. Plant Breed 130:192–194. https://doi.org/10.1111/j.1439-0523.2010.01793.x

Kiviharju E, Moisander S, Laurila J (2005) Improved green plant regeneration rates from oat anther culture and the agronomic performance of some DH lines. Plant Cell Tissue Organ Cult 81:1–9. https://doi.org/10.1007/s11240-004-1560-0

Kleiber D, Prigge V, Melchinger AE, Burkard F, San Vicente F, Palomino G, Andrés Gordillo G (2012) Haploid fertility in temperate and tropical maize germplasm. Crop Sci 52:623–630. https://doi.org/10.2135/cropsci2011.07.0395

Kochevenko A, Jiang Y, Seiler C, Surdonja K, Kollers S, Reif JC et al (2018) Identification of QTL hot spots for malting quality in two elite breeding lines with distinct tolerance to abiotic stress. BMC Plant Biol 18:1–17. https://doi.org/10.1186/s12870-018-1323-4

Köhler F, Wenzel G (1985) Regeneration of isolated barley microspores in conditioned media and trials to characterise the responsible factor. J Plant Physiol 121:181–191. https://doi.org/10.1016/S0176-1617(85)80044-4

Komeda N, Chaudhary HK, Suzuki G, Mukai Y (2007) Cytological evidence for chromosome elimination in wheat × *Imperata cylindrica* hybrids. Genes Genet Syst 82:241–248. https://doi.org/10.1266/ggs.82.241

Krolow KD (1970) Investigations on compatibility between wheat and rye. Z Pflanzenzuchtung 64: 44–72

Kwon YH, Kabange NR, Lee JY, Lee SM, Cha JK, Shin DJ et al (2021) Novel QTL associated with shoot branching identified in doubled haploid rice (*Oryza sativa* L.) under low nitrogen cultivation. Genes (Basel) 12:745. https://doi.org/10.3390/genes12050745

Lange W (1971) Crosses between *Hordeum vulgare* L. and *H. bulbosum* L. I. Production, morphology and meiosis of hybrids, haploids and dihaploids. Euphytica 20:14–29. https://doi.org/10.1007/BF00146769

Laurie DA (1989) The frequency of fertilization in wheat × pearl millet crosses. Genome 32:1063–1067. https://doi.org/10.1139/g89-554

Laurie DA, Bennett MD (1986) Wheat × maize hybridization. Can J Genet Cytol 28:313–316

Laurie DA, Bennett MD (1988a) Cytological evidence for fertilization in hexaploid wheat × sorghum crosses. Plant Breed 100:73–82

Laurie DA, Bennett MD (1988b) The production of haploid wheat plants from wheat × maize crosses. Theor Appl Genet 76:393–397. https://doi.org/10.1007/BF00265339

Laurie DA, Bennett MD (1989) The timing of chromosome elimination in hexaploid wheat × maize crosses. Genome 32:953–961. https://doi.org/10.1139/g89-537

Lazar MD, Schaffer GW, Baenziger PS (1985) The physical environment in relation to high frequency callus and plantlet development in anther cultures of wheat (*Triticum aestivum* L.) cv. Chris J Plant Physiol 121:103–109. https://doi.org/10.1016/S0176-1617(85)80034-1

Lee YT, Lim MS, Kim HS, Shin HT, Kim CH, Bae SH, Cho CI (1989) An anther derived new high-quality variety with disease and insect resistance 'Hwacheong byeo'. Res Rep Rural Dev Admin Rice 31(2):27–23

Lee SY, Lee JH, Kwon TO (2003) Selection of salt-tolerant doubled haploids in rice anther culture. Plant Cell Tissue Organ Cult 74:143–149

Li L, Xu X, Jin W, Chen S (2009) Morphological and molecular evidences for DNA introgression in haploid induction via a high oil inducer CAUHOI in maize. Planta 230:367–376. https://doi.org/10.1007/s00425-009-0943-1

Li X, Meng D, Chen S et al (2017) Single nucleus sequencing reveals spermatid chromosome fragmentation as a possible cause of maize haploid induction. Nat Commun 8:991. https://doi.org/10.1038/s41467-017-00969-8

Liu X, Li R, Chang X, Jing R (2013) Mapping QTLs for seedling root traits in a doubled haploid wheat population under different water regimes. Euphytica 189:51–66. https://doi.org/10.1007/s10681-012-0690-4

Liu D, Zhang H, Zhang L, Yuan Z, Hao M, Zheng Y (2014) Distant hybridization: a tool for interspecific manipulation of chromosomes. In: Pratap A, Kumar J (eds) Alien gene transfer in crop plants, volume 1 innovations, methods and risk assessment. Springer, New York, pp 25–42. https://doi.org/10.1007/978-1-4614-8585-8_2

Liu C, Chen B, Ma Y et al (2017a) New insight into the mechanism of hetero fertilization during maize haploid induction. Euphytica 213:174. https://doi.org/10.1007/s10681-017-1957-6

Liu C, Li X, Meng D, Zhong Y, Chen C, Dong X, Xu X, Chen B, Li W, Li L, Tian X (2017b) A 4-bp insertion at *ZmPLA1* encoding a putative phospholipase A generates haploid induction in maize. Mol Plant 10:520–522. https://doi.org/10.1016/j.molp.2017.01.011

Liu C, Zhong Y, Qi X, Chen M, Liu Z, Chen C, Tian X, Li J, Jiao Y, Wang D, Wang Y (2019) Extension of the *in vivo* haploid induction system from diploid maize to hexaploid wheat. Plant Biotechnol J 18:316–318. https://doi.org/10.1111/pbi.13218

Ma H, Li G, Würschum T, Zhang Y, Zheng D, Yang X, Li J, Liu W, Yan J, Chen S (2018) Genome-wide association study of haploid male fertility in maize (*Zea mays* L.). Front. Plant Sci 9:974. https://doi.org/10.3389/fpls.2018.00974

Mahato A, Chaudhary HK (2015) Relative efficiency of maize and *Imperata cylindrica* for haploid induction in *Triticum durum* following chromosomal elimination-mediated approach of doubled haploid breeding. Plant Breed 134:379–383. https://doi.org/10.1111/pbr.12288

Maluszynska J (2003) Cytogenetic tests for ploidy level analyses — chromosome counting. In: Maluszynski M, Kasha KJ, Forster BP, Szarejko I (eds) Doubled haploid production in crop plants. Springer, Dordrecht. https://doi.org/10.1007/978-94-017-1293-4_51

Maraschin SF, Caspers M, Potokina E, Wülfert F, Graner A, Spaink HP et al (2006) cDNA array analysis of stress-induced gene expression in barley androgenesis. Physiol Plant 127:535–550. https://doi.org/10.1111/j.1399-3054.2006.00673.x

Maruthachalam R, Chan SWL (2010) Haploid plants produced by centromere-mediated genome elimination. Nature 464:615–618. https://doi.org/10.1038/nature08842

McClinchey SL, Kott LS (2008) Production of mutants with high cold tolerance in spring canola (*Brassica napus*). Euphytica 162:51–67. https://doi.org/10.1007/s10681-007-9554-8

Mehta I, Chaudhary HK, Sharma P, Manoj NV, Singh K, Sran RS (2020) In vivo colchicine manipulation for enhancing DH production efficiency in *Triticum durum* using *Imperata cylindrica*- mediated chromosome elimination approach. Cereal Res Commun 48:217–224. https://doi.org/10.1007/s42976-020-00018-z

Melchinger AE, Schipprack W, Würschum T et al (2013) Rapid and accurate identification of in vivo-induced haploid seeds based on oil content in maize. Sci Rep 3:2129. https://doi.org/10.1038/srep02129

Melchinger AE, Schipprack W, Friedrich Utz H, Mirdita V (2014) In vivo haploid induction in maize: identification of haploid seeds by their oil content. Crop Sci 54:1497–1504. https://doi.org/10.2135/cropsci2013.12.0851

Melchinger AE, Böhm J, Utz HF et al (2018) High-throughput precision phenotyping of the oil content of single seeds of various oilseed crops. Crop Sci 58:670–678. https://doi.org/10.2135/cropsci2017.07.0429

Mishra R, Rao GJN (2016) In vitro androgenesis in rice: advantages, constraints and future prospects. Rice Sci 23:57–68. https://doi.org/10.1016/J.RSCI.2016.02.001

Mochida K, Tsujimoto H, Sasakuma T (2004) Confocal analysis of chromosome behaviour in wheat × maize zygotes. Genome 47:199–205. https://doi.org/10.1139/g03-123

Molenaar WS, Couto EGO, Piepho HP, Melchinger AE (2019a) Early diagnosis of ploidy status in doubled haploid production of maize by stomata length and flow cytometry measurements. Plant Breed 138:266–276. https://doi.org/10.1111/pbr.12694

Molenaar WS, Schipprack W, Brauner PC, Melchinger AE (2019b) Haploid male fertility and spontaneous chromosome doubling evaluated in a diallel and recurrent selection experiment in maize. Theor Appl Genet 132:2273–2284. https://doi.org/10.1007/s00122-019-03353-w

Mujeeb-Kazi A, Gul A, Ahmed J, Mirza JI (2006) A simplified and effective protocol for production of bread wheat haploids (n=3x=21, ABD) with some application areas in wheat improvement. Pak J Bot 38:393–406

Mujeeb-Kazi A, Gul A, Farooq M, Rizwan S, Ahmad I (2008) Rebirth of synthetic hexaploids with global implications for wheat improvement. Aust J Agric Res 59:391–398. https://doi.org/10.1071/AR07226

Murovec J, Bohanec B (2012) Haploids and doubled haploids in plant breeding. In: Abdurakhmonov I (ed) Plant breeding. INTECH, Rijeka. https://doi.org/10.5772/29982

Nair SK, Molenaar W, Melchinger AE et al (2017) Dissection of a major QTL qhir1 conferring maternal haploid induction ability in maize. Theor Appl Genet 130:1113–1122. https://doi.org/10.1007/s00122-017-2873-9

Nanda DK, Chase SS (1966) An embryo marker for detecting monoploids of maize (Zea mays L.). Crop Sci 6:213–215. https://doi.org/10.2135/cropsci1966.0011183X000600020036x

Niazian M, Shariatpanahi ME (2020) In vitro-based doubled haploid production: recent improvements. Euphytica 216:1–21. https://doi.org/10.1007/s10681-020-02609-7

Nitsch JP, Nitsch C (1969) Haploid plants from pollen grains. Science 163:85–87. https://doi.org/10.1126/science.163.3862.85

Niu Z, Jiang A, Abu Hammad W, Oladzadabbasabadi A, Xu SS, Mergoum M et al (2014) Review of doubled haploid production in durum and common wheat through wheat × maize hybridization. Plant Breed 133:313–320. https://doi.org/10.1111/pbr.12162

Ochatt SJ (2008) Flow cytometry in plant breeding. Cytometry A 73:581–598. https://doi.org/10.1002/cyto.a.20562

Ouyang JW, Hu H, Chuang CC, Tseng CC (1973) Induction of pollen plants from anthers of *Triticum aestivum* L. cultured in vitro. Sci Sinica 16:79–90. https://doi.org/10.1360/ya1973-16-1-79

Park GH, Kim JH, Kim KM (2014) QTL analysis of yield components in rice using a cheongcheong/nagdong doubled haploid genetic map. Am J Plant Sci 05:1174–1180. https://doi.org/10.4236/ajps.2014.59130

Patial M, Pal D, Thakur A, Bana RS, Patial S (2019) Doubled haploidy techniques in wheat (*Triticum aestivum* L.): an overview. Proc Natl Acad Sci India Sect B Biol Sci 89:27–41. https://doi.org/10.1007/s40011-017-0870-z

Patil VD, Nerkar YS, Misal MB, Harkal SR (1997) Parag 401, a semidwarf rice variety developed through anther culture. Int Rice Res Notes 22(2):19

Pauk J, Janeso M, Simon-Kiss I (2009) Rice doubled haploids and breeding A. In: Touraev BP, Forster SMJ (eds) Advances in haploid production in higher plants. Springer, Netherlands, pp 189–197

Pogna NE, Marzetti A (1977) Frequency of two tubes in in vitro germinated pollen grains. Maize Genet Coop News Lett 51:44

Prasanna BM (2012) Doubled haploid technology in maize breeding: an overview. In: Prasanna BM, Chaikam V, Mahuku G (eds) Doubled haploid technology in maize breeding: theory and practice. CIMMYT, Mexico, pp 1–8

Pratap A, Sethi GS, Chaudhary HK (2005) Relative efficiency of different Gramineae genera for haploid induction in triticale and triticale × wheat hybrids through chromosome elimination technique. Plant Breed 124:147–153. https://doi.org/10.1111/j.1439-0523.2004.01059.x

Pratap A, Sethi GS, Chaudhary HK (2006) Relative efficiency of anther culture and chromosome elimination technique for haploid induction in triticale × wheat and triticale × triticale hybrids. Euphytica 150:339–345. https://doi.org/10.1007/s10681-006-9120-9

Prigge V, Sánchez C, Dhillon BS, Melchinger AE (2011) Doubled haploids in tropical maize: I. Effects of inducers and source germplasm on in vivo haploid induction rates. Crop Sci 51:1498–1506. https://doi.org/10.2135/cropsci2010.10.0568

Prigge V, Schipprack W, Mahuku G et al (2012a) Development of in vivo haploid inducers for tropical maize breeding programs. Euphytica 185:481–490. https://doi.org/10.1007/s10681-012-0657-5

Prigge V, Xu X, Li L et al (2012b) New insights into the genetics of in vivo induction of maternal haploids, the backbone of doubled haploid technology in maize. Genetics 190:781–793. https://doi.org/10.1534/genetics.111.133066

Prins R, Pretorius ZA, Bender CM, Lehmensiek A (2011) QTL mapping of stripe, leaf and stem rust resistance genes in a Kariega × Avocet S doubled haploid wheat population. Mol Breed 27:259–270. https://doi.org/10.1007/s11032-010-9428-y

Qiu F, Liang Y, Li Y et al (2014) Morphological, cellular and molecular evidences of chromosome random elimination in vivo upon haploid induction in maize. Curr Plant Biol 1:83–90. https://doi.org/10.1016/j.cpb.2014.04.001

Qu Y, Liu Z, Zhang Y, Yang J, Li H (2021) Improving the sorting efficiency of maize haploid kernels using an NMR-based method with oil content double thresholds. Plant Methods 17:1–15. https://doi.org/10.1186/s13007-020-00703-4

Rahman MH, Krishnaraj S, Thorpe TA (1995) Selection for salt tolerance in vitro using microspore-derived embryos of *Brassica napus* cv Topas, and the characterization of putative tolerant plants. In Vitro Cell Dev Biol Plant 31:116–121. https://doi.org/10.1007/BF02632248

Rajcan I, Boersma JG, Shaw EJ (2011) Plant genetic techniques: plant breeder's toolbox. Elsevier B.V, Second Edi. https://doi.org/10.1016/B978-0-08-088504-9.00252-X

Rakha MT, Metwally EI, Moustafa SA, Etman AA, Dewir YH (2012) Evaluation of regenerated strains from six Cucurbita interspecific hybrids obtained through anther and ovule in vitro cultures. Aust J Crop Sci 6:23–30

Ravi M, Chan SW (2010) Haploid plants produced by centromere-mediated genome elimination. Nature 464:615–618. https://doi.org/10.1038/nature08842

Ravi M, Marimuthu MP, Tan EH, Maheshwari S, Henry IM, Marin-Rodriguez B, Urtecho G, Tan J, Thornhill K, Zhu F, Panoli A (2014) A haploid genetics toolbox for Arabidopsis thaliana. Nat Commun 5:533–534. https://doi.org/10.1038/ncomms6334

Redha A, Talaat A (2008) Improvement of green plant regeneration by manipulation of anther culture induction medium of hexaploid wheat. Plant Cell Tissue Organ Cult 92:141–146. https://doi.org/10.1007/s11240-007-9315-3

Ren J, Wu P, Trampe B, Tian X, Lübberstedt T, Chen S (2017) Novel technologies in doubled haploid line development. Plant Biotechnol J 15:1361–1370. https://doi.org/10.1111/pbi.12805

Ren J, Boerman NA, Liu R, Vanous K, Trampe B, Frei UK, Chen S, Lübberstedt T (2019) Mapping of QTL and identification of candidate genes conferring spontaneous haploid genome doubling in maize (*Zea mays* L.). Plant Sci 293:110337. https://doi.org/10.1016/j.plantsci.2019.110337

Ribeiro CB, Pereira FC, Filho LN, Rezende BA, Dias KOG, Braz GT, Souza JC (2018) Haploid identification using tropicalized haploid inducer progenies in maize. Crop Breed Appl Biotechnol 18:16–23. https://doi.org/10.1590/1984-70332018v18n1a3

Riley R, Chapman V (1967a) Effect of $5B^S$ in suppressing the expression of altered dosage of $5B^L$ on meiotic chromosome pairing in *Triticum aestivum*. Nature 216:60–62. https://doi.org/10.1038/216060a0

Riley R, Chapman V (1967b) The inheritance in wheat of crossability with rye. Genet Res 9:259–267. https://doi.org/10.1017/S0016672300010569

Röber FK, Gordillo GA, Geiger HH (2005) In vivo haploid induction in maize-performance of new inducers and significance of doubled haploid lines in hybrid breeding. Maydica 50:275

Rotarenco VA, Kirtoca IH, Jacota AG (2007) Using oil content to identify kernels with haploid embryos. Maize Genet Coop News Lett 81:11

Rotarenco VA, Dicu G, State D, Fuia S (2010) New inducers of maternal haploids in maize. Maize Genet Coop News Lett 84:21–22

Samantaray S, Ali J, Nicolas KLC (2021) Doubled haploids in rice improvement: approaches, applications, and future prospects. In: Ali J, Wani SH (eds) Rice improvement: physiological, molecular breeding and genetic perspective. Springer, Cham, pp 425–447. https://doi.org/10.1007/978-3-030-66530-2

Sánchez-Díaz RA, Castillo AM, Vallés MP (2013) Microspore embryogenesis in wheat: new marker genes for early, middle and late stages of embryo development. Sex Plant Reprod 26:287–296. https://doi.org/10.1007/s00497-013-0225-8

Sarkar KR, Coe EH Jr (1971) Analysis of events leading to heterofertilization in maize. J Hered 62:118–120. https://doi.org/10.1093/oxfordjournals.jhered.a108136

Scagliusi SM (2014) Establishing isolated microspore culture to produce doubled haploid plants in Brazilian wheat (*Triticum aestivum* L.). Aust J Crop Sci 8:887–894

Scheeren LP, Caetano RV, Caierao E, Silva MS, Nascimento A, Eichelberger L, Miranda MZ, Brammer SP (2014) BRS 328 - Double haploid bread wheat cultivar. Crop Breed Appl Biotechnol 14:65–67. https://doi.org/10.1590/S1984-70332014000100011

Schwarzacher Robinson T, Finch RA, Smith JB, Bennett MD (1987) Genotypic control of centromere positions of parental genomes in *Hordeum* × *Secale* hybrid metaphases. J Cell Sci 87:291–304. https://doi.org/10.1242/jcs.87.2.291

Senadhira D, Zapata-Arias FJ, Gregorio GB, Alejar MS, De La Cruz HC, Padolina TF, Galvez AM (2002) Development of the first salt tolerant rice cultivar through indica/indica anther culture. Field Crops Res 76(2/3):103–110. https://doi.org/10.1016/S0378-4290(02)00032-1

Shariatpanahi ME, Ahmadi B (2016) Isolated microspore culture and its applications in plant breeding and genetics. In: Anis M, Ahmad N (eds) Plant tissue culture: propagation, conservation and crop improvement. Springer, Singapore. https://doi.org/10.1007/978-981-10-1917-3_21

Sharma P, Chaudhary HK, Manoj NV, Kumar P (2019a) New protocol for colchicine induced efficient doubled haploidy in haploid regenerants of tetraploid and hexaploid wheats at in vitro level. Cereal Res Commun 47:356–368. https://doi.org/10.1556/0806.47.2019.09

Sharma P, Chaudhary HK, Manoj NV, Singh K, Relan A, Sood VK (2019b) Haploid induction in triticale × wheat and wheat × rye derivatives following *Imperata cylindrica-* mediated chromosome elimination approach. Cereal Res Commun 47:701–713. https://doi.org/10.1556/0806.47.2019.46

Sidhu PK, Davies PA (2009) Regeneration of fertile green plants from oat isolated microspore culture. Plant Cell Rep 28:571–577. https://doi.org/10.1007/s00299-009-0684-4

Singh RB (1998) Agricultural biotechnology in Asia-Pacific region Agricultural Biotechnology in the Developing World. In: FAO Research & Technology Development Division. Daya Publishing House, New Delhi, pp 53–55

Sitch LA, Snape JW, Firman SJ (1985) Intra chromosomal mapping of crossability genes in wheat (*Triticum aestivum*). Theor Appl Genet 70:309–314. https://doi.org/10.1007/BF00304917

Snape JW, Simpson E, Parker BB et al (1986) Criteria for the selection and used doubled haploid systems in cereal breeding programmes. In: Horn W, Jensen CJ, Odenbach W, Shieder O (eds) Genetic manipulation in plant breeding. Walter de Gruyter, Berlin, pp 217–229

Sserumaga JP, Beyene Y, Pillay K, Kullaya A, Oikeh SO, Mugo S et al (2018) Grain-yield stability among tropical maize hybrids derived from doubled-haploid inbred lines under random drought stress and optimum moisture conditions. Crop Pasture Sci 69:691–702. https://doi.org/10.1071/CP17348

Strigens A, Schipprack W, Reif JC, Melchinger AE (2013) Unlocking the genetic diversity of maize landraces with doubled haploids opens new avenues for breeding. PLoS One 8:e57234. https://doi.org/10.1371/journal.pone.0057234

Subrahmanyam NC, Kasha KJ (1973) Selective chromosomal elimination during haploid formation in barley following interspecific hybridization. Chromosoma 42:111–125. https://doi.org/10.1007/BF00320934

Sugimoto K, Arai T (2002) Stability of characters of a doubled haploid rice variety, Shirayukihime. Breed Sci 52:15–21

Swapna M, Sarkar KR (2011) Anomalous fertilization in haploidy inducer lines in maize (*Zea mays* L). Maydica 56:221–225

Szarejko I, Forster BP (2007) Doubled haploidy and induced mutation. Euphytica 158:359–370. https://doi.org/10.1007/s10681-006-9241-1

Tang F, Tao Y, Zhao T, Wang G (2006) In vitro production of haploid and doubled haploid plants from pollinated ovaries of maize (*Zea mays*). Plant Cell Tissue Organ Cult 84:233–237. https://doi.org/10.1007/s11240-005-9017-7

Tefera AA (2017) Review on concept and impact of double haploid techniques in crop improvement. J Nat Sci Res 7:10–20. www.iiste.org

Thomas WTB, Forster BP, Gertsson B (2003) Doubled haploids in breeding. In: Maluszynski M, Kasha KJ, Forster BP, Szarejko I (eds) Doubled haploid production in crop plants. Springer, Dordrecht. https://doi.org/10.1007/978-94-017-1293-4_47

Tian X, Qin Y, Chen B et al (2018) Hetero-fertilization together with failed egg–sperm cell fusion supports single fertilization involved in *in vivo* haploid induction in maize. J Exp Bot 69(20): 689–4701. https://doi.org/10.1093/jxb/ery177

Touraev A, Indrianto A, Wratschko I, Vicente O, Heberle BE (1996) Efficient microspore embryogenesis in wheat (*Triticum aestivum* L.) induced by starvation at high temperature. Sex Plant Reprod 9:209–215. https://doi.org/10.1007/BF02173100

Touraev A, Pfosser M, Heberle-Bors E (2001) The microspore: a haploid multipurpose cell. Adv Bot Res 35:53–109. https://doi.org/10.1016/S0065-2296(01)35004-8

Trampe B, dos Santos IG, Frei UK, Ren J, Chen S, Lübberstedt T (2020) QTL mapping of spontaneous haploid genome doubling using genotyping-by-sequencing in maize (*Zea mays* L.). Theor Appl Genet 133:2131–2140. https://doi.org/10.1007/s00122-020-03585-1

Tripathy SK, Swain D, Mohapatra PM, Prusti AM, Sahoo B, Panda S, Dash M, Chakma B, Behera SK (2019) Exploring factors affecting anther culture in rice (*Oryza sativa* L.). J Appl Biol Biotechnol 7:87–92. https://doi.org/10.7324/JABB.2019.70216

Tsuwamoto R, Fukuoka H, Takahata Y (2007) Identification and characterization of genes expressed in early embryogenesis from microspores of *Brassica napus*. Planta 225:641–652. https://doi.org/10.1007/s00425-006-0388-8

Vafadar Shamasbi F, Jamali SH, Sadeghzadeh B, Abdollahi Mandoulakani B (2017) Genetic mapping of quantitative trait loci for yield-affecting traits in a barley doubled haploid population derived from clipper × sahara 3771. Front Plant Sci 8:688. https://doi.org/10.3389/fpls.2017.00688

Vanous K, Vanous A, Frei UK, Lubberstedt T (2017) Generation of maize (Zea mays) doubled haploids via traditional methods. Curr Protoc Plant Biol 2:147–157. https://doi.org/10.1002/cppb.20050

Wang JL, Sun JS, Lu TG, Fang R, Cui HR, Cheng SZ, Yang C (1991) Fertilization and embryo development in wheat × maize crosses. Acta Bot Sin 33:674–679

Wang H, Liu J, Xu X et al (2016) Fully-automated high-throughput NMR system for screening of haploid kernels of maize (corn) by measurement of oil content. PLoS One 11:e0159444. https://doi.org/10.1371/journal.pone.0159444

Wang X-Y, Liao W-X, An D, Wei Y-G (2018) Maize haploid identification *via* LSTM-CNN and hyperspectral imaging technology. ArXiv, abs/1805.09105.

Wang S, Jin W, Wang K (2019) Centromere histone H3- and phospholipase-mediated haploid induction in plants. Plant Methods 15:42. https://doi.org/10.1186/s13007-019-0429-5

Wedzony M, Röber FK, Geiger HH (2002) Chromosome elimination observed in selfed progenies of maize inducer line RWS. In: XVIIth International Congress on sex plant reproduction. Maria Curie-Sklodowska University Press, Lublin

Wu P, Li H, Ren J, Chen S (2014) Mapping of maternal QTLs for *in vivo* haploid induction rate in maize (*Zea mays* L.). Euphytica 196:413–421. https://doi.org/10.1007/s10681-013-1043-7

Wu P, Ren J, Tian X et al (2017) New insights into the genetics of haploid male fertility in maize. Crop Sci 57:637–647

Xu Q, Yuan X, Yu H, Wang Y, Tang S, Wei X (2011) Mapping quantitative trait loci for sheath blight resistance in rice using double haploid population. Plant Breed 130:404–406. https://doi.org/10.1111/j.1439-0523.2010.01806.x

Xu X, Li L, Dong X et al (2013) Gametophytic and zygotic selection leads to segregation distortion through *in vivo* induction of a maternal haploid in maize. J Exp Bot 64:1083–1096. https://doi.org/10.1093/jxb/ers393

Yang XR, Fu HH (1989) Hua-03: a high protein indica rice. Intl Rice Res News Lett 14(3):14–15

Yao L, Zhang Y, Liu C, Liu Y, Wang Y, Liang D, Liu J, Sahoo G, Kelliher T (2018) OsMATL mutation induces haploid seed formation in indica rice. Nat Plants 4:530–533. https://doi.org/10.1038/s41477-018-0193-y

Ye JM, Kao KN, Harvey BL, Rossnagel BG (1987) Screening salt tolerant barley genotypes via F1 anther culture in salt stress media. Theor Appl Genet 74:426–429

Yu W, Birchler JA (2016) A green fluorescent protein-engineered haploid inducer line facilitates haploid mutant screens and doubled haploid breeding in maize. Mol Breed 36:1–12. https://doi.org/10.1007/s11032-015-0428-9

Yuan J, Guo X, Hu J, Lv Z, Han F (2015) Characterization of two *CENH 3* genes and their roles in wheat evolution. New Phytol 206:839–851. https://doi.org/10.1111/nph.13235

Zenketler M, Straub J (1979) Cytoembryological study on the process of fertilization and the development of haploid embryo of *Triticum aestivum* (2n = 42) after crossing with *Hordeum bulbosum* (2n = 14). Z Pflanzenzuchtung 82:36–44

Zenkteler M, Nitzsche W (1984) Wide hybridization experiments in cereals. Theor Appl Genet 68: 311–315. https://doi.org/10.1007/BF00267883

Zhahg-Yi Y, Hong-Ru K, Zhahg-Jin W, Li-Zheng Y, Zeng-Qian C (2008) High quality and blast resistance DH lines via anther culture. Southwest China J Agrl Sci 21:75–79

Zhang Z, Qiu F, Liu Y et al (2008) Chromosome elimination and *in vivo* haploid production induced by Stock 6-derived inducer line in maize (*Zea mays* L.). Plant Cell Rep 27:1851–1860. https://doi.org/10.1007/s00299-008-0601-2

Zhang W, Wang K, Lin ZS, Du LP, Ma HL, Xiao LL, Ye XG (2014) Production and identification of haploid dwarf male sterile wheat plants induced by corn inducer. Bot Stud 55:26. https://doi.org/10.1186/1999-3110-55-26

Zheng MY, Konzak CF (1999) Effect of 2,4-dichlorophenoxyacetic acid on callus induction and plant regeneration in anther culture of wheat (*Triticum aestivum* L.). Plant Cell Rep 19:69–73. https://doi.org/10.1007/s002990050712

Zheng YL, Luo MC, Yen C, Yang JL (1992) Chromosome location of a new crossability gene in common wheat. Wheat Inf Serv 75:36–40. https://doi.org/10.1534/genetics.109.107706

Zhong Y, Liu C, Qi X et al (2019) Mutation of *ZmDMP* enhances haploid induction in maize. Nat Plants 5:575–580. https://doi.org/10.1038/s41477-019-0443-7

Zhou H, Konzak CF (1989) Improvement of anther culture methods for haploid production in wheat. Crop Sci 29:817–821

Zhou C, Yang HY (1981) Induction of haploid rice plantlets by ovary culture. Plant Sci Lett 20: 231–237. https://doi.org/10.1016/0304-4211(81)90267-4

Zhu ZC, Wu HS, An QK, Liu ZY (1981) Induction of haploid plantlets from unpollinated ovaries of *Triticum aestivum* cultured in vitro. Acta Genet Sin 8:586–590

Zhu T, Peterson DJ, Tagliani L, St. Clair G, Baszczynski CL, Brown B (1999) Targeted manipulation of maize genes *in vivo* using chimeric RNA/DNA oligonucleotides. Proceeding of National Academy of Sciences 96:8768–8773. https://doi.org/10.1073/pnas.96.15.8768

Żur I, Dubas E, Krzewska M, Sánchez-Díaz RA, Castillo AM, Vallés MP (2014) Changes in gene expression patterns associated with microspore embryogenesis in hexaploid triticale (*Triticosecale* Wittm.). Plant Cell Tissue Organ Cult 116:261–267. https://doi.org/10.1007/s11240-013-0399-7

Rapid Generation Advancement and Fast-Track Breeding Approaches in Wheat Improvement

7

H. Prashanth Babu, Manjeet Kumar, Kiran B. Gaikwad, Ravindra Kumar, Naresh Kumar, Dharmateja Palaparthi, Hemalata Bharti, Kranti Kamre, and Rajbir Yadav

Abstract

The development of homozygous pure lines in wheat requires more than 3 and 6 years with and without offseason facilities. The traditional way of generation advancement is time-consuming, laborious and high-cost task. In recent times, advances in understanding the plant physiology and response of plants to different photoperiod regimes have helped breeders to adopt rapid generation advancement (RGA) protocols. These protocols have enabled the rapid generation of homozygous lines with more number of crop generations per year while enhancing the rate of genetic gain. Breeding strategies such as marker-assisted backcross breeding (MABB) and genomic selection (GS) can be easily integrated with RGA technology to develop stress-resilient modern wheat varieties quickly and efficiently. Generally, standardized protocols of doubled haploid (DH) technology and speed breeding are available and can be employed to reduce the time required to achieve homozygosity and develop a cultivar in wheat. In this chapter, we discuss different RGA protocols, their adaptive costs and limitations for successfully applying these strategies for accelerated breeding and maximization of genetic gain through an increased number of generations per year.

H. Prashanth Babu, Manjeet Kumar and Kiran B. Gaikwad contributed equally with all other contributors.

H. Prashanth Babu (✉) · M. Kumar · K. B. Gaikwad · N. Kumar · D. Palaparthi · H. Bharti · K. Kamre · R. Yadav (✉)
Indian Agricultural Research Institute, New Delhi, India

R. Kumar
Indian Institute of Wheat and Barley Research, Karnal, Haryana, India

M. G. Mallikarjuna et al. (eds.), *Next-Generation Plant Breeding Approaches for Stress Resilience in Cereal Crops*, https://doi.org/10.1007/978-981-19-1445-4_7

Keywords

Wheat · Rapid generation advancement · Doubled haploid · Speed breeding · MABB · GS

7.1 Introduction

According to the United Nations (UN) estimate, the global human population will be approximately 10.0 billion by 2050 (Yadav et al. 2018). The growing population and climate change have escalated global food security concerns. Additionally, land shrinkage for crop production due to environmental reasons and anthropogenic factors such as rapid urban and commercial development warrants the need to produce more crops per unit area. Besides fulfilling the food demand of the growing human population on planet earth, the science of plant breeding plays a pivotal role in adapting cropping systems under changing climate scenarios. The conventional breeding techniques demand colossal time and efforts to result in a successful and popular crop cultivar.

Wheat (*Triticum aestivum* L.) is one of the major food crops grown and consumed worldwide (Yadav et al. 2021). Conventional wheat breeding programmes globally have significantly delivered numbers of improved varieties with superior grain yield and resistance against various stresses in the past 100 years. However, in the present context, the progress achieved through conventional breeding looks slow, owing to lengthy breeding cycles, which often take approximately 10 years from cross to variety release. Despite increased wheat production due to bumper yields of the crop, which is achieved with fertilizer-responsive high-yielding wheat cultivars, the research community is far behind the target as far as wheat improvement is concerned. The current mean annual genetic gain in wheat has been around 1%, whereas the requirement is increasing by 1.7% annually. Therefore, 1.0 billion tonnes of wheat have to produce by 2050 to meet the target wheat requirement (Tadesse et al. 2019). At this juncture, rapid breeding cycles can play a crucial role in enhancing genetic gain by hastening the development and release of wheat cultivars with superior yield, stress resilience and quality traits.

Rapid generation advancement (RGA) techniques have been developed in many crops to quicken the breeding cycles and breeding advancement (Bhattarai et al. 2009; Depauw and Clarke 1976; Gaur et al. 2007; Ishigaki 2010; Rizal et al. 2014; Wang et al. 2011). The RGA technique was first proposed by Goulden (1939). Grafius (1965) suggested some modifications for the existing protocols, and subsequently, the most recent application of it in new form was proposed as 'speed breeding' (Watson et al. 2018). The time requisite in variety development depends on the number of cropping seasons required to create homozygous and stable genotypes followed by crossing two parents. In wheat, if only one crop generation is produced a year, it takes seven to nine cropping seasons or years to create homozygous lines after hybridization. Therefore, production of doubled haploids (DH) and employment of rapid generation advance (RGA) methods are most common contemporary tools being practiced in reducing the number of years required to produce stable homozygous lines and to develop cultivars in a short period. Many crops have utilized these procedures to fast-track the breeding cycles and achieve genetic improvement in less time. RGA approaches also significantly reduce the harvest time of crops to speed up the agricultural research to ensure better food production to deal with the increasing population pressure. Application of new protocols called 'speed breeding' (Atlin et al. 2017), which has been designed to accomplish up to six wheat generations in a single year (Watson et al. 2018). Therefore, furnishing as an advantageous mechanism in minimizing the duration of breeding cycles (Alahmad et al. 2018). This technology involves the generation of complete plants by sowing immature seeds harvested at physiological maturity under controlled conditions. RGA methods enhance the genetic gain rate by reducing the time between the crop to crop and with accelerated selection cycles. In addition, the number of phenotyping methodologies was reshaped to the speed breeding system and has been expanded to permit the characterization and selection for crucial traits in wheat. For example, seminal root phenotyping for drought tolerance (Richard et al. 2015), grain dormancy for tolerance to pre-harvest sprouting and disease resistance traits like adult plant resistance (APR) to leaf rust (Hickey et al. 2011; Alahmad et al. 2018), stripe rust (Riaz et al. 2016), yellow spot (Dinglasan et al. 2016) and crown rot (Alahmad et al. 2018) in bread wheat. In comparison to pedigree/bulk breeding methods, RGA protocols are much easier and effective as they do not involve selection and maintenance at every generation. As most of the RGA protocols are designed to undertake under a controlled environment, thus the risk of local harsh weather conditions could be eliminated in achieving the target of generation advancements. RGA provides additional advantages in terms of space as we employ the single seed descent (SSD) method of generation advancement, which uses a single seed per plant and less time to generate the next immediate generation of seeds. In RGA, selections are not practiced in the early generations (F_2-F_4) of segregating populations, which saves time and avoids any risks of losing valuable genotypes, possessing unfavourable linkages. Here in RGA, the early generations are developed under controlled greenhouse conditions, which are very equipped with breeders. Subsequently, the advanced stable generations (F_5-F_6) are evaluated under field conditions for the agronomic and other physiological traits. The rapid breeding methods help in deploying crop improvement strategies with much more efficiency on a pilot basis at the regional research centres to fasten the breeding progress. Further, the integration of contemporary crop breeding techniques with RGA methods helps to overcome the limitations of varying photoperiods and adverse seasonal changes associated with field conditions with mere or no losses of breeding germplasm.

7.2 Importance of Rapid Breeding Cycles in Wheat

Wheat is affected by various biotic stresses comprising of diseases caused by fungi, bacteria, viruses, nematodes, etc. and insects pests. Among the biotic stresses, diseases caused by fungi are very critical to achieve the potential yield of newly bred wheat cultivars. These diseases pose hurdles in the realization of the maximum yield output. The major diseases are rusts, powdery mildew, foliar blights and upcoming blast disease, whose impact on wheat production is well established (Figueroa et al. 2018). Rust diseases constitute the highest economically important fungal diseases of wheat and have a wide distribution in wheat-growing regions across the globe Babu et al. (2020). Norman E. Borlaug famously and appropriately said, 'rust never sleeps', which means rust pathotypes continuously evolve. A constant vigil on their occurrence at the global level and continuous breeding efforts are needed to be one step ahead of these rust pathogens. These pathogens can transform themselves into new races, biotypes or variants, which can evolve in a short span of time. Therefore, the rapid breeding cycles support the wheat breeding against such type of ever-evolving rust pathogens through rapid generation advancement. Rapid screening under control conditions using speed breeding can be a possibility in a short time. The six generations of disease-resistant material can be achieved in a single year through screening against rust using rapid breeding cycles. Against leaf rust of wheat, 15 genotypes were tested at Wellington under field conditions and speed breeding. There were not many differences in the disease reaction patterns in susceptible and resistant cultivars under both conditions. But these accelerated breeding cycles with the technology of speed breeding hastened the screening process and can enable the development of resistant cultivars.

New biotic stresses in crops have appeared across the globe during the last few decades intimidating food safety and security. Till date, diseases and pathotypes of exotic origin such as the Ug99 race of wheat stem rust and blast of wheat are not reported from India, which can be noted as a remarkable success of our wheat researchers and policymakers working under the aegis of India's National Agricultural Research System. Apart from the Ug99 race of stem rust, the blast is another emerging disease in wheat caused by *Magnaporthe oryzae* pathotype *Triticum* (MoT). Since its first report in 1985 from the Parana state of Brazil, the wheat blast has spread to different countries and reached Bangladesh in Southeast Asia in 2016 (Mottaleb et al. 2018). Although the anticipatory breeding efforts are going on prior to the occurrence of wheat blast disease in the country. Nevertheless, speed breeding that enabled rapid breeding cycles may support the screening of potential genotypes against wheat blast disease under anticipatory breeding programmes at different sites in the countries where this disease exists. The ever-evolving pathogens like *Magnaporthe oryzae* pathotype *Triticum* (MOT) can push back by utilizing rapid breeding cycles, which allow six wheat generations in a single year and hasten the disease resistance breeding. However, handling of pathogenic inoculum, rapid inoculum multiplication, inoculations and latent period required for symptom expression under screening are to be tested on a wider scale for better adaptability of rapid breeding cycles in wheat resistance breeding programmes.

7.3 Maximizing Genetic Gain Through RGA

Genetic gain is an improvement in the mean trait value within a population over breeding cycles as a response to selection (Crespo-Herrera et al. 2017). The disparity in the realized genetic gain across the world could be due to germplasm nature, crop duration, agronomic practices, prevailing weather conditions, soil and many other factors. However, developing advanced crop cultivars through improving several agronomic traits has ever been the major reason to increase the genetic gain via grain yield productivity enhancements. Genetic gains have primarily been studied by systematic evaluation of historical varieties released over different points of time (Beche et al. 2014). In wheat, a shorter duration findings have estimated more than 1% genetic yield gain per year (Underdahl et al. 2008).

More interestingly, the generation cycle (L) is the single parameter in the denominator of the breeder's equation for estimating genetic gain. Thus, the exponential increase in genetic gain would be possible by manipulating the time factor compared to other factors such as additive variance, additive genetic variation within the population(σ_a), selection intensity(i) and selection accuracy (r) in the genetic gain equation. Eberhart (1970) later introduced the 'L' into the denominator as a way to evaluate efficiency by expressing the response to selection as change over time.

$$\text{Genetic gain } (\Delta G) = (\sigma_a)(i)(r)/L$$

This equation keeps its importance in any crop breeding programme. The quantum of improvements is measured in terms of few parameters that the breeder can manipulate to gain maximum in important economic traits. The application of RGA methods in breeding programmes is found to be very much essential in achieving genetic gains, very quickly and efficiently, by reducing the 'L'. Knowing the complexity of crop breeding programmes, in understanding genetic and phenotypic information to carry out selections, there is a need to incentivize the breeding teams for better exploitation of parameters deciding the genetic gains. Among the parameters in the breeder's equation, the generation cycle is the simplest to perceive, economical to deploy and the very potent parameter for enhancing the genetic gain. The generation cycle involves recycling of breeding materials from advanced segregating materials into the crossing block when the breeder determines that the genotype is over-performing than the average breeding value of the individuals in the population. The breeding values are usually estimated as genomic estimated breeding value (GEBV) using advanced estimates predicted from genomic selection models.

The most preferred way to increase the genetic gain is to reduce the generation cycle time without altering the growth and development of the crop plants. In wheat, it takes an average of 9–10 years to come out with a commercial wheat cultivar (Atlin et al. 2017). After the development of the variety, it takes a longer time in its commercialization and spread, which hinders in achieving the maximum genetic gains in any breeding programmes. The vernalization requirement of winter wheat to enter into flowering stage is very well-known phenomenon, and the same adds to the prolonged cycle time of wheat (Davidson et al. 1985; Evans 1987). Many recent studies showed exciting findings that long exposure to cold temperatures has drastically reduced the cycling time in wheat (Watson et al. 2018). This strategy can be readily applied in wheat to reduce the cycle time and enhance the genetic gain as described in the breeder's equation.

As discussed earlier, though accelerating generation cycles will be the best approach to enhance the genetic gains, it has been very much underexploited by global breeding programmes across the world. The plant breeders emphasize the other three parameters of the breeders' equation, namely, heritable additive genetic variance, selection intensity and selection accuracy. Though these are very effective in the first few breeding cycles, they cause diminishing gains, leading to increased costs and decreased efficiency. A linear increase in heritability is an almost impossible task, and it does not increase the genetic gains linearly. The other two factors, selection accuracy and selection intensity, need a larger population size and come with higher cost investments in larger field trials and more replications to effectively reduce the amount of genetic gain achievable from the breeding materials. The impact of short breeding cycles in most breeding programmes is much greater than heritability or reduced selection proportion of breeding materials. Compared to the pedigree method, most RGA methods reduce the cultivar development time to 3–4 years, and there is still an opportunity to reduce the generation time to 1 or 1 0.5 years in many of the cereals and legumes. This can be achieved if the crop of interest is not very much sensitive to photoperiod and does not have specific photoperiod requirements. It is very common in many breeding programmes that the parents are selected only when they are completely stabilized and homozygous, which lengthen the breeding cycles. There are many approaches available wherein breeding values estimated on non-inbred individuals are considered and used as parents to maximize the genetic gains. Any typical breeding programme with an offseason facility would take two seasons a year and generate fixed lines in 3 years before taking up yield trials in the larger plots and multi-locations. Then the parents are selected among these better performing lines, and they would be cycled into the crossing blocks.

More typically, as well as understood phenomena, breeding cycles could be accelerated by carrying out selections in early generations of selfing, and the selected can be used as parents instead of waiting until later stages of fixation of lines. This will not only save time; it will also improve the breeding value of the population to develop better cultivars. As proposed by many workers, the recurrent selection is entirely based on crossing among individuals in the early generations of breeding and developing diverse lines instead of obtaining highly homozygous lines. In most recent times, invention of useful platforms like GS has greatly contributed in selecting parents based on the GEBVs, but still, experiments have to prove its effectiveness in estimating correct breeding values. Overall, recent advancements in achieving more generations in a year and fast-tracking the breeding programmes have shed a ray of hopes in maximizing genetic gains across the crops.

7.4 Accelerated Breeding Technologies in Wheat

7.4.1 Doubled Haploidy (DH)

With the availability of inadequate natural resources, land and water, and climate change-mediated stresses, the yield of staple food crops needs to be increased over time. Continued genetic gain in these major food crops requires innovative breeding technologies like doubled haploid (DH) technique which can significantly shorten the breeding cycles along with maintaining the genetic gain. Using DH strategy, the breeding process can be shortened to about 6–7 years, and rapid development of homozygous lines can be achieved instead of six to ten generations of inbreeding (Fig. 7.1; Prigge et al. 2012) which is a significant innovation to speed up varietal development (Dunwell 2010). Doubled haploids in wheat can be induced through anther culture and wide hybridization. However, it has commonly been experienced in recent years that the wheat x maize system of haploid induction is an effective and versatile tool among the available methods involving chromosome elimination.

Major wheat breeding programmes in the world like CIMMYT, ICARDA and PBI Sydney regularly utilize DH strategy in their wheat breeding programmes for genetic studies of economically important traits like rust resistance, nutritional quality, etc. Several countries, namely, China, Canada, France, Hungary and Romania have already released several wheat varieties developed through DH origin (Tadesse 2013). The wheat x maize DH production strategy is an integral part of the wheat breeding programme of Australia, which is dominated by two major companies, namely, Australian Grain Technology Pty. Ltd. (AGT) and Longreach Plant Breeder (Kuchel et al. 2005). Longreach came up with a wheat variety in 2016 named 'Longreach Reliant', developed through wheat × maize DH strategy. Public institutions in Australia like Plant Breeding Institute, University of Sydney, South Australian Research and Development Institute and Department of Agriculture and Food, Western Australia, have also employed this technique in their wheat breeding programmes for basic and applied research. In USA, wheat varieties 'Bond CL' and 'Gallagher' were developed in 2004 and 2012 by Colorado State University (Haley

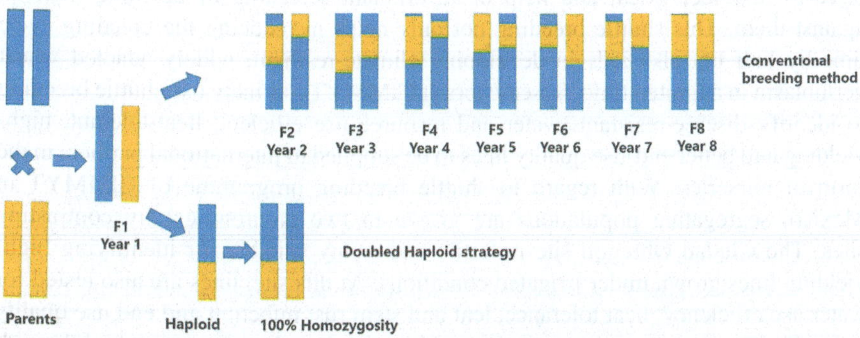

Fig. 7.1 Rapid development of homozygous wheat lines through DH strategy

et al. 2006) and Oklahoma State University using wheat × maize crosses. In Japan, the DH wheat cultivar 'Sanukioyume-2000' was developed through wheat × maize system (Yuichi et al. 2002).

In Canada, 12 years after the release of the first wheat DH variety, 27 wheat varieties were released that were developed through wheat x maize crosses approach at wheat breeding centres of Universities of Saskatchewan, Agriculture and Agri-food Canada and Manitoba (DePauw et al. 2011). Another widely used high protein variety, 'Lillian' having the gene *Gpc-B1/Yr36*, was also a product of DH technology (DePauw et al. 2011). In India, during the year 2015–2016, two DH lines PBW751 and PBW755 developed by Punjab Agricultural University, Ludhiana, were tested under National Initial Varietal Trials in a coordinated programme (Srivastava and Bains 2018). India's first wheat DH cultivar, 'Him Pratham', was bred at Himachal Pradesh Agricultural University, Palampur, through wheat × *Imperata cylindrica* crosses (Chaudhary et al. 2014), and was released in 2013. Integration of the DH technology combined with marker-assisted breeding can also effectively expedite wheat improvement programmes. In conclusion, wheat × maize DH production strategy offers an opportunity for rapid development of homozygous lines through accelerated breeding and therefore is reflected in the release of so many wheat varieties from major wheat breeding programmes across the globe.

7.4.2 Shuttle Breeding

Shuttle breeding refers to raising two or more crop generations in contrasting environments to shorten the breeding cycle and advance the generations. Shuttle breeding was initially employed at the International Maize and Wheat Improvement Centre (CIMMYT) by Norman E. Borlaug (Borlaug 1968). CIMMYT had identified two sites for shuttling their breeding material, namely, the Toluca station (19^{0}N latitude, 3660 m ASL), which is in the state of Mexico, and Ciudad Obregon station (27.5^{0}N latitude, 40.8 m ASL) in the state of Sonora, in Mexico. These Toluca and Obregon stations have diverse climatic conditions with respect to rainfall, temperature and photoperiods. These differences in climatic conditions allow various diseases to infect wheat and help in an efficient screening of breeding material against them. This shuttle breeding not only helps in reducing the breeding cycle time by half but also aids in developing climate-resilient, widely adapted wheat germplasm in a limited time. Researchers at CIMMYT routinely use shuttle breeding to identify disease-resistant, water and resource use efficient, heat-tolerant, high-yielding and better end-use quality lines to be supplied to international partners in the form of nurseries. With regard to shuttle breeding programme of CIMMYT in Mexico, segregating populations are grown in two environmentally contrasting sites. The Ciudad Obregon site is fertile and sunny suitable for identifying high-yielding lines grown under irrigated conditions. At this site, lines are also tested for water use efficiency, heat tolerance, leaf and stem rust infection and end use quality traits. The Toluca station has a cooler and high land environment with high humidity. These climatic conditions favour efficient screening of breeding material against stripe rust, *Septoria tritici* blotch (STB) and fusarium. At Cd. Obregon, planting of wheat is done in the month of November, and harvesting is carried out in the month of April or May. On the other hand, at the Toluca site, planting is done in May/June, and the produce is harvested in the month of October. The segregating material is routinely shuffled between these two sites, making it a useful approach for selecting the lines and studying the inheritance of simplex or complex traits at a relatively low cost. The CIMMYT wheat breeding programme has also extended its shuttle breeding facility at the Njoro station in Kenya to screen wheat germplasm for resistance against Ug99 and other variants of stem rust. Every year thousands of wheat breeding lines from many countries are screened at this site against the deadly Ug 99 stem rust race.

The wheat genetic enhancement programme at International Centre for Agricultural Research in the Dry Areas (ICARDA) also utilizes a shuttle breeding programme for developing wheat germplasm lines suitable for rainfed and irrigated ecologies throughout the world. For spring bread wheat, a shuttle breeding approach involving the winter-summer cycle at Terbol station (34° N; 36° E, 900 m ASL) in Lebanon, winter cycle at Merchouch station (33.6° N; 6.7° W, 430 m ASL) in Morocco, the Sids station (29° N; 31° E, 32.2 m.a.s.l.) in Egypt and the summer cycle at the Kulumsa station (08° N; 39° E, 2220 m.a.s.l) in Ethiopia is being followed (Tadesse et al. 2019). ICARDA has established germplasm phenotyping facilities in partnership with national programmes of the above-mentioned countries. In this context, Merchouch station is utilized for screening against stripe rust, *Septoria*, Hessian fly resistance and drought tolerance. The Sidi Alydi stations in Morocco are being used for terminal drought stress; Sids station in Egypt for yield potential; Izmir station in Turkey for SRT and APR screening against rust; Wadmedani station in Sudan for heat tolerance; and Kulumsa station for stem and stripe rust, Septoria blotch and Fusarium blight (Tadesse et al. 2019).

The Indian wheat breeding programme was initiated around 1905. However, till 1962 the pace in developing improved varieties was slow. The varieties developed during these 60 years were tall with weak stems and were unsuitable for intensive agriculture (Smale et al. 2008). After the onset of the 'Green revolution', semi-dwarf, lodging tolerant, disease-resistant, fertilizer-responsive genotypes were developed. During this period, the regional station of ICAR-Indian Agricultural Research Institute, at Wellington, was effectively utilized for shuttle breeding purpose. The seeds of imported Mexican varieties (Sonora 64, Lerma Roho) were multiplied at IARI-Wellington during 1964–1965 (SMS Tomar personal communication). After this, this station was regularly utilized for generation advancement and screening breeding material against leaf and stem rust. After the harvest of wheat in the month of April/May (winter season), the sowing is immediately taken (May/-June, off-season) at IARI- Wellington station, and the breeding material with one generation advanced is made available in the month of October for planting at main season again. This station is now providing space for many major wheat breeding centres for screening their breeding material against rusts, powdery mildew, fusarium head blight and generation advancement (Fig. 7.2).

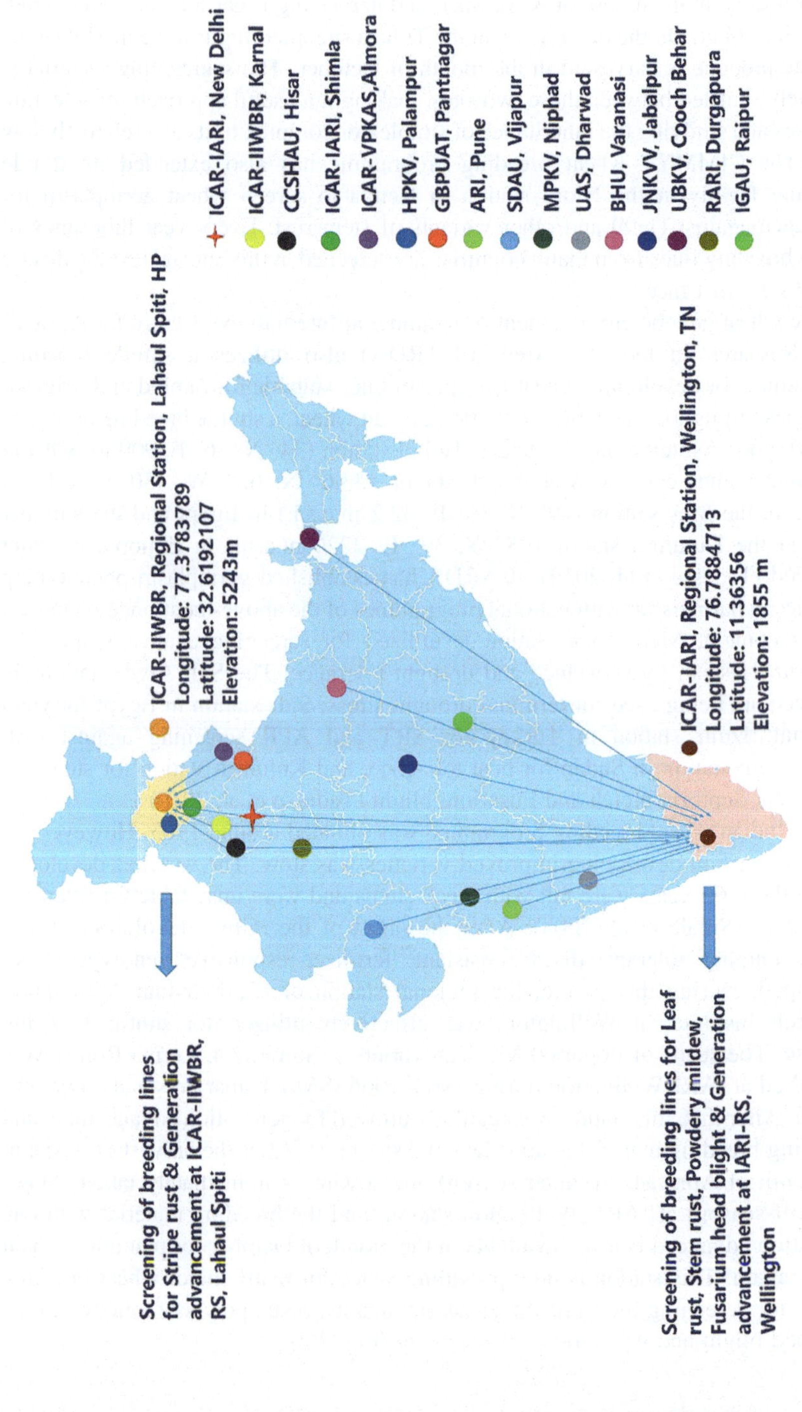

Fig. 7.2 Shuttle breeding programme of wheat operated at ICAR-IIWBR, RS Lahaul Spiti and ICAR-IARI, Wellington

Another important shuttle breeding facility is provided by the regional station of ICAR-Indian Institute of Wheat and Barley Research at Dalang Maidan, Lahaul Spiti, Himachal Pradesh. This station is regularly utilized for screening the breeding material against stripe rust, generation advancement and making corrective crosses. Major institutes working on wheat improvement programmes in North India like ICAR-IARI New Delhi, ICAR-IIWBR Karnal, CCSHAU Hisar, ICAR-IARI Shimal, HPKV Palampur, GBPUAT Pantnagar, etc. are utilizing this station as an off-season nursery (Fig. 7.2). These two stations working with major wheat breeding centres are playing an important role in assisting the breeding of improved wheat varieties with resistance to many biotic stresses.

7.4.3 Speed Breeding

The development of cultivars with conventional generation advancement procedures requires several years after the crossing of selected parental lines. The four to six generations of inbreeding are typically required to have the advanced stable lines for evaluation of grain yield and agronomic traits (Watson et al. 2018). This is even time-consuming for wheat having the off-season generation advancement through shuttle breeding with two generations per year. Globally, shuttle breeding in wheat improvement has been effectively utilized over the past decades, and the pace of yield gain has remained at par with the rising wheat demand. However, by 2050, crop production including wheat needs to double to fulfil the projected good grain requirement resulting from population growth, diet shifts and increasing biofuel consumption (Ray et al. 2013). Globally, the grain yield of wheat was increased by 0.9% per year, non-compounding rates, which is less than the required growth rate, i.e. 2.4% per year to double production by 2050 (Ray et al. 2013). Further, the presence of the narrow genetic base of breeding stocks is also a tremendous challenge to achieve the required growth rate of 2.4% per year. At the current growth rate of wheat production, global production would increase by only 38%, which would fall very short to meet projected demand. Therefore, the accelerated genetic gain of grain yield is the utmost requirement to meet the projected demand and save millions of people from hunger and starvation. The breeders' equation clearly shows that genetic gain can be enhanced by increasing the selection intensity, accuracy and additive genetic variance and shortening the breeding cycle. Tweaking selection accuracy and intensity can lead to minor improvements in genetic gain; however, shortening the breeding cycles per year would be very useful to boost the rate of genetic gain substantially (Li et al. 2018). From the equation, it is well evident that genetic gain can be double if the breeding cycle is reduced to half while maintaining the other factors like selection intensity, heritability and additive genetic variance as such. The shortening of breeding cycles per year can be accomplished by shuttle breeding and speed breeding under controlled artificial conditions.

In speed breeding, environmental conditions for crop growth are artificially manipulated under fully enclosed, controlled environment growth chambers aiming to accelerate flowering and seed set to advance to the next breeding generation as quickly as possible. The research and findings related to speed breeding are not of recent time; the systematic findings had been reported back to the year 1880. Siemens (1880) reported the effects of continuous light on the growth of quick-growing crops such as mustard, carrot, beans, cucumber and melons. Since then, research work on deciphering the effect of artificial environment on plant growth and development and improvement in LED technologies have been carried out (Pfeiffer 1926; Arthur et al. 1930; Bula et al. 1991; Darko et al. 2014; Stutte 2015). The first dwarf wheat variety, 'USU-Apogee' suited for rapid cycling under controlled conditions, was developed by NASA and Utah State University (Bugbee and Koerner 1997). However, the term 'speed breeding' was coined by researchers at the University of Queensland after getting inspiration from NASA in 2003. The systematic and very efficient protocol of speed breeding in wheat was designed by the researchers of the same organization (Watson et al. 2018; Ghosh et al. 2018). This technique requires the crop-specific optimal quality light, light intensity, day length and controlled temperature to speed up photosynthesis and flowering, which is coupled with early harvest of seed to shorten the generation time (Hickey et al. 2019). It is appropriate for variable germplasm and does not necessitate specialized laboratory facilities for in vitro culturing (Hickey et al. 2019). Speed breeding is a highly adaptable platform to achieve rapid generation advancement, where up to six generations per year can be achieved in bread wheat and durum wheat (Watson et al. 2018). The basic purpose of speed breeding is to develop the panel of homozygous lines with sufficient diversity retention after crossing parental lines as early as possible. Therefore, it would be very useful in accelerating the wheat improvement programme through faster generation of populations and adult plant phenotyping for specific traits (Watson et al. 2018) and identification of genomic regions associated with traits of interest. This technique would be very useful in harnessing the diversity present in gene banks by an integrated approach using a combination of speed breeding and genomic selection that could accelerate gene bank mining (Li et al. 2018). Speed breeding can be effectively integrated with novel breeding strategies like genomic selection to enhance the genetic gain per unit of time and multiplex genome editing with CRISPR-Cas9 tool for large-scale genome re-writing (Li et al. 2018) to understand the complex biochemical pathways and/or improvement of trait of interest. In Australia, the first spring wheat variety, i.e. DS Faraday, was released in 2017 after discovering new sources of resistance and genomic regions linked with DNA markers in the Vavilov wheat collection (Riaz et al. 2017).

7.4.4 Marker-Assisted Backcrossing

In marker-assisted backcross breeding (MABB), initially genomic regions associated with a trait of interest are identified with the QTL mapping approach utilizing the mapping population developed either from bi-parental crossing or natural population having sufficient variability for a trait of interest (Gaikwad et al. 2020). MABB is quite effective in transferring traits controlled by few genes and having a large effect on the phenotypic appearance of the target trait. In conventional backcross breeding, at least six to eight backcrosses are required to completely recover the recurrent parent genome (Collard et al. 2005). This typical formula, i.e. $(2^{n+1}-1)/2^{n+1}$, gives the theoretical percentage of the recurrent parent genome after n generations of backcrossing under no genetic drift (Collard et al. 2005). The recurrent parent genome recovery would be 75% in BC1; 87.5% in BC2; and 93.8% in BC3; likewise, four to six generations are needed for almost recovery of recurrent parent genome. However, recurrent parent genome recovery is average recovery over all individuals of the entire population. Some of the individuals might be carrying a more percentage of recurrent parent genome than average genome recovery. For rapid generation advancement and fixation of homozygosity, the selection of these individuals having high parent genome recovery is crucial. In this regard, molecular markers are very useful, by which researchers can select the individuals having the high proportion of recurrent parent genome. As a result, maximum recurrent parent genome recovery within two to three backcrosses and the target trait of interest is possible, making rapid generation advancement feasible. In MABB, tightly linked flanking markers to QTLs or gene(s) of interest and evenly distributed genome-wide markers from other genomic regions of the recurrent parent are employed for selection, the introgression of target QTLs and retrieval of recurrent parental genome (Collard et al. 2005). Although recurrent parent genome recovery in MABB largely depends on the number of markers, population size in each backcross generation and selection strategies (Rai et al. 2018). Frisch et al. (1999) did the computer simulation study to compare selection strategies with respect to proportion of the recurrent parent genome recovery, number of marker data points and population size in each generation for effective introgression. He observed that increasing population sizes from generation BC_1 to BC_3 reduced the required marker data points by 50% without disturbing the proportion of the recurrent parent genome recovery. In BC_1 and BC_2, two-stage selection (1. select individuals carrying the target allele and 2. select one individual which is homozygous for the recurrent parent allele at the maximum number of all markers across the genome) is superior to three- and four-stage selection because it reaches a larger recurrent parent genome proportion with given population size. However, suppose the removal of linkage drag is on high priority. In that case, three-stage (in addition to two-stage, 'Select individuals homozygous for the recurrent parent allele at most flanking markers') and four-stage (in addition to three-stage, 'Select individuals homozygous for the recurrent parent allele at all additional markers on the carrier chromosome') selection should be applied. A four-stage selection approach reduced the required number of marker data points by as much as 75% compared to all markers across the genome. The shortening of backcross generations, i.e. from six to three, with the requirement of moderate population sizes and the number of marker data points, is viable and efficient method to accelerate the breeding program. MABB can also be very effectively integrated with speed breeding for getting the maximum output per unit of time in the wheat improvement program.

7.4.5 Genomic Selection

To enhance wheat productivity, breeders need to use novel breeding strategies which can boost larger genetic gains in shorter times. The capacity of RGS methods to ensure the logistical and cost benefits has its own substantial advantages in crop breeding. The possibility of choosing parents from early generations of breeding cycles (first or second stage of selfing after hybridization) has largely proved their benefits in achieving maximum genetic gains. Though an effective combination of various contemporary approaches in practical crop improvement programmes remains challenging and requires substantial validation. A recent breeding strategy termed 'genomic selection' (Meuwissen et al. 2001) in crop plants has been studied carefully by many workers, and the primary advantage it offers is through reduced cycle time but not the increased accuracy. In its comparison, RGA with much lower cost involvement has the same capacities as genomic selection allows for more quickly realized genetic gains with the reduction in generation cycles. These two tools (GS and RGA) affect important steps in the breeding cycle, including choosing better candidates through selection and the production of seeds for the next generation and reducing the length of the breeding cycle with increased genetic gain per unit of time. Reductions in cycle time due to these methods are useful due to the more accurate selections and evaluation of homozygous, stable lines in replicated yield trials, which is not easier in pedigree and bulk methods without a proper selection and advancement among segregation generations of breeding cycles.

The only major point to be noted when anyone thinks to integrate GS in breeding programmes is the genotyping of fixed lines, which makes it non-economical. It could be employed to enhance the number of selection and individual genotypes with a standard budget provided genotyping is cheap as compared to field phenotyping. The GS has become an integral part of most wheat breeding programmes and has proved its effectiveness in accurately predicting yield, quality and disease resistance traits. Many parametric and nonparametric genomic prediction models with high accuracy have been proposed on specific experimental designs and data sets. Compared to phenotypic selection, GS contributes to higher genetic gain with slightly reduced selection accuracy, which is compensated by lowering breeding cycles (Heffner et al. 2010). Further, in winter wheat, Heffner et al. (2010) employed and trained GS model in the material developed from F_5 lines, which are generated through rapid generation advance (RGA) scheme. The accuracy of genomic predictions relies mainly on the training population, and it needs a larger size of the population with due consideration to the genetic relatedness of the population from which individuals are selected. Moreover, higher epistasis in the total genetic variance would need nonparametric models, and the additive genetic variance can be easily predicted using parametric models.

The higher genetic gains in any breeding programmes are more profound when breeding approaches bring rapid changes in the factors affecting genetic gains. It can be achieved through exploring methods which allow an increase in selection intensity, reduced generation cycles and higher heritability of important traits. The approaches like RGA and GS could enhance the genetic gains achieved per unit area with a positive impact on these factors. GS is very phenomenal in improving the '*i*' component (selection intensity), and RGA shortens the generation time (*L*). In addition to the higher selection intensity and the reduced generation time, these approaches in combination are more effective in reducing the costs of breeding programmes. The GS approach would reduce the large-scale field evaluations of individual lines and allows a larger number of different populations to be tested (Endelman et al. 2014). Therefore, GS increases the genetic gains per unit time with lesser costs in the development of new cultivars. Because of phenotyping of only selected individuals with minimal replication, increase both the accuracy and intensity of selection. Lorenz (2013) and Riedelsheimer and Melchinge (2013) in their simulation studies confirm that the use of genomic prediction usually led to an increased response to selection. Before advanced yield testing, selective screening of genotypes in an early stage of selection with markers linked to few economically important traits like disease resistance would be advantageous in increasing the selection intensity (*i*). But, due to undesirable linkages or correlations between some of the characteristics evaluated under early generations, it may not be feasible to select lines based on the early generation testing compared to more advanced generation testing for traits like yield and quality. The intense selection in early generations can result in reduced response to selection in addition to selection intensity and genetic gain. Hence, parallel testing should be done to study the impact of early generation selection prior to testing of advanced generations on overall genetic gain for important economic traits.

In conclusion, as genomic selection can be carried out on immature seedlings, it could reduce the breeding cycle time to a year or even less, and a new generation cycle can be initiated as soon as the selected candidates reach maturity through rapid generation advancement. Optimization of resources is also possible as GS reduces the size from F_3 generation onwards. It provides a significant advantage compared to traditional breeding through increasing the genetic improvement rate by tenfold. The only prerequisite in the successful integration of RGA and GS is to have an optimized breeding programme and strategies to reduce breeding cycle time in a cost-effective way.

7.5 Procedures and Protocols of Rapid Generation Advancement

7.5.1 Doubled Haploidy (DH)

In wheat, the fixation of target marker loci to stabilize crop yield, stress resilience and other agronomic traits using traditional breeding techniques would take several years for continuous inbreeding and selection. Doubled haploid production in wheat would be the best alternative to facilitate the wheat breeders to achieve line fixation in a single year and deliver lines with cent per cent homozygosity in a very short period. There are several strategies to produce DHs in wheat, including interspecific hybridization, microspore culture and wheat x maize-based crossing system. Among

Fig. 7.3 Production of DH in wheat: (**a**) growing maize as a pollen source, (**b**) wheat in controlled conditions from synchronization, (**c**) the emasculated wheat plants, (**d**) 2,4-D treated ears of wheat, (**e**) the ears with caryopses harvested, (**f**) growing haloid plants under tissue culture, (g) culturing haploid plants under tissue culture facility, (**h**) hardening haploid plants under controlled conditions, (**i**) tillering in haploid plants, (**j**) colchicine-treated haploid plants, (**k**) planting and hardening colchicine-treated plants and (**l**) the production of DH seed on colchicine-treated plants

these, wheat × maize-based system is the easiest and feasible method, which works on the principle that the chromosomes of maize are eliminated on crossing with maize which is followed by an embryo rescue technique and doubling of wheat haploid chromosomes using colchicine (Sadasivaiah et al. 1999; Ushiyama et al. 2007). In wheat doubled haploids were successfully employed in gene/QTL mapping and GWAS, including several genetic studies (Collard et al. 2005; Czembor et al. 2003; Trkulja et al. 2012) (Fig. 7.3; Table 7.1).

7.5.2 Speed Breeding

Speed breeding or accelerated plant breeding is a fast-accepting strategy among plant research groups worldwide to achieve plant generations more rapidly and

Table 7.1 A simple protocol for DH production in wheat is provided below

Steps	Stages	Detailed procedure
Emasculation of wheat spikes	Stage I	Desired ear of the wheat should be medium hard, and anther colour must be light green to green. Central florets of the spikelets are cut off, leaving the primary and secondary florets on the left and right side for better emasculation and seed set
	Stage II	Five to eight spikelets from the central portion of wheat ear must be used for the emasculation and hybridization. Later the emasculated spike is closed with crossing bags to avoid any further pollen contamination
Pollination	Stage I	After 2–3 days of emasculation, a pool of fresh maize pollen is dusted on to the spikelets for pollination. For good fertilization, it may be done two times between 8.30 am to 12 noon and 2.00 pm to 3.30 pm
	Stage II	After the pollination, cover the spike again with a crossing paper bag
Application of 2, 4-D	Stage I	Twenty-four hours after pollination, a drop of 150 ppm 2, 4-D needs to be applied using a plastic pasture pipette at the central portion of two florets.
	Stage II	The spikes should not be covered with crossing bag after 2, 4-D spraying to avoid any fungal growth
Collection of caryopses	Stage I	Wheat ears are cut off from the wheat plants 20–21 days after pollination and are placed in a flask with sterilized water. Later caryopses are scooped out of florets carefully using forceps, sterilized and cultured on to the medium on the same day
	Stage II	Full-strength Milton antibacterial soln. 0.95% w/w sodium hypochlorite equivalent to 1.00% w/w available chlorine is used for sterilization followed by rinsing three times with distilled water
Culturing caryopses	Stage I	The caryopses are cultured on B5 medium in a 30-ml tissue culture vial
	Stage II	Then keep the tubes in a refrigerator for 2 days at 4 °C, followed by shifting to a growth chamber maintained at 20–22 °C with 16 h of light for 3 weeks
Growing haploid plants	Stage I	After 20 days, carefully transfer the haploid plants from tissue culture tubes into small pots containing hardening soil media without fertilizer
	Stage II	After 4–5 days, the plants are shifted to a net house for tillering and establishment
Colchicine treatment	Stage I	The roots of haploid plants are thoroughly washed and treated in 0.15% colchicine solution for 3 h. roots are dipped in the solution up to the crown root level
	Stage II	After 3–4 h, plantlets will be removed from colchicine solution and are washed thoroughly under running tap water
	Stage III	The plants will then be transplanted in larger pots filled with soil
Production of doubled haploids	Stage I	After 2 weeks, plants are transferred to a net house and grown at 22–30 °C for production of seeds

develop varieties at a quicker pace. The procedures followed in undertaking the speed breeding involve simpler protocols and are easily adopted by research groups with minimal facilities with congenial environments. Under speed breeding, plants are grown under controlled greenhouse conditions using optimal light intensity and required day length and temperature. The conditions provided under these controlled conditions can accelerate several physiological activities in plants, particularly photosynthesis and flowering, consequently reducing the generation time. Additionally, speed breeding also allows to accomplish four to six generations per year instead of two to three generations which can be achieved where offseason facilities are available or under the controlled conditions. The speed breeding protocols are well established in some important staple crops and can be referred to Watson et al. (2018).

7.5.3 Marker-Assisted Backcross Breeding

Plant breeders widely use marker-assisted backcross breeding to transfer gene(s) of interest into superior agronomic lines. Commonly, those crops with less than two generations that can be taken in a year would need at least 4 years to develop gene introgressed lines (NILs). The integration of RGA with the MABB method can enable the quick genetic fixation of lines through modifying plant's growth conditions such that early flowering and seed set is achieved as compared to normal field conditions. The major benefits of RGA strategies compared to conventional approaches are speed, technical simplicity, the requirement of fewer resources and reduced costs. A schematic representation of integrating the RGA and MABB for simultaneous characterization and introgression of genes is given in Fig. 7.4.

7.6 Conclusion

The rapid breeding cycles are very important in wheat (*Triticum* spp.), which can be utilized to achieve superior wheat cultivars with better yield, disease resistance and nutritional status within the shortest time required from hybridization to release of varieties. Moreover, these rapid breeding cycles can be employed to get more yield per unit area from better performing wheat cultivars in the near future, which will be pertinent due to unfold urbanization and the development of commercial buildings on agricultural land.

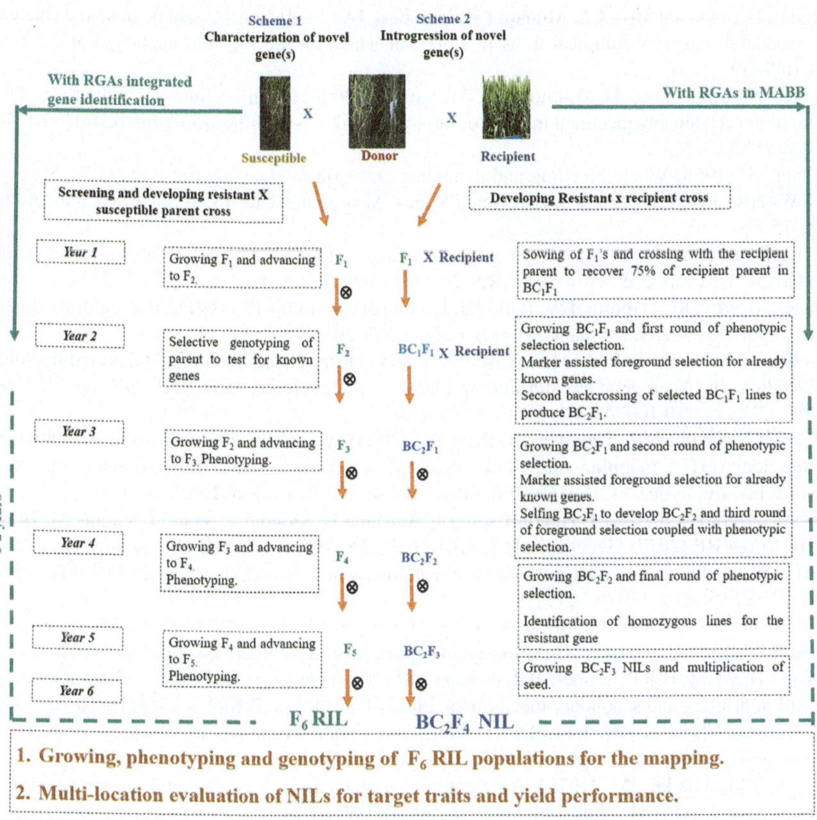

Fig. 7.4 A schematic representation showing the integration of RGA and MABB for simultaneous characterization and introgression of genes for various stress-resilient and agronomic traits

References

Alahmad S, Dinglasan E, Leung KM, Riaz A, Derbal N, Voss-Fels KP, Able JA, Bassi FM, Christopher J, Hickey LT (2018) Speed breeding for multiple quantitative traits in durum wheat. Plant Methods 14:36. https://doi.org/10.1186/s13007-018-0302-y

Arthur JM, Guthrie JD, Newell JM (1930) Some effects of artificial climates on the growth and chemical composition of plants. Am J Bot 17:416–482

Atlin GN, Cairns JE, Das B (2017) Rapid breeding and varietal replacement are critical to adaptation of cropping systems in the developing world to climate change. Glob Food Sec 12: 31–37. https://doi.org/10.1016/j.gfs.2017.01.008

Babu P, Baranwal DK, Harikrishna PD, Bharti H, Joshi P, Thiyagarajan B, Gaikwad KB, Bhardwaj SC, Singh GP, Singh A (2020) Application of genomics tools in wheat breeding to attain durable rust resistance. Front Plant Sci 11:567147. https://doi.org/10.3389/fpls.2020.567147

Beche E, Benin G, da Silva CL, Munaro LB, Marchese JA (2014) Genetic gain in yield and changes associated with physiological traits in Brazilian wheat during the 20th century. Eur J Agron 61:49–59

Bhattarai SP, La PenaRC D, Midmore DJ, Palchamy K (2009) In vitro culture of immature seed for rapid generation advancement in tomato. Euphytica 167:23–30. https://doi.org/10.1007/s10681-008-9855-6

Borlaug NE (1968) Wheat breeding and its impact on world food supply. In: Finlay EW, Sheperd KW (eds) 3rd International Wheat Genet Symp. Australian Academy of Science, Canberra, pp 5–15

Bugbee B, Koerner G (1997) Yield comparisons and unique characteristics of the dwarf wheat cultivar 'USU-apogee'. Adv Space Res 20:1891–1894

Bula RJ, Morrow RC, Tibbitts TW, Barta DJ, Ignatius RW, Martin TS (1991) Light-emitting diodes as a radiation source for plants. Hort Sci 26(2):203–205

Chaudhary HK, Kaila V, Rather SA, Tayeng T (2014) Distant hybridisation and doubled-haploidy breeding. In: Alien gene transfer in crop plants, vol 1. Springer, New York, NY, pp 143–164. https://doi.org/10.1007/978-1-4614-8585-8_6

Collard BC, Jahufer MZ, Brouwer JB, Pang EC (2005) An introduction to markers, quantitative trait loci (QTL) mapping and marker-assisted selection for crop improvement: the basic concepts. Euphytica 142(1):169–196. https://doi.org/10.1007/s10681-005-1681-5

Crespo-Herrera LA, Crossa J, Huerta-Espino J, Autrique E, Mondal S, Velu G, Vargas M, Braun HJ, Singh RP (2017) Genetic yield gains in CIMMYT's international elite spring wheat yield trials by modeling the genotype× environment interaction. Crop Sci 57(2):789–801. https://doi.org/10.2135/cropsci2016.06.0553

Czembor PC, Arseniuk E, Czaplicki A, Song QJ, Cregan PB, Ueng PP (2003) QTL mapping of partial resistance in winter wheat to Stagonospora nodorum blotch. Genome 46:546–554

Darko E, Heydarizadeh P, Schoefs B, Sabzalian MR (2014) Photosynthesis under artificial light: the shift in primary and secondary metabolism. Philos Trans R Soc B Biol Sci 369(1640):20130243

Davidson JL, Christian KR, Jones DB, Bremner PM (1985) Responses of wheat to vernalization and photoperiod. Aus J Agri Res 36(3):347–359

Depauw RM, Clarke JM (1976) Acceleration of generation advancement in spring wheat. Euphytica 25:415–418. https://doi.org/10.1007/BF00041574

DePauw RM, Knox RE, Humphreys DG, Thomas JB, Fox SL, Brown PD, Singh AK, Pozniak C, Randhawa HS, Fowler DB, Graf RJ (2011) New breeding tools impact Canadian commercial farmer fields. Czech J Genet Plant Breed 47(Special Issue):New-breeding. https://doi.org/10.17221/3250-CJGPB

Dinglasan E, Godwin ID, Mortlock MY, Hickey LT (2016) Resistance to yellow spot in wheat grown under accelerated growth conditions. Euphytica 209:693–707. https://doi.org/10.1007/s10681-016-1660-z

Dunwell JM (2010) Haploids in flowering plants: origins and exploitation. Plant Biotechnol J 8(4): 377–424. https://doi.org/10.1111/j.1467-7652.2009.00498.x

Eberhart SA (1970) Factors effecting efficiencies of breeding methods. African Soils 15(1/3):655–680

Endelman JB, Atlin GN, Beyene Y, Semagn K, Zhang X, Sorrells ME, Jannink JL (2014) Optimal design of preliminary yield trials with genome-wide markers. Crop Sci 54(1):48–59

Evans L (1987) Short day induction of inflorescence initiation in some winter wheat varieties. Funct Plant Biol 14:277–286. https://doi.org/10.1071/PP9870277

Figueroa M, Hammond-Kosack KE, Solomon PS (2018) A review of wheat diseases—a field perspective. Mol Plant Pathol 19(6):1523–1536. https://doi.org/10.1111/mpp.12618

Frisch M, Bohn M, Melchinger AE (1999) Minimum sample size and optimal positioning of flanking markers in marker-assisted backcrossing for transfer of a target gene. Crop Sci 39: 967–975. https://doi.org/10.2135/cropsci1999.0011183X003900040003x

Gaikwad KB, Rani S, Kumar M, Gupta V, Babu PH, Bainsla NK, Yadav R (2020) Enhancing the nutritional quality of major food crops through conventional and genomics-assisted breeding. Front Nutrit 7:533453. https://doi.org/10.3389/fnut.2020.533453

Gaur PM, Samineni S, Gowda CL, Rao BV (2007) Rapid generation advancement in chickpea. J SAT Agric Res 3(1)

Ghosh S, Watson A, Gonzalez-Navarro OE, Ramirez-Gonzalez RH, Yanes L, Mendoza-Suárez M, Simmonds J, Wells R, Rayner T, Green P, Hafeez A (2018) Speed breeding in growth chambers and glasshouses for crop breeding and model plant research. Nat Protoc 13(12):2944–2963. https://doi.org/10.1038/s41596-018-0072-z

Goulden CH (1939) Problems in plant selection. In: Burnett RC (ed) Proceeding of the seventh genetics congress. Cambridge University Press, Edinburgh, pp 132–133

Grafius JE (1965) Shortcuts in plant breeding. Crop Sci 5:377

Haley SD, Johnson JJ, Peairs FB, Quick JS, Westra PH, Stromberger JA, Clayshulte SR, Clifford BL, Rudolph JB, Giura A (2006) Registration of 'Bond CL' wheat. 46(2):993. https://doi.org/10.2135/cropsci2005.0031

Heffner EL, Lorenz AJ, Jannink JL, Sorrells ME (2010) Plant breeding with genomic selection: gain per unit time and cost. Crop Sci 50:1681. https://doi.org/10.2135/cropsci2009.11.0662

Hickey LT, Lawson W, Platz GJ, Dieters M, Arief VN, German S, Fletcher S, Park RF, Singh D, Pereyra S, Franckowiak J (2011) Mapping Rph20: a gene conferring adult plant resistance to *Puccinia hordei* in barley. Theor Appl Genet 123:55–68. https://doi.org/10.1007/s00122-011-1566-z

Hickey LT, Hafeez AN, Robinson H, Jackson SA, Leal-Bertioli SC, Tester M, Gao C, Godwin ID, Hayes BJ, Wulff BB (2019) Breeding crops to feed 10 billion. Nat Biotechnol 37(7):744–754. https://doi.org/10.1038/s41587-019-0152-9

Ishigaki Y (2010) Establishment of cultivation technique with rapid generation advancement of *Cyclamen persicum* by sowing seeds right after picking seeds. Bull Gifu Pref Res Inst Agr Sci in Hill Mount Are 6:7–12

Kuchel H, Ye G, Fox R, Jefferies S (2005) Genetic and economic analysis of a targeted marker-assisted wheat breeding strategy. Mol Breed 16(1):67–78. https://doi.org/10.1007/s11032-005-4785-7

Li H, Rasheed A, Hickey LT, He Z (2018) Fast-forwarding genetic gain. Trends Plant Sci 23(3): 184–186. https://doi.org/10.1016/j.tplants.2018.01.007

Meuwissen TH, Hayes BJ, Goddard ME (2001) Prediction of total genetic value using genome-wide dense marker maps. Genetics 157(4):1819–1829. https://doi.org/10.1093/genetics/157.4.1819

Mottaleb KA, Singh PK, Sonder K, Kruseman G, Tiwari TP, Barma NC, Malaker PK, Braun HJ, Erenstein O (2018) Threat of wheat blast to South Asia's food security: an ex-ante analysis. PLoS One 13(5):e0197555. https://doi.org/10.1371/journal.pone.0197555

Pfeiffer NE (1926) Microchemical and morphological studies of effect of light on plants. Bot Gaz 81(2):173–195

Prigge V, Xu X, Li L, Babu R, Chen S, Atlin GN, Melchinger AE (2012) New insights into the genetics of in vivo induction of maternal haploids, the backbone of doubled haploid technology in maize. Genetics 190(2):781–793. https://doi.org/10.1534/genetics.111.133066

Rai N, Bellundagi A, Kumar PK, Kalasapura Thimmappa R, Rani S, Sinha N, Krishna H, Jain N, Singh GP, Singh PK, Chand S (2018) Marker-assisted backcross breeding for improvement of drought tolerance in bread wheat (Triticum aestivum L. em Thell). Plant Breed 137(4):514–526. https://doi.org/10.1111/pbr.12605

Ray DK, Mueller ND, West PC, Foley JA (2013) Yield trends are insufficient to double global crop production by 2050. PLoS One 8(6):e66428. https://doi.org/10.1371/journal.pone.0066428

Riaz A, Periyannan S, Aitken E, Hickey L (2016) A rapid phenotyping method for adult plant resistance to leaf rust in wheat. Plant Methods 12(1):1–10

Riaz A, Athiyannan N, Periyannan S, Afanasenko O, Mitrofanova O, Aitken EA, Lagudah E, Hickey LT (2017) Mining Vavilov's treasure chest of wheat diversity for adult plant resistance to Puccinia triticina. Plant Dis 101(2):317–323. https://doi.org/10.1094/PDIS-05-16-0614-RE

Richard CA, Hickey LT, Fletcher S, Jennings R, Chenu K, Christopher JT (2015) High-throughput phenotyping of seminal root traits in wheat. Plant Methods 11(1):1–1. https://doi.org/10.1186/s13007-015-0055-9

Riedelsheimer C, Melchinge AE (2013) Optimizing the allocation of resources for genomic selection in one breeding cycle. Theor Appl Genet 126(11):835–2848

Rizal G, Karki S, Alcasid M, Montecillo F, Acebron K, Larazo N, Garcia R, Slamet-Loedin IH, Quick WP (2014) Shortening the breeding cycle of sorghum, a model crop for research. Crop Sci 54(2):520–529. https://doi.org/10.2135/cropsci2013.07.0471

Sadasivaiah RS, Orshinsky BR, Kozub GC (1999) Production of wheat haploids using anther culture and wheat x maize hybridization techniques. Cereal Res Commun 27(1):33–40

Siemens CW III (1880) On the influence of electric light upon vegetation, and on certain physical principles involved. Proc R Soc London 30(200–205):210–219. https://doi.org/10.1098/rspl.1879.0108

Smale M, Singh J, Di Falco S, Zambrano P (2008) Wheat breeding, productivity and slow variety change: evidence from the Punjab of India after the green revolution. Aust J Agric Resour Econ 52(4):419–432. https://doi.org/10.1111/j.1467-8489.2008.00435.x

Srivastava P, Bains NS (2018) Accelerated wheat breeding: doubled haploids and rapid generation advance. In: Biotechnologies of crop improvement, vol 1. Springer, Cham, pp 437–461. https://doi.org/10.1007/978-3-319-78283-6_13

Stutte GW (2015) Commercial transition to LEDs: a pathway to high-value products. Hort Sci 50(9):1297–1300. https://doi.org/10.21273/HORTSCI.50.9.1297

Tadesse W (2013) Methods and applications of doubled haploid technology in wheat breeding-A technical manual https://hdl.handle.net/20.500.11766/7530

Tadesse W, Sanchez-Garcia M, Assefa SG, Amri A, Bishaw Z, Ogbonnaya FC, Baum M (2019) Genetic gains in wheat breeding and its role in feeding the world. Crop Breed Genet Genom 1: e190005

Trkulja D, Kondić-Špika A, Brbaklić L, Kobiljski B, Hristov N (2012) Marker-trait associations for spike-related characters in a doubled haploid population of wheat. Rom Agric Res 29:9–16

Underdahl JL, Mergoum M, Ransom JK, Schatz BG (2008) Agronomic traits improvement and associations in hard red spring wheat cultivars released in North Dakota from 1968 to 2006. Crop Sci 48(1):158–166. https://doi.org/10.2135/cropsci2007.01.0018

Ushiyama T, Kuwabara T, Yoshida T (2007) Effects of various phytohormones on haploid wheat production in wheat x maize crosses. Plant Prod Sci 10(1):36–41

Wang X, Wang Y, Zhang G, Ma Z (2011) An integrated breeding technology for accelerating generation advancement and trait introgression in cotton. Plant Breed 130(5):569–573. https://doi.org/10.1111/j.1439-0523.2011.01868.x

Watson A, Ghosh S, Williams MJ, Cuddy WS, Simmonds J, Rey MD, Md Hatta MA, Hinchcliffe A, Steed A, Reynolds D, Adamski NM, Breakspear A, Korolev A, Rayner T, Dixon LE, Riaz A, Martin W, Ryan M, Edwards D, Batley J, Raman H, Carter J, Rogers C, Domoney C, Moore G, Harwood W, Nicholson P, Dieters MJ, DeLacy IH, Zhou J, Uauy C, Boden SA, Park RF, Wulff BBH, Hickey LT (2018) Speed breeding is a powerful tool to accelerate crop research and breeding. Nat Plants 4(1):23–29. https://doi.org/10.1038/s41477-017-0083-8

Yadav R, Gaikwad K, Bhattacharyya R, Bainsla NK, Kumar M, Yadav SS (2018) Breeding new generation genotypes for conservation agriculture in maize-wheat cropping systems under climate change. In: Yadav SS, Redden RJ, Hatfield JL, Ebert AW, Hunter D (eds) Food security and climate change, pp 189–228. https://doi.org/10.1002/9781119180661.ch10

Yadav R, Gupta S, Gaikwad KB, Bainsla NK, Kumar M, Babu P, Ansari R, Dhar N, Dharmateja P, Prasad R (2021) Genetic gain in yield and associated changes in agronomic traits in wheat cultivars developed between 1900 and 2016 for irrigated ecosystems of Northwestern Plain Zone of India. Front Plant Sci 12:719394. https://doi.org/10.3389/fpls.2021.719394

Yuichi H, Takashi O, Tetsuhiro M, Shinji T (2002) Breeding of new wheat cultivar 'Sanukinoyume 2000'. Bull 55:1–8

Integrating Advanced Molecular, Genomic, and Speed Breeding Methods for Genetic Improvement of Stress Tolerance in Rice

8

Suvarna Rani Chimmili, Sruthi Kanneboina, Prashanthkumar S. Hanjagi, Basavaraj P.S., Akshay S. Sakhare, Senguttuvel P., Sudhir Kumar, and Suneetha Kota

Abstract

Conventional plant breeding methods have contributed significantly to improving economic traits toward the development and release of elite cultivars of several crop species, including rice. Generation advancement through conventional methods is time-consuming and requires at least ten cycles to stabilize the variability generated before it is released for commercial cultivation. Recent advances in genomics and molecular tools have accelerated the breeding methods that can handle large datasets more precisely and efficiently, enabling marker-assisted selection, gene pyramiding for multiple stress tolerance, and gene stacking for introgression of multiple traits into an elite background. A comprehensive breeding approach is required which includes novel techniques that allow for rapid inbred line development and evaluation. Recently, a rapid generation advancement technique called "speed breeding" has been proposed to accelerate the generation advancement by shortening the generation cycle in several crops, including rice. Genomic selection (GS) is another novel breeding approach utilizing genome-wide high-throughput, cost-effective molecular markers for genetic evaluation. GS can increase genetic gain for quantitative traits, such as grain yield and stress resistance traits, by reducing the length of the selection cycle. Genomic selection combined with "speed breeding" (SB) could speed up

S. R. Chimmili · S. Kanneboina · A. S. Sakhare · S.P. · S. Kota (✉)
ICAR-Indian Institute of Rice Research, Hyderabad, Telangana, India

P. S. Hanjagi
ICAR-National Institute of Rice Research, Cuttack, Odisha, India

B. P.S.
ICAR-National Institute of Abiotic Stress Management, Baramati, India

S. Kumar
ICAR-Indian Agricultural Research Institute, New Delhi, India

M. G. Mallikarjuna et al. (eds.), *Next-Generation Plant Breeding Approaches for Stress Resilience in Cereal Crops*, https://doi.org/10.1007/978-981-19-1445-4_8

the breeding cycle, even more allowing for rapid generation advancement. Prior to field trials in a target environment, an integrated breeding technique could comprise quick derivation of line, phenotyping of yield-associated variables, and indirect phenotypic and multivariate GS for relevant traits. The present chapter focuses on the genetic improvement of rice through speed breeding methods, genomic selection approaches, and advanced genomic tools available in rice.

Keywords

Speed breeding · Rapid generation advancement · Genomic selection · Rice

8.1 Introduction

Rice, a staple crop, has been deeply imbibed in cultural regimes and is common in everyday diets. The worldwide population will increase by 25% and reach 10 billion over the span of 30 years ahead, with an alarming threat of climate volatility ringing the bells. To mitigate the hunger of the growing population and overcome climatic changes, rapid and improved technologies are required to feed the need with time. Agricultural scientists are put forth with the challenge of developing varieties that are more nutritious, are high yielding, and can withstand emerging pests, diseases, heavy-metal pollutants, and other abiotic stress factors. The transition from conventional breeding to molecular breeding due to rapid advances in second- and third-generation DNA technologies, high-throughput genotyping, and next-generation sequencing platforms made plant breeders and scientists affordably use DNA markers for indirect selections, gene discovery, trait dissection, and predictive plant breeding. Though new technologies have tremendous advantages, plant breeders are under pressure to develop climate-smart crop varieties in a given time frame because a variety in cereal crops like rice will be in the spotlight for 9–10 years using a pedigree method and may be outdated by that time frame (Hickey et al. 2019). The paramount limiting factor for the plant breeders is the time regime for attaining the homozygous fixed lines, which generally takes 3–4 years with one to two generations per year with the pedigree method, and it is exorbitant. A rapid generation advancement is one of the possible solutions, which can be attained by speed breeding. It can be customized for different crops with extended photoperiods and controlled conditions to increase crop generation within a time frame and has shown positive results in crops like wheat, barley, chickpea, and canola by more than half (Ghosh et al. 2018; Watson et al. 2018).

Combining contemporary state-of-the-art tools and methodologies with speed breeding strategies will reinforce the efforts to meet the challenges of plant breeders and will aid in feeding billions of people. Hence, it is indispensable to use the pooled techniques that are inexpensive and shorten the life cycle to release a new variety in optimum time. The basic principle behind speed breeding is that controlled conditions such as optimum light intensity, quality of light, temperature, and day-light increase the rate of photosynthesis, which directly stimulates early flowering and early seed harvest (Sunny Ahmar et al. 2020). Rapid generation advancement (RGA) is a speed breeding approach used in conjunction with the single seed descent (SSD) method and was first reported by Goulden in 1939 and then by Snape and Riggs in 1975; similarly, Japanese rice breeding programs employed controlled environment greenhouse-based rapid generation advancement technique to develop the rice cultivar "Nipponbare" (Koumura 1972a, b). Furthermore, production of experimental breeding lines such as isogenic cultivars, chromosome segment substitution lines (CSSLs), backcrossed inbred lines (BILs), and recombinant inbred lines (RILs) can accelerate the progress of rice genomics using rapid generation advancement method of breeding. Segregating populations were raised in a close-spaced environment with low nitrogen input, high temperature, and short days to reduce growth duration and produce multiple generations per year, as the stress factors hasten flowering and seed set earlier than in normal field conditions. Our aim in the present section is to discuss rice-customized speed breeding methods in combination with advanced genomic tools, as speed breeding in general is suitable for day-neutral and long-day crops with increased day length under controlled greenhouses.

8.2 Plant Breeding Contributions to Rice Crop Improvement and Breeding Methods for Development of Advanced Plant Materials

Plant breeding is considered the basis for human civilization. Without agriculture, civilization could not exist, and without modern cultivars, agriculture could not sustain the civilized world (Breseghello and Coelho 2013). Selection is the most primitive form of plant breeding, where, continuously observed, genetic variation was subjected to selection pressure, leading to significant changes in plant phenotypes. In the 1760s, the early phase of plant breeding started with the first hybridization experiments carried out by Kolreuter; later, with the discovery of Mendel's laws in 1866, the importance of hybridization has been widely recognized. Accumulation of favorable mutations is the major cause of plant domestication and the origin of new cultivated species. Most of the mutations are unfavorable and eliminated from the wild relatives through natural selection. However, some of the mutations may result in desirable plant phenotypes for cultivation or in terms of food/feed quality. Because of the founder effect during the domestication process, many valuable genes for pest and disease resistance were left out of the cultivated gene pool. Introgression of the valuable genes into modern cultivars remains a big challenge for advanced plant breeding and molecular tools. *Oryza sativa* and *Oryza glaberrima* are two cultivated species of rice domesticated from *Oryza rufipogon* and *Oryza barthii*, respectively (Sweeney and McCouch 2007). *Oryza rufipogon* is a perennial and outcrossing species, while *Oryza nivara* is an annual and self-pollinating species like cultivated *Oryza sativa*. According to Vaughan et al. (2008), the genes from both species contributed to the origin of cultivated species. *Indica* and *Japonica* are two species of *Oryza sativa*; still, it is unknown whether both of these subspecies evolved from a single domestication event or from separate domestication events (Kovach et al. 2007). In domestication process, many typical features from wild rice like seed shattering, seed dormancy, awns, dark and hairy hulls, and red pericarp were eliminated in the cultivated form.

Rice landraces originated and evolved in the field as a result of intentional and unintentional selection by farmers over generations for better grain yield and grain shape, being shaped by biotic and abiotic stresses of that region. Landraces are intermediate forms that are well differentiated genetically from wild relatives and still not exploited for cultivar development. Landraces may possess early rice domestication events and contain a specific combination of alleles extremely valuable for the genetic improvement of modern breeding cultivars. Pure-line selection is the earliest plant breeding method for cultivar development with uniformity. In pure-line selection, selection of desirable plant types from landraces of self-pollinating crops like rice is carried out. Subsequently, individual plant progenies are evaluated, and superior progeny are released as a pure-line variety. Pure-line varieties do not have wide adaption and stability than heterogeneous populations for various stresses because of their homogeneity. Later, plant breeding based on hybridization, i.e., pedigree method, became available to breeders to play with parents and combine the parents' best characters. This method is the interest of breeders for elite × elite crosses, where he can put best traits into the single background and easily trace back the pedigree of the new cultivar. However, in the pedigree method, the rate of the genetic gain of quantitative traits such as yield is normally modest, rarely exceeding 1% per year (Breseghello et al. 2011). Rice ideotype breeding is a strategy to improve the rate of genetic gain for quantitative traits, especially for complex traits like yield, and to increase the efficiency of pedigree breeding complex quantitative traits. IRRI and China have used this approach for increasing yield potential (Peng et al. 2008). The new plant type (NPT) model in rice was designed with key plant traits such as few productive tillers, large panicle with more than 200 grains with thick and lodging-resistant stems, thick erect, and intense green color leaves. First-generation NPT lines developed from tropical *japonica*s did not perform well because of poor grain filling and low biomass production. Second-generation NPT lines developed from *indic*a and tropical *japonicas* out-yielded first-generation NPT lines. China remains the leading country with the massive adoption of hybrid rice cultivation across the world, covering about 50% of its irrigated rice area (Janaiah 2002). India is second to China to breed and release the indigenous rice hybrid during 1994. In other prominent rice-growing Asian countries, viz., Vietnam and Bangladesh, the first released rice hybrids were imported and introduced from China (Janaiah and Hossain 2003). Over the last 32 years, miraculous progress has happened in India. So far, 127 hybrids have been released for commercial cultivation; among these, 37 have been released from the public sector, with the remaining 90 from the private sector (AICRIP, 2020 Crop Improvement Report). The QTL mapping approach effectively explains the contrasting traits between parents and understands the genetic control of quantitative traits. However, it is inefficient in explaining the genetic variation of traits present in germplasm. The association mapping offers a great promise to overcome the hindrance of biparental QTL mapping. QTLs explaining large phenotypic variance may directly jump for marker-assisted selection by using the markers lying close to the trait of interest. The advanced molecular tools like genomic selection and speed breeding together can increase the rate of genetic gain, reduce the number of breeding cycles, and accelerate product development. The impact of all these advanced plant breeding methods and genomic tools in the farmer's fields is just in the starting stage. Improved Samba Mahsuri (ISM) is the product of marker-assisted breeding released for bacterial blight resistance carrying three BLB resistance genes *(Xa21, xa13, and xa5*) (Sundaram et al. 2008) and *ISM* introgressed with blast resistance genes (*Pi-2* and *Pi-54*) and BLB resistance gene (*Xa-38*) (Madhavi et al. 2016; Yugander et al. 2018). Pusa Basmati-1 improved for blast two (*Pi2* + *Pi5*) and three *(Pi54 + Pil + Pita)* resistance (Khanna et al. 2015), while Pusa Basmati 1121 and Pusa Basmati 6 improved for blast (*Pi-2* and *Pi-54*) and BLB *(Xa21 and xa13)* resistance (Ellur et al. 2016). Abiotic stress tolerance QTLs were introgressed into various cultivars. Submergence tolerance (*Sub*-1) QTL was introgressed into Swarna (Neeraja et al. 2007); *Saltol* QTL was introgressed into Pusa Basmati 1121 and Pusa Basmati 6 (Waziri ct al. 2016).

8.3 Current Challenges in Plant Breeding

Owing to the development and release of high-yielding varieties for commercial cultivation, the global food production has increased at a faster rate. Plant breeding continuously develops adequate varieties to combat climate change pressures and many other stresses. Plant breeding can provide greater contributions in the near future with the help of supporting sciences like molecular- and genomics-assisted breeding, which are advancing rapidly. To ensure world food security and feed the hungry world, plant breeding must be the highest priority for governments and policymakers. The development of improved varieties for global food security and sustainable agriculture is the major aim of plant breeding. Plant breeding plays a major role in adapting to climate change and contributes to the stable increase of agricultural productivity. Plant breeding involves intensive research. Germplasm diversity is an important factor that influences biodiversity in agriculture. In recent decades, the decreasing trend of genetic diversity of the crop varieties was observed. The major challenge for plant breeding is the development of stable high-yielding stress-tolerant/resilient and resource-efficient varieties with high nutritional values to feed the ever-increasing population in the pace of climate change and with a shrinking natural resource base. Plant breeders are working continuously to integrate the recent advances in molecular biology into their toolbox to increase breeding efficiency by utilizing existing and inducing novel genetic variations. However, the development of new breeding techniques has not led to the complete replacement of older techniques. Based on their breeding goals, the breeders must choose the appropriate tools to accomplish the task more efficiently and in a specified manner.

8.4 Speed Breeding

The concept of growing plants under artificial light was experimented by botanists hundreds of years ago (Singh and Janeja 2021). During the 1980s in the United States, the National Aeronautics and Space Administration (NASA) in collaboration with Utah State University (USU) explored and experimented with the possibilities of rapidly growing and fastening the generations of wheat in the space station. These experimentations resulted the development of new wheat dwarf variety "USU-Apogee." The success of NASA and USU inspired plant scientists at Australia (University of Queensland and University of Sydney) and the United Kingdom (John Innes Centre) to improve the rapid generation advancement method and come up with the new technique. Dr. Lee T. Hickey coined the term "speed breeding" while working with wheat and peanut (Pfeiffer 1926; Bugbee and Koerner 1997). Standardized speed breeding protocols are available for various crop species under controlled conditions of prolonged photoperiod, light intensity, quality of light, and proper temperature with ambient humidity, which enhances the process of photosynthesis, thereby speeding up the growth and reducing the time of harvest. It has additional advantages over other technologies like accelerating the backcrossing, pyramiding traits, transgenic pipelines, etc. The first variety of spring wheat crop, "DS Faraday," was developed in 2017 in Australia by speed breeding methods (Singh and Janeja 2021). However, breeding for short-day plants like rice, which are photosensitive, could not be as beneficial as that of day-neutral plants like wheat, barley, and oats. So, evolved speed breeding methods for rice like biotron breeding system (BBS), simplified biotron breeding system (sBBS), and rapid generation advancement (RGA) will be dealt in the latter part of this section.

8.5 Integrating Speed Breeding with Contemporary Breeding Approaches for Stress Tolerance

Speed breeding is a prospective approach for improving crop varieties within a short time period compared to conventional approaches. Speed breeding employs an artificial environment that expands the light duration to create long-day situation to manipulate the life cycle of photo-insensitive crops.

8.5.1 Speed Breeding, Marker Assisted Breeding and Genomics

The rapid development of climate-smart varieties to sustain rice productivity is the best mitigation strategy to adapt to various stresses (biotic and abiotic) caused by climate change. Genomics-assisted breeding coupled with advanced plant breeding strategies such as speed breeding is the premier strategy for quick product development. Speed breeding is an advanced plant breeding strategy that can easily integrate with high-throughput genotyping and phenotyping techniques such as MAS/GAB and genome editing techniques (Fiyaz et al. 2020). Genomics-assisted breeding allows precise introgression of genes of interest into the popular mega-varieties. New genomic tools and knowledge can change the strategies in crop plant research, and there is a need to integrate advanced genomic tools with plant breeding (Varshney et al. 2005). Rapid selection of best performing inbred lines from a large breeding population by testing the genetic makeup of individuals would be the ultimate goal of a breeder. Peleman and van der Voort (2003) described the in silico designing of genotypes by breeding by design, which is an extension of the whole genomic survey. With the availability of rice sequence data in the public domain and high-throughput genotyping, marker-assisted breeding slowly changed into genomics-assisted breeding (Varshney et al. 2005). The rapid advancement of rice genomics facilitated the precise introgression of beneficial traits with no linkage drag (Ali et al. 2020). Varshney et al. (2021) discussed a fast-track approach for accelerated product development (GAB 2.0). Biotic stress-resistant, abiotic stress-tolerant, and nutritionally rich, superior grain quality varieties can be bred through advanced genomics-assisted breeding 2.0 (GAB 2.0) within a short time. These varieties can be expected to give a stable performance under climate change scenarios with the reduced application of chemical fertilizers and insecticides and enable environmental protection. In conventional breeding, it takes around 7–10 years to develop near-isogenic lines or introgressed lines, while in the case of marker-assisted breeding, it takes around 4–5 years; significant time has been reduced in marker-assisted breeding. However, when speed breeding is integrated with marker-assisted breeding, additional half of the time can be saved (Fig. 8.1). Marker-assisted speed breeding shortens the breeding cycle, even with a minimum of three generations per year; within 2–3 years, homozygous introgressed lines (ILs) can be developed rapidly. Rana et al. (2019) deployed SNP-based marker-assisted selection coupled with biotron-based speed breeding technique to improve salt tolerance (hst-1) in Yukinkomai, a high yielding variety from the Yokohama region of Japan. Fang et al. (2021) successfully advanced four generations in 1 year by adopting speed breeding combined with offsite summer and winter nurseries for generation advancement; fresh seeding method, which can further shorten the 1-month period; and marker-assisted selection for rapid and precise selection. Hickey et al. (2017) employed speed breeding for rapid introgression of multiple disease resistance traits (leaf rust, net and spot forms of net blotch and spot blotch) into barley Scarlett variety.

8.5.2 Speed Breeding and Genomic Selection

Many advanced molecular breeding strategies have been deployed for crop improvement, viz., marker-assisted backcrossing (MAB), marker-assisted recurrent selection (MARS), and genomic selection (GS). Recently, speed breeding (SB) has been added to the list to accomplish the breeding procedure more quickly. The major challenge for rice scientists is to increase rice productivity with shrinking natural resources like agricultural land and water to meet the demand of the rapidly expanding global population. On the other hand, the climate change scenario limits

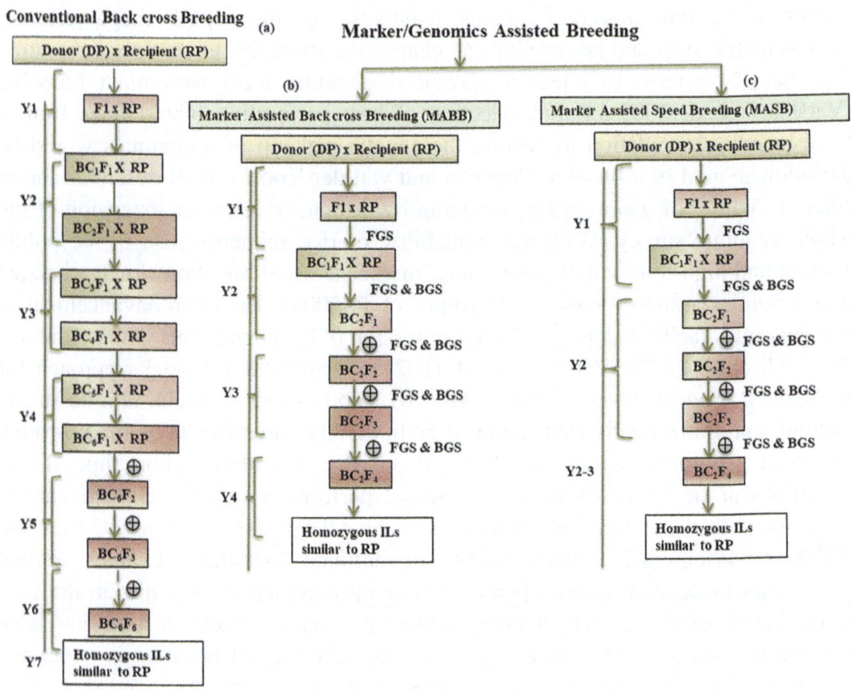

Fig. 8.1 Schematic representation of a comparison of duration between (**a**) conventional backcross breeding, (**b**) marker-assisted backcross breeding, and (**c**) speed breeding integrated with marker-assisted backcross breeding for rapid introgression of multiple traits. *Y* year, *DP* donor parent, *RP* recipient parent, *FGS* foreground selection, *BGS* background selection

the rice yield in terms of biotic and abiotic stress factors. Therefore, rapid development and release of climate-smart varieties and hybrids is the need of the hour. Advances in next-generation sequencing (NGS) technologies made it possible to integrate the genomic selection approach with new plant breeding strategies such as speed breeding to increase the rate of genetic improvement. Genomic selection has the ability to fix the complete genetic variation present in the breeding material and can accurately select the genotypes with higher breeding values without their phenotypic information. This enables the rapid selection and intercrossing of early-generation material with higher breeding values. It is really an exciting opportunity for the rice breeders to combine two advanced plant breeding tools (speed breeding and genomic selection) for rapid development and release of rice varieties or hybrids for the future. To accelerate the breeding procedure and efficiency, applying the predictive-based breeding approaches is a big challenge to the rice breeders. When integrating genomic selection and speed breeding (Fig. 8.2), it is possible to breed a minimum of three generations per year. Here, the genotyping cost is the major limitation; however, under NGS platforms, the high-throughput genotyping costs will decline in the near future, and it will make genomic selection

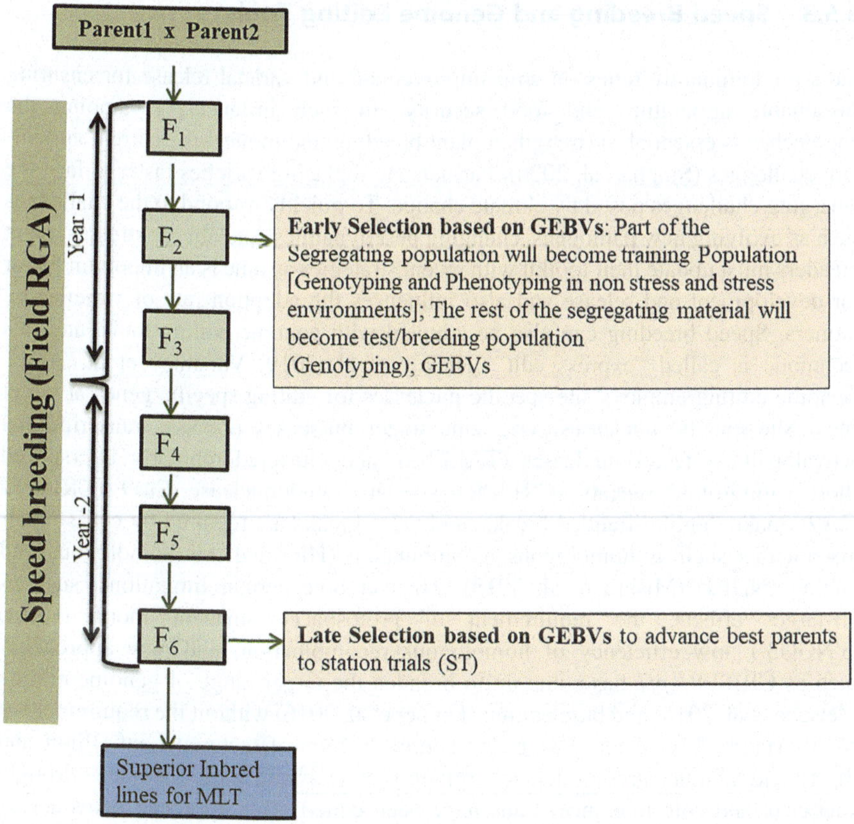

Fig. 8.2 Schematic representation for the integration of speed breeding and genomic selection with the pedigree method for stress tolerance. The training and test populations are from the same generation of a biparental cross

a routine job in crop breeding. Speed breeding and genomic selection together will become a game-changer in plant breeding for rapid breeding of quantitative traits. If resources allow, the selection based on GEBVs proposed twice in the breeding scheme (Spindel et al. 2015), once during early segregating generations for avoiding the elimination of beneficial alleles, and this will also increase the proportion of best-performing lines that are advanced to station trials (ST) and late selection based on GEBVs, can be used to select fixed lines to advance to station trials with highest prediction accuracy, and simultaneously can be used as parents for other crossing programs. For stress tolerance breeding, the phenotyping of the training population can be done in stress and non-stress environments.

8.5.3 Speed Breeding and Genome Editing Tools

India predominantly relies on crop improvement and varietal release for ensuring sustainable agriculture and food security. In such instances, revamping the approaches is essential to strengthen plant breeding and to meet the current agriculture challenges (Singh et al. 2020). Farmers are replacing varieties faster to face the emerging challenges posed by climate change. To quickly respond to the challenges such as evolving new pathogens, changing pest dynamics, and abiotic stresses, plant breeders must update their toolkit with recent strategies as time is an important factor for development and release and also influences the adoption rate of varieties by farmers. Speed breeding can also be coupled with genome editing tools, and this technique is called "express edit" (Hickey et al. 2019; Varshney et al. 2021). Genome editing employs site-specific nucleases for editing specific genes at target sites. Site-specific nucleases, viz., zinc finger nucleases (ZFNs), transcriptional activator-like effector nucleases (TALENs), and clustered regularly interspaced short palindromic repeats (CRISPR)-associated endonuclease Cas9 (CRISPR/*Cas9*), make double-stranded breaks, and these breaks are repaired by cell's repair machineries such as homologous recombination (HR) and non-homologous end joining (NHEJ) (Mishra et al. 2018). To overcome certain limitations, such as off-target effects, the requirement of protospacer adjacent motif (PAM) (5′NGG3′), low efficiency of homologous recombination, and new approaches such as CRISPR/*cpf1* has come up to broaden the target range of genome editing (Zetsche et al. 2015) and base editing (Komer et al. 2016) without the requirement of double-standard breakage. Using this approach, biotic (Bacterial Leaf Blight and Blast) and abiotic stress tolerance (glyphosate-resistant, salt tolerance, drought tolerance, and cold tolerance) traits have been edited. This approach takes a very long time and requires time-consuming tissue culture and sophisticated labs for regeneration. Express edit can overcome the main drawback of tissue culture procedure where the edited plants with *Cas9* gene can be used as a donor with elite lines, and by using marker-assisted speed breeding, the plants without *Cas9* gene can be generated through segregation (Fig. 8.3) (Hickey et al. 2019).

8.6 Cost-Effective Speed Breeding Techniques

The advent of genomic tools and reduced cost for genome sequencing empowered plant research to shift from model plants to crop plants. However, due to large seed-to-seed duration, there is a gap in the development of varieties in relation to climate and biotic challenges (Ghosh et al. 2018). The well-known strategy to shoot up the generations is "shuttle breeding" proposed by Norman Borlaug at CIMMYT during the 1950s, where they could get two generations per year for cereals like wheat and maize. Embryo rescue is another way to narrow down the time for seed maturation, where embryos are separated from seed and grown under culture media with or without phytohormones. Studies showed that with the use of PHs in culture media for embryo rescue, four generations of lentil up to 6.8 generations in faba bean had

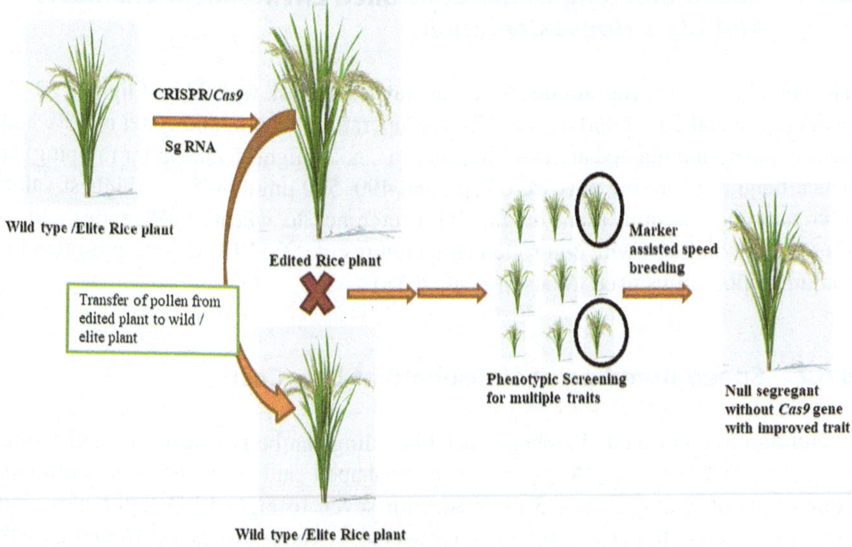

Fig. 8.3 Schematic representation of express editing in rice where speed breeding can be coupled with genome editing tools (CRISPR/*Cas9* system)

been achieved (Bermejo et al. 2016; Mobini et al. 2015). SB could attain six generations per annum for spring wheat, barley, chickpea, and pea and four generations per year for canola (Watson et al. 2018). Double haploid is another method for chromosomal doubling that uses haploid embryos and has been extensively used in breeding programs to reduce the number of generations to attain homozygosity from six or more generations to two, but it is labor-intensive and require high manpower for the removal of the embryo from the seed coat, and it is cumbersome for large population size and rate of haploid generation varies with crop species and varieties. SB can mitigate and accelerate the production of homozygous lines rapidly by speeding up the crosses and reducing the span for a generation in a more economical way (Ghosh et al. 2018).

Speed breeding is a strategy to hasten the breeding cycles of field crops. There is a dire need to increase the number of cycles per year and cut shot the capital requirement for development of a variety across time periods and in economic terms. The benefit-cost ratio varies from one method to another when compared over the year or in the long term. The protocol of SB varies from crop to crop and also depends on the trait of interest; initially, it is a trial-and-error method, and the standardization and stabilization of the protocol is the most important step. Established SB protocols are highly suitable for long-day plants or photoperiod neutral plants, which led to the evolution of RGA for short-day plants. Based on the capability of investing in the initial framework, facilities, and trained, skilled personnel, crop, and the ultimate goal of product development, the SB process has emerged into different speed breeding techniques.

8.6.1 Speed Breeding Under Controlled Environment Chambers and Glass House Condition

The chambers were programmed to mimic dawn and dusk with 22 h of light at 22 °C temperature and 2 h of darkness at 17 °C temperature, with humidity set to 70% and light intensity maintained at 360–380 μmol m^{-2} s^{-1} (highest value after ramping) at bench height, where the pots were kept, and 490–500 μmol m^{-2} s^{-1} (highest value after ramping) at adult plant height (with reference to wheat *cv.* Paragon) and at glasshouse condition with controlled temperatures of 22° C/17 °C from pre-pressure sodium vapor lamps used (Watson et al. 2018).

8.6.2 Speed Breeding at Household at Low Cost

In addition to controlled chambers, speed breeding can be practiced in small rooms in a cost-effective way. A room can be reshaped and designed with optimum dimensions of 3 m x 3 m x 3 m fitted with seven to eight LED lightboxes, and domestic air conditioner was set up as a low-cost alternative to the controlled growth chambers and with light intensity at bench height ranging from 210 to 260 μmol m^{-2} s^{-1} and at 50 cm above the pot from 340 to 590 μmol m^{-2} s^{-1} (Watson et al. 2018). This prototype can accommodate up to 90 pots of 8" diameter and 5 L volume. Hunter 10 Station Irrigation Controller, with one solenoid per room, can be employed to ensure automatic water supply, and ambient humidity conditions are created with 13 mm mainline with spike drippers (one per 8" pot).

Watering was achieved automatically with the Hunter 10 Station Irrigation Controller, with one solenoid per room and a 13 mm mainline with spike drippers (one per 8" pot) with ambient humidity conditions.

8.6.3 Speed Breeding Capsules

Initial setup for controlled growth chambers, screen houses, or glasshouse facilities is a bit expensive, and it was a major setback adoption of speed breeding in many crop improvement programs. There are many incidences of using disused shipping containers and customized built tankers, which were reused by the amateur farmers for hydroponic green production around the world, including most developing counties and underdeveloped countries, which include Kenya and Nigeria. The focus of this is to grow the produce near the point of consumption with very little usage of land and resources (CropBox 2018). In a similar way, shipping containers can be retrofitted with multitier cropping benches with lighting and air condition facilities for 25,000 USD (Saenz 2011); these customized speed breeding capsules can be shipped to any part of the world provided with access to minimum facilities, electricity, and water availability. To further reduce the start-up cost of speed breeding, all internal setup can be provided in a kit form with collapsible multitier growing benches and electrical and solar panel system packed with guiding or training manual. The deployment of a speed breeding capsule encourages the local breeders to customize the protocols to local landraces with important economic traits and further ensure regional food security (Chiurugwi et al. 2019).

8.6.4 Speed Breeding Centers

Established or the proposal for the development of speed breeding centers at potential research institutes like the CGIAR Centres and Research Programmes (worldwide), the African Orphan Crops Consortium (Kenya), West Africa Centre for Crop Improvement (Ghana), the Global Pulse Confederation (UAE), the World Vegetable Center (Taiwan), the BeCa-ILRI Hub (Kenya), and Crops for the Future (Malaysia), is a better approach for wide versatility of speed breeding (Chiurugwi et al. 2019). Researchers at CGIAR institutes, viz., International Institute of Tropical Agriculture (IITA), International Crops Research Institute for the Semi-Arid Tropics (ICRISAT), and the International Centre for Agricultural Research in the Dry Areas (ICARDA) had engaged there to develop facilities for speed breeding to accelerate breeding cycles for mandate crops.

8.6.5 Other Cost-Effective Strategies

Some of the innovative strategies like evaporative cooling systems that use seawater, semitransparent solar panels that selectively transmit wavelengths that promote plant growth, and more efficient lighting systems using laser light; supplementary LED lighting that provides more efficient power usage and reduced heat than other lighting types, such as sodium vapor lamps (Ghosh et al. 2018); and the use of drones for everyday monitoring could reduce the hindrance to adaptability in the future and broaden its use.

8.7 Rapid Generation Advancement (RGA) Methods

New races, pathotypes, and biotypes are emerging day to day in the current climate change scenario. The rice breeders need to adopt faster breeding methods to enhance the rate of genetic gain and to reduce operational costs. Rice breeders can transform their breeding method by adopting RGA as their main breeding scheme with simple reallocation of resources, which is currently spent on expensive pedigree method. As rice is a short-day plant, the response to increasing photoperiod may not be similar to that of long-day plants like wheat, barley, and oats. When rice is grown under speed breeding method with enhanced photoperiod, the vegetative phase will hasten up, but studies have shown that flowering will be initiated when taken outside the controlled condition and short days (Ohnishi et al. 2011; Tanaka et al. 2016).

8.7.1 Rapid Generation Advancement (RGA) Method in Rice

Rapid generation advancement is a flexible and feasible method where plants are grown in a greenhouse or screenhouse facilities to advance the breeding material from F_1 and subsequent generations within a short duration than in normal field conditions. In RGA, after the development of segregating material, as contrary to the pedigree method, selection will be delayed till F_6 generation. Segregating material will be grown in greenhouses, seedling trays, or raised beds of the fields with older seedlings, closer spacing, lower fertilizer, and higher temperature with restriction in root growth resulting in early flowering and maturity, thereby reducing crop duration. Once the target generation had been achieved (e.g., F_5 or F_6) under controlled environment, the panicles were planted in the field (*i.e.*, "panicle rows") to increase seed quantity (i.e., usually $F_{5:6}$ seed). However, planting in field condition also allows checking of lines for genetic uniformity (i.e., check for segregation due to outcrossing or mixtures) and selection. Screening for disease and pest resistance will be carried out at F_6 generation as line stage testing (LST), and the identified elite lines will be forwarded for the varietal release program. The combination of RGA and single seed descent is an efficient approach for plant breeders. It hastens the breeding cycle by increasing the number of favorable genotypes; thereby, breeding cost is reduced. Selection by RGA has limited applicability due to small-statured plants produced, and the ratio of favorable genotypes in the breeding population does not increase. Although there are limitations using RGA, it is feasible to screen and observe the resistance to biotic factors like pests and diseases, physiological parameters such as low-temperature tolerance, high-temperature tolerance, and grain characters that are stable under miniature culture, morphological characters like hairlessness, and characters linked to marker genes.

RGA method can be applied either in the greenhouse or in the field. In the greenhouse, the biotron breeding system can rapidly advance the generations in rice breeding. This system uses a growth chamber where segregating generations will be grown in seedling trays, Minoru trays, or wooden boxes or metal trays with controlled CO_2 and day length by maintaining appropriate root volume, which significantly decreases the days to heading (Tanaka et al. 2016). However, in greenhouses, one cannot accommodate a large number of crosses with a large population, and further, cost-effectiveness, nutrient, and pest management are difficult. In contrary to greenhouse-based RGA, field RGA (FRGA) can be used when greenhouse facilities are not available or not sufficient to screen a large number of crosses. The merits of adopting FRGA over greenhouse-based RGA are artificial microclimate, which can be avoided and often encountered in greenhouses and reduced pest outbreak. There is a chance of genetic drift and possible loss of material due to field pests, and natural disasters are the main concerns of FRGA. In field RGA, adopted by Bangladesh Rice Research Institute (BRRI), seeds can be sown directly or can be transplanted (older seedlings of 36 days) in the selected raised beds with very close spacing (5 cm × 5 cm) and low fertilizer or in other method described by Fahim et al. (1998), where they sow in seedling trays with high density, instead directly into the soil. In both these methods, the selected field area should have access to irrigation. International Rice Research Institute (IRRI) adopted the modified FRGA method by placing the seedling trays directly into the soil, which permits the roots to grow into the ground from the hole of the seedling tray. The major benefits of RGA include its technical simplicity, which allows for easy maintenance of a large population while selfing. Varietal development can be sped up by using RGA methods by rapid fixation of homozygosity in segregating populations. As it needs fewer inputs, it is inexpensive compared to normal pedigree breeding. This method has a higher selection efficiency than pedigree because selection will be practiced on homozygous fixed lines rather than segregating lines. In addition, the genetic gain can be enhanced because of handling a large number of populations with wide variation. Besides, RGA has advantages both in economic and also in the timeframe. The problem of carrying forward poor genotypes through RGA can be mitigated by using the single seed descent method (Fahim et al. 1998). It is estimated that reducing breeding cycle by only 1 year resulted in $19 million and $39 million of extra benefits over 2 years, with a standard discount rate of 5% (Pandey and Rajatasereekul 1999). Fahim et al. (1998) concluded that in comparison with the pedigree method, FRGA was found to be five to ten times cheaper. Using the RGA system, IRRI could advance two lakh plants per batch and can obtain three generations per year, which means six lakhs lines per year (Vergara et al. 1982).

8.7.2 Biotron-Based Speed Breeding System

Biotron breeding method or system (BBS) has been evolved to facilitate rapid and reliable rice cultivation under controlled and equipped conditions. Three major factors conditioned in BBS are regulation of CO_2 levels, removal of tillers, and embryo rescue technique. This system is more advantageous for photoperiod-sensitive rice varieties rather than insensitive ones. By using this biotron breeding system, Ohnishi and co-workers made a study and deliberated that it was possible to shorten the life cycle of Nipponbare by 2 months, and in addition to that embryo rescue method of immature seeds at 7 days after pollination reduces the time for seed maturation and seed dormancy period. This BBS system also allows the selective crossing program during unseasonal flexes, where the crossing program can be scheduled properly; once the plants from the biotron are taken to the outside ambient temperature, it starts flowering and the panicles which were used as females were dipped in hot water of 42 °C for 7 min, and the unopened spikelets will be clipped off; even though the crossed seeds were limited in quantity, the hybridity is more assured. There is a possibility of six generations with the shortened generation of Nipponbare to approximately 2 months. The improved artificial conditions will be useful for transgenic rice where the strict regulation regarding transgenic crops was imposed (Ohnishi et al. 2011). In addition to the biotron breeding system (BBS), with slight modifications, simplified biotron breeding system (sBBS) was proposed by Tanaka and co-workers where the necessity for tiller removal and embryo rescue is not required as it is tedious for larger populations, but with controlled conditions of

Table 8.1 Some of the key cultivars delivered for various stress-resilient and quality traits through RGA in rice (adapted from Collard Bertrand et al. 2017)

Country	Name of variety	Pedigree	Key feature	Release year	Reference
Philippines	IRRI 142	IR68333-R-R-B-22	Grain quality	2008	Ha et al. (2011)
	IRRI 165	IR71896-3R-8-3-1	Salinity tolerance	2011	Gregorio et al. (2013)
Bangladesh	BINA dhan8 (FL449)	IR66946-3R-149-1-1	Salinity tolerance	2010	Gregorio et al. (2013)
	BRRI dhan61	BR7105-4R-2	Salinity tolerance	2013	http://www.brri.gov.bd/
	BRRI dhan62	BR7517-2R-27-3	First Zn-enriched variety	2013	http://www.brri.gov.bd/
	BRRI dhan67	BR7100-R-6-6	Salinity tolerance	2014	http://www.brri.gov.bd/
	BRRI dhan72	BR7527-2R-19-HR10	Zn-enriched variety	2015	http://www.brri.gov.bd/
India	Luna Sankhi (CR Dhan 405)	IR72046-B-3-3-3-1	Salinity tolerance	2012	Gregorio et al. (2013)

CO_2 levels and day length and with appropriate root volume, this sBBS could reduce the interval between two generations of Nipponbare to 3 months without tedious manipulations and could facilitate four crossing cycles in a year (Tanaka et al. 2016). Some of the key product delivered through RGA in rice is presented in Table 8.1.

8.8 The Future Needs and Way Forward

Plant breeding originated when man started selecting the best plants through the process of domestication for human consumption. Because little was understood about the scientific foundation of plant traits, this type of plant breeding was purely based on art, i.e., human skills in judging and selecting superior plants. Over the years, plant breeding developed on sound scientific principles through Mendelian principles. Earlier, plant breeding was considered an art, now completely science-driven. Over the past decades, conventional breeding methods like selection, hybridization, and mutation breeding led to the evolution of superior crop varieties with high grain yield, superior quality, and tolerance to biotic and abiotic stresses. The development and release of high-yielding cultivars and hybrids have played a key role in increasing food grain production and ensuring food security of developing nations through the "green revolution." In the current climate change scenario, farmers and consumer preferences have changed for new, improved crop varieties for domestic consumption and exports. Agriculture and climate change are two processes that are intertwined. Climate change can have a negative impact on agriculture in a variety of ways, including variations in average temperatures (heat and cold stress), rainfall distribution (drought and floods), and the prevalence of biotic stressors (pests and diseases), which have a negative effect on food production and quality. In general, rather than depending on management approaches, stabilizing yield through the development and deployment of climate-resilient (tolerant to biotic and abiotic stresses) rice varieties is a plausible and economical approach. The current rate of varietal development is not sufficient to feed or meet the food demand of the ever-increasing population. Thus, the integration of innovative techniques with conventional breeding will greatly help to accelerate varietal development.

Plant breeders continuously look for new breeding techniques and approaches to bypass the recurring problems in selection and speed up the breeding process per unit time. Therefore, to cope up with the advancing several generations per unit time for achieving the desired genetic fixation of lines, several innovative approaches/technologies, viz., doubled haploidy, embryo culture, marker-assisted selection, transgenic breeding, speed breeding, and genome editing, are devised to supplement/complement the conventional breeding. Classically, plant breeders developed new cultivars by selecting directly or indirectly for yield and yield components in specific environments as most varieties will not perform stable across the environments. Furthermore, these traditional methods will take a longer time period for line development, thereby a longer period for varietal development and release for commercial cultivation. Integration of novel breeding techniques, viz., rapid generation advance (RGA) doubled haploid (DH), shuttle breeding, marker-assisted backcrossing (MABC), marker-assisted recurrent selection (MARS), genomic selection (GS), and speed breeding with traditional breeding methods (pedigree, backcross, and SSD) have great potential in hastening the breeding cycle along with efficient screening for specific biotic and abiotic stresses. Hence, accelerated breeding methods are efficient tools for developing new varieties in a shorter time to suit the stress environments to reduce the impact of climate change. Among the various novel breeding technologies, speed breeding emerged as a powerful technology to shorten the crop breeding cycle and fasten crop improvement through rapid generation advancement. Growing crops in the customized growth chambers of speed breeding helps to speed up research on crops with adult plant phenotyping, crossing, mutant studies, and transformation. Future needs include developing multiple stress-tolerant rice cultivars to safeguard food security in the climate change scenario. To overcome the problem of the rapid evolution of new races of pests and diseases, much emphasis should be laid on identifying new genes and sources of genes and precise incorporation with the aid of novel breeding tools like MAS, MABB, genome editing, etc. Optimization of speed breeding protocols accelerates the breeding programs by way of rapid fixation of lines. In addition, to combat climate change, genomic prediction technologies must be integrated with the routine breeding program for enhanced genetic gain. Furthermore, the integration of genetic engineering and gene editing tools like CRISPR/Cas with speed breeding approaches helps in the creation of novel genetic variability and rapid line fixation and development of new rice varieties in less time.

8.9 Conclusions

With the advent of speed breeding techniques pertaining to various crops and major staples, there is ample scope for fast-tracking and accelerating the development of elite cultivars with the progressive world population dynamics, traits preferred with climate volatilities, and changing preferences of markets. Now, advanced plant breeding and vast genomic tools and techniques have become available for accelerated rice breeding. So far, significant progress has been achieved in genomics-assisted breeding (GAB), and a number of GAB-derived climate-smart products in rice are now available for farmer cultivation. With the availability of whole-genome sequence and emerging new breeding and genomic strategies, the genomic breeding approaches like MAS, marker-assisted backcross breeding (MABC), marker-assisted recurrent selection (MARS), integrating genomic selection along with speed breeding, genome editing, and haplotype-based breeding can be used broadly for the rapid development and designing the future rice varieties. Rationalized and schematic operations of speed breeding reduce the cost, time, space, and manpower, making it more prominent to be integrated with crop improvement programs. For the wider adaptation of speed breeding or rapid generation advancement in the public domain of plant breeding, researchers and technicians should be trained along with the creation of proper infrastructure facilities. Cost of initial investment and installation of the setup should be made economical to mitigate its limitations, and also with foresight information on trending strategies of climate change, population expansion and market, pre-breeding research, and particularly successful dissemination of technology for large-scale adaptation of improved varieties will have the potency for food sufficiency and nutritional security.

References

Ahmar S, Gill RA, Jung KH, Faheem A, Qasim MU, Mubeen M, Zhou W (2020) Conventional and molecular techniques from simple breeding to speed breeding in crop plants: recent advances and future outlook. Int J Mol Sci 21(7):2590. https://doi.org/10.3390/ijms21072590

Ali, JA Mahender, GD Prahalada, Ma Anna, L. Sevilla, A Galang, E J De Asis, MD Paz, CM, Marfori-Nazarea, KL Nicolas, R Vinarao (2020) Rice Breeding Platform, International Rice Research Institute, Philippines, Springer Nature Switzerland AG, Genomic Designing of Climate-Smart Cereal Crops. https://doi.org/10.1007/978-3-319-93381-8

Bermejo C, Gatti I, Cointry E (2016) In vitro embryo culture to shorten the breeding cycle in lentil (*Lens culinaris* Medik). Plant Cell Tiss Org Cult 127:585–590. https://doi.org/10.1007/s11240-016-1065-7

Breseghello F, Coelho ASG (2013) Traditional and modern plant breeding methods with examples in rice (*Oryza sativa* L.). J Agric Food Chem 61:8277–8286. https://doi.org/10.1021/jf305531j

Breseghello F, Morais OP, Pinheiro PV, Silva ACS, Castro EM, Guimaraes EP, Castro AP, Pereira JA, Lopes AM, Utumi MM, Oliveira JP (2011) Results of 25 years of upland rice breeding in Brazil. Crop Sci 51:914–923. https://doi.org/10.2135/cropsci2010.06.0325

Bugbee B, Koerner G (1997) Yield comparisons and unique characteristics of the dwarf wheat cultivar 'USU-Apogee'. Adv Space Res 20(10):1891–1894

Chiurugwi T, Kemp S, Powell W, Hickey LT (2019) Speed breeding orphan crops. Theor Appl Genet 132(3):607–616. https://doi.org/10.1007/s00122-018-3202-7

Collard Bertrand CY, Beredo JC, Lenaerts B, Mendoza R, Santelices R, Lopena V, Verdeprado H, Raghavan C, Gregorio GB, Vial L, Demont M, Biswas PS, Iftekharuddaula KM, Rahman MA, Cobb JN, Islam MR (2017) Revisiting rice breeding methods – evaluating the use of rapid generation advance (RGA) for routine rice breeding. Plant Prod Sci 20(4):337–352. https://doi.org/10.1080/1343943X.2017.1391705

CropBox (2018) A new, smarter way to farm. http://cropbox.com. Accessed 15 Aug 2018.

Ellur RK, Khanna A, Yadav A, Pathania S, Rajashekara H, Singh VK, Krishnan SG, Bhowmick PK, Nagarajan M, Vinod KK, Prakash G (2016) Improvement of Basmati rice varieties for resistance to blast and bacterial blight diseases using marker assisted backcross breeding. Plant Sci 242:330–341. https://doi.org/10.1016/j.plantsci.2015.08.020

Fahim M, Dhanapala MP, Senadhira D, Lawrence MJ (1998) Quantitative genetics of rice. II. A comparison of the efficiency of four breeding methods. Field Crop Res 55:257–266. https://doi.org/10.1016/S0378-4290(97)00090-7

Fang Y, Wang L, Sapey E, Fu S, Wu T, Zeng H, Sun X, Qian S, Khan MAA, Yuan S, Wu C, Hou W, Sun S, Han T (2021) Speed-breeding system in soybean: Integrating off-site generation advancement, fresh seeding, and marker-assisted selection. Front Plant Sci 12:717077. https://doi.org/10.3389/fpls.2021.717077

Fiyaz AR, Ajay BC, Ramya KT, Aravind Kumar J, Sundaram RM, Subba Rao LV (2020) Speed breeding: methods and applications. In: Gosal S, Wani S (eds) Accelerated plant breeding, vol 1. Springer, Cham. https://doi.org/10.1007/978-3-030-41866-3_2

Ghosh S, Watson A, Lee TH (2018) Speed breeding in growth chambers and glasshouses for crop breeding and model plant research. Nat Protoc 13:2944–2963. https://doi.org/10.1038/s41596-018-0072-z

Goulden CH (1939) Problems in plant selection. In Proceedings of the Seventh International Genetics Congress. Cambridge University Press, pp. 132–133.

Gregorio GB, Islam MR, Vergara GV, Thirumeni S (2013) Recent advances in rice science to design salinity and other abiotic stress tolerant rice varieties. SABRAO J Breed Genet 45:31–41

Ha WG, Torollo GV, Kang KH, Lapiz M, Padolina T (2011) Breeding temperate japonica rice MS11 for the tropics. Philippine J Crop Sci 36:111–115

Hickey LT, Germán SE, Pereyra SA, Diaz JE, Ziems LA, Fowler RA, Platz GJ, Franckowiak JD, Dieters MJ (2017) Speed breeding for multiple disease resistance in barley. Euphytica 213(3): 64. https://doi.org/10.1007/s10681-016-1803-2

Hickey LT, Amber NH, Hannah R, Scott AJ, Soraya CM, Leal-Bertioli MT, Caixia G, Ian DG, Ben JH, Brande BHW (2019) Breeding crops to feed 10 billion. Nat Biotechnol 37(7):744–754. https://doi.org/10.1038/s41587-019-0152-9

Janaiah A (2002) Hybrid rice for Indian farmers: myths and reality. Econ Pol Wkly 42(37): 4319–4328. https://doi.org/10.2307/4412746

Janaiah A, Hossain M (2003) Can hybrid rice technology help productivity growth in asian tropics? Farmers' experiences. Econ Pol Wkly 38(25):2492–2501. http://www.jstor.org/stable/4413705

Khanna A, Sharma V, Ellur RK, Shikari AB, Krishnan SG, Singh UD, Prakash G, Sharma TR, Rathour R, Variar M, Prashanthi SK (2015) Development and evaluation of near-isogenic lines for major blast resistance gene(s) in Basmati rice. Theor Appl Genet 128:1243–1259. https://doi.org/10.1007/s00122-015-2502-4

Komer AC, Kim YB, Packer MS, Zuris JA, Liu DR (2016) Programmable editing of a target base in genomic DNA without double stranded DNA cleavage. Nature 533:420–424. https://doi.org/10.1038/nature17946

Koumura T (1972a) Breeding of new rice variety 'Nipponbare' (1). Agric Tech 27:112–116

Koumura T (1972b) Breeding of new rice variety "Nipponbare" (2). Agric Tech 27:159–161

Kovach MJ, Sweeney MT, McCouch SR (2007) New insights into the history of rice domestication. Trends Genet 23:578–587. https://doi.org/10.1016/j.tig.2007.08.012

Madhavi KR, Rambabu R, Abhilash Kumar V, Vijay Kumar S, Aruna J, Ramesh S, Prasad MS (2016) Marker assisted introgression of blast (*Pi*-2 and *Pi*-54) genes in to the genetic background of elite, bacterial blight resistant indica rice variety, improved Samba Mahsuri. Euphytica 212(2):331–342. https://doi.org/10.1007/s10681-016-1784-1

Mishra R, Joshi RK, Zhao K (2018) Genome editing in rice: recent advances, challenges, and future implications. Front Plant Sci 9:1361. https://doi.org/10.3389/fpls.2018.01361

Mobini SH, Lulsdorf M, Warkentin TD, Vandenberg A (2015) Plant growth regulators improve in vitro flowering and rapid generation advancement in lentil and faba bean. In Vitro Cell Dev Biol Plant 51:71–79. https://doi.org/10.1007/s11627-014-9647-8

Neeraja CN, Maghirang-Rodriguez R, Pamplona A, Heuer S, Collard BC, Septiningsih EM, Vergara G, Sanchez D, Xu K, Ismail AM, Mackill DJ (2007) A marker-assisted backcross approach for developing submergence-tolerant rice cultivars. Theor Appl Genet 115(6): 767–776. https://doi.org/10.1007/s00122-007-0607-0

Ohnishi T, Yoshino M, Yamakawa H, Kinoshita T (2011) The biotron breeding system: a rapid and reliable procedure for genetic studies and breeding in rice. Plant Cell Physiol 52(7):1249–1257. https://doi.org/10.1093/pcp/pcr066

Pandey S, Rajatasereekul S (1999) Economics of plant breeding: the value of shorter breeding cycles for rice in Northeast Thailand. Field Crop Res 64:187–197. https://doi.org/10.1016/S0378-4290(99)00059-3

Peleman JD, van der Voort JR (2003) Breeding by design. Trends Plant Sci 8(7):330–334. https://doi.org/10.1016/S1360-1385(03)00134-1

Peng S, Khush GS, Virk P, Tang Q, Zou Y (2008) Progress in ideotype breeding to increase rice yield potential. Field Crop Res 108:32–38. https://doi.org/10.1016/j.fcr.2008.04.001

Pfeiffer NE (1926) Microchemical and morphological studies of effect of light on plants. Bot Gaz 81(2):173–195. https://doi.org/10.1086/333584

Rana MM, Takamatsu T, Baslam M, Kaneko K, Itoh K, Harada N, Sugiyama T, Ohnishi T, Kinoshita T, Takagi H, Mitsui T (2019) Salt tolerance improvement in rice through efficient SNP marker-assisted selection coupled with speed-breeding. Int J Mol Sci 20(10):2585. https://doi.org/10.3390/ijms20102585

Saenz A (2011) Transforming shipping containers into local farms podponics brings produce to the city. In: Singul. Hub. https://singularityhub.com/2011/08/30/transforming-shipping-containers-into-local-farms-podponics-brings-produce-to-the-city/#sm.000100v3z66e9fdyt062k15j3s238. Accessed 7 Aug 2019.

Singh H, Janeja HS (2021) Speed breeding a ray of hope for the future generation in terms of food security: a review. Plant Arch 21(1):155–158

Singh RK, Prasad A, Muthamilarasan M, Parida SK, Prasad M (2020) Breeding and biotechnological interventions for trait improvement: status and prospects. Planta 252(4):1–8. https://doi.org/10.1007/s00425-020-03465-4

Snape JW, Riggs TJ (1975) Genetical consequences of single seed descent in the breeding of self-pollinating crops. Heredity 35(2):211–219. https://doi.org/10.1038/hdy.1975.85

Spindel J, Begum H, Akdemir D, Virk P, Collard B, Redona E, Atlin G, Jannink JL, Mc Couch SR (2015) Correction: genomic selection and association mapping in rice (*Oryza sativa*): effect of trait genetic architecture, training population composition, marker number and statistical model on accuracy of rice genomic selection in elite, tropical rice breeding lines. PLoS Genet 11(6): e1005350. https://doi.org/10.1371/journal.pgen.1005350

Sundaram RM, Vishnupriya MR, Biradar SK, Laha GS, Reddy GA, Rani NS, Sonti RV (2008) Marker assisted introgression of bacterial blight resistance in Samba Mahsuri, an elite indica rice variety. Euphytica 160(3):411–422. https://doi.org/10.1007/s10681-007-9564-6

Sweeney M, McCouch S (2007) The complex history of the domestication of rice. Ann Bot 100: 951–957. https://www.ncbi.nlm.nih.gov/pmc/articles/PMC2759204/#

Tanaka J, Hayashi T, Iwata H (2016) A practical, rapid generation-advancement system for rice breeding using simplified biotron breeding system. Breed Sci 66(4):542–551. https://doi.org/10.1270/jsbbs.15038

Varshney RK, Graner A, Sorrells ME (2005) Genomics-assisted breeding for crop improvement. Trends Plant Sci 10:621–630. https://doi.org/10.1016/j.tplants.2005.10.004

Varshney RK, Bohra A, Jianming Y, Graner A, Zhang Q, Sorrells ME (2021) Designing future crops: genomics-assisted breeding comes of age. Trends Plant Sci 26(6):631–649. https://doi.org/10.1016/j.tplants.2021.03.010

Vaughan DA, Lu BR, Tomooka N (2008) The evolving story of rice evolution. Plant Sci 174:394–408. https://doi.org/10.1016/j.plantsci.2008.01.016

Vergara BS, Patena G, Lopez FSS (1982) Rapid Generation of rice at the International Rice Research Institute, IRRI Research Paper Series No. 84. International Rice Research Institute, Los Baños

Watson A, Ghosh S, Williams MJ, Cuddy WS, Simmonds J, Rey MD, Hickey LT (2018) Speed breeding is a powerful tool to accelerate crop research and breeding. Nat Plants 4(1):23–29. https://doi.org/10.1038/s41477-017-0083-8

Waziri A, Kumar P, Purty RS (2016) Saltol QTL and their role in salinity tolerance in rice. Austin J Biotechnol Bioeng 3(3):1067

Yugander A, Sundaram RM, Singh K, Ladhalakshmi D, Subba Rao LV, Madhav MS, Badri J, Prasad MS, Laha GS (2018) Incorporation of the novel bacterial blight resistance gene *Xa38* into the genetic background of elite rice variety Improved Samba Mahsuri. PLoS One 13(5): e0198260. https://doi.org/10.1371/journal.pone.0198260

Zetsche B, Gootenberg JS, Abudayyeh OO, Slaymaker IM, Makarova KS, Essletzbichler P et al (2015) Cpf1 is a single RNA-guided endonuclease of a class 2 CRISPR-Cas system. Cell 163: 759–771. https://doi.org/10.1016/j.cell.2015.09.038

Varshney RK, Graner A, Sorrells ME (2005) Genomics-assisted breeding for crop improvement. Trends Plant Sci 10(12):621–630. https://doi.org/10.1016/j.tplants.2005.10.004
Varshney RK, Bohra A, Yu J, Graner A, Zhang Q, Sorrells ME (2021) Designing future crops: genomics-assisted breeding comes of age. Trends Plant Sci 26(6):631–649. https://doi.org/10.1016/j.tplants.2021.03.010
Vaughan DA, Lu BR, Tomooka N (2008) The evolving story of rice evolution. Plant Sci 174(4):394–408. https://doi.org/10.1016/j.plantsci.2008.01.016
Vergara BS, Puranik B, Chang TT (1982) Rapid generation of rice at the International Rice Research Institute. IRRI Research Paper Series No. 84. International Rice Research Institute, Los Baños
Watson A, Ghosh S, Williams MJ, Cuddy WS, Simmonds J, Rey MD et al (2018) Speed breeding is a powerful tool to accelerate crop research and breeding. Nat Plants 4(1):23–29. https://doi.org/10.1038/s41477-017-0083-8
Waziri A, Kumar P, Purty RS (2016) Saltol QTL and their role in salinity tolerance in rice. Austin J Biotechnol Bioeng 3(3):1067
Yugander A, Sundaram RM, Singh K, Ladhalakshmi D, Subba Rao LV, Madhav MS, Badri J, Prasad MS, Laha GS (2018) Incorporation of the novel bacterial blight resistance gene Xa38 into the genetic background of elite rice variety Improved Samba Mahsuri. PLoS One 13(5):e0198260. https://doi.org/10.1371/journal.pone.0198260
Zetsche B, Gootenberg JS, Abudayyeh OO, Slaymaker IM, Makarova KS, Essletzbichler P, Volz SE, Joung J, van der Oost J, Regev A, Koonin EV, Zhang F (2015) Cpf1 is a single RNA-guided endonuclease of a class 2 CRISPR-Cas system. Cell 163(3):759–771. https://doi.org/10.1016/j.cell.2015.09.038

CRISPR Genome Editing Brings Global Food Security into the First Lane: Enhancing Nutrition and Stress Resilience in Crops

9

Tanushri Kaul, Sonia Khan Sony, Jyotsna Bharti, Khaled Fathy Abdel Motelb, Rachana Verma, Arulprakash Thangaraj, Rashmi Kaul, Mamta Nehra, and Murugesh Eswaran

Abstract

Crop yield is adversely being influenced very frequently due to biotic and abiotic stresses, globally. The insufficient yield and low nutritional value of the crops are due to the numerous ambient stresses. Which has challenged the nutritional security of people in developing and underdeveloped nations that are already been malnourished. The tremendously changing global climate and the ever-increasing world population are the principal apprehensions guiding towards the adaptation of a neoteric technique that can aid in achieving the sustainable development of agriculture with enriched nutritional value, plant resilience, and improved yield potential. The clustered regularly interspaced short palindromic repeat (CRISPR) and CRISPR-associated (Cas) protein-based genome-editing (CRISPR-Cas) tool is the most valuable technique to boon the modern world and offers an edge over to meganucleases, zinc finger nucleases (ZFNs), and transcription activator-like effector nucleases (TALENs) by being its tremendous potency, accuracy, ease of use, and versatility. Here, we have highlighted the neoteric advancements of the CRISPR-Cas-based approaches that have revolutionized the way of food production in the agriculture industry. It has paved the way for food security by modifying crucial crop attributes by introducing desirable characteristics that employ knockout and/or knockin of targeted genes to generate resistant crop plants with enriched nutritional quality, yield enhancement, and stress resilience. In addition, we have also shed light on different mechanisms, challenges, approaches for the minimization of off-target effects, and future possibilities of these neoteric genome-editing tools.

T. Kaul (✉) · S. K. Sony · J. Bharti · K. F. A. Motelb · R. Verma · A. Thangaraj · R. Kaul · M. Nehra · M. Eswaran
Nutritional Improvement of Crops Group, International Centre for Genetic Engineering and Biotechnology (ICGEB), New Delhi, India
e-mail: tanushri@icgeb.res.in

M. G. Mallikarjuna et al. (eds.), *Next-Generation Plant Breeding Approaches for Stress Resilience in Cereal Crops*, https://doi.org/10.1007/978-981-19-1445-4_9

Furthermore, with the advent of the CRISPR-based platform, the numerous emerging biotechnologies have broadened the basic crop research toolbox and synthetic biotechnology via the incorporation of artificial intelligence (AI) and various bioinformatics frameworks. Eventually, the current global regulatory stratagems and social approval of CRISPR-Cas-based crop trait enhancement have been explored.

Keywords

Climate change · CRISPR-Cas · Genome editing · Nutrition · Stress resilience

9.1 Introduction

9.1.1 Genome Editing: An Introduction to the Plethora of Tools in the Arsenal of Gene Editing

Genetic modification for the development of desired traits in plants utilized for food began before the end of the Pleistocene era about 12,000–11,000 years ago (Larson et al. 2014). Since then, numerous changes have happened due to natural evolutionary processes, which resulted in new crop species that are now genetically different from their ancestors. After establishing Mendel's 'principle of genetics' in 1865, actual genetic modification was started. Plant genetic engineering has designed to generate plants with neoteric attributes that could conquer sustainability goals. Hence, it necessitates introducing advanced genetic engineering strategies, for instance, mutagenesis, transgenic approach, RNAi approaches, genome editing (GEd) via ZFNs, TALENS, CRISPR-Cas approaches. GEd was promoted by introducing double-strand breaks (DSBs) at the targeted locus, which relies on sequence-specific nucleases (SSNs). Recently, the toolkit of GEd comprises four classes of SSNs: meganucleases, ZFNs, TALEN, and CRISPR-Cas systems.

Meganucleases are the naturally available endonucleases, also familiar as homing endonuclease. It is the first-generation SSN and came into the limelight as a self-splicing component of mitochondrial large ribosomal DNA (mtLrDNA) introns of *Saccharomyces cerevisiae* (Colleaux et al. 1988). This can identify a wide range of DNA (14–40 bp) (Orlowski et al. 2007). Due to variation in target recognition and cleavage site, these SSNs can be grouped into six major families, for instance, His-Cys Box, LAGLIDADG, HNH, EDxHD, GIY-YIG, and PD-(D/E)xK (Belfort et al. 2014). I-SecI is the most frequently utilized meganucleases and was first utilized in tobacco plant (Puchta 1999), since then it was being used by plant biologists for GEd. D'Halluin et al. (2007) reported the utilization of meganucleases in maize. Nevertheless, the lack of editing capability of broad target sequences via protein redesign mightily narrows this SSN's applications (Rosen et al. 2006).

The re-programmability lacking of meganucleases was solved with ZFNs. ZFNs comprise multiple zinc finger domain harbouring proteins. Those protein domains are generated from the typical Cys-2-His 2-zinc finger domain (Gaj et al. 2013); after recognition of specific sequences, those protein motifs were fastened to DNA in a sequence-specific manner (Weeks et al. 2016). The C-terminal part of each ZFN motif is responsible for targeted sequence recognition. The composition of each ZFN motif binds a 3-bp DNA sequence and is made up of almost 30 amino acids (Maeder et al. 2008). Therefore, unlike meganucleases, these separate domain arrangements made ZFN simpler. Nevertheless, due to lack of endonuclease activity, they require to be fused with Fok I endonuclease domain for cleavage of DNA at the target site (Kaul et al. 2019). The efficiency of ZFNs-mediated gene editing was first successfully employed in *Arabidopsis* (Lloyd et al. 2005). Similarly, Dicer-like DCL4a and DCL4b gene in soybean was successfully edited utilizing ZFNs (Curtin et al. 2011). ZFN is also used for HDR-mediated gene editing, for instance, the amino acid substitution of SuRa and SuRb gene in tobacco, which conferred resistance to sulphonylurea herbicide (Townsend et al. 2009). However, two separate ZFN motifs target two proximal sites and double the construct size, which may complicate the design of this SSN.

TALENs are specific DNA-binding proteins, large sequences (>30 bp) targets make them more precise (Miller et al. 2011). The TALEs proteins were identified from plant pathogen *Xanthomonas* sp. Unlike ZFNs, these proteins have DNA-binding modular domains, specifically recognizing one single base instead of three (Moscou and Bogdanove 2009). Additionally, like ZFNs, this single DNA identification modules must be fused with Fok I endonuclease domain (Mahfouz et al. 2014). The possibility of TALEN-based editing was realized in 2009, wherein successful gene editing was first reported in yeast (Christian et al. 2013). Over the past few years, TALENs have emerged as a choice of GEd in plants. It has been successfully employed in a variety of crops, for instance, tobacco (Moore et al. 2014), barley (Budhagatapalli et al. 2015), tomato Čermák et al. 2015), *Arabidopsis* (Forner et al. 2015). Genome modification via TALENs is handy in comparison to ZFNs, due to its simplicity in using TALEs repeats for each of the DNA nucleotide recognition.

Amongst all the approaches, recently discovered CRISPR-Cas9-based GEd tools have replaced the ZFNs and TALENs and opened the way to modify plant's genomes with unprecedented precision. This GEd system is revolutionizing the field of plant biology due to its efficiency, specificity, unparalleled flexibility, and target design simplicity. Apart from these, CRISPR-Cas9 has additional advantages over ZFNs and TALENs, including target specificity design, efficiency in incorporating the guide RNA (gRNA), and the RNAs guided Cas9 protein and the ability of multiplexing in a single event. Compared to the previously available techniques, designing a CRISPR-Cas9 vector is easy and efficient with the availability and accessibility of enhanced bioinformatics tools, which can be utilized to find the most selective sequences for designing gRNAs, eliminating the potential for screening libraries to find the most effective target. This technology has been rapidly and widely adopted for a range of applications for instance, multiplex gene knockout, targeted sequence insertion, base editing, prime editing, and so on. A variety of strategies have been developed for optimizing the CRISPR-Cas9 reagents and their delivery systems. This chapter tries to compile a detailed review of the existing GEd approaches, emphasizing the CRISPR-Cas9 technique. We also shed light on a glimpse of information about novel breakthrough and milestone achievements of CRISPR-Cas9 systems and the impact of this system as the next gene tool for crop improvement.

9.2 Era of CRISPR-Cas-Based Genome Editing

The invention of the CRISPR-Cas microbial self-defense mechanism and its ongoing achievement as a genome-editing tool represents the findings of numerous researchers all over the world. Our concise historical era will represent the contributions of different scientists who pushed this GEd field forward from the initial discovery. The clusters of repeats which are separated by spacers were first observed in 1987 during the study of *E. coli* harbouring *jap* gene (Ishino et al. 1987). In 1989, the structure of the CRISPR array was defined but without its functional mechanism (Nakata et al. 1989). Interestingly, similar structures were identified later in numerous bacteria and archaea (Hermans et al. 1991; Mojica et al. 1995; Bult et al. 1996). Francisco Mojica characterizes those sequences for the first time in 1993, what is now known as CRISPR locus, and the potentiality of this locus was shown in 2000 (Mojica et al. 2000). Simultaneously, 45 protein families were identified with clusters of CRISPR-associated genes (Haft et al. 2005). After increasing the volume of prokaryotic sequence data, the crucial breakthrough happened in 2005. It was reported that identified CRISPR sequences showed similarity with some bacteriophage and led to immunity against those infectious bacteriophages (Mojica et al. 2005; Pourcel et al. 2005). Bolotin revealed some anomaly in the CRISPR locus and found a large protein with nuclease activity, which is now known as Cas9 (Bolotin et al. 2005). Although found some viral genes resemble sequences at one end, those are the PAM sequence. In the same year 2005, Jennifer Doudna and Jillian Banfield started their investigation on CRISPR and their functions. In 2006, the hypothetical scheme of the CRISPR-Cas adaptive immunity mechanism was proposed by Koonim (Makarova *et al.* 2006). Later on, this hypothesis was confirmed by Barrangou et al. (2007). After that, scientists started to report the CRISPR-Cas action that how this RNA-mediated system interferes with the invading phage DNA (Brouns *et al.* 2008). It was also demonstrated that this system could target both DNA (Marraffini and Sontheimer 2008) and RNA (Hale et al. 2009). Alongside, PAM sequences are also essential for some systems described simultaneously (Mojica et al. 2009). Details about the transcription mechanisms of crRNAs were also revealed in 2010 (Haurwitz et al. 2010). The classification of the CRISPR-Cas systems was demonstrated in 2011 (Makarova *et al.* 2011). In the same year, 2011, Emmanuelle Charpentier and Jenifer Dounda conjointly started to study the CRISPR-Cas mechanism, and they discovered the function of tracrRNA for the Cas9 system. Moreover, the role of RNase III in pre-crRNA and tracrRNA processing was characterized in 2011 (Deltcheva et al. 2011). In 2012, Siksnys and his team mechanically characterized the mode of adaptation of Cas9 via understanding the cell infection kinetics with the CRISPR-Cas system (Datsenko et al. 2012; Gasiunas et al. 2012). At the same time, in 2012, similar findings were reported by Jennifer Doudna in collaboration with Emmanuelle Charpentier. They demonstrated that synthetic gRNAs could be generated via fusion of the crRNA and the tracrRNA (Jinek et al. 2012; Qi et al. 2021). Finally, the newly invented CRISPR-Cas system was led to use for targeted genome modification in bacteria (Gasiunas et al. 2012), yeast (DiCarlo et al. 2013), human (Cong et al. 2013a, b; Jinek et al. 2013; Mali et al. 2013a, b). In 2013, CRISPR-Cas machinery was successfully employed to engineer plant genomes (Shan et al. 2013). In 2014, CRISPR-Cas9 was demonstrated in primates (Cas9 mRNA and sgRNAs were coinjected in monkey embryo); therefore, Cas9/sgRNA screens were established as a tool for genetic analysis in mammalian cells (Shen 2014). In 2015, US and UK research scientist, Medical Research Council (MRC) declared their support for using GEd strategy for human cells (Charo 2015). After that, the International Summit on human gene editing was met to discuss about the medical and ethical issues (Charo 2016). New protein-Cpf1 was invented in the same year, which made gene editing become simpler (Koonin et al. 2017). Moreover, US scientists reported about the modified CRISPR-Cas9 technique with fewer off-target effects. The first clinical trial of the genetically modified human embryo was approved in 2016 (Cyranoski 2016; Reardon 2016). A new base editing technique was discovered in 2016 by US scientists, offered a new approach where any gene can modify without cleavage of double-stranded DNA as well as without donor DNA template (Rees and Liu 2018; Porto et al. 2020; Bharat et al. 2020). For RNA editing, a new CRISPR approach was identified in 2017 (Cox et al. 2017; Adli 2018). In 2018, Weissman's lab created a new GEd strategy called CRISPRa (for 'activation'), which activate the gene expression, and they also made CRISPRi (for 'interference') technology (Kampmann 2018). In 2018, a group of scientists identified pre-existing Cas9 antibodies in cells, leading to immune issues during gene therapy employing CRISPR-Cas9 (Crudele and Chamberlain 2018; Wagner et al. 2021). Same year another Cas variant Cas14 (a-c) was identified (Harrington et al. 2018) In 2019, Chinese researchers declared their first gene-edited human baby (Wang et al. 2020). Newly developed 'search and replace' tool for GEd known as prime editing was discovered in 2020 (Hampton 2020). In the same year 2020, a Chinese researcher was convicted for employing CRISPR-Cas9 in a human baby (Cyranoski 2020). Moreover, in 2020 for the first time, one patient received gene editing therapy employing the CRISPR-Cas9 approach (He 2020; Ledford 2020). In early 2020, a novel severe acute respiratory syndrome coronavirus 2 (SARS-CoV-2) outbreaks rapidly evolved into a global pandemic. For the detection and quantification of SARS-CoV-2 RNA, a CRISPR-Cas13-based approach was employed (Konwarh 2020; Kumar et al. 2020). In 2020, Jennifer Doudna and Emmanuelle Charpentier were jointly awarded the Nobel Prize in Chemistry for the identification of an efficient method in GEd known as the CRISPR-Cas9 technique (Ledford and Callaway 2020). The CRISPR 'on-off switch'- a new genome-editing approach was discovered by MIT and UCSF researchers in 2021, successfully implicated in Alzheimer disease (James et al. 2021). Any part of the targeted genome can be silent via controlling the gene's expression without altering DNA sequences. Therefore, unlike first-generation GEd tool, the CRISPR-Cas9 technology has empowered researchers with an unprecedented toolbox via breakthrough discoveries and methodological advancements in science.

9.3 CRISPR-Cas: New-Fangled Dawn in Genome Editing

CRISPR-Cas is the most efficacious and ease-to-design editing tool, which generate a buzz in the field of research in current times. This is one of the crucial tools in an endless arms race between bacterial and archaeal hosts and viruses (Newsom et al. 2021). The CRISPR immunity gets triggered when a virus' foreign genetic material (DNA/RNA) is introduced into bacterial cells. The bacterial cell effectively produced specialized molecules (Cas protein) that can recognize the past similarity of foreign DNA and destroy them as antibodies work. The defense mechanism of this system comprises into three-stage process, i.e. (i) Adaptation: small DNA sequences (protospacers) of foreign plasmid are chosen and incorporated into the particular CRISPR locus of the host genome; (ii) Biogenesis of crRNA: multiple gRNA spacers and their repeats are transcribed into a precursor RNA and processed into mature gRNAs. Targeting complexes are produced via fastening of gRNAs with the Cas enzyme, which contain a distinctive spacer sequence resembling foreign target DNA; and (iii) Interference: Cas nuclease starts searching the unique sequences complementary to the gRNA. Cas nucleases fasten up to the gRNA resemble target foreign DNA site via complementary base pairing and cleave the targeted DNA sequences. By utilizing this machinery, bacteria generated the ability to avoid transcribing the matching targeted viral DNA, making its genome resistant to viral invasion. As research gains grounded, numerous CRISPR-Cas systems have been identified for GEd. All these systems have their own attributes, for instance, variation at PAM regions, varying sizes of Cas protein, and different cleavage sites. Amongst all, the type II CRISPR-Cas9 system provided the most simple, versatile precision editing in crop plants. This system required only two key molecules, i.e. Cas9 endonuclease and gRNA: fusion of CRISPR RNA (crRNA, a 20-nucleotide sequence complementary to the target DNA) and trans-activating crRNA (tracrRNA, acts as a binding scaffold for the Cas9 endonuclease). It also should be noted that the gRNA can be expressed as synthetic sgRNA, where the crRNA and tracrRNA are fused into one molecule for ease of expression (Fig. 9.1). Widely accepted *Sp*Cas9 (*Streptococcus pyogenes*) comprises a conserved core with two major big globular recognition lobe, for instance, REC (recognition) and a NUC (nuclease) lobe for nucleic acid binding. Wherein, the REC (functional domain of Cas9) contains bi-partite domain, for instance, REC1 & REC2 and bridge helix cd domain. It was revealed that base pairing between the ligand DNA strand and the seed region of gRNA (up to 8–12 bp) triggers the development of RNA–DNA heteroduplex, which occupied by both NUC and REC lobe (Anders et al. 2014). The small NUC nuclease lobe comprises a highly conserved RuvC- & HNH- and PI-domain (arginine-rich alpha-helical bridge helix) (Hsu et al. 2014). Simultaneously, RuvC and HNH nick the complimentary and non-complimentary strand in the target

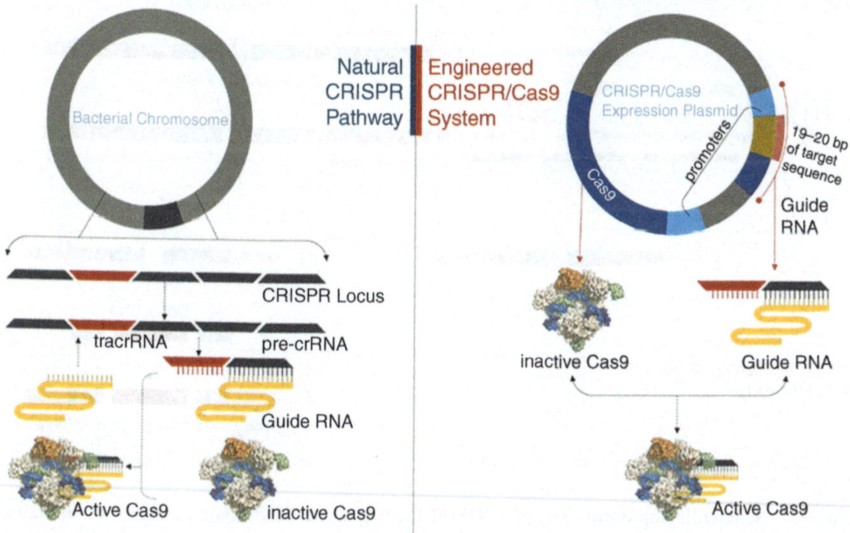

Fig. 9.1 CRISPR mechanism in action: Natural vs Engineered CRISPR system. This system required only two key molecules, i.e. Cas9 endonuclease and gRNA. gRNA is fusion of CRISPR RNA (crRNA, a 20-nucleotide sequence complementary to the target DNA) and trans-activating crRNA (tracrRNA, acts as a binding scaffold for the Cas9 endonuclease). Inactive Cas9 is become active when bind with gRNA. In synthetic sgRNA, the crRNA and tracrRNA are fused into one molecule for ease of expression

sequence, introducing double-strand breaks (DSBs) (Nishimasu et al. 2014). According to previous studies, the PI domain plays a crucial role in the PAM site (5′-NGG-3′) recognition because of having a tryptophan-rich flexible loop (Jinek et al. 2014). At 3 bp prior to PAM sites, the assembled CRISPR-Cas complex created DSBs. DSBs can be repaired at defined positions by integrating numerous alterations utilizing DNA repairing machinery, i.e. HDR and NHEJ (Fig. 9.2).

NHEJ is the primary DSB fixing pathway in plant cells and is comparatively effortless to exploit for GEd (Lieber 2010; Pannunzio et al. 2017). This error-prone pathway generally introduces indel mutations (insertions and/or deletions) by disrupting the targeted DNA, resulting in gene knockout (KO). The CRISPR-Cas9-based KO is utilized in gene function study, and modifying a variety of beneficial traits, for instance, stress resistance (Singh et al. 2020); disease resistance (Schenke and Cai 2020); higher yield (Huang et al. 2018; Ma et al. 2019; Liu et al. 2021; Tabassum et al. 2021); nutritional enhancement (Zhang et al. 2018a; Sanchez-Leon et al. 2018; Ku and Ha 2020; Dong et al. 2020; Huang et al. 2020; Dong et al. 2019; Xu et al. 2020; Yang et al. 2020; Zeng et al. 2020; Kaul et al. 2020a, b; Sashidhar et al. 2020; Tiwari et al. 2020); and male sterility (Chen et al. 2021). For the achievement of successful KO, it is recommended to target early exon because functional activities of a gene will be less if indel mutation is generated in either 3′ end of exon sequences or intron region. Nevertheless, due to alternative splicing if

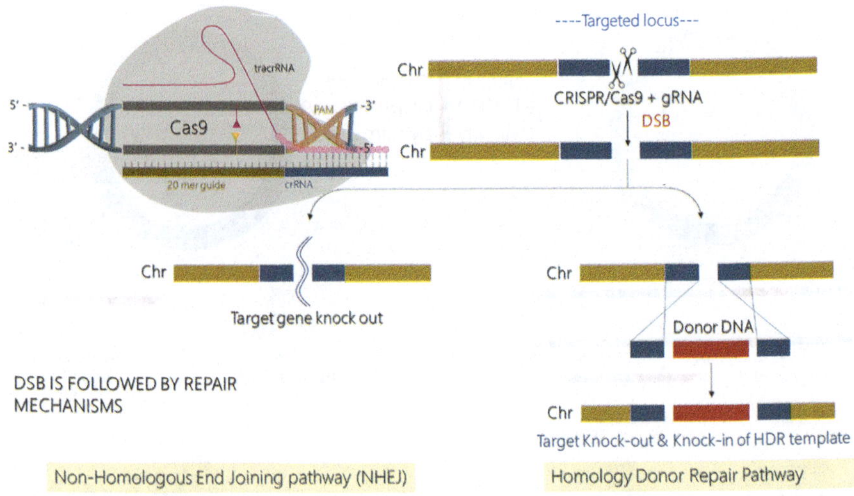

Fig. 9.2 Schematic representation of CRISPR/Cas9-based DSBs repair mechanism, including NHEJ and HDR-mediated repair pathways. The CRISPR-associated endonuclease Cas9 generated DSBs in the target DNA. NHEJ pathway results in random indels via gene disruption at the target site. HDR pathway uses homologous donor DNA sequences for accurate insertions or base substitutions between DSB sites. *DSB* double-strand break; *NHEJ* non-homologous end joining, *HDR* homology donor repair, *Indels* insertions and deletions

target gene enciphers various proteins, then frameshift mutation or stop codon introduction in early exon may not reveal gene KO. In this situation, complete gene deletion can be possible by utilizing the multiplex CRISPR-Cas9 KO strategy by targeting the gene's 3′ and 5′ end. For example, (115–245) kb in size chromosomal deletions were generated via gene cluster deletion in rice (Zhou et al. 2014) employing multiplex CRISPR-based KO strategy. Recently, multiplex CRISPR-Cas9 system utilized for simultaneous KO of multiple genes and revealed de novo domestication of wild tomato (Zsögön et al. 2018; Xie and Liu 2021). Gene KO is extremely difficult in polyploidy species due to its gene functional redundancy. It was successfully utilized in hexaploid wheat to develop fungal-resistant wheat by KO of disease susceptible S-gene (Wang et al. 2014; Wang et al. 2018a). Corteva Agriscience generated amylopectin rich (waxy) corn via KO of *Wx1* gene (DuPont Pioneer 2016). Similarly, two japonica rice varieties (glutinous sticky) were achieved through Waxy (*OsWx*) gene KO (Yunyan et al. 2019). Moreover, amylose-rich rice grain was revealed via targeted modification of the *SBEIIb* gene (Sun et al. 2017). Gaoneng et al. (2017) developed fragrance enriched rice via targeted KO of the *BADH2* gene (negative regulator for aroma production). Additionally, KO of *OsERF922* gene generated blast-resistant rice lines was reported by Wang et al. (2016). Edited rice lines with pale green colour in leaf were generated via KO of chlorophyll biosynthesis regulated gene *OsCAO1* gene (Jung et al. 2021). Interestingly, *OsHOL1* plays a major role in the production of methyl iodide, and KO of this gene abolished methyl iodide emissions from rice plants (Carlessi et al.

2021). Targeted KO of TERMINAL FLOWER 1 (*TF1*) gene in *Brassica napus* altered the flowering time and plant architecture (Sriboon et al. 2020). Targeted KO mutations of *HvHPT* and *HvHGGT* gene rendered a high level of vitamin E (tocopherol) in barley (Zeng et al. 2020). According to Li et al. (2019), KO of numerous genes, i.e. *SVP, AP1*, and *TFL* elicited floral features advancement in *Arabidopsis*. High-oleic acid content was generated in allotetraploid cotton (*Gossypium hirsutum* L.) (Chen et al. 2021) and tobacco (Tian et al. 2020) via KO of *GhFAD2* and *NtFAD2–2* genes, respectively, as well as Monounsaturated Fatty Acid (MUFAs) contents enhancement in Hexaploid *Camelina sativa* seed oil was generated through *FAD2* Gene KO using CRISPR-Cas9 (Lee et al. 2021). Functional KO of *StDND1, StCHL1,* and *StDMR6–1* generated potatoes highly resistance against late blight disease (Kieu et al. 2021).

On the other hand, HDR mechanism introduces specific base pair substitution point mutations via target DNA recombination with complementary HDR template (Reis et al. 2014; Sander and Joung 2014). Knocking-in of targeted and precise sequences has been more challenging. Repairing Cas9-induced DSBs or nicks using HDR-mediated pathway makes GEd more accurate. Thus, unlike NHEJ, in case of knockin, the incision must be embedded precisely, without extra insertions/deletions (indel) mutation. Unfortunately, GEd frequency employing HDR mechanism is relatively low in plants in comparison to NHEJ. Amongst numerous approaches to recurrence, the HDR efficiency in plants, the utilization of mastrevirus (Geminiviridae) vectors for delivery of donor template is the most successful generated so far. This method was first demonstrated in tobacco to develop bean yellow dwarf virus-resistant (Baltes et al. 2014). The HDR template frequency was increased dramatically in nucleus due to replication of donor template, revealing a high editing frequency. Later on, it was employed in tomato ANT1 gene to precisely insert a promoter upstream of the gene, resulted in pigment accumulation in foliage, flowers, and fruits via controlling anthocyanin biosynthesis (Čermák et al. 2015). Moreover, point mutation was introduced in the potato *ALS*1 gene, conferring herbicide resistance (Butler et al. 2016). A viable alternative method is delivering a large copy of the donor template into plant genome employing biolistic approach. This approach was successfully utilized in rice and maize (Baltes et al. 2015; Gil-Humanes et al. 2017; Wang et al. 2017) resulted in higher precise edits. In line with this, numerous advancements had been developed, for instance, in *Arabidopsis* the absence of a repair protein, KU70/80 may lead to a 5–16 fold enhancement in HDR editing frequency via suppressing the NHEJ repair pathway (Endo et al. 2016). Moreover, Lu et al. (2020) discovered a tandem repeat-HDR (TR-HDR) approach for high frequency targeted sequence replacement, wherein the precise editing frequencies ranged from 3.4 to 11.4%. According to Shi et al. (2017), the promoter swapping, for instance, GOS2 promoter by the native ARGOS8 promoter employing CRISPR-Cas based GEd via HDR approach generated drought tolerance in maize. Tomato lines with higher self-life were generated via T317A substitution in the *ALC* gene (Yu et al. 2017). Newly developed RNA-mediated CRISPR/Cpf1-based approach also rendered efficient, targeted gene insertion in tomatoes (Vu et al. 2020). Therefore conjointly, CRISPR-Cas9 and CRISPR/Cpf1 may overcome all difficulties for precise gene knockin via HDR mechanism for crop plant enhancement.

9.4 Novel Technical Breakthrough of Genome Editing in Plants

The necessity to genetically improve crop varieties became the reason for the discovery of target-specific endonucleases (TSENs) and since 2005 there has been a significant improvement and addition of new tools/techniques in the GEd toolkit. The genetic engineering field experienced another boost with the discovery of the CRISPR-Cas9 system which back in 1987 was recognized as the bacterial immune system (Ishino et al. 1987). The last decade has witnessed the evolution of this technique to reduce the bottleneck in terms of efficacy, efficiency, applicability, and other already discussed shortcomings. Substantial diverseness in genes, loci configuration, and action mechanisms of CRISPR-Cas approach made their classification a formidable task. An updated classification was reported by MaKarova et al. (2020), which include 2 classes, 6 types, and 33 subtypes; they identified novel class 2 CRISPR-Cas systems including 3 types and 17 subtypes (Table 9.1). Class 1 systems contain ~90% of all discovered CRISPR-Cas loci constituting type I, III, and IV (Makarova et al. 2015). The Class 2 system contains 10% of CRISPR-Cas loci (Makarova et al. 2015) and clearly differentiating into type II, V, and VI (Makarova et al. 2020). Numerous Cas9 variants are identified in recent years to broaden the opportunity of genome alteration. Thus to greatly expand the range of targets, different orthologs of Cas9 were reviewed, and VQR (5′-NGA-3′) and VRER (5′- NGCG-3′) variants of Cas9 were developed for plants (Hua et al. 2016). In the same line of study, an ortholog from *Francisella novicida* (Fncas9) was engineered to recognize 5′-YG-3′ PAM. FnCas9 is also known to target the RNA substrate, consequently it can be utilized to gain viral resistance in plants (Zhang et al. 2018b). Later to increase the penetrability of the Fn Cas9, proximal CRISPR (proxy-CRISPR) was developed (Chen et al. 2017). To increase the target specificity and reduce the off-target effect, Cas9 nickases mutants (nCas9) came into the picture by introducing point mutations like D10A in RuvC (Jinek et al. 2012) and

Table 9.1 Classification of CRISPR-Cas system

Class	Type	Subtype	Spacer acquisition	crRNA biogenesis	Interference crRNP	Type of nucleic acid targets
1	I	A-G	Cas1, Cas2, Cas4	Cas6/ Cas5d	Cascade	DNA
	III	A-F	Cas1, Cas2	Cas6	Csm/Cmr	DNA/RNA
	IV	A-C	Csf5	Csf	Unknown	DNA
2	II	A-C	Cas1, Cas2, Cas4/Csn2, Cas9	RNase III Cas9	Cas9	DNA
	V	A-I, K	Cas1, Cas2, Cas4	Cas12	Cas12	DNA
	VI	A-D	Cas1, Cas2	Cas13	Cas13	RNA

N863A or H840A in HNH domains (Nishimasu et al. 2014). Along the same line of work, Satomura et al. (2017) designed a 'CRISPR Nickase system' (CNS) to target sequences that were non-editable with the conventional CRISPR-Cas9 tool. Furthermore, the inducible Cas9 or split Cas9 can be used for temporally and spatially restricted Cas9 expression (Zhou et al. 2018; Carlson-Stevermer et al. 2020).

The continuous endeavour led to the discovery of class II type V CRISPR from *Prevotella* and *Francisella* 1- Cpf1/Cas12a was a potential alternative to Cas9 primarily because it could target AT-rich (5′-TTTN-3′) PAM instead of GC rich PAM (Doudna and Charpentier 2014). Besides *cis*-cleavage of the target double strand, it can also cleave non-specific ssDNA in *trans* (Swarts and Jinek 2019) which contributed to the invention of a sensitive nucleic acid detection technique, i.e. DETECTOR (Li et al. 2018a). Recently, Zhang et al. (2021) have contributed exceptionally with the discovery of six highly efficient orthologs (ErCas12a, Lb5Cas12a, BsCas12a, Mb2Cas12a, TsCas12a, and MbCas12a) of Cas12a. Similarly, a related enzymatic activity harbouring Cas12b (C2c1) from *Alicyclobacillus acidiphilus,* i.e. AaCas12b was prospected as a potential add-on to the tool kit. Another class II type VI-A Cas protein, i.e. Cas13a (C2c2) is an effective tool that possesses RNA-guided RNase activity (Abudayyeh et al. 2016). Single strand RNA (ssRNA) targeting LshCas13a (*Leptotrichia shahii*) and other orthologs (b,c,d) have two HEPN (Higher Eukaryotes and Prokaryotes Nucleotide-binding) domain with no requirement of PAM (Bandaru et al. 2020). Cas13a has been utilized for RNA/transcript knockdown and RNA editing (REPAIR and RESCUE; Cox et al. 2017; Abudayyeh et al. 2019). Moreover, Cas13a has been utilized for SHERLOCK, PAC-MAN, and SARS/Covid 19 detection kits (Gootenberg et al. 2018; Joung et al. 2020; Zhang et al. 2020). In plants, it can be utilized to gain viral resistance against specific viral pathogens (Abudayyeh et al. 2017). Another class II type V effector, i.e. the Cas14 family (Cas14a-c: 400–700 amino acids) present in archaea came as a significant discovery (Harrington et al. 2018; Savage 2019). Cas14s can target both ssDNA and dsDNA with no PAM or AT-rich PAM (5′-TTAT-3′) requirement (Karvelis et al. 2019). Due to sensitivity towards mismatching, it can be used for high precision SNP genotyping and because of trans cleavage activity it can be used as a Cas14-DETECTOR and to gain viral resistance in plants (Aquino-Jarquin 2019; Khan et al. 2019a). With continued hustle to discover better alternatives for GEd, the database mining led to the discovery of smaller Cas proteins such as Cas12f and the features closely related to the previously known Cas 14s (Karvelis et al. 2020). The recent classification thus unifies these proteins together Cas12f1 (Cas14a and type V-U3), Cas12f2 (Cas14b), and Cas12f3 (Cas14c, type V-U2 and U4) and expands the utility tools in the GEd artillery (Makarova et al. 2020).

Further expanding the smaller type V effectors family, DpbCasX (Deltaproteobacteria) is one such mini (~980 aa) novel protein (Liu et al. 2019a). CasX (alias Cas12e) is a dual RNA (crRNA and tracrRNA) guided protein (naturally combined into single-guide RNA; sgRNA) targeting dsDNA adjacent to 5′-TTCN-3′ PAM to generate 10 nt staggered break (Yang and Patel 2019). However, it shows the nominal *trans* activity as compared to other type V effectors which highlight structural differences between Cas X and other enzymes. Recently, in Doudna's lab a supercompact CRISPR-CasΦ system encoded by bacteriophage genome has been discovered, where a bacteriophage uses the system to target other competing phages. CasΦ (Cas12j) also has a C-terminal RuvC domain but shares no similarity (<7% amino acid identity) with type V effectors, rather it is remotely related to the TnpB enzymes. The CasΦ locus lacks the spacer acquisition enzymes such as Cas1, 2, and 4 which results in a really compact CRISPR array and the locus also lacks the presence of tracrRNA. CasΦ represents the consolidated form of the CRISPR-Cas system and thus can be utilized to its full potential for genome manipulation (Pausch et al. 2020). With the discovery of such versatile, flexible, and miniature (400–1093 amino acids) effectors, the Cas12 family is expanding and till date, there are 11 subtypes of type V which has been reported, namely Cas12a to k (Li et al. 2021) and a subtype V-U which is more closely related to transposon TnpB. Cas12a, Cas12b, Cas12e, Cas12h, and Cas12i specifically target dsDNA with PAM assisted unwinding (Yan et al. 2019). However, an ortholog Cas12g (thermostable) was reported to initially target ssRNA and then indiscriminately degrade both ssDNA and ssRNA (Chen et al. 2018). Till now, class 2 effectors have dominated the terrain of GEd primarily because it utilizes single subunit protein effectors, whereas class 1 CRISPR-Cas system utilizes multiple subunit protein effectors (Makarova et al. 2018). Type III effectors of class 1 are known to target RNA substrates and hold the potential to be developed into diagnostic tools or to attain tolerance against viruses or mobile genetic elements (MGE) (Samai et al. 2015; Staals et al. 2014) in any system. Type III effectors are divided into III-A (Csm), III-B (Cmr), III-C, III-D subtypes, and the common feature between these subtypes is the presence of Cas10 (Csm1 or Cmr2) (Burmistrz et al. 2020) in the complex. Cas10 predominantly has two domains (Makarova et al. 2018) notably the palm domain (the cyclase activity of palm domain is absent in type III-C effectors) and a nuclease HD-type (unavailable in type III-D effectors) domain (Zhu et al. 2018). Typically, this multi-subunit complex protein effector is composed of two parallel filaments from which the first filament is generally made of six subunits of Cas7 protein and the other is made up of three subunits of Cas11 homolog (Csm2 or Cmr5) protein. The crRNA is stretched in between these filaments (Lintner et al. 2011) and the 5′- end having the handle derived from repeat is capped by Cas10 and Cas5 (Csm4 or Csm3) proteins (Staals et al. 2014), whereas the maturation of crRNA from pre-crRNA is catalysed by the Cas6 (Nickel et al. 2018) protein. Cas7 as a family of protein (members like *Thermofilum pendens* Csc2 protein) has members in both type I-D (commonly present in Archaea and Cyanobacteria) and type III branches of classification (Staals and Brouns 2013; Cai et al. 2013). Moreover, the protein organization in CASCADE (CRISPR-associated interference complex type I) and type III complex is similar, and proteins of type I-D have HD domain fused to the Cas10 (type III protein) and thus Cas7 is considered as the evolutionary link between type I and type III CRISPR-Cas system (Hrle et al. 2014). Exceptionally, the type III system has three different nuclease activities, and primarily it possesses sequence-specific RNase activity where acidic residues of the RNA-recognition motif (RRM) of Cas7 targets a specific RNA sequence (Estrella et al. 2016). The CRISPR-associated Rossman fold (CARF) located at the N-terminal of Csm6 sense the presence of the cyclic oligoadenylate while the C-terminally located higher eukaryotes and prokaryotes nucleotide-binding (HEPN) domain non-specifically cleaves the ssRNA (Niewoehner and Jinek 2016) molecule. Considering the potency of Csm6 protein, it has been included in the SHERLOCKv2 and this resulted in the three fold increase in the efficiency of the technique by improving the reporter signal (Gootenberg et al. 2017; Kellner et al. 2019). With continued exploration and screening over 11 billion protein sequences revealed the existence of a single-protein effector under type III-D2 CRISPR-Cas system, referred as Cas7x3 which have three Cas7 protein fused into a single protein (Özcan et al. 2021). In consonance, a novel breakthrough has resulted in the discovery of a programmable type III RNA targeting single-protein effector termed as Cas7–11, structurally having four Cas7 proteins fused to a putative Cas11 protein (Makarova et al. 2020). *Di*Cas7–11 from *Desulfonema ishimotonii* is a programmable RNase with no reported collateral activity. The discovery of this protein further expands the classification nomenclature by adding a type III-E subtype to the previously known subtypes. Both Cas7x3 and Cas7–11 process their own pre-crRNA into mature crRNA for targeting specific sequence template and not even display any toxic effect in mammalian cells (Özcan et al. 2021). However, they still need to be developed into programmable CRISPR-Cas tools to utilize their full potential for GEd across different systems including plants.

Introducing foreign DNA and generating of DSBs in any system for GEd raised some regulatory concerns which led to the evolution of the DNA-free GEd strategy. Under this, the ribonucleoproteins (RNP) which are pre-assembled Cas nucleases with the target-specific gRNA are delivered into the target system to achieve the desired GEd in plant and animal systems (Woo et al. 2015; Wu et al. 2020). The use of CRISPR-Cas tool in prokaryotic (Qi et al. 2013) and eukaryotic (Gilbert et al. 2013) systems introducing DSB leads to unexpected changes and toxicity. Precision transcriptional regulation without the introduction of any DSBs, i.e. without changing the underlying DNA sequence, with the strategies like CRISPRi and CRISPRa has revolutionized the field of genetic engineering (Liu et al. 2019b). In CRISPRi and CRISPRa, a dCas9 is fused with transcriptional effector to either repress (repressor like Kruppel associated box, or KRAB) or activate (activators like VP64 and p65) the gene expression (Lawhorn et al. 2014; Mali et al. 2013a, b). The newly developed customizable epigenome memory writer 'CRIPSR on-off' technique can alter gene expression by generating heritable epigenome modification. CRISPRoff is a fusion protein with dCas9 with DNA methyltransferase (DNMT1) and KRAB domains to silence the gene expression. However, the modifications are specific, tunable, and reversible as the methylation can be removed (inhibitor of DNMT1, i.e. 5-aza-2′-deoxycytidine (5-aza-dC)) by CRISPRon, and gene expression can be activated via recruitment of the transcriptional machinery. Genome-wide screen helped to find the targetable genes and showed that genes lacking the CpG islands can also be silenced with the CRISPR-off technique (Nunez et al. 2021). 'CRISPR on-off' is a complementary technique to the already existing CRIPSRi, CRISPRa, and CRISPR nuclear approaches.

Base editors (BEs) in conjunction with the CRISPR-Cas tool have been used for precise, specific single base modification with no induction of DSB and as an alternative to HDR-based GEd (Komor et al. 2016). dCas9 or any inactive RNA-guided Cas protein with cytidine base editor (CBE; cytidine deaminase) can catalyse target specific C-to-U (Uracil recognized as T) base substitution which results in C-G to T-A base pair conversion (Rees and Liu 2018) and with adenine base editor (ABE; deoxyadenosine deaminase), it can catalyse the A-to-I (Inosine; recognized as G) base substitution which results in A-T to G-C base pair conversion (Gaudelli et al. 2017), respectively. CRISPR-BE has gone under severe optimization and development and in a recent generation a D10A nCas9 (to induce nick in unedited strand) fused with cytidine deaminase enzyme, i.e. rAPOBEC1 (rat apolipoprotein B mRNA editing enzyme) or to Lamprey cytidine deaminase (pmCDA1) for activation-induced cytidine deaminase (AID) at N-terminal, and two copies of uracil DNA glycosylase inhibitor (UGI) at C-terminal are used for base editing. The fusion of UGI increases the efficiency of editing in the case of C-to-U conversion as it helps in retaining the U in the target sequence till the next cycle of replication by inhibiting the inherent conversion of U-to-C again by uracil DNA glycosylase (UDG) (Abdullaha et al. 2020). Whereas, nCas9 (D10A) was also utilized in conjunction with the TadA (tRNA adenosine deaminase) and TadA* (modified at K157N, I156F, E155V, R152P, D147Y, S146C, H123Y, D108 N, A106V, L84F, R51L, P48A, H36L, W23R) domains connected via varying linker length, for ABE optimization and development (Bharat et al. 2020). Along with DNA base editing, RNA base editing can be achieved with the RNA directed RNA targeting dCas13. RNA editing comprises REPAIR (RNA editing for programmable A-to-I (G) replacement; catalysed by ADARs) and RESCUE (RNA editing for specific C-to-U exchange; catalysed by cytidine deaminase) techniques in plants (Abudayyeh et al. 2019). Although the base editing approach has faced few challenges in terms of off-target, range of editing, and bystander editing (Jeong et al. 2020). These shortcomings have led to the revolutionary discovery of prime editing (PE) which is based on the search and replace ideology and is a template free strategy (Anzalone et al. 2019). PE2 system is dependent on an amalgamation of the nCas9 (H840A), reverse transcriptase (RT; M-MLV from mouse-murine leukaemia virus), and the prime guide RNA (pegRNA). pegRNA have a primer binding site (PBS) sharing sequence complementarity to the sequence of the nicked DNA strand upstream of PAM and a reverse transcriptase template strand (RT strand). The 3′ flap is utilized as the primer to transcribe the desired sequence (written in the RT template), whereas the 5′ flap is cleaved via structure-specific host endogenous flap endonuclease (FEN1; Flap endonuclease *Homo sapiens*). Later the edited strand is ligated after 5′ flap digestion forming a heteroduplex of edited and unedited strands co-exist (Anzalone et al. 2019). The induction of a second nick on the unedited strand 10–12 nt away from the original pegRNA cut on the edited strand resulted in the development of the PE3 system. In PE3 when the unedited strand is repaired after induction of the second nick, it leads to the formation of the homoduplex of the edited dsDNA (Anzalone et al. 2020). In order to avoid incorporation of indel mutation by PE3 while repairing, in PE3b the second nick was introduced after successful completion of the flap resolution and editing (Kantor et al. 2020). PE till now displays really low events of off-target in any system

(Scholefield and Harrison 2021). In a study by Lin et al. (2020), the efficiency of the plant PE system increased at some locus by using the PPE-Ribozyme (PPE-R) system where the PE protein transcript is expressed by Polymerase II (Pol II) and pegRNA is processed by the ribozyme. Prime editing has come as a boon in the field of GEd and has expanded the toolbox for deep genome modification with enhanced efficiency, specificity, and tenacity even in polyploidy genomes such as wheat, as well (Lin et al. 2020).

9.5 Revisiting Challenges and Impediments of CRISPR-Based Approach for Precise Genome Editing

CRISPR is regularly portrayed as 'cut and paste' approach for genes, but the actual procedure is not that easy. However, further research is needed to gain a deeper understanding of the CRISPR-Cas process and its neoteric uses in plants. To date, researchers face umpteen obstacles related to utilizing the CRISPR approach in plant research, including hurdles in GMO regulation. Recently, *r*esearchers have achieved huge achievements utilizing CRISPR in its native and closely related organisms. But, employing CRISPR into bigger genomes containing complex organisms has accompanied its own set of difficulties. Some plants have multiple copies of each chromosome, for instance, hexaploid wheat (6 copies), strawberries (up to 10 copies), which is become strenuous to engineer compared to humans and animals. Subsequently, the probability of getting target gene editing in each copy decreases as the quantity of chromosome copies increases (Yang et al. 2020). Scientists are improving traditional CRISPR-Cas workflow by employing varying modifications so that multiple copies of the identical gene can be altered at once (Wilson et al. 2019; Lin et al. 2020; Jouanin et al. 2020; Smedley et al. 2021). Lamentably, this type of alteration sometimes create off-target mutation/s. Screening of accurate mutation and potential off-target sites is a very sensitive and significant challenge in the field of gene editing.

Earlier PCR/RE strategy was utilized to screen mutation in edited plants (Shan et al. 2014). The T7 endonuclease I (*T7E1*) assay was employed to detect off-target mutations; however, it is neither feasible nor cost-effective for large-scale screening due to its deprived sensitivity. Therefore, RNA-guided endonucleases, i.e. SpCas9- or FnCpf1- based PCR/RNP method for identifying indel/s, overcome the PCR/RE strategy (Liang et al. 2018). Unlike *T7EI*, this PCR/RNP-based technique can differentiate the mutant types, i.e. homozygous, heterozygous, bi-allelic, and mosaic mutants. It is also a SNPs independent mutation detection method essential for polyploidy plants like wheat (Liang et al. 2018). Numerous web-based approaches, for instance, deep sequencing (mutation detection range: 0.01–0.1%), genome-wide, unbiased identification of DSBs facilitated by sequencing (GUIDE-seq), RNA-guided endonucleases (RGEN), had been widely adapted (Wu et al. 2014; Zhang et al. 2015; Tsai and Joung 2016; Kosicki et al. 2018). Consequently, different bioinformatics-based programs (TALE-NT, CAS-OFF Finder, PROGNOS) have been developed to profile off-target mutations via CRISPR-Cas nucleases (Fine et al. 2013; Listgarten et al. 2018; Minkenberg et al. 2019). Recently, genome-wide off-target edit frequencies were identified using the whole-genome resequencing (WGRS) approach in rice, maize, cotton (Tang et al. 2018; Lee et al. 2019; Li et al. 2019).

In addition, the plant regeneration and transformation approach is quintessential for delivering the editing reagents into plant cells for genome editing. Wherein, genotype-dependency is one of the major bottlenecks in completely appearing the incredible capability of genome altering in plant species (Alpeter et al. 2016). The development regulator (DR) genes of maize: *Baby Boom (Bbm)* and *Wuschel2 (Wus2)* in combination with phytohormones lead to enhance the transformation efficiency in plants (Lowe et al. 2016; Maher et al. 2020). Moreover, *Agrobacterium*-mediated transformation is frequently restricted due to the narrow range of genotypes within a species. As well as plant growth conditions, co-incubation time & temperature, pre-treatment with phytohormones, variability of *Agrobacterium tumefaciens* are well-known factors to affect transformation efficiency (Zambre et al. 2003; Gelvin 2006). However, these shortcomings can be overcome by utilizing the biolistic transformation approach due to its efficient and potent high transformation efficiency (Wu et al. 2015; Li et al. 2019; Kaul et al. 2021). The CRISPR-Cas-based genome editing in crop plants can only be manifested by fine-tuning the targeted gene or genetic elements (Kwon et al. 2019; Oliva et al. 2019).

Identifying the targets (quantity) due to an inadequate understanding of biological networks and their interactions with environmental factors is another critical obstacle for CRISPR-Cas-based plant genome editing. Applications of multidisciplinary strategy, for instance, genome-wide, and high-throughput functional genomics strategy for identification of beneficial agronomic traits harbouring targets in both the model and non-model crop plants are crucial for genome editing (Lu et al. 2017; Meng et al. 2017; Araus et al. 2018). Alongside, the achievement of high base substitution efficiency via fragment knockout and knockin of homology donor repair (HDR) is an important implication for crop enhancement. However, precision editing in plants employing an HDR-based approach is a significant challenge due to its lower editing potential. Optimizing the optimal quantity and the effective delivery methods of the donor DNA template might ease the base substitution editing approach (Kaul et al. 2020a, b).

The presence of protein inhibitors of CRISPR-Cas systems, known as anti-CRISPR (Acr) proteins, enables the generation of more precision in CRISPR-Cas-based GEd. More than 50 Acr proteins are currently shown to interact with CRISPR-Cas variants, for instance, Cascade-Cas3, Cas9, Cas12, and Cas13 (Dolgin 2019; Marino et al. 2020). The functional mechanism of ACr proteins is one of three ways: firstly, prevention of DNA binding: Acr either blocks or reduced Cas9's interaction with the PAM recognition site; secondly, prevention of crRNA loading: the interaction of Cas9 may disrupt or prevents the proper integration of the crRNA-Cas complex; and thirdly, and blocking of DNA extraction: Acr binds with HNH endonuclease domain of Cas9 and inhibits its activity (Dong et al. 2017; Zhu et al. 2019). However, Acrs can be used to eliminate allergies in unidentified areas

(Aschenbrenner et al. 2020; Shin et al. 2017), unwanted mutations in unintentional cell types or tissues (Hirosawa et al. 2019; Hoffmann et al. 2020). In addition, Acrs (AcrII4s) can be employed as a ligand biosensor to detect and measure CRISPR-Cas9 RNP affinity reagents (Johnston et al. 2019). Similarly, other alternative approaches also being developed to prevent Cas9 activity, for instance, nucleic acid-base inhibitors and (Barkau et al. 2021) and smaller molecules of inhibitors (Maji et al. 2019). Despite precision genome alteration, Acrs provide a prospect to exploit their ability to inhibit Cas9 and to address other engineering limitations of the Cas9 genome.

Comparative genomic analysis revealed that CRISPR and its associated proteins, especially Cas9, were present in umpteen bacterial phylogenetic groups (Lillestøl et al. 2006; Makarova et al. 2006). Cas9 from *S. pyogenes* showed 23 to 58% and 35% similarity to Cas9 proteins from *Streptococcus thermophilus and Lactobacillus plantarum*, respectively. Those organisms were utilized for various human edible food processing, for instance, yoghurt, cheese, kefir, fermented drinks, and so many (Settachaimongkon et al. 2014; Sidira et al. 2017; Behera et al. 2018). Thus, humans were exposed to Cas9 protein in their diet long before the development of CRISPR-Cas9 genome editing. Additionally, Cas9 from *S. pyogenes* showed 80% sequence similarities with a variety of gram-positive and negative bacteria that present in human body (Qin et al. 2010; Louwen et al. 2014). The above-mentioned findings do not imply that human exposure to Cas9 used in genome-editing planning is insignificant (Pineda et al. 2019). However, the biosafety risk assessment regarding human exposure to Cas9 after consuming GEd plants product requires further testing.

Another hindrance is the adoption of edited crop plants success in natural field conditions. An enormous number of researches on genome editing reported so far, but the majority is only about proof of concepts in the greenhouse environment. The performances uncertainties of the edited plants are still existed due to the lack of field trials. Despite all these challenges and impediments, the CRISPR-Cas9 approach is considered the most promising tool due to its precision editing. This approach incorporates numerous heritable traits in plants, which may produce modified plants similar to those developed through conventional breeding. CRISPR-Cas9 strategy leads towards a progressive change via high yielding crop plant production to meet food security globally.

9.6 Overcoming Challenges for 'Off-Target' Mutations

Alteration of plant genome employing the CRISPR-Cas approach sometimes resulted in off-target effects (alteration of the additional region beyond the target region of the genome), which is a pivotal impediment of this application. However, numerous strategies can be employed to minimize off-target mutations. Till date, above than 30 plant varieties (~100 attribute traits) have been edited successfully employing the CRISPR-Cas9system. The precession binding of Cas9 depends on the 7 8 nucleotides seed sequence and the existence of the PAM close to the target sequence, but unwanted insertions/deletions could happen in the genome

(Hajiahmadi et al. 2019). To improve genome-editing efficiency, scientists devised in vivo/vitro biological analysis and algorithm-based computational methods to uncover and increase gene editing efficiency. Promoters and target genes are essential elements involved in the regulation of gene expression by modifying transcription factors (TFs) via RNA polymerase recognition. The specificity of a promoter is essential for controlling transgenic expression in target tissues or throughout the plant. Over the last few years, constitutive promoters like the cauliflower mosaic virus (CaMV35S) promoter (Paparini and Romano-Spica 2006; McCaw et al. 2021) and the maize ubiquitin (pZmUbi) promoter (Xu et al. 2018; Samalov and Moore 2021) have already been used. In dicot plants, the CaMV35S driven promoter showed a high level of expression, in contrast, monocot plants employ pZmUbi promoter more effectively. In *Arabidopsis*, the promoters of (rd29A and rd29B) genes showed well performance to a variety of stress stimuli, such as salinity and drought (Bihmidine et al. 2013). Salt induces activity in the BADH promoter from *Suaeda liaotungensis* (Zhang et al. 2008). The Rab16A promoter might up-regulate GUS expression in transgenic rice under salt stress (Rai et al. 2009). The TsVP1 promoter from *Thellungiella halophila* is effective in almost all tissues except the seeds, and salt stress in leaves and roots, particularly root tips (Sun et al. 2010). DREB2 coordinated expression of transcription factors will generate successful regulatory activity; thus, monocotyledonous plant promoters' operations are higher in monocots as in dicots. Heat-shock protein 17.5E (Hsp17.5E) gene promoter from soybean (*Glycine max*) has been utilized to direct Cas9 expression in rice for genome editing. Several methods have been described to reduce off-target mutations; primarily, the effect can be minimized by using a highly specific Cas nuclease or a stringent sgRNA design that differs from the other genomic regions by three mismatches, in addition to one mismatch in the PAM proximal region. The designed sgRNAs determine the occurrence of a 'off-target' effect; sgRNAs with more than 50% GC content are competent enough to promote on-target mutagenesis due to strong binding to the target sites (Kim et al. 2015; Ren et al. 2014). Precisely designed sgRNAs enable specific targeting, even if so many homologous loci are present in the studied genome (Baysal et al. 2016). Many recently introduced computer-based innovations, i.e. Cas-OFF Finder that identifies the unique target sequences and possible off-target sites in the genomes of various species minimizes the off-target sites (Cong et al. 2013a, b; Hsu et al. 2013). CRISPR-P enables gRNA design for substantially all plant species with accessible genome sequences, as well as off-target site and restriction enzyme sequence analyses (Lei et al. 2014). Subsequently, CRISPR-PLANT has used a genome-wide platform of highly comprehensive RNAs in more than eight plant species and favours restriction endonuclease analysis of target sites. Various guidelines for sgRNAs design to lower the potential off-target effects which can be beneficial for various crop species have been documented in recent articles. It is critical to avoid using sgRNAs with seeds that are homologous to various other genome loci in order to minimize off-target mutations. Indispensable components of CRISPR-Cas9, the PAM and seed sequence, need to be carefully designed.

The sites cleaved with the genome-editing tool CRISPR-Cas9 system can be both on-target and off-target sites and that need to appropriately balance according to the different experimental purposes. To avoid these events, bioinformatics tools, for instance, E-CRISPR and Cas OT, can promote sgRNA design concerning whole-genome sequence information. A vector performs as a vehicle for delivering an element of interest. The vector only needs two components: the single-guide RNA (sgRNA) sequence and the Cas9 gene, both of which may be expressed from a single vector system. A variable crRNA (approximately 20 bp) and a constant tracrRNA make up the sgRNA. To boost performance and eliminate off-target impacts, various target sequences of the same gene might be introduced. The Cas9 gene encompasses multiple nuclear localization signals (NLS) for nuclear targeting, and Cas9 can be defined in a variety of ways (Heintze et al. 2013). In addition to this, various delivery methods such as agrobacterium-mediated, bombardment or biolistic approach, PEG-mediated protoplast, and floral-dip are widely used in plants to regulate genes properly (Table 9.2). There are widely used transformation mechanisms, but the agrobacterium-mediated method is extensively used for the delivery of various Cas enzymes (Ali et al. 2015). The RNP strategy is another important way to reduce the intended effect when sgRNA and RNP nuclease processes are introduced by biolistic and electroporation into plant protoplasts, showing a small frequency of target changes and successfully reported on various plants such as maize (*Zea mays*), rice (*Oryza sativa*), tomato (*Solanum lycopersicum*), and many others (Woo et al. 2015). Nanoparticle-mediated RNP delivery systems have been successfully adopted in plant species due to the reduction of unwanted changes via the potentiality of RNP. The recommended system is time-effective, affordable, species-independent, and equipment-independent CRISPR-Cas9 vector or ribonucleoprotein complexes. Consequently, the specificity of CRISPR-Cas9 is influenced by several parameters, including the aggregation of the Cas9/sgRNA complex and the characteristics of the off-target sites.

Off-target mutation is a major apprehension in the emergence of the CRISPR-Cas9 system in plants used whole genome sequencing WGS and deep sequencing, respectively, to investigate CRISPR-Cas9 specificity in *Arabidopsis thaliana*. According to their findings, CRISPR-Cas9 is highly specific in plants owing to low Cas9 protein expression levels, which resulted in undetectable levels of off-target alterations. Most CRISPR-Cas9 investigations in plants have reported a low frequency of off-target mutation, which could be attributed to its occurrence in non-coding areas and, as a result, the inability to detect off-target implications (Zhang et al. 2018a). CRISPR-PLANT v2 is a popular tool for predicting off-target mutations in plants. This software has the highest sensitivity of among all off-target prediction tool and can be utilized in the genomes of seven plants, including *Sorghum bicolor, Arabidopsis thaliana, Oryza sativa, Medicago truncatula, Solanum lycopersicum, Glycine max, and Brachypodium distachyon*. However, in eukaryotes, several strategies for off-target recognition have been introduced, including deep sequencing and online prediction software. Although in vitro approaches for investigating potential off-target sites have been established, exact prophecies of the prevalence of undesired mutations in vivo are difficult to

Table 9.2 Novel delivery approaches of CRISPR-Cas based genome editing in agronomically important crop plants

Crop plant	CRISPR-Cas9 ribonucleoprotein complexes (RNP)-based vector	Targeted genes	Delivery method	References
Apple (*Malus domestica*)	Cas9-sgRNA ribonucleoprotein complexes	DIPM-1, 2, 4	PEG-mediated CRISPR-Cas9 components delivery	Malnoy et al. (2016)
Soybean (*Glycine max*)	pCas9-GmU6-sgRNA, pCas9AtU6 sgRNA	Glyma08g02290, Glyma12g37050, Glyma06g14180	PEG-mediated CRISPR-Cas9 components delivery	Sun et al. (2015)
	QC810 and RTW830, QC799 and RTW831	DD20, DD43	Particle bombardment method for CRISPR-Cas9 component delivery	Li et al. (2015)
	p201N Cas9	GFP transgene	*Agrobacterium*-mediated delivery of CRISPR-Cas9 components	Jacobs et al. (2015)
Rice (*Oryza sativum*)	pRGE3, pRGE6, pUC19-*Os*Cas9, pJIT163-2NLSCas9	*Os*MPK5, *Os*SWEET14, *Os*SWEET11, *Os*PDS, *Os*BADH2, crtI, OsPDS1, OsPDS, OsBADH2, OsPDS, OsDEP1	PEG-mediated CRISPR-Cas9 components delivery	Jiang et al. (2013)
	pCam1300-CRISPR-B CRISPR-RNP complex pJIT163-2NLSCas9 p*Os*U3-sgRNA, pJIT163-2NLSCas9, VK005	crtI,,*Os*PDS1, *Os*PDS1, *Os*DEP1	Particle bombardment method for CRISPR-Cas9 component delivery	Banakar et al. (2019)
	VK005	ISA1	*Agrobacterium*-mediated delivery of CRISPR-Cas9 components	Shufen et al. (2019)
	Cas9-sgRNA Ribonuclease	PhACO1	PEG-mediated CRISPR-Cas9	Xu et al. (2020)

(continued)

Table 9.2 (continued)

Crop plant	CRISPR-Cas9 ribonucleoprotein complexes (RNP)-based vector	Targeted genes	Delivery method	References
Petunia (*Petunia hybrida*)	protein complexes (RNPs)		components delivery	
Wheat (*Triticum aestivum*)	pCR8-U6-gRNA	*Ta*EPSPS	PEG-mediated CRISPR-Cas9 components delivery	Arndell et al. (2019)
	pJIT163-ubi	TaMLO-A1, TaMLO B1, TaMLO-D1	Particle bombardment method for CRISPR-Cas9 component delivery	Wang et al. (2014)
	pBI121	Inox, PDS	*Agrobacterium*-mediated delivery of CRISPR-Cas9 components	Upadhyay et al. (2013)
Maize (*Zea mays*)	pZmU3-gRNA, T-nCas9	*Zm*IPK, *Zm*ALS1, *Zm*ALS2	PEG-mediated CRISPR-Cas9 components delivery	Svitashev et al. (2015)
	pSB11-ubi:Cas9	LIG1, Ms26, Ms45, ALS1, ALS2	Particle bombardment method for CRISPR-Cas9 component delivery	Liang et al. (2014)
	pMCG1005	Argonaute 18, Dihydroflavonol-4-reductase Strain- EHA101	*Agrobacterium*-mediated delivery of CRISPR-Cas9 components	Char et al. (2017)
Barley (*Hordeum vulgare*)	pCas9:sgRNA	ENGase	PEG-mediated CRISPR-Cas9 components delivery	Kapusi et al. (2017)
Arabidopsis (*Arabidopsis thaliana*)	pCAMBIA1300	AtPDS3, AtFLS2, RACK1b, RACK1c, BRI1, GAI, JAZ1	*Agrobacterium*-mediated delivery of CRISPR-Cas9 components	Feng et al. (2013)
Tomato (*Solanum lycopersicum*)	pYLCRISPR-Cas9	SGR1, LCY-E, Blc, LCY B1, LCY-B2	1.1. *Agrobacterium*-mediated	Li et al. (2018b)
		SlCCD8		

(continued)

Table 9.2 (continued)

Crop plant	CRISPR-Cas9 ribonucleoprotein complexes (RNP)-based vector	Targeted genes	Delivery method	References
	pENTR-sgRNA: pMR290/Cas9		delivery of CRISPR-Cas9 components	Bari et al. (2019)
	pMDC32	StALS1		
	Cas9-sgRNA ribonucleoprotein complexes (RNPs)	GBSS(GT4)	PEG-mediated CRISPR-Cas9 components delivery	

acquire. Digenome-seq, SITE-seq, and CIRCLE-seq are the most used in vitro genome-wide detection systems and quantifying off-target effects (Cameron et al. 2017). Digested genome sequencing (Digenome-seq) is a reliable, delicate (~ 0.1%), and frequently used for detecting Cas9 and other nucleases for off-target effects in genome-wide. The most prominent strategies established to solve the Digenome-seq difficulties are selective enrichment and identification of tagged genomic DNA ends by sequencing (SITE-Seq), followed by circularization for in vitro reporting of cleavage effects by sequencing (CIRCLE–seq). The SITE-Seq approach could map all of the Cas9 cleavage sites in a genome (Naeem et al. 2020). This study employed sgRNA and Cas9 RNPs in a cell-free environment to cleave purified genomic DNA. Afterwards, both (on- and off-target) cleavage fragments are tagged, and off-target sites are detected using next-generation sequencing (NGS). The total amount of off-target sites has a considerable impact on nuclease concentration. RNPs (low to high) were employed as variable concentrations to recover off-target locations with low and high cleavage sensitivity. When low doses of RNPs are subjected to cell identification, they exhibit a significant proclivity for off-target alterations. SITE-Seq also requires less NGS read depth than Digenome-seq, with some procedural modifications; CIRCLE-Seq has a similar concept. In CIRCLE-Seq, the DNA is first trimmed, then circularized, and finally destroyed. Prior to treatment with (Cas9–sgRNA) RNPs, the degradation phase practically eliminates high background DNA to boost sensitivity, condensing NGS read space that would otherwise be squandered on random reads. Following that, DNA is linearized using Cas9 and then exposed to NGS for off-target detection. CIRCLE-Seq, like SITE-Seq, could be employed in a reference-independent manner to discover off-target cleavage sites, for organisms whose genome sequences are less well-characterized and/or show considerable genetic variability. Several approaches were proposed, including bioinformatics tools for in silico detection of off-target mutations and increased on-target efficiency to mitigate off-target impacts. However, off-target effects might have happened, yet the alterations will be lower than those developed via conventional breeding. Thus, GEd employing the CRISPR-

Cas approach produces a far less off-target effect in comparison to the traditional crop enhancement strategy.

9.7 CRISPR Implementation in Sustainable Agriculture: Climate-Smart and Nutritionally Secure Crops

The global population is assumed to increase 9.2 billion in 2050, and so agronomic production needs to rise by about 70% from existing levels to encounter the increased demand of food, as predicted by Food and Agriculture Organization (accessed on 1 February 2021). Cereal crops such as rice, wheat, and maize are the world's most important sources of energies intended for humans, livestock feed for animals, and raw material for biofuel. Therefore, improving cereal-crop-grain production is critical to meet further demand. For most cereal crops, the annual yield relates to grain production. Until the last decade, the core crop improvement strategies banked upon chemical mutations, hybrids, and expression of trans gene/s (Chari et al. 2017). The shift from the conventional breeding approach, which relied on the occurrence of the naturally relevant variations to the molecular breeding approach, has alleviated some barriers attached with the conventional methods. Now, the targeted traits can be swiftly incorporated into the plant system to generate a new plant variety for food as well as nutritional security. The gradual increase in human population, deteriorating arable land conditions, the drastic climatic changes through uplifted temperature, and escalated pollutants by excessive emission of greenhouse gases (GHG) causes threat to agriculture and food security (Asseng et al. 2014). Therefore, to develop climate-smart crops via sustainable agriculture, the need of the hour is to achieve a 'triple win' by targeting enhanced productivity, improved adaptivity, and GHG mitigation. Targeted GEd made the revolution in molecular biology by discovering programmable SSNs (Chandrasegaran and Carroll 2016). CRISPR-Cas-based GEd has become an essential tool that has effectively caused enormous ripple effects in plant research. Throughout the last decade, we have seen fast development in numerous fields, including plant functional genomics and crop enhancement (>45 genera of plants) in a manner that straightforwardly benefits consumers (Shan et al. 2020). In plant species, the practice of CRISPR-Cas9-mediated genome alteration in diverse crops was successful, for instance, in maize, rice, wheat, maize, and cotton. In 2015, the fourth quarter experienced the employment of DNA-free, pre-assembled RNP complex of CRISPR-Cas9 for genome alteration in model plants such as *Arabidopsis*, rice, lettuce, tobacco, wheat, maize, and so on (Woo et al. 2015). An extremely systematic transgene integration free GEd and most importantly callus-based methodology were introduced for wheat pertaining transitory expression of CRISPR-Cas9 in the form of DNA or RNA (dubbed TECCDNA or TECCRNA, respectively), the technique had the potential to be applied in different crops (Zhang et al. 2016). The crops developed via RNP complex mediated and TECCRNA-based editing techniques are foreign gene integration free, thus they could be spared from GMO regulatory concerns. In a recent study, by altering the sequence of a S gene, namely

SlAGAMOUS-LIKE 6 (SlAGL6) which is linked to enhanced fruit setting even under heat stress, tolerance towards high temperature was attained in tomato (Klap et al. 2017). Further, optimization of method for targeting multiple genes via CRISPR-Cas9 in a single organism was done for numerous crops which include rice, cotton, maize, and wheat (Miao et al. 2013; Gao et al. 2017; Char et al. 2017; Wang et al. 2018b). Thus, CRISPR-Cas9 is a remarkable technique, which is potent enough to develop crop with multiple stress tolerance by choosing concurrently different S genes as a target in exclusively high productive but sensitive cultivars.

The growth of plants is linked to diverse developmental and environmental cues. Plants receive and respond to those cues via cellular signaling cascades, which regulate gene expression at the pre-mRNA level by tuning splicing patterns and controlling the transcript abundance at mature-mRNA level. Alternatively, spliced pre-mRNA represents the genome's coding potential for multi-exon genes and synchronizes gene expression by different mechanisms. In *Solanum tuberosum* (potato), vegetative reproduction (tuberization) is regulated via photoperiod, for example, flowering controlling transcription factor- *StCDF1* (CYCLING DOF FACTOR 1), which regulate the antisense transcript of *StFLORE* to gain drought tolerance. Loss of function mutation in promoter of this *StFLORE* via CRISPR-Cas9 revealed drought tolerance via stomatal size and number regulation (Gonzales et al. 2020).

In agriculture, weed control is critical for a high yield of crop production, which can reduce the phytotoxicity of herbicides to crops, cut off the cost of the weeding, and upgrade the efficiency of the chemical weeding. Consequently, substantial attempts to develop herbicide -resistant crop varieties have been undertaken to contribute the frugal and economic tools to serve farmers for clean and effortless weed management. To develop robust herbicide-resistant crop plants, endogenous genes like cellulose synthase A catalytic subunit 3 (*CESA3*), splicing factor 3B subunit 1 (*SF3B1*) and more commonly acetolactate synthase (*ALS*), 5-enolpyruvylshikimate-3-phosphate synthase (*EPSPS*) are targeted for CRISPR-Cas9-mediated gene editing. The crucial amino acid substitution in *EPSPS* and *ALS* genes in rice employing CRISPR-Cas9 HDR-mediated machinery conferred resistance to glyphosate and sulfonylurea herbicide, respectively (Li et al. 2015; Sun et al. 2016). Similarly, T102I/P106S and T102I/P106A substitution were introduced in *EPSPS* gene of flax (Sauer et al. 2016) and cassava plant (Hummel et al. 2017). To acquire effective gene replacement, CRISPR-Cas9 is employed with target sequence-specific sgRNAs directing the CRISPR-associated RNA endoribonuclease csy4 from *Pseudomonas aeruginosa,* for sequence-specific induction of DSBs (Wang et al. 2021). Till now among the developed crop germplasm specifically resistant to herbicides, crops only resistance towards *ALS*-inhibiting herbicides, ACCase-inhibiting herbicides, and glyphosate has been successfully established. One of the greatest important applications intended for gene editing in agriculture is biotic stress resistance. The genetic mechanisms of the agents that cause biotic stressors in plants can be examined in order to overcome these stresses by GEd (Yin and Qiu 2019; Zafar et al. 2020; Pak et al. 2020). In addition to some crop species like rice, a CRISPR-Cas9 targeted mutation in the ethylene responsive factor,

OsERF922, has been effectively established to improve resistance to *Magnaporthe oryzae* blast disease (Wang et al. 2016). Similarly, *OsMPK5*, a negative regulator of biotic and abiotic stressors in rice, was identified for targeted mutagenesis in rice protoplasts utilizing three gRNAs by using a more precise gRNA design strategy with a low level of off-targets (Xie and Yang 2013). By producing genetically modified resistant crop varieties, which have proven to be a significant effort to fight against biotic stressors. Despite CRISPR-Cas9 inimitable accomplishment, there are substantial trials in incorporating this technology into agricultural research, especially with transformation-resistant crops reproduced asexually. Several projects are presently in progress to fine-tune CRISPR-Cas9-based technologies for precise editing in the plant genome of the target locus.

Consequently, crop improvement now targets not only improving quantity (yield), but quality (nutrition) of the crop product as well. Great quality food grains have a critical and direct impact on human health and well-being, as plants produce numerous molecules with anti-inflammatory, anti-cancerous, and anti-oxidation properties (Liu et al. 2021) that have beneficial effects on human health. Thus, plants are the major source of nutrients and natural dietary products and are considered as 'dietary doctors' as they can cure the prevalent undernourishment (FAO 2020). Thus, crops biofortified with micronutrients and minerals such as iron, zinc, selenium, and iodine can curb the nutrient deficiency in addition making anti-nutrient, such as heavy metals, phytate, and gluten, devoid crops can make the unavailable nutrient available for absorption in human body and protect humans from developing allergies, metabolic disorders, and chronic ailments. Conventional breeding accompanied by technology has saved humanity from the food crisis in the past but now these approaches culminate into no added benefit in enhancing the productivity, whereas new techniques like CRISPR-Cas hold the potential to drive the way towards sustainable food security. Recent breakthroughs (Table 9.3) have paved the way to introduce or manipulate the inherent genes to improve the quality of the majorly consumed crops. Alteration of genes for crop biofortification as well as for removing anti-nutrients have the potential to provide macro and micronutrients and alleviate the 'hidden hunger' (Majumder et al. 2019) condition as well as to cure and prevent the non-infectious, lifestyle related chronic ailments in humans. Although the CRISPR-Cas system is still developing and evolving, the latent potential of this technique has resulted in some benchmark studies, and it will continue to bestow the field of genetic engineering with more novel breakthroughs.

9.8 Amalgamation of MI and CRISPR-Based Genome Editing

Despite being one of the common genetic engineering techniques, CRISPR-Cas9 GEd relies on the accuracy of well-designed guide RNAs as it is an essential aspect of successful target gene editing (Cox et al. 2015). In recent years, various algorithms have been generated for assessing CRISPR activity (on-target) and specificity (off-target) as well as web-based tools for in silico gRNA designing (Henry et al. 2014; Zhu 2015). Machine learning (ML) and Artificial Intelligence

Table 9.3 Recent breakthroughs of CRISPR-Cas based approaches for the quality improvement of majorly consumed crops

Crop	Targeted gene	Gene function	Editing technique	Associated trait	References
Biofortification					
Rice	Starch-branching enzyme I (OsBEI) and starch-branching enzyme IIb (OsBEIIb)	OsBEI is expressed in all tissue, whereas OsBEIIb is expressed in the endosperm. Both genes are involved in the starch synthesis. Downregulation of these genes will result in the pathway direction towards amylose formation	Cas9: Knockout	High amylose content	Sun et al. (2017)
	Phytoene synthase (PSY), carotene desaturase (CrtI)	PSY converts geranylgeranyl-diphosphate (GGPP) to phytoene, and CrtI catalyses the desaturation reaction and introduces four double bond in phytoene and results in the formation of ζ-carotene and ultimately lycopene	Cas9: Gene insertion	High β-carotene content	Dong et al. (2020)
	Glutamate decarboxylase 3 (OsGAD3)	GAD is known to catalyse the conversion of L-glutamate to gamma-aminobutyric acid (GABA). OsGAD3 one of the five GAD genes present in rice is predominantly expressed in the seeds	Cas9: Deletion of C-terminal calmodulin binding domain of OsGAD3	High GABA content	Akama et al. (2020)
	Vacuolar iron transporter (OsVIT2)	VIT is an iron transporter which is expressed in various tissue and downregulation of this transporter can reduce Fe allocation to leaf sheath, nodes, and aleurone and in contrast increase allocation to leaf	Cas9: Knockout	High iron in rice grain	Che et al. (2021)

		blade and grains. High Fe in grains increase the bioavailability of Fe for absorption in human gut			
	Arsenite tolerant 1 (ASTOL1)	ASTOL1 encodes the chloroplast-localized O-acetylserine (thiol)lyase (OAS-TL) which catalyses the condensation of O-acetylserine (OAS) and sulphide to form cysteine (Cys). OAS in turn is produced by the action of serine acetyltransferase (SAT) on serine and acetyl CoA. SAT and OAS-TL form the hetero-oligomeric cysteine synthase complex (CSC)	Cas9: Gene correction	High selenium content	Sun et al. (2021)
	Omega-6 fatty acid desaturase (OsFAD2–1)	FAD2–1 is highly expressed in seeds and catalyses the desaturation of the C18:1 to C18:2 at position sn-2 and conversion of oleic acid to linoleic acid	Cas9: Knockout	High-oleic acid proportion and reduced linolenic acid content	Abe et al. (2018)
	Phospholipase D gene (OsPLDα1)	It converts phosphatidylcholine (PC) to phosphatidic acid (PA) and changes the flux from the biosynthesis of lysophospholipid (LPL) to the accumulation of phytic acid	Cas9: Knockout	Increased LPL content	Khan et al. (2019b)
Banana	Lycopene ε-cyclase (MaLCY-E)	LCY-E catalyses the cyclization of the lycopene to produce α-carotene	Cas9: Knockout	High β-carotene content	Kaur et al. (2020)
Tomato		SlGAD2 and SlGAD3 primarily express at the time of fruit		High GABA content	

(continued)

Table 9.3 (continued)

Crop	Targeted gene	Gene function	Editing technique	Associated trait	References
	Glutamate decarboxylase 2 (SlGAD2), glutamate decarboxylase 3 (SlGAD3)	development and are key enzymes for the biosynthesis of GABA in tomato fruits	Cas9: Introduction of stop codon before autoinhibitory domain		Nonaka et al. (2017)
	Stay-green 1 (SGR1), phytoene desaturase (slyPDS), lycopene β-cyclase (slLCY-B1 & 2), lycopene ε-cyclase (slLCY-E)	SGR1 directly interacts with phytoene synthase 1 gene (PSY1) to regulate lycopene accumulation in ripening fruits. PDS catalyses the desaturation to convert the phytoene to ζ-carotene which further converts to the lycopene, whereas LCY-B1 & 2 and LCY-E catalyses the cyclization of the lycopene to produce β-carotene and α-carotene, respectively	Cas9: Gene replacement of PDS and knockout of SGR1, LCY-B, and LCY-E	Increased lycopene content	Li et al. (2018b)
Rapeseed	Fatty acid desaturase (BnFAD2)	FAD catalyses the desaturation of oleic acid to linoleic acid	Cas9: Knockout	High-oleic acid proportion	Okuzaki et al. (2018)
	BnTT8 (transcription factor involved in flavonoid pathway; TT comes from transparent testa mutant)	Play a role in flavonoid biosynthesis pathway in the seed coat and lead to the accumulation of oxidized form of flavonoids known as proanthocyanidins (PA; condensed tannins) in the endothelial layer of the inner integument which gives dark colour to the seed coat. It also play a role in the fatty acid synthesis pathway	Cas9: Knockout	High oil production and GPC	Zhai et al. (2020)
Camelina	Fatty acid desaturase (CsFAD2)		Cas9: Knockout		

		FAD catalyses the desaturation of oleic acid to linoleic acid		High-oleic acid proportion	Jiang et al. (2017)
Potato	Starch-branching enzyme 1 (StSBE1), starch-branching enzyme 2 (StSBE2)	SBE's genes are involved in the starch synthesisDown regulation of these genes will result in the pathway direction towards amylose formation. It basically reduces the branching frequency which results in compact starch formation	Cas9: Knockout	High amylose content	Tuncel et al. (2019)
Sweet potato	Granule-bound starch synthase I (IbGBSSI), starch-branching enzyme II (IbSBEII)	GBSSI is involved in amylose biosynthesis, and SBEII is involved in amylopectin biosynthesis	Cas9: Knockout	High amylose content	Wang et al. (2019)
Anti-nutrients					
Rapeseed	Inositol tetrakisphosphate kinase (BnITPK)	ITPK catalyses the penultimate step of phytate synthesis in rapeseed	Cas9: Knockout	Low phytic acid content	Sashidhar et al. (2020)
Wheat	α-gliadin genes	α-type gliadin is encoded by Gli-2 locus and is responsible for the gliadin, i.e. gluten protein in the wheat flour. Although there are numerous copies that code for gluten protein thus by targeting conserved region of α-gliadin genes, gluten content can be reduced	Cas9: Knockout	Low gluten content	Sanchez-Leon et al. (2018)
	Inositol 1,3,4,5,6-pentakisphosphate 2-kinase 1 (TaIPK1)	Catalyses the final step of PA biosynthesis by phosphorylating inositol pentaphosphate (IP5) to	Cas9: Knockout	Low phytic acid content and improved accumulation of Fe and Zn	Ibrahim et al. (2021)

(continued)

Table 9.3 (continued)

Crop	Targeted gene	Gene function	Editing technique	Associated trait	References
		phytate (inositol hexakisphosphate/IP6)			
Rice	Natural resistance- associated macrophage protein 5 (OsNramp5)	OsNramp5 located on the exo- and endo-dermis of root and is the major transporter for cd influx in plants	Cas9: Knockout	Low cd accumulations	Tang et al. (2017)
	Low affinity cation transporter (OsLCT1)	LCT regulates the cd transport in rice	Cas9: Knockout	Low cd accumulations	
	Phospholipase D gene (OsPLDα1)	From multidisciplinary actions of OsPLDα1, it has been highlighted that it is involved in lipid-dependent phytic acid biosynthesis pathway by converting phosphatidylcholine (PC) to phosphatidic acid (PA) and changing the flux from the biosynthesis of lysophospholipid (LPL) to the accumulation of phytic acid	Cas9: Knockout	Low phytic acid content	Khan et al. (2019b)
	Inositol 1,3,4-trisphosphate 5/6-kinase (OsITPK6)	IPTK is encoded by six genes in rice, and OsIPTK6 knockout significantly reduces phytic acid content in rice grains. By phosphorylating inositol triphosphate (IP3) at 5 and sixth position, ITPK plays a key role in phytic acid synthesis	Cas9: Knockout	Low phytic acid content	Jiang et al. (2019)
	Arsenite tolerant 1 (ASTOL1)	ASTOL1 encodes the chloroplast-localized O-acetylserine	Cas9: Gene correction	Low arsenic content	Sun et al. (2021)

		(thiol)lyase (OAS-TL) which catalyses the condensation of O-acetylserine (OAS) and sulphide to form cysteine (Cys). OAS in turn is produced by the action of serine acetyltransferase (SAT) on serine and acetyl CoA. SAT and OAS-TL form the hetero-oligomeric cysteine synthase complex (CSC). Cys is the key regulator of production of glutathione (GSH) and phytochelatins (PCs). GSH and PCs are responsible of sequestration of Ar into vacuoles and thus restricting its distribution to grains			
Grape	L-idonate dehydrogenase gene (ldnDH)	IdnDH controls the biosynthesis of tartaric acid (TA) in grape	Cas9: Knockout	Low tartaric acid	Ren et al. (2016)

(AI) offer revolutionary approaches for utilizing the CRISPR-Cas9 technology to analyse edited crop lines with better features, for example, higher nutrient value, palatability, modified root, flower architectures, stress tolerance, and so on. Some examples of CRISPR-based design tools are described in Table 9.4. All of these gRNA design tools, off- and on-target prediction tools have contributed to the success and application of CRISPR genome technology.

Several functions have been shown to be important for target gRNA activity, including secondary structure, sequence composition, thermodynamics, and physicochemical characteristic, but for off-target predictions, this is the size, composition, and combination of discrepancies. Many machines and deep learning methods have been established to represent the activity of CRISPR, which can be broadly divided into two types. (1) Machine learning based, which includes CRISPRscan, sgRNA Scorer, SSC, sgRNA Designer, and CRISPRater. CRISPRScan, CRISPRater, and SSC are trained using simple linear models, and Azimuth2.0 and TUSCAN are trained using general linear models that are logistic regression and random forests, respectively (Listgarten et al. 2018). (2) Deep learning based. CNN_std, DeepCas9, DeepCRISPR, CRISPRpred, and DeepCpf1 predict sgRNA activity builds on automatic recognition of sequence characters using a Convolutional Nuclear Network (CNN). MIT server estimates off-targets based on the distance and number between unpaired nucleotides (Hsu et al. 2013). Subsequently, a cutting frequency determination (CFD) score was developed that predicts off-target scores by reproducing the frequency of bases in gRNA spacer sequence (Doench et al. 2016). Synergizing CRISPR combines the projection results of five different models (CCTop, CFD, CROPIT, MIT, and MIT website) into an input function based on hypothetical and statistical methods (Dobson et al. 2015; Singh et al. 2015). There are currently numerous procedures available to generate accurate sgRNAs using basic rules. Here, a new algorithm called CRISPR target estimation (CRISTA) was introduced as part of ML, which performed the important task of identifying specific genomic regions to be accurately removed via given sgRNAs. The CRISTA predictions have been proven to be more accurate than previously predicted thresholds (Abadi et al. 2017). However, identifying prospective off-target sites required the recognition of short sequence motifs up to 20 bp, besides the PAM with frequent mismatches. In most cases, the aligners first match the seed sequence and extend the seed sequence in a specific direction and then check for a match. Therefore, ML and AI analysed possible regression points that may converge or deviate from on-target and off-target specificity charts.

The precision of these tools for predicting gRNA activity in different species and cell types remains unclear (Chuai et al. 2017). Large variations between species have led to the development of species-specific software (e.g. CRISPR-P for plants, flyCRISPR for fruit flies, CRISPRscan for zebrafish, and EuPaGDT for pathogens). Of these, only CRISPRscan was generated based on ML, and the rest were theoretical software. Since organisms cannot rapidly limit the previous off-target scoring process, researchers wanted to create a new procedure for assessing off-target action called CASPER (Mendoza and Trinh 2018). Although these tools can be selected for prior study when performing experiments by editing them in corresponding species,

Table 9.4 Different types of CRISPR-based designing tools

Tool	Input	PAM	Website	References
Azimuth2.0	DNA sequence	NGG	https://github.com/maximilianh/crisporWebsite/tree/master/bin/Azimuth-2.0	Doench et al. (2016)
Benchling CRISPR gRNA design	Gene ID/genome coordinates	User customizable	https://benchling.com/crispr	Doench et al. (2016)
Cas-designer	DNA sequence	NGG,NRG, NNAGAAW, NNNNGMTT	www.rgenome.net/cas-designer	Park et al. (2015)
Cas-OFFinder	crRNA sequence	20 PAMs (NGG, NRG, NNAGAAW, . . .)	http://www.rgenome.net/cas-offinder/	Bae et al. (2014), Baltes et al. (2014)
CasOT	DNA sequence	NGG, NAG, NNGG	http://eendb.zfgenetics.org/casot/	Xiao et al. (2014)
CASPER	DNA sequence	TTTN, NGG, NGCG	https://github.com/TrinhLab/CASPER	Mendoza and Trinh (2018)
CCTop	DNA sequence	NGG, NRG, NNGRRT, NNNNGATT, NNAGAAW, NAAAAC	https://crispr.cos.uni-heidelberg.de/	Stemmer et al. (2015)
CFD	DNA sequence	NGG, NAG, NCG, NGA	https://broadinstitute.org/rnai/public/software/index	Doench et al. (2016)
ChopChop	RefSeq, genomic region, gene ID	NGG, NGA, NAG, NRG, NNNNGANN, . . .), user customizable	https://chopchop.cbu.uib.no/	Montague et al. (2014)
ChopChop v2	RefSeq gene ID genomic region	User customizable	http://chopchop.cbu.uib.no/	Labun et al. (2016, 2019)
CINDEL	DNA sequence	TTTN, TTTA, TTTC, TTTG, TTTT, TTTV	http://big.hanyang.ac.kr/cindel	Kim et al. (2017)
CNN_std	DNA sequence	NAG, NGT, NTG, NGC, NGA, NGG, NAA, NCG	https://github.com/MichaelLinn/off_target_prediction	Lin and Wong (2018)
COD				

(continued)

Table 9.4 (continued)

Tool	Input	PAM	Website	References
	DNA sequence	NGG, NRG NNAGAAW NNNNGMTT NNGRRT	http://cas9.wicp.netsgRNAcas9	Park et al. (2015)
CrisFlash	DNA sequence	NGG	https://github.com/crisflash	Jacquin et al. (2019)
CRISPick	DNA sequence	NGG, CGGH, CGGT, TGGG	https://portals.broadinstitute.org/gpp/public/analysis-tools/sgrna-design	Doench et al. (2014)
CRISPOR	DNA sequence/genomic region	NGG, NGA, NGCG, NGGNG, NNAGAA, NNGRRT, NNNRRT, NNNNACA, NNNNGMTT, TTTN	http://crispor.tefor.net	Haeussler et al. (2016)
CRISPR finder	DNA sequence	NGG/user customizable	www.crispr.u-psud.fr/server	Kurtz (2003), Doench et al. (2014)
CRISPR MultiTargeter	DNA sequence/gene ID	NGG, user customizable	http://www.multicrispr.net/	Prykhozhij et al. (2015)
CRISPR primer designer	DNA sequence	NGG	http://www.plantsignal.cn/	Yan et al. (2015)
CRISPR-ERA	DNA sequence	NGG	www.CRISPR-ERA.stanford.edu	Liu et al. (2015)
CRISPR-GE	DNA sequence/gene ID	NGG, TTN, TTTN, user customizable	http://skl.scau.edu.cn/	Xie et al. (2017)
CRISPR-P	DNA sequence/gene locus/genome coordinates	NGG, NAG	http://crispr.hzau.edu.cn/cgi-bin/CRISPR/CRISPR	Lei et al. (2014)
CRISPR-P 2.0	DNA sequence/gene locus/genome coordinates	14 PAMs (NGG, NNAGAAW, NNNNGMTT, TTTN, . . .)	http://crispr.hzau.edu.cn/CRISPR2/	Liu et al. (2017)
Crispr-plant	Gene locus/genome coordinates	NGG	https://www.genome.arizona.edu/crispr/	Minkenberg et al. (2019)
CRISPRater		NGG		

(continued)

Table 9.4 (continued)

Tool	Input	PAM	Website	References
	DNA sequence		https://crispr.cos.uni-heidelberg.de/	Labuhn et al. (2018)
CRISPRdirect	DNA sequence genome coordinates	NNN, user customizable	http://crispr.dbcls.jp/	Naito et al. (2015)
CRISPRoff	DNA sequence	NGG, NAG, NGA	https://rth.dk/resources/crispr/	Alkan et al. (2018)
CRISPRpred	DNA sequence	NGG	https://github.com/khaled-buet/CRISPRpred	Rahman and Rahman (2017)
CRISPRscan	DNA sequence	NGG	www.crisprscan.org	Moreno-Mateos et al. (2015)
CRISPRseek	DNA sequence	NRG, NGG, user customizable	http://www.bioconductor.org/packages/release/bioc/html/CRISPRseek.html	Zhu et al. (2014)
CRISTA	DNA sequence	NGG	http://crista.tau.ac.il/pair_score.html	Abadi et al. (2017)
CROPIT	DNA sequence	NGG, NNG, GGG	http://cheetah.bioch.virginia.edu/AdliLab/CROP-IT/homepage.html	Singh et al. (2015)
CT-finder	DNA sequence	NGG	http://bioinfolab.miamioh.edu/ct-finder	Zhu et al. (2016)
DeepCas9	DNA sequence	NGG	https://github.com/lje00006/DeepCas9	Xue et al. (2019)
DeepCpf1	DNA sequence	TTTN	http://deepcrispr.info/	Luo et al. (2019)
DeepCRISPR	sgRNA sequence	NGG, NGT, NGA, NAG, NGC, NCG, NTG, NAA	http://www.deepcrispr.net/	Chuai et al. (2018)
E-CRISP	Gene ID/DNA sequence	NGG, user customizable	http://www.e-crisp.org/E-CRISP/	Heigwer et al. (2014), MacPherson and Scherf (2015)
Elevation	Gene ID transcript		https://crispr.ml/	Listgarten et al. (2018)

(continued)

Table 9.4 (continued)

Tool	Input	PAM	Website	References
	ID genomic region	NAG, NGA, NCG, NGC, NGG, NTG, NGT		
Elevation-search/ dsNickFury	DNA sequence	NGG, NCG, NAG, NGA, NGG, NGC, NTG, NGT	https://github.com/michael-weinstein/dsNickFury3PlusOrchid	Listgarten et al. (2018)
EuPaGDT	DNA sequence	NGG, NAG, NGA	http://grna.ctegd.uga.edu/	Peng and Tarleton (2015)
FlashFry	DNA sequence	NGG	http://aaronmck.github.io/FlashFry/	McKenna and Shendure (2018)
FlyCRISPR	DNA sequence	NGG	www.tools.flycrispr.molbio.wisc.edu/targetFinder	Gratz et al. (2014)
Ge-CRISPR	DNA sequence	NGG	http://bioinfo.imtech.res.in/manojk/gecrispr/	Kaur et al. (2016)
GT-scan	DNA sequence	User customizable	https://gt-scan.csiro.au/	O'Brien and Bailey (2014)
Off-spotter	DNA sequence	NGG, NAG, NNGRRT, NNNNACA (R is Aor G)	https://cm.jefferson.edu/Off-Spotter/	Pliatsika and Rigoutsos (2015)
Optimized CRISPR design	DNA sequence	NGG, NAG	https://crispr.mit.edu	Hsu et al. (2013)
Predict CRISPR	DNA sequence	NGG	https://github.com/penn-hui/OfftargetPredic	Peng et al. (2018)
Protospacer workbench	Gene ID/DNA sequence	NGG	www.protospacer.com	MacPherson and Scherf (2015)
sgRNA designer	DNA sequence, gene ID, transcript ID	NGG	https://portals.broadinstitute.org/gpp/public/analysistools/sgrna-design	Doench et al. (2014)
sgRNA scorer	DNA sequence	NGG, NAG, NNNNGMTT, NNAGAAW	https://crispr.med.harvard.edu/sgRNAScorerV2/	Chari et al. 92,015)
sgRNAcas9	DNA sequence	NGG, NAG	www.biootools.com	Xie et al. (2014)
SSC		NGG		

(continued)

Table 9.4 (continued)

Tool	Input	PAM	Website	References
	DNA sequence		www.crispr.dfci.harvard.edu/SSC/	Xu et al. (2015)
SSFinder	DNA sequence	NGG	https://code.google.com/archive/p/ssfinder/	Upadhyay and Sharma (2014)
SynergizingCRISPR	DNA sequence	NGG	https://github.com/Alexzsx/CRISPR	Zhang et al. (2019)
Synthego design tool	DNA sequence	NGG	https://design.synthego.com/#/	Roginsky (2018)
TUSCAN	DNA sequence	NGG	https://github.com/BauerLab/TUSCAN	Wu et al. (2014)
uCRISPR	DNA sequence	NGG, NAG, NGA	https://github.com/Vfold-RNA/uCRISPR	Zhang et al. (2019)
WGE	DNA sequence	NGG	www.sanger.ac.uk/htgt/wge	Hodgkins et al. (2015)
WU-CRISPR	RNA sequence	NGG	http://crispr.wustl.edu/	Wong et al. (2015)
ZiFiT	DNA sequence	NGG	http://zifit.partners.org/ZiFiT	Sander et al. (2010)

their prediction of sgRNA efficiency and target in various cell types is debatable. However, these tools have been proved in the laboratory using mouse cell lines, and human or both, major and cross-species variations have not yet been testified. Therefore, ML-based learning approaches can effectively predict lethal sgRNA interactions and characterize target regions in specific gene combinations. However, there is a large amount of work to be tested and optimized for utilizing CRISPR-Cas9 gene editing in plant systems. In the future, genome-wide engineering crops will include trained data sets, including variants and orthologs.

9.9 Regulatory Aspects of Genome Edited Crops

GEd technology has proved its potential uses in a broad array of industries, notably human and animal health, food, agriculture, and others, in a relatively short period of time. GEd innovations, the same as any other new technology, have dual-use prospective and so raise both safety and security concerns. Novel GEd techniques, particularly CRISPR-Cas9, have a unified mechanism for the insertion of elite attributes in crops plants, allowing unconstrained base substitutions, additions, deletions, and gene introduction or replacement. The offspring produced are similar to those produced by random mutagenesis, natural genetic variants, and traditional

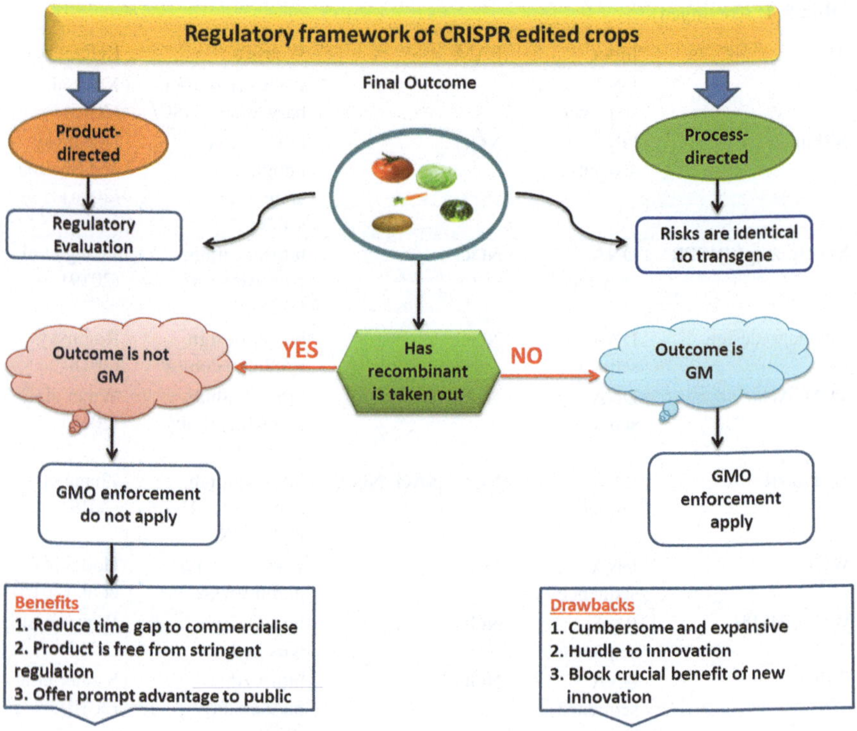

Fig. 9.3 Illustrates the regulatory roadmap for the CRISPR/Cas-based genome editing (GEd), including genome edited crops. Here, showing product/process based regulatory policy for GEd crops. *GM* genetically modified, *GMO* genetically modified organism

breeding. The Cartagena Protocol governs the regulation of genetically modified organisms (GMOs), which is part of the worldwide regulatory framework for living modified organisms (LMOs). LMOs, according to their definition, are living organisms with a unique combination of genetic material that has been improved via the use of contemporary technological methods. In contrast to GMO, the integration site is pre-decided, precise and without an insertion of foreign DNA in GEd organisms. The Cartagena Protocol is based on international terms and conditions that each state and its government must adhere to when enacting biosafety legislation. In addition, the lack of clear perception mentioned in the protocol has been a subject of an argument to date.

Universally, there are differing perspectives on how to harmonize genome edited product/process-based policy in every region (Fig. 9.3). One argument is that GEd species do not need to be regulated because there is no trace of genetic engineering in particular categories, and they resemble organisms that have evolved naturally. The opposing point of view is that GEd organisms must be regulated, but they do not have to go through the same stringent biosafety regulatory process as all GMOs/ LMOs. Such divergent viewpoints reflect the rules and regulations that govern the regulation of GE organisms and products in each country. The insertion of considerable modifications to the genomes of GE crop plants generated using gene editing or Site-Directed Nuclease (SDN) technologies showed genetic differences. There are three types of SDN technology: SDN-1: These were made by cleaving double-stranded DNA in the existing genome without involving of foreign DNA particles, as a result the end products characteristics are almost similar to what arose from natural plant mechanisms and or artificial mutation . SDN-2: involves a short homologous DNA fragment that contains few base pair different from the targeted DNA template. Double strand cut is recognized by the host repair system and simultaneously repaired with the help of donor DNA fragment and introduces predetermined mutations. Lastly, SDN-3: requires a DNA repair donor template longer than 20 bp for incorporation into the target area, which is accomplished by a DSBs nick in the gene that is accomplished by a fragment carrying a gene or other genetic material template. The first and second SDN approaches lack foreign DNA insertions or recombinant DNA because they do not produce new plant varieties. SDN-3, on the other hand, would be subject to GMO regulation if newly created plant types comprised more than 20 bp foreign DNA insertions, showing the same outcome as the classic recombinant DNA technique (Pauwels et al. 2014). Mutation breeding (induced random mutagenesis) or CRISPR-Cas9 (gene editing technology) can be used to create crop features with similar phenotypes, and they will fall into the same category. The change to genetic modifications is appealing due to the possibility for developers to use SDN technology to build superior crops that could bypass the cumbersome regulatory assessments associated with GE crop adoption (Arora and Narula 2017; Yin et al. 2017; Pacher and Puchta 2017; Kumlehn et al. 2018; Sedeek et al. 2019). Policymaker laws that facilitate the commercialization of gene-edited crops could reduce the time between the lab and the farmer even more. Globally, the countries that have welcomed GM crop production and export policy have a planned structure that is quick, simple to comprehend and follow, and enforced (Levin 1994). Notwithstanding their various process or product-based techniques, Argentina, Brazil, Chile, Costa Rica, Honduras, Mexico, and Uruguay were among the first Latin American countries to give GM agricultural permits (Ishii and Araki 2017; Rosado and Craig 2017). This day, these nations are fast forward in cultivating biotech crops and thus, their economic success could be explained by something other than the GMO framework (Table 9.5) (Rosado and Craig 2017). SDN-1 products are almost universally regarded as non-GMO, and the final product would go through the same legislative framework as classically produced plant species (Schmidt et al. 2020).

Divergence re-emerges, however, when it comes to SDN-2 techniques: Australia and Japan have taken a cautious approach, determining that organisms modified with the SDN-2 technology will be classified as GMOs (Thygesen 2019; Tsuda et al. 2019). Plants that have undergone a genetic modification requiring an initial assessment on the basis of their creation using NBTs are characterized as gene-edited organisms, with the exception of those that have been modified without a template or with a modest template. This is not always a negative attitude; in fact, it is one of the key causes driving the formation of biosafety regulation in the first place: it upholds

Table 9.5 Worldwide regulation status of genome edited crops

Country	Regulation status	Remarks
Australia	Deregulated	Edited crops are deregulated when modification occurs via NHEJ-mediated repair pathway (SDN-1), wherein regulated, if donor template or foreign genetic material inserted for alteration of genes
USA	Deregulated	Edited crops cannot be considered as GM crops when any foreign DNA is absent there
Europe	Regulated	Genome edited crops must have regulated via assessment rules designed for the GM crops release
Japan	Deregulated	Edited crops can be reassessed any time, if insufficient information is provided
Brazil	Under existing GMO regulations	Case to case assessment of edited crops, crops are deregulated if they don't carry any transgene
India	Regulation guidelines released	Department of Biotechnology (DBT) under the ministry of science and technology released the much-awaited regulatory guidelines for GE organisms. Edited (SDN-1)/KO crops are in the pipeline of deregulation.
Canada	Deregulated	Edited crops are deregulated, those are regarded as fast version of conventional breeding
Chile	Under existing GMO regulations	Edited crops are deregulated if they have not any transgene
New Zealand	Regulated	Genome edited crops must have defined regulated policy as designed for the release GM crops
Argentina	Under existing GMO regulations	Edited crops become deregulated due to absence of any transgene

societal ideals of risk assessment and risk management with the ultimate goal of safeguarding human, animal, and environmental health.

9.10 Conclusion

Implementation of Noble Prize winner CRISPR-Cas GEd technique for plant GEd and regulation has revolutionized the field of genetic engineering and advanced the plant molecular breeding aspect for crop improvement. Recent advances in genome sequencing (reading) and DNA editing or engineering (writing) techniques have led to an era where we can read and write or even re-write the complex genome of plants. With novel breakthroughs of CRISPR-Cas system, we have witnessed the rise of genetic engineering 2.0 which has contributed enormously to the development of practical, valuable, applicable, and multifaceted tools. These tools are the arsenal for future gene editing, genome modification, metabolic engineering avenues via gene knockout, knockin, replacement, point mutations, fine-tuning of gene regulation, and other modifications at any gene locus. This comprehensive review highlights the successful implementation of the CRISPR-Cas system in plant GEd as well as aids in the documentation of some novel events in the field of plant genetic engineering. In addition, it addresses the technical limitations & shortcomings of the tools, and how to overcome those challenges. Additionally, grieve regulatory concerns and applicability of the machine learning approach i achieve the next-generation engineering or breeding technique. However, it encourages the utilization of the new addition of the CRISPR tool kit for their development into programmable nucleases for efficient, precise, and easy to achieve plant GEd tools. While public acceptance will always be a great concern but there has been a shift in the general notion of disapproval of the genome edited crops however not completely. But even a pitch positive turn with the subsisting endeavours of the scientific community and government ministries of INDIA will be a great achievement, and a way forward to the release of the CRISPR generated new robust variety in the global market.

Acknowledgements The authors thank the ICAR scheme-National Agricultural Science Fund (NASF/GTR-7025/2018-19) for financially supporting this research. We also thank Department of Biotechnology (DBT) Funds (BT/PR38339/GET/119/334/2020) for additional support towards the completion of this work.

References

Abadi S, Yan WX, Amar D, Mayrose I (2017) A machine learning approach for predicting CRISPR-Cas9 cleavage efficiencies and patterns underlying its mechanism of action. PLoS Comput Biol 13:1–24. https://doi.org/10.1371/journal.pcbi.1005807

Abdullaha JZ, Honga X, Zhanga S, Yao R, Xiao Y (2020) CRISPR base editing and prime editing: DSB and template-free editing systems for bacteria and plants. Synth Syst Biotechnol 5:277–292. https://doi.org/10.1016/jsynbio.2020.08.003

Abe K, Araki E, Suzuki Y, Toki S, Saika H (2018) Production of high oleic/low linoleic rice by genome editing. Plant Physiol Biochem 131:58–62

Abudayyeh OO, Gootenberg JS, Konermann S, Joung J, Slaymaker IM, Cox DB, Shmakov S, Makarova KS, Semenova E, Minakhin L, Severinov K, Regev A, Lander ES, Koonin EV, Zhang F (2016) C2c2 is a single-component programmable RNA-guided RNA-targeting CRISPR effector. Science 353(6299):5573. https://doi.org/10.1126/science.aaf5573

Abudayyeh OO, Gootenberg JS, Essletzbichler P, Han S, Joung J, Belanto JJ, Verdine V, Cox DBT, Kellner MJ, Regev A, Lander ES, Voytas DF, Ting AY, Zhang F (2017) RNA targeting with CRISPR-Cas13. Nature 550(7675):280–284

Abudayyeh G, Gootenberg JS, Franklin B, Koob J, Kellner MJ, Ladha A, Joung J, Kirchgatterer P, Cox DBT, Zhang F (2019) A cytosine deaminase for programmable single-base RNA editing. Science 365(6451):382–386

Adli M (2018) The CRISPR tool kit for genome editing and beyond. Nat Commun 9:1911. https://doi.org/10.1038/s41467-018-04252-2

Akama K, Akter N, Endo H, Kanesaki M, Endo M, Toki S (2020) An in vivo targeted deletion of the calmodulin-binding domain from rice glutamate decarboxylase 3 (OsGAD3) increases γ-aminobutyric acid content in grains. Rice 13:20

Ali Z, Abulfaraj A, Idris A, Ali S, Tashkandi M, Mahfouz MM (2015) CRISPR/Cas9-mediated viral interference in plants. Genome Biol 16:238

Alkan F, Wenzel A, Anthon C, Havgaard JH, Gorodkin J (2018) CRISPR-Cas9 off-targeting assessment with nucleic acid duplex energy parameters. Genome Biol 19:1–13. https://doi.org/10.1186/s13059-018-1534-x

Anders C, Niewoehner O, Duerst A, Jinek M (2014) Structural basis of PAM-dependent target DNA recognition by the Cas9 endonuclease. Nature 513:569–573. https://doi.org/10.1038/nature13579

Anzalone AV, Randolph PB, Davis JR, Sousa AA, Koblan LW, Levy JM, Chen PJ, Wilson C, Newby GA, Raguram A, Liu DR (2019) Search-and-replace genome editing without double-strand breaks or donor DNA. Nature 576:149–157

Anzalone AV, Koblan LW, Liu DR (2020) Genome editing with CRISPR–Cas nucleases base editors transposases and prime editors. Nat Biotechnol 38:824–844. https://doi.org/10.1038/s41587-020-0561-9

Aquino-Jarquin G (2019) CRISPR-Cas14 is now part of the artillery for gene editing and molecular diagnostic. Nanomedicine 18:428–431

Arndell T, Sharma N, Langridge P, Baumann U, Watson-Haigh NS, Whitford R (2019) gRNA validation for wheat genome editing with the CRISPR-Cas9 system. BMC Biotechnol 19(1): 1–12

Arora L, Narula A (2017) Gene editing and crop improvement using CRISPR-Cas9 system. Front Plant Sci 8:1932. https://doi.org/10.3389/fpls.2017.01932

Aschenbrenner S, Kallenberger SM, Hoffmann MD, Huck A, Eils R, Niopek D (2020) Coupling Cas9 to artificial inhibitory domains enhances CRISPR- Cas9 target specificity. Sci Adv 6(6): 0187

Asseng S, Ewert F, Martre P, Rötter RP, Lobell DB, Cammarano D, Kimball BA, Ottman MJ, Wall GW, White JW, Reynolds MP, Alderman PD, Prasad PVV, Aggarwal PK, Anothai J, Basso B, Biernath C, Challinor AJ, De Sanctis G, Doltra J, Fereres E, Garcia-Vila M, Gayler S, Hoogenboom G, Hunt LA, Izaurralde RC, Jabloun M, Jones CD, Kersebaum KC, Koehler AK, Müller C, Naresh KS, Nendel C, O'Leary G, Olesen JE, Palosuo T, Priesack E, Eyshi Rezaei E, Ruane AC, Semenov MA, Shcherbak I, Stöckle C, Stratonovitch P, Streck T, Supit I, Tao F, Thorburn PJ, Waha K, Wang E, Wallach D, Wolf J, Zhao Z, Zhu Y (2014) Rising temperatures reduce global wheat production. Nat Clim Change 5(2):143–147. https://doi.org/10.1038/nclimate2470

Bae S, Park J, Kim JS (2014) Cas-OFFinder: a fast and versatile algorithm that searches for potential off-target sites of Cas9 RNA-guided endonucleases. Bioinformatics 30:1473–1475. https://doi.org/10.1093/bioinformatics/btu048

Baltes NJ, Gil-Humanes J, Cermak T, Atkins PA, Voytas DF (2014) DNA replicons for plant genome engineering. Plant Cell 26:151–163. https://doi.org/10.1105/tpc.113.119792

Baltes NJ, Hummel AW, Konecna E, Cegan R, Bruns AN, Bisaro DM, Voytas DF (2015) Conferring resistance to geminiviruses with the CRISPR–Cas prokaryotic immune system. Nat Plants 1:15145. https://doi.org/10.1038/nplants.2015.145

Banakar R, Eggenberger AL, Lee K, Wright DA, Murugan K, Zarecor S, Lawrence-Dill CJ, Sashital DG, Wang K (2019) High-frequency random DNA insertions upon co-delivery of CRISPR-Cas9 ribonucleoprotein and selectable marker plasmid in rice. Sci Rep 9(1):1–13

Bari VK, Nassar JA, Kheredin SM, Gal-On A, Ron M, Britt A, Steele D, Yoder J, Aly R (2019) CRISPR/Cas9-mediated mutagenesis of *CAROTENOID CLEAVAGE DIOXYGENASE 8* in tomato provides resistance against the parasitic weed *Phelipanche aegyptiaca*. Sci Rep 9:11438

Barkau CL, Reilly D, Eddington SB, Damha MJ, Gagnon KT (2021) Small nucleic acids and the path to the clinic for anti-CRISPR. Biochem Pharmacol 189:114492

Barrangou R, Fremaux C, Deveau H, Richards M, Boyaval P, Moineau S, Romero DA, Horvath P (2007) CRISPR provides acquired resistance against viruses in prokaryotes. Science 315:1709–1712. https://doi.org/10.1126/science.1138140

Baysal C, Bortesi L, Zhu C, Farré G, Schillberg S, Christou P (2016) CRISPR/Cas9 activity in the rice OsBEIIb. Mol Breed 36:1–11

Behera SS, Ray RC, Zdolec N (2018) Lactobacillus plantarum with functional properties: an approach to increase safety and shelf-life of fermented foods. Biomed Res Int 2018:9361614. https://doi.org/10.1155/2018/9361614

Belfort M, Roberts RJ (1997) Homing endonucleases: keeping the house in order. Nucleic Acids Res 25(17):3379–3388. https://doi.org/10.1093/nar/25.17.3379

Bharat SS, Li S, Li J, Yan L, Xia L (2020) Base editing in plants: current status and challenges. Crop J 8(3):384–395. https://doi.org/10.1016/j.cj.2019.10.002

Bihmidine S, Lin J, Stone JM, Awada T, Specht JE, Clemente TE (2013) Activity of the Arabidopsis RD29A and RD29B promoter elements in soybean under water stress. Planta 237(1):55–64. https://doi.org/10.1007/s00425-012-1740-9

Bolotin A, Quinquis B, Sorokin A, Ehrlich SD (2005) Clustered regularly interspaced short palindrome repeats (CRISPRs) have spacers of extrachromosomal origin. Microbiology 151: 2551–2561. https://doi.org/10.1099/mic028048-0

Brouns SJ, Jore MM, Lundgren M, Westra ER, Slijkhuis RJ, Snijders AP, Dickman MJ, Makarova KS, Koonin EV, Van der Oost J (2008) Small CRISPR RNAs guide antiviral defense in prokaryotes. Science 321:960–964. https://doi.org/10.1126/science.1159689

Budhagatapalli N, Rutten T, Gurushidze M, Kumlehn J, Hensel G (2015) Targeted modification of gene function exploiting homology-directed repair of TALEN-mediated double-strand breaks in Barley. G3 (Bethesda) 5(9):1857–1863. https://doi.org/10.1534/g3.115.018762

Bult CJ, White O, Olsen GJ, Zhou L, Fleischmann RD, Sutton GG, Blake JA, FitzGerald LM, Clayton RA, Gocayne JD, Kerlavage AR, Dougherty BA, Tomb JF, Adams MD, Reich CI, Overbeek R, Kirkness EF, Weinstock KG, Merrick JM, Glodek A, Scott JL, Geoghagen NS, Venter JC (1996) Complete genome sequence of the methanogenic archaeon *Methanococcus jannaschii*. Science 273:1058–1073. https://doi.org/10.1126/science.273.5278.1058

Burmistrz M, Krakowski K, Krawczyk-Balska A (2020) RNA-targeting CRISPR–Cas systems and their applications. Int J Mol Sci 21:1122. https://doi.org/10.3390/ijms21031122

Butler NM, Baltes NJ, Voytas DF, Douches DS (2016) Geminivirus-mediated genome editing in potato (*Solanum tuberosum* L) using sequence-specific nucleases. Front Plant Sci 7:1045. https://doi.org/10.3389/fpls.2016.01045

Cai F, Axen SD, Kerfeld CA (2013) Evidence for the widespread distribution of CRISPR-Cas system in the Phylum Cyanobacteria. RNA Biol 10:687–693

Cameron P, Fuller CK, Donohoue PD, Jones BN, Thompson MS, Carter MM, Gradia S, Vidal B, Garner E, Slorach EM, Lau E, Banh LM, Lied AM, Edwards LS, Settle AH, Capurso D, Llaca V, Deschamps S, Cigan M, Young JK, May AP (2017) Mapping the genomic landscape of CRISPR-Cas9 cleavage. Nat Methods 14(6):600–606. https://doi.org/10.1038/nmeth.4284

Carlessi M, Mariotti L, Giaume F, Fornara F, Perata P, Gonzali S (2021) Targeted knockout of the gene *OsHOL1* removes methyl iodide emissions from rice plants. Sci Rep 11:17010. https://doi.org/10.1038/s41598-021-95198-x

Carlson-Stevermer J, Kelso R, Kadina A, Joshi S, Rossi N, Walker J, Stoner R, Maures T (2020) CRISPRoff enables spatio-temporal control of CRISPR editing. Nat Commun 11:5041. https://doi.org/10.1038/s41467-020-18853-3

Čermák T, Baltes NJ, Čegan R, Zhang Y, Voytas DF (2015) High-frequency precise modification of the tomato genome. Genome Biol 16:232. https://doi.org/10.1186/s13059-015-0796-9

Chandrasegaran S, Carroll D (2016) Origins of programmable nucleases for genome engineering. J Mol Biol 428:963–989. https://doi.org/10.1016/j.jmb.2015.10.014

Char SN, Neelakandan AK, Nahampun H, Frame B, Main M, Spalding MH, Becraft PW, Meyers BC, Walbot V, Wang K, Yang B (2017) An agrobacterium-delivered CRISPR/Cas9 system for high-frequency targeted mutagenesis in maize. Plant Biotechnol 15:257–268

Chari R, Mali P, Moosburner M, Church GM (2015) Unraveling CRISPR-Cas9 genome engineering parameters via a library-on-library approach. Nat Methods 12:823–826. https://doi.org/10.1038/nmeth.3473

Charo RA (2015) Yellow lights for emerging technologies. Science 349(6246):384–385. https://doi.org/10.1126/science.aab3885

Charo RA (2016) The legal and regulatory context for human gene editing. Issues Sci Technol 32(3):39–45

Che J, Yamaji N, Ma JF (2021) Role of a vacuolar iron transporter OsVIT2 in the distribution of iron to rice grains. New Phytol 230:1049–1062

Chen F, Ding X, Feng Y, Seebeck T, Jiang Y, Davis GD (2017) Targeted activation of diverse CRISPR-Cas systems for mammalian genome editing via proximal CRISPR targeting. Nat Commun 8:14958. https://doi.org/10.1038/ncomms14958

Chen JS, Ma E, Harrington LB, Da Costa M, Tian X, Palefsky JM, Doudna JA (2018) CRISPR-Cas12a target binding unleashes indiscriminate single-stranded DNase activity. Science 360(6387):436–439. https://doi.org/10.1126/science.aar6245

Chen Y, Fu M, Li H et al (2021) High-oleic acid content nontransgenic allotetraploid cotton (*Gossypium hirsutum* L) generated by knockout of GhFAD2 genes with CRISPR/Cas9 system. Plant Biotechnol J 19(3):424–426. https://doi.org/10.1111/pbi.13507

Christian M, Qi Y, Zhang Y, Voytas DF (2013) Targeted mutagenesis of Arabidopsis thaliana using engineered TAL effector nucleases. G3(Bethesda) 3:1697–1705

Chuai G, Hui WQL, Liu Q (2017) In silico meets in vivo: towards computational CRISPR-based sgRNA design. Trends Biotechnol 35:12–21. https://doi.org/10.1016/jtibtech201606008

Chuai G, Ma H, Yan J, Chen M, Hong N, Xue D et al (2018) DeepCRISPR: optimized CRISPR guide RNA design by deep learning. Genome Biol 19:1–18. https://doi.org/10.1186/s13059-018-1459-4

Colleaux L, D'Auriol L, Galibert F, Dujon B (1988) Recognition and cleavage site of the intron-encoded omega transposase. Proc Natl Acad Sci U S A 85(16):6022–6026. https://doi.org/10.1073/pnas.85.16.6022

Cong L, Ran FA, Cox D, Lin S, Barretto R, Habib N, Hsu PD, Wu X, Jiang W, Marraffini LA, Zhang F (2013a) Multiplex genome engineering using CRISPR/Cas systems. Science 339:819–823. https://doi.org/10.1126/science1231143

Cong L, Ran FA, Cox D, Lin S, Barretto R, Habib N, Zhang F (2013b) Multiplex genome engineering using CRISPR/Cas systems. Science 339:819–823

Cox DBT, Platt RJ, Zhang F (2015) Therapeutic genome editing: prospects and challenges. Nat Med 21:121–131. https://doi.org/10.1038/nm.3793

Cox DBT, Gootenberg JS, Abudayyeh OO, Franklin B, Kellner MJ, Joung J, Zhang F (2017) RNA editing with CRISPR-Cas13. Science 358(6366):1019–1027. https://doi.org/10.1126/science.aaq0180

Crudele JM, Chamberlain JS (2018) Cas9 immunity creates challenges for CRISPR gene editing therapies. Nat Commun 9:3497. https://doi.org/10.1038/s41467-018-05843-9

Curtin SJ, Zhang F, Sander JD, Haun WJ, Starker C, Baltes NJ, Reyon D, Dahlborg EJ, Goodwin MJ, Coffman AP, Dobbs D, Joung JK, Voytas DF, Stupar RM (2011) Targeted mutagenesis of duplicated genes in soybean with zinc-finger nucleases. Plant Physiol 156(2):466–473. https://doi.org/10.1104/pp.111.172981

Cyranoski D (2016) CRISPR gene-editing tested in a person for the first time. Nature 539:479. https://doi.org/10.1038/nature.2016.20988

Cyranoski D (2020) What CRISPR-baby prison sentences mean for research. Nature 577:154–155. https://doi.org/10.1038/d41586-020-00001-y

D'Halluin K, Vanderstraeten C, Stals E, Cornelissen M, Ruiter R (2007) Homologous recombination: a basis for targeted genome optimization in crop species such as maize. Plant Biotechnol 6(1):93–102. https://doi.org/10.1111/j1467-7652200700305x

Datsenko K, Pougach K, Tikhonov A, Wanner BL, Severinov K, Semenova E (2012) Molecular memory of prior infections activates the CRISPR/Cas adaptive bacterial immunity system. Nat Commun 3:945. https://doi.org/10.1038/ncomms1937

DiCarlo JE, Norville JE, Mali P, Rios X, Aach J, Church GM (2013) Genome engineering in Saccharomyces cerevisiae using CRISPR-Cas systems. Nucleic Acids Res 41(7):4336–4343. https://doi.org/10.1093/nar/gkt135

Dobson L, Reményi I, Tusnády GE (2015) CCTOP: a consensus constrained topology prediction web server. Nucleic Acids Res 43:408–412. https://doi.org/10.1093/nar/gkv451

Doench JG, Hartenian E, Graham DB, Tothova Z, Hegde M, Smith I, Sullender M, Ebert BL, Xavier RJ, Root DE (2014) Rational design of highly active sgRNAs for CRISPR-Cas9-mediated gene inactivation. Nat Biotechnol 32(12):1262–1267. https://doi.org/10.1038/nbt.3026

Doench JG, Fusi N, Sullender M, Hegde M, Vaimberg EW, Donovan KF, Smith I, Tothova Z, Wilen C, Orchard R, Virgin HW, Listgarten J, Root DE (2016) Optimized sgRNA design to maximize activity and minimize off-target effects of CRISPR-Cas9. Nat Biotechnol 34:184–191. https://doi.org/10.1038/nbt.3437

Dolgin E (2019) Finding the CRISPR off-switch. Nature 577:309

Dong D, Guo M, Wang S, Zhu Y, Wang S, Xiong Z, Yang J, Xu Z, Huang Z (2017) Structural basis of CRISPR–SpyCas9 inhibition by an anti-CRISPR protein. Nature 546(7658):436–439

Dong L, Qi X, Zhu J, Liu C, Zhang X, Cheng B, Mao L, Xie C (2019) Supersweet and waxy: meeting the diverse demands for specialty maize by genome editing. Plant Biotechnol 17:1853–1855. https://doi.org/10.1111/pbi.13144

Dong OX, Yu S, Jain R, Zhang N, Duong PQ, Butler C, Li Y, Lipzen A, Martin JA, Barry KW, Schmutz J, Tian L, Ronald PC (2020) Marker-free carotenoid-enriched rice generated through targeted gene insertion using CRISPR-Cas9. Nat Commun 11:1178. https://doi.org/10.1038/s41467-020-14981-y

Doudna JA, Charpentier E (2014) The new frontier of genome engineering with CRISPR-Cas9. Science 346:1258096

DuPont Pioneer (2016) DuPont Announces Intentions to Commercialize First CRISPR/Cas Product Press Release. https://www.pioneercom/home/site/about/news-media/newreleases/template

Endo A, Masafumi M, Kaya H, Toki S (2016) Efficient targeted mutagenesis of rice and tobacco genomes using Cpf1 from *Francisella novicida*. Sci Rep 6:38169

Estrella MA, Kuo FT, Bailey S (2016) RNA-activated DNA cleavage by the Type III-B CRISPR-Cas effector complex. Genes Dev 30:460–470

Feng Z, Zhang B, Ding W, Liu X, Yang DL, Wei P, Cao F, Zhu S, Zhang F, Mao Y, Zhu JK (2013) Efficient genome editing in plants using a CRISPR/Cas system. Cell Res 23(10):1229

Fine EJ, Cradick TJ, Zhao CL, Lin Y, Bao G (2013) An online bioinformatics tool predicts zinc finger and TALE nuclease off-target cleavage. Nucleic Acids Res 42(6):2014

Forner J, Pfeiffer A, Langenecker T, Manavella PA, Lohmann JU (2015) Correction: germline-transmitted genome editing in Arabidopsis thaliana using TAL-effector-nucleases. PLoS One 10(7):0133945

Gaj T, Gersbach CA, Barbas CF (2013) ZFN TALEN and CRISPR/Cas-based methods for genome engineering. Trends Biotechnol 31(7):397–405. https://doi.org/10.1016/j.tibtech.2013.04.004

Gao W, Long L, Tian X, Xu F, Liu J, Singh PK, Botella JR, Song C (2017) Genome editing in cotton with the CRISPR/Cas9 system. Front Plant Sci 8:1364

Gaoneng S, Lihong GJ, Xiangjin W, Zhonghua S, Shaoqing T, Peisong HU (2017) CRISPR/CAS9-mediated editing of the fragrant gene Badh2 in rice. Chin J Rice Sci 31:216–222

Gasiunas G, Barrangou R, Horvath P, Siksnys V (2012) Cas9-crRNA ribonucleoprotein complex mediates specific DNA cleavage for adaptive immunity in bacteria. Proc Natl Acad Sci U S A 109:2579–2586. https://doi.org/10.1073/pnas.1208507109

Gaudelli NM, Komor AC, Rees HA, Packer MS, Badran AH, Bryson DI, Liu DR (2017) Programmable base editing of A• T to G• C in genomic DNA without DNA cleavage. Nature 551:464–471

Gelvin SB (2006) Agrobacterium virulence gene induction. Methods Mol Biol 343:77–84. https://doi.org/10.1385/1-59745-130-4:77

Gilbert LA, Larson MH, Morsut L, Liu Z, Brar GA, Torres SE, Stern-Ginossar N, Brandman O, Whitehead EH, Doudna JA, Lim WA, Weissman JS, Qi LS (2013) CRISPR-mediated modular RNA-guided regulation of transcription in eukaryotes. Cell 154(2):442–451

Gil-Humanes JY, Wang Z, Liang Q, Shan CV, Ozuna S, Sánchez-León NJ, Baltes C, Starker F, Barro C, Gao DF, Voytas DF (2017) High-efficiency gene targeting in hexaploid wheat using DNA replicons and CRISPR/Cas9. Plant J 89(6):1251–1262

Gonzales LR, Li S, Bergonzi S, Oortwijn M, Bachem C (2020) Potato cycling Dof Factor1 and its lncRNA counterpart StFLORE link tuber development and drought response. Plant J 105:855–869. https://doi.org/10.1111/tpj.15093

Gootenberg JS, Abudayyeh OO, Lee JW, Essletzbichler P, Dy AJ, Joung J, Verdine V, Donghia N, Daringer NM, Freije CA, Myhrvold C, Bhattacharyya RP, Livny J, Regev A, Koonin EV, Hung DT, Sabeti PC, Collins JJ, Zhang F (2017) Nucleic acid detection with CRISPR-Cas13a/C2c2. Science 356(6336):438–442. https://doi.org/10.1126/science.aam9321

Gratz SJ, Ukken FP, Rubinstein CD, Thiede G, Donohue LK, Cummings AM, O'Connor-Giles KM (2014) Highly specific and efficient CRISPR/Cas9-catalyzed homology-directed repair in Drosophila. Genetics 196:961–971. https://doi.org/10.1534/genetics.113.160713

Haeussler M, Schönig K, Eckert H, Eschstruth A, Mianné J, Renaud JB, Schneider-Maunoury S, Shkumatava A, Teboul L, Kent J, Joly JS, Concordet JP (2016) Evaluation of off-target and on-target scoring algorithms and integration into the guide RNA selection tool CRISPOR. Genome Biol 17:1–12. https://doi.org/10.1186/s13059-016-1012-2

Haft DH, Selengut J, Mongodin EF, Nelson KE (2005) A guild of 45 CRISPR-associated (Cas) protein families and multiple CRISPR/Cas subtypes exist in prokaryotic genomes. PLoS Comput Biol 1:60

Hajiahmadi Z, Shirzadian-Khorramabad R, Kazemzad M, Sohani MM (2019) Enhancement of tomato resistance to Tuta absoluta using a new efficient mesoporous silica nanoparticle-mediated plant transient gene expression approach. Sci Hortic 243:367–375

Hale CR, Zhao P, Olson S, Duff MO, Graveley BR, Wells L, Terns RM, Terns MP (2009) RNA-guided RNA cleavage by a CRISPR RNA-Cas protein complex. Cell 139:945–956

Hampton T (2020) DNA prime editing: a new CRISPR-based method to correct most disease-causing mutations. JAMA 323(5):405–406. https://doi.org/10.1001/jama.2019.21827

Harrington LB, Burstein D, Chen JS, Paez-Espino D, Ma E, Witte IP, Cofsky JC, Kyrpides NC, Banfield JF, Doudna JA (2018) Programmed DNA destruction by miniature CRISPR-Cas14 enzymes. Science 362:839–842

Haurwitz RE, Jinek M, Wiedenheft B, Zhou K, Doudna JA (2010) Sequence- and structure-specific RNA processing by a CRISPR endonuclease. Science 329(5997):1355–1358. https://doi.org/10.1126/science.1192272

He S (2020) The first human trial of CRISPR-based cell therapy clears safety concerns as new treatment for late-stage lung cancer. Sig Transduct Target Ther 5:168. https://doi.org/10.1038/s41392-020-00283-8

Heigwer F, Kerr G, Boutros M (2014) E-CRISP: fast CRISPR target site identification. Nat Methods 11:122–123. https://doi.org/10.1038/nmeth.2812

Henry VJ, Bandrowski AE, Pepin AS, Gonzalez BJ, Desfeux A (2014) OMICtools: an informative directory for multi-omic data analysis. Database (Oxford) 2014:bau069. https://doi.org/10.1093/database/bau069

Hermans PW, Van Soolingen D, Bik EM, De Haas PE, Dale JW, Van Embden JD (1991) Insertion element IS*987* from *Mycobacterium bovis* BCG is located in a hot-spot integration region for insertion elements in *Mycobacterium tuberculosis* complex strains. Infect Immun 59:2695–2705

Hirosawa M, Saito H (2021) Cell-type-specific CRISPR-Cas9 system with miRNAs In CRISPR-Cas methods. Humana, New York, NY, pp 265–279

Hodgkins A, Farne A, Perera S, Grego T, Parry-Smith DJ, Skarnes WC, Iyer V (2015) WGE: a CRISPR database for genome engineering. Bioinformatics 31:3078–3080. https://doi.org/10.1093/bioinformatics/btv308

Hoffmann M, Kleine-Weber H, Schroeder S et al (2020) SARS-CoV-2 cell entry depends on ACE2 and TMPRSS2 and is blocked by a clinically proven protease inhibitor. Cell 181(2):271–280. https://doi.org/10.1016/j.cell.2020.02.052

Hrle A, Maier L, Sharma K, Ebert J, Basquin C, Urlaub H, Marchfelder A, Conti E (2014) Structural analyses of the CRISPR protein Csc2 reveal the RNA-binding interface of the type I-D Cas7 family. RNA Biol 11(8):1072–1082. https://doi.org/10.4161/rna.29893

Hsu PD, Scott DA, Weinstein JA, Ran FA, Konermann S, Agarwala V, Li Y, Fine EJ, Wu X, Shalem O, Cradick TJ, Marraffini LA, Bao G, Zhang F (2013) DNA targeting specificity of RNA-guided Cas9 nucleases. Nat Biotechnol 31:827–832. https://doi.org/10.1038/nbt.2647

Hua K, Tao X, Han P, Wang R, Zhu JK (2019) Genome engineering in rice using Cas9 variants that recognize NG PAM sequences. Mol Plant 12:1003–1014

Huang J, Li J, Zhou J, Wang L, Yang S, Hurst LD, Li WH, Tian D (2018) Identifying a large number of high-yield genes in rice by pedigree analysis whole-genome sequencing and CRISPR-Cas9 gene knockout. Proc Natl Acad Sci USA 115:7559–7567. https://doi.org/10.1073/pnas.18061101

Huang L, Li Q, Zhang C, Chu R, Gu Z, Tan H, Zhao D, Fan X, Liu Q (2020) Creating novel Wx alleles with fine-tuned amylose levels and improved grain quality in rice by promoter editing using CRISPR/Cas9 system. Plant Biotechnol 18(11):2164–2166. https://doi.org/10.1111/pbi.13391

Hummel AW, Chauhan RD, Cermak T, Mutka AM, Vijayaraghavan A, Boyher A, Starker CG, Bart R, Voytas DF, Taylor NJ (2017) Allele exchange at the EPSPS locus confers glyphosate tolerance in cassava. Plant Biotechnol 16:1275–1282

Ibrahim S, Saleem B, Rehman N, Zafar SA, Naeem MK, Khan MR (2021) CRISPR/Cas9 mediated disruption of Inositol pentakisphosphate 2-kinase 1(TaIPK1) reduces phytic acid and improves iron and zinc accumulation in wheat grains. J Adv Res 37:33–41. https://doi.org/10.1016/j.jare.2021.07.006

Ishii T, Araki M (2017) A future scenario of the global regulatory landscape regarding genome-edited crops GM. Crop Food 8:44–56. https://doi.org/10.1080/21645698.2016.1261787

Ishino Y, Shinagawa H, Makino K, Amemura M, Nakata A (1987) Nucleotide sequence of the *iap* gene responsible for alkaline phosphatase isozyme conversion in *Escherichia coli* and identification of the gene product. J Bacteriol 169:5429–5433. https://doi.org/10.1128/jb.169.12.5429-5433.1987

Jacob H, Christin L, Robin K (2013) A CRISPR CASe for high-throughput silencing. Front Plant Sci 4:193. https://doi.org/10.3389/fgene.2013.00193

Jacobs TB, LaFayette PR, Schmitz RJ, Parrott WA (2015) Targeted genome modifications in soybean with CRISPR/Cas9. BMC Biotechnol 15:16

Jacquin ALS, Odom DT, Lukk M (2019) Crisflash: open-source software to generate CRISPR guide RNAs against genomes annotated with individual variation. Bioinformatics 35:3146–3147. https://doi.org/10.1093/bioinformatics/btz019

James K, Nuñez JC, Greg C, Pommier JZC, Joseph MR, Carmen A, Gokul NR, Quanming S, King L, Avi JS, Angela NP, James YSK, Amanda C, Manuel DL, Howard YC, Martin K, Bradley EB, Hovestadt V, Luke AG, Jonathan SW (2021) Genome-wide programmable transcriptional memory by CRISPR-based epigenome editing. Cell 184(9):2503–2519.e17. https://doi.org/10.1016/j.cell.2021.03.025

Jeong YK, Song B, Bae S (2020) Current status and challenges of DNA base editing tools. Mol Ther 28:91938–91952

Jiang W, Zhou H, Bi H, Fromm M, Yang B, Weeks DP (2013) Demonstration of CRISPR/Cas9/sgRNA-mediated targeted gene modification in Arabidopsis tobacco sorghum and rice. Nucleic Acids Res 41(20):188–188

Jiang WZ, Henry IM, Lynagh PG, Comai L, Cahoon EB, Weeks DP (2017) Significant enhancement of fatty acid composition in seeds of the allohexaploid *Camelina sativa* using CRISPR/Cas9 gene editing. Plant Biotechnol J 15:648–646

Jiang M, Liu Y, Liu Y, Tan Y, Huang J, Shu Q (2019) Mutation of inositol 134-trisphosphate 5/6-kinase6 impairs plant growth and phytic acid synthesis in rice. Plan Theory 8:114. https://doi.org/10.3390/plants8050114

Jinek M, Chylinski K, Fonfara I, Hauer M, Doudna JA, Charpentier E (2012) A programmable dual-RNA-guided DNA endonuclease in adaptive bacterial immunity. Science 337(6096):816–821. https://doi.org/10.1126/science.1225829

Jinek M, East A, Cheng A, Lin S, Ma E, Doudna J (2013) RNA-programmed genome editing in human cells. Elife 2:00471

Jinek M, Jiang F, Taylor DW, Sternberg SH, Kaya E, Ma E, Kaplan M (2014) Structures of Cas9 endonucleases reveal RNA-mediated conformational activation. Science 343:1247997

Johnston RK, Seamon KJ, Saada EA, Podlevsky JD, Branda SS, Timlin JA, Harper JC (2019) Use of anti-CRISPR protein AcrIIA4 as a capture ligand for CRISPR/Cas9 detection. Biosens Bioelectron 141:111361. https://doi.org/10.1016/jbios.2019.111361

Joung J, Ladha A, Saito M, Segel M, Bruneau R, Huang MW, Kim NG, Yu X, Li J, Walker BD, Greninger AL, Jerome KR, Gootenberg JS, Abudayyeh OO, Zhang F (2020) Point-of-care testing for COVID-19 using SHERLOCK diagnostics. medRxiv. https://doi.org/10.1101/2020.05.04.20091231

Jung YJ, Lee HJ, Yu J, Bae S, Cho YG, Kang KK (2021) Transcriptomic and physiological analysis of OsCAO1 knockout lines using the CRISPR/Cas9 system in rice. Plant Cell Rep 40:1013–1024. https://doi.org/10.1007/s00299-020-02607-y

Kampmann M (2018) CRISPRi and CRISPRa screens in mammalian cells for precision biology and medicine. ACS Chem Biol 13(2):406–416. https://doi.org/10.1021/acschembio.7b00657

Kantor A, McClements ME, MacLaren RE (2020) CRISPR-Cas9 DNA base-editing and prime-editing. Int J Mol Sci 21:6240. https://doi.org/10.3390/ijms21176240

Kapusi E, Corcuera-Gomez M, Melnik S, Stoger E (2017) Heritable genomic fragment deletions and small indels in the putative ENgase gene induced by CRISPR/Cas9 in barley. Front Plant Sci 8:540. https://doi.org/10.3389/fpls.2017.00540

Karvelis T, Bigelyte G, Young JK, Hou Z, Zedaveinyte R, Pociute K, Silanskas A, Venclovas C, Siksnys V (2019) PAM recognition by miniature CRISPR-Cas14 triggers programmable double-stranded DNA cleavage. BioRxiv Preprint. https://doi.org/10.1101/654897

Karvelis T, Bigelyte G, Young JK, Hou Z, Zedaveinyte R, Budre K, Paulraj S, Djukanovic V, Gasior S, Silanskas A, Venclovas Č, Siksnys V (2020) PAM recognition by miniature CRISPR-Cas12f nucleases triggers programmable double-stranded DNA target cleavage. Nucleic Acids Res 48(9):5016–5023. https://doi.org/10.1093/nar/gkaa208

Kaul T, Raman NM, Eswaran M, Thangaraj A, Verma R, Sony SK, Sathelly KM, Kaul R, Yadava P, Agrawal PK (2019) Data mining by pluralistic approach on CRISPR gene editing in plants. Front Plant Sci 10:801. https://doi.org/10.3389/fpls.2019.00801

Kaul T, Sony SK, Raman NM, Eswaran M, Verma R, Thangaraj A, Bharti J, Motelb KFA, Kaul R (2020a) How crisp is CRISPR? CRISPR/Cas mediated crop improvement with special focus on nutritional traits. In: Tuteja N, Tuteja R, Passricha N, Saifi S (eds) Advancement in crop improvement techniques, vol 20. Woodhead Printing, New Delhi, pp 159–197

Kaul T, Sony SK, Verma R, Motelb KFA, Thangaraj A, Eswaran M, Bharti J, Nehra M, Kaul R (2020b) Revisiting CRISPR–Cas mediated crop improvement:special focus on nutrition. J Biosci 45:137

Kaul T, Sony SK, Raman NM, Motelb KFA, Bharti J (2021) Genotype–independent regeneration and transformation protocol for rice cultivars. In: Bandyopadhyay A, Thilmony R (eds) Rice genome engineering and gene editing. Methods in Molecular Biology. Humana, New York, NY, p 2238

Kaur K, Gupta AK, Rajput A, Kumar M (2016) Ge-CRISPR - An integrated pipeline for the prediction and analysis of sgRNAs genome editing efficiency for CRISPR/Cas system. Sci Rep 6:1–12. https://doi.org/10.1038/srep30870

Kaur N, Alok A, Shivani KP, Kaur N, Awasthi P, Chaturvedi S, Pandey P, Pandey A, Pandey AK, Tiwari S (2020) CRISPR/Cas9 directed editing of lycopene epsilon-cyclase modulates metabolic flux for beta-carotene biosynthesis in banana fruit. Metab Eng 59:76–86. https://doi.org/10.1016/j.ymben.2020.01.008

Kellner MJ, Koob J, Gootenberg JS, Abudayyeh OO, Zhang F (2019) SHERLOCK: nucleic acid detection with CRISPR nucleases. Nat Protoc 14(10):2986–3012. https://doi.org/10.1038/s41596-019-0210-2

Khan MZ, Haider S, Mansoor S, Amin I (2019a) Targeting plant ssDNA viruses with engineered miniature CRISPR-Cas14a. Trends Biotechnol 37(8):800–804

Khan MSS, Basnet R, Islam SA, Shu Q (2019b) Mutational analysis of OsPLD1 reveals its involvement in phytic acid biosynthesis in rice grains. J Agric Food Chem 67:11436–11443

Kieu NP, Lenman M, Wang ES, Petersen BL, Andreasson E (2021) Mutations introduced in susceptibility genes through CRISPR/Cas9 genome editing confer increased late blight resistance in potatoes. Sci Rep 11:4487. https://doi.org/10.1038/s41598-021-83972-w

Kim D, Bae S, Park J, Kim E, Kim S, Yu HR, Kim JS (2015) Digenome-seq: genome-wide profiling of CRISPR/Cas9 off-target effects in human cells. Nat Methods 12:237

Kim HK, Song M, Lee J, Menon AV, Jung S, Kang YM (2017) In vivo high-throughput profiling of CRISPR-Cpf1 activity. Nat Methods 14:153–159. https://doi.org/10.1038/nmeth4104

Klap C, Yeshayahou E, Bolger AM, Arazi T, Gupta SK, Shabtai S, Usadel B, Salts Y, Barg R (2017) Tomato facultative parthenocarpy results from SlAGAMOUS-LIKE 6 loss of function. Plant Biotechnol J 15:634–647

Komor AC, Kim YB, Packer MS, Zuris JA, Liu DR (2016) Programmable editing of a target base in genomic DNA without double-stranded DNA cleavage. Nature 533:420–424. https://doi.org/10.1038/nature17946

Konwarh R (2020) Can CRISPR/Cas technology be a felicitous stratagem against the COVID-19 fiasco? Prospects and hitches. Front Mol Biosci 7:557377. https://doi.org/10.3389/fmolb2020557377

Koonin EV, Makarova KS, Zhang F (2017) Diversity classification and evolution of CRISPR-Cas systems. Curr Opin Microbiol 37:67–78. https://doi.org/10.1016/j.mib.2017.05.008

Kosicki M, Tomberg K, Bradley A (2018) Repair of double-strand breaks induced by CRISPR–Cas9 leads to large deletions and complex rearrangements. Nat Biotechnol 36:765–771

Ku HK, Ha SH (2020) Improving nutritional and functional quality by genome editing of crops: status and perspectives. Front Plant Sci 11:577313. https://doi.org/10.3389/fpls.2020.577313

Kumar P, Malik YS, Ganesh B, Rahangdale S, Saurabh S, Natesan S, Srivastava A, Sharun K, Yatoo MI, Tiwari R, Singh RK, Dhama K (2020) CRISPR-Cas system: an approach with potentials for COVID-19 diagnosis and therapeutics. Front Cell Infect Microbiol 10:576875. https://doi.org/10.3389/fcimb.2020.576875

Kumlehn J, Pietralla J, Hensel G, Pacher M, Puchta H (2018) The CRISPR/Cas revolution continues: From efficient gene editing for crop breeding to plant synthetic biology. J Integr Plant Biol 60:1127–1153. https://doi.org/10.1111/jipb.12734

Kurtz S (2003) The Vmatch large scale sequence analysis software Ref Type. Computer Program:4–12

Labuhn M, Adams FF, Ng M, Knoess S, Schambach A, Charpentier EM, Schwarzer A, Mateo JL, Klusmann JH, Heckl D (2018) Refined sgRNA efficacy prediction improves largeand small-scale CRISPR-Cas9 applications. Nucleic Acids Res 46:1375–1385. https://doi.org/10.1093/nar/gkx1268

Labun K, Montague TG, Gagnon JA, Thyme SB, Valen E (2016) CHOPCHOP v2: a web tool for the next generation of CRISPR genome engineering. Nucleic Acids Res 44:272–276. https://doi.org/10.1093/nar/gkw398

Labun K, Montague TG, Krause M, Torres CYN, Tjeldnes H, Valen E (2019) CHOPCHOP v3: expanding the CRISPR web toolbox beyond genome editing. Nucleic Acids Res 47:171–174. https://doi.org/10.1093/nar/gkz365

Larson G, Piperno DR, Allaby RG, Purugganan MD, Andersson L, Arroyo-Kalin M, Barton L, Climer VC, Denham T, Dobney K, Doust AN, Gepts P, Gilbert MTP, Gremillion KJ, Lucas L, Lukens L, Marshall FB, Olsen KM, Pires JC, Richerson PJ, Rubio CR, Sanjur OI, Thomas MG, Fuller DQ (2014) Current perspectives and the future of domestication studies. Proc Natl Acad Sci 111(17):6139–6146. https://doi.org/10.1073/pnas1323964111

Lawhorn IE, Ferreira JP, Wang CL (2014) Evaluation of sgRNA target sites for CRISPR-mediated repression of TP53. PLoS One 9:113232

Ledford H (2020) CRISPR treatment inserted directly into the body for first time. Nature 579:185. https://doi.org/10.1038/d41586-020-00655-8

Ledford H, Callaway E (2020) Pioneers of revolutionary CRISPR gene editing win chemistry Nobel. Nature 586:346–347. https://doi.org/10.1038/d41586-020-02765-9

Lee K, Zhang Y, Kleinstiver BP et al (2019) Activities and specificities of CRISPR/Cas9 and Cas12a nucleases for targeted mutagenesis in maize. Plant Biotechnol J 17(2):362–372. https://doi.org/10.1111/pbi.12982

Lei Y, Lu L, Liu HY, Li S, Xing F, Chen LL (2014) CRISPR-P: a web tool for synthetic single-guide RNA design of CRISPR-system in plants. Mol Plant 7:1494–1496. https://doi.org/10.1093/mp/ssu044

Levin M (1994) The role of risk assessment in developing statutes and regulations. In: Krattiger AF, Rosemarin A (eds) Biosafety for sustainable agriculture: sharing biotechnology regulatory experiences of the western hemisphere. ISAAA and Stockholm Environment Institute, Ithaca/Stockholm, pp 127–137

Li Z, Liu ZB, Xing A, Moon BP, Koellhoffer JP, Huang L, Ward RT, Clifton E, Falco SC, Cigan AM (2015) Cas9- guide RNA directed genome editing in Soybean. Plant Physiol 169:960–970

Li C, Liu C, Qi X, Wu Y, Fei X, Mao L, Cheng B, Li X, Xie C (2017) RNA-guided Cas9 as an in vivo desired-target mutator in maize. Plant Biotechnol J 15:1566–1576

Li S, Cheng Q, Liu J, Nie X, Zhao G, Wang J (2018a) CRISPR-Cas12a has both cis- and trans-cleavage activities on single-stranded DNA. Cell Res 28:1–3. https://doi.org/10.1038/s41422-018-0022-x

Li X, Wang Y, Chen S, Tian H, Fu D, Zhu B, Luo Y, Zhu H (2018b) Lycopene is enriched in tomato fruit by CRISPR/Cas9-mediated multiplex genome editing. Front Plant Sci 9:559

Li S, Li J, He Y, Xu M, Zhang J, Du W, Zhao Y, Xia L (2019) Precise gene replacement in rice by RNA transcript templated homologous recombination. Nat Biotechnol 37:445–450

Li Z, Zhang H, Xiao R, Han R, Chang L (2021) Cryo-EM structure of the RNA-guided ribonuclease Cas12g. Nat Chem Biol 17(4):387–393. https://doi.org/10.1038/s41589-020-00721-2

Liang Z, Zhang K, Chen K, Gao C (2014) Targeted mutagenesis in *Zea mays* using TALENs and the CRISPR/Cas system. Genet Genomics 41:63–68

Liang Z, Chen K, Yan Y, Zhang Y, Gao C (2018) Genotyping genome-edited mutations in plants using CRISPR ribonucleoprotein complexes. Plant Biotechnol J 16(12):2053–2062. https://doi.org/10.1111/pbi.12938

Lieber MR (2010) The mechanism of double-strand DNA break repair by the nonhomologous DNA end-joining pathway. Annu Rev Biochem 79:181–211. https://doi.org/10.1146/annurev.biochem.052308.093131

Lillestøl RK, Redder P, Garrett RA, Brügger KA (2006) A putative viral defence mechanism in archaeal cells. Archaea 2(1):59–72. https://doi.org/10.1155/2006/542818

Lin J, Wong KC (2018) Off-target predictions in CRISPR-Cas9 gene editing using deep learning. Bioinformatics 34:656–663. https://doi.org/10.1093/bioinformatics/bty.554

Lin Q, Zong Y, Xue C, Wang S, Jin S, Zhu Z, Wang Y, Anzalone AV, Raguram A, Doman JL, Liu DR, Gao C (2020) Prime genome editing in rice and wheat. Nat Biotechnol 38:582–585. https://doi.org/10.1038/s41587-020-0455-x

Lintner NG, Kerou M, Brumfield SK, Graham S, Liu H, Naismith JH, Sdano M, Peng N, She Q, Copie V et al (2011) Structural and functional characterization of an archaeal clustered regularly interspaced short palindromic repeat (CRISPR)-associated complex for antiviral defense (CASCADE). J Biol Chem 286:21643–21656. https://doi.org/10.1074/jbcM111238485

Listgarten J, Weinstein M, Kleinstiver BP, Sousa AA, Joung JK, Crawford J, Gao K, Hoang L, Elibol M, Doench JG, Fusi N (2018) Prediction of off-target activities for the end-to-end design of CRISPR guide RNAs. Nat Biomed Eng 2:38–47. https://doi.org/10.1038/s41551-017-0178-6

Liu H, Wei Z, Dominguez A, Li Y, Wang X, Qi LS (2015) CRISPR-ERA: a comprehensive design tool for CRISPR-mediated gene editing repression and activation. Bioinformatics 31:3676–3678. https://doi.org/10.1093/bioinformatics/btv423

Liu H, Ding Y, Zhou Y, Jin W, Xie K, Chen LL (2017) CRISPR-P 20: an improved CRISPR-Cas9 tool for genome editing in plants. Mol Plant 10:530–532. https://doi.org/10.1016/jmolp.2017.01.003

Liu JJ, Orlova N, Oakes BL, Ma E, Spinner HB, Baney KLM, Chuck J, Tan D, Knott GJ, Harrington LB, Al-Shayeb B, Wagner A, Brötzmann J, Staahl BT, Taylor KL, Desmarais J, Nogales E, Doudna JA (2019a) CasX enzymes comprise a distinct family of RNA-guided genome editors. Nature 566(7743):218–223. https://doi.org/10.1038/s41586-019-0908-x

Liu Y, Wan X, Wang B (2019b) Engineered CRISPRa enables programmable eukaryote-like gene activation in bacteria. Nat Commun 10:3693

Liu L, Gallagher J, Arevalo ED, Chen R, Skopelitis T, Wu Q, Bartlett M, Jackson D (2021) Enhancing grain-yield-related traits by CRISPR-Cas9 promoter editing of maize CLE genes. Nat Plants 7(3):287–294. https://doi.org/10.1038/s41477-021-00858-5

Lloyd A, Plaisier CL, Carroll D, Drews GN (2005) Targeted mutagenesis using zinc-finger nucleases in Arabidopsis. Proc Natl Acad Sci U S A 102:232–2237

Louwen R, Staals RH, Endtz HP, Van Baarlen P, van der Oost J (2014) The role of CRISPR-Cas systems in virulence of pathogenic bacteria. Microbiol Mol Biol Rev 78(1):74–88. https://doi.org/10.1128/MMBR00039-13

Lu Y, Tian Y, Shen R, Yao Q, Wang M, Chen M, Dong J, Zhang T, Li F, Lei M, Zhu JK (2020) Targeted efficient sequence insertion and replacement in rice. Nat Biotechnol 38(12):1402–1407. https://doi.org/10.1038/s41587-020-0581-5

Luo J, Chen W, Xue L, Tang B (2019) Prediction of activity and specificity of CRISPR-Cpf1 using convolutional deep learning neural networks. BMC Bioinformatics 20:1–10. https://doi.org/10.1186/s12859-019-2939-6

Ma X, Feng F, Zhang Y, Elesawi IE, Xu K, Li T, Mei H, Liu H, Gao N, Chen C, Luo L, Yu S (2019) A novel rice grain size gene OsSNB was identified by genome-wide association study in natural population. PLoS Genet 15:1008191. https://doi.org/10.1371/journal.pgen.1008191

MacPherson CR, Scherf A (2015) Flexible guide-RNA design for CRISPR applications using Protospacer Workbench. Nat Biotechnol 33:805–806. https://doi.org/10.1038/nbt.3291

Maeder ML, Thibodeau-Beganny S, Osiak A, Wright DA, Anthony RM, Eichtinger M, Jiang T, Foley JE, Winfrey RJ, Townsend JA, Unger-Wallace E, Sander JD, Müller-Lerch F, Fu F, Pearlberg J, Göbel C, Dassie JP, Pruett-Miller SM, Porteus MH, Sgroi DC, Iafrate AJ, Dobbs D, McCray PB, Cathomen T, Voytas DF, Joung JK (2008) Rapid "open-source" engineering of customized zinc-finger nucleases for highly efficient gene modification. Mol Cell 31(2):294–301. https://doi.org/10.1016/j.molcel.2008.06.016

Mahfouz MM, Piatek A, Stewart CN (2014) Genome engineering via TALENs and CRISPR/Cas9 systems: challenges and perspectives. Plant Biotechnol 12(8):1006–1014. https://doi.org/10.1111/pbi.12256

Maji B, Gangopadhyay SA, Lee M, Shi M, Wu P, Heler R, Mok B, Lim D, Siriwardena SU, Paul B, Dančík V (2019) A high-throughput platform to identify small-molecule inhibitors of CRISPR-Cas9. Cell 177(4):1067–1079

Makarova KS, Grishin NV, Shabalina SA, Wolf YI, Koonin EV (2006) A putative RNA-interference-based immune system in prokaryotes: computational analysis of the predicted enzymatic machinery functional analogies with eukaryotic RNAi and hypothetical mechanisms of action. Biol Direct 1:7. https://doi.org/10.1186/1745-6150-1-7

Makarova KS, Haft DH, Barrangou R, Brouns SJ, Charpentier E, Horvath P, Moineau S, Mojica FJ, Terns RM, Terns MP, White MF, Yakunin AF, Garrett RA, Van der Oost J, Backofen R, Koonin EV (2011) Evolution and classification of the CRISPR-Cas systems. Nat Rev Microbiol 9:467–477. https://doi.org/10.1038/nrmicro2577

Makarova KS, Wolf YI, Alkhnbashi OS, Costa F, Shah SA, Saunders SJ (2015) An updated evolutionary classification of CRISPR-Cas systems. Nat Rev Microbiol 13:722–736. https://doi.org/10.1038/nrmicro3569

Makarova KS, Wolf YI, Koonin EV (2018) Classification and nomenclature of CRISPR-Cas systems: where from here? Cris J 1:325–336

Makarova KS, Wolf YI, Iranzo J, Shmakov SA, Alkhnbashi OS, Brouns SJJ, Charpentier E, Cheng D, Haft DH, Horvath P, Moineau S, Mojica FJM, Scott D, Shah SA, Siksnys V, Terns MP, Venclovas Č, White MF, Yakunin AF, Yan W, Zhang F, Garrett RA, Backofen R, van der Oost J, Barrangou R, Koonin EV (2020) Evolutionary classification of CRISPR–Cas systems: a burst of class 2 and derived variants. Nat Rev Microbiol 18:67–83. https://doi.org/10.1038/s41579-019-0299-x

Mali P, Aach J, Stranges PB, Esvelt KM, Moosburner M, Kosuri S, Yang L, Church GM (2013a) CAS9 transcriptional activators for target specificity screening and paired nickases for cooperative genome engineering. Nat Biotechnol 31:833–838

Mali P, Yang L, Esvelt KM, Aach J, Guell M, DiCarlo JE, Norville JE, Church GM (2013b) RNA-guided human genome engineering via Cas9. Science 339:823–826. https://doi.org/10.1126/science.1232033

Malnoy M, Viola R, Jung MH, Koo OJ, Kim S, Kim JS, Velasco R, Nagamangala KC (2016) DNA-free genetically edited grapevine and apple protoplast using Crispr/Cas9 ribonucleoproteins. Front Plant Sci 7:1904

Marino ND, Pinilla-Redondo R, Csörgő B, Bondy-Denomy J (2020) Anti- CRISPR protein applications: natural brakes for CRISPR-Cas technologies. Nat Methods 17(5):471–479

Marraffini LA, Sontheimer EJ (2008) CRISPR interference limits horizontal gene transfer in staphylococci by targeting DNA. Science 322:1843–1845. https://doi.org/10.1126/science.1165771

McCaw ME, Lee K, Kang M, Zobrist JD, Azanu MK, Birchler JA, Wang K (2021) Development of a transformable fast-flowering mini-maize as a tool for maize gene editing. Front Genome Ed 2: 622227. https://doi.org/10.3389/fgeed.2020.622227

McKenna A, Shendure J (2018) FlashFry: a fast and flexible tool for large-scale CRISPR target design. BMC Biol 16:4–9. https://doi.org/10.1186/s12915-018-0545-0

Mendoza BJ, Trinh CT (2018) Enhanced guide-RNA design and targeting analysis for precise CRISPR genome editing of single and consortia of industrially relevant and non-model organisms. Bioinformatics 34:16–23. https://doi.org/10.1093/bioinformatics/btx564

Miao J, Guo D, Zhang J, Huang Q, Qin G, Zhang X, Wan J, Gu H, Qu LJ (2013) Targeted mutagenesis in rice using CRISPR–Cas system. Cell Res 23:1233–1236

Miller JC, Tan S, Qiao G, Barlow KA, Wang J, Xia DF, Meng X, Paschon DE, Leung E, Hinkley SJ, Dulay GP, Hua KL, Ankoudinova I, Cost GJ, Urnov FD, Zhang HS, Holmes MC, Zhang L, Gregory PD, Rebar EJ (2011) A TALE nuclease architecture for efficient genome editing. Nat Biotechnol 29(2):143–148

Minkenberg B, Zhang J, Xie K, Yang Y (2019) CRISPR-PLANT v2: an online resource for highly specific guide RNA spacers based on improved off-target analysis. Plant Biotechnol J 17:5–8. https://doi.org/10.1111/pbi.13025

Mojica FJ, Ferrer C, Juez G, Rodríguez-Valera F (1995) Long stretches of short tandem repeats are present in the largest replicons of the archaea *Haloferax mediterranei* and *Haloferax volcanii* and could be involved in replicon partitioning. Mol Microbiol 17:85–93. https://doi.org/10.1111/j1365-29581995mmi_17010085x

Mojica FJ, Díez-Villaseñor C, Soria E, Juez G (2000) Biological significance of a family of regularly spaced repeats in the genomes of archaea bacteria and mitochondria. Mol Microbiol 36:244–246. https://doi.org/10.1046/j1365-2958200001838x

Mojica FJM, Díez-Villaseñor C, García-Martínez J, Soria E (2005) Intervening sequences of regularly spaced prokaryotic repeats derive from foreign genetic elements. J Mol Evol 60: 174–182. https://doi.org/10.1007/s00239-004-0046-3

Mojica FJ, Diez-Villasenor C, Garcia-Martinez J, Almendros C (2009) Short motif sequences determine the targets of the prokaryotic CRISPR defence system. Microbiology 155:733–740

Montague TG, Cruz JM, Gagnon JA, Church GM, Valen E (2014) CHOPCHOP: a CRISPR/Cas9 and TALEN web tool for genome editing. Nucleic Acids Res 42:401–407. https://doi.org/10.1093/nar/gku410

Moore R, Chandrahas A, Bleris L (2014) Transcription activator-like effectors: a toolkit for synthetic biology. ACS Synth Biol 3(10):708–716. https://doi.org/10.1021/sb400137b

Moreno-Mateos MA, Vejnar CE, Beaudoin JD, Fernandez JP, Mis EK, Khokha MK, Giraldez AJ (2015) CRISPRscan: designing highly efficient sgRNAs for CRISPR-Cas9 targeting in vivo. Nat Methods 12:982–988. https://doi.org/10.1038/nmeth3543

Moscou MJ, Bogdanove AJ (2009) A simple cipher governs DNA recognition by TAL effectors. Science 326(5959):1501

Naeem M, Majeed S, Hoque MZ, Ahmad I (2020) Latest developed strategies to minimize the off-target effects in CRISPR-Cas-mediated genome editing. Cell 9(7):1608. https://doi.org/10.3390/cells9071608

Naito Y, Hino K, Bono H, Ui-Tei K (2015) CRISPRdirect: Software for designing CRISPR/Cas guide RNA with reduced off-target sites. Bioinformatics 31:1120–1123. https://doi.org/10.1093/bioinformatics/btu743

Nakata A, Amemura M, Makino K (1989) Unusual nucleotide arrangement with repeated sequences in the *Escherichia coli* K-12 chromosome. J Bacteriol 171:3553–3556. https://doi.org/10.1128/jb.171.6.3553-3556.1989

Newsom S, Parameshwaran HP, Martin L, Rajan R (2021) The CRISPR-Cas mechanism for adaptive immunity and alternate bacterial functions fuels diverse biotechnologies. Front Cell Infect Microbiol 10:619763. https://doi.org/10.3389/fcimb.2020.619763

Nickel L, Ulbricht A, Alkhnbashi OS, Förstner KU, Cassidy L, Weidenbach K, Backofen R, Schmitz RA (2018) Cross-cleavage activity of Cas6b in crRNA processing of two different CRISPR-Cas systems in *Methanosarcina mazei* Gö1. RNA Biol 16:492–503

Niewoehner O, Jinek M (2016) Structural basis for the endoribonuclease activity of the type III-A CRISPR-associated protein Csm6. RNA 22:318–329

Nishimasu H, Ran FA, Hsu PD, Konermann S, Shehata SI, Dohmae N, Ishitani R, Zhang F, Nureki O (2014) Crystal structure of Cas9 in complex with guide RNA and target DNA. Cell 156:935–949. https://doi.org/10.1016/jcell201402001

Nonaka S, Arai C, Takayama M, Matsukura C, Ezura H (2017) Efficient increase of γ-aminobutyric acid (GABA) content in tomato fruits by targeted mutagenesis. Sci Rep 7:7057

Nunez JK, Chen J, Pommier GC, Cogan JZ, Replogle JM, Adriaens C, Ramadoss GN, Shi Q, Hung KL, Samelson AJ, Pogson AN, Kim JYS, Chung A, Leonetti MD, Chang HY, Kampmann M, Bernstein BE, Hovestadt V, Gilbert LA, Weissman JS (2021) Genome-wide programmable transcriptional memory by CRISPR-based epigenome editing. Cell 184(9):2503–2519. https://doi.org/10.1016/jcell202103025

O'Brien A, Bailey TL (2014) GT-Scan: Identifying unique genomic targets. Bioinformatics 30: 2673–2675. https://doi.org/10.1093/bioinformatics/btu354

Okuzaki A, Ogawa T, Koizuka C, Kaneko K, Inaba M, Imamura J, Koizuka N (2018) CRISPR/Cas9-mediated genome editing of the fatty acid desaturase 2 gene in *Brassica napus*. Plant Physiol Biochem 131:63–69

Orlowski J, Boniecki M, Bujnicki JM (2007) I-Ssp6803I: the first homing endonuclease from the PD-(D/E)XK superfamily exhibits an unusual mode of DNA recognition. Bioinformatics 23(5): 527–530

Özcan A, Krajeski R, Ioannidi E, Lee B, Gardner A, Makarova KS, Koonin EV, Abudayyeh OO, Gootenberg JS (2021) Programmable RNA targeting with the single-protein CRISPR effector Cas7-11. Nature 597(7878):720–725. https://doi.org/10.1038/s41586-021-03886-5

Pacher M, Puchta H (2017) From classical mutagenesis to nuclease-based breeding –directing natural DNA repair for a natural end-product. Plant J 90:819–833. https://doi.org/10.1111/tpj.13469

Pak H, Wang H, Kim Y, Song U, Tu M, Wu D, Jiang L (2020) Creation of male-sterile lines that can be restored to fertility by exogenous methyl jasmonate for the establishment of a two-line system for the hybrid production of rice (*Oryza sativa* L). Plant Biotechnol J 19:365–374. https://doi.org/10.1111/pbi.13471

Pannunzio NR, Watanabe G, Lieber MR (2017) Nonhomologous DNA end joining for repair of DNA double-strand breaks. J Biol Chem:117000374. https://doi.org/10.1074/jbcTM117000374

Paparini A, Romano-Spica V (2006) Gene transfer and cauliflower mosaic virus promoter 35S activity in mammalian cells. J Environ Sci Health B41(4):437–449. https://doi.org/10.1080/03601230600616957

Park J, Bae S, Kim JS (2015) Cas-designer: a web-based tool for choice of CRISPR-Cas9 target sites. Bioinformatics 31:4014–4016. https://doi.org/10.1093/bioinformatics/btv537

Pausch P, Al-Shayeb B, Bisom-Rapp E, Tsuchida CA, Li Z, Cress BF, Knott GJ, Jacobsen SE, Banfield JF, Doudna JA (2020) CRISPR-CasΦ from huge phages is a hypercompact genome editor. Science 369(6501):333–337. 10.1126/science.abb1400

Pauwels K, Podevin N, Breyer D, Carroll D, Herman P. 2014 Engineering nucleases for gene targeting: safety and regulatory considerations. N Biotechnol 31(1):18-27. https://doi.org/10.1016/j.nbt.2013.07.001

Peng D, Tarleton R (2015) EuPaGDT: a web tool tailored to design CRISPR guide RNAs for eukaryotic pathogens. Microb Genomics 1:1–7. https://doi.org/10.1099/mgen.0.000033

Peng H, Zheng Y, Blumenstein M, Tao D, Li J (2018) CRISPR/Cas9 cleavage efficiency regression through boosting algorithms and Markov sequence profiling. Bioinformatics 34:3069–3077. https://doi.org/10.1093/bioinformatics/bty298

Pineda M, Lear A, Collins JP, Kiani S (2019) Safe CRISPR: challenges and possible solutions. Trends Biotechnol 37:389–401. https://doi.org/10.1016/jtibtech.2018.09.010

Pliatsika V, Rigoutsos I (2015) "Off-Spotter": very fast and exhaustive enumeration of genomic lookalikes for designing CRISPR/Cas guide RNAs. Biol Direct 10:1–10. https://doi.org/10.1186/s13062-015-0035-z

Porto EM, Komor AC, Slaymaker IM, Yeo GW (2020) Base editing: advances and therapeutic opportunities. Nat Rev Drug Discov 19:839–859. https://doi.org/10.1038/s41573-020-0084-6

Pourcel C, Salvignol G, Vergnaud G (2005) CRISPR elements in *Yersinia pestis* acquire new repeats by preferential uptake of bacteriophage DNA and provide additional tools for evolutionary studies. Microbiology 151:653–663. https://doi.org/10.1099/mic.0.27437-0

Prykhozhij SV, Rajan V, Gaston D, Berman JN (2015) CRISPR multitargeter: a web tool to find common and unique CRISPR single guide RNA targets in a set of similar sequences. PLoS One 10:1–18. https://doi.org/10.1371/journal.pone.0119372

Puchta H (1999) Double-strand break-induced recombination between ectopic homologous sequences in somatic plant cells. Genetics 152(3):1173–1181

Qi LS, Larson MH, Gilbert LA, Doudna JA, Weissman JS, Arkin AP, Lim WA (2021) Repurposing CRISPR as an RNA-guided platform for sequence-specific control of gene expression. Cell 184(3):844

Qin J, Li R, Raes J, Arumugam M, Burgdorf KS, Manichanh C, Nielsen T, Pons N, Levenez F, Yamada T, Mende DR, Li J, Xu J, Li S, Li D, Cao J, Wang B, Liang H, Zheng H, Xie Y, Wang J (2010) A human gut microbial gene catalogue established by metagenomic sequencing. Nature 464(7285):59–65. https://doi.org/10.1038/nature.08821

Rahman MK, Rahman MS (2017) CRISPRpred: a flexible and efficient tool for sgRNAs on-target activity prediction in CRISPR/Cas9 systems. PLoS One 12:1–14. https://doi.org/10.1371/journal.pone.0181943

Rai M, He C, Wu R (2009) Comparative functional analysis of three abiotic stress-inducible promoters in transgenic rice. Transgenic Res 18:787–799

Reardon S (2016) First CRISPR clinical trial gets green light from US panel. Nature. https://doi.org/10.1038/nature.2016.20137

Rees HA, Liu DR (2018) Base editing: precision chemistry on the genome and transcriptome of living cells. Nat Rev Genet 19(12):770–788. https://doi.org/10.1038/s41576-018-0059-1

Reis A, Hornblower B, Robb B, Tzertzinis G (2014) CRISPR/Cas9 & targeted genome editing: new era in molecular biology. Biolabs

Ren X, Yang Z, Xu J, Sun J, Mao D, Hu Y, Yang SJ, Qiao HH, Wang X, Hu Q, Deng P, Liu LP, Ji JY, Li JB, Ni JQ (2014) Enhanced specificity and efficiency of the CRISPR/Cas9 system with optimized sgRNA parameters in Drosophila. Cell Rep 9:1151–1162

Ren C, Liu X, Zhang Z, Wang Y, Duan W, Li S, Liang Z (2016) CRISPR/Cas9-mediated efficient targeted mutagenesis in Chardonnay (*Vitis vinifera* L). Sci Rep 6:1–9

Roginsky J (2018) Analyzing CRISPR editing results. Genet Eng Biotechnol News 38:24–26. https://doi.org/10.1089/gen.381113

Rosado A, Craig W (2017) Biosafety regulatory systems overseeing the use of genetically modified organisms in the Latin America and Caribbean region. AgBioForum 20:120–132

Rosen LE, Morrison HA, Masri S, Brown MJ, Springstubb B, Sussman D, Stoddard BL (2006) Seligman LM: homing endonuclease I-CreI derivatives with novel DNA target specificities. Nucleic Acids Res 34(17):4791–4800

Samai P, Pyenson N, Jiang W, Goldberg GW, Hatoum-Aslan A, Marraffini LA (2015) Co-transcriptional DNA and RNA cleavage during type III CRISPR–Cas immunity. Cell 161: 1164–1174

Samalov M, Moore I (2021) The steroid-inducible pOp6/LhGR gene expression system is fast, sensitive and does not cause plant growth defects in rice (*Oryza sativa*). BMC Plant Biol 21:461. https://doi.org/10.1186/s12870-021-03241-w

Sanchez-Leon S, Gil-Humanes J, Ozuna CV, Gimenez MJ, Sousa C, Voytas DF, Barro F (2018) Low-gluten nontransgenic wheat engineered with CRISPR/Cas9. Plant Biotechnol 16:902–910. https://doi.org/10.1111/pbi.12837

Sander JD, Joung JK (2014) CRISPR-Cas systems for editing regulating and targeting genomes. Nat Biotechnol 32:347. https://doi.org/10.1038/nbt.2842

Sander JD, Maeder ML, Reyon D, Voytas DF, Joung JK, Dobbs D (2010) ZiFiT (Zinc Finger Targeter): an updated zinc finger engineering tool. Nucleic Acids Res 38:462–468. https://doi.org/10.1093/nar/gkq.319

Sashidhar N, Harloff HJ, Potgieter L, Jung C (2020) Gene editing of three BnITPK genes in tetraploid oilseed rape leads to significant reduction of phytic acid in seeds. Plant Biotechnol 18(11):2241–2250. https://doi.org/10.1111/pbi.13380

Satomura A, Nishioka R, Mori H, Sato K, Kuroda K, Ueda M (2017) Precise genome-wide base editing by the CRISPR Nickase system in yeast. Sci Rep 7:2095. https://doi.org/10.1038/s41598-017-02013-7

Sauer NJ, Narvaez-Vasquez J, Mozoruk J, Miller RB, Warburg ZJ, Woodward MJ, Mihiret YA, Lincoln TA, Segami RE, Sanders SL, Walker KA, Beetham PR, Schöpke CR, Gocal GF (2016) Oligonucleotide-mediated genome editing provides precision and function to engineered nucleases and antibiotics in plants. Plant Physiol 170:1917–1928

Savage DF (2019) Cas14: big advances from small CRISPR proteins. Biochem 58:1024–1025

Schenke D, Cai D (2020) Applications of CRISPR/Cas to improve crop disease resistance: beyond inactivation of susceptibility factors. iScience 23(9):101478. https://doi.org/10.1016/j.isci.2020.101478

Schmidt SM, Belisle M, Frommer WB (2020) The evolving landscape around genome editing in agriculture. EMBO Rep 21:19–22. https://doi.org/10.15252/embr.202050680

Scholefield J, Harrison PT (2021) Prime editing – an update on the field. Gene Ther 28:396–401. https://doi.org/10.1038/s41434-021-00263-9

Sedeek KEM, Mahas A, Mahfouz M (2019) Plant genome engineering for targeted improvement of crop traits. Front Plant Sci 10:114. https://doi.org/10.3389/fpls.2019.00114

Settachaimongkon S, Nout MJR, Antunes FEC, Van Hooijdonk Toon CM, Zwietering Marcel H, Smid Eddy J, Van Valenberg Hein JF (2014) The impact of selected strains of probiotic bacteria on metabolite formation in set yoghurt. Int Dairy J 38(1):1–10. https://doi.org/10.1016/j.idairyj.2014.04.002

Shan Q, Wang Y, Li J, Zhang Y, Chen K, Liang Z, Zhang K, Liu J, Xi JJ, Qiu JL, Gao C (2013) Targeted genome modification of crop plants using a CRISPR-Cas system. Nat Biotechnol 31(8):686–688

Shan Q, Wang Y, Li J, Gao C (2014) Genome editing in rice and wheat using the CRISPR/Cas system. Nat Protoc 9:2395–2410

Shan S, Soltis PS, Soltis DE, Yang B (2020) Considerations in adapting CRISPR/Cas9 in nongenetic model plant. Appl Plant Sci 8:11314. https://doi.org/10.1002/aps3.11314

Shen H (2014) First monkeys with customized mutations born. Nature. https://doi.org/10.1038/nature.2014.14611

Shi J, Gao H, Wang H, Lafitte HR, Archibald RL, Yang M, Hakimi SM, Mo H, Habben JE (2017) ARGOS 8 variants generated by CRISPR/Cas9 improve maize grain yield under field drought stress conditions. Plant Biotechnol J 15:207–216

Shin J, Jiang F, Liu JJ, Bray NL, Rauch BJ, Baik SH, Nogales E, Bondy-Denomy J, Corn JE, Doudna JA (2017) Disabling Cas9 by an anti-CRISPR DNA mimic. Sci Adv 3:1701620

Shufen C, Yicong C, Baobing F, Guiai J, Zhonghua S, Ju L, Shaoqing T, Jianlong W, Peisong H, Xiangjin W (2019) Editing of rice isoamylase gene ISA1 provides insights into its function in starch formation. Ric Sci 26(2):77–87. https://doi.org/10.1016/j.rsci.2018.07.001

Sidira MV, Santarmaki M, Kiourtzidis AA, Argyri OS, Papadopoulou N, Chorianopoulos C, Tassou S, Kaloutsas A, Galanis KY (2017) Evaluation of immobilized *Lactobacillus plantarum* 2035 on whey protein as adjunct probiotic culture in yoghurt production. Lebensm Wiss Technol 75:137–146

Singh R, Kuscu C, Quinlan A, Qi Y, Adli M (2015) Cas9-chromatin binding information enables more accurate CRISPR off-target prediction. Nucleic Acids Res 43:1–8. https://doi.org/10.1093/nar/gkv575

Singh A, Roychowdhury R, Singh T, Wang W, Yadav D, Kumar A, Modi A, Rai AC, Ghughe S, Kumar A, Kumar Singh P (2020) Improvement of crop's stress tolerance by gene editing CRISPR/CAS9 system. In: Choudhury RR, Hasanuzzaman S, Srivastava M (eds) Sustainable agriculture in the era of climate change. Springer, Cham

Sriboon S, Li H, Guo C et al (2020) Knock-out of *TERMINAL FLOWER 1* genes altered flowering time and plant architecture in *Brassica napus*. BMC Genet 21:52. https://doi.org/10.1186/s12863-020-00857-z

Staals RH, Brouns SJ (2013) In: Barrangou R, van de Oost J (eds) Distribution and mechanism of the type I CRISPR-Cas systems. Springer, Berlin, Heidelberg, pp 115–144

Staals RH, Zhu Y, Taylor DW, Kornfeld JE, Sharma K, Barendregt A, Koehorst JJ, Vlot M, Neupane N, Varossieau K, Sakamoto K, Suzuki T, Dohmae N, Yokoyama S, Schaap PJ, Urlaub H, Heck AJ, Nogales E, Doudna JA, Shinkai A, van der Oost J (2014) RNA targeting by the type III-A CRISPR–Cas Csm complex of *Thermus thermophiles*. Mol Cell 56:518–530

Stemmer M, Thumberger T, Del SKM, Wittbrodt J, Mateo JL (2015) CCTop: an intuitive flexible and reliable CRISPR/Cas9 target prediction tool. PLoS One 10:1–11. https://doi.org/10.1371/journal.pone.0124633

Sun Q, Gao F, Zhao L, Li K, Zhang J (2010) Identification of a new 130 bp *cis*-acting element in the *TsVP1* promoter involved in the salt stress response from *Thellungiella halophile*. BMC Plant Biol 10:90

Sun X, Hu Z, Chen R, Jiang Q, Song G, Zhang H, Xi Y (2015) Targeted mutagenesis in soybean using the CRISPR-Cas9 system. Sci Rep 5:10342

Sun Y, Zhang X, Wu C, He Y, Ma Y, Hou H, Guo X, Du W, Zhao Y, Xia L (2016) Engineering herbicide-resistant rice plants through CRISPR/Cas9-mediated homologous recombination of acetolactate synthase. Mol Plant 9:628–631

Sun Y, Jiao G, Liu Z, Zhang X, Li J, Guo X, Du WDJ, Francis F, Zhao Y, Xia L (2017) Generation of high amylose rice through CRISPR/Cas9-mediated targeted mutagenesis of starch branching enzymes. Front Plant Sci 8:298

Sun SK, Xu XJ, Tang Z, Tang Z, Huang XY, Wirtz M, Hell R, Zhao FJ (2021) A molecular switch in sulfur metabolism to reduce arsenic and enrich selenium in rice grain. Nat Commun 12:1392

Svitashev S, Young JK, Schwartz C, Gao H, Falco SC, Cigan AM (2015) Targeted mutagenesis precise gene editing and site-specific gene insertion in maize using Cas9 and guide RNA. Plant Physiol 169:931–945

Swarts DC, Jinek M (2019) Mechanistic insights into the Cis- and Trans-acting deoxyribonuclease activities of Cas12a. Mol Cell 73:589–600. https://doi.org/10.1016/j.molcel.2018.11.021

Tabassum J, Ahmad S, Hussain B, Mawia AM, Zeb A, Ju L (2021) Applications and potential of genome-editing systems in rice improvement: current and future perspectives. Agronomy 11: 1359. https://doi.org/10.3390/agronomy11071359

Tang L, Mao B, Li Y, Lv Q, Zhang L, Chen C, He H, Wang W, Zeng X, Shao Y, Pan Y, Hu Y, Peng Y, Fu X, Li H, Xia S, Zhao B (2017) Knockout of OsNramp5 using the CRISPR/Cas9 system produces low Cd-accumulating indica rice without compromising yield. Sci Rep 7: 14438

Tang T, Yu X, Yang H, Gao Q, Ji H, Wang Y, Yan G, Peng Y, Luo H, Liu K, Li X, Ma C, Kang C, Dai C (2018) Development and validation of an effective CRISPR/Cas9 vector for efficiently isolating positive transformants and transgene-free mutants in a wide range of plant species. Front Plant Sci 9:1533. https://doi.org/10.3389/fpls.2018.01533

Thygesen P (2019) Clarifying the regulation of genome editing in Australia: situation for genetically modified organisms. Transgenic Res 28:151–159. https://doi.org/10.1007/s11248-019-00151-4

Tian Y, Chen K, Li X, Zheng Y, Chen F (2020) Design of high-oleic tobacco (*Nicotiana tabacum* L) seed oil by CRISPR-Cas9-mediated knockout of *NtFAD2–2*. BMC Plant Biol 20:233. https://doi.org/10.1186/s12870-020-02441-0

Tiwari M, Trivedi P, Pandey A (2020) Emerging tools and paradigm shift of gene editing in cereals fruits and horticultural crops for enhancing nutritional value and food security. Food Ener Secur 10(1):e258. https://doi.org/10.1002/fes.3258

Townsend JA, Wright DA, Winfrey RJ, Fu F, Maeder ML, Joung JK, Voytas DF (2009) High-frequency modification of plant genes using engineered zinc-finger nucleases. Nature 459:442–445

Tsai SQ, Joung JK (2016) Defining and improving the genome-wide specificities of CRISPR-Cas9 nucleases. Nat Rev Genet 17(5):300–312. https://doi.org/10.1038/nrg.2016.28

Tsuda M, Watanabe KN, Ohsawa R (2019) Regulatory status of genome edited organisms under the Japanese Cartagena act. Front Bioeng Biotechnol 7:387. https://doi.org/10.3389/fbioe.2019.00387

Tuncel A, Corbin KR, Ahn-Jarvis J, Harris S, Hawkins E, Smedley MA, HarwoodW WFJ, Patron NJ, Smith AM (2019) Cas9-mediated mutagenesis of potato starch-branching enzymes generates a range of tuber starch phenotypes. Plant Biotechnol J 17:2259–2271

Upadhyay SK, Sharma S (2014) SSFinder: high throughput CRISPR-Cas target sites prediction tool. Biomed Res Int. https://doi.org/10.1155/2014/742482

Upadhyay SK, Kumar J, Alok A, Tuli R (2013) RNA guided genome editing for target gene mutations in wheat. G3 (Bethesda) 3:2233–2238

Vu TV, Sivankalyani V, Kim EJ, Doan DTH, Tran MT, Kim J, Kim JY (2020) Highly efficient homology-directed repair using CRISPR/Cpf1-geminiviral replicon in tomato. Plant Biotechnol J 18(10):2133–2143. https://doi.org/10.1111/pbi.13373

Wagner DL, Peter L, Schmueck-Henneresse M (2021) Cas9-directed immune tolerance in humans—a model to evaluate regulatory T cells in gene therapy? Gene Ther 28(9):549–559. https://doi.org/10.1038/s41434-021-00232-2

Wang Y, Cheng X, Shan Q, Zhang Y, Liu J, Gao C, Qiu JL (2014) Simultaneous editing of three homoeoalleles in hexaploid bread wheat confers heritable resistance to powdery mildew. Nat Biotechnol 32:947–951

Wang F, Wang C, Liu P, Lei C, Hao W, Gao Y, Liu YG, Zhao K (2016) Enhanced rice blast resistance by CRISPR/Cas9-targeted mutagenesis of the ERF transcription factor gene OsERF922. PLoS One 11(4):0154027. https://doi.org/10.1371/journal.pone.0154027

Wang M, Lu Y, Botella JR, Mao Y, Hua K, Zhu JK (2017) Gene targeting by homology-directed repair in rice using a geminivirus-based CRISPR/Cas9 system. Mol Plant 10(7):1007–1010. https://doi.org/10.1016/jmolp.2017.03.002

Wang W, Simmonds J, Pan Q, Davidson D, He F, Battal A, Akhunova A, Trick HN, Uauy C, Akhunov E (2018a) Gene editing and mutagenesis reveal inter-cultivar differences and additivity in the contribution of TaGW2 homoeologues to grain size and weight in wheat. Theor Appl Genet 131:2463–2475

Wang M, Wang S, Liang Z, Shi W, Gao C, Xia G (2018b) From genetic stock to genome editing: gene exploitation in wheat. Trends Biotechnol 36:160–172

Wang H, Wu Y, Zhang Y, Yang J, Fan W, Zhang H, Zhao S, Yuan L, Zhang P (2019) CRISPR/Cas9-based mutagenesis of starch biosynthetic genes in sweet potato (*Ipomoea Batatas*) for the improvement of starch quality. Int J Mol Sci 20:4702

Wang D, Wang K, Cai Y (2020) An overview of development in gene therapeutics in China. Gene Ther 27:338–348

Weeks DP, Spalding MH, Yang B (2016) Use of designer nucleases for targeted gene and genome editing in plants. Plant Biotechnol 14:483–495

Wong N, Liu W, Wang X (2015) WU-CRISPR: characteristics of functional guide RNAs for the CRISPR/Cas9 system. Genome Biol 16:1–8. https://doi.org/10.1186/s13059-015-0784-0

Woo JW, Kim J, Kwon SI, Corvalán C, Cho SW, Kim H, Kim S, Kim S, Choe S, Kim J (2015) DNA-free genome editing in plants with preassembled CRISPR-Cas9 ribonucleoproteins. Nat Biotechnol 33(11):1162–1164. https://doi.org/10.1038/nbt.3389

Wu X, Scott DA, Kriz AJ, Chiu AC, Hsu PD, Dadon DB (2014) Genome-wide binding of the CRISPR endonuclease Cas9 in mammalian cells. Nat Biotechnol 32:670–676. https://doi.org/10.1038/nbt.2889

Wu H, Awan FS, Vilarinho A, Zeng Q, Kannan B, Phipps T (2015) Transgene integration complexity and expression stability following biolistic or *Agrobacterium*-mediated transformation of sugarcane. In Vitro Cell Devel Biol Plant 51:603–611

Wu S, Zhu H, Liu J, Yang Q, Shao X, Bi F, Hu C, Huo H, Chen K, Yi G (2020) Establishment of a PEG-mediated protoplast transformation system based on DNA and CRISPR/Cas9 ribonucleoprotein complexes for banana. BMC Plant Biol 20:425. https://doi.org/10.1186/s12870-020-02609-8

Xiao A, Cheng Z, Kong L, Zhu Z, Lin S, Gao G et al (2014) CasOT: a genome-wide Cas9/gRNA off-target searching tool. Bioinformatics 30:1180–1182. https://doi.org/10.1093/bioinformatics/btt.764

Xie X, Liu YG (2021) *De novo* domestication towards new crops. Nat Sci Rev 8(4):033. https://doi.org/10.1093/nsr/nwab.033

Xie K, Yang Y (2013) RNA-guided genome editing in plants using a CRISPR-Cas system. Mol Plant 6:1975–1983. https://doi.org/10.1093/mp/sst.119

Xie S, Shen B, Zhang C, Huang X, Zhang Y (2014) SgRNAcas9: a software package for designing CRISPR sgRNA and evaluating potential off-target cleavage sites. PLoS One 9:1–9. https://doi.org/10.1371/journal.pone.0100448

Xie X, Ma X, Zhu Q, Zeng D, Li G, Liu YG (2017) CRISPR-GE: a convenient software toolkit for CRISPR-based genome editing. Mol Plant 10:1246–1249. https://doi.org/10.1016/j.molp.2017.06.004

Xu H, Xiao T, Chen CH, Li W, Meyer CA, Wu Q, Wu D, Cong L, Zhang F, Liu JS, Brown M, Liu XS (2015) Sequence determinants of improved CRISPR sgRNA design. Genome Res 25:1147–1157. https://doi.org/10.1101/gr.191452.115

Xu R, Qin R, Li H, Li J, Yang J, Wei P (2018) Enhanced genome editing in rice using single transcript unit CRISPR-LbCpf1 systems. Plant Biotechnol J 17:553–555. https://doi.org/10.1111/pbi.13028

Xu J, Kang BC, Naing AH, Bae SJ, Kim JS, Kim H, Kim CK (2019) CRISPR/Cas9-mediated editing of 1-aminocyclopropane-1-carboxylate oxidase1 enhances Petunia flower longevity. Plant Biotechnol J 18(1):287–297. https://doi.org/10.1111/pbi.13197

Xu Y, Lin Q, Li X, Wang F, Chen Z, Wang J, Li W, Fan F, Tao Y, Jiang Y, Wei X, Zhang R, Zhu QH, Bu Q, Yang J, Gao C (2020) Fine-tuning the amylose content of rice by precise base editing of the Wx gene. Plant Biotechnol 19(1):11–13. https://doi.org/10.1111/pbi.13433

Xue L, Tang B, Chen W, Luo J (2019) Prediction of CRISPR sgRNA activity using a deep convolutional neural network. J Chem Inf Model 59:615–624. https://doi.org/10.1021/acs.jcim.8b00368

Yan L, Wei S, Wu Y, Hu R, Li H, Yang W, Xie Q (2015) High-efficiency genome editing in arabidopsis using YAO promoter-driven CRISPR/Cas9 system. Mol Plant 8:1820–1823. https://doi.org/10.1016/j.molp.2015.10.004

Yan WX, Hunnewell P, Alfonse LE, Carte JM, Keston-Smith E, Sothiselvam S, Garrity AJ, Chong S, Makarova KS, Koonin EV, Cheng DR, Scott DA (2019) Functionally diverse type V CRISPR-Cas systems. Science 363(6422):88–91. https://doi.org/10.1126/science.aav7271

Yang H, Patel DJ (2019) CasX: a new and small CRISPR gene-editing protein. Cell Res 29:345–346. https://doi.org/10.1038/s41422-019-0165-4

Yang Q, Zhong X, Li Q, Lan J, Tang H, Qi P, Ma J, Wang J, Chen G, Pu Z, Li W, Lan X, Deng M, Harwood W, Li Z, Wei Y, Zheng Y, Jiang Q (2020) Mutation of the d-hordein gene by RNA-guided Cas9 targeted editing reducing the grain size and changing grain compositions in barley. Food Chem 311:125892. https://doi.org/10.1016/j.foodchem.2019.125892

Yin K, Qiu JL (2019) Genome editing for plant disease resistance: applications and perspectives. Biol Sci 374(1767):20180322. https://doi.org/10.1098/rstb.2018.0322

Yin K, Gao C, Qiu JL (2017) Progress and prospects in plant genome editing. Nat Plants 3:17107. https://doi.org/10.1038/nplants.2017.107

Yu Q, Wang B, Li N, Tang Y, Yang S, Yang T, Xu J, Guo YP, Wang Q, Asmutola P (2017) CRISPR/Cas9- induced targeted mutagenesis and gene replacement to generate long-shelf life tomato lines. Sci Rep 7:11874

Yunyan F, Jie Y, Fangquan W, Fangjun F, Wenqi L, Jun W, Yang X, Jinyan Z, Weigong Z (2019) Production of two elite glutinous rice varieties by editing Wx Gene. Ric Sci 26:118–124

Zafar SA, Zaidi SS, Gaba Y, Singla-Pareek SL, Dhankher OP, Li X, Mansoor SP (2020) A engineering abiotic stress tolerance via CRISPR/Cas-mediated genome editing. J Exp Bot 71(2):470–479. https://doi.org/10.1093/jxb/erz476

Zambre M, Terryn N, De CJ, De BS, Dillen W, Van MM, Van DSD, Angenon G (2003) Light strongly promotes gene transfer from *Agrobacterium tumefaciens* to plant cells. Planta 216(4): 580–586. https://doi.org/10.1007/s00425-002-0914-2

Zeng Z, Han N, Liu C, Buerte B, Zhou C, Chen J, Wang M, Zhang Y, Tang Y, Zhu M, Wang J, Yang Y, Bian H (2020) Functional dissection of HGGT and HPT in barley vitamin E biosynthesis via CRISPR/Cas9-enabled genome editing. Ann Bot 126(5):929–942. https://doi.org/10.1093/aob/mcaa115

Zhai Y, Yu K, Cai S, Hu L, Amoo O, Xu L, Yang Y, Ma B, Jiao C, Zhang C (2020) Targeted mutagenesis of BnTT8 homologs controls yellow seed coat development for effective oil production in *Brassica napus* L. Plant Biotechnol J 18:1153–1168

Zhang F (2019) Development of CRISPR-Cas systems for genome editing and beyond. Q Rev Biophys 52. https://doi.org/10.1017/S0033583519000052

Zhang Y, Yin H, Li D, Zhu W, Li Q (2008) Functional analysis of BADH gene promoter from *Suaeda liaotungensis* K. Plant Cell Rep 27(3):585–592. https://doi.org/10.1007/s00299-007-0459-8

Zhang XH, Tee LY, Wang XG, Huang QS, Yang SH (2015) Off-target effects in CRISPR/Cas9-mediated genome engineering. Mol Ther Nucl Acids 4(11):264. https://doi.org/10.1038/mtna.2015.37

Zhang Y, Liang Z, Zong Y, Wang Y, Liu J, Chen K, Qiu JL, Gao C (2016) Efficient and transgene-free genome editing in wheat through transient expression of CRISPR/Cas9 DNA or RNA. Nat Commun 7:12617

Zhang J, Zhang H, Botella JR, Zhu JK (2018a) Generation of new glutinous rice by CRISPR/Cas9-targeted mutagenesis of the Waxy gene in elite rice varieties. J Integr Plant Biol 60:369–375. https://doi.org/10.1111/jipb.12620

Zhang T, Zheng Q, Yi X, An H, Zhao Y, Ma S, Zhou G (2018b) Establishing RNA virus resistance in plants by harnessing CRISPR immune system. Plant Biotechnol J 16(8):1415–1423. https://doi.org/10.1111/pbi.12881

Zhang S, Li X, Lin Q, Wong KC (2019) Synergizing CRISPR/Cas9 off-target predictions for ensemble insights and practical applications. Bioinformatics 35:1108–1115. https://doi.org/10.1093/bioinformatics/bty748

Zhang F, Abudayyeh OO, Gootenberg JS (2020) A protocol for detection of COVID-19 using CRISPR diagnostics Broad Institute. Bioarchive

Zhang Y, Ren Q, Tang X, Liu S, Malzahn AA, Zhou J, Wang J, Yin D, Pan C, Yuan M, Huang L, Yang H, Zhao Y, Fang Q, Zheng X, Tian L, Cheng Y, Le Y, McCoy B, Franklin L, Selengut JD, Mount SM, Que Q, Zhang Y, Qi Y (2021) Expanding the scope of plant genome engineering with Cas12a orthologs and highly multiplexable editing systems. Nat Commun 12:1944. https://doi.org/10.1038/s41467-021-22330-w

Zhou H, Liu B, Weeks DP, Spalding MH, Yang B (2014) Large chromosomal deletions and heritable small genetic changes induced by CRISPR/Cas9 in rice. Nucleic Acids Res 42(17): 10903–10914. https://doi.org/10.1093/nar/gku806

Zhou XX, Zou X, Chung HK, Gao Y, Liu Y, Qi LS, Lin MZ (2018) A single-chain photoswitchable CRISPR-Cas9 architecture for light-inducible gene editing and transcription. ACS Chem Biol 13:443–448. https://doi.org/10.1021/acschembio.7b00603

Zhu LJ (2015) Overview of guide RNA design tools for CRISPR-Cas9 genome editing technology. Front Biol (Beijing) 10:289–296. https://doi.org/10.1007/s11515-015-1366-y

Zhu LJ, Holmes BR, Aronin N, Brodsky MH (2014) CRISPRseek: a bioconductor package to identify target-specific guide RNAs for CRISPR-Cas9 genome-editing systems. PLoS One 9(9): e108424. https://doi.org/10.1371/journal.pone.0108424

Zhu H, Misel L, Graham M, Robinson ML, Liang C (2016) CT-finder: a web service for CRISPR optimal target prediction and visualization. Sci Rep 6:1–8. https://doi.org/10.1038/srep25516

Zhu Y, Klompe SE, Vlot M, van der Oost J, Staals RHJ (2018) Shooting the messenger: RNA-targetting CRISPR-Cas systems. Biosci Rep 38(3):BSR20170788

Zhu Y, Gao A, Zhan Q, Wang Y, Feng H, Liu S, Gao G, Serganov A, Gao P (2019) Diverse mechanisms of CRISPR/Cas9 inhibition by type IIC anti-CRISPR proteins. Mol Cell 74:296–309

Zsögön A, Čermák T, Naves E et al (2018) De novo domestication of wild tomato using genome editing. Nat Biotechnol 36:1211–1121

Genome Editing for Stress Tolerance in Cereals: Methods, Opportunities, and Applications

10

H. M. Mamrutha, Kapil Deswal, Zeenat Wadhwa, Rajender Singh, Rakesh Kumar, Ratan Tiwari, and G. P. Singh

Abstract

Many advanced technologies were used along with conventional breeding to develop novel varieties, which increases the productivity of major cereal crops. Regardless of this progress, continuous increase in biotic and abiotic stresses imposes challenges for crop scientists to ensure the future food security to growing population. Recently, the availability of whole-genome sequence information and the advances in precise genome editing technology have revolutionized the crop breeding domain. The genome editing methods are becoming more accurate, simple, and highly efficient with time. The genome editing applications have been successfully proved in several cereals, viz., rice, wheat, maize, and barley, and produced various stress-tolerant crops. The current chapter compiles information on the advantages of using genome editing tools like zinc finger nucleases (ZFNs), transcription activator-like effector nucleases (TALENs), clustered regularly interspaced short palindromic repeats (CRISPR)/CRISPR-associated (Cas) (CRISPR/Cas9), and base editing and their application in cereals to enhance stress resilience. It also includes different steps involved in genome editing approaches in cereal crops. The emerging genome editing technologies can provide non-transgenic stress-resilient cultivars in less time to

H. M. Mamrutha (✉) · Z. Wadhwa · R. Singh · R. Tiwari · G. P. Singh
ICAR—Indian Institute of Wheat and Barley Research, Karnal, India
e-mail: Mamrutha.M@icar.gov.in

K. Deswal
ICAR—Indian Institute of Wheat and Barley Research, Karnal, India

CCSHAU, Hisar, India

R. Kumar
Tel AVIV University, Ramat AVIV, Tel AVIV, Israel

M. G. Mallikarjuna et al. (eds.), *Next-Generation Plant Breeding Approaches for Stress Resilience in Cereal Crops*, https://doi.org/10.1007/978-981-19-1445-4_10

cope with rapidly changing climatic conditions. Furthermore, the ethical and regulatory policies to produce new cultivars through genome editing approaches are updated with respect to the global and national context.

Keywords

Genome editing · Cereals · Abiotic stress · Biotic stress · Crops

10.1 Introduction

Changing environment and increasing human population are the two major concerns that raised questions of worldwide food security, which enforce the present improvement of important crops to meet future requirement. Domestication and natural breeding processes have taken more than 10,000 years to produce varieties from landraces. To meet the human needs and to adapt local environment, modern crop varieties have various better agronomic traits. However, it takes a long time and a lot of effort to improve present elite germplasm. On the other hand, linkage drag and the transmission of detrimental genetic material associated to favorable features make it difficult to introduce helpful traits into an elite variety. However, introgression breeding also involves numerous rounds of backcrossing and selection to reestablish the elite genotypic background, which takes a long time and is inconvenient too.

Therefore, the slow pace of improvement via traditional breeding is assumed to be due to longer generation duration, random nature of recombination, and undirected mutagenesis of crop plants. The emergence of advanced breeding tools such as genome editing brought about a paradigm shift in biological and agricultural research, providing plant breeders with an open opportunity in de novo domestication to produce genetic variation for breeding in a unique approach to reduce generation time and develop elite varieties. Indeed, for improving characteristics in crop plants, new plant breeding techniques (NPBTs) have developed as alternative to traditional breeding and transgenic methods. The genome editing technique allows for the modification of endogenous genes in crops to improve the target qualities without having to allocate transgene crossway species boundaries. In particular, clustered regularly interspaced short palindromic repeats (CRISPR)/CRISPR-associated (Cas) has emerged as the most popular technology for editing the genome of crops, with rapidly expanding agricultural applications in cereals such as rice, wheat, maize, and other food security crops. Cereals are staple food crops in our diet and provide ample primary sources of energy in the form of carbohydrates, minerals, fibers, niacin, riboflavin, thiamine, etc. Hence, cereals have huge significance for worldwide food security. Given its prominence, genome editing methods are commonly used in the genetic improvement of cereal crops to produce elite cultivars that are resistant to stress.

This chapter compiles the information on different genome editing tools—ZFN, TALENs, and CRISPR/Cas9—and also briefly sums up recent advancements in genome editing, i.e., base editing, that have revolutionized the crop improvement program that allows effective and specific gene editing to single base level. Furthermore, various steps in genome editing are described, as well as current applications of genome editing in cereals, with a focus on its prospective for genetic enhancement of crops in terms of abiotic and biotic stress. Additionally, various aspects of challenges and opportunities in cereals and regulatory issues related to genome-edited crops are also discussed.

10.2 Types of Genome Editing Tools

ZFNs, TALENs, and CRISPR-cas9 are the keystones of gene editing tools, which are theoretically well-defined as deliberate alteration of gene sequences by means of molecular scissors by opening the novel way of targeted genome editing (Fig. 10.1). The introduction of double-stranded break (DSBs)at the target regions is the typical feature of these genome editing tools. Endogenous DNA repair mechanism such as non-homologous end joining (NHEJ) or by homology directed repair (HDR) (Gallagher and Haber 2018; Sander and Joung 2014) repairs these DSBs, resulting in DNA alteration such as insertion or deletions (indels) at the DSB sites. Indel frequency, on the other hand, has been used to assess the complete activity and preciseness (off-target) of genome editing tools.

10.2.1 Zinc Finger Nucleases (ZFNs)

The first genome editing tool utilized programmable nucleases, zinc finger nucleases (ZFNs), resulting in a breakthrough in genome engineering (Chandrasegaran and Carroll 2016). By taking the advantage of endogenous DNA repair mechanism, the reagents of the DNA repair mechanism can be utilized to accurately modify the genomes of higher species that lead to both targeted mutagenesis and gene replacement remarkably at higher frequency. ZFNs are the targetable DNA cleavage reagent made by fusing the DNA-binding zinc finger protein (ZFP) domain at the amino terminus with the *Fok I* nuclease cleavage domain at the carboxyl terminus, resulting in a target-specific desired sequence. ZF domain comprised of eukaryotic transcription factor and tandem array of Cys2His2 zinc finger in each unit of approximately 30 amino acids bound to a single atom of zinc that each recognizes 3 bp of DNA (Wolfe et al. 2000). To dimerize and cleave DNA, standard ZFNs merge the cleavage domain at the C terminus of every single zinc finger domain, and then two distinct ZFNs must bind opposite strands of DNA with their C termini at a particular distance away from each other. Zinc finger nucleases act as heterodimer because *Fok I* must dimerize to cut target DNA sequence (Bitinaite et al. 1998). However, monomeric is not active, so cleavage does not occur at single binding sites. Cytotoxicity will result from poor targeting and a high number of off-target effects. Thus, the usage of ZFNs is limited as compared to other programmable nucleases.

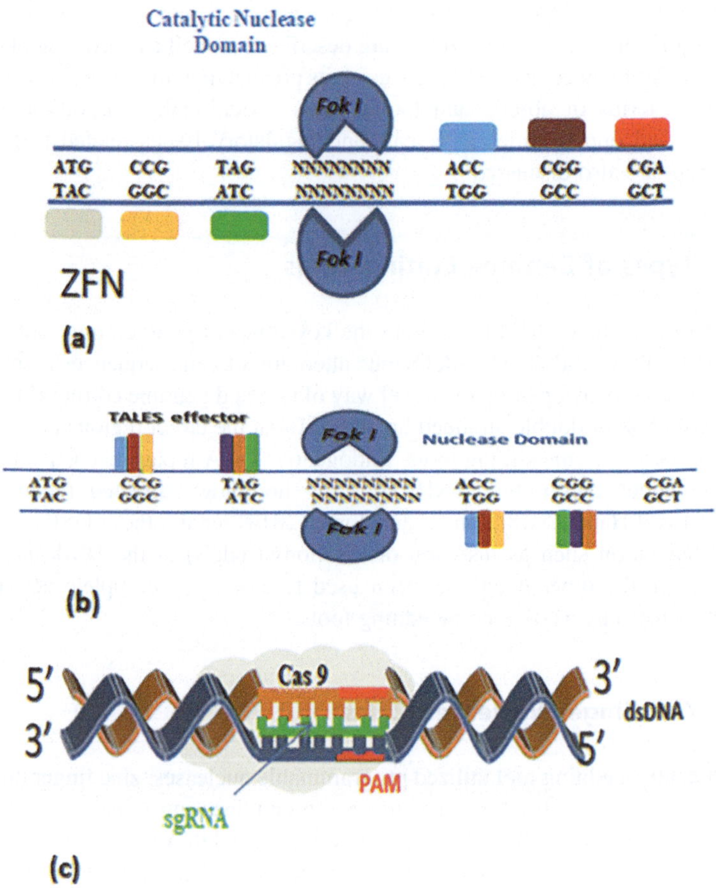

Fig. 10.1 Tools of genome editing: (**a**) zinc finger nucleases (ZFNs), (**b**) transcription activator-like effector nucleases (TALENs), (**c**) clustered regularly interspaced short palindromic repeats (CRISPR)/CRISPR-associated (Cas)

10.2.2 Transcription Activator-Like Effector Nucleases (TALENs)

The next uprising in gene editing history is transcription activator-like effector nucleases (TALENs), generated by the fusion of DNA-binding domain protein called TALEs and deduced from transcription activators like effectors of *Xanthomonas* (Miller et al. 2011) to DNA cleavage *Fok* I nuclease domain. The TALEN DNA-binding domain is defined by order and number of four repeated domains, which has extremely preserved 33–34 amino acid (aa) repetitive domain with divergent 12th and 13th aa, well known as repeat variable di-residue (RVD): NG, HD, NI, and NN/NH/NK to mark T, C, A, G nucleotide, respectively. Selecting a combination of repeat segments including RVDs makes engineering a specific binding domain simple. Like ZFNs, *Fok I* function as a heterodimer with unique

DNA-binding domains for locations in the marked genome that are properly oriented and spaced. TALENs have a far more rigorous protein DNA coding for targeting, and it can also recognize a single base rather than triplet, giving it more versatility than ZFNs. Scientists all across the world are interested in TALENs because of its apparent advantages, such as greater precision and cleavage efficiency when introducing mutations over ZFN. However, as compared to ZFNs, the use of TALENs is limited due to higher amount of the encoding cDNA (3 kb).

10.2.3 Clustered Regularly Interspaced Short Palindromic Repeats/CRISPR-Associated (CRISPR/Cas9)

In 2012, French and American scientists Emmanuelle Charpentier and Jennifer Doudna discovered CRISPR/Cas9, a third-generation genome editing tool. Due to its apparent benefits like more precision and cleavage efficiency to introduce mutation over ZFN and TALENs, CRISPR/Cas9 shows great attention of scientists around the world. The CRISPR/Cas9 system is a straightforward tool for site-specific mutagenesis, where genes can be knocked out or precisely altered by harnessing the different repair mechanisms. In bacteria and archea, CRISPR/Cas system is an RNA-mediated acquired immune response. It comprises CRISPR spacer arrays and Cas protein, which is naturally evolved to provide defense against phages. On the basis of Cas genes and interference complex, CRISPR/Cas system has been divided into two classes which have been further subdivided into six types. For interference, Class 1 CRISPR/Cas systems (types I, III, and IV) use multi-Cas protein complexes, whereas Class 2 systems (types II, V, and VI) use single effector proteins in complex with CRISPR RNAs (crRNAs) (Koonin et al. 2017).The *Streptococcus pyogenes* type II CRISPR/Cas9 was the first one to accurately break DNA in eukaryotic cells. It has two main components, namely, Cas9 nuclease and non-coding single guide RNA (sgRNA). Cas9 is a dual RNA-guided DNA endonuclease with 5′-NGG-3′ sequence as PAM (protospacer-adjacent motif) following the gRNA with 20 bp as target that creates blunt ends 3 nt upstream of the protospacer-adjacent motif. The Cas9 consists of HNH nuclease and RuV C domain, each cleaving one strand of the target. The other component is two short non-coding RNA that comprises of crRNA, which is composed of 20 nt target-specific sequence that establishes the uniqueness of this system and a *trans* activating crRNA (tracrRNA), which interacts with crRNA to mediate endonuclease activity of the CRISPR/Cas9 complex (Wiedenheft et al. 2012). When all of the components are carried to a target cell, three base-pair NGG (PAM) on the target DNA strand direct the Cas9 endonuclease to cut 3 bp upstream to PAM sequence (Jinek et al. 2013). As a result, the Cas9-gRNA complex images the protospacer-adjacent motif region and generates complementary base pairing with 20 nucleotides of the target DNA. This configuration allows the endonuclease to cut site-specific target DNA. Lastly, the cell repair DSB's internal DNA repair mechanism makes the appropriate alterations. Schunder et al. (2013) discovered another kind of Cas Cpf1 (also known as Cas12) in *Francisella* spp., which is a CRISPR type V endonuclease that identifies and cleaves protospacer-adjacent motif 50-TTN, which will be more prevalent in the genome that cleaves the target DNA, creating 5 nt 50 overhang 18–23 bases away from protospacer-adjacent motif. There are many Cas9 variants developed, i.e., nickase Cas9 (nCas9) and dead Cas9 (dCas9), to overcome the limitations of Cas9, especially with respect to off-target mutations and indel formation (Certo et al. 2011; Brookhouser et al. 2017). As nCas9 produces single-stranded binding (SSBs), a pair of nCas9 can be used to produce paired nicks in its place of DSB, reducing off-target cleavage, whereas dCas9 functions as a site-specific DNA-binding vehicle that can combine with other effectors to modify target DNA sites with higher specificity and efficacy than nCas9 (Guilinger et al. 2014). In addition to mutagenesis, CRISPR/Cas9 can be used to repress or induce gene expression by combining repressor or transcriptional activator with a catalytically inactive Cas9 (dCas9) (Bortesi and Fischer 2015). As a result, it has the potential to replace standard traditional methods of gene overexpression and silencing. These advances have greatly aided to the broader adaptability of this technique among the eukaryotes.

10.2.4 Base Editing

To bypass the limitation of CRISPR/Cas, a modern evolution of a single base pair editing system has been devised based on CRISPR/Cas-based technologies. The base editing system directly creates point mutation in targeted DNA without inducing DSB. Furthermore, compared to non-DSB-mediated genome editing in plants, base editing approach improves gene modification efficiency by lowering off-target and random mutations in the DNA, multiplex, or whole-gene editing. This approach enables the programmed conversion of single bases into another (e.g., A/T to G/C, C/G to T/A) and allows four transitions. Aside from base editing (BE), prime editing (PE) allows for non-double-stranded break and template-free random sequence addition, removal, or nucleotide replacement. On the other hand, PE was created to allow base-to-base transitions, which facilitates targeted deletion and insertion.

Base editors are chimeric complexes that contain catalytically inactive CRISPR/Cas domain and cytosine or adenosine domain that creates desirable point alterations in the target region, allowing for precision genome editing. Cytosine base editors (CBEs) and adenine base editors (ABEs) are the two main types of DNA base editors that have been described (Fig. 10.2a, b). The two essential components of DNA base editors are a Cas enzyme for customizable DNA binding and a ssDNA-modifying enzyme for selective nucleotide alteration. The CBE systems consist of cytidine deaminase coupled to nCas9 and a uracil glycosylase inhibitor that converts targeted cytosine to uracil in genomic DNA (Komor et al. 2016). Cytosine (C) in DNA is converted to uracil (U), and subsequently U is replaced by T during DNA replication using cytidine deaminase. Through this process, uracil glycosylase inhibitors attach to and inhibit uracil DNA glycosylase, thus blocking uridine excision, resulting in the base excision repair pathway and enhanced base editing efficacy. The CBE system consists of a uracil glycosylase inhibitor and cytidine deaminase fused with

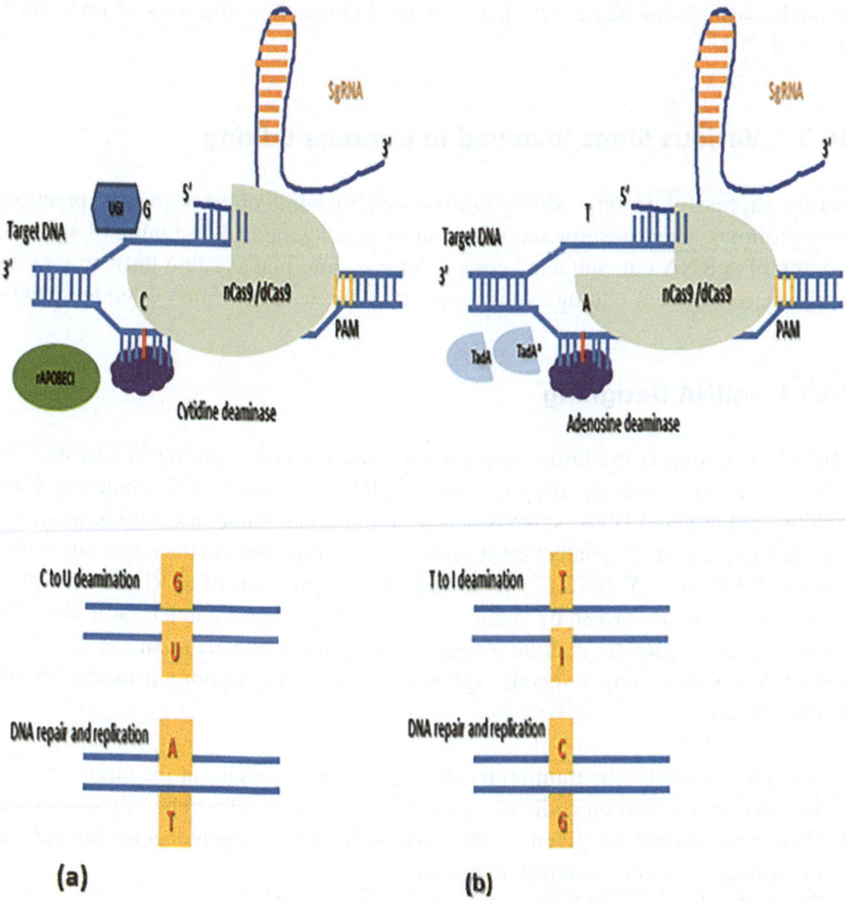

Fig. 10.2 Mechanism of DNA-based base editors: (**a**) cytosine base editing mechanism, conversion of C to T, (**b**) adenine base editing mechanism, conversion of A to G

Cas9, and changes targeted C to U in genomic DNA. The human APOBEC3A-based plant CBE has been employed in rice, wheat, and potato to efficiently convert Cs to Ts (Li et al. 2018; Zong et al. 2018).

Liu's group later produced ABEs, which used to facilitate the alteration of A to G in genomic DNA. In contrast to CBEs, ABEs do not require DNA glycosylase inhibitors. A deoxyadenosine deaminase (TadA*) and TadA-TadA* heterodimer was produced from modified *E. coli* transfer RNA adenosine deaminase (TadA) and linked with nCas9 (D10A) (Gaudelli et al. 2017). With excellent efficiency and product purity, the seventh-generation ABEs (7.10) were employed to convert A to G in extensive range of targets (Gaudelli et al. 2017). Rice and wheat ABE systems have also been optimized. In rice and wheat, the practice of improved sgRNAs [sgRNA(F + E)] in combination with three replicas of nuclear localization sequences at the C terminus of nCas9 resulted in A to G conversion efficacies of up to 60% (Li et al. 2018).

10.3 Various Steps Involved in Genome Editing

For the successful genome editing in plant system, some of the sequential practices are as follows: identification and selection of target gene and designing of sgRNA, cloning of sgRNA into suitable vectors, delivery into plant system through various methods, selection of editing events in plants, and characterization of edited plants.

10.3.1 sgRNA Designing

sgRNA designing is the initial step for successful genome editing in cereals. The editing capability mainly depends upon sgRNA structure, GC contents, Cas9 codons, and targeted DNA. sgRNA acts as a functional guide in CRISPR-mediated editing and contains 20 nucleotide complementary sequences to the target site with a specific PAM site (5′-NGG-3′) at 3′ end. The expression of sgRNAs into plant system is commonly driven by small nuclear RNA gene promoters (U3 or U6). The transcription of sgRNAs is done through RNA polymerase III (Jiang et al. 2013). Before designing of target-specific sgRNA, the following important factors should be considered:

1. sgRNA size should be minimum (18–21 nucleotide length) at the target site.
2. Identify most common coding sequence of all isomers of a gene in the genome.
3. Preference should be given to the first exon of the targeted gene for sgRNA designing for loss of function mutation.
4. There should be a PAM site in the end of a target site.
5. There should be suitable restriction enzyme (RE) sites at each side of sgRNA for cloning work.
6. Designed sgRNAs should exhibit minimum off-target effects in the targeted organism.

The sgRNA designing can be done by using an online bioinformatics tool that allows identifying new target sites (Stemmer et al. 2015). Various online tools are now available with plant databases that will enable the identification of new target sites for sgRNA designing (Stemmer et al. 2015). The following were mainly used for many cereal crops: wheatCRISPR, CasOT, E-CRISP, biotools, Cas-OFFinder, CRISPRdirect, etc. In addition, a CRISPR Design tool was developed by Zhang and colleagues (http: //www.genome-engineering.org) and another by Xie and co-workers in 2014, CRISPR-PLANT, to get efficient sgRNA constructs which are used in genome editing events (Xie et al. 2014). Similarly, some novel web tools were developed for designing sgRNA for every plant whose genome sequence is available (Lei et al. 2014).

Fig. 10.3 Flow diagram depicting the different steps involved in sgRNA designing

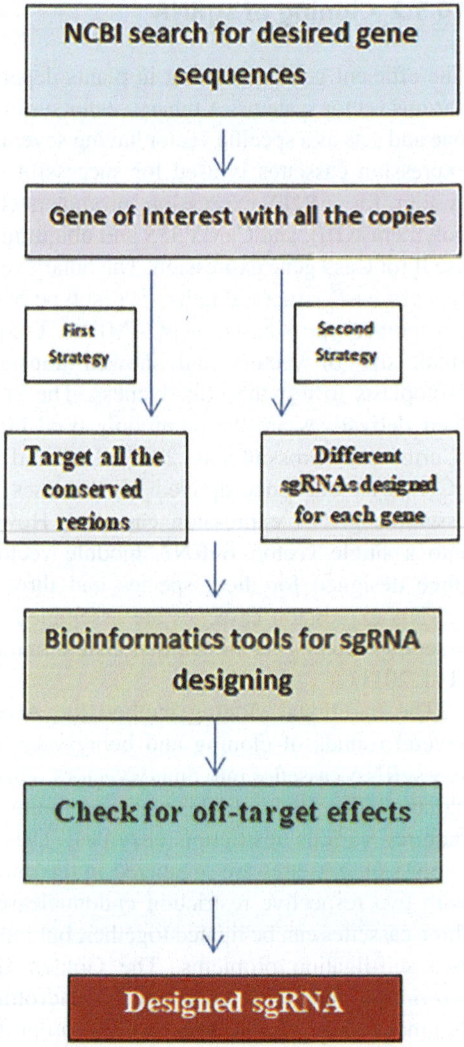

The designing criteria for efficient sgRNA in plant systems are as follows:

(a) Designed sgRNA should show G/C content range between 30% and 80%.
(b) Designed sgRNA should contain intact secondary structures except for stem-loop 1.
(c) sgRNA contains not more than 12 total base pairs and no more than 7 consecutive base pairs between guide sequence and the other sequence.
(d) Not more than six internal base pairs (IBPs) (http://www.genome-engineering.org/). The complete process of sgRNA designing is shown in Fig. 10.3.

10.3.2 Cloning of sgRNA

The efficient genome editing in plants depends on the cloning of sgRNA through various vector systems. A binary vector system that utilizes features of two vectors in one and acts as a specific vector having several sgRNAs and cas9 proteins along with expression cassettes is used for successful genome editing events. In this vector system, for sgRNA expression, promoters (U6/U3) are designed (driven by RNA polymerase III), and CaMV35S and ubiquitin promoters are used by RNA polymerase II for Cas9 gene expression. The binary vectors utilized in cloning work used two types of basic structural units: (i) first type of structural unit is based on pGreen, and (ii) another type is based on pCAMBIA. The pGreen vectors were used owing to the small size of vectors and showed transient Cas9 and sgRNA expression in protoplasts to test the effectiveness. The vectors pCAMBIA1300/2300/3300 and their derivatives are the commonly used binary vectors for various plant species (Curtis and Grossniklaus 2003; Lee and Gelvin 2008). Improvement in the pCAMBIA backbone-derived vectors uses the BsaI site in the pVS1 region to assemble gRNA expression cassettes. However, for multiple sgRNA insertions into a single vector, 6gRNA module vectors are constructed, which consist of three designed for dicot species and three designed for monocot plants. More sgRNA expression cassettes are assembled into one vector either through Gibson assembly cloning or the Golden Gate cloning methods (Engler et al. 2008; Weber et al. 2011).

The traditional cloning method has several disadvantages, like the need for several rounds of cloning and being very time-consuming for the expression of few sgRNA cassettes into binary vectors. This cloning method is also called "regular cloning" (Fig. 10.4a). In this method, cloning of several sgRNA expression cassettes requires various restriction enzymes. The U3/U6 promoter (Pr)-driven sgRNA expression cassettes are organized in the middle vector and recovered by digestion with two respective restriction endonucleases. Generally in a binary vector, only three cassettes can be ligated together, but more than three fragments cause competitive self-ligation problems. The Golden Gate cloning method uses restriction enzymes (type II) to create non-palindromic sticky ends among multiple DNA fragments (Fig. 10.4b). This method can proficiently ligate several DNA fragments in a particular procedure (Engler et al. 2008). In the same method, two sets of vector systems have been developed in a single round of cloning to make CRISPR/Cas9 binary constructs with the help of PCR-amplified sgRNA expression cassettes (Ma et al. 2015; Xing et al. 2014). The expression cassettes are digested with a restriction endonuclease (RE type II)(BsaI) to create cohesive ends and joined all together to a binary vector for cloning (Ma et al. 2015). The Gibson cloning can capably join multiple DNA fragments.

The Gibson assembly technique can join several DNA fragments with homologous termini using the collective work of the Taq DNA ligase,T5 exonuclease, and the Phusion (DNA) polymerase (Gibson et al. 2009). The sgRNA expression cassettes ready in vitro by PCR and ligated to a binary vector shown (Fig. 10.4c) in the Gibson assembly method. Another strategy, the PTG which is polycistronic

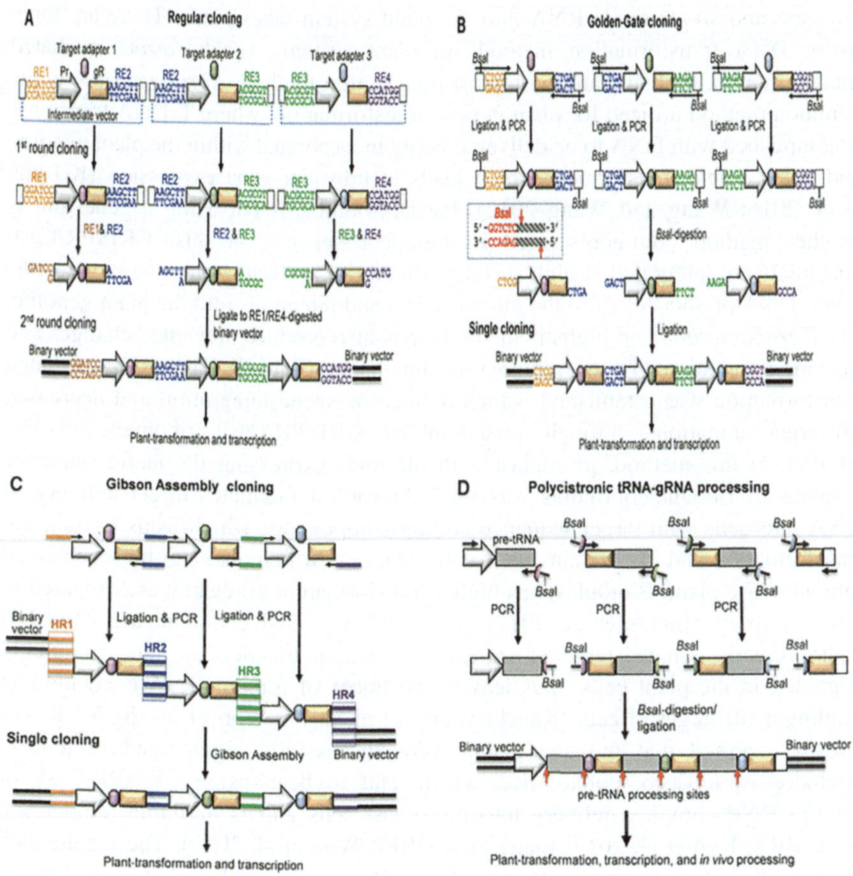

Fig. 10.4 Different methods of sgRNA cloning into a binary vector. (**a**) Regular cloning. (**b**) Golden Gate cloning. (**c**) Gibson assembly cloning. (**d**) Polycistronic tRNA-gRNA cloning. Adopted from Ma et al. (2016)

tRNA-gRNA system has been utilized by flanking the sgRNAs with a tRNA precursor sequence, and multiple sgRNAs are generated with different target sequences (Fig. 10.4d). The multiple sgRNAs with U3/U6 promoter is linked to pre-tRNA/sgRNA scaffolds using Golden Gate ligation.

10.3.3 Transformation into Plant System

Genetic transformation and regeneration processes are the major steps of gene editing. CRISPR/Cas9-mediated genome editing requires effective delivery of editing reagents, including sgRNAs and Cas9 nucleases that perform the actual targeted genome modification in most plant cells. The CRISPR/Cas system is able to make a cut in both strands of the DNA when the efficient transformation of Cas9 nuclease and single guide RNA into the plant system takes place. There are three major DNA transformation methods in plant system: *Agrobacterium*-mediated transformation, biolistic, and protoplast transfection method. *Agrobacterium* is the common method utilized for plant genetic transformation, where T-DNA transfer is accomplished with DNA to be delivered being incorporated within the plant genome and being stable transformed, which leads to transient gene expression (Krishna et al. 2016; Wang and Wang 2012). Particle bombardment using a gene gun is another method commonly used in monocot species. In this, CRISPR/Cas9 constructs are integrated at high speeds with carrier molecules into the target cells. Later, DNA dissociates from the microcarriers and integrates into the plant genome. The *Agrobacterium* and biolistic methods may also produce unwanted changes and off-target mutation. To overcome this drawback, CRISPR/Cas9 RNP-mediated transformation was established, which avoids transgene integration and decreased off-target mutations through preassembled CRISPR/Cas9 ribonucleoproteins (RNPs). In this method, protoplast with plasmids expressing the target sequence reagents or ribonucleoproteins was used. Protoplast facilitates direct delivery of DNA into cells with target sequence editing components, which leads to transient transformation and also retains their cell identities, which also can be regenerated into an entire plant. Protoplast has higher transformation efficiency as compared to other methods (Baltes et al. 2017). These RNA-guided endonuclease (RGENs) RNPs directly edit the targeted sequences just after transfection and are quickly degraded in the plant cells, thus leaving no traces of foreign DNA elements and minimum off-target effects (Kanchiswamy et al. 2017; Woo et al. 2015). It was already reported that mutated plants were successfully regenerated in lettuce, Arabidopsis, tobacco, potato, rice, wheat, and soybean using CRISPR/Cas9 or Cas12a RNP complex delivery into protoplast cells and is heritable (Andersson et al. 2018; Kim et al. 2017; Liang et al. 2017; Woo et al. 2015). The regenerated plants would likely be exempted from the regulatory process because of no integration of any foreign DNA into the targeted plants (Clasen et al. 2016; Haun et al. 2014). Another simplest and frequently used method in plant transformation is the floral dip using agroinfection, which replaces the lengthy tissue culture procedures (Zlobin et al. 2020).

10.3.4 Characterization of Edited Plants

The main practices used for screening CRISPR/Cas system-induced mutants include qPCR assay, surveyor nuclease (T7EI assays), high-resolution melting analysis (HRMA)-based assay, high-throughput tracking of mutations (Hi-TOM), and whole-genome sequencing (WGS) to detect it in targeted sequences. The qPCR is used to identify mutated DNA sequences by amplifying the locus and sequencing of PCR products, which is mainly used to differentiate between heterozygous and homozygous mutations. This method is widely used as it is a highly effective, rapid, and simple method to detect induced mutations and already validated in Arabidopsis, sorghum, maize, and rice (Peng et al. 2017).

The T7EI assays or surveyor nuclease are widely used and considered suitable for any target sequence. The CEL family of mismatch-specific nucleases includes surveyor nuclease. These nucleases recognize mismatches and create cut in heteroduplex DNA sequences. It targets mismatch sequences by cleaving both DNA and identifies mutations of up to 12 nt (Qiu et al. 2004). However, detection sensitivity of T7EI assays method is considerably less than PCR assays and are more labor- and time-consuming and cleave several double-stranded DNA molecules if their structure is curved and bend (Cong et al. 2013; Declais et al. 2006).

The high-resolution melting analysis (HRMA) technique entails DNA sequence amplification by qPCR covering (about 90–200 bp) genomic target, with fluorescent dye after that amplicon melt curve analysis (Wang et al. 2015). The nondestructive nature of this method requires less than 2 h for the whole procedure, from genomic DNA preparation to mutation detection. After that, amplicons could be studied using sequencing and gel electrophoresis. The advantage of this method is that they are a simple and more delicate technique and has a high-throughput screening format. However, this assay has some limitations: assay is not able to identify larger mutations and the cost is also higher for operating the assay. However, higher cost of this method can be decreased by coupling HRMA with online HRMA software (Talbot and Amacher 2014). The Hi-TOM assay is an online tool that can be used for quantitative and precise mutation detection induced by the CRISPR/Cas9 system, without any extra-complex parameter configuration and data analysis. The advantage of assay is it is easy and user-friendly, and does not require a skilled person for bioinformatics tools or next-generation sequencing (NGS). The Hi-TOM online tool has suitable high-throughput detection methodology for CRISPR system-induced mutations because of its convenience to use (Liu et al. 2018). A non-denaturing PAGE gel is also used for detecting the genome editing-induced mutations in the target genomes. In this method, variations in the single-stranded sequences can alter their conformations and show different migration rates in PAGE gel, and it is also called "single-strand conformation polymorphism" (SSCP) (Zheng et al. 2016). Sanger sequencing can be used to study the amplicon containing targeted sequences. The Sanger sequencing approach is suitable for identifying mutations; however, the real problem is that it is expensive and tedious.

10.4 Application of Genome Editing in Cereals for Abiotic and Biotic Stress Tolerance

The major losses in crop production are caused more by abiotic stresses, for instance, drought, salinity, and high temperature than by biotic stresses. The genome editing tools can contribute significantly toward creating novel plant types having tolerance to abiotic and biotic stresses in cereal crops as summarized in Table 10.1.

Table 10.1 Promising genes edited for abiotic/biotic stress tolerance in cereals

Crop	Gene	Stress	Reference
Abiotic stress			
Oryza sativa	*OsDST*	Drought and salt tolerance	Kumar et al. (2020)
Zea mays	*ARGOS8*	Drought tolerance	Shi et al. (2017)
Triticum aestivum	*TaDREB2, TaERF3*	Drought tolerance	Kim et al. (2018)
Triticum aestivum	*TaCer9*	Drought tolerance	Liang et al. (2018)
Oryza sativa	*OsNramp5*	Salinity tolerance	Tang et al. (2017)
Oryza sativa	*OsRR22*	Salinity tolerance	Zhang et al. (2019)
Zea mays	*ZmCLCg*	Salinity tolerance	Luo et al. (2021)
Biotic stress			
Triticum aestivum	*TaLpx-1, TaMLO*	*Fusarium graminearum* and powdery mildew	Wang et al. (2018)
Oryza sativa	*OsSEC3A*	Bacterial blast	Ma et al. (2017)
Triticum aestivum	*TaMLO-A1*	Resistance to powdery mildew	Wang et al. (2014)
Triticum aestivum	*TaLox2*	*Fusarium head blight*	Shan et al. (2013)
Oryza sativa	*OsERF922*	Blast resistance	Wang et al. (2016)
Oryza sativa	*eIF4G*	Rice tungro spherical virus resistance	Macovei et al. (2018)
Oryza sativa	*SWEET11, SWEET13, SWEET14*	Bacterial blight resistance	Oliva et al. (2019)
Triticum aestivum	*TaABCC6, TansLTP9.4, TaNFXL1*	Fusarium head blight (FHB) resistance	Cui et al. (2019)
Hordeum vulgare	*HvMORC1*	Resistant to powdery mildew and *Fusarium graminearum*	Kumar et al. (2018)
Triticum aestivum	*TaEDR1*	Powdery mildew resistance	Zhang et al. (2017)

10.4.1 Drought Stress

In the future, increase in frequency and severity in drought stress are expected in many regions. Hence, there is a need to develop genotypes tolerant to decreased precipitation and increased evaporation. The rice *OsDST* gene encodes a zinc finger transcription factor, and the loss of DST protein function in rice resulted in drought and salt tolerance (Huang et al. 2009). The mutation in the gene enhanced the leaf width, reduced stomatal density, and enhanced stomatal closure through modulation of H_2O_2 homeostasis (Huang et al. 2009). TheCRISPR/Cas9-induced mutation in the DST gene of indica rice enhanced the leaf water retention under dehydration stress. In the seedling stage, the Cas9-free DST mutant showed moderate resistance to osmotic stress and excellent tolerance to salt stress (Kumar et al. 2020). Under drought conditions, targeted editing of two abiotic stress-responsive transcription factors in wheat, dehydration response element-binding protein 2 (TaDREB2), and ethylene responsive factor 3 (TaERF3) revealed increased expression of both genes in seedlings (Kim et al. 2018). The precise genomic DNA modification of a negative regulator of ethylene responses, *ARGOS8* (auxin-regulated gene involved in organ size), *in* maize through CRISPR/Cas technology generated novel variants of *ARGOS8*, showing increased grain yield under drought stress at flowering stage (Shi et al. 2017).

10.4.2 Salt Stress

To improve salt tolerance in rice, CRISPR/Cas9 system has been used for target site genome editing. The salinity tolerance of T_2 homozygous mutant lines of *OsRR22*gene was engineered through a Cas9-OsRR22-gRNA expressing vector significantly enhanced as compared to wild-type plants (Zhang et al. 2019). CRISPR/Cas9 editing system was used to construct mutation alleles in drought and salt tolerance (DST) gene in indica rice cv. MTU1010. In the seedling stage, the Cas9-free DST mutant showed a high amount of salt stress (Kumar et al. 2020). Maize yield and quality is significantly affected by salt stress, and the genome-wide association analysis identified several QTLs for maize seedling salt tolerance. The functional validation of the candidate gene *ZmCLCg* in salt tolerance was done by generating gene knockout mutations through CRISPR/Cas9 technology. Under 100 mm NaCl treatment, three *ZmCLCg* mutants demonstrated higher decrease in root length, root fresh weight, shoot length, and shoot fresh weight than that of the wild type (Luo et al. 2021).

10.4.3 Biotic Stress

A major fungal disease powdery mildew is affected by *Blumeria graminis* f. sp. *tritici* in wheat and causes considerable yield losses. In bread wheat, different genome editing tools are utilized to introduce targeted mutations in the three homoeoalleles that encode mildew resistance locus (MLO) to provide powdery mildew resistance, a characteristic that is not seen in natural populations (Wang et al. 2014). The enhanced disease resistance1 (EDR1), a Raf-like mitogen-activated protein kinase kinase kinase (MAPKKK), reported as a negative regulator of powdery mildew resistance in *Arabidopsis thaliana* (Frye et al. 2001). The single guide RNA (T-EDR1) targeting a highly conserved region in the fourth exon was used in CRISPR/Cas9 technology to generate frameshift mutations for three homologs of wheat *EDR1*. The edited plants were resistant to powdery mildew and did not observe mildew-induced cell death (Zhang et al. 2017). The mutations were introduced through CRISPR/Cas9-mediated genome editing in three sucrose transporter genes, namely, *SWEET11*, *SWEET13*, and *SWEET14* of rice. The mutations in the promoter regions of these genes resulted in broad-spectrum resistance to bacterial blight caused by *Xanthomonas oryzae* pv. *oryzae* (Oliva et al. 2019). The confined screenhouse experiments indicated that most of the lines performed similarly to wild type for key agronomic characters such as plant height, panicle length, number of reproductive tillers, and fertility rate. Resistance to rice tungro spherical virus was acquired by a mutation in the eukaryotic translation initiation factor 4 gamma gene (eIF4G) in indica rice cv.IR64, a commonly planted variety across tropical Asia (Macovei et al. 2018).

The ethylene responsive factors are associated in imparting stress tolerance to abiotic and biotic stresses. The rice ethylene-responsive factor gene *OsERF922* is a negative regulator of blast disease produced by *Magnaporthe oryzae* in rice (Liu et al. 2012). The CRISPR/Cas9-targeted knockout of the ERF transcription factor gene *OsERF922* in a *japonica* rice variety improved the blast resistance. The mutant lines showed a significant reduction in blast lesion after pathogen infection. However, there were no significant variations in agronomic attributes between mutant lines and natural plants. Multiple sites within the *OsERF922* gene were also targeted to induce two or more mutations using Cas9/multi-target sgRNAs (Wang et al. 2016). Microrchidia (MORC) proteins act as negative regulators of immunity in barley. The barley genome contains seven MORC genes. *Hv*MORC1 was knocked out by *Streptococcus pyogenes* Cas9 (*Sp*Cas9) to generate loss-of-function alleles by targeting upstream of the ATPase domain of the gene. The hvmorc1-KO mutants showed increased resistance to fungal pathogens *Blumeria graminis* f. sp. *Hordei*, which bases powdery mildew and mycotoxin-producing fungus *Fusarium graminearum* (Kumar et al. 2018).

10.5 Challenges and Opportunities of Genome Editing in Cereals

The focus of genome editing is refining crops through improving yield and its associated traits. However, despite the significant attainments in crop improvement, there are few key obstacles that want to be focused while using genome editing technologies, such as polyploidy, off-target mutations, delivery methods, etc.

10.5.1 Polyploidy

Polyploidy is one of the biggest challenges to achieve the desired mutation, owing to additional complete chromosome sets within an organism. Due to the dosage impact of paralogous gene copies, this may not result in phenotypic alterations, especially during gene knockdown or knockout. However, occasionally desired trait alteration necessitates editing of all paralogs, which significantly reduces efficacy. In many polyploid crops such as sugarcane, wheat, etc., polyploidy is the major issue, as the targeted locus is present in many copies. Therefore, it is more exciting for GE tools to achieve homozygous plant that contains multiple target loci in polyploid crops. The key success for editing the multigene family members is to avoid the conserved sequence and select a unique gene sequence. Another major problem in polyploid crops is to screen a large number of edited plant population. To overcome the current problem, high-throughput phenotyping facility development is considered as a potential strategy. But successful implementation of this strategy again requires special skills and high-tech instrumentation. So far, the CRISPR/Cas-mediated homology direct repair (HDR) has been reported in many plant species such as Arabidopsis, tomato, rice, and maize (Li et al. 2021). Furthermore, the prime plant editors were generated in wheat protoplasts to attain specific point mutations at seven exogenous gene targets with single-nucleotide substitutions; the frequency observed was 1.4% (Lin et al. 2020). This indicates that more efficient plant prime editors require further improvement, mainly in polyploid species (Li et al. 2020).

10.5.2 Transformation Methods

In CRISPR/Cas9, another greatest challenge is the effective delivery method to the proper tissue and subsequent regeneration or expression of viable plants. The main difficulties during transformation are the time-consuming process, lower transformation frequency, lower titer of DNA, less precision, many crops recalcitrant to regeneration, and random somatic mutations (Gao 2018).This challenge creates an urgent need to improve the delivery system to obtain high efficiency of genome editing by using regeneration, use of booster to enable tissue culture in recalcitrant crops, or direct delivery to apical meristem to get edited plants without tissue culture. With progression in ribonucleoproteins (RNPs), viral delivery and nanoparticle systems offer other transformation methods for a tissue culture-free GE system. These technologies not only enhance the efficiency of genome editing but also decrease the regeneration period of edited plants. This prime obstacle offers an opportunity to augment plant transformation and regeneration responses by targeting an extensive range of tissues and genotypes.

10.5.3 Off-Target Effect

Another major concern is an off-target effect that impedes the potential application of the CRISPR/Cas9 system, where Cas9 cleaves genomic DNA sites that are imperfect complements of sgRNA. RGEN (RNA-guided endonuclease)-induced mutations has high frequency in off-target activity ($\geq$50%) at sites other than the intended on-target site. Unwanted cleavage and undesired chromosomal rearrangements due to higher number of off-targets can induce cellular toxicity. Inversion, translocations, and deletions triggered by the repair of these off-target DSBs can be damaging to plants. Cas9's sensivity is related to the 20 nucleotide sgRNA guide sequences and the PAM sequence. Many studies have found off-target DNA cleavage in sgRNA sequences with 1–5 bp mismatches. It has also been suggested the PAM sequence involved in the binding of Cas9. At the same time, 3′ end is essential for target identification, R-loop formation, and nuclease activation in Cas9 (Sander and Joung 2014). Potential off-target effects of the CRISPR/Cas system in the target sequence can be overcome by truncated Cas9 (Ran et al. 2013). An alternative approach is to use truncated gRNAs to minimize off-site targeting of the CRISPR/Cas system (Fu et al. 2014; Pattanayak et al. 2013). Further, ribonucleoprotein (RNP)-mediated genome editing has helped in the reduction of the off-target effects in wheat and maize (Liang et al. 2017; Svitashev et al. 2016). In silico prediction, HTGTS (high-throughput genome-wide translocation sequencing), ChIP-seq (chromatin immunoprecipitation), T7E1 (T7 endonuclease1) assay, fluorescence in situ hybridization, deep sequencing tools, etc. have been reported to examine off-target events. Off-target detection approaches like GUIDE-seq and Digenome-seq have been advanced with 0.1% sensitivity. The composition and structure of guide RNAs are the primary determinants of on-target and off-target cleavages. The off-target events can be reduced by manipulating the structure and composition of sgRNA (Manghwar et al. 2020). Another approach is using different Cas9 variants to reduce the off-target effects in various crops, and different cas9 variants are developed, merging dCas9 with FokI nuclease to develop fCas9 (Guilinger et al. 2014) and three to four amino acid replacements in Cas9, which lead to no detectable off-targets (Kleinstiver et al. 2016). However, various tools are being developed such as PEM-seq, CCTop, CHOPCHOP, CRISPR-PLANT v2, CT-Finder, CROP-IT, CFD (Cutting Frequency Determination) Score, CRISPOR, CRISPR-GE, etc. for sgRNA finding evaluation and predicting.

10.6 Regulatory Issues of Genome Edited Crops

Globally, the legal status of genome editing is not decided yet or is still under discussion. In Europe, the Court of Justice of the European Union (CJEU) rule on directed mutagenesis is that the genome edited products are subjected to current legal GMO framework without any exemptions, and genome editing are subject to the legal framework applicable to release, marketing, labelling, and traceability of GMOs (Menz et al. 2020). In Israel, during 2017, the National Committee for Transgenic Plants stated that genome-edited plants with no insertion of foreign DNA and with deletion of nucleotides will not be considered as transgenic and thus not to be subjected to the GE seed regulation. Moreover, the genome-edited plants with foreign DNA incorporated will be subjected to regulations and guidelines found in the GE seed regulation (USDA FAS 2020).

Different countries adopted different approaches of regulations to approve the genome editing crops. Mainly, two regulatory approaches are adopted by different countries, known as process-based and product-based regulations. The process-based regulations are adopted in the EU and Norway, and Canada is an example of product-based regulation. However, the United States is described as a combined regulatory approach (Ishii and Araki 2017; Zetterberg and Bjornberg 2017). The gene editing plants are considered under the GMO regulations in New Zealand and

European countries, while GMO regulations are removed by the United States from gene-edited plants (Gupta et al. 2021).

In India, the Department of Biotechnology under the Ministry of Science &Technology drafted guidelines after expert consultations and invited comments from researchers, institutions, and other stakeholders. According to draft guidelines, the GEd Group I (SDN-1, ODM) and GEd Group II (SDN-2) would be assessed mainly to confirm targeted edit(s) as well as an absence of any biologically significant off-target genomic changes. In addition, they would be subjected to phenotypic equivalence analysis. The GEd Group II would also be used for trait efficacy through appropriate contained and/or confined field trials. However, GEd Group III (SDN-3) harboring large or foreign DNA may represent similar biosafety concerns as that of genetically engineered (GE) organisms (DBT, India 2020).

10.7 Conclusion

The progress made through conventional breeding for food security is remarkable. However, climate change offers new challenges for further yield improvement in cereal crops. The genome editing tools offer a novel opportunity for designing the crop with the preferred trait(s). There is an urgent need for human resource development in these emerging technologies. The method also needs to be standardized for each crop for greater harvesting of products through genome editing in cereal improvement.

References

Andersson M, Turesson H, Olsson N, Fält AS, Ohlsson P, Gonzalez MN, Samuelsson M, Harvinder P (2018) Genome editing in potato via CRISPR-Cas9 ribonucleoprotein delivery. Physiol Plant 164(4):378–384. https://doi.org/10.1111/ppl.12731

Baltes NJ, Gil-Humanes J, Voytas DF (2017) Chapter one – Genome engineering and agriculture: opportunities and challenges. Prog Mol Biol Transl Sci 149:1–26

Bitinaite J, Wah DA, Aggarwal AK, Schildkraut I (1998) FokI dimerization is required for DNA cleavage. Proc Natl Acad Sci U S A 95:10570–10575. https://doi.org/10.1073/pnas.95.18.10570

Bortesi L, Fischer R (2015) The CRISPR/Cas9 system for plant genome editing and beyond. Biotechnol Adv 33(1):41–52. https://doi.org/10.1016/j.biotechadv.2014.12.006

Brookhouser N, Raman PC, Brafman DA (2017) May i cut in? Gene editing approaches in human induced pluripotent stem cells. Cell 6(1):5. https://doi.org/10.3390/cells6010005

Certo MT, Ryu BY, Annis JE, Garibov M, Jarjour J, Rawlings DJ, Scharenberg AM (2011) Tracking genome engineering outcome at individual DNA breakpoints. Nat Methods 8:671–676. https://doi.org/10.1038/nmeth.1648

Chandrasegaran S, Carroll D (2016) Origins of programmable nucleases for genome engineering. J Mol Biol 428(5):963–989. https://doi.org/10.1016/j.jmb.2015.10.014

Clasen BM, Stoddard TJ, Luo S, Demorest ZL, Li J, Cedrone F, Tibebu R, Davison S, Ray EE, Daulhac A, Coffman A (2016) Improving cold storage and processing traits in potato through targeted gene knockout. Plant Biotechnol J 14(1):169–176. https://doi.org/10.1111/pbi.12370

Cui X, Balcerzak M, Schernthaner J, Babic V, Datla R, Brauer EK, Labbel N, Subramaniam R, Ouellet T (2019) An optimised CRISPR/Cas9 protocol to create targeted mutations in homoeologous genes and an efficient genotyping protocol to identify edited events in wheat. Plant Methods *15*(1):1–12. https://doi.org/10.1186/s13007-019-0500-2

Curtis MD, Grossniklaus U (2003) A gateway cloning vector set for high-throughput functional analysis of genes in planta. Plant Physiol *133*(2):462–469. https://doi.org/10.1104/pp.103.027979

DBT (2020) Draft Document on Genome Edited Organisms: Regulatory Framework and Guidelines for Risk Assessment, Department of Biotechnology, Ministry of Science & Technology, Government of India. https://dbtindia.gov.in/sites/default/files/Draft_Regulatory_Framework_Genome_Editing_9jan2020a.pdf. Accessed 13 May 2021

Declais AC, Liu J, Freeman AD, Lilley DM (2006) Structural recognition between a four-way DNA junction and a resolving enzyme. J Mol Biol *359*(5):1261–1276. https://doi.org/10.1016/j.jmb.2006.04.037

Engler C, Kandzia R, Marillonnet S (2008) A one pot, one step, precision cloning method with high throughput capability. PLoS One *3*(11):e3647. https://doi.org/10.1371/journal.pone.0003647

Frye CA, Tang D, Innes RW (*2001*) Negative regulation of defence responses in plants by a conserved MAPKK kinase. Proc Natl Acad Sci *98*(1):373–378. https://doi.org/10.1073/pnas.98.1.373

Fu Y, Sander JD, Reyon D, Cascio VM, Joung JK (2014) Improving CRISPR-Cas nuclease specificity using truncated guide RNAs. Nat Biotechnol *32*(3):279–284. https://doi.org/10.1038/nbt.2808

Gallagher DN, Haber JE (2018) Repair of a site-specific DNA cleavage: old-school lessons for Cas9-mediated gene editing. ACS Chem Biol 13:397–405. https://doi.org/10.1021/acschembio.7b00760

Gao C (2018) The future of CRISPR technologies in agriculture. Nat Rev Mol Cell Biol *19*(5):275–276. https://doi.org/10.1038/nrm.2018.2

Gaudelli NM, Komor AC, Rees HA, Packer MS, Badran AH, Bryson DI, Liu DR (2017) Programmable base editing of T to G C in genomic DNA without DNA cleavage. Nature 551:464–471. https://doi.org/10.1038/nature24644

Gibson DG, Young L, Chuang RY, Venter JC, Hutchison CA, Smith HO (2009) Enzymatic assembly of DNA molecules up to several hundred kilobases. Nat Methods *6*(5):343–345. https://doi.org/10.1038/nmeth.1318

Guilinger JP, Thompson DB, Liu DR (2014) Fusion of catalytically inactive Cas9 to foki nuclease improves the specificity of genome modification. Nat Biotechnol 32:577–582. https://doi.org/10.1038/nbt.2909

Gupta S, Kumar A, Patel R, Kumar V (2021) Genetically modified crop regulations: scope and opportunity using the CRISPR-Cas9 genome editing approach. Mol Biol Rep 1-13. https://doi.org/10.1007/s11033-021-06477-9

Huang XY, Chao DY, Gao JP, Zhu MZ, Shi M, Lin HX (2009) A previously unknown zinc finger protein, DST, regulates drought and salt tolerance in rice via stomatal aperture control. Genes Dev 23:1805–1817. https://doi.org/10.1101/gad.1812409

Ishii T, Araki M (2017) A future scenario of the global regulatory landscape regarding genome-edited crops. GM Crops Food *8*(1):44–56. https://doi.org/10.1080/21645698.2016.1261787

Jiang W, Zhou H, Bi H, Fromm M, Yang B, Weeks DP (2013) Demonstration of CRISPR/Cas9/sgRNA-mediated targeted gene modification in Arabidopsis, tobacco, sorghum and rice. Nucleic Acids Res *41*(20):e188–e188. https://doi.org/10.1093/nar/gkt780

Jinek M, East A, Cheng A, Lin S, Ma E, Doudna J (2013) RNA-programmed genome editing in human cells. Elife 2:e00471. https://doi.org/10.7554/eLife.00471

Kanchiswamy CN, Malnoy M, Velasco R, Kim JS, Viola R (2017) Non-GMO genetically edited crop plants. Trends Biotechnol *33*(9):489–491. https://doi.org/10.1016/j.tibtech.2015.04.002

Kim H, Kim ST, Ryu J, Kang BC, Kim JS, Kim SG (2017) CRISPR/Cpf1-mediated DNA-free plant genome editing. Nat Commun 8(1):1–7. https://doi.org/10.1038/ncomms14406

Kim D, Alptekin B, Budak H (2018) CRISPR/Cas9 genome editing in wheat. Funct Integr Genomics 18:31–41. https://doi.org/10.1007/s10142-017-0572-x

Kleinstiver BP, Pattanayak V, Prew MS, Tsai SQ, Nguyen NT, Zheng Z, Joung JK (2016) High-fidelity CRISPR–Cas9 nucleases with no detectable genome-wide off-target effects. Nature 529(7587):490–495. https://doi.org/10.1038/nature16526

Komor AC, Kim YB, Packer MS, Zuris JA, Liu DR (2016) Programmable editing of a target base in genomic DNA without double-stranded DNA cleavage. Nature 533:420–424. https://doi.org/10.1038/nature17946

Koonin EV, Makarova KS, Zhang F (2017) Diversity, classification and evolution of CRISPR-Cas systems. Curr Opin Microbiol 37:67–78. https://doi.org/10.1016/j.mib.2017.05.008

Krishna H, Alizadeh M, Singh D, Singh U, Chauhan N, Eftekhari M, Sadh RK (2016) Somaclonal variations and their applications in horticultural crops improvement. 3Biotech 6(1):54. https://doi.org/10.1007/s13205-016-0389-7

Kumar N, Galli M, Ordon J, Stuttmann J, Kogel K-H, Imani J (2018) Further analysis of barley MORC1 using a highly efficient RNA-guided Cas9 gene-editing system. Plant Biotechnol J 16: 1892–1903. https://doi.org/10.1111/pbi.12924

Kumar VVS, Verma RK, Yadav SK, Yadav P, Watts A, Rao MV, Chinnusamy V (2020) CRISPR-Cas9 mediated genome editing of drought and salt tolerance *(OsDST)* gene in indica mega rice cultivar MTU1010. Physiol Mol Biol Plants 26:1099–1110. https://dx.doi.org/10.1007%2Fs12298-020-00819-w

Lee LY, Gelvin SB (2008) T-DNA binary vectors and systems. *Plant Physiol 146*(2):325–332. https://doi.org/10.1104/pp.107.113001

Lei Y, Lu L, Liu HY, Li S, Xing F, Chen LL (2014) CRISPR-P: a web tool for synthetic single-guide RNA design of CRISPR-system in plants. Mol Plant 7(9):1494–1496. https://doi.org/10.1093/mp/ssu044

Li C, Zong Y, Wang Y, Jin S, Zhang D, Gao C (2018) Expanded base editing in rice and wheat using a Cas9-adenosine deaminase fusion. Genome Biol 9:59. https://doi.org/10.1186/s13059-018-1443-z

Li J, Li H, Chen J, Yan L, Xia L (2020) Toward precision genome editing in crop plants. Mol Plant 13(6):811–813. https://doi.org/10.1016/j.molp.2020.04.008

Li S, Zhang C, Li J, Yan L, Wang N, Xia L (2021) Present and future prospects for wheat improvement through genome editing and advanced technologies. Plant Commun 2, 100211 (4). https://doi.org/10.1016/j.xplc.2021.100211

Liang Z, Chen K, Li T, Zhang Y, Wang Y, Zhao Q, Gao C (2017) Efficient DNA-free genome editing of bread wheat using CRISPR/Cas9 ribonucleoprotein complexes. Nat Commun *8*(1): 1–5. https://doi.org/10.1038/ncomms14261

Liang Z, Chen K, Zhang Y, Liu J, Yin K, Qiu JL (2018) Genome editing of bread wheat using biolistic delivery of CRISPR/Cas9 in vitro transcripts or ribonucleoproteins. Nat Protoc 13:413–430. https://doi.org/10.1038/nprot.2017.145

Lin Q, Zong Y, Xue C, Wang S, Jin S, Zhu Z, Gao C (2020) Prime genome editing in rice and wheat. Nat Biotechnol *38*(5):582–585. https://doi.org/10.1038/s41587-020-0455-x

Liu DF, Chen XJ, Liu JQ, Ye JC, Guo ZJ (2012) The rice ERF transcription factor OsERF922 negatively regulates resistance to *Magnaportheoryzae* and salt tolerance. J Exp Bot 63:3899–3912. https://doi.org/10.1093/jxb/ers079

Liu Q, Liu R, Ma Y, Song J (2018) Physiological and molecular evidence for Na+ and Cl– exclusion in the roots of two Suaeda salsa populations. Aquat Bot 146:1–7. https://doi.org/10.1016/j.aquabot.2018.01.001

Luo M, Zhang Y, Li J, Zhang P, Chen K, Song W, Wang X, Yang J, Lu X, Lu B, Zhao Y, Zhao J (2021) Molecular dissection of maize seedling salt tolerance using a genome-wide association analysis method. Plant Biotechnol J 19(10):1937–1951. https://doi.org/10.1111/pbi.13607

Ma X, Zhang Q, Zhu Q, Liu W, Chen Y, Qiu R, Liu YG (2015) A robust CRISPR/Cas9 system for convenient, high-efficiency multiplex genome editing in monocot and dicot plants. Mol Plant *8*(8):1274–1284. https://doi.org/10.1016/j.molp.2015.04.007

Ma X, Zhu Q, Chen Y, Liu YG (2016) CRISPR/Cas9 platforms for genome editing in plants: developments and applications. Mol Plant *9*(7):961–974. https://doi.org/10.1016/j.molp.2016.04.009

Ma J, Chen J, Wang M, Ren Y, Wang S, Lei C, Cheng Z (2017) Disruption of OsSEC3A increases the content of salicylic acid and induces plant defence responses in rice. J Exp Bot 69(5): 1051–1064. https://doi.org/10.1093/jxb/erx458

Macovei A, Sevilla NR, Cantos C, Jonson GB, Slamet-Loedin I, Čermák T, Voytas DF, Choi IR, Chadha-Mohanty P (2018) Novel alleles of rice eIF4G generated by CRISPR/Cas9-targeted mutagenesis confer resistance to Rice tungro spherical virus. Plant Biotechnol J 16:918–1927. https://doi.org/10.1111/pbi.12927

Manghwar H, Li B, Ding X, Hussain A, Lindsey K, Zhang X, Jin S (2020) CRISPR/Cas systems in genome editing: methodologies and tools for sgRNA design, off-target evaluation, and strategies to mitigate off-target effects. Adv Sci *7*(6):1902312. https://doi.org/10.1002/advs.201902312

Menz J, Modrzejewski D, Hartung F, Wilhelm R, Sprink T (2020) Genome edited crops touch the market: a view on the global development and regulatory environment. Front Plant Sci 11: 586027. https://doi.org/10.3389/fpls.2020.586027

Miller JC, Tan S, Qiao G, Barlow KA, Wang J, Xia DF, Meng X, Paschon DE, Leung E, Hinkley SJ, Dulay GP (2011) ATALE nuclease architecture for efficient genome editing. Nat Biotechnol 29:143–148. https://doi.org/10.1038/nbt.1755

Oliva R, Ji C, Atienza-Grande G, Huguet-Tapia JC, Perez-Quintero A, Li T, Eom JS, Li C, Nguyen H, Liu B, Auguy F, Sciallano C, Luu VT, Dossa GS, Cunnac S, Schmidt SM, Slamet-Loedin IH, Cruz CV, Szurek B, Frommer WB, White FF, Yang B (2019) Broad-spectrum resistance to bacterial blight in rice using genome editing. Nat Biotechnol 37:1344–1350. https://doi.org/10.1038/s41587-019-0267-z

Pattanayak V, Lin S, Guilinger JP, Ma E, Doudna JA, Liu DR (2013) High-throughput profiling of off-target DNA cleavage reveals RNA-programmed Cas9 nuclease specificity. Nat Biotechnol *31*(9):839–843. https://doi.org/10.1038/nbt.2673

Peng A, Chen S, Lei T, Xu L, He Y, Wu L, Yao L, Zou X (2017) Engineering canker-resistant plants through CRISPR/Cas9-targeted editing of the susceptibility gene Cs LOB 1 promoter in citrus. Plant Biotechnol J 15(12):1509–1519. https://doi.org/10.1111/pbi.12733

Qiu P, Shandilya H, D'Alessio JM, O'Connor K, Durocher J, Gerard GF (2004) Mutation detection using Surveyor™ nuclease. Biotechniques *36*(4):702–707. https://doi.org/10.2144/04364PF01

Ran FA, Hsu PD, Lin CY, Gootenberg JS, Konermann S, Trevino AE, Zhang F (2013) Double nicking by RNA-guided CRISPR Cas9 for enhanced genome editing specificity. Cell *154*(6): 1380–1389. https://doi.org/10.1016/j.cell.2013.08.021

Sander JD, Joung JK (2014) CRISPR-Cas systems for editing, regulating and targeting genomes. Nat Biotechnol 32:347–355. https://doi.org/10.1038/nbt.2842

Shan Q, Wang Y, Li J, Zhang Y, Chen K, Liang Z, Gao C (2013) Targeted genome modification of crop plants using a CRISPR-Cas system. Nat Biotechnol *31*(8):686–688. https://doi.org/10.1038/nbt.2650

Shi J, Gao H, Wang H, Lafitte HR, Archibald RL, Yang M, Hakimi SM, Mo H, Habben JE (2017) ARGOS8 variants generated by CRISPR-Cas9 improve maize grain yield under field drought stress conditions. Plant Biotechnol J 15:207–216. https://doi.org/10.1111/pbi.12603

Stemmer M, Thumberger T, del Sol KM, Wittbrodt J, Mateo JL (2015) CCTop: an intuitive, flexible and reliable CRISPR/Cas9 target prediction tool. PLoS One 10(4):e0124633. https://doi.org/10.1371/journal.pone.0124633

Svitashev S, Schwartz C, Lenderts B, Young JK, Cigan AM (2016) Genome editing in maize directed by CRISPR–Cas9 ribonucleoprotein complexes. Nat Commun *7*(1):1–7. https://doi.org/10.1038/ncomms13274

Talbot JC, Amacher SL (2014) A streamlined CRISPR pipeline to reliably generate zebrafishframeshifting alleles. Zebrafish 11(6):583–585. https://doi.org/10.1089/zeb.2014.1047

Tang L, Mao B, Li Y, Lv Q, Zhang L, Chen C, He H, Wang W, Zeng X, Shao Y, Pan Y (2017) Knockout of OsNramp5 using the CRISPR/Cas9 system produces low Cd-accumulating indica rice without compromising yield. Sci Rep 7(1):1–12. https://doi.org/10.1038/s41598-017-14832-9

USDA FAS (2020) Agricultural Biotechnology Annual 2019. Israel's Agricultural Biotechnology Regulations Remain Unchanged GAIN Report Number: IS2019-0006. USDA FAS. http://agriexchange.apeda.gov.in/marketreport/Reports/Agricultural_Biotechnology_Annual_Sarajevo_Bosnia_and_Herzegovina_10-20-2019.pdf

Wang QM, Wang L (2012) An evolutionary view of plant tissue culture: somaclonal variation and selection. Plant Cell Rep *31*(9):1535–1547. https://doi.org/10.1007/s00299-012-1281-5

Wang Y, Cheng X, Shan Q, Zhang Y, Liu J, Gao C, Qiu JL (2014) Simultaneous editing of three homoeoalleles in hexaploid bread wheat confers heritable resistance to powdery mildew. Nat Biotechnol 32:947. https://doi.org/10.1038/nbt.2969

Wang K, Mei DY, Liu QN, Qiao XH, Ruan WM, Huang T, Cao GS (2015) Research of methods to detect genomic mutations induced by CRISPR/Cas systems. J Biotechnol *214*:128–132. https://doi.org/10.1016/j.jbiotec.2015.09.029

Wang F, Wang C, Liu P, Lei C, Hao W, Gao Y, Liu YG, Zhao K (2016) Enhanced rice blast resistance by CRISPR/Cas9-targeted mutagenesis of the ERF transcription factor gene *OsERF922*. PLoS One 11:e0154027. https://doi.org/10.1371/journal.pone.0154027

Wang W, Pan Q, He F, Akhunova A, Chao S, Trick H, Akhunov E (2018) Transgenerational CRISPRCas9 activity facilitates multiplex gene editing in allopolyploid wheat. CRISPR J 1:65–74. https://doi.org/10.1089/crispr.2017.0010

Weber E, Engler C, Gruetzner R, Werner S, Marillonnet S (2011) A modular cloning system for standardized assembly of multigene constructs. PLoS One *6*(2):e16765. https://doi.org/10.1371/journal.pone.0016765

Wiedenheft B, Sternberg SH, Doudna JA (2012) RNA-guided genetic silencing systems in bacteria and archaea. Nature 482:331–338. https://doi.org/10.1038/nature10886

Wolfe SA, Nekludova L, Pabo CO (2000) DNA recognition by Cys2His2 zinc finger proteins. Annu Rev Biophys Biomol Struct 29:183–212. https://doi.org/10.1146/annurev.biophys.29.1.183

Woo JW, Kim J, Kwon SI, Corvalan C, Cho SW, Kim H, Kim JS (2015) DNA-free genome editing in plants with preassembled CRISPR-Cas9 ribonucleoproteins. Nat Biotechnol *33*(11):1162–1164. https://doi.org/10.1038/nbt.3389

Xie K, Zhang J, Yang Y (2014) Genome-wide prediction of highly specific guide RNA spacers for CRISPR–Cas9-mediated genome editing in model plants and major crops. Mol Plant *7*(5):923–926. https://doi.org/10.1093/mp/ssu009

Xing HL, Dong L, Wang ZP, Zhang HY, Han CY, Liu B, Chen QJ (2014) A CRISPR/Cas9 toolkit for multiplex genome editing in plants. BMC Plant Biol *14*(1):1–12. https://doi.org/10.1186/s12870-014-0327-y

Zetterberg C, Bjornberg KE (2017) Time for a new EU regulatory framework for GM crops? J Agric Environ Ethics *30*(3):325–347. https://doi.org/10.1007/s10806-017-9664-9

Zhang Y, BaiY WG, Zou S, Chen Y, Gao C, Tang D (2017) Simultaneous modification of three homoeologs of TaEDR1 by genome editing enhances powdery mildew resistance in wheat. Plant J 91:714–724. https://doi.org/10.1111/tpj.13599

Zhang A, Liu Y, Wang F, Li T, Chen Z, Kong D, Bi J, Zhang F, Luo X, Wang J, Tang J, Yu X, Liu G, Luo L (2019) Enhanced rice salinity tolerance via CRISPR/Cas9-targeted mutagenesis of the *OsRR22* gene. Mol Breeding 39:47. https://doi.org/10.1007/s11032-019-0954-y

Zheng X, Yang S, Zhang D, Zhong Z, Tang X, Deng K, Zhang Y (2016) Effective screen of CRISPR/Cas9-induced mutants in rice by single-strand conformation polymorphism. Plant Cell Rep *35*(7):1545–1554. https://doi.org/10.1007/s00299-016-1967-1

Zlobin NE, Lebedeva MV, Taranov VV (2020) CRISPR/Cas9 genome editing through in planta transformation. Crit Rev Biotechnol *40*(2):153–168. https://doi.org/10.1080/07388551.2019.1709795

Zong Y, Song Q, Li C, Jin S, Zhang D, Wang Y, Qiu JL, Gao C (2018) Efficient C-to-T base editing in plants using a fusion of nCas9 and human APOBEC3A. Nat Biotechnol 36:950–953. https://doi.org/10.1038/nbt.4261

Editing Plant Genome with CRISPR/Cas: A Sustainable Strategy for Disease Management

11

Shilpi Bansal, Alexander Balamurugan, V. Mohan Murali Achary, Aundy Kumar, Malireddy K. Reddy, and Ganesan Prakash

Abstract

The world's food and nutritional security is adversely affected by the plant diseases which in turn affects the economic growth and environmental sustainability at the global level. The threat in climate change combined with increasing food production demand poses growing stress on the agro-ecosystems. Worldwide, due to pests and diseases, around 20–40% of crop losses occur in agricultural productivity every year. Basic understanding on the infection process of plant pathogens and their interaction with the host during disease establishment, virulence/avirulence/effector genes in plant pathogens and resistance genes in hosts becomes mandatory for developing disease-resistant cultivars in a sustainable manner. Different management strategies, viz. host plant resistance, use of agrochemicals, cultural practices, and biocontrol agents, have been practiced for reducing the losses caused by plant pathogens. But the sustainability is a question mark because of the rapid occurrence of new pathotypes/ strains/ races of plant pathogens which are infecting the major crops of economic value. With the development of novel genome editing techniques, it is feasible to generate pathogen-resistant crops which are durable with less time. One such powerful genome editing tool is CRISPR/Cas due to its high precision, robustness, minimal off-target effects, and ability to edit multiple targets.

S. Bansal · A. Balamurugan · A. Kumar · G. Prakash (✉)
Division of Plant Pathology, ICAR-Indian Agricultural Research Institute, New Delhi, India

V. M. M. Achary · M. K. Reddy
Crop Improvement Group, International Centre for Genetic Engineering and Biotechnology, New Delhi, India

M. G. Mallikarjuna et al. (eds.), *Next-Generation Plant Breeding Approaches for Stress Resilience in Cereal Crops*, https://doi.org/10.1007/978-981-19-1445-4_11

Keywords

Genome editing · CRISPR/Cas · Plant pathogens · Cereals crops · Disease resistance

11.1 Introduction

Technological advancement is the utmost requirement, especially in the agricultural science field to feed the rapidly increasing population. According to the present scenario, the world population is approximately 7.8 billion, which by 2050 may undergo a 34% increase and reach up to 9.1 billion (https://www.un.org/development/desa/en/news/population/world-population-prospects-2019.html). The increase in population can be associated with an increase in workforce, intellectual minds, technology developments, but it is equally associated with a decrease in resources available to an individual. Further, a large population consumes enormous renewable and non-renewable resources, which are often difficult to replenish. Another biggest challenge is to feed this huge population. According to United Nations' Food and Agriculture Organization (FAO), there has been a global increase in hunger affected people since 2014.

Malnutrition is still very much existing, and the target to achieve zero hunger is far from what was expected by 2030 (http://www.fao.org/publications/sofi/2020/en). The health and socio-economic impacts of the COVID-19 pandemic have further deteriorated the food security of the most vulnerable population. Considering the population explosion and pandemic disease outbreaks, we need alternative solid steps to meet the differences between the supply and demand of food (Aday and Aday 2020). Solution for the problem can only be determined after properly understanding the problem itself and its causative agents. Agricultural production is reduced due to over-exploitation of land reserves and forest areas by humans, which leads to reduced land area under cultivation, loss in soil fertility, soil erosion, pollution, etc. (Sreekanth et al. 2017). Other than human intervention, the major reason for decreased agricultural production is yield loss by abiotic and biotic factors. Abiotic stress factors include climatic conditions such as erratic rainfall and temperature. At present, the mean temperature change is a function of global warming and affects the yield and quality of the crop (Minhas et al. 2017). Other challenges the plants face are drought, heat, cold, soil salinity, soil acidity, low fertility of soil, water logging, etc. These conditions lead to a series of abnormalities in morphology, physiology, metabolism, reproduction, seed viability, pollination, accumulation of toxic chemicals, depletion of nutrients from the soil, and so on (Yadav et al. 2020).

Biotic factors comprise an array of crop destruction elements like weeds (monocots and dicots), pests (insects, mites, nematodes, birds, etc.), and pathogens (fungi, bacteria, viruses). This chapter mainly deals with plant pathogens and their sustainable management through employing Clustered Regularly Interspaced Short Palindromic Repeats associated Cas protein (CRISPR/Cas) editing technology.

Breeding methods for disease resistance mainly depend on broad genetic variation in the elite primary gene pool of target crop species. It takes years to introduce the desirable alleles through the breeding method (Darwin and Murray 2010; Scheben et al. 2017). Moreover, the traditional breeding methods are laborious and non-durable because of the emergence of plant pathogens of major crops with new genetic structure. Therefore, under this current scenario, advanced breeding tools like genome editing technologies are very promising. The technology of genome editing has witnessed wide applications in humans and animals (e.g. genetic studies, gene therapy, drug development, disease diagnosis) (Gori et al. 2015; Sánchez-Rivera et al. 2014) and plants (e.g. biotic stress resistance, abiotic stress resistance, nutrition improvement, herbicide resistance, and yield improvement) (El-Mounadi et al. 2020). Genetic manipulations made in the food crops mark one way to ensure food security. In crops, site-specific genome modifications can be made so that they are tolerant to the biotic stresses without any off or non-target effects. Genome editing can advance the plant breeding process by manipulating the genes involved in host plant recognition during the infection process, genes involved in plant susceptibility, and silencing the essential pathogen genes (Andolfo et al. 2016; Zaidi et al. 2018; Oliva et al. 2019). The present chapter covers the various genome editing techniques with their advantages and disadvantages, with reference to CRISPR/Cas and its application in developing pathogen-resistant plants.

11.2 Genetic Modifications in Plants: From Meganuclease to CRISPR/Cas System

Improvement of crops against biotic stresses has always posed a challenge for the research community. Numerous techniques have been employed till date which plays with the genome of plants, including deletions, insertions, and mutations of single nucleotide or fragments from which wild or dominant varieties to elite varieties. The most primitive approaches include random mutagenesis, naturally occurring mutations, classical breeding, hybridization, etc., which have been commonly used to develop disease-resistant cultivars (Acquaah 2015). The random nature of mutations mediated through physical (X-rays, radiation), chemical (ethyl methane sulphonate), and biological (transposons) agents necessitates the large-scale screening of mutagenized populations (Sikora et al. 2011). Often, these techniques work on the concept of hit and trial and mostly lack any control over the integration site and the type of mutations they cause (Daboussi et al. 2015). Further pyramiding of multiple disease resistance genes takes about 8–10 years via conventional breeding, and this long duration is enough for pathogens to escape the resistance barrier. Additionally, high variability and mutations in pathogens result in resistance break down in genes pyramided cultivars (Fuchs 2017).

The recombinant DNA technology formed the next level of modifications through which the concept of cloning, transfection, and transformation came into existence. Various methods, such as polyethylene glycol mediated, electroporation, particle bombardment, etc., were devised to introduce a foreign gene into plants

(Ortiz-Matamoros et al. 2018). However, the *Agrobacterium* mediated genetic transformation technique gained maximum popularity (Narusaka et al. 2012). *Agrobacterium* is a gram-negative soil bacterium that has the natural capability to transfer foreign genes into the plant system (Gelvin 2003). Similar to other methods, the transgenic approach is also associated with certain constraints as the integration of foreign gene(s) is not site-specific, and the occurrence of undesired traits is prevalent. Many people deny eating transgenic food crops since the foreign DNA can be of any origin; thus, social acceptance of transgenics is challenging. Furthermore, chances of insertion of some fragments of bacterial DNA along with target DNA is also there. Critics of Genetically Modified (GM) crops have also warned against the incidences of cross contamination with wild type varieties, harmful effects on non-target organisms, thereby disrupting the natural ecosystem. Due to above stated factors, large scale commercialization of GM crops has come to the back-foot as the per-unit involvement of time and infrastructure the returns are less (Prakash et al. 2011).

Further, as science progressed, the concepts of new plant breeding techniques were developed to overcome the disadvantages offered by conventional and transgenic breeding methods. These techniques include oligonucleotide directed mutagenesis (ODM), cisgenesis/transgenesis, reverse breeding, grafting, RNA dependent methylation, agrofiltration, RNAi mediated gene silencing, etc. In brief, in ODM, a site-specific oligonucleotide having a single point mutation but otherwise having an identical sequence to plant DNA is repaired by plant machinery, which leads to alteration in the genome (Sauer et al. 2016). In cisgenesis, the fragment is completely unchanged, while in transgenesis new combination of DNA is introduced. Except for the T-DNA border sequences, all the DNA used in the modification are from same or cross compatible species (Hou et al. 2014). Reverse breeding is a technique where both the original and hybrid plants are genetically similar; here, just the steps involved in hybrid production are reversed (Hou et al. 2014). Attachment of a non-transgenic scion on genetically modified rootstock results in grafting (Nawaz et al. 2016). Agrofiltration technique is mediated by *Agrobacterium* where plant part is infected with *Agrobacterium* harbouring the desired construct and its effect is studied. Majorly the response generated by transformation in this case is transient in nature (Schaart and Visser 2009; Lusser and Davies 2013). Some of these mediate stable integration of gene(s); however, strategies like RNAi, agrofiltration suffer from some disadvantages like each of the independent transgenic lines behaves independently in terms of gene expression and hence screening of a large population is required. Sometimes, transgene event is unstable and incomplete, resulting in partial suppression of target gene expression (Dietz-Pfeilstetter 2010).

Thus, it is mandatory to search for genome modification strategies which are more specific and precise. One such approach is the introduction of double-stranded breaks (DSBs), which on repairing by non-homologous end joining (NHEJ) or homologous recombination (HR) generates either insertion or deletions in the target sequences. These DSBs are induced by molecular scissors, which are engineered nucleases generating breaks at a specific site. At present, mainly four nucleases are being used: (1) Transcription activator-like effector nucleases (TALENs);

(2) Meganucleases; (3) Zinc finger nucleases (ZFN); and (4) CRISPR/Cas9 (Saha et al. 2019).

11.2.1 Meganucleases

Meganucleases can be categorized as the first sequence-specific nuclease utilized to create double-stranded breaks. Meganucleases are also called homing endonucleases and can recognize up to 12–40 bp target DNA (Gallagher et al. 2014). Compared to other nucleases such as ZFNs these are less toxic in cells. However, its limitation is less availability of naturally occurring meganucleases, thereby creating a need to design sequence-specific enzymes, which is expensive, laborious, and time-consuming (Prieto et al. 2007).

11.2.2 Zinc Finger Nucleases

ZFN comprises two domains: DNA binding domain at an amino terminal and DNA cleavage domain at carboxyl terminal (Osakabe and Osakabe 2015). The DNA cleavage domain is from FokI endonuclease, while the DNA binding domain is made from 3 to 4 zinc finger arrays where each zinc finger binds to 3 bp of DNA. This FokI endonuclease containing DNA cleavage domain on fusion with DNA binding domain forms a monomer. ZFN acts in dimer as the FokI nuclease domain needs dimerization to be active and cleave DNA (Carroll 2016). The C-terminal end of each of the ZFN monomers in a dimer binds to opposite DNA strands of the target site. Amino acids present at −1,+1, +2,+3,+4,+5,+6 relative to the initiation site of zinc finger α-helix are responsible for sequence-specific interaction with DNA. Any changes in these amino acids facilitate specific binding to DNA sequences (Pavletich and Pabo 1991; Elrod-Erickson and Pabo 1999); thus, ZFN has the capability to cleave a large stretch of double-stranded DNA (Durai et al. 2005). However, the ZFN usage is restricted due to poor targeting resulting in various off-target effects.

11.2.3 Transcription Activator-like Effector Nucleases

The TALENs gene editing system came into existence when transcription activator-like effectors were identified from a pathogenic bacterium, *Xanthomonas*. The bacterium causes uncontrolled growth of plant cells mediated by these TALE proteins (Boch and Bonas 2010). These proteins are a part of DNA binding protein family which regulates the target genes expression by coupling with activation or repressor proteins. This domain contains monomeric repeats of 34 amino acids, with each monomer binding to one nucleotide in the target sequence. Out of the 34 amino acids, two amino acids placed at 12 and 13 positions are highly variable and called as repeat variable di-residue (RVD). These RVDs channelize the recognition of target specific nucleotide (Weeks et al. 2016). A set of codes determines the pairing of

RVDs with a particular nucleotide base. Similar to ZFNs, TALENs also have FokI nuclease domain. The combination of this DNA binding domain and cleavage domain leads to generation of DSBs. TALENs also function in dimer form and are more specific due to larger target sites. TALENs require a thymidine residue at the 5′ end of target residue for efficient binding, and this property limits its use for genome editing (Miller et al. 2015). In addition, the large number of amino acids involved in TALEN monomers (15–20 RVDs per monomer) further restricts its usage (Razzaq et al. 2019).

11.3 CRISPR/Cas-Mediated Genome Modification

11.3.1 History

The roots of today's CRISPR technique, a popular gene editing tool can be found back in late 1980s when Ishino et al. (1987) reported some repetitive elements in *E. coli* followed by extensive experimentation since 2005. The credit of the success and continuous advancement in CRISPR/Cas9 microbial adaptive immune system goes to the efforts of researchers working continually across the globe (Arora and Narula 2017). The series of historical events is in Fig. 11.1.

CRISPR/Cas9 mechanism is an endogenous technique developed by prokaryotes as an adaptive immunity against viruses and plasmids. It requires two components Cas 9; a monomeric DNA endonuclease and a single guide RNA sequence which can be modified according to our requirement to bind a specific target DNA. Using this two-unit system, numerous objectives such as determination of gene function, activation/inactivation of the gene, development of stress tolerant plant varieties, etc. can be achieved (Hsu et al. 2014).

CRISPR-Cas system can be categorized into two classes (Class 1 and Class 2), and each of these classes is further divided into types and subtypes. The classification is based on the number of effector proteins and sequence divergence between

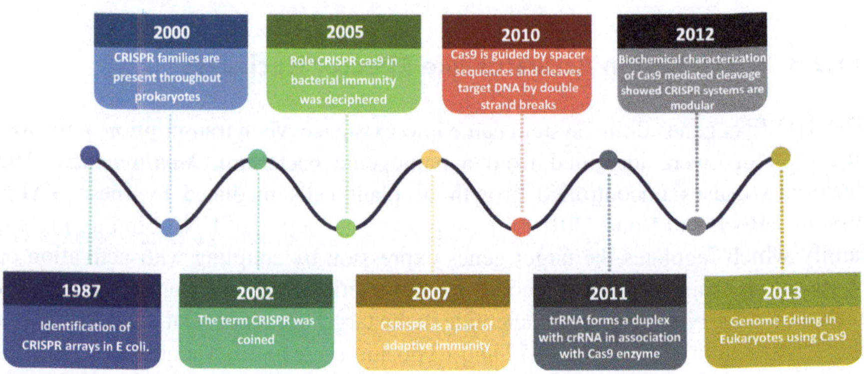

Fig. 11.1 The chronology of historical developments in the CRISPR/Cas system

these effector modules. In class 1, multiple Cas proteins (Cas3, Cas5-8, Cas10-11) in various combinations and permutations are utilized. In class 2, single Cas protein (Cas9, Cas12 or Cas13) is sufficient to act (Makarova et al. 2015, 2018).

11.3.2 Mechanism of CRISPR/Cas System

11.3.2.1 Adaptation

On infection by a foreign organism, a fragment from it called protospacer is incorporated as a new spacer into the CRISPR array of the host chromosome with the help of acquisition assembly. As a part of adaptive immunity, the host will remember this foreign material and will show immunity against it during future infections (Barrangou et al. 2007). This step is facilitated by Cas1 and the structural subunit of Cas2 proteins in all the types of CRISPR-Cas systems (Makarova et al. 2020). Mutation in Cas1 nuclease inhibits adaptation events (Yosef et al. 2012). However, apart from these Cas1 and Cas2 nucleases, different types and subtypes have additional requirement of Cas proteins such as Cas4 nuclease [type I-B] (Li et al. 2014), Csn2 [type II-A] (Vorontsova et al. 2015), reverse transcriptase fused to Cas1 protein [type III-B] (Silas et al. 2016). Cas1 and Cas2 nucleases form a complex and mediate the acquisition of spacers in a pattern similar to integrases and transposases. At the leader repeat boundary of the Cas protein array, a new spacer is introduced with the first repeat being duplicated (Barrangou et al. 2007; Yosef et al. 2012; Wei et al. 2015).

11.3.2.2 Biogenesis

After integration, the new spacer along with other spacers is co-transcribed to form a long precursor CRISPR RNA (pre-crRNA) which is processed into mature guide crRNAs (Carte et al. 2008; Haurwitz et al. 2010). Similar to adaptation in biogenesis also; as the CRISPR type and subtypes varies the Cas proteins which catalyses the processing of pre-crRNA also varied. In all the type I and III systems except type I-C, the Cas6 family leads to processing generating 5′tagged intermediate crRNAs. In type I-C, Cas5d protein mimics the role of Cas6 and generates intermediate crRNAs flanked by 11 bp 5′tag (Garside et al. 2012; Nam et al. 2012; Richter et al. 2012). This intermediate crRNA is cleaved at its 3′end by a nuclease to yield mature crRNA. The mature crRNA contains a spacer at 5′end, short RNA fragment complementary to foreign genetic material sequence, and CRISPR repeat sequence at 3′ end. The spacer and the complementary foreign material forms a hairpin structure and triggers the cleavage event (Hale et al. 2009; Garneau et al. 2010). Type II system recruits tracer RNA for processing. This tracrRNA first transcribes separately and then anneals to each of the pre-crRNA repeats leaving the guide spacer region free to form RNA duplex stabilized by Cas9. This RNA duplex is recognized and acted upon by RNase III to yield intermediate crRNA, which is converted to mature small guide RNA by an unknown mechanism and activates Cas9 (Deltcheva et al. 2011; Jiang and Doudna 2017). In type V and type VI systems, catalytic centre of a large effector protein regulates processing (Makarova et al. 2020).

11.3.2.3 Interference

Mature crRNAs (guide sequence) in association with Cas proteins form crRNA–effector complexes, which recognize the target DNA and cleave at the complementary sequence. A 2–5 bp short, conserved motif called PAM is a pre-requisite for this event. It was observed that due to mutations in PAM sequence AGAA located downstream the protospacer in *S. thermophilus*, the phage escaped the CRISPR defence (Deveau et al. 2008; Garneau et al. 2010). In a type II CRISPR system, mature crRNA–tracrRNA structure engages a single Cas9 endonuclease to recognize the target via PAM. Identification of PAM initiates unwinding of dsDNA, enabling invasion of crRNA and base pairing with the target (Sternberg et al. 2014). This leads to the formation of R-loop, which moves away from PAM (Szczelkun et al. 2014) and promotes base pairing between crRNA and target DNA. Cas 9 then cleaves the dsDNA with each strand being cleaved by a separate nuclease domain HNH or RuvC. The combination of crRNA and tracrRNA by sgRNA to form a single transcript simplifies the system and maintains the activity of Cas9 (Jinek et al. 2012). PAM sequences are required to identify the target sequences to avoid any self-targeting or off-targets. All the types of class2 possess this PAM sequence; wherein in type II, it is located downstream of the protospacer sequence, while in type I and V, it is located upstream. In type III systems, 5′tag of mature crRNA shields from non-self-targeting. In type I systems, invading DNA is localized in a crRNA dependent fashion by the cascade followed by Cas3 nuclease recruitment, which generates a cut on the foreign DNA leading to target DNA degradation (Westra et al. 2012). To target both the nucleic acids, i.e. DNA and RNA, an important signal transduction pathway for type III system has been characterized, which recruits Cas10-Csm and Cas10-Cmr corresponding to types III-A & III-D and types III-B & III-C, respectively. The subunit Cas10 cleaves the DNA, while Csm3 cleaves the transcribed mRNA in type III-A (Staals et al. 2014; Samai et al. 2015) and Cmr4 (Tamulaitis et al. 2014) cleaves the transcribed mRNA in type III-B CRISPR-Cas systems. Except for type V-A, which needs only crRNA for identifying a target and degrading it, rest all type V systems act in a way similar to type II (Hille and Charpentier 2016). It is this guide RNA sequence which is employed to select and aim at any sequence and generate a DSB with the aid of Cas9. During the repair of these DSB, the creation of various insertions and deletions leads to genome modification (Fig. 11.2).

11.3.3 Genome-Editing Applications Using Cas9 or Modified Nuclease Variants

The present technique has gained immense popularity due to its widespread applications involving gene knockouts, replacement of genes, gene function determination, gene mutation, epigenome editing, multiplex editing (Kim and Kim 2014; Khatodia et al. 2016; Song et al. 2016). Since its inception, CRISPR/Cas9 has been utilized majorly in animal and bacterial systems, but its application in plants is also

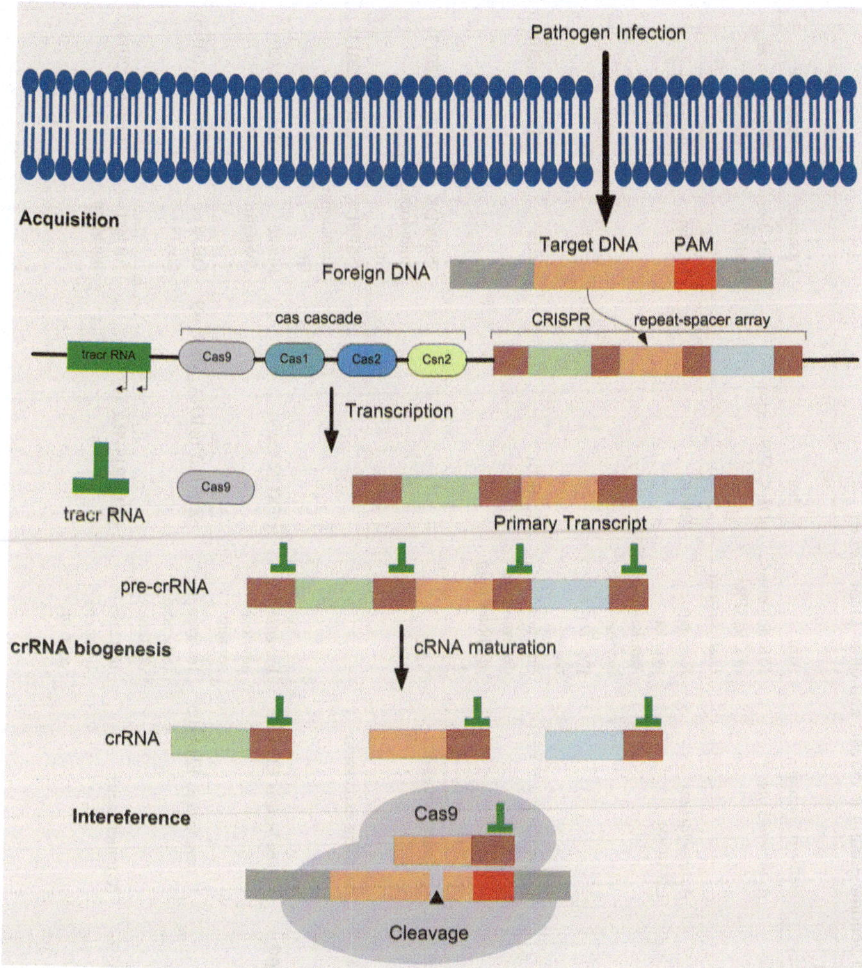

Fig. 11.2 An overview of CRISPR/Cas9 mediated genome editing. (Source-Jiang and Doudna 2017)

evident. Using this technique, the rice genes modulating grain number (42.5%), panicle architecture (27.5%), and grain size (57.5%) were mutated. The genes regulating these traits were Gn1a, DEP1, GS3, and IPA1; mutation in which led to increase in yield (Li et al. 2016a, b). Table 11.1 enlisted some of the selected examples, where CRISPR/Cas9 has led to the change in the genome of plants.

Multiplex genome editing involves either targeting multiple sites in a genome or incorporating multiple traits in a genome or different members of a gene in a genome (Yin et al. 2017). In rice, it was observed that editing of REC 8, PAIR 1, and OSD

Table 11.1 Application of CRISPR/Cas9 mediated genome editing in cereals and other plant species

Plant	Gene	Mutation	Editing efficiency	Delivery system	Aim	Vector	Promoter	Reference
Apple	PDS, TFL1	Target mutation	84–90%	*A. tumefaciens*	Efficacy check of CRISPR/Cas9 system	pDE-CAS9 vector	MdU3 or MdU6	Charrier et al. (2019)
Rice	Os8N3	Gene knockout	–	*A. tumefaciens* LBA4404	Resistance against *Xanthomonas oryzae* pv. *Oryzae* (Xoo)	PHAtC	OsU6a	Kim et al. (2019)
Kalanchoe fedtschenkoi	Phototropin 2	Gene knockout	–	*A. tumefaciens* GV3101	Function determination of phototropin 2	pKSE401	AtU6	Liu et al. (2019)
A. thaliana	PDS, FLS	Multiplex editing	1.1–5.6%	Transient expression	–	–	35SPPDK Arabidopsis U6	Li et al. (2013)
N. benthamiana	PDS	Multiplex editing	37–38%	Transient expression	–	–	35SPPDK Arabidopsis U6	Li et al. (2013)
Rice	Starch branching enzyme	–	26.7–40%	*A. tumefaciens* EHA105	Increase in amylose content	pCXUN-Cas9	Maize ubiquitin	Sun et al. (2017)
Rice	FAD	Gene knockout		*A. tumefaciens* EHA105	Increase in fatty acid content	pZH_OsU6gRNA_MMCas9	OsU6 or OsU3	Abe et al. (2018)
Tomato	*LCY-E*, *LCY-B1*, and *LCY-B2*, as well as *Blc*. *LCY-E*	Multiplex editing	0–95.83%s	A. tumefaciens	Increase in lycopene content	pYLCRISPR/Cas9	AtU6-1 or AtU6-29	Li et al. (2018a)

Maize	ARGOS8	Target mutation	60–98%	Biolistic method	Increased grain yield under drought stress	CPB	GOS2	Shi et al. (2017)
A. thaliana	MIR169a, MIR827a, TFL1	Gene deletion/ replacement	0.8%	*A. tumefaciens* GV3301	Establishment of gene replacement technique		At U6-26	Zhao et al. (2016)
Tomato	Pds	Knockout mutation	61.8–68.4%	–	Checking the efficacy of CRISPR system	pYLCRISPR/Cas9	–	Li et al. (2018b)
Tomato	*GABA-TP1*, *GABA-TP2*, *GABA-TP3*, *CAT9* and *SSADH*	Multiplex genome editing	0–56.82	–	GABA gene function determination	pYLCRISPR/Cas9	LacZ-AtU3d, AtU3d, AtU3b, AtU3b, AtU6-1 or AtU6-29	Li et al. (2018b)
Potato	GBBS	Targeted mutagenesis	Up to 67%	Transient transformation	Modulation in starch content	pEX-A2	AtU6	Andersson et al. (2017)
Cucumber	Elf4E	Base editing	–	*A. tumefaciens* GV3301	Resistance against virus	pRCS binary vector	AtU6	Chandrasekaran et al. (2016)
A. thaliana	ALS	Cytosine base editing	14.3–66.7%	*A. tumefaciens* GV3101	Imidazolone herbicide resistance	pHEE901	AtU6-26	Dong et al. (2020)
Petunia	NR	Site directed mutagenesis	17.8–21%	Direct delivery through RNP's	Nitrogen metabolism regulation	–	–	Subburaj et al. (2016)
Lotus japonicus	SYMRK	Targeted mutagenesis	35	*A. tumefaciens*EHA105, *A. rhizogenes*LBA1334	Symbioses regulation in legumes	pCAMBIA1300	LjU6	Wang et al. (2016)

(continued)

Table 11.1 (continued)

Plant	Gene	Mutation	Editing efficiency	Delivery system	Aim	Vector	Promoter	Reference
Grape	MLO-7	Targeted mutagenesis	0.1%	Direct delivery through RNP's	Improved tolerance against powdery mildew	–	–	Malnoy et al. (2016)
Orange	LOB1	Targeted mutagenesis	11.5–64.7	*A. tumefaciens*EHA105	Resistance against canker disease	pCas9-GN	AtU6	Peng et al. (2017)

1 together led to the production of heterozygous F1 hybrid rice while editing of another gene MATRILINEAL results in haploid seeds. Editing all the four genes together enabled clonal propagation of rice plants via seeds (Wang et al. 2019). Any novel technology always has a scope of improvement, which also serves true for CRISPR/Cas9 where the various number of nuclease variants have been established, which helps in subsiding the shortcomings of CRISPR/Cas9 and also increase its effectiveness. One such variant is dCas 9 or dead Cas 9. This variant is developed by inactivation of nuclease activity while retaining the recognition activity to allow identification of target DNA by guide RNA (Gilbert et al. 2013). Gene regulation via CRISPR/Cas recruits dCas9 for silencing and activation of genes. Cas9 and sgRNA complex binds to the target DNA and blocks the passage for the movement of RNA polymerase, thereby interfering with the transcript elongation (Khatodia et al. 2016). dCas9 targets transcription initiation and elongation and restricts change in target sequences or cell death by genome disruption (Larson et al. 2013; Cho et al. 2018). When dCas9 is fused with transcriptional activator domain or effector such as VP16, VP64, it either enhance or repress the gene expression. In the case of tobacco, the fusion of dCas9 with EDLL and TAL effectors (activation) and SRDX (repressor) led to strong induction of activation of Bs3::uidA gene and repression of endogenous, respectively (Piatek et al. 2015). dCas9 is also associated with epigenome editing. Epigenetics causes modification in DNA which are also heritable but does not alter the DNA sequence. DNA methylation and demethylation, histone modification, hydroxyl-methylation, gene imprinting, chromatin remodeling, ubiquitination, and noncoding RNA are some modes to bring out epigenetics (Xie et al. 2018). Epigenome editing plays with epigenetic markers either by modifying them at a locus to determine the underlying mechanism involved in the interaction between the changes and its subsequent effect on transcriptional regulation and phenotype or by epigenetic treatment of the problem causing gene. DNA methylation is an epigenetic modification that adds a methyl group to the cytosine ring by DNA methyltransferase at the fifth position (Mercé et al. 2020). In Arabidopsis, targeted methylation and demethylation have been reported recently (Gallego-Bartolomé et al. 2018). Attempts have been made to further improve the epigenome editing efficiency by incorporation of SunTag to the dCas9 epi-effector complex. SunTag is a repeating peptide array that has a potential to bind with multiple protein copies simultaneously and has been recently employed in regulating flowering by targeting the FWA gene in *Arabidopsis* (Papikian et al. 2019). Since CRISPR/Cas9 involves site repairing either by homologous DNA direct repair (HDR) or NHEJ, chances of gene replacement and knockout is always prevalent. In rice, the 5-enolpyruvylshikimate-3-phosphate synthase gene was replaced by herbicide glyphosate tolerant protein with an editing efficiency of 2.0% (Li et al. 2016a, b) (Fig. 11.3).

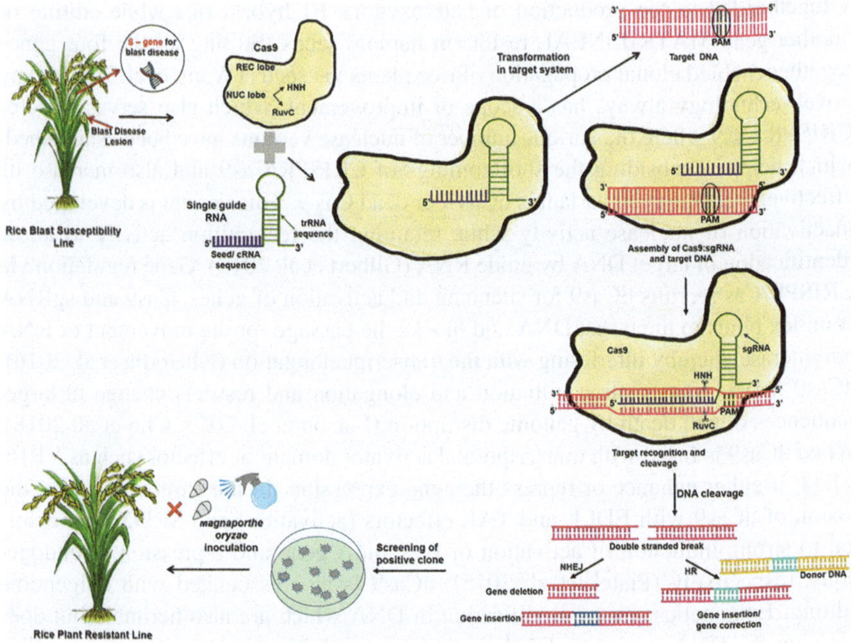

Fig. 11.3 A brief overview of mechanism involved in the development of pathogen-resistant line in rice via CRISPR/Cas 9 system. *NUC* nuclease lobe, *REC* recognition lobe, *HNH and RuvC* are the nuclease domains of NUC, *sgRNA* single guide RNA, *PAM* protospacer adjacent motif, *NHEJ* non-homologous end joining, *HR* homologous recombination.

11.4 Understanding the Host–Pathogen Interaction Events during Susceptibility and Resistance

Plant diseases are the result of interaction between two organisms, which is pathogen and host plant. Upon host–pathogen interactions, there are two kinds of successful events, i.e. degree of susceptibility and degree of resistance controlled by segment of its DNA making up the genes. In resistant reaction, the microbial molecules called '*elicitors*' are released from the pathogen, and correspondingly, recognition of elicitor by receptor molecules triggers the defence responses and makes the host resistant. Whereas in the susceptible reaction, there was no specific pathogen mediated elicitor molecules released from the host, and or no receptor site present in the host plant or both the elicitor and receptor site when absent in the host plants makes the plant susceptible; hence no defence reaction was triggered on it. In both the reactions, elicitors and receptor molecules play a key role in which resistance genes (*R*-genes) operate functions in resistant host. In contrast, virulence genes operate in the susceptible host (Agrios 2005). The mechanisms of molecular interactions between the pathogens and host systems are diagrammatically presented in Fig. 11.4.

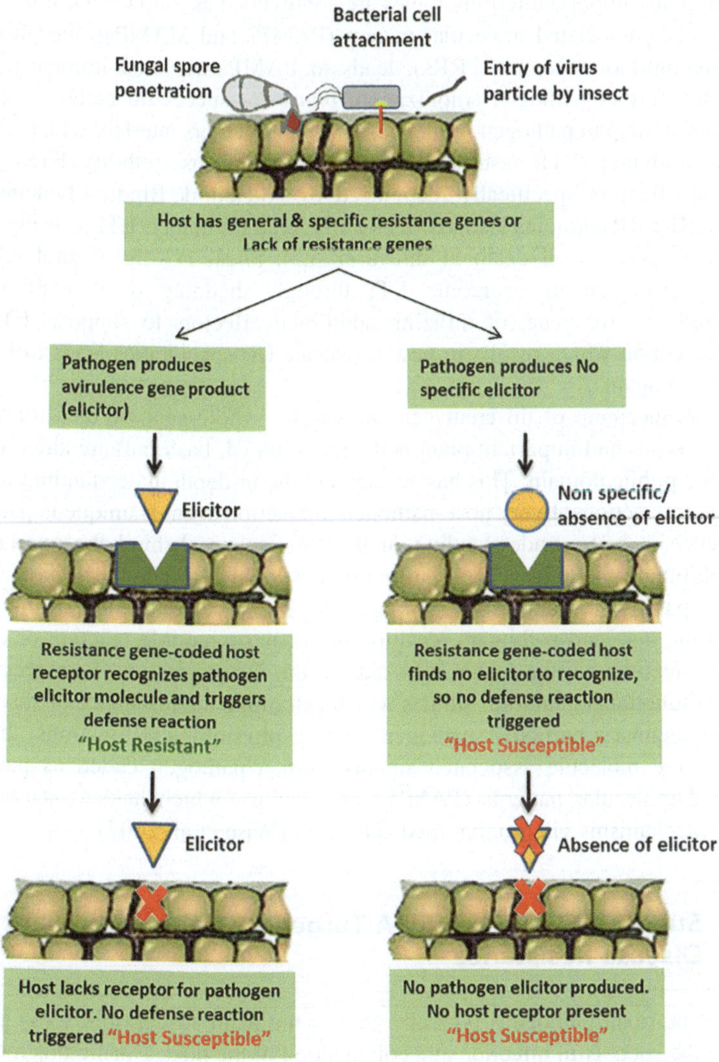

Fig. 11.4 Understanding of host–pathogen interaction during susceptibility and resistance reactions

During pathogen infection, host plants respond to two different branched innate-immune system in which the first type branch responds and recognizes the microbial mediated signals, including non-pathogenic organisms. In the second branch, host plants respond to the invading pathogen secreted virulence factors by direct or their effects on the host target. At present, there are four phased plant immune systems which are clearly reviewed by Jones and Dangl (2006), also called as 'zigzag' model, in which many important terminologies are abbreviated as follows: *phase-I*—pathogen-microbes associated molecular pattern (PAMPs and MAMPs), recognized by pattern-recognition-receptor (PRRs), leads to PAMPs-triggered immunity which restrict the further pathogen colonization; *phase-II*—successful pathogen uses its effectors that help in pathogen virulence, in which effectors interfere with pathogen-triggered immunity (PTI), resulting to effector-triggered susceptibility (ETS); *phase-III*—these effectors specifically recognized by Nucleotide-Binding-Leucine Rich Repeat (NB-LRR) proteins lead to effector-triggered immunity (ETI) causing hypersensitive response (cell death at infection site); *phase-IV*—the natural selection functions pathogen to overcome ETI through shedding or diversifying the recognized effector gene, or utilizing additional effectors to suppress ETI. The natural selection which results in new resistance (*R*) specificities leads to ETI can be triggered again.

The advancement of different genomics tools, whole-genome data for various important crops and important plant pathogens (fungal, bacterial) are already available in the public domain. This has accelerated the in-depth understanding of host–pathogen interactions. Every host–pathogen interaction event is unique in pathogenesis. Hence, a better understanding of the mechanisms behind the resistance or susceptibility is of utmost importance for developing disease resistance. During the host–pathogen relationship, the specific event activates signal transaction between the host and pathogen, resulting in either susceptible reaction or defence reaction. Genome editing technology aids in the determination of key regulatory genes via functional genomics studies which determine the resistance or susceptible functions against a particular pathogen. During physiological functions, the plant perceives the molecule associated signals from a pathogen called as pathogen-associated molecular patterns (PAMPs) or effectors, which lead to activation of defence mechanisms via programmed cell death (Wise et al. 2007).

11.5 Susceptibility Gene(S): A Target for Enhancing Plant Disease Resistance

As an event, plant pathogens typically exploit the plant susceptibility (*S*) genes to facilitate the successful infection and colonization of the host. During this molecular mechanism between the host and pathogen, three main interactions take place in the susceptible genotypes such as: (i) basic compatibility (assist for recognition and penetration); (ii) sustained compatibility (assist for pathogen proliferation and spread within the host); and (iii) negative regulation of immune signals (assisting to control immune suppression in the host) (van Schie and Takken 2014). For genome editing of crop plants, both resistant and susceptible genes are the options in which resistant genes are dominant and resistance against disease is provided by manipulating these genes in genotypes with susceptible genes, which are recessive and associated with a fitness cost. Whereas susceptible gene-mediated resistance is broad-spectrum and pathogen non-specific, and in which impaired pathways are implicated, thereby infection process including pre-penetration, penetration, and post-penetration are restricted in the plant systems. Therefore, susceptible gene-based genome editing can ensure broad-spectrum resistance against plant pathogens with durable and constitute effects. Various susceptible genes classes and their molecular mechanism are reported in many crops (Zaidi et al. 2018). For example, in rice *OsMPK5* gene (Xie and Yang 2013) and *OsSWEET11, OsSWEET14* (Jiang et al. 2013) genes were used against *Magnaporthe oryzae*, *Burkholderia glumae,* and *Xanthomonas oryzae* pv. *oryzae,* respectively.

11.6 Genome Editing for Resistance against Fungal Pathogens in Cereal Crops

Among all the phytopathogens, the fungal pathogens are the most dominant in causing various plant diseases worldwide and possess great negative impact on agriculture food production. Due to broad geographical distribution and high genetic diversity, they produce enormous infective spores, and can quickly colonize the new hosts, and can break the resistance gene (*R* gene) mediated resistance, and display resistance to various fungicides, thus always making challenges for plant disease control (Doehlemann et al. 2017; Yin and Qiu 2018). Recently, genome editing technology has addressed this challenge by modifying the susceptible host gene (S gene). Wang et al. (2014) used TALEN and CRISPR based technology to edit wheat gene*Mlo* (*mildew resistance locus o*)and confirmed that the edited wheat plants showed resistance to *Blumeria graminis* f. sp. *tritici*, the causal agent of powdery mildew disease when simultaneous mutation was carried out in all six copies of *TaMlo*. The results of TALEN and CRISPR-based genome editing technology showed that genome editing is a superior tool in disturbing or modifying the targets within polyploidy fungal genomes. Rice blast caused by *Magnaporthe oryzae* is considered top first fungal plant pathogens worldwide and also cause devastating effects in rice production worldwide (Dean et al. 2012). The modification of targeted gene *OsERF922*by using CRISPR/Cas9 tool was generated rice *Oserf922* knockout mutants (Liu et al. 2012). These null mutants decreased pathogen virulence and lead to enhanced blast resistance without changing the other plant physiological functions. These results suggest us that the targeted gene editing of negative regulators and/or susceptibility genes through genome editing represents a powerful tool for plant disease management. Further, the *EDR1* is highly conserved across plant species similar to *Mlo,* and by adopting CRISPR/Cas9 technology the *Taedr1*gene region was edited in wheat plant system by targeting three homologs of *EDR1*. The results found that the *Taedr1* mutant plants were exhibited resistant to *Blumeria graminis* f. sp. *tritici* but without mildew-induced cell death (Table 11.2).

Table 11.2 Key examples of genome editing for improving the crop resistance to various plant pathogens

Crops	Targeted gene	Genome editing technique	Molecular function related to disease	References
Rice	*11 N3/SWEET14*	TALENs	Bacterial blight	Li et al. (2012)
	SWEET11 and *SWEET14*	CRISPR/ Cas9	Bacterial blight	Jiang et al. (2013)
	OsMPK5	CRISPR/ Cas9	Blast	Xie and Yang (2013)
	SWEET13	TALENs	Bacterial blight	Zhou et al. (2015)
	ERF922	CRISPR/Cas 9	Blast	Wang et al. (2016)
	SWEET14	TALENs	Bacterial blight	Blanvillain-Baufumé et al. (2017)
	09 g29100	TALENs	Bacterial leaf streak	Cai et al. (2017)
	USTA ustiloxin and UvSLT2 MAP kinase	CRISPR/ Cas9	False smut	Liang et al. (2017)
	SEC3A	CRISPR/ Cas9	Blast	Ma et al. (2018)
	Xa10-Ni and Xa23-Ni	TALENs	Bacterial blight	Wang et al. (2017)
	ALB1, SDI1 and RSY1	CRISPR/ Cas9 (RNP)	Blast	Foster et al. (2018)
	BSR-k1	CRISPR/ Cas9	Bacterial blight	Zhou et al. (2018)
	eIF4G	CRISPR/ Cas9	Rice tungro spherical virus (RTSV)	Macovei et al. (2018)
	SWEET11, SWEET13 and *SWEET14*	CRISPR/ Cas9	Bacterial blight	Oliva et al. (2019)
	TMS5, Pi21, and Xa13	CRISPR/ Cas9	Bacterial blight	Li et al. (2019)
Wheat	TaMLO/exon	TALEN	Powdery mildew	Wang et al. (2014)
	TaEDR1/exon	CRISPR/ Cas9	Powdery mildew	Zhang et al. (2017)
	TaMLO	CRISPR/ Cas9	Powdery mildew	Shan et al. (2014)

11.7 Genome Editing for Resistance Against Bacterial Pathogens in Cereal Crops

In nature, bacteria are diverse, omnipresent, and play beneficial and harmful roles in the plant living system (Yin and Qiu 2018). Plant pathogenic bacteria possess an important role and cause significant important threat to agriculture, by causing several diseases such as leaf spots, blight, vascular wilts, soft rots, tumours, and galls, which result in severe yield loss (Vale et al. 2001; Zeng et al. 2010). Once the occurrence of these bacterial diseases in an area which form epidemic and are difficult to control because the bacteria multiply rapidly and can spread quickly in several ways (Yin and Qiu 2018). There are some reports which have been conducted through the CRISPR-Cas9 system to develop a transgenic plant against bacterial pathogens that were listed in Table 11.2. Through CRISPR-Cas9 tool, Zhou et al. (2014) developed a rice resistance line against bacterial blight pathogen *Xanthomonas oryzae* pv. *oryzae*, by mutating the *OsSWEET13*gene which is involved to host susceptibility in sucrose transportation during pathogenesis. After protein involvement, i.e. *PthXo2,* the expression of *OsSWEET13* is induced by the *X. oryzae* effector and thus boost the host susceptibility, therefore knocked out of this *OsSWEET13* gene will lead to creation of null mutant, which resulted in improved resistance against *X. oryzae* (Zhou et al. 2015).

11.8 Genome Editing for Resistance against Viral Pathogens in Cereal Crops

Most of the viral diseases transmitted by insect vectors are difficult to manage. Plant viruses are obligate parasites and multiply rapidly only in the plant living systems. Plant viruses with a larger host range infect almost all agricultural crop plants. Management through chemicals is difficult due to the typical intercellular nature of host invasion. Most plant viruses are transmitted by diverse insect vectors; hence, pesticide application is not advisable in every situation. So, pesticide application is not advisable; keeping the environment pollution and diverse vectors plays a key role in transmitting the virus on the new hosts. The development of resistant cultivars by traditional approach is well known, and many resistant cultivars are developed against various virus diseases. Over the time, such varieties have become susceptible once the virus pathogenesis systems evolve very quickly (Ansari et al. 2020). Widely viral nucleocapsids were used to develop virus resistance plants via transgenic expression which is called as pathogen-derived resistance. Recently, RNAi-mediated plant virus resistance is considered as the most efficient method (Ding and Voinnet 2007). Therefore, the genome editing approach is one we can look for development resistant cultivar with durable and long-lasting effects. Macovei et al. (2018) recently developed a new resistance cultivar against rice tungro spherical virus (RTSV). Rice tungro spherical virus resistant plants were obtained through transgene free T2 plants and not showed any mutational effects in the off-target sites. By altering host genes, the infection spread of viruses is checked effectively, generating resistance against RTSV.

11.9 Future Perspectives

In the present scenario, ensuring food security to all is questionable due to insufficient food availability caused by various environmental factors, especially biotic factors, and the application of chemicals to plants for increased production has hazardous effects on health. Here, CRISPR/Cas9 poses a promising solution to the problem. Owing to an increased interest of the scientific community in understanding the agronomic traits, plant defence mechanism and easy accessibility of various bioinformatics tools, a substantial amount of information on genes, mode of action, and their function is available on the public databases from which target gene can be selected, and vice versa CRISPR/Cas can be used for determination of genes. The question is how this information available can be exploited to the best of our benefit. It has been already discussed previously that pathogens have the potential to escape from resistance, and these problems in conventional breeding have been dealt with stacking of genes or incorporating NBS-LRR receptor kinases. Apart from generating allelic mutants, CRISPR/Cas can also be used to stack up genes by encompassing multiple sgRNA and that too without employing the nearby genetic regions. Targeting a single gene for generating resistance against pathogens or stacking up of multiple genes is an efficient way to curb the spread of pathogens. CRISPR/Cas technique facilitates both single target editing and multiplex editing. At the gene level, this is achieved by targeted DNA break via Cas9 and DNA repair by NHEJ. However, when alleles of a gene and not gene alone is involved, HDR is more suitable for repair mechanism. HDR allows the entry of synthetically designed sequences and thus can broaden the range of mutations in any genome. However, due to its low efficacy, HDR is less preferred. Thus, there is a need to work upon it. Another point that needs attention is targeted mutagenesis itself. It creates indels which are random and usually cause loss of function, due to which always there is a possibility of damage to the host system also. An example of it is elongation initiation factor 4E that mediates the translation of RNA into proteins. Mutation in genes provided resistance against the virus in various crop plants; however, it also affects the host plant as the gene is important for both virus and plant. If a base mutation can be induced in such a manner that only the virus and not the host plant is affected, then the purpose can be solved. Base editing or genome editing is more influential than targeted mutagenesis. The transition from C to T and A to G has come into existence, but its application at a wider scale still needs to be established. Despite its specificity, off-target effects have been reported, which can be resolved by developing more efficient bioinformatics tools to design specific sgRNA and modifying nuclease. Some already available nuclease variants such as Cas9 nickase, dCas9 can be employed more frequently. Variations in promoter for cas9 and sgRNA can also be brought about that are specific for gene expression regulation or are constitutive ones or are tissue or developmental stage-specific or can induce the expression of a gene.

For successful editing events, the correct choice of Cas9 and sgRNA is equally important as the efficient regeneration and transformation protocol in target plants. However, in many plant systems, the regeneration and transformation are unsuccessful, and further screening of mutants is difficult. In such cases, alternate culture free systems like ribo-nucleoproteins, nanoparticles, viral delivery, etc. can be employed, which could further enhance the efficacy of the editing event. The major concern associated with genome editing is the regulatory concerns posed by the ethical committees. Court of Justice of the European Union issued a judgement that genome edited organisms are to be categorized under genetically modified organisms. Also, despite genome-edited lines being T-DNA free; the problem of its acceptance by the public needs to be addressed. In this regard, it is mandatory to ensure that the lincs are transgene free for which genetic segregation followed by genotypic validation is required.

11.10 Conclusion

Presently, more reliable techniques are needed to edit the crops genome to ensure food safety and security. Over the years, tremendous progress in the development of genome editing tools has been marked. CRISPR/Cas 9 in recent years has gained immense popularity owing to its features like simplicity, efficiency, less off-target effects, etc. CRISPR/Cas 9 provides a platform by which one can easily edit the genome according to our requirements and generate plants with desired traits. The development of pathogen-resistant cultivars is the pre-requisite for the agricultural industry. This technique gives us the chance to design a toolbox that could repair most of the concerns of the plant sector. Modification in nucleases, promoters, and target site selection is in continual progress to improve editing efficiency. None of the scientific inventions has been perfect; each contains certain drawbacks and limitations. It is up to us how we overcome these limitations and exploit the techniques to the maximal; similar holds for CRISPR/Cas9, which offers a range of advantages but the scope for improvement remains open.

Acknowledgement The authors are highly thankful to the project on **"Editing rice genes through CRISPR/Cas9 technology for enhanced and durable blast resistance in rice"** (Grant No. BT/PR32125/AGIII/103/1147/2019) sponsored by the Department of Biotechnology (DBT), India. We are also grateful to Directors, ICAR-IARI, New Delhi and ICGEB, New Delhi for support and encouragement.

References

Abe K, Araki E, Suzuki Y, Toki S, Saika H (2018) Production of high oleic/low linoleic rice by genome editing. Plant Physiol Biochem 131:58–62. https://doi.org/10.1016/j.plaphy.2018.04.033

Acquaah G (2015) Conventional plant breeding principles and techniques. In: Al-Khayri J, Jain S, Johnson D (eds) Advances in plant breeding strategies: breeding, biotechnology and molecular tools. Springer, Cham. https://doi.org/10.1007/978-3-319-22521-0

Aday S, Aday MS (2020) Impact of COVID-19 on the food supply chain. Food Qual Saf 4(4): 167–180

Agrios GN (2005) Genetics of plant disease. In: Plant pathology, Elsevier, 5th Edn, pp 139–142

Andersson M, Turesson H, Nicolia A, Fält AS, Samuelsson M, Hofvander P (2017) Efficient targeted multiallelic mutagenesis in tetraploid potato (*Solanum tuberosum*) by transient CRISPR-Cas9 expression in protoplasts. Plant Cell Rep 36(1):117–128. https://doi.org/10.1007/s00299-016-2062-3

Andolfo G, Iovieno P, Frusciante L, Ercolano MR (2016) Genome-editing technologies for enhancing plant disease resistance. Front Plant Sci 7:1813. https://doi.org/10.3389/fpls.2016.01813

Ansari WA, Chandanshive SU, Bhatt V, Nadaf AB, Vats S, Katara JL, Deshmukh R (2020) Genome editing in cereals: approaches, applications and challenges. Int J Mol Sci 21(11):4040. https://doi.org/10.3390/ijms21114040

Arora L, Narula A (2017) Gene editing and crop improvement using CRISPR-Cas9 system. Front Plant Sci 8:1932. https://doi.org/10.3389/fpls.2017.01932

Barrangou R, Fremaux C, Deveau H, Richards M, Boyaval P, Moineau S, Romero DA, Horvath P (2007) CRISPR provides acquired resistance against viruses in prokaryotes. Science 315 (5819):1709–1712. https://doi.org/10.1126/science.1138140

Blanvillain-Baufumé S, Reschke M, Solé M, Auguy F, Doucoure H, Szurek B, Meynard D, Portefaix M, Cunnac S, Guiderdoni E, Boch J (2017) Targeted promoter editing for rice resistance to Xanthomonas oryzae pv. oryzae reveals differential activities for SWEET 14-inducing TAL effectors. Plant Biotechnol J 15(3):306–317. https://doi.org/10.1111/pbi.12613

Boch J, Bonas U (2010) Xanthomonas AvrBs3 family-type III effectors: discovery and function. Annu Rev Phytopathol 48:419–436. https://doi.org/10.1146/annurev-phyto-080508-081936

Cai L, Cao Y, Xu Z, Ma W, Zakria M, Zou L, Cheng Z, Chen G (2017) A transcription activator-like effector Tal7 of *Xanthomonasoryzae*pv. *oryzicola* activates rice gene Os09g29100 to suppress rice immunity. Sci Rep 7(1):5089. https://doi.org/10.1038/s41598-017-04800-8

Carroll D (2016) The development and use of zinc-finger nucleases. Ingenome editing. Springer, New York, NY, pp 15–28

Carte J, Wang R, Li H, Terns RM, Terns MP (2008) Cas6 is an endoribonuclease that generates guide RNAs for invader defense in prokaryotes. Genes Dev 22(24):3489–3496. https://doi.org/10.1101/gad.1742908

Chandrasekaran J, Brumin M, Wolf D, Leibman D, Klap C, Pearlsman M, Sherman A, Arazi T, Gal-On A (2016) Development of broad virus resistance in non-transgenic cucumber using CRISPR/Cas9 technology. Mol Plant Pathol 17:1140–1153. https://doi.org/10.1111/mpp.12375

Charrier A, Vergne E, Dousset N, Richer A, Petiteau A, Chevreau E (2019) Efficient targeted mutagenesis in apple and first time edition of pear using the CRISPR-Cas9 system. Front Plant Sci 10:40. https://doi.org/10.3389/fpls.2019.00040

Cho S, Shin J, Cho BK (2018) Applications of CRISPR/Cas system to bacterial metabolic engineering. Int J Mol Sci 19(4):1089. https://doi.org/10.3390/ijms19041089

Gilbert LA, Larson MH, Morsut L, Liu Z, Brar GA, Torres SE, Stem-Ginossar N, Brandman O, Whitehead EH, Doudna JA, Lim WA (2013) CRISPR-mediated modular RNA-guided regulation of transcription in eukaryotes. Cell 154:442–451. https://doi.org/10.1016/j.cell.2013.06.044

Daboussi F, Stoddard TJ, Zhang F (2015) Engineering meganuclease for precise plant genome modification. InAdvances in new technology for targeted modification of plant genomes. Springer, New York, NY, pp 21–38. https://doi.org/10.1007/978-1-4939-2556-8_2

Darwin C, Murray J. (2010) The variation of animals and plants under domestication. London: John Murray. 1, 12

Dean R et al (2012) The Top 10 fungal pathogens in molecular plant pathology. Mol Plant Pathol 13 (4):414–430. https://doi.org/10.1111/j.1364-3703.2011.00783.x

Deltcheva E, Chylinski K, Sharma CM, Gonzales K, Chao Y, Pirzada ZA, Eckert MR, Vogel J, Charpentier E (2011) CRISPR RNA maturation by trans-encoded small RNA and host factor RNase III. Nature 471(7340):602–607. https://doi.org/10.1038/nature09886

Deveau H, Barrangou R, Garneau JE, Labonté J, Fremaux C, Boyaval P, Romero DA, Horvath P, Moineau S (2008) Phage response to CRISPR-encoded resistance in Streptococcus thermophilus. J Bacteriol 190(4):1390–1400. https://doi.org/10.1128/JB.01412-07

Dietz-Pfeilstetter A (2010) Stability of transgene expression as a challenge for genetic engineering. Plant Sci 79(3):164–167. https://doi.org/10.1016/j.plantsci.2010.04.015

Ding SW, Voinnet O (2007) Antiviral immunity directed by small RNAs. Cell 130:413–426. https://doi.org/10.1016/j.cell.2007.07.039

Doehlemann G, Ökmen B, Zhu W, Sharon A (2017) Plant pathogenic fungi. In: Heitman J, Howlett B, Crous P, Stukenbrock E, James T, Gow N (eds) The fungal kingdom. ASM Press, Washington, DC, pp 703–726

Dong H, Wang D, Bai Z, Yuan Y, Yang W, Zhang Y, Ni H, Jiang L (2020) Generation of imidazolinone herbicide resistant trait in Arabidopsis. PLoS One 15(5):e0233503. https://doi.org/10.1371/journal.pone.0233503

Durai S, Mani M, Kandavelou K, Wu J, Porteus MH, Chandrasegaran S (2005) Zinc finger nucleases: custom-designed molecular scissors for genome engineering of plant and mammalian cells. Nucleic Acids Res 33(18):5978–5990. https://doi.org/10.1093/nar/gki912

Elrod-Erickson M, Pabo CO (1999) Binding Studies with Mutants of Zif268: contribution of individual side chains to binding affinity and specificity in the Zif268 zinc finger-DNA complex. J Biol Chem 274(27):19281–19285. https://doi.org/10.1074/jbc.274.27.19281

El-Mounadi K, Morales-Floriano ML, Garcia-Ruiz H (2020) Principles, applications, and biosafety of plant genome editing using CRISPR-Cas9. Front Plant Sci 13(11):56. https://doi.org/10.3389/fpls.2020.00056

Foster AJ, Martin-Urdiroz M, Yan X, Wright HS, Soanes DM, Talbot NJ (2018) CRISPR-Cas9 ribonucleoprotein-mediated co-editing and counterselection in the rice blast fungus. Sci Rep 8(1):1–2. https://doi.org/10.1038/s41598-018-32702-w

Fuchs M (2017) Pyramiding resistance-conferring gene sequences in crops. Curr Opin Virol 26:36–42. https://doi.org/10.1016/j.coviro.2017.07.004

Gallagher RR, Li Z, Lewis AO, Isaacs FJ (2014) Rapid editing and evolution of bacterial genomes using libraries of synthetic DNA. Nat Protoc 9(10):2301. https://doi.org/10.1038/nprot.2014.082

Gallego-Bartolomé J, Gardiner J, Liu W, Papikian A, Ghoshal B, Kuo HY, Zhao JM, Segal DJ, Jacobsen SE (2018) Targeted DNA demethylation of the Arabidopsis genome using the human TET1 catalytic domain. Proc Natl Accad Sci USA 115(9):E2125–E2134. https://doi.org/10.1073/pnas.1716945115

Garneau JE, Dupuis MÈ, Villion M, Romero DA, Barrangou R, Boyaval P, Fremaux C, Horvath P, Magadán AH, Moineau S (2010) The CRISPR/Cas bacterial immune system cleaves bacteriophage and plasmid DNA. Nature 468(7320):67–71. https://doi.org/10.1038/nature09523

Garside EL, Schellenberg MJ, Gesner EM, Bonanno JB, Sauder JM, Burley SK, Almo SC, Mehta G, MacMillan AM (2012) Cas5d processes pre-crRNA and is a member of a larger family of CRISPR RNA endonucleases. RNA 18(11):2020–2028. https://doi.org/10.1261/rna.033100.112

Gelvin SB (2003) Agrobacteriummediated plant transformation: the biology behind the "gene-jockeying" tool. Microbiol Mol Biol Rev 67(1):16–37. https://doi.org/10.1128/mmbr.67.1.16-37.2003

Gori JL, Hsu PD, Maeder ML, Shen S, Welstead GG, Bumcrot D (2015) Delivery and specificity of CRISPR/Cas9 genome editing technologies for human gene therapy. Hum Gene Ther 26(7):443–451. https://doi.org/10.1089/hum.2015.074

Hale CR, Zhao P, Olson S, Duff MO, Graveley BR, Wells L, Terns RM, Terns MP (2009) RNA-guided RNA cleavage by a CRISPR RNA-Cas protein complex. Cell 139(5):945–956. https://doi.org/10.1016/j.cell.2009.07.040

Haurwitz RE, Jinek M, Wiedenheft B, Zhou K, Doudna JA (2010) Sequence-and structure-specific RNA processing by a CRISPR endonuclease. Science 329(5997):1355–1358. https://doi.org/10.1126/science.1192272

Hille F, Charpentier E (2016) CRISPR-Cas: biology, mechanisms and relevance. Philos Trans R Soc B 371(1707):20150496. https://doi.org/10.1098/rstb.2015.0496

Hou H, Atlihan N, Lu ZX (2014) New biotechnology enhances the application of cisgenesis in plant breeding. Front Plant Sci 5:389. https://doi.org/10.3389/fpls.2014.00389

Hsu PD, Lander ES, Zhang F (2014) Development and applications of CRISPR-Cas9 for genome engineering. Cell 157(6):1262–1278. https://doi.org/10.1016/j.cell.2014.05.010

Ishino Y, Shinagawa H, Makino K, Amemura M, Nakata A (1987) Nucleotide sequence of the *iap* gene, responsible for alkaline phosphatase isozyme conversion in *Escherichia coli*, and identification of the gene product. J Bacteriol 169(12):5429–5433. https://doi.org/10.1128/jb.169.12.5429-5433.1987

Jiang F, Doudna JA (2017) CRISPR–Cas9 structures and mechanisms. Annu Rev Biophys 46:505–529. https://doi.org/10.1146/annurev-biophys-062215-010822

Jiang W, Zhou H, Bi H, Fromm M, Yang B, Weeks DP (2013) Demonstration of CRISPR/Cas9/sgRNA-mediated targeted gene modification in arabidopsis, tobacco, sorghum and rice. Nucleic Acids Res 41(20):e188. https://doi.org/10.1093/nar/gkt780

Jinek M, Chylinski K, Fonfara I, Hauer M, Doudna JA, Charpentier E (2012) A programmable dual-RNA–guided DNA endonuclease in adaptive bacterial immunity. Science 337(6096):816–821. https://doi.org/10.1126/science.1225829

Jones JD, Dangl JL (2006) The plant immune system. Nature 444(7117):323–329. https://doi.org/10.1038/nature05286

Khatodia S, Bhatotia K, Passricha N, Khurana SM, Tuteja N (2016) The CRISPR/Cas genome-editing tool: application in improvement of crops. Front Plant Sci 7:506. https://doi.org/10.3389/fpls.2016.00506

Kim H, Kim JS (2014) A guide to genome engineering with programmable nucleases. Nat Rev Genet 15(5):321–334. https://doi.org/10.1038/nrg3686

Kim YA, Moon H, Park CJ (2019) CRISPR/Cas9-targeted mutagenesis of Os8N3 in rice to confer resistance to *Xanthomonasoryzae*pv. *oryzae*. Rice 12(1):1–3. https://doi.org/10.1186/s12284-019-0325-7

Larson MH, Gilbert LA, Wang X, Lim WA, Weissman JS, Qi LS (2013) CRISPR interference (CRISPRi) for sequence-specific control of gene expression. Nat Protoc 8(11):2180–2196. https://doi.org/10.1038/nprot.2013.132

Li T, Liu B, Spalding MH, Weeks DP, Yang B (2012) High-efficiency TALEN-based gene editing produces disease-resistant rice. Nat Biotechnol 30(5):390. https://doi.org/10.1038/nbt.2199

Li JF, Norville JE, Aach J, McCormack M, Zhang D, Bush J, Church GM, Sheen J (2013) Multiplex and homologous recombination–mediated genome editing in Arabidopsis and *Nicotianabenthamiana* using guide RNA and Cas9. Nat Biotechnol 31(8):688–691. https://doi.org/10.1038/nbt.2654

Li J, Meng X, Zong Y, Chen K, Zhang H, Liu J, Li J, Gao C (2016a) Gene replacements and insertions in rice by intron targeting using CRISPR–Cas9. Nat Plants 2(10):1–6. https://doi.org/10.1038/nplants.2016.139

Li M, Li X, Zhou Z, Wu P, Fang M, Pan X, Lin Q, Luo W, Wu G, Li H (2016b) Reassessment of the four yield-related genes Gn1a, DEP1, GS3, and IPA1 in rice using a CRISPR/Cas9 system. Front Plant Sci 7:377. https://doi.org/10.3389/fpls.2016.00377

Li R, Li R, Li X, Fu D, Zhu B, Tian H, Luo Y, Zhu H (2018a) Multiplexed CRISPR/Cas9 mediated metabolic engineering of γ aminobutyric acid levels in *Solanumlycopersicum*. Plant Biotechnol J 16(2):415–427. https://doi.org/10.1111/pbi.12781

Li X, Wang Y, Chen S, Tian H, Fu D, Zhu B, Luo Y, Zhu H (2018b) Lycopene is enriched in tomato fruit by CRISPR/Cas9-mediated multiplex genome editing. Front Plant Sci 9:559. https://doi.org/10.3389/fpls.2018.00559

Liang Z, Chen K, Li T, Zhang Y, Wang Y, Zhao Q, Liu J, Zhang H, Liu C, Ran Y, Gao C (2017) Efficient DNA-free genome editing of bread wheat using CRISPR/Cas9 ribonucleoprotein complexes. Nat Commun 8:14261. https://doi.org/10.1038/ncomms14261

Li M, Wang R, Zhao D, Xiang H (2014) Adaptation of the Haloarcula hispanica CRISPR-Cas system to a purified virus strictly requires a priming process. Nucleic Acids Res 42(4):2483–2492. https://doi.org/10.1093/nar/gkt1154

Li S, Shen L, Hu P, Liu Q, Zhu X, Qian Q, Wang K, Wang Y (2019) Developing disease-resistant thermosensitive male sterile rice by multiplex gene editing. J Integr Plant Biol 61(12):1201–1205. https://doi.org/10.1111/jipb.12774

Liu D, Chen X, Liu J, Ye J, Guo Z (2012) The rice ERF transcription factor OsERF922 negatively regulates resistance to *Magnaportheoryzae* and salt tolerance. J Exp Bot 63:3899–3911. https://doi.org/10.1093/jxb/ers079

Liu D, Chen M, Mendoza B, Cheng H, Hu R, Li L, Trinh CT, Tuskan GA, Yang X (2019) CRISPR/Cas9-mediated targeted mutagenesis for functional genomics research of crassulacean acid metabolism plants. J Exp Bot 70(22):6621–6219. https://doi.org/10.1093/jxb/erz415

Lusser M, Davies HV (2013) Comparative regulatory approaches for groups of new plant breeding techniques. New Biotechnol 30(5):437–446

Ma J, Chen J, Wang M, Ren Y, Wang S, Lei C, Cheng Z (2018) Disruption of OsSEC3A increases the content of salicylic acid and induces plant defense responses in rice. J Exp Bot 69(5):1051–1064. https://doi.org/10.1093/jxb/erx458

Macovei A, Sevilla NR, Cantos C, Jonson GB, Slamet-Loedin I, Čermák T, Voytas DF, ChoiI R, Chadha-Mohanty P (2018) Novel alleles of rice eIF4G generated by CRISPR/Cas9-targetedmutagenesis confer resistance to Rice tungro spherical virus. Plant Biotechnol J16 (11):1918–1927. https://doi.org/10.1111/pbi.12927

Makarova KS, Wolf YI, Alkhnbashi OS, Costa F, Shah SA, Saunders SJ, Barrangou R, Brouns SJ, Charpentier E, Haft DH, Horvath P (2015) An updated evolutionary classification of CRISPR–Cas systems. Nat Rev Microbiol 13(11):722–736. https://doi.org/10.1038/nrmicro3569

Makarova KS, Wolf YI, Koonin EV (2018) Classification and nomenclature of CRISPR-Cas systems: where from here? CRISPR J 1(5):325–336. https://doi.org/10.1089/crispr.2018.0033

Makarova KS, Wolf YI, Iranzo J, Shmakov SA, Alkhnbashi OS, Brouns SJ, Charpentier E, Cheng D, Haft DH, Horvath P, Moineau S (2020) Evolutionary classification of CRISPR–Cas systems: a burst of class 2 and derived variants. Nat Rev Microbiol 18(2):67–83. https://doi.org/10.1038/s41579-019-0299-x

Malnoy M, Viola R, Jung MH, Koo OJ, Kim S, Kim JS, Velasco R, NagamangalaKanchiswamy C (2016) DNA-free genetically edited grapevine and apple protoplast using CRISPR/Cas9 ribonucleoproteins. Front Plant Sci 7:1904. https://doi.org/10.3389/fpls.2016.01904

Mercé C, Bayer PE, Tay Fernandez C, Batley J, Edwards D (2020) Induced methylation in plants as a crop improvement tool: progress and perspectives. Agronomy 10(10):1484. https://doi.org/10.3390/agronomy10101484

Miller JC, Zhang L, Xia DF, Campo JJ, Ankoudinova IV, Guschin DY, Babiarz JE, Meng X, Hinkley SJ, Lam SC, Paschon DE (2015) Improved specificity of TALE-based genome editing using an expanded RVD repertoire. Nat Methods 12(5):465–471. https://doi.org/10.1038/nmeth.3330

Minhas PS, Rane J, Pasala RK (2017) Abiotic stresses in agriculture: an overview. Abiot Stress Manage Resil Agric 3-8. https://doi.org/10.1007/978-981-10-5744-1_1

Nam KH, Haitjema C, Liu X, Ding F, Wang H, DeLisaMP KA (2012) Cas5d protein processes pre-crRNAand assembles into a cascade-like interference complex in subtype I-C/Dvulg CRISPR-Cas system. Structure 20:1574–1584. https://doi.org/10.1016/j.str.2012.06.016

Narusaka Y, Narusaka M, Yamasaki S, Iwabuchi M (2012) Methods to transfer foreign genes to plants. In: Agricultural and Biological Sciences. "Transgenic Plants-Advances and Limitations", vol 7. In Tech Publishing, pp 173–188

Nawaz MA, Imtiaz M, Kong Q, Cheng F, Ahmed W, Huang Y, Bie Z (2016) Grafting: a technique to modify ion accumulation in horticultural crops. Front Plant Sci 7:1457. https://doi.org/10.3389/fpls.2016.01457

Oliva R, Ji C, Atienza-Grande G, Huguet-Tapia JC, Perez-Quintero A, Li T, Eom JS, Li C, Nguyen H, Liu B, Auguy F (2019) Broad-spectrum resistance to bacterial blight in rice using genome editing. Nat Biotechnol 37(11):1344–1350. https://doi.org/10.1038/s41587-019-0267-z

Ortiz-Matamoros MF, Villanueva MA, Islas-Flores T (2018) Genetic transformation of cell-walled plant and algae cells: delivering DNA through the cell wall. Brief Funct Genomics 17(1):26–33. https://doi.org/10.1093/bfgp/elx014

Osakabe Y, Osakabe K (2015) Genome editing with engineered nucleases in plants. Plant Cell Physiol 56(3):389–400. https://doi.org/10.1093/pcp/pcu170

Papikian A, Liu W, Gallego-Bartolomé J, Jacobsen SE (2019) Site-specific manipulation of Arabidopsis loci using CRISPR-Cas9 SunTag systems. Nat Commun 10:729. https://doi.org/10.1038/s41467-019-08736-7

Pavletich NP, Pabo CO (1991) Zinc finger-DNA recognition: crystal structure of a Zif268-DNA complex at 2.1 A. Science 252(5007):809–817. https://doi.org/10.1126/science.202825

Peng A, Chen S, Lei T, Xu L, He Y, Wu L, Yao L, Zou X (2017) Engineering canker resistant plants through CRISPR/Cas9-targeted editing of the susceptibility gene Cs LOB 1 promoter in citrus. Plant Biotechnol J 15(12):1509–1519. https://doi.org/10.1111/pbi.12733

Piatek A, Ali Z, Baazim H, Li L, Abulfaraj A, Al SS, Aouida M, Mahfouz MM (2015) RNA guided transcriptional regulation in planta via synthetic dC as9based transcription factors. Plant Biotechnol J 13(4):578–589. https://doi.org/10.1111/pbi.12284

Prakash D, Verma S, Bhatia R, Tiwary BN (2011) Risks and precautions of genetically modified organisms. Int Sch Res Notices. https://doi.org/10.5402/2011/369573

Prieto J, Redondo P, Padro D, Arnould S, Epinat JC, Pâques F, Blanco FJ, Montoya G (2007) The C-terminal loop of the homing endonuclease I-CreI is essential for site recognition, DNA binding and cleavage. Nucleic Acids Res 35(10):3262–3271. https://doi.org/10.1093/nar/gkm183

Razzaq A, Saleem F, Kanwal M, Mustafa G, Yousaf S, Imran Arshad HM, Hameed MK, Khan MS, Joyia FA (2019) Modern trends in plant genome editing: an inclusive review of the CRISPR/Cas9 toolbox. Int J Mol Sci 20(16):4045. https://doi.org/10.3390/ijms20164045

Richter H, Zoephel J, Schermuly J, Maticzka D, Backofen R, Randau L (2012) Characterization of CRISPR RNA processing in *Clostridiumthermocellum* and *Methanococcusmaripaludis*. Nucleic Acids Res 40(19):9887–9896. https://doi.org/10.1093/nar/gks737

Saha SK, Saikot FK, Rahman MS, Jamal MA, Rahman SK, Islam SR, Kim KH (2019) Programmable molecular scissors: applications of a new tool for genome editing in biotech. Mol Ther Nucleic Acids 14:212–238. https://doi.org/10.1016/j.omtn.2018.11.016

Samai P, Pyenson N, Jiang W, Goldberg GW, Hatoum-Aslan A, Marraffini LA (2015) Co-transcriptional DNA and RNA cleavage during type III CRISPR-Cas immunity. Cell 161(5):1164–1174. https://doi.org/10.1016/j.cell.2015.04.027

Sánchez-Rivera FJ, Papagiannakopoulos T, Romero R, Tammela T, Bauer MR, Bhutkar A, Joshi NS, Subbaraj L, Bronson RT, Xue W, Jacks T (2014) Rapid modelling of cooperating genetic events in cancer through somatic genome editing. Nature 516(7531):428–431. https://doi.org/10.1038/nature13906

Sauer NJ, Mozoruk J, Miller RB, Warburg ZJ, Walker KA, Beetham PR, Schöpke CR, Gocal GF (2016) Oligonucleotide-directed mutagenesis for precision gene editing. Plant Biotechnol J 14(2):496–502. https://doi.org/10.1111/pbi.12496

Schaart JG, Visser RGF (2009) Novel plant breeding techniques. Consequences of new genetic modification-based plant breeding techniques in comparison to conventional plant breeding. Cogem. https://edepot.wur.nl/137009

Scheben A, Wolter F, Batley J, Puchta H, Edwards D (2017) Towards CRISPR/Cas crops–bringing together genomics and genome editing. New Phytol 216(3):682–698. https://doi.org/10.1111/nph.14702

Shan Q, Wang Y, Li J, Gao C (2014) Genome editing in rice and wheat using the CRISPR/Cas system. Nat Protoc 9:2395. https://doi.org/10.1038/nprot.2014.157

Shi J, Gao H, Wang H, Lafitte HR, Archibald RL, Yang M, Hakimi SM, Mo H, Habben JE (2017) ARGOS 8 variants generated by CRISPR-Cas9 improve maize grain yield under field drought stress conditions. Plant Biotechnol J 15(2):207–216. https://doi.org/10.1111/pbi.12603

Sikora P, Chawade A, Larsson M, Olsson J, Olsson O (2011) Mutagenesis as a tool in plant genetics, functional genomics, and breeding. Int J Plant Genomics 2011:1–13. https://doi.org/10.1155/2011/314829

Silas S, Mohr G, Sidote DJ, Markham LM, Sanchez-Amat A, Bhaya D, Lambowitz AM, Fire AZ (2016) Direct CRISPR spacer acquisition from RNA by a natural reverse transcriptase–Cas1 fusion protein. Science 351(6276). https://doi.org/10.1126/science.aad4234

Song G, Jia M, Chen K, Kong X, Khattak B, Xie C, Li A, Mao L (2016) CRISPR/Cas9: a powerful tool for crop genome editing. Crop J 4(2):75–82. https://doi.org/10.1016/j.cj.2015.12.002

Sreekanth M, Hakeem AH, Peer QJ, Rashid I (2017) Low productivity of Indian agriculture with special reference on cereals. J Pharmacogn Phytochem 6(5):239–243

Staals RH, Zhu Y, Taylor DW, Kornfeld JE, Sharma K, Barendregt A, Koehorst JJ, Vlot M, Neupane N, Varossieau K, Sakamoto K (2014) RNA targeting by the type III-A CRISPR-CasCsm complex of *Thermusthermophilus*. Mol Cell 56(4):518–530. https://doi.org/10.1016/j.molcel.2014.10.005

Sternberg SH, Redding S, Jinek M, Greene EC, Doudna JA (2014) DNA interrogation by the CRISPR RNA-guided endonuclease Cas9. Nature 507(7490):62–67. https://doi.org/10.1038/nature13011

Subburaj S, Chung SJ, Lee C, Ryu SM, Kim DH, Kim JS, Bae S, Lee GJ (2016) Site-directed mutagenesis in Petunia× hybrida protoplast system using direct delivery of purified recombinant Cas9 ribonucleoproteins. Plant Cell Rep 35(7):1535–1544. https://doi.org/10.1007/s00299-016-1937-7

Sun Y, Jiao G, Liu Z, Zhang X, Li J, Guo X, Du W, Du J, Francis F, Zhao Y, Xia L (2017) Generation of high-amylose rice through CRISPR/Cas9-mediated targeted mutagenesis of starch branching enzymes. Front Plant Sci 8:298. https://doi.org/10.3389/fpls.2017.00298

Szczelkun MD, Tikhomirova MS, Sinkunas T, Gasiunas G, Karvelis T, Pschera P, Siksnys V, Seidel R (2014) Direct observation of R-loop formation by single RNA-guided Cas9 and Cascade effector complexes. PNAS 111(27):9798–9803. https://doi.org/10.1073/pnas.1402597111

Tamulaitis G, Kazlauskiene M, Manakova E, Venclovas Č, Nwokeoji AO, Dickman MJ, Horvath P, Siksnys V (2014) Programmable RNA shredding by the type III-A CRISPR-Cas system of *Streptococcus thermophilus*. Mol Cell 56(4):506–517. https://doi.org/10.1016/j.molcel.2014.09.027

Vale FXRD, Parlevliet JE, Zambolim L (2001) Concepts in plant disease resistance. J Fitopatol Bras 26:577–589. https://doi.org/10.1590/S0100-41582001000300001

van Schie CC, Takken FL (2014) Susceptibility genes101: how to be a good host. Annu Rev Phytopathol 52:551–581. https://doi.org/10.1146/annurev-phyto-102313-045854

Vorontsova D, Datsenko KA, Medvedeva S, Bondy-Denomy J, Savitskaya EE, Pougach K, Logacheva M, Wiedenheft B, Davidson AR, Severinov K, Semenova E (2015) Foreign DNA acquisition by the IF CRISPR–Cas system requires all components of the interference machinery. Nucleic Acids Res 43(22):10848–10860. https://doi.org/10.1093/nar/gkv1261

Wang Y, Cheng X, Shan Q, Zhang Y, Liu J, Gao C, Qiu JL (2014) Simultaneous editing of threehomoeoalleles in hexaploid bread wheat confers heritable resistance to powdery mildew. Nat Biotechnol 32:947–951. https://doi.org/10.1038/nbt.2969

Wang L, Wang L, Tan Q, Fan Q, Zhu H, Hong Z, Zhang Z, Duanmu D (2016) Efficient inactivation of symbiotic nitrogen fixation related genes in *Lotus japonicus* using CRISPR-Cas9. Front Plant Sci 7:1333. https://doi.org/10.3389/fpls.2016.01333

Wang M, Mao Y, Lu Y, Tao X, Zhu JK (2017) Multiplex gene editing in rice using the CRISPR-Cpf1system. Mol Plant 10(7):1011–1013. https://doi.org/10.1016/j.molp.2017.03.001

Wang C, Liu Q, Shen Y, Hua Y, Wang J, Lin J, Wu M, Sun T, Cheng Z, Mercier R, Wang K (2019) Clonal seeds from hybrid rice by simultaneous genome engineering of meiosis and fertilization genes. Nat Biotechnol 7(3):283–286. https://doi.org/10.1038/s41587-018-0003-0

Weeks DP, Spalding MH, Yang B (2016) Use of designer nucleases for targeted gene and genome editing in plants. Plant Biotechnol J 14(2):483–495. https://doi.org/10.1111/pbi.12448

Wei Y, Chesne MT, Terns RM, Terns MP (2015) Sequences spanning the leader-repeat junction mediate CRISPR adaptation to phage in Streptococcus thermophilus. Nucleic Acids Res 43 (3):1749–1758. https://doi.org/10.1093/nar/gku1407

Westra ER, van Erp PB, Künne T, Wong SP et al (2012) CRISPR immunity relies on the consecutive binding and degradation of negatively supercoiled invader DNA by Cascade and Cas3. Mol Cell 46(5):595–605. https://doi.org/10.1016/j.molcel.2012.03.018

Wise RP, Moscou MJ, Bogdanove AJ, Whitham SA (2007) Transcript profiling in host–pathogen interactions. Annu Rev Phytopathol 45:329–369. https://doi.org/10.1146/annurev.phyto.45.011107.143944

Xie K, Yang Y (2013) RNA-guided genome editing in plants using a CRISPR–Cas system. Mol Plant 6(6):1975–1983. https://doi.org/10.1093/mp/sst119

Xie N, Zhou Y, Sun Q, Tang B (2018) Novel epigenetic techniques provided by the CRISPR/Cas9 system. Stem Cells Int 2018:1–12. https://doi.org/10.1155/2018/7834175

Yadav S, Modi P, Dave A, Vijapura A, Patel D, Patel M (2020) Effect of abiotic stress on crops. sustainable crop production. https://doi.org/10.5772/intechopen.88434

Yin K, Qiu JL (2018) Genome editing for plant disease resistance: applications and perspectives. Philos Trans R Soc Lond B Biol Sci 374:1767. https://doi.org/10.1098/rstb.2018.0322

Yin K, Gao C, Qiu JL (2017) Progress and prospects in plant genome editing. Nat Plants 3(8):1–6. https://doi.org/10.1038/nplants.2017.107

Yosef I, Goren MG, Qimron U (2012) Proteins and DNA elements essential for the CRISPR adaptation process in Escherichia coli. Nucleic Acids Res 40(12):5569–5576. https://doi.org/10.1093/nar/gks216

Zaidi SSEA, Mukhtar MS, Mansoor S (2018) Genome editing: targeting susceptibility genes for plant disease resistance. Trends Biotechnol 36(9):898–906. https://doi.org/10.1016/j.tibtech.2018.04.005

Zeng W, Melotto M, He SY (2010) Plant stomata: a checkpoint of host immunity and pathogen virulence. Curr Opin Biotechnol 21(5):599–603. https://doi.org/10.1016/j.copbio.2010.05.006

Zhang Y, Bai Y, Wu G, Zou S, Chen Y, Gao C, Tang D (2017) Simultaneous modification of threehomoeologs of TaEDR1 by genome editing enhances powdery mildew resistance in wheat. Plant J 91:714–724. https://doi.org/10.1111/tpj.13599

Zhao Y, Zhang C, Liu W, Gao W, Liu C, Song G, Li WX, Mao L, Chen B, Xu Y, Li X (2016) An alternative strategy for targeted gene replacement in plants using a dual-sgRNA/Cas9 design. Sci Rep 6(1):1–1. https://doi.org/10.1038/srep23890

Zhou H, Liu B, Weeks DP, Spalding MH, Yang B (2014) Large chromosomal deletions and heritable small genetic changes induced by CRISPR/Cas9 in rice. Nucleic Acids Res 42(17):10903–10914. https://doi.org/10.1093/nar/gku806

Zhou J, Peng Z, Long J, Sosso D, Liu B, Eom J-S, Huang S, Liu S, Vera Cruz C, Frommer WB et al (2015) Gene targeting by the TAL effector PthXo2 reveals cryptic resistance gene for bacterial blight of rice. Plant J 82(4):632–643. https://doi.org/10.1111/tpj.12838

Zhou X, Liao H, Chern M, Yin J, Chen Y, Wang J, Zhu X, Chen Z, Yuan C, Zhao W, Wang J (2018) Loss of function of a rice TPR-domain RNA-binding protein confers broad-spectrum disease resistance. Proc Natl Acad Sci 115(12):3174–3179. https://doi.org/10.1073/pnas.1705927115

CRISPR/Cas for Improved Stress Tolerance in Rice

12

Khondokar Nowshin Islam, Mohammad Mahmood Hasan, and Mohammad Nurul Islam

Abstract

The discovery of clustered regularly interspaced short palindromic repeats (CRISPR) and CRISPR-associated protein (Cas) system has revolutionized genome editing technology. Though in nature it is found in bacteria and archaea as a defense mechanism against viruses, it has been successfully repurposed as an effective and robust genome editing tool in all forms of life, e.g., bacteria, plants, animals, and humans. The utilization of this ingenious tool in agriculture is increasing day by day as it can be used to introduce the gene of interest in a specific site within the genome and to eliminate the expression of a gene of choice through knockout at DNA or RNA level. To date, this technology has effectively installed resistance against both abiotic and biotic stresses in different crops. In this chapter, we have discussed the basic mechanisms of CRISPR/Cas and its latest classification. Further, we discuss the recent successes of this tool in rice breeding, which is the staple food for billions of people around the world. Finally, we highlight the prospects of CRISPR/Cas technology in providing resistance against stresses in rice.

K. N. Islam
Department of Microbiology, Miami University, Oxford, OH, USA

M. M. Hasan
Department of Microbiology and Cell Science, University of Florida, Gainesville, FL, USA

M. N. Islam (✉)
Plant Breeding and Biotechnology Laboratory, Department of Botany, University of Dhaka, Dhaka, Bangladesh
e-mail: mnurul@du.ac.bd

M. G. Mallikarjuna et al. (eds.), *Next-Generation Plant Breeding Approaches for Stress Resilience in Cereal Crops*, https://doi.org/10.1007/978-981-19-1445-4_12

Keywords

Abiotic stress · Biotic stress · Climate change · Genome editing · Rice · Resistance breeding

12.1 Introduction

12.1.1 Introduction to CRISPR/Cas

In the late 2019, readers of *The Irish Times* declared gene editing as the innovation of the last decade (O'Connell 2019) because it has opened a new dimension to scientists in the field of biological science by allowing them to alter an organism's (humans, plants, microbes, etc.) DNA and consequently enabling them to develop stress-resilient crop varieties (Zafar et al. 2019), treat inherited diseases (Rajeev Rai and Cavazza 2021), understand a gene's function (Martin et al. 2016), and sometimes even detect unknown species in the environment (Baerwald et al. 2020). Among several types of gene-editing tools such as ZFN (zinc finger nucleases), TALEN (transcription activator-like effector nucleases), CRISPR/Cas, etc., the latter one is most conveniently applied due to its robustness, editing efficiency, simplicity, and most importantly, flexibility (Adli 2018). The clustered regularly interspaced short palindromic repeats (CRISPR)/Cas (CRISPR-associated proteins) is a microbial adaptive immune system that cleaves foreign genetic elements by using RNA-guided nucleases and can be utilized to facilitate efficient genome engineering in eukaryotic cells (Ran et al. 2013). This technology involves the engineering of a single guide RNA (sgRNA) and base-pairing between the sgRNA and the target DNA site that remains adjacent to the protospacer adjacent motif (PAM) followed by a double-stranded breakage (DSB) on the genome by Cas endonucleases (Wang et al. 2017a).

The DSB repair machinery and the outcome of the process play a key role in determining the nature of a genome edit (van Overbeek et al. 2016). There are two common pathways, i.e., non-homologous end joining (NHEJ) and homology-directed repair (HDR), either of which may facilitate the DSB repair (Barman et al. 2020). In higher eukaryotes, NHEJ is the leading pathway for DSB repair that may result in either deletions or insertions or substitutions (commonly termed as indels) at the break site (Shen et al. 2017b). On the other hand, the HDR repair pathway induces specific genetic changes to the DSB by the introduction of a homologous DNA repair template and results in precise point mutations, gene deletions, or insertions of genes of interest (Fig. 12.1) (Salsman and Dellaire 2017).

12.1.2 CRISPR/Cas in Agriculture

The global agricultural production needs to be enhanced drastically as, by 2050, the world's population will be around 9.6 billion, increasing the demand for staple crops

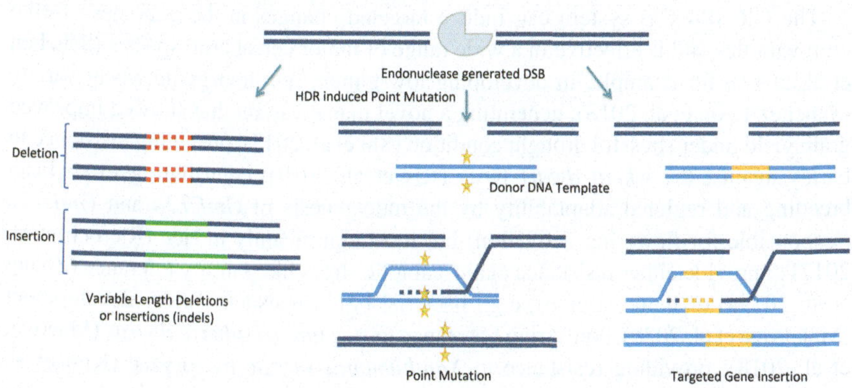

Fig. 12.1 Double-stranded breakage induced by CRISPR/Cas followed by the repair mechanism through non-homologous end joining (NHEJ) or homology-directed repair (HDR). The NHEJ may result in either deletion or insertion (indels) and HDR could be used to introduce a point mutation or insert a gene of interest; a donor template is required to be delivered in cells with CRISPR/Cas machinery for HDR

by 60%. However, the conventional crop breeding techniques alone cannot accomplish this objective as these methods are often time-consuming, laborious, and complicated. Hence, CRISPR/Cas, a rapid and more reliable technology, has been widely used in improving several crop characteristics (yield, quality, disease resistance, herbicide resistance, etc.) in recent years (Zhu et al. 2020). Since its first application in 2012, advancement in CRISPR/Cas technology has revolutionized research in the field of life sciences (Gao 2018) especially in the fields of functional genomics and crop improvement by allowing researchers to develop novel plant varieties with either deletion of harmful traits or addition of desired characteristics (Arora and Narula 2017). CRISPR/Cas has been employed in influencing the genome of different plant species including *Arabidopsis*, *Medicago truncatula*, tomato, potato, wheat, corn, rice, etc. (Afzal et al. 2020).

CRISPR/Cas was first introduced in agriculture in 2013, and since then, it has been successfully implemented in several crop species. One study reported targeted mutagenesis in the tomato *PMR4* gene could generate higher resistance in comparison to RNAi-silenced transgenic plants (Santillán Martínez et al. 2020). In another report, CRISPR/Cas9 gene editing was employed to knock out the *BBL* genes that are responsible for nicotine production in tobacco plants resulting in the development of nicotine-free, non-transgenic plants, thus reducing the risk of death from tobacco use (Schachtsiek and Stehle 2019). CRISPR/Cas9 engineering also demonstrated potential for genetic modification of potato that has high nutritional value and is considered one of the major starch-producing crops. Successful knockout of the *GBSS* gene (responsible for the synthesis of amylose enzyme) of tetraploid potato (*Solanum tuberosum*) by transient CRISPR-Cas9 expression in protoplasts resulted in mutations in all four alleles without stable integration of DNA (Andersson et al. 2017).

The CRISPR/Cas system can induce targeted changes in the genomes of elite crop varieties and is effective in a wide range of major cereal crop species (Scheben et al. 2017), for example, in developing low-gluten, non-transgenic wheat variety (Sánchez-León et al. 2018); generating a novel maize variant that showed improved grain yield under stressful drought condition (Shi et al. 2017); providing resistance in barley against the *wheat dwarf virus* (Kis et al. 2019); contributing to soybean breeding and regional adaptability by the mutagenesis of *GmFT2a* and *GmFT5a* (responsible for flowering activation) but most significantly in rice (Ricroch et al. 2017) to provide either resistance against abiotic stress like salinity tolerance (Zhang et al. 2019a) or biotic stresses, e.g., bacterial blight-resistant variety development (Makarova et al. 2020); conferring resistance to *rice tungro spherical virus* (Macovei et al. 2018); providing resistance to *Xanthomonas oryzae* pv. oryzae (Kim et al. 2019), etc.

12.1.3 Economic Importance of Rice and Production Constraints

Rice is the most common cereal crop and acts as the staple food for approximately half of the world's population. In Asia alone, over 2 billion people obtain 80% of their energy from the consumption of rice. Not only that, reports have been made that rice contains lesser antioxidant molecules along with several other medicinal properties in comparison to other cereal crops, making it an ideal contender for a natural source of antioxidants and exploitation in the pharmaceutical industry (Chaudhari et al. 2018). In addition to providing calories, rice is a potential source of magnesium, phosphorus, manganese, iron, folic acid, selenium, thiamine, and niacin. It also contains low fiber and fat (Fukagawa and Ziska 2019). Moreover, a recent study investigated the antidiabetic activity of purple rice bran and discovered the great potential for its application to improve hepatic insulin signaling and in decreasing hepatic gluconeogenesis (Hlaing et al. 2019).

However, due to various constraints such as biotic factors (insects and pests, weeds, diseases, etc.) and abiotic factors (scarcity of good quality water, salt stress, nutrient imbalance, climatic factors, etc.), overall rice productivity has been harmfully affected (Fahad et al. 2019). A quantification study in Tamil Nadu reported a total loss of 2.73 million tonnes of rice due to various constraints that are about 39.45% of the total production (Shanmugam et al. 2006). Another report suggested that drought alone can be responsible for as much as 40% loss in rice production, reducing income up to 58% in South and Southeast Asia. Furthermore, it has been reported that owing to several rice diseases, more than 40% yield of the total harvest is lost in South Asia where it represents the first source of caloric intake (Savary et al. 2012).

12.1.4 CRISPR/Cas in Rice

It is estimated that by 2030, rice production will need to be increased by one-fourth percent of the current production to meet the demand of the expanding global population (Ansari et al. 2015). Hence, anticipating the immense importance of rice in the present and upcoming future, numerous measures have been undertaken in the previous years to ensure adequate rice production by improving yield (Khan et al. 2015), developing tolerance against biotic (Sreewongchai et al. 2010), and abiotic (Singh et al. 2010) stresses, etc. Since its first application in 2013 (Shan et al. 2013), CRISPR/Cas-mediated genome editing has demonstrated immense potential in rice breeding towards an improved production because of its ease of use, economic nature, and efficiency (Bandyopadhyay et al. 2019).

One of the most prominent applications of CRISPR/Cas technology in rice is the construction of a genome-wide mutant library that can be utilized to find out gene functions, genetic improvement, and functional characterization of unknown genes. Another significant use could be the precise elimination of selective marker genes in transgenic plants to further improve breeding techniques. Besides, CRISPR/Cas can also be applied to rice to provide biotic and abiotic stress tolerance, improving grain yield, replacing alleles efficiently, and thereby, hastening the crop improvement process through the induction of point mutation or base editing or multiplex genome editing or insertion or deletion of a precise region of a gene (Romero and Gatica-Arias 2019). In this chapter, we'll discuss various stresses of rice production and applications of different types of CRISPR/Cas technologies to overcome the hurdles. We'll also discuss different delivery methods of CRISPR/Cas that have previously been employed on rice. Additionally, we'll highlight the success of CRISPR/Cas-mediated gene editing in providing stress tolerance/resistance to date in rice and its future implications for increasing overall rice productivity for a sustainable future.

12.2 Stresses Hampering Rice Production

The global rice production has declined in the past years owing to various constraints that include limited yield potential of high-yielding varieties, pressure from abiotic and biotic stresses, socioeconomic concerns, increasing production costs, etc., and if appropriate measures are not immediately carried out to address and reduce the effects of these factors on the overall rice productivity, then serious food scarcity may occur shortly in different parts of the world (Van Nguyen and Ferrero 2006). Some of the major abiotic and biotic stresses in rice are described below along with stating their potential threats to loss of rice yield and productivity.

12.2.1 Abiotic Stresses in Rice

Abiotic stresses are induced by abiotic factors, i.e., unfavorable environmental conditions (extreme temperature, cold stress, drought stress, salt stress, etc.), causing significant variation in the ideal production environment of plants and thus resulting in the visible decline of their growth, development, and production (Fahad et al. 2019). Abiotic stresses not only play a crucial role in the yield loss of rice but also are responsible for lower grain quality which in turn causes decreased consumer acceptance (Fahad et al. 2019).

12.2.1.1 High-Temperature Stress

Due to global warming and the greenhouse effect, the global air temperature has significantly increased in the past few decades. This high temperature is responsible for increasing floods, storms, and other adverse calamities worldwide and eventually affecting overall food production. Not only that, but an increase in temperature is also accountable for amplifying the atmospheric CO_2 concentration that possesses a severe threat to crop production (Wheeler et al. 2000). Temperature is a key aspect in controlling several features in rice such as germination, seedling growth, leaf emergence, tillering, heading, plant height, dark respiration, grain filling, grain quality, yield, etc. Therefore, a temperature rise may cause serious alteration in rice structure and a decline in overall productivity. Additionally, high temperature disturbs the water, ion, and organic solute movement across plant membranes and thereby affects photosynthesis and respiration causing a reduction in yield performance (Krishnan et al. 2011).

In many tropical and subtropical countries, high day temperature has caused significant loss of rice production (Fahad et al. 2019). Asia is the biggest global rice contributor accounting for about 87% of the global rice production. Rice exports from Asian regions especially from China and India (about 49% of the world's rice producers) play a key role in maintaining global food security (Bandumula 2018). However, due to climate change and increasing temperature, several reports on yield reduction have been made in major rice-producing regions of Asia such as China (Lv et al. 2018), India (Setiyono et al. 2018), Malaysia (Vaghefi et al. 2011), and Vietnam (Thuy and Saitoh 2017). Similar outcomes of yield reduction with increasing temperature have been reported in Africa, another leading rice-producing region of the world (Adhikari et al. 2015; van Oort and Zwart 2018).

12.2.1.2 Cold Stress

Cold stress concludes both chilling injuries (under 20 °C) and freezing injuries (under 0 °C) and is one of the most significant abiotic stresses reducing production and yield of several major crops such as rice, maize, soybean, and cotton (Thakur et al. 2010). Cold stress possesses an immense threat to rice production affecting both vegetative and reproductive phases of its life cycle with the latter one being more prominent. As a result, abnormal development of anthers, low spikelet fertility, and eventually notable yield losses are observed (Bai et al. 2015).

Recent studies on cold stress on rice production have suggested that low temperatures during long, cold springs in several low-altitude and high-altitude regions of China, Japan, Korea, and other parts of the world can result in inhibition of germination and restrict early seedling growth (Zhang et al. 2014). One study in Southern China recorded the cold stress effect on rice from the year 1981 to 2009 and represented production loss with decreasing temperature (Zhang et al. 2016). It is indicated that if the temperature drops below 15 ° C, the germination and seedling growth of rice will be severely affected (Lv et al. 2019).

12.2.1.3 Drought Stress

Drought is termed as the inadequacy of water for a while due to insignificant rainfall, unavailability of water in the soil, and lack of moisture in the air that in turn causes continuous loss of water from plants through excessive evaporation and transpiration (Singh et al. 2018). Drought is one of the primary stresses in rice, because of which plant growth and development are severely hampered eventually resulting in reduced grain yield of rice (Sharifunnessa and Islam 2017). Severe drought stress in rice causes economic yield loss in both reproductive (48–94%) and grain-filling stage (60%) (Kim et al. 2020).

Drought negatively affects rice production and yield stability and causes rigorous yield loss in many rainfed areas in many Asian countries but most significantly in Eastern India and adjacent parts of Nepal (more than 17 million hectares of rainfed area) which is considered as the largest drought-prone regions of Asia (Palanog et al. 2014). In 1987 and 2002/2003, owing to severe drought, 50% of the total cropped area of India was affected and more than 300 million people had to suffer across the country. Thailand encountered adverse effects of drought in the year 2004 affecting 20% of rice lands and more than eight million people (Wassmann et al. 2009).

12.2.1.4 Salt Stress

Normal salt pH ranges between 4.5 and 7.5 making the soil most favorable for nutrient availability and plant growth. Salt stress is the condition when the soil contains a high concentration of soluble salts such as sodium (Na^+), magnesium (Mg^{2+}), calcium (Ca^{2+}), chloride (Cl^-), and sulfate (SO_4^{2-}), etc. creating an environment that negatively impacts plant growth (Hussain et al. 2017). More than 90% of the world's rice is grown in Asia where approximately 60% of the earth's population resides. One of the harshest abiotic stresses that affect rice plants in their early seedling stage and cause serious yield and production loss in Asia is soil salinity (Kumar et al. 2013).

A study on the impact of salt stress on the growth and yield of some native rice cultivars of Kerala in India reported significant height reduction in two cultivars among the seven studied. Further, tiller production was seen to have decreased in three cultivars due to salt stress. Additionally, panicle length, spikelet per panicle, and fertility percentage were found to have reduced due to salinity (Joseph 2013). Rice production and yield in Pakistan also declined due to increased salinity in approximately ten million ha of irrigated land (Zaman et al. 2018).

12.2.2 Biotic Stresses in Rice

Biotic stress refers to the infection of different pathogens as well as herbivore pests in plants under natural conditions that possesses an enormous threat to plant productivity and yield by causing many diseases (Suzuki et al. 2014). Rice is the second most important cereal crop in the world in terms of productivity and is considered the principal food of most developing countries (Molla et al. 2019). Among various constraints of rice production, diseases are the major factors behind low yields of rice throughout the world. To date, more than 70 diseases caused by bacteria, fungi, viruses, and nematodes have been reported in rice (Singh et al. 2020).

Bacterial leaf blight, primarily reported in Japan during 1884–1885 and then in many other major rice-growing regions of the world (Gnanamanickam 2009), is responsible for serious damage of rice plant and can result in a yield reduction of 20–40% and 50% if infection occurs during tillering stage and early stage, respectively (Chukwu et al. 2019). Bacterial leaf streak can reduce yields of rice up to 8–17% in the wet season and 1–3% in the dry season. However, it can be more severe in certain areas; for instance, studies in India reported yield loss of rice to reach as high as 30% due to the bacterial leaf streak (Kumar et al. 2017). Another major bacterial disease frequently occurring in rice is bacterial sheath rot causing grain sterility and notable yield loss (Rostami et al. 2005). In Indonesia, an extreme yield loss of 72.2% was recorded due to the disease, whereas the highest yield loss in Malaysia was 20%.

Rice blast is a major fungal disease in rice, usually causing 30% yield loss which could be fed to 60 million people, and not only that, the disease has even been regarded as capable of causing 100% yield loss. Some reports on yield loss of rice due to blast disease have been reported in India (5–10%), Korea (8%), China (14%), and the Philippines (50–85%) (Fahad et al. 2019). Another most significant fungal disease of rice is brown spot that can cause serious damage to rice production (up to 90%) and was responsible for the great Bengal famine during 1942. The percentage of yield loss varies depending on the rice cultivar and stage of infection, mostly ranging between 18.75% and 22.50% (Sunder et al. 2014). Rice production has been acutely damaged in tropical regions especially in South Asia due to sheath blight, a fungal disease, whose infection is favored by warm temperature and high humidity. The disease was first reported in Japan in 1910 and can cause a yield loss of up to 45% (Singh et al. 2019).

Among the viral diseases of rice, tungro disease is highly significant as it possesses a great deal of economic and social consequences on the rice production of Asia and Southeast Asia (more than 90% of the world's rice producer and consumer) and is estimated to cause an annual yield loss of 5% to 10% as well as an economic loss of approximately US $1.5 billion (Dai and Beachy 2009). Africa is the second-largest rice importer in the world representing 25% of the world's rice importation (Woin et al. 2010). Rice yellow mottle virus was first reported in Kenya (Bakker 1974), and since then, it has been reported in many countries in East and West Africa. This disease is one of the most damaging diseases of rice in Africa and can cause a yield loss of 10–100% depending on plant age and susceptibility of the rice variety (Kouassi et al. 2005).

Here, Table 12.1 represents the major bacterial, fungal, and viral diseases that are responsible for high quantities of annual yield loss along with their causal organisms, disease symptoms, and commonly occurring regions in the world.

Table 12.1 Different major diseases of rice caused by bacteria, fungi, and bacteria, their causal organisms, typical symptoms, and occurrence

Pathogen	Disease name	Causal organism	Disease symptoms	Occurrence
Bacteria	Bacterial leaf blight	*Xanthomonas oryzae* pv. oryzae	Kresek phase (sudden wilting and death of plant), leaf wilting and upward rolling, leaf color change from grayish-green to yellow, water-soaked lesions, milky bacterial ooze, plant drying	Southeast Asia (Zhou et al. 2013), Africa (Verdier et al. 2012), Australia, Latin America, and the Caribbean (Mew 1989)
	Bacterial leaf streak	*Xanthomonas oryzae* pv. *oryzicola*	Dark-green and water-soaked streaks on interveins, streaks soon turn yellow or orange-brown, infection in the florets and seeds discoloration and death of ovary, stamens, browning of glumes	Southern and Central China, Southeast Asia, and Africa (Wu et al. 2019)
	Brown sheath rot	*Pseudomonas fuscovaginae*	Symptoms on the flag leaf sheath and the panicle, a systemic discoloration spreading to the midrib or veins of the leaves, yellow to brown discoloration of seedlings, necrosis, grain discoloration	Temperate regions of Asia, Africa, South America, and Australia (Kakar et al. 2014)
Fungi	Rice blast	*Magnaporthe oryzae*	White to gray-green lesions or spots, lesions can enlarge and coalesce to kill the entire leaf	China, India, Japan, South Korea, Indonesia (Wang and Valent 2009)
	Brown spot	*Bipolaris oryzae*	Small, circular, yellow-brown or brown lesions, spikelet and floret infection, grain-filling disruption, and grain quality reduction and discoloration	Japan, China, Burma, Sri Lanka, Bangladesh, Iran, Africa, South America, Russia, North America, Philippines, Saudi Arabia, Australia, Malaya, and Thailand (Sunder et al. 2014)

(continued)

Table 12.1 (continued)

Pathogen	Disease name	Causal organism	Disease symptoms	Occurrence
	Sheath blight	*Rhizoctonia solani*	Greenish gray lesions on the leaf sheaths, sclerotia	Africa, Bangladesh, Brazil, Burma, Colombia, China, Cuba, Germany, Fiji, Formosa, India, Indonesia, Iran, Korea, Liberia, Madagascar, Malaya, Malaysia, Netherland, Nigeria, Papua New Guinea, Philippines, Russia, Senegal, Sri Lanka, Surinam, Taiwan, Thailand, Trinidad, Tobago, UK, USA, Venezuela, and Vietnam (Singh et al. 2016)
Virus	Tungro disease	Combine infection of *rice tungro bacilliform virus* (RTBV) and *rice tungro spherical virus* (RTSV)	Presence of leafhoppers, leaf (yellow or orange-yellow) discoloration, leaves show mottled or striped appearance, interveinal necrosis	Philippines, India, Indonesia, Malaysia, Bangladesh, Nepal, Pakistan, Sri Lanka, Vietnam, China, and Japan (Dai and Beachy 2009)
	Rice yellow mottle disease	*Rice yellow mottle virus* (RYMV)	Yellow-green spots on the base of the youngest leaves, mottled and twisted leaves, discoloration and poor panicle exertion, reduced tillers, sterile spikelets, and eventually death	Kenya, Liberia, Nigeria, Sierra Leone, Tanzania, Nigeria, Burkina Faso, Mali, Malawi, Rwanda, Madagascar, Gambia, Guinea Bissau, Senegal, Mauritania, Zanzibar, Cameroon, and Chad (Kouassi et al. 2005)

12.3 Structure of CRISPR/Cas

The CRISPR/Cas system mainly consists of a CRISPR array that includes short direct repeats spaced by short variable DNA sequences termed as a spacer. This CRISPR array is flanked by a variety of *cas* genes. In front of the CRISPR array,

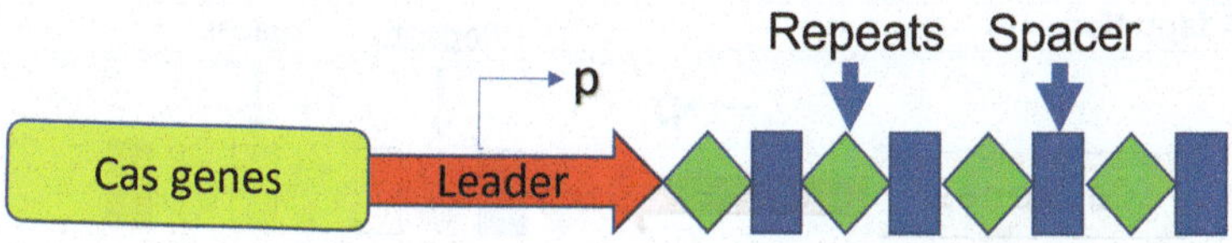

Fig. 12.2 Diagrammatic representation of different components of CRISPR/Cas system. The *cas* genes produce the Cas proteins required for acquiring new spacers from invader DNA, crRNA biogenesis, and interference. The CRISPR array contains palindromic repeats and spacers that transcribe into pre-crRNA. In the upstream of the CRISPR array, there is a leader sequence that contains a promoter for the expression of pre-crRNA

there is a leader sequence that contains the promoters required to transcribe the CRISPR array (Fig. 12.2). There is no open reading frame present in the CRISPR array. The repeats are identical direct repeats in sequence and they can be 21 to 50 nucleotides long. The number of repeats varies from organism to organism ranging from 2 to several hundred (mostly around 50). The spacers, on the other hand, are highly variable in the sequence of similar size ranging from 20 to 84 nucleotides long. These sequences can be identical to sequences from bacteriophages, plasmids, or rarely from chromosomes. The leader sequence is always present upstream of the CRISPR array. It also does not contain any open reading frame and could be several hundred nucleotides long (AT-rich); however, it contains all the necessary promoters and protein binding sites required for the biogenesis of crRNAs (Amitai and Sorek 2016).

12.4 Mechanism of CRISPR/Cas System

The overall mechanism of CRISPR/Cas can be divided into three major steps, e.g., adaptation, crRNA biogenesis, and interference (Fig. 12.3). These major steps are discussed in brief as below.

12.4.1 Adaptation

This is the first step of the CRISPR/Cas mechanism where a complex of Cas proteins functions together to bind to a target DNA, mostly upon recognizing a distinct, short motif called "protospacer-adjacent motif" of simply PAM presented on the upstream or downstream of the target DNA. After binding to the target DNA, the Cas proteins cleave a portion of that target DNA (known as protospacer) and insert it into the spacer of the CRISPR array which is then called the "spacer" (Amitai and Sorek 2016). There are other types of CRISPR/Cas system that acquire such spacer from RNAs through the reverse transcription procedure governed by reverse transcriptase enzymes encoded by the CRISPR/Cas locus (Makarova et al. 2020). Such spacer sequences are stored in the CRISPR array as the immunological memory. The spacer is integrated between the leader sequence and the first repeat of the CRISPR array

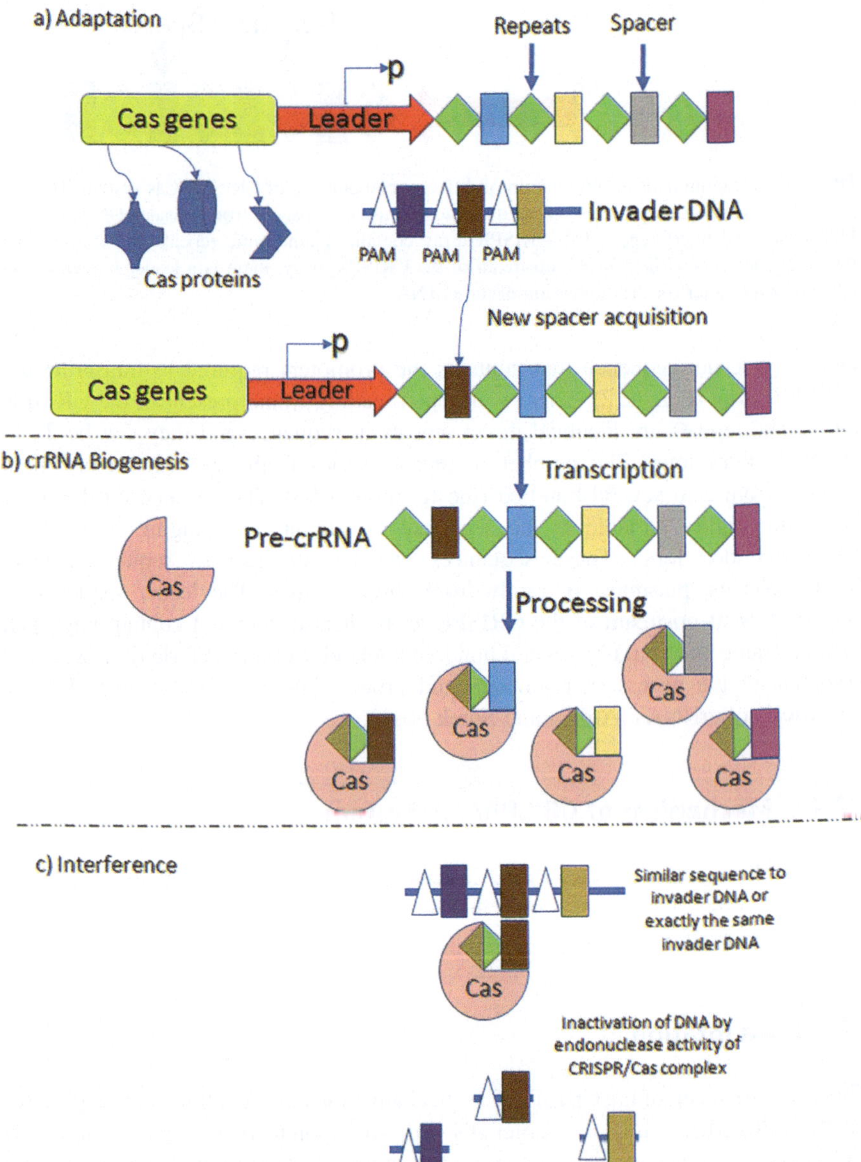

Fig. 12.3 A simplified graphical representation of CRISPR/Cas mechanism. (**a**) The adaptation step: the cas protein complexes recognize an invader DNA acquiring a new "spacer" into the CRISPR array. (**b**) Expression/crRNA biogenesis: the *cas* genes express Cas protein with endonuclease activity and the CRISPR array is expressed into pre-crRNA through transcription using the promoter sequence present in the upstream leader sequence of the CRISPR array. The pre-crRNA is further processed into mature crRNA by Cas protein to form the CRISPR/Cas complex. (**c**) Interference: when an invader DNA is present in the cell having complementary sequence with the CRISPR/Cas complex, the complex binds with the invader DNA at the complementary site and cleaves it into pieces, thus inactivating the invading DNA

and is accompanied by a duplication of the repeat. Several protospacers could be added to the array (each with its repeat) from a single invader to enhance the resistance level.

12.4.2 Expression/crRNA Biogenesis

This step starts with the transcription of the CRISPR array that is driven by a promoter situated in the leader sequence and produces a long precursor CRISPR-RNA (pre-crRNA) that is processed further by the distinct subunit of multiprotein Cas complex or by a single multidomain Cas protein, depending on the variant, into small mature crRNAs. Sometimes, this processing also involves accessory factors, such as non-Cas host RNases (Wimmer and Beisel 2020). Each crRNA contains a part of a repeat on its 5′ side, a part or all spacer, and sometimes also a part of the repeat on the 3′ side.

12.4.3 Interference

Interference takes place when the same foreign nucleic acid tries again to invade the cell itself or the daughter cells. The crRNAs form a complex (Cas-crRNA complex) with the one or multiple Cas protein. These complexes scan the invading nucleic acid and find the protospacer sequence (with the help of the PAM and seed sequences; if this is a system that uses PAM). crRNP inactivates this DNA or RNA by silencing or degradation. Most CRISPR-Cas systems recognize and attack DNA. Some systems attack ssRNA or both DNA and mRNA during transcription.

12.4.4 Distinguishing Between Target and Genomic CRISPR Array

To eliminate the possibility of harming self-DNA, CRISPR/Cas systems need to distinguish between the target DNA and its own CRISPR array. The PAM sequence is found responsible for such safety mechanisms in type I and type II CRISPR/Cas systems. The PAM sequence is situated adjacent to the protospacer sequence that is essential for recognition and cleavage of target DNA. Oppositely, it is absent in the spacer sequence of the CRISPR array. Therefore, the CRISPR/Cas system never cleaves its own CRISPR array. However, in type III system, such self-discrimination does not depend on the PAM. Rather, it utilizes the 5′ handle of the crRNA that interacts with the repeat sequence in the CRISPR locus followed by the inhibition of nuclease recruitment. Thus, the cleavage of self-DNA is prevented (Burmistrz and Pyrć 2015).

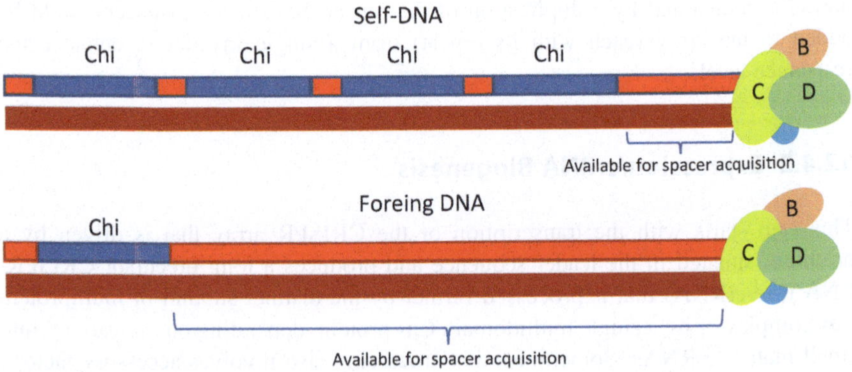

Fig. 12.4 Richness in Chi-site in *E. coli* genome prevents the degradation activity by RecBCD having only a small portion available for spacer acquisition, whereas the foreign DNA has fewer Chi-sites resulting in long-range DNA degradation by RecBCD and having many materials for spacer acquisition

12.4.5 Negligence of Acquiring Spacer from Self-DNA

In nature, the acquisition of spacer from the chromosomal DNA of the organism instead of the invader DNA is detrimental as it leads towards the breakage of self-DNA by the interference mechanism of the CRISPR/Cas system. Such an event leads to CRISPR/Cas autoimmunity. Organisms adopt different changes in their genome to prevent such autoimmunity when accidentally the CRISPR/Cas system obtains a spacer from its own chromosomal DNA. Such changes involve inactivation of the Cas genes through mutation, bringing changes in the sequence of the repeats next to the self-derived spacer or changing the PAM sequence. However, such mutational changes might not always be favorable. Therefore, the CRISPR/Cas system should eliminate the chances of acquiring self-DNA and avoid any unwanted impacts on the organism itself. It has been observed that the CRISPR/Cas system is indeed fond of acquiring spacer from foreign invading DNA rather than from its own chromosomal DNA. Such preference is believed to be based on the RecBCD machinery and Chi-sites. For instance, the *E. coli* genome is high in Chi-sites (once in every 4.6 kb, on average); as a result, even if there is a DSB the RecBCD only degrades a short length of the self-DNA, and the degradation is halted by the adjacent Chi-site. In comparison to the self-DNA in *E. coli*, the Chi-sites in the exogenous DNA are less densely distributed, e.g., around one Chi-site in every 65 kb. As a result, the RecBCD degrades long-length DNA that ultimately generates ample substrates for new spacers (Fig. 12.4). Even if the system acquires spacer from its own genomic DNA, the system may mutate or delete the specific spacer, mutate or delete flanking repeat, mutate or delete the PAM sequence, mutate or delete the *cas* genes, and even sometimes delete the whole system to avoid any kind of detrimental impact (Wimmer and Beisel 2020).

12.5 Classification of CRISPR/Cas and Frequently Used Systems in Rice Genome Editing

To date, there have been six types of CRISPR/Cas system reported which were further classified into two major classes (Makarova et al. 2020). The classification of CRISPR-Cas systems is based primarily on Cas protein composition differences and sequence divergence between the effector modules. The *cas* genes of different types of CRISPR/Cas system can be broadly divided into four categories based on their function; however, some of them might have an overlapping role. The first category, the adaptation module, includes enzymes that are involved in spacer acquisition. Cas1 and Cas2 are common in all the types and Cas4 is seen in Types I, II, and V. In Type III, a reverse transcriptase (RT) enzyme is involved in this function. In all the types of Class 1, the pre-crRNA processing is done by Cas6 enzyme. In Class 2 large effector Cas proteins, Cas9, Cas12, and Cas13 play this role in Types II, V, and VI, respectively. However, in Type II, this function is accompanied by a non-Cas protein, bacterial RNaseIII. In Class 1 systems, the effector module consists of multiple *cas* genes, e.g., Cas3, Cas5–8, Cas10, and Cas11, in different combinations. On the other hand, in Class 2 systems, the effector module is represented by large single Cas proteins – Cas9, Cas12, and Cas13. Besides these Cas proteins, there are several other genes involved in signal transduction or ancillary functions in some of the systems (Table 12.2).

Among all the systems, CRISPR/Cas9 is the most widely used for genome editing in all forms of organisms including rice (Mishra et al. 2018). The CRISPR/Cas12 system has also been used for different purposes in rice (Mishra et al. 2018). The CRISPR/Cas13, however, is being least used in rice. Since this system targets RNA molecule instead of DNA, it has been demonstrated in rice that it can be used to eliminate rice RNA viruses (Yue et al. 2020).

12.5.1 Use of CRISPR/Cas9 System in Rice

This system has two main components, namely, Cas9 and gRNA. The gRNA is around 100-nucleotide (nt) long that contains two parts, e.g., CRISPR RNA (crRNA), a 17–20 nt sequence that is complementary to the target DNA, and tracrRNA that functions as the binding scaffold with the Cas9. Sometimes a single RNA strand contains both the crRNA and tracrRNA and is called single guide RNA (sgRNA). Cas9 is an RNA-dependent DNA endonuclease enzyme that induces double-stranded breakage (DSB). The crRNA is required to recruit Cas9 with the target DNA. This system also requires another sequence in the target DNA called protospacer adjacent motif (PAM). In the Cas9 system, the PAM is situated at the 3′ end of the target DNA. The PAM sequence usually varies in the originating organism, for instance, in *Streptococcus pyogenes*, 5′-NGG-3′ is recognized as the PAM, whereas, in *Staphylococcus aureus* it is 5′-NNGRRT-3′ (Wada et al. 2020).

In rice, the CRISPR/Cas9 has been shown as an effective tool for targeted mutagenesis and functional genomics studies (Char et al. 2019). The orthologue of

Table 12.2 Different types of CRISPR/Cas systems and function of different Cas proteins

		Adaptation	Expression	Interference		Signal transduction/ancillary			
Class	Type	Spacer integration	Pre-crRNA processing	Effector module (crRNA and target binding)	Target cleavage	CoA synthesis	Sensor effector	Ring nuclease	Helper, unknown role
Class 1	Type I	Cas1, Cas2, Cas4	Cas6	Cas7, Cas5, SS, Cas8/LS	Cas3″, Cas3′				
	Type III	Cas1, Cas2, RT	Cas6	Cas7, Cas5, SS, Cas10, or LS	Cas10 or LS	Cas10 or LS	CARF#, HEPN#	CARF#	
	Type IV	Cas1, Cas2	Cas6	Cas7, Cas5, SS, Csf1/LS	–				DinG
Class 2	Type II	Cas1, Cas2, Cas4	RNaseIII, Cas9	Cas9	Cas9			Csn2	Csn2
	Type V	Cas1, Cas2, Cas4	Cas12	Cas12	Cas12				
	Type VI	Cas1, Cas2	Cas13	Cas13	Cas13				

an early developmental gene *EPFL9* (epidermal patterning factor like-9), a positive regulator of stomatal development of *Arabidopsis* in rice, has been knocked out using the CRISPR/Cas9 system to elucidate its function (Yin et al. 2017). Targeted knockout of rice-dominant *Waxy* gene that controls the amylose content showed low amylose content and glutinous characteristics in rice grain (Zhang et al. 2018). Insertion of a 5.2-kb carotenoid biosynthesis cassette in the rice genome is also achieved using the CRISPR/Cas9 system by HDR. This system has been found effective to confer resistance or tolerance against different stresses which is discussed in the later section of this chapter.

12.5.2 Use of CRISPR/Cas12a or Cpf1 in Rice

The CRISPR/Cas12a (previously known as Cpf1) is a comparatively new genome-editing tool than the Cas9 system. There are several benefits of using the Cas12a system over the Cas9. First of all, the PAM sequence in Cas9 is G-rich that sometimes makes it difficult for organisms rich in AT, whereas this system recognized the PAM of AT-rich, e.g., 5′-TTTN-3′ and 5′-TTN-3′ (Mishra et al. 2018). Besides, this system cleaves the DNA in a staggered manner leaving 4–5 nucleotide sticky overhangs; in contrast, the Cas9 creates blunt ends during DSB that makes the Cas12a system more specific and less error-prone. Also, Cas12 cuts DNA as a distal end to the PAM allowing repeated targeting. Unlike Cas9, this system can target and cleave the DNA without the need for tracrRNA. The Cas12a also has RNase activity along with endonuclease activity; thus, it can process its own crRNAs (Zetsche et al. 2015).

CRISPR/Cas12a system was demonstrated as a tool for targeted mutagenesis in rice through mutating *OsPDS* and *OsBEL* genes (Xu et al. 2017). Multiplex gene editing in rice using this tool was established targeting six sites of three endogenous genes *5-enolpyruvylshikimate 3-phosphate synthase* (*OsEPSPS*, LOC_Os06g04280), *bentazon-sensitive lethal* (*OsBEL*, LOC_Os03g55240), and *phytoene desaturase* (*OsPDS*, LOC_Os03g08570) (Wang et al. 2017c). A separate experiment successfully edited eight genes at a time using this system (Wang et al. 2018). Targeted mutation of *EPFL9* (*epidermal patterning factor*) orthologue in rice through CRISPR/Cas12a was shown to be a heritable change (Yin et al. 2019; Yin et al. 2017). Several other studies are also present where Cpf1 has been reported successful for genome editing (Li et al. 2018). In a comparative study, where the rice *phytoene desaturase* (*PDS*) gene was targeted using both CRISPR/Cas9 and CRISPR/Cas12a, the latter one was found more effective by achieving higher targeted mutagenesis frequency (Banakar et al. 2020).

12.5.3 Use of CRISPR/Cas13 in Rice

The use of CRISPR/Cas was extended from DNA to RNA with the discovery of CRISPR/Cas13 system that belongs to the Type VI of Class 2 (Mahas et al. 2018).

To date, three families of CRISPR/Cas13 have been discovered, namely, Cas13a (earlier termed as C2c2), Cas13b, and Cas13c. The use of this system ranges from RNA knockdown, transcript tracking to editing tools in both animal and plant cells (Abudayyeh et al. 2017; Cox et al. 2017). Recently it is reported that Cas13a could be used as a nucleic acid detection tool (Gootenberg et al. 2017) too. Unlike other Cas proteins, Cas13 contains two higher eukaryotes and prokaryotes nucleotide-binding domains (HEPN) with exclusive RNase activity (Anantharaman et al. 2013). This system has been demonstrated as a successful plant RNA virus controlling tool (Aman et al. 2018) including the *rice stripe mosaic virus* (RSMV) (Zhang et al. 2019b).

12.6 Delivery Methods of CRISPR/Cas System in Rice

Successful delivery of the CRISPR/Cas system into plant cells is the prerequisite of successful genome editing in plants. Scientists have been using mainly two types of genetic transformation methods in plants, namely, (1) direct and (2) indirect methods. The direct method involved the delivery of CRISPR/Cas components directly through physical or chemical means, whereas the indirect method involves stable expression of transgenes using Ti-plasmid-based vectors or modified plant virus-based vectors (Ran et al. 2017).

12.6.1 Indirect Methods

12.6.1.1 *Agrobacterium*-Mediated Transformation

The *Agrobacterium tumefaciens* is a soil-borne plant pathogenic bacterium that causes crown gall disease in dicot plants. This bacterium has a special ability to transfer a portion of its DNA, called T-DNA, of Ti (tumor inducing) plasmid into the plant genome. The Ti plasmid contains virulence (*vir*) genes that are being activated by phenolic compounds, e.g., acetosyringone (Engström et al. 1987). Upon activation, the *vir* genes produce proteins necessary to transfer the T-DNA portion as a single-stranded DNA (ssDNA) to the plant cell and integrate into the plant genome. The T-DNA is flanked by two directly repeated orientations of 25-bp-long highly homologous sequence, termed as the left border (LB) and right border (RB). These sequences are necessary for recognition by the Vir protein and successful transfer and integration of the T-DNA into the plant genome. The T-DNA naturally contains oncogenes that upon integration in the plant genome produce excessive auxin and cytokinin ultimately result in tumor formation (Gelvin 2003). These pathogenic genes of T-DNA can be removed keeping the LB and RB and expression cassette (s) can be integrated to produce heterologous proteins in the plant. Such Ti plasmids without oncogenes are called disarmed Ti plasmids (Tzfira and Citovsky 2006). Another *Agrobacterium* species, *Agrobacterium rhizogenes*, is also used for the same purpose. This bacterium possesses root inducing (Ri) plasmid harboring T-DNA and this species results in hairy root formation (Ream 2009).

Ti-plasmid-based *Agrobacterium*-mediated genetic transformation has been widely used for delivering CRISPR/Cas components in rice for genome editing. In a study, a single binary Ti plasmid harboring single guide RNA (sgRNA) and Cas9 was delivered in rice using *Agrobacterium* where successful silencing of the rice *bentazon-sensitive lethal* (*BEL, LOC_Os03g0760200*) gene was achieved (Xu et al. 2014). Other studies also being successful to enrich amylose content (Sun et al. 2017), understanding the role of *Isoamylase 1* (*ISA1*) in starch synthesis and endosperm development (Shufen et al. 2019), developed low Cd-accumulating *indica* rice by knocking out *OsNramp5* (Tang et al. 2017) where the CRISPR/Cas components were delivered into rice using *Agrobacterium tumefaciens*. Multiplexed targeting of three genes, namely, *GW2*, *GW5*, and *TGW6*, by CRISPR/Cas9 delivered by *Agrobacterium* is also reported successful where grain weight and size of rice have been increased rapidly (Xu et al. 2016). Apart from enhanced nutritional quality and yield increase and understanding the function of genes, *Agrobacterium*-mediated transformation of CRISPR/Cas components is also used to confer resistance/tolerance in rice, e.g., blast resistance (Wang et al. 2016; Zhou et al. 2018), bacterial leaf blight (Zhou et al. 2018), herbicide tolerance (Sun et al. 2016), and cold tolerance (Shen et al. 2017a).

12.6.1.2 Agroinfiltration

One of the major drawbacks of *Agrobacterium*-mediated genetic transformation is that it integrates the gene of interest(s) into the plant genome and plants generated from explants become transgenic, i.e., stable expression of the transgene. Since transgenic plants are always controversial, another way to use *Agrobacterium* to deliver CRISPR/Cas components is agroinfiltration, where *Agrobacterium* harboring T-DNA with CRISPR/Cas components are injected into plant leaves directly that results in only transient expression of sgRNA and Cas protein. However, this method leads to chimeric rather than systematic expression. During writing this book chapter agroinfiltration of CRISPR/Cas components in several citrus plants has been reported but no study in rice was reported yet (Jia et al. 2019; Jia and Wang 2014; Jia et al. 2017).

12.6.1.3 Viral Vector-Based Transformation of CRISPR/Cas Components

The benefit of using genetically modified plant viruses as a vector to transiently express foreign proteins in the plant over agroinfiltration is that it has the ability to systematically infect the whole plant. The plant viruses have a wide range of host specificity and have their own replicating mechanisms. Till date, many viruses have been adopted for genome engineering in a wide range of crops including rice (Zaidi and Mansoor 2017). The use of plant virus as a vector was first demonstrated using *tobacco mosaic virus* (TMV) for virus-induced gene silencing (VIGS) of carotenoid biosynthesis in *Nicotiana benthamiana* (Kumagai et al. 1995). The use of viral vectors for genome engineering was first reported using geminivirus (Baltes et al. 2014). The geminivirus is a single-stranded DNA (ssDNA) virus that can infect a wide range of dicot and monocot plants. Upon infecting plant cells, it requires only one protein, Rep, to initiate DNA replication through rolling circle amplification

(RCA). However, geminiviruses are not a good option to deliver large DNA fragments due to their smaller genome size (2.5–3.0 kb); therefore, it is recommended to use geminiviruses for the production of an increased amount of sgRNA (Yin et al. 2015). A modified genome of geminiviruses expressing guide RNA could be integrated into T-DNA and delivered into transgenic plants expressing the Cas gene through agroinfiltration.

Another class of viral vector that is extensively used to express alien genes is single-stranded RNA viruses of the family Virgaviridae. Viruses of this family can infect more than 400 plant species belonging to 50 families. *Tobacco rattle virus* (TRV) is the most frequently used of this class of viral vector. Though it can carry more foreign DNA than geminiviruses, it is still not suitable to express Cas protein. Therefore, this vector can also be used to guide RNA delivery (Kuluev et al. 2019).

Recently, the geminivirus-based CRISPR/Cas system has been optimized for rice to knock in a gene of interest (Wang et al. 2017b). This study designed expression cassettes based on the *wheat dwarf virus* (WDV) to express gRNA, *ACT1*, and *GST*. Transgenic rice calli expressing Cas9 were used and successful knock-in of *ACT1* and *GST* has been reported. In this study, wild-type rice calli were also used and another expression cassette to express both Cas9 and gRNA along with knocking in cassettes of *ACT1* and *GST*; however, the success rate was lower in the latter approach.

12.6.2 Direct Methods

12.6.2.1 Biolistic or Particle Gun Bombardment

Biolistic or particle gun bombardment method of genetic transformation requires a special machine called "gene gun" or "biolistic gun." In this method particles of gold, silver, or tungsten coated with DNA are used to transfer DNA into plant explants by applying high pressure. After the successful integration of foreign DNA, the explants are regenerated on selective media, i.e., it is a stable expression of a transgene. This method was used to knock out *OsPDS* and *OsBADH2* where rice calli were bombarded with Cas9 plasmid and sgRNA expression plasmid (Shan et al. 2013). CRISPR/Cas9-based targeted insertion of a 52-Kb carotenoid biosynthesis expression cassette in the targeted site in the rice genome has also been achieved through the particle gun bombardment method (Dong et al. 2020). This method has also been adopted to deliver CRISPR reagents as ribonucleoproteins (RNPs). Integration of the CRISPR/Cas system in the plant genome raises ethical concerns and biosafety issues as it causes continuous genome editing and off-target effects in next generations; therefore, DNA-free delivery of CRISPR reagents is the most desirable as RNPs have limited half-life (Liang et al. 2019). However, a recent study showed that biolistic delivery of CRISPR reagents either in the form of DNA or RNPs results in the insertion of random DNA fragments in the targeted site which was not observed in the case of *Agrobacterium*-mediated delivery (Banakar et al. 2019).

12.6.2.2 Protoplast Transfection

Upon enzymatic removal of plant cell walls, DNA, proteins, and other reagents can be directly transformed into naked protoplasts of plants employing electroporation of polyethylene glycol (PEG) treatment. After successful transformation, the protoplast is regenerated into plants on suitable culture media. For dicot plants, mesophyll protoplasts are used where embryonic callus-derived protoplasts are more preferable for monocots (Ran et al. 2017). Transfection of CRISPR RNPs has been demonstrated in protoplasts of rice (Woo et al. 2015) and in vitro derived zygote of rice (Toda et al. 2019). Protoplast transfection is preferred for the delivery of CRISPR/Cas RNPs. The delivery of CRISPR/Cas RNPs has several benefits over the DNA-based delivery as it does not integrate any DNA in the crop genome. The mutated plants derived through transfecting plant protoplasts by CRISPR RNPs are non-transgenic in nature, thus getting rid of the controversies regarding GM crops. Nonetheless, DNA-based delivery of CRISPR/Cas9 reagents in rice has also been achieved to confer blast disease resistance by mutagenesis of the *ERF transcription factor* gene *OsERF922* (Wang et al. 2016). Besides electroporation and PEG, lipoinfection-mediated delivery of Cas9/gRNA RNPs has proven effective as well but not being tried in rice protoplast yet (Liu et al. 2020).

12.7 CRISPR/Cas for Enhancing Resistance Against Biotic Stresses in Rice

12.7.1 CRISPR/Cas in Providing Resistance Against Rice Bacterial Diseases

Bacterial leaf blight (BLB), caused by *Xanthomonas oryzae* pv. oryzae (Xoo), is one of the most common bacterial diseases of rice frequently observed mostly in Asia and Africa. Many attempts have already been taken to tackle this disease following genome engineering technology including CRISPR/Cas. Immediately after attack by bacteria the plant recognized a pattern in the pathogen, e.g., bacterial flagellin, that triggers immunity in plants, called pathogen-associated molecular patterns (PAMPs)-triggered immunity (PTI) (Zipfel 2014). However, the pathogen also injects molecules called "effector" that bypasses the PTI system and leads towards effector-triggered susceptibility (ETS). If the plants can recognize the effector molecules, then a second layer of immunity is boosted called effector-triggered immunity (ETI) (Spoel and Dong 2012). The Xoo contains T3SS effectors, transcription activator-like effectors (TALEs), which are inducing the expression of *OsSWEET* family of putative sugar transporter genes that leads towards susceptibility. Disruption of two susceptible genes to TALEs, *OsSWEET11* and *OsSWEET14*, in rice cv. Kitaake showed broad-spectrum resistance against most of the Xoo (Xu et al. 2019). TALEs attacks the host nucleus and binds with the specific promoters and activates their expression, altering the transcriptome of the plant. Rice *OS8N*, also called *OsSWEET11*, was edited in another study that was found to be providing enhanced significant resistance against Xoo in homozygous mutants

(Kim et al. 2019). A similar result was also obtained in another rice cultivar Zhonghua 11 (CR-S14) where a codon region of *OsSWEET14* was edited using CRISPR/Cas9 to disrupt its function without any yield penalty (Zeng et al. 2020a). All other studies being carried out to confer resistance against Xoo strains in different rice cultivars were successful using CRISPR/Cas9 technology when the expression of *OsSWEET11*, *OsSWEET13*, and *OsSWEET14* was disrupted either targeting the promoter or directly the coding region (Blanvillain-Baufumé et al. 2017; Oliva et al. 2019; Varshney et al. 2019; Zafar et al. 2020). Interference or knockdown of another rice gene, *Xa13*, is reported to provide resistance against leaf blight; however, this gene is also involved in anther development. Therefore, disruption of this gene results in a penalty for fertility. Recently, selective deletion of the promoter region of this gene using the CRISPR/Cas9 system has been proven effective in both japonica and indica rice varieties without losing the expression of the gene, i.e., plants remain fertile (Li et al. 2020).

12.7.2 CRISPR/Cas in Providing Resistance Against Rice Fungal Diseases

Among the different fungal pathogens of rice, the blast disease caused by the filamentous Ascomycetes *Magnaporthe oryzae* is the most devastating one. Worldwide this pathogen results in up to 30% yield loss which would be enough to feed 60 million people (Nalley et al. 2016). There are about 100 resistant (R) genes that have been found in rice that confer resistance against this disease; among them 30 are already cloned at the molecular level (Xiao et al. 2019). These major R genes have been deployed to develop resistant rice lines worldwide through resistance breeding and transgenic approaches. Genome editing techniques like CRISPR/Cas9 have also been used for this purpose (Mishra et al. 2021). In the plant cells, the pattern recognition receptors (PRRs) recognize PAMPs and trigger the PTI. Upon triggering of PTI plants produce different hormones like jasmonic acid, salicylic acid, and ethylene that are involved in defense mechanisms. Plant *ethylene-responsive factor*, a subfamily of the *APETELA/ethylene response factor* (*AP2*/*ERF*) transcription factor superfamily, has been reported to provide resistance in plants. In rice, genes of this family, e.g., *OsBIERF1*, *OsBIERF3*, and *OsBIERF4*, are involved in providing resistance against *M. oryzae* (Cao et al. 2006). Blast resistance was improved in Kuiku131, a japonica type rice widely cultivated in Northern China, following CRISPR/Cas9-mediated knockout of *ERF transcription factor* gene *OsERF922* (Wang et al. 2016). Around 42% mutation was observed in T_0 in the targeted gene site and a stable heredity of this mutation was also observed in T1 and T2. Besides, no significant loss in agronomic performances was observed, revealing that proper editing using CRISPR/Cas technology does not compromise other desired traits. *Pi21* is another broad-spectrum resistance that encodes a proline-rich protein that contains a putative heavy metal binding domain and a putative protein-protein interaction domain. Wild-type *Pi21* shows susceptibility in plants against *M. oryzae*. Using CRISPR/Cas9 this gene was mutated and enhanced stability and non-race-specific resistance against *M. oryzae* were obtained (Nawaz et al. 2020). The mutation rate was 66%, among which 26% biallelic, 22% homozygous, 12% heterozygous, and 3% chimeric were obtained. They were also able to achieve transgene-free mutants with enhanced resistance. Mutation of the *Pi21* gene did not compromise the agronomic characteristics.

The plant R genes usually encode for proteins with nucleotide-binding site-leucine-rich repeat (NLR) domains. Using CRISPR/Cas9, a study showed that Ptr, an R gene, provides broad-spectrum resistance against blast disease. Using CRISPR/Cas technology this gene can be knocked in in other susceptible rice varieties to develop resistance against blast disease (Zhao et al. 2018). Another study found another potential gene that plays a role in resistance against *M. oryzae*, the *OsSEC3A*, which is an important unit of the exocyst complex of the rice. Through disrupting this gene using CRISPR/Cas9, enhanced resistance against *M. oryzae* was obtained. Such disruption is linked with enhanced salicylic acid synthesis, thus providing more resistance against *M. oryzae*. However, this came with the penalty of dwarf structure and lesion-mimic phenotype (Ma et al. 2018). Further optimization on the mutating of this gene may result in better resistance against blast disease without any unwanted penalty.

12.7.3 CRISPR/Cas in Providing Resistance Against Rice Viral Diseases

The plant viruses cause a severe economic loss around the world and greatly alter the agronomic traits and physiological functions in crops (Nicaise 2014). Controlling plant viruses greatly depends on controlling their vectors through applying synthetic pesticides. Identification of different resistant (R) genes has also made it possible to control plant viruses through molecular breeding though it is a time-consuming technique (Khan et al. 2018). Besides, several transgenic approaches have been proven effective including the RNAi mechanism. However, such promising techniques of controlling plant viruses come with many hurdles, such as, for DNA viruses, RNAi mechanism can suppress the expression of genes at the post-transcription level rather than eliminating the virus itself (Voinnet 2005). It is proven that the viruses also develop a counter-mechanism of plants' RNAi mechanism (Pumplin and Voinnet 2013). CRISPR/Cas technology can be used as an alternative to resistance breeding or RNAi mechanism to control viral diseases of rice.

The major viral disease of rice is rice tungro disease (RTD), a severe production constraint mainly in tropical Asia. Two viruses act jointly to cause this disease, e.g., the single-stranded RNA virus, namely, *rice tungro spherical virus* (RTSV) and the double-stranded DNA virus named *rice tungro bacilliform virus* (RTBV) (Chancellor et al. 2006). In a recent study, a susceptible (S) gene, *eIF4G*, in rice was mutated using CRISPR/Cas9 and achieved resistance against this disease in an RTD susceptible rice variety, IR-64 (Macovei et al. 2018). They obtained a mutation frequency of 36 to 86% and no potential off-target issue was observed. Analyzing the sequences in mutant lines, it was observed that among all the obtained mutations, the resistance was conferred by the in-frame mutation in the SVLFPNLAGKS residues (mainly NL), nearby the YUV residues. Such technology could be used to develop rice varieties suitable for cultivating in RTD-prone areas to achieve the targeted yield. SNP in the codon for Val$^{1060-1061}$ of the *eIF4G* gene in rice is also reported to be associated with resistance against RTSV (Lee et al. 2010). Recently, the CRISPR/Cas system has also been optimized for base editing to introduce point mutations (Kantor et al. 2020).

Another study obtained RTD resistant lines using RNAi technology targeting the ORF IV of RTBV (Valarmathi et al. 2016). Alternatively, CRISPR/Cas13 could be used to control RNA viruses in rice at the post-transcriptional level by inhibiting translation (Cao et al. 2020). The benefit of using CRISPR/Cas13 over CRISPR/Cas9 or CRISPR/Cas12 is that it targets RNAs rather than DNA that turns this into an effective tool to control plant RNA viruses (Khan et al. 2018). *Southern rice black-streaked dwarf virus* (SRBSDV), a major virus infecting rice plants in several East Asian countries, was successfully controlled using CRISPR/Cas13 (Zhang et al. 2019b). This study designed three crRNAs to target the double-stranded RNA genome. They also targeted a single-stranded RNA virus of rice, named *rice stripe mosaic virus* (RMSV). Therefore, stable expression of CRISPR/Cas13 in rice can be a better option to control RNA viruses, e.g., it can also be implemented to control RTD.

Other approaches to using CRISPR/Cas are base editing and prime editing. For DNA base editing a Cas enzyme for programmable DNA binding is required with a single-stranded DNA-modifying enzyme for single nucleotide base alteration. There are two types of base editor present, e.g., cytosine base editors (C → T, T → C) and adenine base editors (A → G, and G → A) (Kantor et al. 2020). Recently, C → G transversion using the CRISPR/Cas system is also being reported (Kurt et al. 2021). The limitations of these base editors have been omitted very recently with another technique that does not require any DSB or donor DNA. This method directly alters the DNA information in a specific site of a targeted DNA using a catalytically impaired Cas9 (Cas9 nickase-nCas9) protein coupled with an engineered reverse transcriptase, programmed with prime editing guide RNA (pegRNA) that both specifies the target site and encodes the desired edit (Anzalone et al. 2019). This prime genome editing technology has also been adopted for rice too (Lin et al. 2020).

12.7.4 CRISPR/Cas in Providing Tolerance Against Abiotic Stresses of Rice

Several abiotic stresses result in rice yield reduction including salinity, cold stress, drought, and so on. Among them, the development of saline tolerant rice lines is a promising approach to increase the rice yield as it will permit the cultivation of rice in areas that are saline-prone. Usually, rice is a saline-sensitive crop and cannot be grown in saline-prone land; however, many saline-tolerant genes have been cloned, namely, *SKC1*, *DST*, *OsRR22*, *OsHAL3*, *P5CS*, *SNAC2*, and *OsNAP*. Among them, *OsRR22* gene is reported to be linked with enhanced saline tolerance in rice when its natural function is disrupted. With this aim, CRISPR/Cas9 was used to mutate this gene and the saline tolerance increased significantly from the seedling stage in rice (Zhang et al. 2019a). Sequence analysis of this study reported six mutation types at the target sites that are proven to be linked with saline tolerance.

Another important environmental factor that limits rice production is cold stress and developing cold stress-tolerant lines would allow cultivating rice in areas with low temperature. Several cold stress tolerance genes have been identified and cloned, such as *COLD1*, *OsSRFP1*, *SGD1*, and *OsMYB30*. Several cold stress-mutant rice lines with improved agronomic characteristics have been developed using CRISPR/Cas9 targeting *OsPIN5b* (a panicle length gene), *GS3* (a grain size gene), and *OsMYB30* (cold tolerance gene) genes in Nipponbare, a japonica rice (Zeng et al. 2020b).

CRISPR/Cas9 technology has been employed to develop herbicide resistance in rice (Sun et al. 2016). Targeting the rice ALS gene, this study used two guide RNAs, a Cas9 enzyme and a 476-bp donor template, to bring several point mutations in the targeted gene (W548L and S627I substitutions). The donor DNA also had some other features like several synonymous substitutions that did not change the amino acid sequence but restricted the Cas9 enzyme from further targeting the gene. They used the particle gun bombardment method to deliver the donor DNA and CRISPR reagents. Herbicide tolerance in rice has also been achieved via prime editing (Butt et al. 2020). In this study, three different gene loci were targeted, termed as *ACETOLACTATE SYNTHASE (OsALS)*, *IDEAL PLANT ARCHITECTURE 1 (OsIPA1)*, and *TEOSINTE BRANCHED 1 (OsTB1)*; however, the authors suggested further studies on using this technology.

12.8 Conclusion and Future Implications

The production of rice is estimated to increase by 1% annually to meet the demand of the growing population (Normile 2008) and the total production must be increased by 40% by 2050 (Milovanovic and Smutka 2017). Biotic and abiotic stresses are the major constraints of lower rice yield worldwide and these issues must be addressed to meet the targeted yield (Stallworth et al. 2020). The development of resistant or tolerant rice lines has always been a continuous process worldwide in rice improvement projects. Developing resistance or tolerance in crops has always been carried out through bringing changes in the genetic constitution by natural mutation followed by selection, hybridization, mutation breeding, or genetic engineering. Conventional breeding is always time-consuming and somewhat becomes static due to the unavailability of genetic variation and loss of genetic diversity due to crop domestication. Natural mutation is a slow process and always needs to rely on fate, and artificial mutagenesis by means of physical (e.g., radiation) and chemical (e.g., ethyl methylsulfonate) mutagens is random, time-consuming, and labor intensive. In contrast, the genetic engineering approach is always less labor intensive, time-saving, and precise and allows transferring of genes from distantly related organisms, for instance, Golden Rice, a genetically modified rice that contains precursor genes to produce β-carotene that is naturally lacking in rice grains (Paine et al. 2005). With the discovery of CRISPR/Cas system and its establishment as a genome editing tool, it has become the most powerful tool for crop genetic modification because of its high efficiency, preciseness, easiness, and cost-effectiveness (Manghwar et al. 2019). However, this system is not free from limitations. The biggest limitation of the CRISPR/Cas system is the continuous expression of Cas proteins and potential off-targets that may be detrimental to changes in organisms. Though new techniques are being implemented to address this issue, in our opinion DNA-free delivery method of CRISPR/Cas ribonucleoprotein complex in rice zygote or protoplast is the most effective option (Banakar et al. 2019; Toda et al. 2019). Such methods avoid the integration of CRISPR machinery into the genome of the organism and eliminate the potential risks of off-target. More recently, another technique has been developed to address the off-target issue using light-induced degradation of sgRNA named as CRISPRoff (Carlson-Stevermer et al. 2020). This CRISPRoff sgRNA was synthesized artificially using solid-phase synthesis, and photocleavable residues containing o-nitrobenzyl groups were incorporated at specific positions that undergo degradation when exposed to UV light. Upon exposure to UV light, these sgRNAs did not form any complex with Cas9 protein. The authors also demonstrated that these sgRNAs were cleaved within cells when exposed to light. In contrast, the cell that was kept in dark had an abundance of sgRNA. They also successfully showed that sgRNAs could be cleaved in specific tissues via selective illumination. However, this study is yet confined within human cells and no study has been published on plant cells. Development and optimization of CRISPRoff for plant cells including rice have great potential in genome editing for crop improvement.

Another major limitation of the CRISPR/Cas system is PAM. Due to the high specificity of the PAM requirement in successful interference, it limits the number of targets. For instance, the CRISPR/Cas9 system only recognized GC-rich PAMs that limit its application in AT-rich genomes. Through the discovery of the CRISPR/Cas12a system, AT-rich genomes are now also available to be edited using CRISPR/Cas. Besides, new variants and modified systems of CRISPR/Cas are being discovered and developed regularly that are increasing the range of the target sequences (Mishra et al. 2018). Another CRISPR/Cas system-based technique that must be mentioned that can address all the limitations and issues regarding genome editing is "prime editing" which is PAM independent and DSB-free. This method is precise enough to introduce point mutations. Recently, another Cas protein has been discovered, namely, Cas14, that can be used for targeted cleavage of single-stranded DNA (ssDNA) and independent of PAM requirement. In addition, this system is also compact (950 to 1400 amino acids; half of the Cas12a protein) (Harrington et al. 2018). This compact-sized Cas protein can also be used in viral-based vectors. As the viral-based vectors cannot carry large size gene fragments, expression of Cas9 or Cas12a protein has remained impossible and is only limited to CRISPR array expression. Therefore, there is the scope of investigating Cas14 as a potential genome editing tool in rice as well as using viral-based vectors for its transient expression.

Most of the resistance or tolerance is generally achieved through gain-of-function mutation; therefore, knock-in of the desired gene to provide resistance or tolerance against biotic and abiotic stresses is most desired. However, due to the low efficiency of HR in plants, CRISPR/Cas-based knock-in of a gene of interest in plants has remained a major challenge. Therefore, a better delivery method of donor DNAs along with CRISPR/Cas systems will always be admired.

Apart from all the technical issues, other major issues that limit the genome-edited crops being cultivated in the farmers' fields are the regulatory issue and biosafety issue. Genetically modified crops are always a topic of controversy and a great political issue. Contradictory opinions regarding GM crops among different nations and different regulation policies have always made it difficult to be cultivated. Therefore, a proper and unified regulation is required for the whole world that how CRISPR/Cas-edited crops will be treated. Also, identification of a method through which a genome-edited crop will be free of all kind of controversies is a must.

In conclusion, with the advancement of sequencing techniques, more and more information on the function of genes is being revealed. By achieving such genomic data and with advancement of CRISPR/Cas-based genome editing techniques, developing stress resistance and tolerance in rice is becoming easier day by day. Through this, food security could be achieved for the ever-growing population of the world. Therefore, we must be prepared and address all the issues regarding genome editing techniques and regulations to successfully achieve the "Zero Hunger" sustainable development goal of the United Nations.

References

Abudayyeh OO, Gootenberg JS, Essletzbichler P, Han S, Joung J, Belanto JJ, Verdine V, Cox DBT, Kellner MJ, Regev A, Lander ES, Voytas DF, Ting AY, Zhang F (2017) RNA targeting with CRISPR-Cas13. Nature 550:280–284

Adhikari U, Nejadhashemi AP, Woznicki SA (2015) Climate change and eastern Africa: a review of impact on major crops. Food Energy Secur 4:110–132

Adli M (2018) The CRISPR tool kit for genome editing and beyond. Nat Commun 9:1911

Afzal S, Sirohi P, Singh NK (2020) A review of CRISPR associated genome engineering: application, advances and future prospects of genome targeting tool for crop improvement. Biotechnol Lett 42:1611–1632

Aman R, Ali Z, Butt H, Mahas A, Aljedaani F, Khan MZ, Ding S, Mahfouz M (2018) RNA virus interference via CRISPR/Cas13a system in plants. Genome Biol 19:1

Amitai G, Sorek R (2016) CRISPR–Cas adaptation: insights into the mechanism of action. Nat Rev Microbiol 14:67–76

Anantharaman V, Makarova KS, Burroughs AM, Koonin EV, Aravind L (2013) Comprehensive analysis of the HEPN superfamily: identification of novel roles in intra-genomic conflicts, defense, pathogenesis and RNA processing. Biol Direct 8:15

Andersson M, Turesson H, Nicolia A, Fält A-S, Samuelsson M, Hofvander P (2017) Efficient targeted multiallelic mutagenesis in tetraploid potato (Solanum tuberosum) by transient CRISPR-Cas9 expression in protoplasts. Plant Cell Rep 36:117–128

Ansari MR, Shaheen T, Bukhari S, Husnain T (2015) Genetic improvement of rice for biotic and abiotic stress tolerance. Turk J Bot 39:911–919

Anzalone AV, Randolph PB, Davis JR, Sousa AA, Koblan LW, Levy JM, Chen PJ, Wilson C, Newby GA, Raguram A, Liu DR (2019) Search-and-replace genome editing without double-strand breaks or donor DNA. Nature 576:149–157

Arora L, Narula A (2017) Gene editing and crop improvement using CRISPR-Cas9 system. Front Plant Sci 8:1932

Baerwald MR, Goodbla AM, Nagarajan RP, Gootenberg JS, Abudayyeh OO, Zhang F, Schreier AD (2020) Rapid and accurate species identification for ecological studies and monitoring using CRISPR-based SHERLOCK. Mol Ecol Resour 20:961–970

Bai B, Wu J, Sheng W-T, Zhou B, Zhou L-J, Zhuang W, Yao D-P, Deng Q-Y (2015) Comparative analysis of anther transcriptome profiles of two different Rice male sterile lines genotypes under cold stress. Int J Mol Sci 16:11398–11416

Bakker W (1974) Characterization and ecological aspects of rice yellow mottle virus in Kenya. Pudoc

Baltes NJ, Gil-Humanes J, Cermak T, Atkins PA, Voytas DF (2014) DNA replicons for plant genome engineering. Plant Cell 26:151–163

Banakar R, Eggenberger AL, Lee K, Wright DA, Murugan K, Zarecor S, Lawrence-Dill CJ, Sashital DG, Wang K (2019) High-frequency random DNA insertions upon co-delivery of CRISPR-Cas9 ribonucleoprotein and selectable marker plasmid in rice. Sci Rep 9:19902

Banakar R, Schubert M, Collingwood M, Vakulskas C, Eggenberger AL, Wang K (2020) Comparison of CRISPR-Cas9/Cas12a ribonucleoprotein complexes for genome editing efficiency in the Rice phytoene desaturase (OsPDS) gene. Rice 13:4

Bandumula N (2018) Rice production in Asia: key to global food security. Proc Natl Acad Sci India Sect B Biol Sci 88:1323–1328

Bandyopadhyay A, Yin X, Biswal A, Coe R, Quick WP (2019) CRISPR-Cas9-mediated genome editing of rice towards better grain quality. In: Sreenivasulu N (ed) Rice grain quality: methods and protocols. Springer, New York, NY, pp 311–336

Barman A, Deb B, Chakraborty S (2020) A glance at genome editing with CRISPR–Cas9 technology. Curr Genet 66:447–462

Blanvillain-Baufumé S, Reschke M, Solé M, Auguy F, Doucoure H, Szurek B, Meynard D, Portefaix M, Cunnac S, Guiderdoni E, Boch J, Koebnik R (2017) Targeted promoter editing for rice resistance to Xanthomonas oryzae pv. oryzae reveals differential activities for SWEET14-inducing TAL effectors. Plant Biotechnol J 15:306–317

Burmistrz M, Pyrć K (2015) CRISPR-Cas systems in prokaryotes. Pol J Microbiol 64

Butt H, Rao GS, Sedeek K, Aman R, Kamel R, Mahfouz M (2020) Engineering herbicide resistance via prime editing in rice. Plant Biotechnol J 18:2370–2372

Cao Y, Song F, Goodman RM, Zheng Z (2006) Molecular characterization of four rice genes encoding ethylene-responsive transcriptional factors and their expressions in response to biotic and abiotic stress. J Plant Physiol 163:1167–1178

Cao Y, Zhou H, Zhou X, Li F (2020) Control of plant viruses by CRISPR/Cas system-mediated adaptive immunity. Front Microbiol:11, 2613

Carlson-Stevermer J, Kelso R, Kadina A, Joshi S, Rossi N, Walker J, Stoner R, Maures T (2020) CRISPRoff enables spatio-temporal control of CRISPR editing. Nat Commun 11:5041

Chancellor TC, Holt J, Villareal S, Tiongco ER, Venn J (2006) Spread of plant virus disease to new plantings: a case study of rice tungro disease. Adv Virus Res 66:1–29

Char SN, Li R, Yang B (2019) CRISPR/Cas9 for mutagenesis in rice. Methods Mol Biol 1864:279–293

Chaudhari PR, Tamrakar N, Singh L, Tandon A, Sharma D (2018) Rice nutritional and medicinal properties: A. J Pharmacogn Phytochem 7:150–156

Chukwu S, Rafii M, Ramlee S, Ismail S, Hasan M, Oladosu Y, Magaji U, Akos I, Olalekan K (2019) Bacterial leaf blight resistance in rice: a review of conventional breeding to molecular approach. Mol Biol Rep 46:1519–1532

Cox DBT, Gootenberg JS, Abudayyeh OO, Franklin B, Kellner MJ, Joung J, Zhang F (2017) RNA editing with CRISPR-Cas13. Science 358:1019–1027

Dai S, Beachy RN (2009) Genetic engineering of rice to resist rice tungro disease. In Vitro Cell Dev Biol Plant:45–517

Dong OX, Yu S, Jain R, Zhang N, Duong PQ, Butler C, Li Y, Lipzen A, Martin JA, Barry KW, Schmutz J, Tian L, Ronald PC (2020) Marker-free carotenoid-enriched rice generated through targeted gene insertion using CRISPR-Cas9. Nat Commun 11:1178

Engström P, Zambryski P, Van Montagu M, Stachel S (1987) Characterization of agrobacterium tumefaciens virulence proteins induced by the plant factor acetosyringone. J Mol Biol 197:635–645

Fahad S, Adnan M, Hassan S, Saud S, Hussain S, Wu C, Wang D, Hakeem KR, Alharby HF, Turan V, Khan MA, Huang J (2019) Chapter 10. Rice responses and tolerance to high temperature. In: Hasanuzzaman M, Fujita M, Nahar K, Biswas JK (eds) Advances in rice research for abiotic stress tolerance. Woodhead Publishing, pp 201–224. ISBN: 978-0-12-814332-2 (print), ISBN: 978-0-12-814333-9 (online), Copyright © 2019 Elsevier Inc., Woodhead Publishing, Cambridge, UK

Fukagawa NK, Ziska LH (2019) Rice: importance for global nutrition. J Nutr Sci Vitaminol 65:S2–S3

Gao C (2018) The future of CRISPR technologies in agriculture. Nat Rev Mol Cell Biol 19:275–276

Gelvin SB (2003) Agrobacterium-mediated plant transformation: the biology behind the "gene-jockeying" tool. Microbiol Mol Biol Rev 67:16–37. table of contents

Gnanamanickam SS (2009) An overview of progress in biological control. Biological control of rice diseases. Springer, Dordrecht, pp 43–51

Gootenberg JS, Abudayyeh OO, Lee JW, Essletzbichler P, Dy AJ, Joung J, Verdine V, Donghia N, Daringer NM, Freije CA, Myhrvold C, Bhattacharyya RP, Livny J, Regev A, Koonin EV, Hung DT, Sabeti PC, Collins JJ, Zhang F (2017) Nucleic acid detection with CRISPR-Cas13a/C2c2. Science 356:438–442

Harrington LB, Burstein D, Chen JS, Paez-Espino D, Ma E, Witte IP, Cofsky JC, Kyrpides NC, Banfield JF, Doudna JA (2018) Programmed DNA destruction by miniature CRISPR-Cas14 enzymes. Science 362:839

Hlaing EE, Rungcharoenarrichit S, Lailerd N, Roytrakul S, Piamrojanaphat P (2019) Chapter 25 - purple rice bran improves hepatic insulin signaling via activation of Akt and stabilization of IGF in diabetic rats. In: Watson RR, Preedy VR (eds) Dietary interventions in liver disease. Academic Press, pp 297–312

Hussain S, Zhang J h, Zhong C, Zhu L-f, Cao X-c, Yu S-m, Allen Bohr J, Hu J-j, Jin Q-y (2017) Effects of salt stress on rice growth, development characteristics, and the regulating ways: a review. J Integr Agricl 16:2357–2374

Jia H, Wang N (2014) Targeted genome editing of sweet orange using Cas9/sgRNA. PLoS One 9: e93806–e93806

Jia H, Zhang Y, Orbović V, Xu J, White FF, Jones JB, Wang N (2017) Genome editing of the disease susceptibility gene CsLOB1 in citrus confers resistance to citrus canker. Plant Biotechnol J 15:817–823

Jia H, Orbović V, Wang N (2019) CRISPR-LbCas12a-mediated modification of citrus. Plant Biotechnol J 17:1928–1937

Joseph A (2013) A study on the effect of salinity stress on the growth and yield of some native rice cultivars of Kerala State of India. Agric For Fisher 2:141

Kakar KU, Duan YP, Nawaz Z, Sun G, Almoneafy AA, Hassan MA, Elshakh A, Li B, Xie G-L (2014) A novel rhizobacterium Bk7 for biological control of brown sheath rot of rice caused by pseudomonas fuscovaginae and its mode of action. Eur J Plant Pathol 138:819–834

Kantor A, McClements ME, MacLaren RE (2020) CRISPR-Cas9 DNA base-editing and prime-editing. Int J Mol Sci 21(17):6240

Khan MH, Dar ZA, Dar SA (2015) Breeding strategies for improving rice yield-a review. Agric Sci 05:12

Khan MZ, Amin I, Hameed A, Mansoor S (2018) CRISPR-Cas13a: prospects for plant virus resistance. Trends Biotechnol 36:1207–1210

Kim Y-A, Moon H, Park C-J (2019) CRISPR/Cas9-targeted mutagenesis of Os8N3 in rice to confer resistance to Xanthomonas oryzae pv. oryzae. Rice 12:67

Kim Y, Chung YS, Lee E, Tripathi P, Heo S, Kim K-H (2020) Root response to drought stress in rice (Oryza sativa L.). Int J Mol Sci 21:1513

Kis A, Hamar É, Tholt G, Bán R, Havelda Z (2019) Creating highly efficient resistance against wheat dwarf virus in barley by employing CRISPR/Cas9 system. Plant Biotechnol J 17:1004–1006

Kouassi N, N'guessan P, Albar L, Fauquet C, Brugidou C (2005) Distribution and characterization of rice yellow mottle virus: a threat to African farmers. Plant Dis 89:124–133

Krishnan P, Ramakrishnan B, Reddy KR, Reddy VR (2011) Chapter three. High-temperature effects on rice growth, yield, and grain quality. In: Sparks DL (ed) Advances in agronomy. Academic Press, pp 87–206. ISBN: 978-0-12-387689-8 © Copyright 2011 Elsevier Inc. Academic Press, San Diego, CA, USA

Kuluev BR, Gumerova GR, Mikhaylova EV, Gerashchenkov GA, Rozhnova NA, Vershinina ZR, Khyazev AV, Matniyazov RT, Baymiev AK, Baymiev AK, Chemeris AV (2019) Delivery of CRISPR/Cas components into higher plant cells for genome editing. Russ J Plant Physiol 66: 694–706

Kumagai MH, Donson J, Della-Cioppa G, Harvey D, Hanley K, Grill LK (1995) Cytoplasmic inhibition of carotenoid biosynthesis with virus-derived RNA. Proc Natl Acad Sci U S A 92: 1679–1683

Kumar K, Kumar M, Kim S-R, Ryu H, Cho Y-G (2013) Insights into genomics of salt stress response in rice. Rice 6:27

Kumar S, Meshram S, Sinha A (2017) Bacterial diseases in rice and their eco-friendly management. Agric Sci Res 7:31–42

Kurt IC, Zhou R, Iyer S, Garcia SP, Miller BR, Langner LM, Grünewald J, Joung JK (2021) CRISPR C-to-G base editors for inducing targeted DNA transversions in human cells. Nat Biotechnol 39:41–46

Lee J-H, Muhsin M, Atienza GA, Kwak D-Y, Kim S-M, De Leon TB, Angeles ER, Coloquio E, Kondoh H, Satoh K, Cabunagan RC, Cabauatan PQ, Kikuchi S, Leung H, Choi I-R (2010) Single nucleotide polymorphisms in a gene for translation initiation factor (eIF4G) of rice (Oryza sativa) associated with resistance to rice tungro spherical virus. Mol Plant-Microbe Interact 23:29–38

Li S, Zhang X, Wang W, Guo X, Wu Z, Du W, Zhao Y, Xia L (2018) Expanding the scope of CRISPR/Cpf1-mediated genome editing in rice. Mol Plant 11:995–998

Li C, Li W, Zhou Z, Chen H, Xie C, Lin Y (2020) A new rice breeding method: CRISPR/Cas9 system editing of the Xa13 promoter to cultivate transgene-free bacterial blight-resistant rice. Plant Biotechnol J 18:313–315

Liang Z, Chen K, Gao C (2019) Biolistic delivery of CRISPR/Cas9 with ribonucleoprotein complex in wheat. Methods Mol Biol 1917:327–335

Lin Q, Zong Y, Xue C, Wang S, Jin S, Zhu Z, Wang Y, Anzalone AV, Raguram A, Doman JL, Liu DR, Gao C (2020) Prime genome editing in rice and wheat. Nat Biotechnol 38:582–585

Liu W, Rudis MR, Cheplick MH, Millwood RJ, Yang J-P, Ondzighi-Assoume CA, Montgomery GA, Burris KP, Mazarei M, Chesnut JD, Stewart CN (2020) Lipofection-mediated genome editing using DNA-free delivery of the Cas9/gRNA ribonucleoprotein into plant cells. Plant Cell Rep 39:245–257

Lv Z, Zhu Y, Liu X, Ye H, Tian Y, Li F (2018) Climate change impacts on regional rice production in China. Clim Chang 147:523–537

Lv Y, Hussain MA, Luo D, Tang N (2019) Current understanding of genetic and molecular basis of cold tolerance in rice. Mol Breed 39:159

Ma J, Chen J, Wang M, Ren Y, Wang S, Lei C, Cheng Z (2018) Disruption of OsSEC3A increases the content of salicylic acid and induces plant defense responses in rice. J Exp Bot 69:1051–1064

Macovei A, Sevilla NR, Cantos C, Jonson GB, Slamet-Loedin I, Čermák T, Voytas DF, Choi I-R, Chadha-Mohanty P (2018) Novel alleles of rice eIF4G generated by CRISPR/Cas9-targeted mutagenesis confer resistance to rice tungro spherical virus. Plant Biotechnol J 16:1918–1927

Mahas A, Neal Stewart C Jr, Mahfouz MM (2018) Harnessing CRISPR/Cas systems for programmable transcriptional and post-transcriptional regulation. Biotechnol Adv 36:295–310

Makarova KS, Wolf YI, Iranzo J, Shmakov SA, Alkhnbashi OS, Brouns SJJ, Charpentier E, Cheng D, Haft DH, Horvath P, Moineau S, Mojica FJM, Scott D, Shah SA, Siksnys V, Terns MP, Venclovas Č, White MF, Yakunin AF, Yan W, Zhang F, Garrett RA, Backofen R, van der Oost J, Barrangou R, Koonin EV (2020) Evolutionary classification of CRISPR–Cas systems: a burst of class 2 and derived variants. Nat Revi Microbiol 18:67–83

Manghwar H, Lindsey K, Zhang X, Jin S (2019) CRISPR/Cas system: recent advances and future prospects for genome editing. Trends Plant Sci 24:1102–1125

Martin A, Serano Julia M, Jarvis E, Bruce Heather S, Wang J, Ray S, Barker Carryn A, O'Connell Liam C, Patel Nipam H (2016) CRISPR/Cas9 mutagenesis reveals versatile roles of Hox genes in crustacean limb specification and evolution. Curr Biol 26:14–26

Mew T (1989) An overview of the world bacterial blight situation. Proceedings of the International Workshop on Bacterial Blight of Rice. International Rice Research Institute, 12, p 7

Milovanovic V, Smutka L (2017) Asian countries in the global rice market. Acta Univ Agric Silvic Mendel Brun 65:679–688

Mishra R, Joshi RK, Zhao K (2018) Genome editing in rice: recent advances, challenges, and future implications. Front Plant Sci 9:1361

Mishra R, Zheng W, Joshi RK, Kaijun Z (2021) Genome editing strategies towards enhancement of rice disease resistance. Rice Sci 28:133–145

Molla KA, Azharudheen TPM, Ray S, Sarkar S, Swain A, Chakraborti M, Vijayan J, Singh ON, Baig MJ, Mukherjee AK (2019) Novel biotic stress responsive candidate gene based SSR (cgSSR) markers from rice. Euphytica 215:17

Nalley L, Tsiboe F, Durand-Morat A, Shew A, Thoma G (2016) Economic and environmental impact of rice blast pathogen (Magnaporthe oryzae) alleviation in the United States. PLoS One 11:e0167295

Nawaz G, Usman B, Peng H, Zhao N, Yuan R, Liu Y, Li R (2020) Knockout of Pi21 by CRISPR/Cas9 and iTRAQ-based proteomic analysis of mutants revealed new insights into M. oryzae resistance in elite rice line. Genes 11:735

Nicaise V (2014) Crop immunity against viruses: outcomes and future challenges. Fron Plant Sci:5, 660

Normile D (2008) Agricultural research. Reinventing rice to feed the world. Science 321:330–333

O'Connell C (2019) Why gene editing is decade's most significant innovation. The Irish Times

Oliva R, Ji C, Atienza-Grande G, Huguet-Tapia JC, Perez-Quintero A, Li T, Eom J-S, Li C, Nguyen H, Liu B, Auguy F, Sciallano C, Luu VT, Dossa GS, Cunnac S, Schmidt SM, Slamet-Loedin IH, Vera Cruz C, Szurek B, Frommer WB, White FF, Yang B (2019) Broad-spectrum resistance to bacterial blight in rice using genome editing. Nat Biotechnol 37:1344–1350

Paine JA, Shipton CA, Chaggar S, Howells RM, Kennedy MJ, Vernon G, Wright SY, Hinchliffe E, Adams JL, Silverstone AL, Drake R (2005) Improving the nutritional value of Golden Rice through increased pro-vitamin A content. Nat Biotechnol 23:482–487

Palanog AD, Swamy BPM, Shamsudin NAA, Dixit S, Hernandez JE, Boromeo TH, Cruz PCS, Kumar A (2014) Grain yield QTLs with consistent-effect under reproductive-stage drought stress in rice. Field Crops Res 161:46–54

Pumplin N, Voinnet O (2013) RNA silencing suppression by plant pathogens: defence, counter-defence and counter-counter-defence. Nat Rev Microbiol 11:745–760

Rajeev Rai AJT, Cavazza A (2021) Gene editing for the treatment of primary immunodeficiency diseases. Hum Gene Ther 32:43–51

Ran FA, Hsu PD, Wright J, Agarwala V, Scott DA, Zhang F (2013) Genome engineering using the CRISPR-Cas9 system. Nat Protoc 8:2281–2308

Ran Y, Liang Z, Gao C (2017) Current and future editing reagent delivery systems for plant genome editing. Sci China Life Sci 60:490–505

Ream W (2009) Agrobacterium tumefaciens and A. rhizogenes use different proteins to transport bacterial DNA into the plant cell nucleus. Microb Biotechnol 2:416–427

Ricroch A, Clairand P, Harwood W (2017) Use of CRISPR systems in plant genome editing: toward new opportunities in agriculture. Emerging Top Life Sci 1:169–182

Romero FM, Gatica-Arias A (2019) CRISPR/Cas9: development and application in rice breeding. Rice Sci 26:265–281

Rostami M, Rahimian H, Ghasemi A (2005) Occurrence of bacterial sheath rot of rice caused by pseudomonas syringae in North of Iran. Iran J Plant Path 41:205–206

Salsman J, Dellaire G (2017) Precision genome editing in the CRISPR era. Biochem Cell Biol 95: 187–201

Sánchez-León S, Gil-Humanes J, Ozuna CV, Giménez MJ, Sousa C, Voytas DF, Barro F (2018) Low-gluten, nontransgenic wheat engineered with CRISPR/Cas9. Plant Biotechnol J 16:902–910

Santillán Martínez MI, Bracuto V, Koseoglou E, Appiano M, Jacobsen E, Visser RGF, Wolters A-MA, Bai Y (2020) CRISPR/Cas9-targeted mutagenesis of the tomato susceptibility gene PMR4 for resistance against powdery mildew. BMC Plant Biol 20:284

Savary S, Ficke A, Aubertot J-N, Hollier C (2012) Crop losses due to diseases and their implications for global food production losses and food security. Food Secur 4:519–537

Schachtsiek J, Stehle F (2019) Nicotine-free, nontransgenic tobacco (Nicotiana tabacum l.) edited by CRISPR-Cas9. Plant Biotechnol J 17:2228–2230

Scheben A, Wolter F, Batley J, Puchta H, Edwards D (2017) Towards CRISPR/Cas crops – bringing together genomics and genome editing. New Phytol 216:682–698

Setiyono T, Barbieri M, Prasadini P, Maunahan A, Gatti L (2018) Spatial assessment of heat stress impact on rice production in two districts of Andhra Pradesh, India. World J Agric Res 6:10–14

Shan Q, Wang Y, Li J, Zhang Y, Chen K, Liang Z, Zhang K, Liu J, Xi JJ, Qiu J-L, Gao C (2013) Targeted genome modification of crop plants using a CRISPR-Cas system. Nat Biotechnol 31: 686–688

Shanmugam T, Sendhil R, Thirumalvalavan V (2006) Quantification and prioritization of constraints causing yield loss in rice (Oryza sativa) in India. Agric Trop Subtrop 39:194–201

Sharifunnessa M, Islam MT (2017) Effect of drought stress at different growth stages on yield and yield components of six rice (Oryza sativa L.) genotypes. Fundam Appl Agric 2:285–289

Shen C, Que Z, Xia Y, Tang N, Li D, He R, Cao M (2017a) Knock out of the annexin gene OsAnn3 via CRISPR/Cas9-mediated genome editing decreased cold tolerance in rice. J Plant Biol 60: 539–547

Shen H, Strunks GD, Klemann BJPM, Hooykaas PJJ, de Pater S (2017b) CRISPR/Cas9-induced double-strand break repair in arabidopsis nonhomologous end-joining mutants. G3 (Bethesda) 7:193–202

Shi J, Gao H, Wang H, Lafitte HR, Archibald RL, Yang M, Hakimi SM, Mo H, Habben JE (2017) ARGOS8 variants generated by CRISPR-Cas9 improve maize grain yield under field drought stress conditions. Plant Biotechnol J 15:207–216

Shufen C, Yicong C, Baobing F, Guiai J, Zhonghua S, Ju L, Shaoqing T, Jianlong W, Peisong H, Xiangjin W (2019) Editing of rice Isoamylase gene ISA1 provides insights into its function in starch formation. Rice Sci 26:77–87

Singh RK, Redoña E, Refuerzo L (2010) Varietal improvement for abiotic stress tolerance in crop plants: special reference to salinity in rice. In: Pareek A, Sopory SK, Bohnert HJ (eds) Abiotic stress adaptation in plants: physiological, molecular and genomic foundation. Springer, Dordrecht, pp 387–415

Singh R, Sunder S, Kumar P (2016) Sheath blight of rice: current status and perspectives. Indian Phytopathol 69:340–351

Singh S, Prasad S, Yadav V, Kumar A, Jaiswal B, Kumar A, Khan N, Dwivedi D (2018) Effect of drought stress on yield and yield components of rice (Oryza sativa L.) genotypes. Int J Curr Microbiol Appl Sci 7:2752–2759

Singh P, Mazumdar P, Harikrishna JA, Babu S (2019) Sheath blight of rice: a review and identification of priorities for future research. Planta 250:1387–1407

Singh P, Verma RL, Singh RS, Singh RP, Singh HB, Arsode P, Kumar M, Singh PK (2020) Biotic stress management in rice (Oryza sativa L.) through conventional and molecular approaches. In: Rakshit A, Singh HB, Singh AK, Singh US, Fraceto L (eds) New frontiers in stress management for durable agriculture. Singapore, Springer, pp 609–644

Spoel SH, Dong X (2012) How do plants achieve immunity? Defence without specialized immune cells. Nat Rev Immunol 12:89–100

Sreewongchai T, Toojinda T, Thanintorn N, Kosawang C, Vanavichit A, Tharreau D, Sirithunya P (2010) Development of elite indica rice lines with wide spectrum of resistance to Thai blast isolates by pyramiding multiple resistance QTLs. Plant Breed 129:176–180

Stallworth S, Schumaker B, Fuller M, Tseng TM (2020) Consequences and mitigation strategies of biotic and abiotic stress in rice (Oryza sativa L.). In: Hossain A (ed) Plant stress physiology. IntechOpen

Sun Y, Zhang X, Wu C, He Y, Ma Y, Hou H, Guo X, Du W, Zhao Y, Xia L (2016) Engineering herbicide-resistant rice plants through CRISPR/Cas9-mediated homologous recombination of acetolactate synthase. Mol Plant 9:628–631

Sun Y, Jiao G, Liu Z, Zhang X, Li J, Guo X, Du W, Du J, Francis F, Zhao Y, Xia L (2017) Generation of high-amylose rice through CRISPR/Cas9-mediated targeted mutagenesis of starch branching enzymes. Front Plant Sci 8:298

Sunder S, Singh R, Agarwal R (2014) Brown spot of rice: an overview. Indian Phytopathol 67:201–215

Suzuki N, Rivero RM, Shulaev V, Blumwald E, Mittler R (2014) Abiotic and biotic stress combinations. New Phytol 203:32–43

Tang L, Mao B, Li Y, Lv Q, Zhang L, Chen C, He H, Wang W, Zeng X, Shao Y, Pan Y, Hu Y, Peng Y, Fu X, Li H, Xia S, Zhao B (2017) Knockout of OsNramp5 using the CRISPR/Cas9 system produces low cd-accumulating indica rice without compromising yield. Sci Rep 7: 14438–14438

Thakur P, Kumar S, Malik JA, Berger JD, Nayyar H (2010) Cold stress effects on reproductive development in grain crops: An overview. Environ Exper Bot 67:429–443

Thuy TL, Saitoh K (2017) Responses of fourteen Vietnamese Rice (Oryza sativa L.) cultivars to high temperatures during grain filling period under field conditions. Agronomy 7:57

Toda E, Koiso N, Takebayashi A, Ichikawa M, Kiba T, Osakabe K, Osakabe Y, Sakakibara H, Kato N, Okamoto T (2019) An efficient DNA- and selectable-marker-free genome-editing system using zygotes in rice. Nat Plants 5:363–368

Tzfira T, Citovsky V (2006) Agrobacterium-mediated genetic transformation of plants: biology and biotechnology. Curr Opin Biotechnol 17:147–154

Vaghefi N, Shamsudin MN, Makmom A, Bagheri M (2011) The economic impacts of climate change on the rice production in Malaysia. Int J Agric Res 6:67–74

Valarmathi P, Kumar G, Robin S, Manonmani S, Dasgupta I, Rabindran R (2016) Evaluation of virus resistance and agronomic performance of rice cultivar ASD 16 after transfer of transgene against Rice tungro bacilliform virus by backcross breeding. Virus Genes 52:521–529

Van Nguyen N, Ferrero A (2006) Meeting the challenges of global rice production. Paddy Water Environ 4:1–9

van Oort PAJ, Zwart SJ (2018) Impacts of climate change on rice production in Africa and causes of simulated yield changes. Glob Change Biol 24:1029–1045

van Overbeek M, Capurso D, Carter Matthew M, Thompson Matthew S, Frias E, Russ C, Reece-Hoyes John S, Nye C, Gradia S, Vidal B, Zheng J, Hoffman Gregory R, Fuller Christopher K,

May Andrew P (2016) DNA repair profiling reveals nonrandom outcomes at Cas9-mediated breaks. Mol Cell 63:633–646
Varshney RK, Godwin ID, Mohapatra T, Jones JDG, McCouch SR (2019) A SWEET solution to rice blight. Nat Biotechnol 37:1280–1282
Verdier V, Vera Cruz C, Leach JE (2012) Controlling rice bacterial blight in Africa: needs and prospects. J Biotechnol 159:320–328
Voinnet O (2005) Induction and suppression of RNA silencing: insights from viral infections. Nat Rev Genet 6:206–220
Wada N, Ueta R, Osakabe Y, Osakabe K (2020) Precision genome editing in plants: state-of-the-art in CRISPR/Cas9-based genome engineering. BMC Plant Biol 20:234
Wang X, Valent B (2009) Advances in genetics, genomics and control of rice blast disease. Springer. ISBN 978-1-4020-9499-6, e-ISBN 978-1-4020-9500-9, © Springer Science+Business Media B.V. 2009, Van Godewijckstraat, Dordrecht, Netherlands
Wang F, Wang C, Liu P, Lei C, Hao W, Gao Y, Liu Y-G, Zhao K (2016) Enhanced rice blast resistance by CRISPR/Cas9-targeted mutagenesis of the ERF transcription factor gene OsERF922. PLoS One 11:e0154027
Wang H-X, Li M, Lee CM, Chakraborty S, Kim H-W, Bao G, Leong KW (2017a) CRISPR/Cas9-based genome editing for disease modeling and therapy: challenges and opportunities for nonviral delivery. Chem Rev 117:9874–9906
Wang M, Lu Y, Botella JR, Mao Y, Hua K, Zhu JK (2017b) Gene targeting by homology-directed repair in rice using a geminivirus-based CRISPR/Cas9 system. Mol Plant 10:1007–1010
Wang M, Mao Y, Lu Y, Tao X, Zhu JK (2017c) Multiplex gene editing in rice using the CRISPR-Cpf1 system. Mol Plant 10:1011–1013
Wang M, Mao Y, Lu Y, Wang Z, Tao X, Zhu JK (2018) Multiplex gene editing in rice with simplified CRISPR-Cpf1 and CRISPR-Cas9 systems. J Integr Plant Biol 60:626–631
Wassmann R, Jagadish SVK, Sumfleth K, Pathak H, Howell G, Ismail A, Serraj R, Redona E, Singh RK, Heuer S (2009) Chapter 3. Regional vulnerability of climate change impacts on Asian rice production and scope for adaptation. Advances in agronomy. Academic Press, pp 91–133. ISBN: 978-0-12-374818-8, Copyright © 2009 Elsevier Inc., Academic Press, San Diego, CA, USA
Wheeler TR, Craufurd PQ, Ellis RH, Porter JR, Vara Prasad PV (2000) Temperature variability and the yield of annual crops. Agric Ecosyst Environ 82:159–167
Wimmer F, Beisel CL (2020) CRISPR-Cas systems and the paradox of self-targeting spacers. Front Microbiol 10:3078
Woin N, Sadou I, Bourou S, Bebom T (2010) Potential for biological control of Rice yellow mottle virus vectors. Afr Crop Sci J 15:15
Woo JW, Kim J, Kwon SI, Corvalán C, Cho SW, Kim H, Kim SG, Kim ST, Choe S, Kim JS (2015) DNA-free genome editing in plants with preassembled CRISPR-Cas9 ribonucleoproteins. Nat Biotechnol 33:1162–1164
Wu T, Peng C, Li B, Wu W, Kong L, Li F, Chu Z, Liu F, Ding X (2019) OsPGIP1-mediated resistance to bacterial leaf streak in rice is beyond responsive to the polygalacturonase of xanthomonas oryzae pv. oryzicola. Rice 12:90
Xiao W, Yang Q, Huang M, Guo T, Liu Y, Wang J, Yang G, Zhou J, Yang J, Zhu X, Chen Z, Wang H (2019) Improvement of rice blast resistance by developing monogenic lines, two-gene pyramids and three-gene pyramid through MAS. Rice 12:78
Xu R, Li H, Qin R, Wang L, Li L, Wei P, Yang J (2014) Gene targeting using the agrobacterium tumefaciens-mediated CRISPR-Cas system in rice. Rice 7:5
Xu R, Yang Y, Qin R, Li H, Qiu C, Li L, Wei P, Yang J (2016) Rapid improvement of grain weight via highly efficient CRISPR/Cas9-mediated multiplex genome editing in rice. J Genet Genomics 43:529–532
Xu R, Qin R, Li H, Li D, Li L, Wei P, Yang J (2017) Generation of targeted mutant rice using a CRISPR-Cpf1 system. Plant Biotechnol J 15:713–717

Xu Z, Xu X, Gong Q, Li Z, Li Y, Wang S, Yang Y, Ma W, Liu L, Zhu B, Zou L, Chen G (2019) Engineering broad-spectrum bacterial blight resistance by simultaneously disrupting variable TALE-binding elements of multiple susceptibility genes in rice. Mol Plant 12:1434–1446

Yin K, Han T, Liu G, Chen T, Wang Y, Yu AYL, Liu Y (2015) A geminivirus-based guide RNA delivery system for CRISPR/Cas9 mediated plant genome editing. Sci Rep 5:14926

Yin X, Biswal AK, Dionora J, Perdigon KM, Balahadia CP, Mazumdar S, Chater C, Lin HC, Coe RA, Kretzschmar T, Gray JE, Quick PW, Bandyopadhyay A (2017) CRISPR-Cas9 and CRISPR-Cpf1 mediated targeting of a stomatal developmental gene EPFL9 in rice. Plant Cell Rep 36:745–757

Yin X, Anand A, Quick P, Bandyopadhyay A (2019) Editing a stomatal developmental gene in Rice with CRISPR/Cpf1. Methods Mol Biol 1917:257–268

Yue J-J, Hong C-Y, Wei P, Tsai Y-C, Lin C-S (2020) How to start your monocot CRISPR/Cas project: plasmid design, efficiency detection, and offspring analysis. Rice 13:9

Zafar SA, Zaidi SS, Gaba Y, Singla-Pareek SL, Dhankher OP, Li X, Mansoor S, Pareek A (2019) Engineering abiotic stress tolerance via CRISPR/Cas-mediated genome editing. J Exp Bot 71: 470–479

Zafar K, Khan MZ, Amin I, Mukhtar Z, Yasmin S, Arif M, Ejaz K, Mansoor S (2020) Precise CRISPR-Cas9 mediated genome editing in super basmati rice for resistance against bacterial blight by targeting the major susceptibility. Gene 11:575

Zaidi SS, Mansoor S (2017) Viral vectors for plant genome engineering. Front Plant Sci 8:539

Zaman M, Shahid SA, Heng L (2018) Guideline for salinity assessment, mitigation and adaptation using nuclear and related techniques. Springer Nature

Zeng X, Luo Y, Vu NTQ, Shen S, Xia K, Zhang M (2020a) CRISPR/Cas9-mediated mutation of OsSWEET14 in rice cv. Zhonghua11 confers resistance to Xanthomonas oryzae pv. Oryzae without yield penalty. BMC Plant Biol 20:313

Zeng Y, Wen J, Zhao W, Wang Q, Huang W (2020b) Rational improvement of Rice yield and cold tolerance by editing the three genes OsPIN5b, GS3, and OsMYB30 with the CRISPR–Cas9 system. Front Plant Sci 10:1663

Zetsche B, Gootenberg JS, Abudayyeh OO, Slaymaker IM, Makarova KS, Essletzbichler P, Volz SE, Joung J, van der Oost J, Regev A, Koonin EV, Zhang F (2015) Cpf1 is a single RNA-guided endonuclease of a class 2 CRISPR-Cas system. Cell 163:759–771

Zhang Q, Chen Q, Wang S, Hong Y, Wang Z (2014) Rice and cold stress: methods for its evaluation and summary of cold tolerance-related quantitative trait loci. Rice 7:24

Zhang S, Tao F, Zhang Z (2016) Changes in extreme temperatures and their impacts on rice yields in southern China from 1981 to 2009. Field Crops Res 189:43–50

Zhang J, Zhang H, Botella JR, Zhu JK (2018) Generation of new glutinous rice by CRISPR/Cas9-targeted mutagenesis of the waxy gene in elite rice varieties. J Integr Plant Biol 60:369–375

Zhang A, Liu Y, Wang F, Li T, Chen Z, Kong D, Bi J, Zhang F, Luo X, Wang J, Tang J, Yu X, Liu G, Luo L (2019a) Enhanced rice salinity tolerance via CRISPR/Cas9-targeted mutagenesis of the OsRR22 gene. Mol Breed 39:47

Zhang T, Zhao Y, Ye J, Cao X, Xu C, Chen B, An H, Jiao Y, Zhang F, Yang X, Zhou G (2019b) Establishing CRISPR/Cas13a immune system conferring RNA virus resistance in both dicot and monocot plants. Plant Biotechnol J 17:1185–1187

Zhao H, Wang X, Jia Y, Minkenberg B, Wheatley M, Fan J, Jia MH, Famoso A, Edwards JD, Wamishe Y, Valent B, Wang G-L, Yang Y (2018) The rice blast resistance gene Ptr encodes an atypical protein required for broad-spectrum disease resistance. Nat Commun 9:2039

Zhou G, Xu D, Xu D, Zhang M (2013) Southern rice black-streaked dwarf virus: a white-backed planthopper-transmitted fijivirus threatening rice production in Asia. Front Microbiol 4:270–270

Zhou X, Liao H, Chern M, Yin J, Chen Y, Wang J, Zhu X, Chen Z, Yuan C, Zhao W, Wang J, Li W, He M, Ma B, Wang J, Qin P, Chen W, Wang Y, Liu J, Qian Y, Wang W, Wu X, Li P, Zhu L, Li S, Ronald PC, Chen X (2018) Loss of function of a rice TPR-domain RNA-binding protein confers broad-spectrum disease resistance. Proc Natl Acad Sci U S A 115:3174–3179

Zhu H, Li C, Gao C (2020) Applications of CRISPR–Cas in agriculture and plant biotechnology. Nat Rev Mol Cell Biol 21:661–677

Zipfel C (2014) Plant pattern-recognition receptors. Trends Immunol 35:345–351

Harnessing CRISPR/Cas Tools for Installing Virus Resistance in Cereals: An Overview

13

Ashwini Talakayala, Srinivas Ankanagari, and Mallikarjuna Garladinne

Abstract

The plant viruses constantly threaten crop productivity and food security globally. The genetic improvement in plants is always a sustainable way to meet the food demand of growing population. The viruses are obligatory intracellular pathogens, which can't complete their life cycle without cellular machinery of host. Introduction of mutations in the host disease susceptibility genes or virus genome can restrict the compatibility between the plant host and viruses. In current chapter, we focus on updates of clustered regularly interspaced short palindromic repeats (CRISPR)/CRISPR-associated protein 9 (Cas9) tool for the improvement of molecular immunity in cereals against multifarious eukaryotic viruses, including DNA and RNA viruses. In addition, recent progress made on CRISPR/Cas9 and Cas13a approach for producing viral resistance in model plants and crops is discussed. The prospects and limitations of this tool are also presented.

Keywords

Cereals · CRISPR/Cas9 · Genome editing · Resistance · Viruses

A. Talakayala
Plant Molecular Biology Laboratory, Agri Biotech Foundation, Hyderabad, India

Department of Genetics, Osmania University, Hyderabad, India

S. Ankanagari
Department of Genetics, Osmania University, Hyderabad, India

M. Garladinne (✉)
Plant Molecular Biology Laboratory, Agri Biotech Foundation, Hyderabad, India

M. G. Mallikarjuna et al. (eds.), *Next-Generation Plant Breeding Approaches for Stress Resilience in Cereal Crops*, https://doi.org/10.1007/978-981-19-1445-4_13

13.1 Introduction

Plant viral diseases are serious constraints on agriculture productivity and food security globally. About 1500 plant viral species cause almost half of the plant diseases, and estimated crop loss is more than US$30 billion annually (Nicaise 2014; Cao et al. 2020). Viruses are obligate parasites completely dependent on host for progressing its life cycle. Controlling viral diseases with chemicals and antibiotics is not feasible. Moreover, plant viruses evolve continuously by mutating its genome; thus, it is a challenging task to control them. The impact of viruses and their vectors in cereals including rice, wheat, maize, barley, and oats poses a serious challenge. Indeed, developing plant immunity or virus-resistant plants through molecular breeding plays significant role to combat these obligate parasites that ultimately lead to enhanced crop production. For a long time, the researchers have employed several strategies like RNAi, antisense RNA, and siRNA successfully to control viral diseases in crop plants. However, incomplete resistance and extensive off-target gene silencing hindered the practical applications of these approaches. Thus, to tackle such challenges, durable and efficient resistance strategies are needed.

13.2 Major Viral Diseases in Cereals and Viral Genome Features

Viruses are the smallest obligate parasites replicating in the host cells, and transmission occurs via insects, vectors, aphids, fungi, soil, etc. Cereals are generally affected by numerous viruses including *wheat streak mosaic virus* (WSMV), *soil-borne wheat mosaic virus* (SBWMV), *cereal yellow dwarf virus* (CYDV), *rice tungro virus* (RTV), and *barley yellow dwarf virus* (BYDV). Generally, the viruses are classified into DNA and RNA viruses. DNA viruses replicate within the nucleus, while RNA viruses, particularly +/−−sense ssRNA, replicate in the cytoplasm (Lucas 2006). DNA viruses lack enzymes for its replication; hence, it utilizes the DNA polymerase from host metabolic machinery. RNA polymerase is being used in case of RNA viral replication, and mRNA transcribes to synthesize different proteins in the cytoplasm. The *turnip mosaic virus (TuMV)* viral replication is associated with the globular structures, where viruses move from one cell to neighboring cells through plasmodesmata by symplasmic channels, which occurs via actin microfilaments (Grangeon et al. 2013). Plant viruses and viral particles are transmitted to the new cells by insect vectors, aphids, and whiteflies. The transmission of viruses is taken place by mechanical or non-circulative, and they are absorbed by the insect vector at the time of feeding and attachment to the cuticle or salivary canals of the mouthparts (Ng and Falk 2006). During circulative mode of transmission, the virus is attached with plant cellular contents together and sucked up by the insect vector. Then, the virus can be spread into new host plants via insect saliva (Hogenhout et al. 2008). The virus completes its life cycle during propagative transmission within the vector's body; however, it will replicate within the salivary glands and gut (Hogenhout et al. 2008).

The plants have evolved dynamically with resistance genes involved in acquiring immunity to viruses. The plant defense generally started after entry of specific virus into the host. Plants have developed two varieties of defense mechanisms such as constitutive and inducible way of defenses. Constitutive defenses include epidermal cuticle, cell walls, and bark, which protect plants from invasion by their strength and rigidity. Inducible defense factors include the production of highly toxic chemicals and pathogen-degrading proteins (Freeman and Beattie 2008). Plant defenses against viral pathogens include hypersensitive responses, antimicrobial synthesis through secondary metabolites, and enzyme production where it can fight against viruses. Generally, the pathogen interaction with host plants can aim to induce protective mechanism in the damaged tissues by synthesis of pathogen-related (PR) proteins and enzymes. Lignin and phenolic compounds are major secondary metabolites providing plant disease resistance (Kumar 2020). Most of the plant viruses are transmitted through two ways: firstly, by seed-borne, where plant viruses are transmitted through seeds. In fact, the plant viruses are microscopic, submicroscopic, and infectious particles with protein coat and nucleic acid composition (Sastry 2013), and, secondly, by soilborne where viruses always infect their plant hosts through underground roots (Roberts 2014). Different seed-borne and soilborne viral diseases and the mode of transmission in cereals were listed in Table 13.1.

13.3 Evolution of Viral Diseases in Cereals

Plant viruses are obligate parasites, which enter into the plants via two component systems such as protein coat and nucleic acid form (DNA or RNA). Once the virus makes an entry, it exploits the host machinery for multiplication. Viruses spread very easily through infected seeds, wind, grafting, dripping sap, and pollination. Further, the viral population quickly evolves, the emanation of new variants of viruses with virulence; thus, viruses are switching to other host in cultivated crops (Rakotondrafara et al. 2017). Plant virus transmission is confined by insects, which act as a vector group (i.e., aphids, whiteflies, hoppers, and thrips), nematodes, and mites.

When the capsid of virus opens, the viral genome is released into the plant cells. Further, it will be translated into protein that results in the formation of new viruses from membranes of the endoplasmic reticulum and other organelles. Viral RNAs are replicated and exported into cytoplasm, and viral particles move to other cells through plasmodesmata, which can be expanded by virus-encoded movement proteins (Asher 2018).

Cereals crops affected by viruses are mainly wheat (*Triticum aestivum, T. compactum, and T. durum*), barley (*Hordeum vulgare* and *H. distichon*), oats (*Avena sativa* and *A. byzantina*), rye (*Secale cereale*), maize (*Zea mays*), rice (*Oryza sativa*), sorghum (*Sorghum vulgare*), and millets (Slykhuis 1976). The viral-infected plants showed the symptoms like discoloration of leaf, deformations, and stunting and ultimately loss the entire crop.

Table 13.1 Different viral diseases caused in cereal crops

S. no.	Virus	Genus	Genome	Transmission	Infected crops	Causes	Described by
	Seed-borne plant viruses						
1	*Barley stripe mosaic viruses* (BSMV)	*Hordevirus*	Tripartite positive-sense RNA	Mechanically	Barley, wheat, oat, rye,maize, millets, sorghum	Mosaic, striping	McKinney (1951)
2	*Barley mosaic viruses* (BMV)	*Hordevirus*	Tripartite positive-sense RNA	Aphid	Barley, wheat, oat	Mosaic, chlorosis stunting	Dhanraj and Raychaudhuri (1969)
	Soilborne plant viruses						
1	*Wheat (soilborne) mosaic virus* (WSBMV)	*Furovirus*	Positive-sense RNA	Transmission by fungus, *Polymyxa graminis*	Barley, wheat, and rye	Light green to yellow mosaic or rosetting with stunting	McKinney (1925)
2	*Barley yellow mosaic virus* (BYMV)	*Bymovirus*	Positive-sense RNA	Transmission by fungus, *Polymyxa graminis*	Barley	Chlorotic mottle	Miyamoto (1958)
3	*Oat mosaic virus* (OMV)	*Bymovirus*	Positive-sense RNA	Transmission by fungus, *Polymyxa graminis*	Oats	Chlorotic mottling, streaking, or spotting of oats	McKinney (1946)
4	*Wheat spindle streak mosaic virus* (WSSMV)	*Bymovirus*	Positive-sense RNA	Transmission by wheat roots via fungus, *Polymyxa graminis*	Wheat	Light green to yellow spindle-shaped streaks and necrosis of leaves, reduced heading, and slight stunt	Slykhuis (1970)
5	*Rice necrosis mosaic virus* (RN MV)	*Bymovirus*	Positive-sense RNA	Transmission by wheat roots via fungus, *Polymyxa*, mechanical inoculation	Rice	Necrotic lesions	Fujii et al. (1967)
6	*Brome mosaic virus* (BMV)	*Bromovirus*	Positive-sense RNA	Transmitted by soil-inhabiting nematodes	Barley, wheat, oats, rye, maize	Mosaic necrosis	McKinney et al. (1942)

13.4 Novel CRISPR-Cas and its Variants in Plant Genome Editing

Clustered regularly interspaced short palindromic repeats (CRISPR)/CRISPR-associated protein 9 (Cas9) is an adaptive immune system present in bacteria, which acts as defensive system against intruding pathogens. This system possesses two components including guide RNA (gRNA) and Cas9 endonuclease protein. Cas9 is an endonuclease that can edit the targeted 20 bp dsDNAs in the genome. It consists of trans-activating small RNA (tracrRNA) and a small mature CRISPR RNA (crRNA) (Bhaya et al. 2011). *Streptococcus pyogenes* Cas9 (SpCas9) is guided by gRNA, which is complimentary to the first 17–20 nucleotides of target host DNA attached with NGG protospacer adjacent motif (PAM). The Cas9 contains HNH and RuvC-like nuclease domains, which can cleave both strands of DNA at definite −3 positions upstream of PAM, resulting in the formation of blunt-ended DNA double-strand breaks (Jinek et al. 2012; Hsu et al. 2014).

The CRISPR/Cas13a is another class-II-type VI-A ribonuclease capable of targeting and cleaving the ssRNA (Gosavi et al. 2020). Cas13 proteins target RNA rather than DNA molecules, and it binds to two distinct HEPN RNase domains and combines with single crRNA to form a crRNA-guided protein complex for targeting RNA. Here, 5′ and 3′ protospacer flanking site (PFS) is required for redirecting the Cas13/crRNA complex. Cas13a provides immunity to a bacteriophage in *E. coli* by interfering with MS2 lytic ssRNA phage. It is guided by a crRNA that comprises a 28 nt spacer sequence and cleaves target ssRNAs with PFS of A, U, or C (Aman et al. 2018).

Several researchers have made tremendous contribution by exploiting transgenic tools including expression of viral and nonviral proteins, host-resistant (R) genes, and gene silencing through RNA interference (Ahmad et al. 2020; Zhao et al. 2020). Recently, CRISPR tool emerged as an alternative and safe breeding tool. CRISPR-Cas9 is a type II adaptive immune system, initially identified in *Streptococcus pyogenes*, that serves immunity against invading phages or invaders (Jinek et al. 2012), CRISPR tool completely and permanently silences the gene at the DNA and RNA level by Cas9 and Cas13a endonucleases. CRISPR tool has emerged as an important tool that targets both DNA and RNA plant viruses by generating site-specific double-strand breaks based on the gRNA-defined sequences (Jinek et al. 2012). Recently, genome editing has emerged as an alternative weapon against DNA- and RNA-type viruses. In CRISPR system, the Cas9 confers viral interference against DNA viruses in plants. Cas13 system interferes RNA viruses as well as transcriptome in mammalian and plant cells (Mahas et al. 2019).

CRISPR variants are important editing system such as *Streptococcus pyogenes* Cas9 (SpCas9) and *Staphylococcus aureus* Cas9 (SaCas9) from the type II system and *Lachnospiraceae bacterium* ND2006 (LbCas12a) and also LbCpf1 from type V system (Jinek et al. 2012). Cas13a is an efficient RNA targeting and editing tool to engineer resistance against plant RNA viruses. CRISPR/Cas9 system utilizes 20 bp DNA target (guide RNA), followed by a short, trinucleotide (5’-NGG-3′) protospacer adjacent motif (PAM) in the host DNA (Xie et al. 2014). The sgRNA directs the activity of Cas9 nuclease, thereby creating DSBs and mutation at target sites. DNA binding and cleavages that are associated with PAM sequence in the host are recognized. CRISPR/Cas13a is a type II VI-A ribonuclease capable of targeting and cleaving ssRNA molecules of the phage genome. Cas13a protein has a nucleotide-binding domain (HEPN), which is associated with RNase activity by cleaving ssRNAs with protospacer flanking sequence (PFS) of A, U, or C (Aman et al. 2018).

CRISPR system has emerged as an efficient gene silencing and gene editing tool in hosts, where it uses only a 20 nt sgRNA as target nucleotides of DNA and RNA plant viruses in order to develop resistance against viral pathogens. By using CRISPR tool, desired modifications are made by Homologous directed repair (HDR)- and Non Homologous end-joining repair (NHEJ)-based repair system. Gene silencing is also possible by suppressing or silencing the targeted region to repress its transcription. Knock in/down of the genes by HDR is made to develop the virus-resistant plants (Barrangou and Doudna 2016). DNA virus resistance is achieved by designing sgRNA and mainly directing Cas9 to bind to the dsDNA inside the nucleus in eukaryotic cells, where viruses replicate; therefore, it targets the different conserved regions of the DNA viruses to achieve resistance (Das et al. 2019), whereas RNA viral resistance in plants is achieved by targeting ssRNA in the cytoplasm, which targets RNA virus genome mainly the initiation factors. The Cas13a system having an RNases activity to cleave viral RNA established that the plants acquire potent defense against viral infection in both dicot and monocot plants by showing high efficiency in generating RNA virus-resistant plants (Zhang et al. 2019).

13.5 Applications of Viral Genome Editing for Plant Resistance in Cereals

13.5.1 Editing DNA Viruses

The viral pathogens and their diseases are prevented either by editing viral genome or by susceptible host plant genes. The editing of viral genome may anticipate the suppression of viral replication and reduction in viral load in infected plants. On the other side, editing host-susceptible genes/factors would lead to enhancement of plant immunity and inhibition of viral growth in the host plants.

Geminiviridae is one of the largest families that threatens the productivity in almost all crops. These viruses comprises single-strand DNA-A and B genomes, which can replicate to double strand through rolling circle mode in the nucleus of host plants (Amin et al. 2021). Several researchers have employed zinc finger nucleases (ZFNs) and transcription activator-like effector nucleases (TALENs) for viral genome editing or plant host genome (Chen et al. 2014; Cheng et al. 2015). However, manipulation of DNA by employing ZFNs and TALENs is expensive and laborious. Further, the CRISPR/Cas tools have emerged as promising and alternative approach for the manipulation and engineering of virus-resistant crops (Fig. 13.1).

In the year 2015, the first experiment was conducted to engineer CRISPR-mediated resistance against *beet severe curly top virus* (BSCTV) and *yellow dwarf*

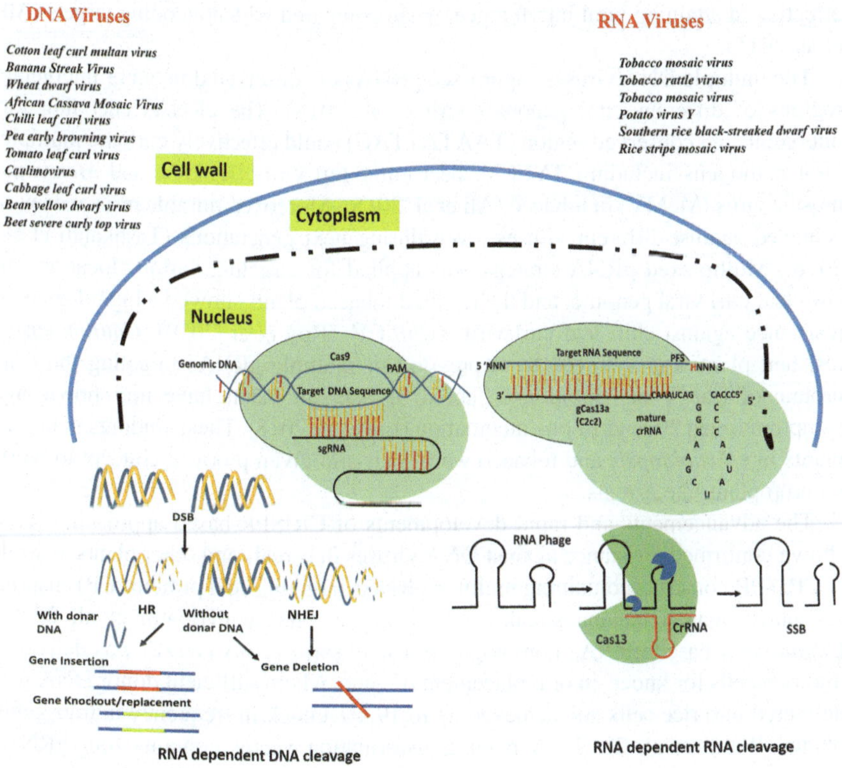

Fig. 13.1 CRISPR tool for editing of viral genome (i.e., DNA-Cas9 and RNA-Cas13a) viruses for enhancing resistance against viral pathogens in plants

virus (BeYDV) in *Nicotiana benthamiana* and *Arabidopsis thaliana* (Liang et al. 2016; Ji et al. 2015). The transgenic *Arabidopsis* and tobacco expressing gRNA-Cas9 construct resulted in the attenuation of viral disease. The overexpression of gRNAs coding for replication and cell mobility coding regions of yellow dwarf virus (YDV) in tobacco showed enhanced resistance against multiple Geminiviruses (Baltes et al. 2015). Cabbage leaf curl virus (*CaLCuV*) of Geminiviridae family has an ability to deliver gRNA molecules and induce systemic gene mutations in *N. benthamiana* by virus-induced gene silencing (VIGS) where it downregulate its gene expression and viral efficiency (Yin et al. 2015). The transgenic tobacco expressing dual gRNA construct targeting C1 (replication-associated protein) and IR regions of *Cotton leaf curl Multan virus* (CLCuMuV) conferred completed resistance to viral infections (Yin et al. 2019). In another study, CRISPR construct harboring gRNAs homologues to noncoding and coding regions of tomato yellow leaf curl virus (TYLCV) was transformed into tobacco through *Agrobacterium*. The resultant plants showed enhanced resistance against viral variants and inhibited viral transmission (Ali et al. 2015). In contrast, gRNAs targeting coding regions were not effective in attaining viral interference, when compared with noncoding targets (Ali et al. 2015).

The multiple DNA viruses suppressed gRNAs of conserved domain or intergenic regions of different viral genomes (Ali et al. 2018). The gRNAs encoding for intergenic and conserved region (TAATATTAC) could effectively curb the multiple viral pathogens including TYLCV, beet curly top virus (BCTV), and *Merremia* mosaic virus (MeMV) in tobacco (Ali et al. 2015). Moreover, durable resistance was achieved against different viruses, even during next generations (Tashkandi et al. 2018). Multiplexed gRNA strategy was applied for targeting various locations of chili leaf curl viral genome, and the resulted tobacco plants showed a high degree of resistance against chili leaf curl virus (*ChiLCV*) (Roy et al. 2019). *Caulimovirus*-resistant plant was achieved by expression of multiple gRNAs targeting the coat protein of CaMV in *Arabidopsis*; about 85–90% of plants have not shown any symptoms even 20 days of postinoculation (Liu et al. 2018). These findings in model plants like *Arabidopsis* and tobacco were basis and given proof of concept to work on crop plant like cereals.

The advancements and rapid developments of CRISPR-based approaches have shown confirmed resistance against DNA viruses in cereals and other plants as well. A CRISPR construct containing multiple gRNAs targeting coat protein (CP) ensured resistance against banana streak virus (BSV) in banana (Tripathi et al. 2019). *Geminivirus*-based gRNAs homologue to *wheat dwarf virus* (*WDV*) was delivered into rice cells for knock-in or replacement of gene, where sufficient donor DNA was delivered into rice cells and achieved up to 19.4% knock-in frequency in transgenic plants (Wang et al. 2017). A plant transformation vector contains four gRNAs encoding for MP, CP in coding region, LIR, WDV, Rep/Rep-A protein of N-terminus and C-terminus in WDV under the regulation of three different monocotyledon-specific small nuclear RNA promoters, and is transferred into barley through *Agrobacterium*-mediated transformation. The resultant plants showed resistance to WDV and did not show any symptoms (Kis et al. 2019). The list of DNA viral genomes editing through CRISPR/Cas approach in model plants and cereals is mentioned in Table 13.2.

13.5.2 Editing RNA Viruses

The single stranded RNA (ssRNA) viruses were also successfully targeted using CRISPR/Cas tools through programmable RNA-guided ssRNA ribonucleases including *FnCas9* (a nuclease from *Francisella novicida*) and CRISPR/Cas13a (a nuclease from *Leptotrichia shahii*) (Price et al. 2015; Aman et al. 2018). Cas13a is a class 2, type VI protein that possesses two RNase domains, i.e., higher eukaryotes and prokaryotes nucleotide-binding domain (HEPN), that can cleave SsRNA molecules (Abudayyeh et al. 2016). Indeed, majority of the plant viruses contains RNA genomes; hence, Cas13 may give new hope to control pathogenic plant viruses. Zhang et al. (2018a, b) developed CRISPR-mediated resistance to RNA viruses, i.e., *cucumber mosaic virus* (CMV) and *tobacco mosaic virus* (TMV)

Table 13.2 Viral genome editing through CRISPR/Cas-DNA viruses in model plants and cereals

S. no.	Target virus/gene	Viral target	Promoter used	Plant species	Effector protein	Strategy of transformation	Trait improvement	Reference
1	*CLCuMuV* C1 (rep) and IR	ssDNA	35S, U6	*Nicotiana benthamiana*	SpCas9	*AgrobacteriumGV2260*	Resistance to *cotton leaf curl Multan virus*	Yin et al. (2019)
2	*ChiLCV*	ssDNA	U3,U6 and CaMV35S	*Nicotiana benthamiana*	SpCas9	*Agrobacterium transient assay GV3103*	Resistant to *ChiLCV-reduced viral accumulation*	Roy et al. (2019)
3	gRNA1 and gRNA2 of tobacco rattle virus (*TRV*) and pea early browning virus (*PEBV*)	ssDNA	pPEBV	*Tobacco and Arabidopsis*	SpCas9	*Agrobacterium GV3101*	Increased the targeted genome editing	Ali et al. (2018)
4	Coat protein (*CP*) or replicase (*rep*)	ssDNA	U6	*Tomato and tobacco*	SpCas9	*Agrobacterium GV3101*	Low accumulation of *TYLCV* in tomato and tobacco transgenic plants	Tashkandi et al. (2018)
5	*IR, CP, rep-CLCuKoV, TYLCV, TYLCSV, MeMV, BCTV*-Logan, *BCTV*-Worland	ssDNA	PEBV promoter	*Nicotiana benthamiana*	SpCas9	*Agrobacterium GV3103*	Disease resistance not confirmed but resistance to Geminiviruses-*CLCuKoV*	Ali et al. (2018)
6	*CP* region of *Caulimovirus*	ssDNA	AtU6	*Arabidopsis*	SpCas9	*Agrobacterium floral dip recA strain*	Resistance to *CaMV*	Liu et al. (2018)
7	gRNAs-pCVA and pCVB of cabbage leaf curl virus (*CaLCuV*)	ssDNA	CaMV35S, AtU6	*Nicotiana benthamiana/ Arabidopsis*	SpCas9	*Agrobacterium* infiltration *GV3101*	Demonstrated geminiviral-mediated VIGE for genome editing	Yin et al. (2015)

(continued)

Table 13.2 (continued)

S. no.	Target virus/gene	Viral target	Promoter used	Plant species	Effector protein	Strategy of transformation	Trait improvement	Reference
8	*TYLCV*-IR, RCA region/IR	ssDNA	PEBV pea early browning virus promoter	*Nicotiana benthamiana*	SpCas9	*Agrobacterium*	Resistance to leaf curl disease	Yin et al. (2015)
9	*BeYDV* (short intergenic region, transacting replication-initiation protein	ssDNA	CaMV35S, AtU6	*Nicotiana benthamiana*	SpCas9	*Agrobacterium*	Resistance against bean yellow dwarf virus and observed mild symptoms due to minimal expression of Cas9	Baltes et al. (2015)
10	SgRNAs targeting *beet severe curly top virus* (*BSCTV*)	ssDNA	AtU6	*Nicotiana and Arabidopsis*	SpCas9	*Agrobacterium*	Reduction in viral load of 97% and severe relief from leaf curly symptoms	Ji et al. (2015)
11	Four *WDV*-specific sgRNAs: *MP, CP, rep/rep, IR/WDV*	ssDNA	ZmUbi1 CaMV35S	*Barley*	SpCas9	*Agrobacterium C58C1*	Resistant to *wheat dwarf virus*	Kis et al. (2019)
12	sgRNA-rep of *wheat dwarf virus*	ssDNA	ZmUbi	*Rice*	SpCas9	*Agrobacterium* and biolistics	To increase gene targeting	Wang et al. (2017)

by expressing a construct in pCAMBIA vector containing gRNAs and FnCas9 in *N. benthamiana* and *A. thaliana*. Engineering of sgRNAs is targeting tobacco rattle virus (TRV) and pea early browning virus (PEBV) delivered into tobacco and *Arabidopsis* (Ali et al. 2018). The CRISPR/Cas13a induced cleavage at ssRNA molecules of the genome of *Potyvirus, turnip mosaic virus* (*TuMV*), in tobacco and showed interference to RNA virus (Aman et al. 2018).

Rice plants expressing LshCas13a serves the editing of genomic RNA of TMV that confers resistance to *southern rice black-streaked dwarf virus* (SRBSDV) and *rice stripe mosaic virus* in tobacco (RSMV) (Zhang et al. 2019). In sum, all these studies signify that targeted editing of the viral genome mediated by CRISPR/Cas is a powerful approach for developing viral disease resistance in plants including cereals (Table 13.3).

13.6 Applications of Plant Host Gene Editing for Viral Resistance in Cereals

New plant-breeding technologies (NPBT) have recently emerged as alternative approaches for improving plant immunity to viruses. Since the viruses are obligatory parasites, they utilize host metabolic pathways to complete their life cycle. In turn, the plant also possesses the susceptibility/recessive loci, which are the regions hijacked by viruses for multiplication and infection to the neighboring cells. Precise editing of such recessive genes will arrest replication and reduce the availability of host machinery for viral growth. The recessive genes can be further edited by CRISPR/Cas9 system for generating transgene-free virus-resistant plants and also backcrossed to eliminate the Cas9 protein in next generations. Recently, researchers employed different CRISPR constructs expressing gRNAs and Cas proteins in plants.

Plant viruses require host machinery to sustain their life cycle. Identification and editing of novel host-susceptible genes by CRISPR serve as an essential tool to engineer plant virus resistance. Several "S" genes have been validated in different crop plants. Eukaryotic translational initiation factors and coilin genes are important host-susceptible factors aided in viral infection and multiplication process. Plant eIF4E, a cap-binding protein, is involved in host susceptibility for viral infection and viral existence in plants. Hence, these "S" genes are known as natural recessive genes in all kinds of plant species. These genes have emerged as an alternative source of resistance (Zaidi et al. 2018). The "S" genes can regulate plant disease resistance mechanisms, suppress immune response, and stimulate pathogen growth in plants (Langner et al. 2018). The studies on plant virus interactions have generated many host genes associated with viral resistance or susceptibility in several plant species (Kang et al. 2005; Gómez et al. 2009).

The eIF4E binds to 5′ 7-methyl guanosine (5'm7GppN) cap structures of mRNA during translational initiation process, with viral protein-associated genome interaction with eIF4E as proof of concept. The recessive host genes (eIF4E and eIF4E (Iso)) have been silenced in several species using CRISPR/Cas for developing viral

Table 13.3 Viral genome editing through CRISPR /Cas-RNA viruses in cereals and model plants

S. no.	Virus and gene modified	Viral target	Promoter used	Target plant	Effector Protein	Transformation Strategy	Target Trait	Reference
1	*TuMV*	ssRNA	CaMV35S	*Nicotiana benthamiana*	LshCas13a	*Agrobacterium GV3101*	Resistance to leaf curl disease	Aman et al. (2018)
2	Tobacco rattle virus (*TRV*) and pea early browning virus (*PEBV*)	ssRNA	pPEBV	*Tobacco and Arabidopsis*	FnCas9	*Agrobacterium GV3101*	Increased the targeted	Ali et al. (2018)
3	Tobacco mosaic virus (*TMV)*	ssRNA	CaMV35S at U6	*Tobacco and Arabidopsis*	*FnCas9*	*Agrobacterium EHA105*	Resistance to tobacco mosaic virus	Zhang et al. (2018a, b)
4	Tobacco mosaic virus (*TMV*), southern rice black-streaked dwarf virus (*SRBDSV*), rice stripe mosaic virus (*RSMV*)	ssRNA	CaMV35S OsU6 at U6	*Tobacco and Oryza sativa*	LshCas13a	–	Resistance to viruses, i.e., TMV, SRBDSV, RSMV	Zhang et al. (2019)

resistance. A viral protein of potyviruses directly binds to eIF4E and completes its life cycle. Mutated eIF4E diminishes the viral ability to interact with host proteins and ceases the translation of the viral genome. Site-specific DSB through CRISPR/Cas has opened up a new dimension in targeting eIF4E for achieving complete resistance against RNA-based turnip mosaic virus (TuMV) in *Arabidopsis* (Pyott et al. 2016). Furthermore, a susceptible eIF4E1 allele N176K was undergoing single base pair (C-G) editing through CRISPR-nCas9-cytidine deaminase, and resultant plants showed resistance against RNA-based *Potyvirus*, i.e., clover yellow vein virus (CYVV) in *Arabidopsis* (Bastet et al. 2019).

It is well known that rice tungro disease (RTD) is a serious problem in rice, which is caused by a combination of two viral interactions such as rice tungro spherical virus (RTSV) and rice tungro bacilliform virus (RTBV). Mutation in native eIF4G through CRISPR/Cas9 conferred resistance to rice tungro streak spherical viral disease in susceptible IR64 rice variety. As a result, the frameshift mutations were transmitted to the further generations without phenotypic aberrations and exhibited higher yield than wild type under greenhouse conditions in rice (Macovei et al. 2018). CRISPR/Cas-mediated genome editing laid a way for understanding the plant virus interactions and development of immunity against single or multiple viruses in plants. However, CRISPR-mediated resistance in plants needs to be evaluated under natural field conditions. The list of the plant host-susceptible genes edited for enhancing plant resistance is listed in Table 13.4.

13.7 Advantages of CRISPR Tool against Protein-Based Editing Tools and RNAi Approaches in Cereals

Plant diseases are generally caused by the incidence of viral pathogens on the crops. The viral diseases can be targeted by the implementation of new technologies like RNA interference and CRISPR system. RNAi is a powerful tool and is a gene-silencing process where it contains siRNA and shRNA, mediating gene knockdown by promoting the degradation of mRNA. However, the process of knockdown is unpredictable. On the other hand, CRISPR consists of tracrRNA made up of a longer stretch of bases that are constant as stem loop structure bound by the CRISPR nuclease; thus, RNA components hybridize to form gRNA, which is programmable and targets CRISPR nucleases to DNA/RNA sequences depending on the complementarity of the crRNA and the presence of other features, i.e., PAM and PFS sequences recognized by the nucleases. This technology implies manipulating and changing ability to delete, insert, and modify DNA. Gene repair can be done by HDR, which is a “knock-in” and “knockout” method. It is more useful because the off-targets can be eliminated by designing the best sgRNA and choosing the low off-target score and high on-target score, to assure a stable heritable mutation (Doench et al. 2014, 2016).

Table 13.4 Plant host-susceptible genes edited for enhancing plant resistance for viruses in model plants

S. no.	Gene target	Promoter	Target plant/crop	Effector protein	Strategy of transformation	Trait improvement	Reference
1	*eIF4E* gene	AtU626	*Arabidopsis thaliana*	Sp-nCas9-cytidine deaminase	*Agrobacterium*	Transgene-free resistant plants against *Potyvirus*, i.e., *clover yellow vein virus*	Bastet et al. (2019)
2	*eIF (iso) 4E*	AtU6	*Arabidopsis thaliana*	SpCas9	*Agrobacterium*	Resistance to turnip mosaic virus (*TuMV*), Potyvirus	Pyott et al. (2016)
3	*eIF4G*	ZmUbi6, TaU6, CaMV35S	*Oryza sativa* L. *Japonica*	SpCas9	*Agrobacterium LBA4404*	Rice tungro spherical virus *(RTSV)*	Macovei et al. (2018)

13.8 Concerns and Regulations

The creation mutations or modifications in endogenous genes would alter the function of a gene. Despite the advantage of viral genome editing via CRISPR, still, there is a major concern that new viral mutant variants or species may evolve as byproducts of this approach. It would anticipate the virus evolution. Thus, the plants may lose resistance to the viruses.

Apart from technological improvements, regulatory developments are also needed for producing and marketing CRISPR crops. The EU regulatory framework considers CRISPR crops as GMO and potentially increases the time and labor cost for producing them to release into market (Custers 2017; Globus and Qimron 2018). The viral genome-edited CRISPR tomato plants were expressing Cas9 stably, and these tomato plants showed minimal viral occurrence and minimal viral load observed continuously for three generations (Pyott et al. 2016; Macovei et al. 2018). However, continuous expression of Cas protein may inculcate the abnormalities in the next generations. Cas9 nuclease expressing crops may treat as genetically modified organisms (GMO).

Developing RNP-mediated transgene-free CRISPR-Cas9 may overcome limitations. The usage and application of CRISPR-Cas9 and its variants are elevated for developing disease-resistant crops. The first-generation transgenic crops were easily apparent from conventional-bred varieties because they invariably carried nonhost DNA sequences that conferred traits not found in nature in the crop species. Transformation technique implies as random insertion of transgenes into the host genome, thereby leaving a specific fingerprint of each event, event-specific, where it's called for testing. However, some of the new varieties being developed through gene editing are practically indistinguishable from those that could be obtained by conventional/mutation breeding. Some countries like the USA and Canada are treating such gene-edited crops as comparable to conventionally bred varieties and hence are out of the ambit of regulation. On the other hand, European Court of Justice has held that products of gene editing are comparable to first-generation transgenic crops and should be regulated. As per Indian regulation, gene-edited crops are GMOs and need to be regulated. Thus, there is considerable debate among scientists as to how new-generation GE crops should be treated, and this has created confusion among various stakeholders (Urnov et al. 2018).

To resolve the issue, various countries have now come with a proposal to classify gene-edited crops into three different categories, namely, site-directed nuclease-1 (SDN1), SDN2, and SDN3. SDN1 are those that carry point mutations, SDN2 with small deletions/insertions, and SDN3 with larger replacements/additions. The first two types will either be regarded as non-GMOs or be subjected to minimal regulation (Jones 2015).

Irrespective of how they are treated with regard to regulation, there is considerable negative perception about GMOs due to protracted battle between vested interested groups. Therefore, fresh efforts have been initiated to inform and educate the stakeholders about the new technologies. Given that food preferences are highly personal and countries differ with regard to food accessibility and needs, the outcomes are going to be different. Considering the need for accelerated breeding of new varieties of cereals to face the challenges of climate change and food and nutritional security, it is imperative that novel breeding techniques should be harnessed at the earliest. Therefore, excessive regulation of gene-edited crops could prove a major deterrent to developers of technology and deprive countries to meet their food needs. Hence, it is prudent that potential technologies should not be sacrificed at the alter of imaginary fears. The innovations of genome editing may ensure the world's food security. However, it all depends on how the public perceive and accept the GE products for consumption or usage (Wolt et al. 2016).

13.9 Prospects and Conclusions

The genome-editing technologies, CRISPR and its variants, have become alternative to classical or molecular breeding for combating and prevention of viruses causing severe diseases in crops. The post emergence of editing technologies would accelerate the development of new resistant resources in the plants. Usage of viral-based vector systems is proven to enhance resistance to viruses. With the advantage of CRISPR systems, host-susceptible genes/recessive factors would be edited or silenced for acquiring antiviral immunity to produce transgene-free non-GMO crops with negligible growth defects. Targeting host susceptibility of genes via CRISPR/Cas9 for insect vectors would cease the spreading of viruses and their diseases. Targeted editing of multiple gRNAs for various plant viruses is achieved. Moreover, transgene-free plants can be produced by employing primer editor, CRISPR-Act. 2.0, and pre-assembled CRISPR-Cas RNPs into protoplasts; site-specific integration; and removal of Cas9 through segregation in CRISPR plants (McCarty et al. 2020; Lowder et al. 2018; Liang et al. 2018; Woo et al. 2015; Baltes et al. 2014; Demirer et al. 2019). These technologies should be promoted at grassroot level with proper understanding, and reaching it to the end is very much needed. Indeed, crops and the plant products that developed through genetic engineering are always a major concern for the public. It is essential to eliminate the obstacles to promote the implementation of GE tools for the crop improvement with future prospects. The crop plants generated through editing tools should be treated as nongenetically modified organism for consumer acceptance. Exploitation of these technologies in agriculture would boost the generation of viral-resistant varieties, which lead to enhanced crop productivity. In crops where plants are sterile, this technology provides an opportunity as an alternative breeding approach to develop resistant varieties (Li et al. 2019). Viral genomes and plant host-susceptible genes have been edited in cereals and in plants by design and assessment of gRNA and delivery of Cas variants and gRNA for genome editing to gain disease-resistant plants to various plant viral pathogens. Finally, genome editing definitely plays a promising role to control plant diseases caused by viruses.

Acknowledgments We sincerely thank the Council for Scientific and Industrial Research for funding (CSIR-EMRII-38 (1450)/18/EMRII) awarded to GM. We thank the executive director of Agri Biotech Foundation, Hyderabad, for the facilities and support.

References

Abudayyeh OO, Gootenberg JS, Konermann S, Joung J, Slaymaker IM, Cox DB, Shmakov S, Makarova KS, Semenova E, Minakhin L, Severinov K (2016) C2c2 is a single-component programmable RNA-guided RNA-targeting CRISPR effector. Science 353:6299

Ahmad S, Wei X, Sheng Z, Hu P, Tang S (2020) CRISPR/Cas9 for development of disease resistance in plants: recent progress, limitations and future prospects. Brief Funct Genom 19(1):26–39

Ali Z, Abulfaraj A, Idris A, Ali S, Tashkandi M, Mahfouz MM (2015) CRISPR/Cas9-mediated viral interference in plants. Genome Biol 16(1):1–1

Ali Z, Eid A, Ali S, Mahfouz MM (2018) Pea early-browning virus-mediated genome editing via the CRISPR/Cas9 system in Nicotiana benthamiana and Arabidopsis. Virus Res 15(244): 333–337

Aman R, Ali Z, Butt H, Mahas A, Aljedaani F, Khan MZ, Ding S, Mahfouz M (2018) RNA virus interference via CRISPR/Cas13a system in plants. Genome Biol 19(1):1–9

Amin I, Ahmed N, Kamal H, Mansoor S (2021) Geminiviruses and their interaction with host proteins. In: Plant virus-host interaction, vol 1. Academic Press, pp 191–229

Asher C (2018) Viruses vs. Plants. Scientist 32(2):33–39

Baltes NJ, Gil-Humanes J, Cermak T, Atkins PA, Voytas DF (2014) DNA replicons for plant genome engineering. Plant Cell 26(1):151–163

Baltes NJ, Hummel AW, Konecna E, Cegan R, Bruns AN, Bisaro DM, Voytas DF (2015) Conferring resistance to geminiviruses with the CRISPR–Cas prokaryotic immune system. Nat Plants 1(10):1–4

Barrangou R, Doudna JA (2016) Applications of CRISPR technologies in research and beyond. Nat Biotechnol 34(9):933–941

Bastet A, Zafirov D, Giovinazzo N, Guyon-Debast A, Nogué F, Robaglia C, Gallois JL (2019) Mimicking natural polymorphism in eIF 4E by CRISPR-Cas9 base editing is associated with resistance to potyviruses. Plant Biotechnol J 17(9):1736–1750

Bhaya D, Davison M, Barrangou R (2011) CRISPR-Cas systems in bacteria and archaea: versatile small RNAs for adaptive defense and regulation. Annu Rev Genet 15(45):273–297

Cao Y, Zhou H, Zhou X, Li F (2020) Control of plant viruses by CRISPR/Cas system-mediated adaptive immunity. Front Microbiol 11:2613

Chen L, Tang L, Xiang H, Jin L, Li Q, Dong Y, Wang W, Zhang G (2014) Advances in genome editing technology and its promising application in evolutionary and ecological studies. Gigascience 3(1):2047–217X

Cheng W, Song XS, Li HP, Cao LH, Sun K, Qiu XL, Xu YB, Yang P, Huang T, Zhang JB, Qu B (2015) Host-induced gene silencing of an essential chitin synthase gene confers durable resistance to fusarium head blight and seedling blight in wheat. Plant Biotechnol J 13(9): 1335–1345

Custers R (2017) The regulatory status of gene-edited agricultural products in the EU and beyond. Emerging topics. Life Sci 1(2):221–229

Das A, Sharma N, Prasad M (2019) CRISPR/Cas9: a novel weapon in the arsenal to combat plant diseases. Front Plant Sci 9:2008

Demirer GS, Zhang H, Matos JL, Goh NS, Cunningham FJ, Sung Y, Chang R, Aditham AJ, Chio L, Cho MJ, Staskawicz B (2019) High aspect ratio nanomaterials enable delivery of functional genetic material without DNA integration in mature plants. Nat Nanotechnol 14(5): 456–464

Dhanraj KS, Raychaudhuri SP (1969) Note on barley mosaic in India. Plant disease reporter

Doench JG, Hartenian E, Graham DB, Tothova Z, Hegde M, Smith I, Sullender M, Ebert BL, Xavier RJ, Root DE (2014) Rational design of highly active sgRNAs for CRISPR-Cas9–mediated gene inactivation. Nat Biotechnol 32(12):1262–1267

Doench JG, Fusi N, Sullender M, Hegde M, Vaimberg EW, Donovan KF, Smith I, Tothova Z, Wilen C, Orchard R, Virgin HW (2016) Optimized sgRNA design to maximize activity and minimize off-target effects of CRISPR-Cas9. Nat Biotechnol 34(2):184–191

Freeman BC, Beattie GA (2008) An overview of plant defenses against pathogens and herbivores. The Plant Health Instructor

Fujii S (1967) Necrosis mosaic a new rice disease. Shokubutsu Boeki 21:188–190

Globus R, Qimron U (2018) A technological and regulatory outlook on CRISPR crop editing. J Cell Biochem 119(2):1291–1298

Gómez P, Rodríguez-Hernández AM, Moury B, Aranda MA (2009) Genetic resistance for the sustainable control of plant virus diseases: breeding, mechanisms and durability. Eur J Plant Pathol 125(1):1–22

Gosavi G, Yan F, Ren B, Kuang Y, Yan D, Zhou X, Zhou H (2020) Applications of CRISPR technology in studying plant-pathogen interactions: overview and perspective. Phytopathol Res 2(1):1–9

Grangeon R, Jiang J, Wan J, Agbeci M, Zheng H, Laliberté JF (2013) 6K2-induced vesicles can move cell to cell during turnip mosaic virus infection. Front Microbiol 4(4):351

Hogenhout SA, Ammar ED, Whitfield AE, Redinbaugh MG (2008) Insect vector interactions with persistently transmitted viruses. Annu Rev Phytopathol 8(46):327–359

Hsu PD, Lander ES, Zhang F (2014) Development and applications of CRISPR-Cas9 for genome engineering. Cell 157(6):1262–1278

Ji X, Zhang H, Zhang Y, Wang Y, Gao C (2015) Establishing a CRISPR–Cas-like immune system conferring DNA virus resistance in plants. Nat Plants 1(10):1–4

Jinek M, Chylinski K, Fonfara I, Hauer M, Doudna JA, Charpentier E (2012) A programmable dual-RNA–guided DNA endonuclease in adaptive bacterial immunity. Science 337(6096): 816–821

Jones HD (2015) Regulatory uncertainty over genome editing. Nat Plants 8(1):1–3

Kang BC, Yeam I, Jahn MM (2005) Genetics of plant virus resistance. Annu Rev Phytopathol 28(43):581–621

Kis A, Hamar É, Tholt G, Bán R, Havelda Z (2019) Creating highly efficient resistance against wheat dwarf virus in barley by employing CRISPR/Cas9 system. Plant Biotechnol J 17(6):1004

Kumar D (2020) Plant immune response strategies against pathogens. Plant Archiv 20(1): 1169–1174

Langner T, Kamoun S, Belhaj K (2018) CRISPR crops: plant genome editing toward disease resistance. Annu Rev Phytopathol 25(56):479–512

Li S, Shen L, Hu P, Liu Q, Zhu X, Qian Q, Wang K, Wang Y (2019) Developing disease-resistant thermosensitive male sterile rice by multiplex gene editing. J Integr Plant Biol 61(12):1201–1205. https://doi.org/10.1111/jipb.12774

Liang G, Zhang H, Lou D, Yu D (2016) Selection of highly efficient sgRNAs for CRISPR/Cas9-based plant genome editing. Sci Rep 6(1):1–8

Liang Z, Chen K, Zhang Y, Liu J, Yin K, Qiu JL, Gao C (2018) Genome editing of bread wheat using biolistic delivery of CRISPR/Cas9 in vitro transcripts or ribonucleoproteins. Nat Protoc 13(3):413

Liu H, Soyars CL, Li J, Fei Q, He G, Peterson BA, Meyers BC, Nimchuk ZL, Wang X (2018) CRISPR/Cas9-mediated resistance to cauliflower mosaic virus. Plant Direct 2(3):e00047

Lowder LG, Zhou J, Zhang Y, Malzahn A, Zhong Z, Hsieh TF, Voytas DF, Zhang Y, Qi Y (2018) Robust transcriptional activation in plants using multiplexed CRISPR-Act2. 0 and mTALE-act systems. Mol Plant 11(2):245–256

Lucas WJ (2006) Plant viral movement proteins: agents for cell-to-cell trafficking of viral genomes. Virology 344(1):169–184

Macovei A, Sevilla NR, Cantos C, Jonson GB, Slamet-Loedin I, Čermák T, Voytas DF, Choi IR, Chadha-Mohanty P (2018) Novel alleles of rice eIF4G generated by CRISPR/Cas9-targeted mutagenesis confer resistance to Rice tungro spherical virus. Plant Biotechnol J 16(11): 1918–1927

Mahas A, Aman R, Mahfouz M (2019) CRISPR-Cas13d mediates robust RNA virus interference in plants. Genome Biol 20(1):1–6

McCarty NS, Graham AE, Studená L, Ledesma-Amaro R (2020) Multiplexed CRISPR technologies for gene editing and transcriptional regulation. Nat Commun 11(1):1–3

McKinney HH (1925) Mosaic disease of winter wheat and winter rye

McKinney HH (1942) Studies on the virus of brome grass mosaic. Phytopathology 34:993

McKinney HH (1946) Mosaics of winter oats induced by soil borne viruses. Phytopathology 36:359–369

McKinney HH (1951) A seed-borne virus causing false-stripe in barley. Phytopathology:563–564

Miyamoto Y (1958) Studies on soil-borne cereal mosaics. IV. On the barley yellow-mosaic virus. (Part 2). Jpn J Phytopathol 23(4):199–206

Ng JC, Falk BW (2006) Virus-vector interactions mediating nonpersistent and semipersistent transmission of plant viruses. Annu Rev Phytopathol 8(44):183–212

Nicaise V (2014) Crop immunity against viruses: outcomes and future challenges. Front Plant Sci 21(5):660

Price AA, Sampson TR, Ratner HK, Grakoui A, Weiss DS (2015) Cas9-mediated targeting of viral RNA in eukaryotic cells. Proc Natl Acad Sci 112(19):6164–6169

Pyott DE, Sheehan E, Molnar A (2016) Engineering of CRISPR/Cas9-mediated potyvirus resistance in transgene-free Arabidopsis plants. Mol Plant Pathol 17(8):1276–1288

Roberts AG (2014) Plant viruses: soil-borne. In: eLS. John Wiley & Sons, Ltd, Chichester. https://doi.org/10.1002/9780470015902.a0000761.pub3

Rakotondrafara AM, Byamukama E, Plumb RT (2017) Virus diseases of cereals. eLS:1–2

Roy A, Zhai Y, Ortiz J, Neff M, Mandal B, Mukherjee SK, Pappu HR (2019) Multiplexed editing of a begomovirus genome restricts escape mutant formation and disease development. PLoS One 14(10):e0223765

Sastry KS (2013) Plant virus transmission through vegetative propagules (asexual reproduction). In: Seed-borne plant virus diseases. Springer, India, pp 285–305

Slykhuis JT (1970) Factors determining the development of wheat spindle streak mosaic caused by a soil-borne virus in Ontario. Phytopathology 60(31):9–331

Slykhuis JT (1976) Virus and virus-like diseases of cereal crops. Annu Rev Phytopathol 14(1): 189–210

Tashkandi M, Ali Z, Aljedaani F, Shami A, Mahfouz MM (2018) Engineering resistance against tomato yellow leaf curl virus via the CRISPR/Cas9 system in tomato. Plant Signal Behav 13(10):e1525996

Tripathi JN, Ntui VO, Ron M, Muiruri SK, Britt A, Tripathi L (2019) CRISPR/Cas9 editing of endogenous banana streak virus in the B genome of Musa spp. overcomes a major challenge in banana breeding. Commun Biol 2:46. https://doi.org/10.1038/s42003-019-0288-7

Urnov FD, Ronald PC, Carroll D (2018) A call for science-based review of the European court's decision on gene-edited crops. Nat Biotechnol 36(9):800–802

Wang M, Lu Y, Botella JR, Mao Y, Hua K, Zhu JK (2017) Gene targeting by homology-directed repair in rice using a geminivirus-based CRISPR/Cas9 system. Mol Plant 10(7):1007–1010

Wolt JD, Wang K, Yang B (2016) The regulatory status of genome-edited crops. Plant Biotechnol J 14(2):510–518

Woo JW, Kim J, Kwon SI, Corvalán C, Cho SW, Kim H, Kim SG, Kim ST, Choe S, Kim JS (2015) DNA-free genome editing in plants with preassembled CRISPR-Cas9 ribonucleoproteins. Nat Biotechnol 33(11):1162–1164

Xie K, Zhang J, Yang Y (2014) Genome-wide prediction of highly specific guide RNA spacers for CRISPR–Cas9-mediated genome editing in model plants and major crops. Mol Plant 7(5): 923–926

Yin K, Han T, Liu G, Chen T, Wang Y, Yu AY, Liu Y (2015) A geminivirus-based guide RNA delivery system for CRISPR/Cas9 mediated plant genome editing. Sci Rep 5(1):1

Yin K, Han T, Xie K, Zhao J, Song J, Liu Y (2019) Engineer complete resistance to cotton leaf curl multan virus by the CRISPR/Cas9 system in nicotiana benthamiana. Phytopathol Res 1(1):1–9

Zaidi SS, Mukhtar MS, Mansoor S (2018) Genome editing: targeting susceptibility genes for plant disease resistance. Trends Biotechnol 36(9):898–906

Zhang H, Si X, Ji X, Fan R, Liu J, Chen K, Wang D, Gao C (2018a) Genome editing of upstream open reading frames enables translational control in plants. Nat Biotechnol 36(9):894–898

Zhang T, Zheng Q, Yi X, An H, Zhao Y, Ma S, Zhou G (2018b) Establishing RNA virus resistance in plants by harnessing CRISPR immune system. Plant Biotechnol J 16(8):1415–1423

Zhang T, Zhao Y, Ye J, Cao X, Xu C, Chen B, An H, Jiao Y, Zhang F, Yang X, Zhou G (2019) Establishing CRISPR/Cas13a immune system conferring RNA virus resistance in both dicot and monocot plants. Plant Biotechnol J 17(7):1185–1187

Zhao Y, Yang X, Zhou G, Zhang T (2020) Engineering plant virus resistance: from RNA silencing to genome editing strategies. Plant Biotechnol J 18(2):328–336

Genomic and Bioinformatic Resources for Next-Generation Breeding Approaches Towards Enhanced Stress Tolerance in Cereals

14

Rinku Sharma, Mallana Gowdra Mallikarjuna, K. R. Yathish, Chikkappa G. Karjagi, and H. C. Lohithaswa

Abstract

Breeding stress-resilient cereals is a prerequisite to sustaining the food and nutritional security in the era of climate change. The number of cereal genome sequences and the number of genomic resources are growing with the advent of a technological revolution in genomics and molecular biology. These genomic innovations resulted in advanced next-generation cereal breeding methods to improve the genetic gain per unit time. Many contemporary next-generation breeding methods necessitate the genetic and genomic resources for target trait improvement. Therefore, the cereal genetic resources are a substantial opportunity for stress-resilient cereal breeding, preservation, and maintenance of genetic diversity. Interestingly, broad genetic variability for stress-resilient traits in cultivated varieties, landraces, and wild relatives is preserved and actively used in the breeding pipeline. Further, contemporary cereal breeding techniques, viz. genome-wide association mapping, genomic selection, and genome editing, have resulted in various sequence resources, databases, and software packages. These genomic resources are assisting the translational cereal genomics in addition to

R. Sharma · M. G. Mallikarjuna (✉)
Division of Genetics, ICAR-Indian Agricultural Research Institute, New Delhi, India
e-mail: MG.Mallikarjuna@icar.gov.in

K. R. Yathish
Winter Nursery Centre, ICAR-Indian Institute of Maize Research, Hyderabad, India

C. G. Karjagi
Unit Office, ICAR-Indian Institute of Maize Research, New Delhi, India

H. C. Lohithaswa (✉)
Department of Genetics and Plant Breeding, University of Agricultural Sciences, GKVK, Bengaluru, India

M. G. Mallikarjuna et al. (eds.), *Next-Generation Plant Breeding Approaches for Stress Resilience in Cereal Crops*, https://doi.org/10.1007/978-981-19-1445-4_14

the promotion of comparative genomics to accelerate discovery and functional analyses of genes in cereals and model plants. Additionally, various bioinformatic platforms and associated databases for stress-resilient cereal breeding are expected to play key roles in designing effective breeding strategies to make the best use of genomic resources in cereals. Here, we summarised genomic resources, bioinformatic tools, and databases made available in cereal omics towards their possible utility in the next-generation cereal breeding for stress resilience.

Keywords

Bioinformatic tools · Cereals · Climate resilience · Genetic resources · Genomic resources · Next-generation breeding · Stress tolerance

14.1 Introduction

Cereals are the prominent source of global food and nutritional security contributing approximately 60% of global food and energy requirements (FAO 2015). Extensive scientific cereal breeding for more than 100 years resulted in improved cereal varieties and hybrids with higher grain and fodder yields. Further, there is a need to enhance the cereal crop productivity to ensure food and nutrition security for a continuously growing population that is expected to cross nine billion by the year 2050 (http://population.un.org/wpp/). However, the journey is more challenging owing to frequent occurrences of climate change-induced environmental constraints. The occurrences of various abiotic and biotic stresses are causing 30–90% of grain yield losses in cereals depending on the crop, growth stage, and geographical location (Boyer 1982; James 2002; Bahri et al. 2011; Karjagi et al. 2017; Serfling et al. 2017; Jeevan et al. 2020). The tolerant and resistant cultivars to abiotic and biotic stresses, respectively, are the most effective techniques to manage and sustain the cereal grain yield. Conventional breeding delivered substantially to food and nutritional security, especially during and post-green revolution period. Subsequently, the technological advances in the cereal omics research area resulted in high-throughput methods for uncovering the genome sequences, profiling the expressions of thousands of genes, dissection of regulatory elements and pathways in trait expression, and interactions in the proteome.

The cereal genomic revolution started with the availability of draft genome sequences from rice (Goff 2002; Yu 2002) followed by sorghum (Paterson et al. 2009), maize (Schnable et al. 2009), pearl millet (Varshney et al. 2017), barley (Mayer et al. 2012; Mascher et al. 2017), and wheat (Appels et al. 2018). Subsequently, genome sequence availability along with various genomic resources, advanced mapping populations, and panels has accelerated genomics-assisted breeding in cereals (Mochida and Shinozaki 2010). Further, these advancements also helped in investigating gene function in association with the target phenotype. The ultimate goal of cereal genomics is to improve the breeders' ability to identify the genotypes with optimal agronomic and stress-resilient traits to improve the yield with stress tolerance. The next-generation breeding tools in cereals have been employed since the publication of a draft sequence of rice genome and enormously helped in trait mapping and cultivar development with tolerance to various stresses. However, the feasibility of modern breeding tools utility in cereal breeding mainly depends on the availability of genetic resources and genomic tools. Therefore, fair knowledge and access to publicly available information have a direct impact on the application of next-generation breeding approaches in cereals for stress resilience.

14.2 Genetic Resources in Cereals for Stress Tolerance

Cereal genetic resources comprise the basic raw material for genetic improvement of cereals for various stress tolerances. The genetic variability for stress-adaptive traits in addition to grain yield is a key criterion for the exploitation of cereal genetic resources in the breeding programmes. Globally, various efforts have been directed towards conserving, characterising, and evaluating cereal genetic resources. In addition to the utilisation of cereal genetic resources, specialised germplasm sets such as mini-core collections and reference sets to capture the existing variability for stress tolerance facilitate the identification of single and multi-stress-tolerant cereal germplasm, mapping for adaptive traits and mining allele for various biotic and abiotic stresses and agronomic traits. Existing cereal germplasm for stress-resilient breeding is discussed as follows.

14.2.1 Cereal Genebanks and Germplasm Portals

Genebanks harbour the cereal germplasm with tolerance to various abiotic and biotic stress tolerances in addition to valuable agronomic traits. Globally, more than three million cereal germplasm are conserved in the genebanks (FAO 2010). The present status of cereal germplasm in important genebanks is summarised in Fig. 14.1. The major accessions of cereals are catalogued and conserved in CGIAR (Consultative Group on International Agricultural Research) establishments, viz. International Rice Research Institute (IRRI, rice); International Maize and Wheat Improvement Center, Mexico (CIMMYT, wheat and maize); and the International Crops Research Institute for the Semi-Arid Tropics, India (ICRISAT, sorghum and pearl millet). In addition to these international efforts, there are various national genebanks that captured the local cereal diversity. The National Plant Germplasm System (NPGS) of USDA-ARS, USA, is a public germplasm collection represented by various genebanks of the USA. NPGS of the USA preserves the cereal seed samples, epitomising the global cereal diversity in crops like barley, maize, oat, rice, rye, triticale, wheat, and several wild relatives (Jaradat 2013; Byrne et al. 2018). The Lieberman Germplasm Bank of Institute for Cereal Crops Improvement, Tel Aviv, Israel, holds the most extensive collection of wild relatives of wheat, barley, and oat falling under the genera *Aegilops* (7520), *Hordeum* (4610), *Triticum* (3282), and

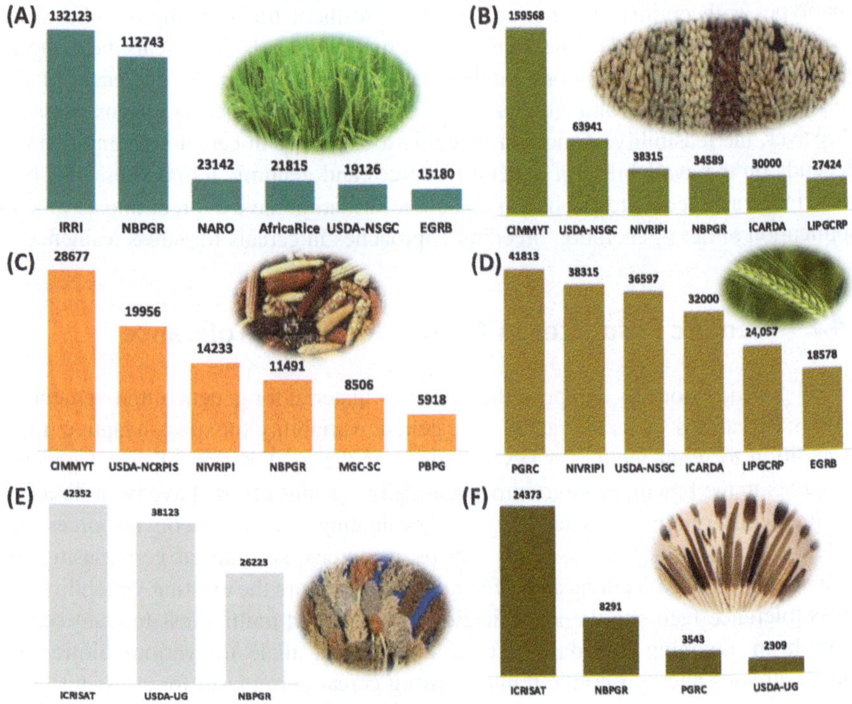

Fig. 14.1 Overview of (**a**) rice, (**b**) wheat, (**c**) maize, (**d**) barley, (**f**) sorghum, and (**g**) pearl millet germplasm status in the major global and national genebanks. The accession numbers are collected from Genesys, Genebank Platform (https://www.genebanks.org), and respective databases as on July 2021. *CIMMYT* International Maize and Wheat Improvement Center, Mexico, *EGRB* Embrapa Genetic Resources & Biotechnology, Brazil, *ICARDA* International Center for Agricultural Research in the Dry Areas, Lebanon, *ICRISAT* International Crops Research Institute for the Semi-Arid Tropics, India, *IRRI* International Rice Research Institute, Philippines, *LIPGCRP* Genebank, Leibniz Institute of Plant Genetics and Crop Plant Research, Germany; *MGC-SC* Maize Genetics Cooperation Stock Center, USA; *NARO* National Agriculture and Food Research Organization, Japan; *NBPGR* ICAR-National Bureau of Plant Genetic Resources, India, *NIVIRIPI* N.I. Vavilov Research Institute of Plant Industry, Russia, *PBPG* Portuguese Bank of Plant Germplasm, Portugal; *PGRC* Plant Gene Resources of Canada, Saskatoon Research and Development Centre, Canada, *USDA-NCRPIS* North Central Regional Plant Introduction Station, USDA-ARS, USA, *USDA-NSGC* National Small Grains Collection, USDA-ARS, USA, *USDA-UG* Plant Genetic Resources Conservation Unit, Southern Regional Plant Introduction Station, University of Georgia, USDA-ARS, USA

Avena (1592) (https://en-lifesci.tau.ac.il/icci). Likewise, the Plant Gene Resources of Canada, Saskatoon Research and Development Centre, Canada, houses major barley accessions.

Knowledge and access to global cereal diversity are most crucial for their utilisation in breeding programmes. The availability of cereal germplasm on a single platform eases the breeders and plant scientists to find the desirable accession. Furthermore, these portals also assist the genebanks with standardised practices and documentation systems for efficient germplasm management, research, and genetic resource exchange. The FAO's WIEWS portal on Plant Genetic Resources (PGR) provides the information on the identification and analysis of global cereal landraces (http://apps3.fao.org/wiews/wiews.jsp). Similarly, the European Cooperative Programme for Plant Genetic Resources (ECPGR) is a collective outcome of European nations. The ECPGR provides procedures and documentation systems for long-term conservation and facilitates PGR utilisation in Europe (http://www.ecpgr.cgiar.org/).

Additionally, GENESYS and GRIN are the two important portals associated with linking plant genetic resources across the globe. The Genesys portal comprises the data given by three international project partners such as the USDA-ARS National Genetic Resources Program (https://www.genesys-pgr.org/), the European Cooperative Programme for Plant Genetic Resources (ECPGR-EURISCO), and the System-wide Genetic Resources Programme (SGRP-SINGER) of the CGIAR. Another global portal, GRIN (Germplasm Resources Information Network, www.ars-grin.gov), was developed by the Global Crop Diversity Trust, USDA, and Bioversity International. The GRIN portal provides origin, passport information, availability, and quantity of accessions that can be distributed to scientists and plant breeders worldwide (Postman et al. 2010). Some of the cereal germplasm portals and databases are given in Table 14.1.

14.2.2 Stress-Resilient Cereal Landraces and Wild Relatives

The present cereal cultivars are the products of continued selection for several agronomical traits and show a relatively narrow genetic base. However, the present-day cereal cultivars are mostly sensitive to various abiotic and biotic stresses. On the contrary, these cereals' landraces and wild relatives have evolved mostly in environments exposed to various stresses and low nutrient availability and pose broad genetic bases. Compared to improved cultivars, these cereal landraces and wild relatives withstand the stresses and provide appreciable variation for low fertiliser input cropping systems (Newton et al. 2010). Thus, landraces and wild relatives serve as potential allelic donors for stress-resilient breeding in cereals. Many of the previously explained genebanks captured the cereal landrace diversity. It is reported that nearly 24% of 856,168 wheat accessions conserved in 229 institutes are landraces (FAO 2010). Likewise, more than 30% of the CIMMYT wheat accessions are landraces (de Carvalho et al. 2012). In the case of barley, landraces occupied 23% of 470,470 accessions from 204 genebanks across the globe. The leading barley genebank Plant Gene Resources of Canada (PGRC) of Canada holds more than 40% of landraces from a total of ~40,000 accessions (FAO 2010). The CIMMYT has conserved 176 and 2930 wild accessions of maize and wheat, respectively, whereas IRRI and AfricaRice conserved 3718 and 40 wild rice germplasm, respectively (www.genebanks.org). The landraces can be used traditionally as a donor source for the improvement of stress resilience. Alternatively, novel breeding approaches like genome editing allow editing the genome for agronomic

Table 14.1 Cereal genetic resource databases and portals for stress tolerance and other agronomic traits

S. n6.o.	Database	URL	Remarks	Reference
1	Genesys	https://www.genesys-pgr.org/	Provide information on crop genetic resources, including all the cereals conserved in genebanks	–
2	GBIS	http://gbis.ipk-gatersleben.de/	German genebank information system. Allows fetching germplasm information from German ex situ collections and placing the request for the same. Among cereals, barley and wheat are included	Oppermann et al. (2015)
3	EURISCO	https://eurisco.ipk-gatersleben.de/	Gives information on more than two million ex situ preserved crop accessions, including cereals and their wild relatives, by ~400 institutes in Europe	Weise et al. (2017)
4	GRIN	https://www.grin-global.org	Provides immediate access to information on PGR, including cereals conserved across the global databases	Postman et al. (2010)
5	MGCSC	http://maizecoop.cropsci.uiuc.edu	Collect, maintain, and distribute maize genetic stocks along with detailed information about stocks and mutations	–
6	NARO	https://www.gene.affrc.go.jp	Coordinating institute for conservation of plants, microorganisms, and animal germplasm to agriculture in Japan. Contain core collections of cereal crops	–
7	NPGS	https://npgsweb.ars-grin.gov/	Collect, catalogue, and distribute the PGRs to scientists, educators, producers, and research and education institutes	Byrne et al. (2018)
8	PGR portal	https://pgrportal.nbpgr.ernet.in/	Facilitate availability of information on conserved germplasm in the National Bureau of plant genetic resources (NBPGR), India. PGR portal is accessible to researchers, farmers, students, and policymakers	Tyagi (2016); Singh et al. (2020)

traits by keeping the broad genetic base of landraces intact. Editing of *semi-dwarf1* (*SD1*) gene in Kasalath and TeTePu landrace background showed tolerance to low-phosphorus, broad-spectrum resistance to several diseases and insects and semi-dwarf phenotypes in rice (Hu et al. 2019). Similarly, in the African rice landrace, Kabre was edited for *HTD1*, *GS3*, *GW2*, and *GN1A* genes to reduce plant height and improve seed yield (Lacchini et al. 2020).

14.3 Sequence Resources in Cereals

Crop genomics metamorphosed with the delivery of genome sequences of model and crop plants, expressed sequence tag (EST) databases, and rapid gene discovery techniques. These developments further transformed both plant genomics and crop bioinformatics. The accessibility of sequencing information of crops in the public domain enhances the knowledge on the genome, gene discovery, and trait improvement in addition to traditional breeding approaches. With the onset of next-generation sequencing technology in 2005, genome sequencing has become more economical in terms of cost and time, leading to a drastic increment in the number of sequenced plant genomes. These genomes have consistently contributed a variety of important genetic and genomic resources, such as huge compendia of molecular markers, high-throughput sequencing methodologies, high-density genetic maps, novel breeding panels and populations, etc.

The high-resolution reference genome sequences of the cereals offer an exhaustive list of genes, the regulatory elements that regulate gene expression, and the quantum of existing genomic variations. The sequence knowledge is essential for understanding the function of genes in plant growth and development that helps to decipher the mechanisms at the biological system level and efficiently exploit an organism's natural and induced genetic diversity. We have been able to uncover and extract key genes and determine their functions in controlling grain yield, biomass, and tolerance to a variety of environmental pressures due to advancements in plant genomics.

14.3.1 Whole-Genome Sequencing Projects in Cereals

The significant motives behind genome sequencing projects of major crops were identification, classification, tagging, and exploitation of individual alleles for target traits. In addition, genome sequencing allows the development of allele/gene-specific molecular markers, which could be employed to monitor the inheritance of desired alleles in selective breeding programmes. With these motives, several genome sequencing projects of major cereals were implemented, such as the International Rice Genome Sequencing Project (IRGSP; http://rgp.dna.affrc.go.jp/IRGSP/index.html), Maize Genome Sequencing Project (Chandler and Brendel 2002), International Wheat Genome Sequencing Consortium (IWGSC) (http://www.wheatgenome.org/), Sorghum Genome Sequencing Team (Andrew et al. 2009), International Pearl Millet Genome Sequencing Consortium (Varshney et al. 2017), Rye Genome Sequencing Consortium (rye) (Li et al. 2021), and International Barley Genome Sequencing Consortium (barley) (Mayer et al. 2012). These projects utilised new sequencing technologies in conjunction with traditional methods to develop and validate high-quality sequences and cost-effective designs. As a result, by providing essential data for comparative and functional genomics research, these whole-genome sequencing efforts considerably impacted global food security and bio-energy advancement.

Table 14.2 List of genome sequence databases for cereals

Database	URLs	Reference
General		
NCBI genome	http://www.ncbi.nlm.nih.gov/genome/	Sayers et al. (2019)
Phytozome V12.1	http://phytozome.jgi.doe.gov/pz/portal.html	Goodstein et al. (2012)
PLAZA	http://plaza.psb.ugent.be/	Van Bel et al. (2018)
PlantGDB	http://www.plantgdb.org/	Dong et al. (2004)
Ensembl plants	http://plants.ensembl.org/index.html	Bolser et al. (2016)
ChloroplastDB	http://chloroplast.cbio.psu.edu/	Cui et al. (2006)
KEGG	http://www.genome.jp/kegg/	Kanehisa et al. (2017)
JGI GOLD	https://gold.jgi-psf.org/	Mukherjee et al. (2017)
CoGepedia	https://genomevolution.org/wiki/index.php/	–
Species-specific sequence database		
RGAP V.7 (rice)	http://rice.plantbiology.msu.edu/	Ouyang et al. (2007)
RAP-DB (rice)	http://rapdb.dna.affrc.go.jp/	Sakai et al. (2013)
Gramene (Gramineae)	http://www.gramene.org/	Tello-Ruiz et al. (2021)
GrainGenes (Triticeae and *Avena*)	http://wheat.pw.usda.gov/GG3/	Matthews et al. (2003)
MaizeGDB (maize)	http://www.maizegdb.org/	Lawrence et al. (2004)

As the quantity of whole-genome sequences grows at an exponential rate, well-organised and annotated DNA databases are becoming extremely relevant. The three largest well-systematised databases are GenBank, EMBL, and DNA Data Bank of Japan. These databases, which contain millions of plant DNA sequences, are widely considered as the gold standard for publicly accessible annotated DNA sequence collections around the world. For example, according to RefSeq Growth Statistics, the NCBI Genome database has a total of 15,552,676 plant accession entries in May 2021 (www.ncbi.nlm.nih.gov/refseq/statistics/). Back in May 2020, there were 13,991,642 entries, which show a drastic increase in entries per year. Further, this increase is more prominent in almost all the cereal crop datasets, viz. rice, wheat, maize, oat, barley, sorghum, pearl millet, etc. Other public databases that contain additional information on cereal genomes include CoGepedia, ChloroplastDB, Ensembl Plants, Genomes Online Database (GOLD), KEGG, Phytozome, PlantGDB, etc. which are summarised in tabular form (Table 14.2).

Other databases focusing on specific plant species, such as the Rice Genome Annotation Project (RGAP) (Kawahara et al. 2013), the Rice Annotation Project (RAP-DB) (Sakai et al. 2013), Gramene (Ware et al. 2002), GrainGenes (Matthews et al. 2003), MaizeGDB (Lawrence et al. 2004), etc., have recently become available (Table 14.2).To analyse genomic sequences, these databases typically feature a collection of analytical, visualisation, and query tools, such as BLAST for discovering sequence similarity in big datasets.

14.3.2 Expressed Sequence Tags (ESTs), cDNA Clones, and Full-Length cDNA

The availability of a crop's entire genome sequence is beneficial for plant breeding, but for some species with vast and complicated genomes, obtaining a draft can take years. Transcriptome sequencing has been a less expensive alternative to whole-genome sequencing. The genes expressed in a defined spatio-temporal condition are identified using expressed sequence tags (ESTs) generated through cDNA sequencing. The wide and rapid accumulation of cDNA clones, as well as massive datasets containing associated sequence tags, has become a useful resource for functional genomics. Gene discovery could be significantly aided by ESTs obtained from various tissues, including tissues from different stages of development or stress-treated ones. Gene structural annotation, large-scale expression profiles, genome-scale intra- and inter-specific comparative study of expressed genes, and the development of expressed gene-oriented molecular markers and probes for microarrays are all conceivable with ESTs. Although non-coding sequences are not included in EST sequencing efforts, identifying all genes and transcript variations is difficult even with multiple libraries. Despite these drawbacks, EST collections have shown to be highly beneficial to breeders. Several data resources, such as NCBI-UniGene (now available online through FTP), PlantGDB, HarvEST, and TIGR Plant Gene Indices, provide EST datasets of plants (Fig. 14.2).

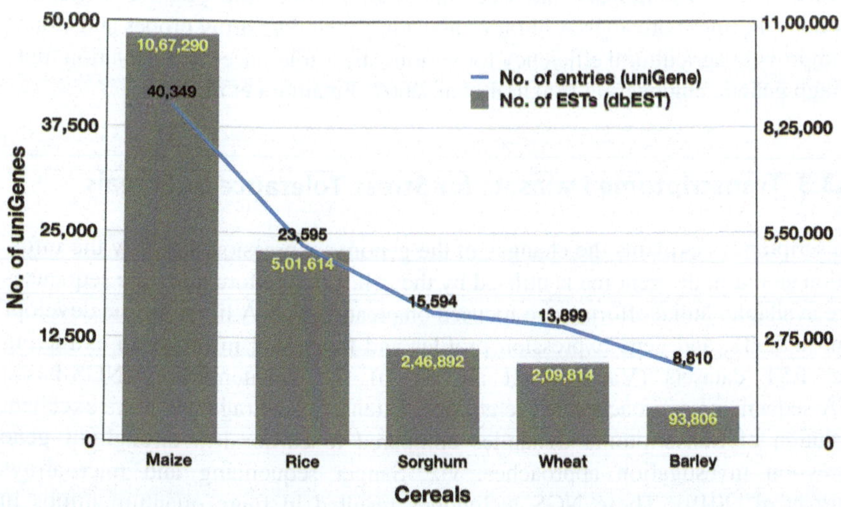

Fig. 14.2 The dbEST and unigene statistics for various cereals (NCBI-UniGene now available online through NCBI FTP site)

Table 14.3 Large-scale collections of full-length cDNA clones in cereals

Plant	Database	Reference
Barley	http://www.shigen.nig.ac.jp/barley/ or https://harvest.ucr.edu/	Sato et al. (2009)
Rice (japonica)	http://cdnaOl.dna.aftrc.go.jp/cDNA/	Kikuchi et al. (2003)
Rice (indica)	http://www.ncgr.ac.cn/ricd	Liu et al. (2007)
Wheat	http://tritldb.psc.riken.jp/	Ogihara et al. (2004); Kawaura et al. (2009)
Maize	http://www.maizecdna.org/	Soderlund et al. (2009)
Sorghum	https://mycocosm.jgi.doe.gov/Sorbi1/Sorbi1.home.html	Tello-Ruiz et al. (2021)

Though partial cDNAs are valuable for rapid creation of expressed gene catalogues, they are ineffective for determining the gene's functional characteristics. As a result, full-length cDNA and large-scale clone sequence resources have become useful tools for several functional characterisation of genes. Full-length cDNA sequence databases can also help identify transcribed areas in complete or draft genome sequences. Furthermore, full-length cDNAs were utilised to detect structural elements of a gene, viz. transcription units, transcription start sites (TSSs), and transcriptional variations in *Arabidopsis* and rice (Seki et al. 2002; Itoh et al. 2007). Additionally, full-length cDNAs were utilised in developing bioinformatic pipelines, which helped in gene discovery in plant species where only the draft genome is available (Nanjo et al. 2007; Umezawa et al. 2008) (Table 14.3). In addition, by comparing target sequences with full-length cDNA libraries of model organisms like *Arabidopsis* and rice, one can discover the putative biological functions. Furthermore, cDNA libraries assist in developing array probes and clones for improving agricultural efficiency for various stress tolerance and adaptation traits through genetic engineering (Sakurai et al. 2007; Futamura et al. 2008).

14.3.3 Transcriptome Datasets for Stress Tolerance in Cereals

Transcriptomics explains the changes in the genome expression and how the information stored in the genome is utilised by the cell. Long before genomic sequences were available, initial efforts were focused on creating cDNA libraries, the development of ESTs, and gene expression profiles and functional information extracting from EST datasets (Varshney et al. 2009b). The contemporary NGS-based RNA-sequencing approaches delivered more extensive coverage and more excellent resolution of transcriptome dynamics compared to earlier low-throughput gene expression investigation approaches, viz. Sanger sequencing and microarrays (Garg et al. 2019). These NGS techniques resulted in huge quantum jumps in stress-resilient cereal improvement programmes. Furthermore, NGS technology made it possible to identify epigenetic alterations on native DNA and allowed sequencing of whole transcripts independent of availability of genome assembly (van Dijk et al. 2018). Numerous stress-responsive expression profiles were generated in cereals using high-throughput expression platforms by various researchers (Cho et al. 2008; Cai et al. 2012; Cal et al. 2013; do Amaral et al. 2016; Aravind et al. 2017; Arora et al. 2017; Thirunavukkarasu et al. 2017; Zeng et al. 2019; Zhou et al. 2019; Liu et al. 2020; Mallikarjuna et al. 2020). Following the creation of minimum information about a microarray experiment (MIAME) standard, gene expression data obtained using high-throughput techniques such as microarray and RNA-seq were archived in public repositories. In a MIAME-compliant way, these are the Gene Expression Omnibus (GEO; https://www.ncbi.nlm.nih.gov/geo/) (Edgar 2002) and EBI ArrayExpress (AE; https://www.ebi.ac.uk/arrayexpress/) (Parkinson et al. 2007) of NCBI and EBI, respectively. Plant genome responses to diverse stressors and developmental phases are studied via transcriptome profiling. Additionally, transcriptome sequencing was employed to investigate genes, annotations, and the finding of non-coding RNA in functional genomics (Morozova and Marra 2008). The transcriptome assemblies for key cereals, viz. rice (Tian et al. 2015), wheat (Jia et al. 2018), and maize (Shan et al. 2013), were given to elucidate the regulation of candidate genes for diverse traits at various plant growth and developmental phases under stress and control conditions.

The expression data generated from transcriptome assemblies allowed the identification of potential genes linked to various target traits and stress responses. Finding potential genes for agronomic traits and stress tolerance requires an understanding of underlying genomic information about particular phenotypes at crucial developmental phases. Additionally, the gene expression atlases (GEAs) enable a comprehensive examination of complete transcriptional profile, illuminating genome-wide gene activity in various model and agricultural plants' tissues. Such gene expression atlases were developed for major cereal crops like rice which encompases 39 tissues of 2 cultivars (Zhenshan 97 and Minghui 63) for whole life cycle (Wang et al. 2010) and maize from 18 tissues. The availability of expression atlases in major cereals like rice and maize helping researchers to identify and characterise various pathways and genes pertaining to developmental stages (Sekhon et al. 2013). These transcriptomic resources shed light on molecular mechanisms underlying numerous phases of plant development and traits like yield, stress tolerance, etc., which in turn assist in the development of improved cereal cultivars.

Additionally, several secondary plant-specific expression databases containing raw expression datasets for cereals with other important processed information are available. PLAnt co-EXpression database (PLANEX; http://planex.plantbioinformatics.org) is a new Internet-based database for plant gene analysis. PLANEX includes publicly accessible GeneChip data from the Gene Expression Omnibus (GEO) of NCBI. It is a genome-wide co-expression database that enables functional gene identification from a wide range of experimental designs (Yim et al. 2013). The EBI Expression Atlas is a valuable resource that contains information on gene and protein expression in a variety of species and conditions, including tissue, developmental stage, disease, and cell type, as well as over 900 plant transcriptome experiments (Papatheodorou et al. 2018; http://www.ebi.ac.uk/gxa). Finally, the AgriSeqDB (https://expression.latrobe.edu.au/agriseqdb) hosts a collection of

RNA-seq data of *Arabidopsis* and crop species from researchers at La Trobe University and other researchers (Robinson et al. 2018).

14.3.4 Small RNA and Micro-RNA Databases

Heterochromatic small interfering RNAs (heterochromatic siRNAs or hc-siRNAs), microRNAs (miRNAs), and phased small interfering RNAs (phased siRNAs or phasiRNAs) are essential regulatory molecules in plants. Plant microRNAs (miRNAs) are a well-studied subclass of sRNAs (Fei et al. 2013). They are endogenous, single-stranded, short (21–23 nt) non-coding regulatory RNA molecules that repress translation and modify messenger RNA (mRNA) degradation by binding complementary sites in target mRNAs' protein-encoding or 3′-untranslated regions. They also perform crucial roles in controlling several biological processes such as stress responses, floral organ identification, cell signalling, and so on (Voinnet 2009; Li and Zhang 2016). PhasiRNAs (phased interfering small RNAs) are sRNAs produced through various pathways and regulated by regulatory cascades or modules. They are usually produced as a result of a miRNA-transcript targeting event, which results in the generation of additional phased short RNAs that can influence gene expression in both cis and trans. phasiRNAs have lately appeared as important regulatory RNAs in practically every phase of plant development and growth. Heterochromatic siRNAs, on the other hand, are the most common kind of sRNA in plants. phasiRNAs are usually involved in DNA methylation, which is a crucial step in transcriptional control (Fei et al. 2013; Liu et al. 2018).

Fortunately, valuable public resources have become increasingly available for mechanistic studies on plant sRNAs in the form of several databases. The miRbase houses 38,589 miRNA entries across diverse species including plants (http://www.mirbase.org/). Additionally, it provides a set of precursor and mature miRNAs discovered in various plant species (Kozomara and Griffiths-Jones 2014). The CSRDB database hosts a collection of maize and rice miRNAs (Johnson et al. 2007; http://sundarlab.ucdavis.edu/smrnas/), and Rfam gives miRNA precursors secondary structures in many plants (Griffiths-Jones et al. 2003; http://rfam.sanger.ac.uk/). Similarly, the two databases PASmiR and miRNEST provide comprehensive literature-curated databases for stress-responsive miRNA and target genes (Zhang et al. 2013; Szczes'niak and Makabowska 2014). Additionally, the mega-database miRNEST houses miRNAs of broad-range species such as plant, animals, and virus miRNAs. miRNEST provides miRNA data based on computational predictions and high-throughput sequencing. It contains miRNAs from 22 viruses and more than 270 plant species that also have 2041 degradome data entries.

The plant miRNA database (PMRD) is another important miRNA database in plants, providing information on plant-miRNA sequences and their targets (Zhang et al. 2010). PMRD also includes sequencing data, secondary structure, targets, expression profiles, and a genomic browser that includes over 8400 miRNA entries for over 120 plant species, including cereals (http://bioinformatics.cau.edu.cn/PMRD/). PMTED (Plant miRNA Target Expression Database) is miRNA database exclusively dedicated to plants that may be explored to study miRNA functions by gathering target gene expression profiles from a global microarray data. PMTED additionally includes tools for searching for miRNA targets and retrieving expression information (Sun et al. 2013a; http://pmted.agrinome.org/). sRNAanno (www.plantsRNAs.org) is another repository that holds all major types of sRNAs (miRNAs, phasiRNAs, and hcsiRNAs) for over 140 plant genomes. These broad annotations were made possible by analysing over 1600 sRNA datasets with well-established compute pipelines that met tight criteria (Chen et al. 2021). Plant miRNA Encyclopedia (PmiREN) is a comprehensive database of functional plant miRNA. PmiREN houses 38,186 high confidence new miRNA loci from 179 species. The database provides various tools and information on the precursor sequence, secondary structure and expression pattern, genome clusters, and synteny. Additionally, it also gives quality miRNA-target pairs validated through parallel analysis of RNA ends (PARE) sequencing (Guo et al. 2021; http://www.pmiren.com/). These small RNA-related resources available in the public domain could help in accelerating the research on understanding the role of small RNAs in cereal plant growth, development, and stress response mechanisms.

14.3.5 Databases for Stress-Regulatory Elements Databases

The expression of a gene is an important aspect of its function and determining its expression profile is crucial to obtaining a complete functional description of each gene. Thus, gene expression regulation is central to all the cellular processes in organisms. Many components in the plant genome regulate gene expression via DNA and regulatory protein interactions at all phases of genetic information flow. *Cis* sequences and *trans* factors are two important categories of gene expression regulators. Non-coding DNA linear nucleotide regions are known as cis-regulatory sequences. Genes differ in the position and direction of cis-regulatory elements (Venter and Botha 2010). The regulatory sequences either can be found directly in the transcribed DNA strand, such as enhancers, promoters, silencers, and insulators, or inserted as post-transcriptional alterations, such as signal sequences 5′ cap, poly-A tail, etc. (Vaughn et al. 2012). *Trans* factors, a type of regulatory protein, interact with *cis* sequences and other proteins to create active complexes. Transcription factors are one of the most significant trans elements in networks, as they influence the expression of other regulatory proteins, which can moderate the expression of genes coding for structural proteins or regulatory trans elements (Biłas et al. 2016). Identifying stress-regulatory networks in which transcription factors and other regulatory elements affect the temporal and spatial expression of all genes in an organism under specific environment is an emerging area in gene expression studies. An evolving topic in expression analysis is identifying stress-regulatory networks in which transcription factors affect the temporal and spatial expression of all genes in an organism. The identification of all TFs and their corresponding cis-regulatory areas in all gene promoters is a first step in achieving this goal (Canales et al. 2014; Sharma et al. 2020).

To investigate for putative regulatory elements in grains, several resources are accessible. PlantCare houses the plant *cis*-acting regulatory elements and allows in silico analyses of promoter sequences for the presence of *cis* elements (Lescot et al. 2002). The PlantPromDB is a non-redundant library of annotated proximal promoter sequences for RNA polymerase II with experimentally determined transcription start sites (TSS) from a variety of plant species (Zuo and Li 2011). PLACE is a collection of motifs discovered in plant cis-acting elements based on published research and exclusively dedicated to vascular plant species (Higo et al. 1999). PlantTFDB stands for Plant Transcription Factor Database that provides a web interface for accessing nearly complete sets of transcription factors from a diverse species of plants, including *Arabidopsis thaliana* (thale cress) and cereals (Jin et al. 2017). The PRI-CAT (Plant Research International ChIP-seq Analysis Tool), providing a web-based workflow to manage and analyse the ChIP-seq results, was developed by Plant Research International. Users can submit their sequencing data directly to the PRI-CAT portal for automated analysis (Muiño et al. 2011). Transfac also includes eukaryotic transcription factor database. This contains DNA-binding profiles and genomic binding locations for transcription factors (Matys et al. 2006). Plant Promoter Analysis Navigator (PlantPAN) is a convenient resource for identifying transcription factor binding sites (TFBSs), respective TFs, and other significant regulatory features such as tandem repeats and CpG islands in promoter sequences. Presently, the PlantPAN comprises 16,960 transcription factors (TFs) and 1143 TF binding site matrices from 76 plant species (Chow et al. 2019; http://plantpan.itps.ncku.edu.tw/). The DoOP (Databases of Orthologous Promoters) comprises orthologous clusters of promoters from *Arabidopsis*, human, rice etc. (Barta et al. 2005). Furthermore, PPDB (Plant Promoter Database) offers transcription start sites (TSS) and other structural information of *Arabidopsis* and rice promoters (Shahmuradov et al. 2003). The availability of these open access resources could help understand the stress-regulatory mechanism operating under vivid biological scenarios in cereals.

14.3.6 Stress-Specific Databases for Cereals

Numerous web-based public platforms and databases specialised to stress reactions in cereals and other crops are also facilitating crop scientists. These resources are vital for organising output from genomic studies and other relevant knowledge from various phases of research and making it available to the scientific community working in the plant stress field. The Plant Stress Gene Database (Prabha et al. 2011) contains 259 genes engaged in stress conditions from 11 plant species. It is possible to query information using the web page by species, gene ID, or function and it contains paralog or ortholog gene information. Another stress-specific database of genes, DroughtDB (Alter et al. 2015), contains manually curated genes implicated in drought stress response, as well as thorough information on calculated ortholog genes in model and crop plants.

Similarly, the Plant Stress Protein Database (PSPDB) is an open access database that includes 2064 carefully curated stress proteins of 134 plant species from UniProt and their functional roles in the influence of 30 various abiotic and biotic stresses. PlantPReS, which houses 20 k plus curated items from 456 articles and more than 10,000 plus distinct stress-associated proteins, is another resource in proteomics for plant responses to various biotic and abiotic stresses (Mousavi et al. 2016). There are several stress-specific regulatory gene databases, such as Stress-Responsive Transcription Factor Database (STIFDB v.2) (Naika et al. 2013), created with stress-responsive transcription factor (STIF) algorithm based on HMM model. STIFDB v.2 contains over 38,000 relations of stress signals, stress-associated genes, and transcription factor binding sites. RiceSRTFDB is an exclusive database which contains information about transcription factors with comprehensive expression and cis-acting element (Priya and Jain 2013). PASmiR (Zhang et al. 2013) is a database of stress-responsive miRNAs that contains information on regulation of plant miRNAs under abiotic stresses which is compiled from approximately 200 publications. It contains 1038 interactions among 682 miRNAs for 35 types of abiotic stresses in 30 plus plant species. Further, PhytoPath provides plant pathogen genomic and phenotypic data. The PhytoPath is also linked to the genes from PHI-base database, which is an expertly curated catalogue of genes for host-pathogen interactions. The genes in the databases are experimentally validated for pathogenicity and the databases utilise Ensembl for data visualisation and analysis (Pedro et al. 2016; www.phytopathdb.org).

14.3.7 Database on Molecular Markers in Cereals

Plant breeders can utilise molecular markers as selection tools since they are DNA segments related with agronomically relevant features. Many crop genomes have been decoded using high-throughput approaches, allowing the development of molecular markers for trait mapping, selection, and genetic enhancement (Bohra et al. 2020). However, the ever-increasing amount of genetic and genomic data needs data management to make the data organised, open, and reachable to scientific community. This activity necessitates the creation of specialised visualisation tools and bioinformatics systems to associate genomics with phenomics.

Many molecular marker databases are freely available in the public domain, with information on a wide range of markers for a variety of species. Some databases, on the other hand, focus on a particular type of marker. dbSNP (http://www.ncbi.nlm.nih.gov/projects/SNP/) is the largest single nucleotide polymorphism database. The dbSNP database mostly contains SNP data for humans and other vertebrates, with some plant data thrown in for good measure (Sherry et al. 2001). For grasses, there are various databases. Many kinds of rice, maize, and other cereal markers are located in the Gramene (Ware et al. 2002; http://www.gramene.org/). Users can use this website's search engine to look for specific markers and marker information, including database cross-references and CMap coordinates related to chromosomes (Ware et al. 2002). The public SSR or microsatellite resources are included in the

International Rice Genome Sequencing Project (Sasaki and Burr 2000), International Rice Microsatellite Initiative (IRMI) (McCouch et al. 2002), MaizeGDB (http://www.maizegdb.org), and the Cornell SSR library (Lawrence et al. 2004). GrainGenes provides information on various markers such as SSR, RFLP, and SNP for Triticeae and *Avena* spp. (Matthews et al. 2003). GrainGenes database also has comparative map views for barley, oats, rye, and wheat based on the CMap tool (http://wheat.pw.usda.gov/cgi-bin/graingenes/).

Further, the maize-specific MaizeGDB database has been given with search option to fetch ESTs, AFLPs, RAPD, and sequence data. Another database for maize and teosinte is Panzea database, which explains the genetic architecture of complex characteristics. A marker search interface for SNP and SSR is also available in this database. The search results provide a list of markers and their location on various chromosomes (Zhao et al. 2006; http://www.panzea.org/). Molecular marker databases for Triticeae may include TriMEDB (Triticeae mapped EST database), which delivers mapped cDNA markers for barley and wheat (Mochida et al. 2008). The Wheatgenome.info database gives wheat genome visualisation based on GBrowse2, CMap, and CMap3D (Lai et al. 2012). The HarvEST, a barley database (http://harvest.ucr.edu/), offers various tools for searching markers using their name and chromosome location and displaying retrieved marker, related linkage maps, chromosome number, map positions, and primer pairs for PCR. The Rice Genome Annotation Project database provides information about putative simple sequence repeats through user-friendly web interface and displays predicted markers filtered by type or chromosome in GBrowser view (http://rice.plantbiology.msu.edu/; Sakai et al. 2013). The autoSNPdb database is built on an early SNP discovery pipeline based on EST sequence data. It offers a user-friendly interface for searching for SNPs within known genes in various crops, viz. rice, barley, and wheat. In this database, SNP identification process was created using sequence variations associated with particular genes that is discovered using a keyword search or sequence resemblance (Duran et al. 2009) (Table 14.4). There are several software and tools available for molecular marker studies, viSGSautoSNP (second-generation sequencing autoSNP) software (Lorenc et al. 2012), SSR Locator (http://fl ora.acpfg.com.au/ssrprimer2/)(da Maia et al. 2008), etc. These tools may help plant breeders in molecular marker discovery, primer design, and PCR simulations.

14.4 Genotyping Platforms for Next-Generation Breeding Approaches in Cereals

Crop improvement involves a larger number of lines with some extent of genetic variation; hence, genotyping is an important part of the process. In the last three decades, the development and implementation of DNA marker technology in crop genetics received much devotion. From low-throughput first marker system, restriction fragment length polymorphisms (RFLPs) to NGS based single nucleotide polymorphism (SNP) markers were made available at breeder's disposal (Tanksley et al. 1989; Varshney et al. 2009b). The contemporary genomic landscapes of crops

Table 14.4 Examples of molecular marker databases for cereals

Database	URLs	Reference
autoSNPdb	http://autosnpdb.appliedbioinformatics.com.au/	Duran et al. (2009)
GenBank dbSNP	http://www.ncbi.nlm.nih.gov/projects/SNP/	Sherry et al. (2001)
GrainGenes	http://wheat.pw.usda.gov/cgi-bin/graingenes/browse.cgi?class-marker	Matthews et al. (2003)
Gramene	http://www.gramene.org/db/markers/marker_view	Ware et al. (2002); Tello-Ruiz et al. (2021)
MaizeGDB	http://www.maizegdb.org/probe.php	Lawrence et al. (2004)
Panzea	http://www.panza.org/db/searches/webform/marker_search	Zhao et al. (2006)
Rice genome annotation project	http://rice.plantbiology.msu.edu/annotation_pseduo_putativessr.shtml	Sakai et al. (2013)
SSR primer 2	http://flora.acpfg.com.au/ssrprimer2/	#
SSR taxonomy tree	http://appliedbioinformatics.com.au/projects/ssrtaxonomy/php/	Jewell et al. (2006)
Triticale mapped EST database ver. 2.0 (TriMEDB)	http://trimedb.psc.riken.jp/index.pl	Mochida et al. (2008)
Wheat genome information	http://www.wheatgenome.info	Lai et al. (2012)

were metamorphosed owing to NGS techniques, which provide a wealth of sequencing data with genome-wide coverage, speed, and lower cost (Bevan and Uauy 2013). These techniques make it much easier to construct chip-based marker platforms for ultra-high-throughput genotyping and genotyping-by-sequencing (GBS). In the following sections, these approaches are examined in greater depth.

14.4.1 Marker-Based Genotyping

Genetic markers are the characteristics governed by allelic variants of genes that can be passed down through generations and utilised as experimental probes or tags to track a specific trait(s). In genetics and plant breeding, classical and DNA markers are two major categories of genetic markers (Varshney et al. 2009a). Classical markers comprise of morphological, cytological, and biochemical indicators, whereas DNA markers are described as a piece of DNA sequence variant that can detect polymorphism between different genotypes or alleles of a gene in a population or gene pool. These fragments are linked to a precise position in the genome and identified as such as RFLP, AFLP, RAPD, SSR, SNP, etc. (Collard et al. 2005). These variations can be detected using appropriate molecular technology. For instance, southern blotting is based on nuclear acid hybridisation principle, and polymerase chain reaction and DNA sequencing are based on hybridisation followed by amplification. The genetic markers are employed to genotype the plants using existing technologies such as low-throughput gel-based approaches like cleaved amplified polymorphic sequence (CAPS) marker (Thiel et al. 2004); PCR-based fluorescently labelled high-throughput methods like HRM curve analysis, TaqMan, and KASPTM assays (Martino et al. 2010); and fixed array systems such as Illumina Infinium (Mason et al. 2017 (Thomson 2014). Genotyping was primarily based on phenotypic variances at the start of Mendelian genetics, which limited the realisation of plant breeding impacts. With the advent of polymerase chain reaction (PCR) method, genotyping technology has evolved dramatically. Molecular markers such as random amplification length polymorphism (RAPD) and amplified fragment length polymorphism (AFLP) were heralded on PCR technology. Many studies used RAPD and AFLP marker types because they do not require nucleotide sequence information and are inexpensive; nevertheless, the marker information is not consistent between populations. Sequence-based PCR markers have dominated genotyping approaches from first-generation sequencing since SSR markers became readily accessible and relatively affordable, ample on plant genomes, and very informative than last PCR-based markers (Poland and Rife 2012; Kim et al. 2020). SSRs' most potent feature was combined with the development of ESTs, which carries actively expressed genes. Scientists used to prefer EST-SSRs because they may relate marker information to genes associated with target traits. Despite this, we are unable to classify SSR as a tool for high-throughput genotyping for four reasons. Firstly, identifying accurate information in terms of multi alleles per locus is difficult. Secondly, integrating or comparing SSR profiles from diverse platforms or populations is problematic. Thirdly, compared to SNPs, SSR repeats are limited in number and are not equally distributed across a genome. Finally, gel-based genotyping SSRs is time-consuming and labour-intensive, whereas automated fragment analysis systems save time and effort. Single nucleotide polymorphisms (SNPs), abundant in plant genomes, are the most recent molecular markers. From the 1990s until the early 2000s, scientists preferred SSRs to SNPs since SNP discovery and genotyping with DNA sequencing were exceedingly pricy and complex. Nevertheless, tremendous versatility, speed, and cost-effectiveness of NGS technologies have made SNPs as primary choice markers in many breeding experiments (Poland and Rife 2012; Kim et al. 2016). Due to nucleotide sequence similarities, SNP markers can be utilised globally for genotyping from various sources, allowing for an integrated analysis across species. As a result, SNPs are the most preferred in contemporary genotyping experiments, despite few unclear interpretations in some polyploidy species due to their biallelic nature (You et al. 2018). For SNP genotyping pipelines, a range of concepts and methodologies have been adopted and applied. In crop breeding, two types of high-throughput SNP genotyping platforms are employed, namely, array- and NGS-based platforms. When the quantity of samples is minimal, these high-throughput genotyping approaches may not be cost-effective. Because of their high degrees of multiplexing, array- or PCR-based genotyping technologies can substantially reduce the cost per data point if the number of samples is large enough. Genotyping systems based on arrays or PCR require prior nucleotide sequence information, in contrast to NGS-based technique. As a result, NGS-based platforms are widely being employed for species without a reference genome. However, it is less precise than array- or PCR-based technologies and is not repeatable between trials. TaqMan (Applied Biosystems), SNPlex (Applied Biosystems), BioMark HD (Fluidigm), KASPar (LGC), Axiom Biobank (Affymetrix), Infinium II (Illumina), GoldenGate (Illumina), and iPlex (Sequenome) are now commercially available array- or PCR-based SNP genotyping platforms (Kim et al. 2020). Restriction association DNA sequencing (RAD-seq), multiplex shotgun genotyping (MSG), and genotyping-by-sequencing (GBS) are few NGS-based pipelines that are widely applied to plant sciences (Poland and Rife 2012).

14.4.2 Next-Generation Sequencing

GBS (genotyping-by-sequencing) is the latest genotyping technology and widely utilised crop improvement programme that uses high-throughput sequencing to detect SNPs and other sequence variations. With the progresses in NGS technology, GBS is a cost-effective and preferred genotyping technique in cereal breeding. The various GBS approaches utilised in crop plants are summarised in Table 14.5 (Rasheed et al. 2017), each with its own set of characteristics and merits. This includes GBS (Elshire et al. 2011), DArT-seq (Cruz et al. 2013), and sequence-based genotyping (SBG) (Truong et al. 2012), restriction fragment sequencing (REST-seq) (Stolle and Moritz 2013), and restriction enzyme site comparative analysis (RESCAN) (Kim and Tai 2013). The most extensively utilised platforms in agricultural genomics are GBS and DArT-seq approaches. Restriction enzyme digestion is used in GBS, followed by adapter ligation, PCR, and sequencing. Another step forward in lowering GBS cost is the development of repeat amplification sequencing (rAmpSeq). rAmpSeq combines ground-breaking computation and robust genotyping to score thousands of markers for < $5 per sample (Buckler et al. 2016). In spite of the various benefits, viz. low cost and the ability to genotype mutations in low copy intervening sequences, it has limited usage owing to generation of fewer markers than standard GBS and necessitates reference genome sequence in constructing a quality test.

GBS was initially intended for high-resolution association analyses. Subsequently, it expanded to various modifications such as RAD-seq in the variety of species having complicated genomes. GBS is a cost-effective tool for finding and genotyping SNPs in crop species and populations especially with available reference genomes. It a technically very simple and amenable for multiplexing that has been widely used in large crop genomes to saturate mapping and breeding populations with thousands to millions of SNPs for evolutionary studies, molecular profiles of genotypes for genetic characterisation, and selection in breeding experiments (Poland et al. 2012). Additionally, the GBS approach is proved to be an excellent platform for a wide range of coverages, i.e. from a gene to whole-genome coverage (Poland and Rife 2012). The GBS allowed genotyping for GWAS, diversity and linkage analyses, marker discovery, and genomic predictions. The GBS approach has been robust across a range of species, and SNP finding and genotyping are

Table 14.5 Array- and NGS-based genotyping platforms for high-throughput genotyping in cereals (modified from: Rasheed et al. 2017)

Crop	Size	Technology	Information/ resource	Reference
Wheat	9 K	Illumina Infinium BeadChip	Wheat 9 K iSelect	Cavanagh et al. (2013)
Wheat	90 K	Illumina Infinium BeadChip	Wheat 90 K select	Wang et al. (2014)
Wheat	660 K	Affymetrix axiom	Wheat 660 K axiom	Sun et al. (2020)
Wheat	820 K	Affymetrix axiom	Wheat HD genotyping array	Winfield et al. (2016)
Wheat	35 K	Affymetrix axiom	Wheat Breeder's genotyping array	Allen et al. (2017)
Rye	600 K	Affymetrix axiom	Rye6OOK	Bauer et al. (2017)
Ryegrass	9 K	Illumina Infinium BeadChip		Blackmore et al. (2015)
Crop specific	Scalable	Exome capture	Exome sequencing	Allen et al. (2013)
De novo (applicable to multiple crops)	50–300 K	GBS	Genotyping-by-sequencing	Elshire et al. (2011)
	~50 K	DArT-seq	DArTsequencing	http://www.diversityarrays.com
	1–2 K	rAmpSeq	Repeat amplification sequencing	Buckler et al. (2016)
	Depend on genome size, sequencing depth, and technology	SLAF-seq	Specific length amplified sequencing	Sun et al. (2013b)
		RAD-seq	Restriction site-associated DNA sequencing	Bérard et al. (2009)
		Two-enzyme GBS		Poland et al. (2012)
		ddRAD	Double-digest RAD	Peterson et al. (2012)
		SBG	Sequencing-based genotyping	Truong et al. (2012)
		REST-seq	Restriction fragment sequencing	Stolle and Moritz (2013)

(continued)

Table 14.5 (continued)

Crop	Size	Technology	Information/ resource	Reference
		RAD capture	Rapture	Ali et al. (2016)
		MSG	Multiplexed shotgun genotyping	Andolfatto et al. (2011)
		ezRAD		Toonen et al. (2013)

conducted simultaneously. Thus, no prior knowledge of the species genomes is required (Elshire et al. 2011; Poland and Rife 2012). Additionally, NGS platforms offer cost-effective genotyping, particularly in orphan crops with minimal genomic information, since the genotyping and SNP discovery happen simultaneously with limited genetic background bias. As a regular limitation of GBS, genotyping errors caused by poor NGS read coverage result in misidentification of homozygotes from heterozygotes. The extent of genotyping errors is quite high with polyploidy and lacks a reference genome, as paralogs might be misconstrued for identical readings when their similarities are extreme. To overcome these obstacles, unusual cutters to increase sequence depth are used, the number of multiplexed samples in the library preparation step is lowered, or the library with latest NGS technology is sequenced. GBS data is typically squandered, resulting in a large number of NGS reads. Allele dropout, which is caused by a variation in the restriction enzyme recognition site, prevents enzyme activation, and leads to genotyping errors, is another disadvantage (Scheben et al. 2017). Another common issue is differences in coverage produced by amplification bias towards fragments with shorter lengths and greater GC content. GBS's frequent use of methylation-sensitive enzymes, which contain almost half of trait-associated SNPs, could lead to ascertainment bias (Hindorff et al. 2009). Other difficulties are prolonged library preparation step, high percentage of missing data points, and lack of adequate computation facility for data imputation analysis and storage facility. Despite these drawbacks, GBS has grown in prominence in agricultural genetics. This is attributable to major advances in sequencing chemistry, the accessibility of long-read sequencing technologies, and enrichments in the existing reference genomes, which have resulted in genotyping approaches that make better use of this technology.

14.4.3 SNP Arrays

SNP array is a genotyping assay with high-throughput, low-cost, and automatic results. It has been utilised extensively in crop genetic investigations, including genome-wide association studies (GWAS) (McCouch et al. 2016), linkage map creation (Felcher et al. 2012), genomic selection (Clarke et al. 2016), population structure analysis (Wang et al. 2016), and gene mapping (Dalton-Morgan et al.

2014). SNP array technology, like many other genotyping tools, offers advantages and disadvantages. SNP arrays have various advantages over conventional genotyping techniques for high-throughput genotyping. First, compared to data obtained using NGS-based approaches, SNP array data is comparatively simple to evaluate, especially when it comes to the time-consuming preparation of NGS libraries and downstream in silico data analysis (GARVIN et al. 2010). Affymetrix or Illumina can call and deliver genotypes for SNP markers, or researchers can call genotypes using the genotype calling pipelines. However, calling genotypes using NGS-based data is more complex and time-consuming. Read trimming, read alignment, SNP genotyping, SNP filtering, etc., are all part of the SNP genotype calling process (Clevenger et al. 2015). Secondly, SNPs from target genomic location can be integrated on the array with precision. Furthermore, interesting SNP numbers located on array are adjustable in Illumina and Affymetrix platforms. Thirdly, despite substantial cost savings associated with NGS, the SNP array is believed to have low to moderate per sample expenses (Peng et al. 2017). SNP arrays and NGS can be used in tandem. Despite its clear advantages in producing sequence and identifying variations, NGS genotyping in polyploids remains a challenge. The SNP array remains a viable genotyping platform based on the rising number of variations revealed by NGS approaches. The SNP array, on the other hand, has some limitations, including the requirement for prior genomic knowledge, the ability to genotype only known SNP locations, and manual dosage scoring (Vos et al. 2015). Furthermore, array design and subsequent standardisation require quite longer time. Further, the ascertainment bias is a usual problem with arrays owing to selective sampling of polymorphisms in the target population (Heslot et al. 2013) or a few samples were used in SNP detection panels (Albrechtsen et al. 2010). To eliminate ascertainment bias, efforts have been made to use whole-genome sequencing with high coverage, update SNP array markers, and combine markers from various arrays. Currently, SNP arrays for many plant species with different capacities are available (Table 14.5).

14.5 Software and Databases for Next-Generation Breeding Tools

14.5.1 Genome-Wide Association Study (GWAS)

GWAS examines the relationships between the single nucleotide polymorphisms (SNPs) or haplotypes with the target phenotype. The linkage disequilibrium (LD) in crop genome is a basis for quantifiable assessment of GWAS. In general, a GWAS infers the cause and variant links employing a hypothesis test with relevant statistical tests like F-test, the exact test of Fisher, Pearson's $\chi 2$ test, or a regression model under statistical assumptions of null hypothesis of no marker-trait associations.

The following three graphs are commonly used to visualise the GWAS results:

1. Manhattan plot: In Manhattan plots the p-values in the –log10 (p) scale are plotted against the genomic physical locations of the SNPs, on respective chromosomes in scatter plot. Large peaks matching to small *p*-values imply a substantial relationship between the trait and the related genomic region. The plot is known as a Manhattan plot because it mimics New York's Manhattan skyscrapers.
2. Quantile-quantile (Q-Q) plot: Q-Q plots depict the divergence of observed p-values from null hypothesis. Here Q-Q plot provides an agreement or discrepancy or theoretical p-values with experimental p-values. In Q-Q plots, the negative logarithms of the p-values from the GWAS fitted models are plotted against the predicted value with a null hypothesis of no association. The SNPs that deviate from the diagonal in the upper right area of the graph are most likely to be linked to the trait of interest. The p-value should follow a uniform distribution over the range [0, 1], if SNPs are fitting with the null hypothesis. However, the presence of any correlations deviates the p-values from the uniform distribution.
3. Principal component (PC) plot: In GWAS, PCA analysis provides population structure estimates effect on multivariate data in terms of the data's covariance structure.

A measure of association or statistical dependency of an SNP with the phenotype is produced in a typical (univariate) genome-wide association study. With the null hypothesis of no association, the chance of getting an association of similar strength among the SNPs and the target phenotype is computed for each association score. If the chance factor or *p*-value is less than a predetermined threshold value (0.01 or 0.05), it indicates the association among the SNPs and the target phenotype. In spite of the strong statistical evidence against the null hypothesis, there is a cent percent chance that the lower significant p-values are due to arbitrary associations; hence, the identified relationship is due to random chance owing to various population parameters. Therefore, one of the most challenging aspects of GWAS is avoiding false-positive results (Bush and Moore 2012; Gumpinger et al. 2018).

Many software programmes have included various statistical analysis approaches and genomic pre-processing stages; among these, GAPIT (Lipka et al. 2012) and PLINK (Purcell et al. 2007) are publicly available and frequently employed in GWAS analyses. In addition, several other tools are also available for GWAS and genome selection (Table 14.6).

Shaun Purcell and co-workers at Harvard University developed C/C++ – based open-source whole-genome association analysis toolset PLINK. The PLINK toolset provides five major utilities, viz. management of datasets, summary statistics, population stratification, association analysis, and inferred ancestry. gPLINK, software built on JAVA programme, was recently developed to make it a user-friendly tool to biologists. GCTA tool is designed for complex trait genome-wide association analysis (Yang et al. 2011). It enables users to do several analyses on SNP data such as two-sample and linear regression tests for univariate GWAS, set-based tests, and epistasis screenings. Furthermore, several software such as FaST-LMM, EMMAX, GEMMA, GRAMMAR-Gamma, BOLT-LMM, etc. implement different

Table 14.6 List of tools/software for genome-wide association analysis

S. no.	Name	Description	Reference
1	AlphaDrop	Simulate genomic data and phenotypes to perform genomic selection and GWAS. It can handle sequence data, SNP data, and pedigrees and also can calculate QTL effects and breeding values	Hickey and Gorjanc (2012)
2	BLINK	BLINK (Bayesian-information and linkage-disequilibrium iteratively nested keyway) employs Bayes and linkage disequilibrium information to dissect the underlying SNPs or genomic regions controlling target phenotypes	Huang et al. (2019)
3	FarmCPU	FarmCPU (fixed and random model circulating probability unification) performs GWAS analysis by GLM framework. Further, through single-marker regression, FarmCPU allocates the data points into bins and sorts out optimal markers set as covariates in the subsequent iteration	Neves et al. (2012)
4	FaST-LMM	FaST-LMM (factored spectrally transformed linear mixed models) package suitable for extensive GWAS studies	Listgarten et al. (2012)
5	G2P	G2P (a genome-wide-association-study simulation tool for genotype simulation, phenotype simulation, and power evaluation) is designed to simulate genotypes in GWAS for genotype and phenotype data along with statistical threshold estimation	Tang and Liu (2019)
6	GEMMA	GEMMA (genome-wide efficient mixed-model association) tests the marker-trait association through Wald statistics	Zhou and Stephens (2012)
7	GWAS pipeline	GWAS pipeline performs data filtration, generates a kinship matrix and covariate files, runs EMMAX, and creates graphical displays such as Manhattan and QQ plots Further, GWAS pipeline also provides the functions to calculate the summary of significant SNPs with allele effect contribution on target phenotype	McCouch et al. (2016)
8	IPGWAS	IPGWAS performs the combined GWAS and quality control analyses. Additionally, IPGWAS provides Manhattan and Q-Q plots and conversion formats for genetic analyses, genotype phasing, and imputation	Fan and Song (2012)
9	OmicABEL	OmicABEL provides the rapid mixed-model-based GWAS analysis for both single and multiple traits	Aulchenko et al. (2014)
10	Wtest	Wtest is an integrated R package for analysing the cause-variants association of principal effects and pairwise and high-order interactions in GWAS data. It also analyses the cis-regulation of SNPs and CpG sites, genome- and epigenome-wide, respectively	Sun et al. (2019)

approaches in association testing with linear mixed models (Purcell et al. 2007). GAPIT is an R package for GWAS and genomic selection. GAPIT was created at Cornell University's Institute for Genomic Diversity. Several statistical approaches for testing of associations among genetic variations and traits, viz. MLM, population parameters previously determined (P3D), and efficient mixed-model association (EMMA), can be used with it (Storey and Tibshirani 2003). GAPIT has key advantages in managing larger datasets (SNPs and genotypes) and is user-friendly.

TASSEL (Trait Analysis by Association, Evolution, and Linkage) is a JAVA-based statistical package for association analyses. It is most popular among GWAS analyses in plants. Like GAPIT, TASSEL can evaluate linkage disequilibrium, trait associations, and evolutionary patterns. TASSEL (TASSEL 5.0) also allows analyses of genetic diversity and SNP calling from GBS datafiles. In addition, TASSEL also has several visualisation tools to picturise the genetic analyses and GWAS results such as LD, Manhattan plot, PCA scatter plots, a phylogenetic tree using Archaeopteryx and genetic distance heat map, and phenotypic variance explained by markers (R^2). The latest TASSEL (TASSEL 5.0) version also provides helpful data summaries to get quick overviews of individuals, markers, missing data, and chromosome-wise marker numbers. Presently, the TASSEL package is freeware and get installed from Buckler Lab website (Bradbury et al. 2007; http://www.maizegenetics.net/tassel).

GenStat is a paid statistical software for performing GWAS analyses using biallelic and multi-allelic markers (https://www.vsni.co.uk/software/genstat/). It performs analysis using GLM model or MLM model with PCA or kinship values to control genetic relatedness and population structure. In addition, GenStat allows graphical visualisation of LD decay plot and provides an option for defining the threshold of significance –log10(p) along with Bonferroni function.

Many databases and web servers which store curated information of genome-wide variant-trait associations for various species may help perform analysis. GWAS Atlas encompasses manually curated data resources of genome-wide variant-trait associations for different species (https://bigd.big.ac.cn/gwas/). Presently, it has 96,141 causal variant and phenotype associations for 614 traits across 7 cultivated plants, including major cereals such as maize, rice, and sorghum (Tian et al. 2020). Similarly, GWASpro is a high-performing freely available web server (https://www.zhaolab.org/GWASPRO/), mainly designed for large-scale GWAS analyses in plants. GWASpro is optimised to handle ten million markers and 10,000 samples from replicable genotypes.

14.5.2 Genomic Selection

In the past, QTLs based on bi-parental mapping populations were used to study the genetics of traits. This method was giving a low-resolution snapshot of marker-trait associations. Subsequently, based on the extent of genetic variability in panel, the researchers were able to dissect the high-resolution association up to 30–40 alleles through association mapping approach. Presently, for quantitative traits, there is a transition in analysing a few loci to studying the entire genome. Because the genome-wide selection models hypothesise that all markers of genome contribute to the expression of target phenotype, either positively or negatively (Meuwissen et al. 2001); therefore, GS models incorporate small-effect marker loci in the

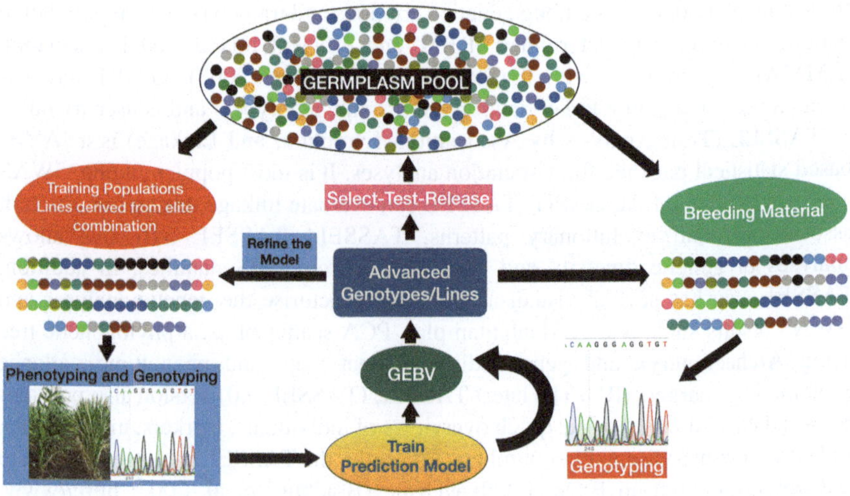

Fig. 14.3 Overview of the general genomic selection scheme followed in cereal crops

selection models (Guo et al. 2012). By integrating SNPs from genetically diverse populations, GS methods can produce lines with the best SNP combinations. The trait expression is governed by the cumulative effects of SNPs, termed as genomic estimated breeding values (GEBVs). GS analysis is mainly made up of two parts; the first part is to predict the GEBVs and second part uses the predicted GEBVs in the selection (Fig. 14.3).

The GEBVs are predicted with GS models using genome-wide distributed SNPs and comprehensive phenotypic data points. The individuals in the training population are stringently phenotyped and genotyped for traits of interest using dense, genome-wide markers such as SNPs to build a genomic prediction model. Genomic estimated breeding values are computed using a model that predicts the in-toto additive genetic effects of genome-wide dispersed alleles on an individual's target phenotype. In selection cycles or breeding populations or testing set, the individuals are genotyped with the identical marker set employed in the training population. Later, the genomic prediction model developed in training population is employed to forecast the individual GEBVs in testing set or population. Superior performers are chosen and progressed to the next selection cycle based on their GEBVs. As a result, during selection cycles or testing population (Wang et al. 2018), the phenotyping stage of similar qualities evaluated in the training population is skipped, which saves time and resources of breeding programmes. Validating the accuracy of the GS model before picking candidates based on GEBVs is a crucial stage in the fitting process. The most popular method is calculating the GEBV correlation with observed phenotypes in the validation group of individuals (Ornella et al. 2014). A validation set can consist of 10% to 30% of the training set's random individuals (Endelman 2011; Ornella et al. 2014). The validation set's GEBVs are calculated using a model built from the rest of the training population. Each time the validation set contains different individuals, ten or more cross-validation tests are performed.

GS is a data-intensive method, and it provides a computational difficulty in terms of data storage infrastructure. Therefore, data administration, statistical analysis workflow, result accessibility, and data sharing are in high demand. Furthermore, breeders lack a thorough understanding of the complexities of GS statistical analysis (Nakaya and Isobe 2012). Flexible GS databases and user-friendly statistical tools would let breeding programmes use GS more effectively. SolGS is a web-based platform designed to help GS researchers with bioinformatics and statistics. It allows the plant breeders to choose and construct prediction models and is employed to predict GEBVs of individuals in selection set. It uses a browser to display data graphically and interactively. It stores phenotype and genotype data and experimental information in an organism-agnostic database structure (Tecle et al. 2014). The ridge regression best linear unbiased predictor (rrBLUP) R package (Endelman 2011) is used for statistical modelling, and the GBLUP (genomic relationship matrix) approach is employed to predict GEBVs. Another pipeline is BWGS (Charmet et al. 2020), an R tool that makes it simple to compute GEBVs for GS. BreedWheat Genomic Selection (BWGS) was built with private-public partnership project BreedWheat (https://breedwheat.fr). The BWGS allows two major functions: (1) perform simulated random cross-validations in the genotypic and phenotypic data points of training population and (2) prediction of GEBVs for a set of genotyped-only lines. Additionally, BWGS has several other methods such as missing data imputation, marker and training set selection, and genomic prediction with 15 parametric and semi-parametric statistical models. Another R-based package BGLR is also widely used for genome-wide regression and prediction analyses.

14.5.3 Genome Editing

Genome editing is a technology that allows manipulating the DNA bases in organisms, which allows the researcher to delete, add to, or replace the genomic region. Editing genomic regions can change the nucleic acid base composition and target trait. The arrival of very adaptable genome-editing tools has given the capability to researchers to rapidly and cost-effectively incorporate the sequence-specific alterations into the genomes. Three most important technologies now utilised as genome editing tools are (1) zinc-finger nucleases (ZFNs) (Cas9), (2) transcription activator-like effector nucleases (TALENs), and (3) clustered regularly interspaced short palindromic repeats (CRISPR)-CRISPR-associated protein9. Among the above genome editing techniques, CRISPR/Cas9 is the most promising approach being employed in plant systems for trait improvement (Kamburova et al. 2017). Genome editing technologies offer a wide range of uses in the current genomics era, including developing novel crop cultivars that are better yielding, showing resistance to biotic and abiotic stresses, and having a better nutritional value. To achieve this, genome editing systems are targeted in plant breeding to modify the genomic sequences governing the target trait expression through (1) creation of point mutations like natural SNPs, (2) creating small changes to gene

Table 14.7 Web server and database links for CRISPR/Cas 9 genome editing method

Tool	URL	Reference
WGE	http://www.sanger.ac.uk/htgt/wge/	Hodgkins et al. (2015)
Cas database	http://www.rgenome.net/cas-database/	Park et al. (2016)
CrisprGE	http://crdd.osdd.net/servers/crisprge/.	Kaur et al. (2015)
CRISPRdb	http://crispr.u-psud.fr/crispr	Grissa et al. (2007)
COSMID	https://crispr.bme.gatech.edu/	Cradick et al. (2014)
PhytoCRISP-ex	http://www.phytocrispex.biologie.ens.fr/CRISP-Ex/	Rastogi et al. (2016)
GT-scan	http://gt-scan.braembl.org.au/gt-scan/	O'Brien and Bailey (2014)

function, (3) introducing the foreign genes/sequences, (4) gene knockout and pyramiding, (5) regulating the gene expression, and (6) epigenetic editing (Jacobs et al. 2015; Kamburova et al. 2017).

The CRISPR-Cas9 system mainly includes single guide (sgRNA) sequence with 19–22 bases and a CRISPR-associated Cas protein that have endonuclease activity. The sgRNA comprises crRNA and tracrRNA. The crRNA is characterised by two important regions: the spacer sequence which mediates the editing complex to the target sequence in the genome and a region which binds to tracrRNA. Many software and web servers are available for designing sgRNA for CRISPR/Cas9 experiments (Table 14.7) (Grissa et al. 2007; Cradick et al. 2014; O'Brien and Bailey 2014; Kaur et al. 2015; Hodgkins et al. 2015; Rastogi et al. 2016; Park et al. 2016), and many software and tools are also available for CRISPR/Cas 9 genome editing method (Table 14.8) (Pliatsika and Rigoutsos 2015; Winter et al. 2016; Oliveros et al. 2016; Concordet and Haeussler 2018).

14.6 Prospects

The changing climate is challenging the assured cereal production to provide food and nutritional requirements of the increasing population. The availability of genetic resources, modern breeding, and genomics-based decision supporting tools is a great asset to achieve the desired genetic gains in cereals. However, despite well-catalogued availability of broad genetic base of cereals and genomic databases, very little has been achieved in integrating the genomic resources with existing breeding pipelines. Further, to speed up the genetic gain per unit time through efficient utilisation of genetic and genomic resources of cereals, active collaboration among cereal geneticists and breeders, bioinformaticians, and molecular biologists should be encouraged. Moreover, the available genetic and genomic databases need to be maintained through timely update to ensure their proper utility. Although vast amounts of transcriptomics and functional genomics datasets are available in cereals for various stresses, there are very few databases available to understand the systems biology and regulatory aspects of stress tolerance mechanisms in cereals. Therefore,

Table 14.8 List of software/tools for CRISPR/Cas9-based genome editing

Software/tool	Category	Tool description	URLs	Reference
Off-spotter	Genome editing	Off-spotter designs the ideal guide RNAs (gRNAs) through providing various PAM motif choices to given input sequence. The off-spotter has an input limitation of single 1000 nucleotide sequence or less than 20 CR-separated 20-mers	https://cm.jefferson.edu/Off-Spotter/	Pliatsika and Rigoutsos (2015)
CrispRVariants	Genome editing	CrispRVariants delivers various tools for studying the CRISPR-Cas9 mutagenesis sequences of variants or other investigations where variants for precise genomic region(s) are required. In the sequence data CrispRVariants locates the specific mutant alleles for the endonuclease cut site and variant alleles. This allows generating summary plots for variant allele and table of counts	https://bioconductor.org/packages/release/bioc/html/CrispRVariants.html	Lindsay et al. (2016)
CRISPOR	Genome editing	The web page CRISPOR aids in the design, assessment, and cloning of the guide sequences for CRISPR/Cas9 applications in 120 genomes, comprising model organisms and plants	http://crispor.tefor.net/	Concordet and Haeussler (2018)
Breaking-CAS	Genome editing	Breaking-CAS is developed to design the putative gRNA and detect the putative sgRNA off-targets for CRISPR/CAS applications. It works with all eukaryotic genomes available in Ensembl and Ensembl genomes databases. The user can provide one to	https://bioinfogp.cnb.csic.es/tools/breakingcas/index.php	Oliveros et al. (2016)

(continued)

Table 14.8 (continued)

Software/tool	Category	Tool description	URLs	Reference
		many sequences up to 20,000 bps in FASTA format. Breaking-CAS also generates interactive web page displaying thorough information on candidate oligos, on-targets and off-targets. Additionally, it also gives score, coordinates, and overlapping genes and generates the results in tabular form		
CaRpools (CRISPR-AnalyzeR for pooled screens caRpools)	High-throughput sequencing	CaRpools is an R package designed to generate the standardised analysis reports of NGS read counts from pooled CRISPR screens. Additionally, CaRpools is a user-friendly and open virtual appliance which does not require prior programming knowledge to perform analysis	https://github.com/boutroslab/caRpools	Winter et al. (2016)

there is a need to enhance the stress-associated genetic and genomic data generation, sharing, and creation of breeder-friendly databases to ensure data-driven cereal breeding for stress resilience.

Acknowledgements MGM is grateful to the 'Early Career Research Award scheme' (ECR/2017/000675), Science and Engineering Research Board (SERB), Government of India, and the network project on 'Computational Biology and Agricultural Bioinformatics (Agril.Edn.14(44)/2014-A&P)' for financial support to research work on the topic. The funding agencies had no role in designing the study, data collection and analyses, the decision to publish, or the preparation of the manuscript.

References

Albrechtsen A, Nielsen FC, Nielsen R (2010) Ascertainment biases in SNP chips affect measures of population divergence. Mol Biol Evol 27:2534–2547. https://doi.org/10.1093/molbev/msq148

Ali OA, O'Rourke SM, Amish SJ et al (2016) RAD Capture (Rapture): Flexible and Efficient Sequence-Based Genotyping. Genetics 202:389–400. https://doi.org/10.1534/genetics.115.183665

Allen AM, Barker GLA, Wilkinson P et al (2013) Discovery and development of exome-based, co-dominant single nucleotide polymorphism markers in hexaploid wheat (*Triticum aestivum* L.). Plant Biotechnol J 11:279–295. https://doi.org/10.1111/pbi.12009

Allen AM, Winfield MO, Burridge AJ et al (2017) Characterization of a wheat breeders' array suitable for high-throughput SNP genotyping of global accessions of hexaploid bread wheat (*Triticum aestivum*). Plant Biotechnol J 15:390–401. https://doi.org/10.1111/pbi.12635

Alter S, Bader KC, Spannagl M et al (2015) DroughtDB: an expert-curated compilation of plant drought stress genes and their homologs in nine species. Database 2015:1–7. https://doi.org/10.1093/database/bav046

Andolfatto P, Davison D, Erezyilmaz D et al (2011) Multiplexed shotgun genotyping for rapid and efficient genetic mapping. Genome Res 21:610–617. https://doi.org/10.1101/gr.115402.110

Andrew H, John E, Rémy B et al (2009) The Sorghum bicolor genome and the diversification of grasses. Nature 457:551–556. https://doi.org/10.1038/nature07723

Appels R, Eversole K, Stein N et al (2018) Shifting the limits in wheat research and breeding using a fully annotated reference genome. Science 361:eaar7191. https://doi.org/10.1126/science.aar7191

Aravind J, Rinku S, Pooja B et al (2017) Identification, characterization, and functional validation of drought-responsive microRNAs in subtropical maize inbreds. Front Plant Sci 8:941. https://doi.org/10.3389/fpls.2017.00941

Arora K, Panda KK, Mittal S et al (2017) RNAseq revealed the important gene pathways controlling adaptive mechanisms under waterlogged stress in maize. Sci Rep 7:10950. https://doi.org/10.1038/s41598-017-10561-1

Aulchenko Y, Fabregat-Traver D, Sharapov SZ et al (2014) High-performance mixed models based genome-wide association analysis with omicABEL software. F1000R:3. https://doi.org/10.12688/f1000research.4867.1

Bahri B, Shah SJA, Hussain S et al (2011) Genetic diversity of the wheat yellow rust population in Pakistan and its relationship with host resistance. Plant Pathol 60:649–660. https://doi.org/10.1111/J.1365-3059.2010.02420.X

Barta E, Sebestyén E, Pálfy TB et al (2005) DoOP: databases of orthologous promoters, collections of clusters of orthologous upstream sequences from chordates and plants. Nucleic Acids Res 33:D86–D90. https://doi.org/10.1093/nar/gki097

Bauer E, Schmutzer T, Barilar I et al (2017) Towards a whole-genome sequence for rye (*Secale cereale* L.). Plant J 89:853–869. https://doi.org/10.1111/tpj.13436

Bérard A, Le Paslier MC, Dardevet M et al (2009) High-throughput single nucleotide polymorphism genotyping in wheat (*Triticum spp.*). Plant Biotechnol J 7:364–374. https://doi.org/10.1111/j.1467-7652.2009.00404.x

Bevan MW, Uauy C (2013) Genomics reveals new landscapes for crop improvement. Genome Biol 14:206. https://doi.org/10.1186/gb-2013-14-6-206

Biłas R, Szafran K, Hnatuszko-Konka K, Kononowicz AK (2016) Cis-regulatory elements used to control gene expression in plants. Plant Cell Tiss Org Cult 127:269–287. https://doi.org/10.1007/s11240-016-1057-7

Blackmore T, Thomas I, McMahon R et al (2015) Genetic–geographic correlation revealed across a broad European ecotypic sample of perennial ryegrass (*Lolium perenne*) using array-based SNP genotyping. Theor Appl Genet 128:1917–1932. https://doi.org/10.1007/s00122-015-2556-3

Bohra A, Chand Jha U, Godwin ID, Kumar Varshney R (2020) Genomic interventions for sustainable agriculture. Plant Biotechnol J 18:2388–2405. https://doi.org/10.1111/pbi.13472

Bolser D, Staines DM, Pritchard E, Kersey P (2016) Ensembl plants: integrating tools for visualizing, mining, and analyzing plant genomics data BT. In: Edwards D (ed) Plant bioinformatics: methods and protocols. Springer, New York, NY, pp 115–140

Boyer JS (1982) Plant productivity and environment. Science 218:443–448. https://doi.org/10.1126/science.218.4571.443

Bradbury PJ, Zhang Z, Kroon DE et al (2007) TASSEL: software for association mapping of complex traits in diverse samples. Bioinformatics 23:2633–2635. https://doi.org/10.1093/bioinformatics/btm308

Buckler E, Ilut D, Wang X et al (2016) rAmpSeq: Using repetitive sequences for robust genotyping. bioRxiv:096628. https://doi.org/10.1101/096628

Bush WS, Moore JH (2012) Genome-wide association studies. PLOS Comp Biol 8:e1002822. https://doi.org/10.1371/journal.pcbi.1002822

Byrne PF, Volk GM, Gardner C et al (2018) Sustaining the future of plant breeding: the critical role of the USDA-ARS national plant germplasm system. Crop Sci 58:451–468. https://doi.org/10.2135/cropsci2017.05.0303

Cai H, Lu Y, Xie W et al (2012) Transcriptome response to nitrogen starvation in rice. J Biosci 37:731–747. https://doi.org/10.1007/s12038-012-9242-2

Cal AJ, Liu D, Mauleon R et al (2013) Transcriptome profiling of leaf elongation zone under drought in contrasting rice cultivars. PLoS One 8. https://doi.org/10.1371/journal.pone.0054537

Canales J, Moyano T, Villarroel E, Gutiérrez R (2014) Systems analysis of transcriptome data provides new hypotheses about Arabidopsis root response to nitrate treatments. Front Plant Sci 5:22. https://doi.org/10.3389/fpls.2014.00022

Cavanagh CR, Chao S, Wang S et al (2013) Genome-wide comparative diversity uncovers multiple targets of selection for improvement in hexaploid wheat landraces and cultivars. Proc Natl Acad Sci 110:8057–8062. https://doi.org/10.1073/pnas.1217133110

Chandler VL, Brendel V (2002) The maize genome sequencing project. Plant Physiol 130:1594–1597. https://doi.org/10.1104/pp.015594

Charmet G, Tran LG, Auzanneau J et al (2020) BWGS: A R package for genomic selection and its application to a wheat breeding programme. PLoS One 15:1–20. https://doi.org/10.1371/journal.pone.0222733

Chen C, Li J, Feng J et al (2021) sRNAanno—a database repository of uniformly annotated small RNAs in plants. Horticul Res 8:45. https://doi.org/10.1038/s41438-021-00480-8

Cho K, Shibato J, Agrawal GK et al (2008) Integrated transcriptomics, proteomics, and metabolomics analyses to survey ozone responses in the leaves of rice seedling. J Proteome Res 7:2980–2998. https://doi.org/10.1021/pr800128q

Chow C-N, Lee T-Y, Hung Y-C et al (2019) PlantPAN3.0: a new and updated resource for reconstructing transcriptional regulatory networks from ChIP-seq experiments in plants. Nucleic Acids Res 47:D1155–D1163. https://doi.org/10.1093/nar/gky1081

Clarke WE, Higgins EE, Plieske J et al (2016) A high-density SNP genotyping array for Brassica napus and its ancestral diploid species based on optimised selection of single-locus markers in the allotetraploid genome. Theor Appl Genet 129:1887–1899. https://doi.org/10.1007/s00122-016-2746-7

Clevenger J, Chavarro C, Pearl SA et al (2015) Single nucleotide polymorphism identification in polyploids: A review, example, and recommendations. Mol Plant 8:831–846. https://doi.org/10.1016/j.molp.2015.02.002

Collard BCY, Jahufer MZZ, Brouwer JB, Pang ECK (2005) An introduction to markers, quantitative trait loci (QTL) mapping and marker-assisted selection for crop improvement: The basic concepts. Euphytica 142:169–196. https://doi.org/10.1007/s10681-005-1681-5

Concordet J-P, Haeussler M (2018) CRISPOR: intuitive guide selection for CRISPR/Cas9 genome editing experiments and screens. Nucleic Acids Res 46:W242–W245. https://doi.org/10.1093/nar/gky354

Cradick TJ, Qiu P, Lee CM et al (2014) COSMID: a web-based tool for identifying and validating CRISPR/CAS off-target sites. Mol Ther Nucleic Acids 3:e214. https://doi.org/10.1038/mtna.2014.64

Cruz VMV, Kilian A, Dierig DA (2013) Development of DArT marker platforms and genetic diversity assessment of the U.S. collection of the new oilseed crop Lesquerella and related species. PLoS One 8:e64062

Cui L, Veeraraghavan N, Richter A et al (2006) ChloroplastDB: the chloroplast genome database. Nucleic Acids Res 34:692–696. https://doi.org/10.1093/nar/gkj055
da Maia LC, Palmieri DA, de Souza VQ et al (2008) SSR Locator: tool for simple sequence repeat discovery integrated with primer design and PCR simulation. Int J Plant Genom 2008:412696. https://doi.org/10.1155/2008/412696
Dalton-Morgan J, Hayward A, Alamery S et al (2014) A high-throughput SNP array in the amphidiploid species Brassica napus shows diversity in resistance genes. Funct Integr Genomics 14:643–655. https://doi.org/10.1007/s10142-014-0391-2
de Carvalho MAAP, Bebeli PJ, Bettencourt E et al (2012) Cereal landraces genetic resources in worldwide GeneBanks. A review. Agron Sustain Dev 33:177–203. https://doi.org/10.1007/S13593-012-0090-0
do Amaral MN, Arge LWP, Benitez LC et al (2016) Comparative transcriptomics of rice plants under cold, iron, and salt stresses. Funct Integr Genomics 16:567–579. https://doi.org/10.1007/s10142-016-0507-y
Dong Q, Schlueter SD, Brendel V (2004) PlantGDB, plant genome database and analysis tools. Nucleic Acids Res 32:354–359. https://doi.org/10.1093/nar/gkh046
Duran C, Appleby N, Clark T et al (2009) AutoSNPdb: an annotated single nucleotide polymorphism database for crop plants. Nucleic Acids Res 37:D951–D953. https://doi.org/10.1093/nar/gkn650
Edgar R (2002) Gene expression omnibus: NCBI gene expression and hybridization array data repository. Nucleic Acids Res 30:207–210. https://doi.org/10.1093/nar/30.1.207
Elshire RJ, Glaubitz JC, Sun Q et al (2011) A robust, simple genotyping-by-sequencing (GBS) approach for high diversity species. PLoS One 6:e19379
Endelman JB (2011) Ridge regression and other kernels for genomic selection with R package rrBLUP. Plant, Genome 4(3). https://doi.org/10.3835/plantgenome2011.08.0024
Fan YH, Song YQ (2012) IPGWAS: An integrated pipeline for rational quality control and association analysis of genome-wide genetic studies. Biochem Biophys Res Commun 422: 363–368. https://doi.org/10.1016/j.bbrc.2012.04.117
FAO (2010) The second report on the state of the world's plant genetic resources for food and agriculture. Commission on Genetic Resources for Food and Agriculture, Food and Agriculture Organization of the United Nations.
Fei Q, Xia R, Meyers BC (2013) Phased, secondary, small interfering RNAs in posttranscriptional regulatory networks. Plant Cell 25:2400–2415. https://doi.org/10.1105/tpc.113.114652
Felcher KJ, Coombs JJ, Massa AN et al (2012) Integration of two diploid potato linkage maps with the potato genome sequence. PLoS One 7:e36347
Futamura N, Totoki Y, Toyoda A et al (2008) Characterization of expressed sequence tags from a full-length enriched cDNA library of Cryptomeria japonica male strobili. BMC Genomics 9: 383. https://doi.org/10.1186/1471-2164-9-383
Garg V, Khan AW, Kudapa H et al (2019) Integrated transcriptome, small RNA and degradome sequencing approaches provide insights into Ascochyta blight resistance in chickpea. Plant Biotechnol J 17:914–931. https://doi.org/10.1111/pbi.13026
Garvin MR, Saitoh K, Gharrett AJ (2010) Application of single nucleotide polymorphisms to non-model species: a technical review. Mol Ecol Resour 10:915–934. https://doi.org/10.1111/j.1755-0998.2010.02891.x
Goff SA (2002) A draft sequence of the rice genome (*Oryza sativa* L. ssp. *japonica*). Science 296: 92–100. https://doi.org/10.1126/science.1068275
Goodstein DM, Shu S, Howson R et al (2012) Phytozome: a comparative platform for green plant genomics. Nucleic Acids Res 40:1178–1186. https://doi.org/10.1093/nar/gkr944
Griffiths-Jones S, Bateman A, Marshall M et al (2003) Rfam: an RNA family database. Nucleic Acids Res 31:439–441. https://doi.org/10.1093/nar/gkg006
Grissa I, Vergnaud G, Pourcel C (2007) The CRISPRdb database and tools to display CRISPRs and to generate dictionaries of spacers and repeats. BMC Bioinformatics 8:172. https://doi.org/10.1186/1471-2105-8-172

Gumpinger AC, Roqueiro D, Grimm DG, Borgwardt KM (2018) Methods and tools in genome-wide association studies BT. In: von Stechow L, Santos Delgado A (eds) Computational cell biology: methods and protocols. Springer, New York, NY, pp 93–136

Guo Z, Tucker DM, Lu J et al (2012) Evaluation of genome-wide selection efficiency in maize nested association mapping populations. Theor Appl Genet 124:261–275. https://doi.org/10.1007/s00122-011-1702-9

Guo Z, Kuang Z, Wang Y et al (2021) PmiREN: a comprehensive encyclopedia of plant miRNAs. Nucleic Acids Res 48:D1114–D1121. https://doi.org/10.1093/nar/gkz894

Heslot N, Rutkoski J, Poland J et al (2013) Impact of marker ascertainment bias on genomic selection accuracy and estimates of genetic diversity. PLoS One 8:e74612

Hickey JM, Gorjanc G (2012) Simulated data for genomic selection and genome-wide association studies using a combination of coalescent and gene drop methods. G3 2:425–427. https://doi.org/10.1534/g3.111.001297

Higo K, Ugawa Y, Iwamoto M, Korenaga T (1999) Plant cis-acting regulatory DNA elements (PLACE) database: 1999. Nucleic Acids Res 27:297–300. https://doi.org/10.1093/nar/27.1.297

Hindorff LA, Sethupathy P, Junkins HA et al (2009) Potential etiologic and functional implications of genome-wide association loci for human diseases and traits. Proc Natl Acad Sci 106:9362–9367. https://doi.org/10.1073/pnas.0903103106

Hodgkins A, Farne A, Perera S et al (2015) WGE: a CRISPR database for genome engineering. Bioinformatics 31:3078–3080. https://doi.org/10.1093/bioinformatics/btv308

Hu X, Cui Y, Dong G et al (2019) Using CRISPR-Cas9 to generate semi-dwarf rice lines in elite landraces. Sci Rep 9(1):1–7. https://doi.org/10.1038/s41598-019-55757-9

Huang M, Liu X, Zhou Y et al (2019) BLINK: a package for the next level of genome-wide association studies with both individuals and markers in the millions. Gigascience 8:1–12. https://doi.org/10.1093/gigascience/giy154

Itoh T, Tanaka T, Barrero RA et al (2007) Curated genome annotation of *Oryza sativa* ssp. *japonica* and comparative genome analysis with *Arabidopsis thaliana*: the rice annotation project. Genome Res 17:175–183. https://doi.org/10.1101/gr.5509507

Jacobs TB, LaFayette PR, Schmitz RJ, Parrott WA (2015) Targeted genome modifications in soybean with CRISPR/Cas9. BMC Biotechnol 15:16. https://doi.org/10.1186/s12896-015-0131-2

James C (2002) Global review of commercialized transgenic crops: 2001 Feature: Bt Cotton. ISAAA Briefs No 26 1–44.

Jaradat A (2013) Wheat Landraces: A mini review. Emir J Food Agric 25:20. https://doi.org/10.9755/ejfa.v25i1.15376

Jeevan B, Gogoi R, Sharma D et al (2020) Genetic analysis of maydis leaf blight resistance in subtropical maize (*Zea mays* L.) germplasm. J Genet 99:89. https://doi.org/10.1007/s12041-020-01245-3

Jewell E, Robinson A, Savage D et al (2006) SSRPrimer and SSR taxonomy tree: biome SSR discovery. Nucleic Acids Res 34:W656–W659. https://doi.org/10.1093/nar/gkl083

Jia M, Guan J, Zhai Z et al (2018) Wheat functional genomics in the era of next generation sequencing: an update. Crop J 6:7–14. https://doi.org/10.1016/j.cj.2017.09.003

Jin J, Tian F, Yang DC et al (2017) PlantTFDB 4.0: Toward a central hub for transcription factors and regulatory interactions in plants. Nucleic Acids Res 45:D1040–D1045. https://doi.org/10.1093/nar/gkw982

Johnson C, Bowman L, Adai AT et al (2007) CSRDB: a small RNA integrated database and browser resource for cereals. Nucleic Acids Res 35:4–7. https://doi.org/10.1093/nar/gkl991

Kamburova VS, Nikitina EV, Shermatov SE et al (2017) Genome editing in plants: an overview of tools and applications. Int J Agron 2017:7315351. https://doi.org/10.1155/2017/7315351

Kanehisa M, Furumichi M, Tanabe M et al (2017) KEGG: New perspectives on genomes, pathways, diseases and drugs. Nucleic Acids Res 45:D353–D361. https://doi.org/10.1093/nar/gkw1092

Karjagi CG, Sekhar JC, Lakshmi SP et al (2017) breeding for resistance to insect pests in maize. In: Arora R, Sandhu S (eds) Breeding insect resistant crops for sustainable agriculture. Springer, Singapore, pp 201–229. https://doi.org/10.1007/978-981-10-6056-4_7

Kaur K, Tandon H, Gupta AK, Kumar M (2015) CrisprGE: a central hub of CRISPR/Cas-based genome editing. Database 2015. https://doi.org/10.1093/database/bav055

Kawahara Y, de la Bastide M, Hamilton JP et al (2013) Improvement of the *Oryza sativa* Nipponbare reference genome using next generation sequence and optical map data. Rice 6: 1–10. https://doi.org/10.1186/1939-8433-6-1

Kawaura K, Mochida K, Enju A et al (2009) Assessment of adaptive evolution between wheat and rice as deduced from full-length common wheat cDNA sequence data and expression patterns. BMC Genomics 10:1–11. https://doi.org/10.1186/1471-2164-10-271

Kikuchi S, Satoh K, Nagata T et al (2003) Collection, mapping, and annotation of over 28,000 cDNA clones from japonica rice. Science 301:376–379. https://doi.org/10.1126/science.1081288

Kim S-I, Tai TH (2013) Identification of SNPs in closely related temperate japonica rice cultivars using restriction enzyme-phased sequencing. PLoS One 8:e60176

Kim C, Guo H, Kong W et al (2016) Application of genotyping by sequencing technology to a variety of crop breeding programs. Plant Sci 242:14–22. https://doi.org/10.1016/j.plantsci.2015.04.016

Kim KD, Kang Y, Kim C (2020) Application of genomic big data in plant breeding: past, present, and future. Plants 9(11):1454. https://doi.org/10.3390/plants9111454

Kozomara A, Griffiths-Jones S (2014) miRBase: annotating high confidence microRNAs using deep sequencing data. Nucleic Acids Res 42:D68–D73

Lacchini E, Kiegle E, Castellani M et al (2020) CRISPR-mediated accelerated domestication of African rice landraces. PLoS One 15:e0229782. https://doi.org/10.1371/JOURNAL.PONE.0229782

Lai K, Berkman PJ, Lorenc MT et al (2012) WheatGenome.info: an integrated database and portal for wheat genome information. Plant Cell Physiol 53:e2–e2. https://doi.org/10.1093/pcp/pcr141

Lawrence CJ, Dong Q, Polacco ML et al (2004) MaizeGDB, the community database for maize genetics and genomics. Nucleic Acids Res 32:D393–D397. https://doi.org/10.1093/nar/gkh011

Lescot M, Déhais P, Thijs G et al (2002) PlantCARE, a database of plant cis-acting regulatory elements and a portal to tools for in silico analysis of promoter sequences. Nucleic Acids Res 30: 325–327. https://doi.org/10.1093/nar/30.1.325

Li C, Zhang B (2016) MicroRNAs in control of plant development. J Cell Physiol 231:303–313. https://doi.org/10.1002/jcp.25125

Li G, Wang L, Yang J et al (2021) A high-quality genome assembly highlights rye genomic characteristics and agronomically important genes. Nat Genet 53:574–584. https://doi.org/10.1038/s41588-021-00808-z

Lindsay H, Burger A, Biyong B et al (2016) CrispRVariants charts the mutation spectrum of genome engineering experiments. Nat Biotechnol 34:701–702. https://doi.org/10.1038/nbt.3628

Lipka AE, Tian F, Wang Q et al (2012) GAPIT: genome association and prediction integrated tool. Bioinformatics 28:2397–2399. https://doi.org/10.1093/bioinformatics/bts444

Listgarten J, Lippert C, Kadie CM et al (2012) Improved linear mixed models for genome-wide association studies. Nat Methods 9:525–526. https://doi.org/10.1038/nmeth.2037

Liu X, Lu T, Yu S et al (2007) A collection of 10,096 indica rice full-length cDNAs reveals highly expressed sequence divergence between *Oryza sativa indica* and *japonica* subspecies. Plant Mol Biol 65:403–415. https://doi.org/10.1007/s11103-007-9174-7

Liu L, Ren S, Guo J et al (2018) Genome-wide identification and comprehensive analysis of microRNAs and phased small interfering RNAs in watermelon. BMC Genomics 19:111. https://doi.org/10.1186/s12864-018-4457-8

Liu H, Able AJ, Able JA (2020) Integrated analysis of small RNA, transcriptome, and degradome sequencing reveals the water-deficit and heat stress response network in durum wheat. Int J Mol Sci 21:1–28. https://doi.org/10.3390/ijms21176017

Lorenc MT, Hayashi S, Stiller J et al (2012) Discovery of single nucleotide polymorphisms in complex genomes using SGSautoSNP. Biology 1:370–382. https://doi.org/10.3390/biology1020370

Mallikarjuna MG, Thirunavukkarasu N, Sharma R et al (2020) Comparative transcriptome analysis of iron and zinc deficiency in maize (*Zea mays* L.). Plants 9:1812. https://doi.org/10.3390/plants9121812

Martino A, Mancuso T, Rossi AM (2010) Application of high-resolution melting to large-scale, high-throughput SNP genotyping: A comparison with the TaqMan® method. J Biomol Screen 15:623–629. https://doi.org/10.1177/1087057110365900

Mascher M, Gundlach H, Himmelbach A et al (2017) A chromosome conformation capture ordered sequence of the barley genome. Nature 544:427–433. https://doi.org/10.1038/nature22043

Mason AS, Higgins EE, Snowdon RJ et al (2017) A user guide to the Brassica 60 K Illumina Infinium™ SNP genotyping array. Theor Appl Genet 130:621–633. https://doi.org/10.1007/s00122-016-2849-1

Matthews DE, Carollo VL, Lazo GR, Anderson OD (2003) GrainGenes, the genome database for small-grain crops. Nucleic Acids Res 31:183–186. https://doi.org/10.1093/nar/gkg058

Matys V, Kel-Margoulis OV, Fricke E et al (2006) TRANSFAC® and its module TRANSCompel®: transcriptional gene regulation in eukaryotes. Nucleic Acids Res 34:D108–D110. https://doi.org/10.1093/nar/gkj143

Mayer KFX, Waugh R, Langridge P et al (2012) A physical, genetic and functional sequence assembly of the barley genome. Nature 491:711–716. https://doi.org/10.1038/nature11543

McCouch SR, Teytelman L, Xu Y et al (2002) Development and mapping of 2240 new ssr markers for rice (*Oryza sativa* L.). DNA Res 9:199–207. https://doi.org/10.1093/dnares/9.6.199

McCouch SR, Wright MH, Tung C-W et al (2016) Open access resources for genome-wide association mapping in rice. Nat Commun 7:10532. https://doi.org/10.1038/ncomms10532

Meuwissen THE, Hayes BJ, Goddard ME (2001) Prediction of total genetic value using genome-wide dense marker maps. Genetics 157:1819–1829. https://doi.org/10.1093/genetics/157.4.1819

Mochida K, Shinozaki K (2010) Genomics and bioinformatics resources for crop improvement. Plant Cell Physiol 51:497–523. https://doi.org/10.1093/pcp/pcq027

Mochida K, Saisho D, Yoshida T et al (2008) TriMEDB: A database to integrate transcribed markers and facilitate genetic studies of the tribe Triticeae. BMC Plant Biol 8:72. https://doi.org/10.1186/1471-2229-8-72

Morozova O, Marra MA (2008) Applications of next-generation sequencing technologies in functional genomics. Genomics 92:255–264. https://doi.org/10.1016/j.ygeno.2008.07.001

Mousavi SA, Pouya FM, Ghaffari MR et al (2016) PlantPReS: a database for plant proteome response to stress. J Proteomics 143:69–72. https://doi.org/10.1016/j.jprot.2016.03.009

Muiño JM, Hoogstraat M, van Ham RCHJ, van Dijk ADJ (2011) PRI-CAT: a web-tool for the analysis, storage and visualization of plant ChIP-seq experiments. Nucleic Acids Res 39:W524–W527. https://doi.org/10.1093/nar/gkr373

Mukherjee S, Stamatis D, Bertsch J et al (2017) Genomes online database (GOLD) v.6: data updates and feature enhancements. Nucleic Acids Res 45:D446–D456. https://doi.org/10.1093/nar/gkw992

Naika M, Shameer K, Mathew OK et al (2013) STIFDB2: an updated version of plant stress-responsive transcription factor database with additional stress signals, stress-responsive transcription factor binding sites and stress-responsive genes in Arabidopsis and rice. Plant Cell Physiol 54:1–15. https://doi.org/10.1093/pcp/pcs185

Nakaya A, Isobe SN (2012) Will genomic selection be a practical method for plant breeding? Ann Bot 110:1303–1316. https://doi.org/10.1093/aob/mcs109

Nanjo T, Sakurai T, Totoki Y et al (2007) Functional annotation of 19,841 Populus nigra full-length enriched cDNA clones. BMC Genomics 8:448. https://doi.org/10.1186/1471-2164-8-448

Neves HHR, Carvalheiro R, Queiroz SA (2012) A comparison of statistical methods for genomic selection in a mice population. BMC Genet 13:1–24. https://doi.org/10.1186/1471-2156-13-100

Newton AC, Akar T, Baresel JP et al (2010) Cereal landraces for sustainable agriculture. A review. Agron Sustain Dev 30:237–269. https://doi.org/10.1051/agro/2009032

O'Brien A, Bailey TL (2014) GT-scan: identifying unique genomic targets. Bioinformatics 30: 2673–2675. https://doi.org/10.1093/bioinformatics/btu354

Ogihara Y, Mochida K, Kawaura K et al (2004) Construction of a full-length cDNA library from young spikelets of hexaploid wheat and its characterization by large-scale sequencing of expressed sequence tags. Genes Genet Syst 79:227–232. https://doi.org/10.1266/ggs.79.227

Oliveros JC, Franch M, Tabas-Madrid D et al (2016) Breaking-Cas—interactive design of guide RNAs for CRISPR-Cas experiments for ENSEMBL genomes. Nucleic Acids Res 44:W267–W271. https://doi.org/10.1093/nar/gkw407

Oppermann M, Weise S, Dittmann C, Knüpffer H (2015) GBIS: the information system of the German Genebank. Database 2015. https://doi.org/10.1093/DATABASE/BAV021

Ornella L, Pérez P, Tapia E et al (2014) Genomic-enabled prediction with classification algorithms. Heredity 112:616–626. https://doi.org/10.1038/hdy.2013.144

Ouyang S, Zhu W, Hamilton J et al (2007) The TIGR rice genome annotation resource: improvements and new features. Nucleic Acids Res 35:8–11. https://doi.org/10.1093/nar/gkl976

Papatheodorou I, Fonseca NA, Keays M et al (2018) Expression Atlas: gene and protein expression across multiple studies and organisms. Nucleic Acids Res 46:D246–D251. https://doi.org/10.1093/nar/gkx1158

Park J, Kim J-S, Bae S (2016) Cas-Database: web-based genome-wide guide RNA library design for gene knockout screens using CRISPR-Cas9. Bioinformatics 32:2017–2023. https://doi.org/10.1093/bioinformatics/btw103

Parkinson H, Kapushesky M, Shojatalab M et al (2007) ArrayExpress - a public database of microarray experiments and gene expression profiles. Nucleic Acids Res 35(Database Issue): D747–D750. https://doi.org/10.1093/nar/gkl995

Paterson AH, Bowers JE, Bruggmann R et al (2009) The *Sorghum bicolor* genome and the diversification of grasses. Nature 457:551–556. https://doi.org/10.1038/nature07723

Pedro H, Maheswari U, Urban M et al (2016) PhytoPath: an integrative resource for plant pathogen genomics. Nucleic Acids Res 44:D688–D693. https://doi.org/10.1093/nar/gkv1052

Peng Z, Fan W, Wang L et al (2017) Target enrichment sequencing in cultivated peanut (Arachis hypogaea L.) using probes designed from transcript sequences. Mol Genet Genomics 292:955–965. https://doi.org/10.1007/s00438-017-1327-z

Peterson BK, Weber JN, Kay EH et al (2012) Double Digest RADseq: an Inexpensive method for de novo SNP discovery and genotyping in model and non-model species. PLoS One 7:e37135

Pliatsika V, Rigoutsos I (2015) "Off-Spotter": very fast and exhaustive enumeration of genomic lookalikes for designing CRISPR/Cas guide RNAs. Biol Direct 10:4. https://doi.org/10.1186/s13062-015-0035-z

Poland JA, Rife TW (2012) Genotyping-by-sequencing for plant breeding and genetics. Plant. Genome 5(3). https://doi.org/10.3835/plantgenome2012.05.0005

Poland JA, Brown PJ, Sorrells ME, Jannink J-L (2012) Development of high-density genetic maps for barley and wheat using a novel two-enzyme genotyping-by-sequencing approach. PLoS One 7:e32253

Postman J, Hummer K, Bretting P et al (2010) GRIN-Global: an international project to develop a global plant genebank information management system. Acta Hortic 859:49–56. https://doi.org/10.17660/ACTAHORTIC.2010.859.4

Prabha R, Ghosh I, Singh DP (2011) Plant stress gene database: a collection of plant genes responding to stress condition. ARPN J Sci Technol 1:28–31

Priya P, Jain M (2013) RiceSRTFDB: a database of rice transcription factors containing comprehensive expression, cis-regulatory element and mutant information to facilitate gene function analysis. Database 2013:1–7. https://doi.org/10.1093/database/bat027

Purcell S, Neale B, Todd-Brown K et al (2007) PLINK: a tool set for whole-genome association and population-based linkage analyses. Am J Hum Genet 81:559–575. https://doi.org/10.1086/519795

Rasheed A, Hao Y, Xia X et al (2017) Crop breeding chips and genotyping platforms: progress, challenges, and perspectives. Mol Plant 10:1047–1064. https://doi.org/10.1016/j.molp.2017.06.008

Rastogi A, Murik O, Bowler C, Tirichine L (2016) PhytoCRISP-Ex: a web-based and stand-alone application to find specific target sequences for CRISPR/CAS editing. BMC Bioinformatics 17:261. https://doi.org/10.1186/s12859-016-1143-1

Robinson AJ, Tamiru M, Salby R et al (2018) AgriSeqDB: an online RNA-Seq database for functional studies of agriculturally relevant plant species. BMC Plant Biol 18:200. https://doi.org/10.1186/s12870-018-1406-2

Sakai H, Lee SS, Tanaka T et al (2013) Rice annotation project database (RAP-DB): an integrative and interactive database for rice genomics. Plant Cell Physiol 54:e6–e6. https://doi.org/10.1093/pcp/pcs183

Sakurai T, Plata G, Rodríguez-Zapata F et al (2007) Sequencing analysis of 20,000 full-length cDNA clones from cassava reveals lineage specific expansions in gene families related to stress response. BMC Plant Biol 7:66. https://doi.org/10.1186/1471-2229-7-66

Sasaki T, Burr B (2000) International rice genome sequencing project: the effort to completely sequence the rice genome. Curr Opin Plant Biol 3:138–142. https://doi.org/10.1016/S1369-5266(99)00047-3

Sato K, Shin-I T, Seki M et al (2009) Development of 5006 full-length cDNAs in barley: a tool for accessing cereal genomics resources. DNA Res 16:81–89. https://doi.org/10.1093/dnares/dsn034

Sayers EW, Agarwala R, Bolton EE et al (2019) Database resources of the national center for biotechnology information. Nucleic Acids Res 47:D23–D28. https://doi.org/10.1093/nar/gky1069

Scheben A, Batley J, Edwards D (2017) Genotyping-by-sequencing approaches to characterize crop genomes: choosing the right tool for the right application. Plant Biotechnol J 15:149–161. https://doi.org/10.1111/pbi.12645

Schnable PS, Ware D, Fulton RS et al (2009) The B73 maize genome: complexity, diversity, and dynamics. Science 326:1112–1115. https://doi.org/10.1126/science.1178534

Sekhon RS, Briskine R, Hirsch CN et al (2013) Maize gene atlas developed by RNA sequencing and comparative evaluation of transcriptomes based on Rna sequencing and microarrays. PLoS One 8:e61005

Seki M, Narusaka M, Kamiya A et al (2002) Functional annotation of a full-length Arabidopsis cDNA collection. Science 296:141–145. https://doi.org/10.1126/science.1071006

Serfling A, Kopahnke D, Habekuss A et al (2017) Wheat diseases: an overview, pp 263–294

Shahmuradov IA, Gammerman AJ, Hancock JM et al (2003) PlantProm: a database of plant promoter sequences. Nucleic Acids Res 31:114–117. https://doi.org/10.1093/nar/gkg041

Shan X, Li Y, Jiang Y et al (2013) Transcriptome profile analysis of maize seedlings in response to high-salinity, drought and cold stresses by deep sequencing. Plant Mol Biol Rep 31:1485–1491. https://doi.org/10.1007/s11105-013-0622-z

Sharma R, Upadhyay S, Bhat B et al (2020) Abiotic stress induced miRNA-TF-gene regulatory network: a structural perspective. Genomics 112:412–422. https://doi.org/10.1016/j.ygeno.2019.03.004

Sherry ST, Ward M-H, Kholodov M et al (2001) dbSNP: the NCBI database of genetic variation. Nucleic Acids Res 29:308–311. https://doi.org/10.1093/nar/29.1.308

Singh K, Gupta K, Tyagi V, Rajkumar S (2020) Plant genetic resources in India: management and utilization. Vavilov J Genet Breed 24:306. https://doi.org/10.18699/VJ20.622

Soderlund C, Descour A, Kudrna D et al (2009) Sequencing, mapping, and analysis of 27,455 maize full-length cDNAs. PLoS Genet 5(11):e1000740. https://doi.org/10.1371/journal.pgen.1000740

Stolle E, Moritz RFA (2013) RESTseq – Efficient benchtop population genomics with RESTriction Fragment SEQuencing. PLoS One 8:e63960

Storey JD, Tibshirani R (2003) Statistical significance for genomewide studies. Proc Natl Acad Sci 100:9440–9445. https://doi.org/10.1073/pnas.1530509100

Sun X, Dong B, Yin L et al (2013a) PMTED: A plant microRNA target expression database. BMC Bioinformatics 14. https://doi.org/10.1186/1471-2105-14-174

Sun X, Liu D, Zhang X et al (2013b) SLAF-seq: an efficient method of large-scale de novo SNP discovery and genotyping using high-throughput sequencing. PLoS One 8:e58700

Sun R, Xia X, Chong KC et al (2019) Wtest: an integrated R package for genetic epistasis testing. BMC Med Genomics 12:1–6. https://doi.org/10.1186/s12920-019-0638-9

Sun C, Dong Z, Zhao L, Ren Y, Zhang N, Chen F (2020) The wheat 660 K SNP array demonstrates great potential for marker-assisted selection in polyploid wheat. Plant Biotechnol J 18(6): 1354–1360. https://doi.org/10.1111/pbi.13361

Szczes'niak MW, Makabowska I (2014) miRNEST 2.0: a database of plant and animal microRNAs. Nucleic Acids Res 42:74–77. https://doi.org/10.1093/nar/gkt1156

Tang Y, Liu X (2019) G2P: a genome-wide-association-study simulation tool for genotype simulation, phenotype simulation and power evaluation. Bioinformatics 35:3852–3854. https://doi.org/10.1093/bioinformatics/btz126

Tanksley SD, Young ND, Paterson AH, Bonierbale MW (1989) RFLP mapping in plant breeding: new tools for an old science. Nat Biotechnol 7:257–264. https://doi.org/10.1038/nbt0389-257

Tecle IY, Edwards JD, Menda N et al (2014) solGS: a web-based tool for genomic selection. BMC Bioinformatics 15:1–9. https://doi.org/10.1186/s12859-014-0398-7

Tello-Ruiz MK, Naithani S, Gupta P et al (2021) Gramene 2021: harnessing the power of comparative genomics and pathways for plant research. Nucleic Acids Res 49:D1452–D1463. https://doi.org/10.1093/nar/gkaa979

Thiel T, Kota R, Grosse I et al (2004) SNP2CAPS: a SNP and INDEL analysis tool for CAPS marker development. Nucleic Acids Res 32:e5–e5. https://doi.org/10.1093/nar/gnh006

Thirunavukkarasu N, Sharma R, Singh N et al (2017) Genomewide expression and functional interactions of genes under drought stress in maize. Int J Genom 2017:2568706. https://doi.org/10.1155/2017/2568706

Thomson M (2014) High-throughput SNP genotyping to accelerate crop improvement. Plant Breed Biotechnol 2:195–212

Tian X, Long Y, Wang J et al (2015) De novo transcriptome assembly of common wild rice (*Oryza rufipogon* Griff.) and discovery of drought-response genes in root tissue based on transcriptomic data. PLoS One 10:e0131455

Tian D, Wang P, Tang B et al (2020) GWAS atlas: a curated resource of genome-wide variant-trait associations in plants and animals. Nucleic Acids Res 48:D927–D932. https://doi.org/10.1093/nar/gkz828

Toonen RJ, Puritz JB, Forsman ZH et al (2013) ezRAD: a simplified method for genomic genotyping in non-model organisms. PeerJ 1:e203. https://doi.org/10.7717/peerj.203

Truong HT, Ramos AM, Yalcin F et al (2012) Sequence-based genotyping for marker discovery and co-dominant scoring in germplasm and populations. PLoS One 7:e37565

Tyagi R (2016) Conservation of PGR for effective utilisation: new initiatives at NBPGR. Indian J Plant Genet Resour 29:272. https://doi.org/10.5958/0976-1926.2016.00044.9

Umezawa T, Sakurai T, Totoki Y et al (2008) Sequencing and analysis of approximately 40,000 soybean cDNA clones from a full-length-enriched cDNA library. DNA Res 15:333–346. https://doi.org/10.1093/dnares/dsn024

Van Bel M, Diels T, Vancaester E et al (2018) PLAZA 4.0: an integrative resource for functional, evolutionary and comparative plant genomics. Nucleic Acids Res 46:D1190–D1196. https://doi.org/10.1093/nar/gkx1002

van Dijk EL, Jaszczyszyn Y, Naquin D, Thermes C (2018) The third revolution in sequencing technology. Trends Genet 34:666–681. https://doi.org/10.1016/j.tig.2018.05.008

Varshney RK, Hoisington DA, Nayak SN, Graner A (2009a) Molecular plant breeding: methodology and achievements BT. In: Gustafson JP, Langridge P, Somers DJ (eds) Plant genomics: methods and protocols. Humana Press, Totowa, NJ, pp 283–304

Varshney RK, Nayak SN, May GD, Jackson SA (2009b) Next-generation sequencing technologies and their implications for crop genetics and breeding. Trends Biotechnol 27:522–530. https://doi.org/10.1016/j.tibtech.2009.05.006

Varshney RK, Shi C, Thudi M et al (2017) Pearl millet genome sequence provides a resource to improve agronomic traits in arid environments. Nat Biotechnol 35:969–976. https://doi.org/10.1038/nbt.3943

Vaughn JN, Ellingson SR, Mignone F, Von Arnim A (2012) Known and novel post-transcriptional regulatory sequences are conserved across plant families. RNA 18:368–384. https://doi.org/10.1261/rna.031179.111

Venter M, Botha FC (2010) Synthetic Promoter Engineering BT. In: Pua EC, Davey MR (eds) Plant developmental biology - biotechnological perspectives: volume 2. Springer, Berlin, Heidelberg, pp 393–414

Voinnet O (2009) Origin, biogenesis, and activity of plant microRNAs. Cell 136:669–687. https://doi.org/10.1016/j.cell.2009.01.046

Vos PG, Uitdewilligen JGAML, Voorrips RE et al (2015) Development and analysis of a 20 K SNP array for potato (Solanum tuberosum): an insight into the breeding history. Theor Appl Genet 128:2387–2401. https://doi.org/10.1007/s00122-015-2593-y

Wang L, Xie W, Chen Y et al (2010) A dynamic gene expression atlas covering the entire life cycle of rice. Plant J 61:752–766. https://doi.org/10.1111/j.1365-313X.2009.04100.x

Wang Y, Cheng X, Shan Q et al (2014) Simultaneous editing of three homoeoalleles in hexaploid bread wheat confers heritable resistance to powdery mildew. Nat Biotechnol 32:947–951. https://doi.org/10.1038/nbt.2969

Wang J, Chu S, Zhang H et al (2016) Development and application of a novel genome-wide SNP array reveals domestication history in soybean. Sci Rep 6:20728. https://doi.org/10.1038/srep20728

Wang X, Xu Y, Hu Z, Xu C (2018) Genomic selection methods for crop improvement: current status and prospects. Crop J 6:330–340. https://doi.org/10.1016/j.cj.2018.03.001

Ware D, Jaiswal P, Ni J et al (2002) Gramene: a resource for comparative grass genomics. Nucleic Acids Res 30:103–105. https://doi.org/10.1093/nar/30.1.103

Weise S, Oppermann M, Maggioni L et al (2017) EURISCO: the European search catalogue for plant genetic resources. Nucleic Acids Res 45:D1003–D1008. https://doi.org/10.1093/NAR/GKW755

Winfield MO, Allen AM, Burridge AJ et al (2016) High-density SNP genotyping array for hexaploid wheat and its secondary and tertiary gene pool. Plant Biotechnol J 14:1195–1206. https://doi.org/10.1111/pbi.12485

Winter J, Breinig M, Heigwer F et al (2016) caRpools: an R package for exploratory data analysis and documentation of pooled CRISPR/Cas9 screens. Bioinformatics 32:632–634. https://doi.org/10.1093/bioinformatics/btv617

Yang J, Lee SH, Goddard ME, Visscher PM (2011) GCTA: a tool for genome-wide complex trait analysis. Am J Hum Genet 88:76–82. https://doi.org/10.1016/j.ajhg.2010.11.011

Yim WC, Yu Y, Song K et al (2013) PLANEX: the plant co-expression database. BMC Plant Biol 13:83. https://doi.org/10.1186/1471-2229-13-83

You Q, Yang X, Peng Z et al (2018) Development and applications of a high throughput genotyping tool for polyploid crops: Single nucleotide polymorphism (SNP) array. Front Plant Sci 9:104. https://doi.org/10.3389/fpls.2018.00104

Yu J (2002) A draft sequence of the rice genome (*Oryza sativa* L. ssp. *indica*). Science 296:79–92. https://doi.org/10.1126/science.1068037

Zeng H, Zhang X, Ding M, Zhu Y (2019) Integrated analyses of miRNAome and transcriptome reveal zinc deficiency responses in rice seedlings. BMC Plant Biol 19:1–18. https://doi.org/10.1186/s12870-019-2203-2

Zhang Z, Yu J, Li D et al (2010) PMRD: plant microRNA database. Nucleic Acids Res 38:D806–D813. https://doi.org/10.1093/nar/gkp818

Zhang S, Yue Y, Sheng L et al (2013) PASmiR: a literature-curated database for miRNA molecular regulation in plant response to abiotic stress. BMC Plant Biol 13:33

Zhao W, Canaran P, Jurkuta R et al (2006) Panzea: a database and resource for molecular and functional diversity in the maize genome. Nucleic Acids Res 34:D752–D757. https://doi.org/10.1093/nar/gkj011

Zhou X, Stephens M (2012) Genome-wide efficient mixed-model analysis for association studies. Nat Genet 44:821–824. https://doi.org/10.1038/ng.2310

Zhou M, Zheng S, Liu R et al (2019) Comparative analysis of root transcriptome profiles between low- and high-cadmium-accumulating genotypes of wheat in response to cadmium stress. Funct Integr Genomics 19:281–294. https://doi.org/10.1007/s10142-018-0646-4

Zuo Y-C, Li Q-Z (2011) Identification of TATA and TATA-less promoters in plant genomes by integrating diversity measure, GC-Skew and DNA geometric flexibility. Genomics 97:112–120. https://doi.org/10.1016/j.ygeno.2010.11.002

The manufacturer's authorised representative in the EU is Springer Nature Customer Service Centre GmbH, Europaplatz 3, 69115 Heidelberg, Germany. If you have any concerns regarding our products, please contact ProductSafety@springernature.com

Printed and bound by CPI Group (UK) Ltd, Croydon, CR0 4YY
30/09/2026
02225239-0020